T0213716

Lecture Notes in Computer Science   10306

Commenced Publication in 1973
Founding and Former Series Editors:
Gerhard Goos, Juris Hartmanis, and Jan van Leeuwen

More information about this series at http://www.springer.com/series/7407

Ignacio Rojas · Gonzalo Joya
Andreu Catala (Eds.)

Advances in Computational Intelligence

14th International Work-Conference
on Artificial Neural Networks, IWANN 2017
Cadiz, Spain, June 14–16, 2017
Proceedings, Part II

 Springer

Editors
Ignacio Rojas
Universidad de Granada
Granada
Spain

Andreu Catala
Polytechnic University of Catalonia
Vilanova i la Geltrú, Barcelona
Spain

Gonzalo Joya
University of Malaga
Malaga
Spain

ISSN 0302-9743   ISSN 1611-3349   (electronic)
Lecture Notes in Computer Science
ISBN 978-3-319-59146-9   ISBN 978-3-319-59147-6   (eBook)
DOI 10.1007/978-3-319-59147-6

Library of Congress Control Number: 2017940386

LNCS Sublibrary: SL1 – Theoretical Computer Science and General Issues

Printed on acid-free paper

This Springer imprint is published by Springer Nature
The registered company is Springer International Publishing AG
The registered company address is: Gewerbestrasse 11, 6330 Cham, Switzerland

Preface

We are proud to present the set of final accepted papers for the 13th edition of IWANN – the International Work-Conference on Artificial Neural Networks – held in Cadiz, Spain, during June 14–16, 2017.

IWANN is a biennial conference that seeks to provide a discussion forum for scientists, engineers, educators, and students about the latest ideas and realizations in the foundations, theory, models, and applications of hybrid systems inspired by nature (neural networks, fuzzy logic, and evolutionary systems) as well as in emerging areas related to these areas. As in previous editions of IWANN, this year's event also aimed to create a friendly environment that could lead to the establishment of scientific collaborations and exchanges among attendees. The proceedings include all the presented communications to the conference. The publication of an extended version of selected papers in a special issue of several specialized journals (such as *Neurocomputing, Soft Computing,* and *Neural Proccesing Letters*) is also foreseen.

Since the first edition in Granada (LNCS 540, 1991), the conference has evolved and matured. The list of topics in the successive call for papers has also evolved, resulting in the following list for the present edition:

1. Mathematical and theoretical methods in computational intelligence. Mathematics for neural networks. RBF structures. Self-organizing networks and methods. Support vector machines and kernel methods. Fuzzy logic. Evolutionary and genetic algorithms.
2. Neurocomputational formulations. Single-neuron modelling. Perceptual modelling. System-level neural modelling. Spiking neurons. Models of biological learning.
3. Learning and adaptation. Adaptive systems. Imitation learning. Reconfigurable systems. Supervised, non-supervised, reinforcement and statistical algorithms.
4. Emulation of cognitive functions. Decision-making. Multi-agent systems. Sensor mesh. Natural language. Pattern recognition. Perceptual and motor functions (visual, auditory, tactile, virtual reality, etc.). Robotics. Planning motor control.
5. Bio-inspired systems and neuro-engineering. Embedded intelligent systems. Evolvable computing. Evolving hardware. Microelectronics for neural, fuzzy and bioinspired systems. Neural prostheses. Retinomorphic systems. Brain–computer interfaces (BCI). Nanosystems. Nanocognitive systems.
6. Advanced topics in computational intelligence. Intelligent networks. Knowledge-intensive problem-solving techniques. Multi-sensor data fusion using computational intelligence. Search and meta-heuristics. Soft computing. Neuro-fuzzy systems. Neuro-evolutionary systems. Neuro-swarm. Hybridization with novel computing paradigms.
7. Applications. Expert systems. Image and signal processing. Ambient intelligence. Biomimetic applications. System identification, process control, and manufacturing. Computational biology and bioinformatics. Parallel and distributed computing. Human–computer interaction, Internet modelling, communication and networking.

Intelligent systems in education. Human–robot interaction. Multi-agent systems. Time series analysis and prediction. Data mining and knowledge discovery.

At the end of the submission process, and after a careful peer review and evaluation process (each submission was reviewed by at least two, and on average 2.8, Program Committee members or additional reviewers), 126 papers were accepted for oral or poster presentation, according to the recommendations of the reviewers and the authors' preferences.

It is important to note, that for the sake of consistency and readability of the book, the presented papers are not organized as they were presented in the IWANN 2017 sessions, but classified under 21 chapters. The organization of the papers is in two volumes, arranged according to the topics list included in the call for papers. The first volume (LNCS 10305), entitled "Advances in Computational Intelligence. IWANN 2017. Part I" is divided into nine main parts and includes the contributions on:

1. Bio-inspired Computing
2. E-Health and Computational Biology
3. Human–Computer Interaction
4. Image and Signal Processing
5. Mathematics for Neural Networks
6. Self-Organizing Networks
7. Spiking Neurons
8. Artificial Neural Networks in Industry, ANNI 2017 (Special Session, organized by: Dr. Ahmed Hafaifa, Dr. Kouzou Abdellah, and Dr. Guemana Mouloud)
9. Machine Learning for Renewable Energy Applications (Dr. Sancho Salcedo Sanz, and Dr. Pedro Antonio Gutiérrez)

In the second volume (LNCS 10306), entitled "Advances in Computational Intelligence. IWANN 2017. Part II" is divided into 12 main parts and includes the contributions on:

1. Computational Intelligence Tools and Techniques for Biomedical Applications (Special Session, organized by: Dr. Miguel Atencia, Dr. Leonardo Franco, and Dr. Ruxandra Stoean)
2. Assistive Rehabilitation Technology (Special Session, organized by: Dr. Oresti Baños and Dr. Jose A. Moral-Muñoz)
3. Computational Intelligence Methods for Time Series (Special Session, organized by: Dr. German Gutierrez and Dr. Héctor Pomares)
4. Machine Learning Applied to Vision and Robotics (Special Session, organized by: Dr. José García-Rodríguez, Dr. Enrique Dominguez, Mauricio Zamora, and Dr. Eldon Caldwel)
5. Human Activity Recognition for Health and Well-Being Applications (Special Session, organized by: Dr. Daniel Rodríguez-Martín and Dr. Albert Samà)
6. Software Testing and Intelligent Systems (Special Session, organized by: Dr. Manuel Núñez and Pablo Cerro Cañizares)
7. Real-World Applications of BCI Systems (Special Session, organized by: Dr. Ricardo Ron and Dr. Ivan Volosyak)

8. Machine Learning in Imbalanced Domains (Special Session, organized by: Dr. Jaime S. Cardoso and Dr. María Pérez Ortíz)
9. Surveillance and Rescue Systems and Algorithms for Unmanned Aerial Vehicles (Special Session, organized by: Dr. Wilbert Aguilar)
10. End-User development for Social Robotics (Special Session, organized by: Igor Zubrycki, Hoang-Long Cao, and Dr. Emilia Barakova)
11. Artificial Intelligence and Games (Special Session, organized by: Dr. Antonio J. Fernández-Leiva, Dr. Antonio Mora-García, and Dr. Pablo García Sánchez)
12. Supervised, Non-supervised, Reinforcement and Statistical Algorithms

In this edition of IWANN 2017, we were honored to have the following invited speakers:

– Dr. Matthias Rauterberg, Technische Universiteit Eindhoven, The Netherlands: "How to Design for the Unconscious"
– Prof. Ulrich Rückert, Bielefeld University, Germany: "Cognitronics: Resource-efficient Architectures for Cognitive Systems"
– Prof. Le Lu, U.S. National Institutes of Health, USA: "Towards Big Data, Weak Label and True Clinical Impact on Medical Image Diagnosis: The Roles of Deep Label Discovery and Open-Ended Recognition"

The 14th edition of the IWANN conference was organized by the University of Granada, University of Malaga, Polytechnical University of Catalonia, together with the Spanish Chapter of the IEEE Computational Intelligence Society. We wish to thank to the University of Cadiz for their support and grants.

We would also like to express our gratitude to the members of the different committees for their support, collaboration, and good work. We especially thank the local Organizing Committee, Program Committee, the reviewers, invited speakers, and special session organizers. Finally, we want to thank Springer, and especially Alfred Hofmann and Anna Kramer for their continuous support and cooperation.

June 2017
Ignacio Rojas
Gonzalo Joya
Andreu Catala

Organization

Program Committee

Leopoldo Acosta	University of La Laguna, Spain
Vanessa Aguiar-Pulido	RNASA-IMEDIR, University of A Coruña, Spain
Arnulfo Alanis Garza	Instituto Tecnologico de Tijuana, Mexico
Ali Fuat Alkaya	Marmara University, Turkey
Amparo Alonso-Betanzos	University of A Coruña, Spain
Juan Antonio Alvarez-García	University of Seville, Spain
Jhon Edgar Amaya	University of Tachira (UNET), Venezuela
Gabriela Andrejkova	University of Pavol Jozef Safarik Kosice, Slovakia
Cesar Andres	Universidad Complutense de Madrid, Spain
Miguel Angel Lopez	Lopez University of Cádiz, Spain
Anastassia Angelopoulou	University of Westminster, UK
Plamen Angelov	Lancaster University, UK
Davide Anguita	University of Genoa, Italy
Cecilio Angulo	Universitat Politecnica de Catalunya, Spain
Javier Antich	Universitat de les Illes Balears, Spain
Angelo Arleo	CNRS, Pierre and Marie Curie University of Paris VI, France
Corneliu Arsene	SC IPA SA
Miguel Atencia	University of Malaga, Spain
Jorge Azorín-López	University of Alicante, Spain
Davide Bacciu	University of Pisa, Italy
Javier Bajo	Universidad Politécnica de Madrid, Spain
Juan Pedro Bandera Rubio	ISIS Group, University of Malaga, Spain
Cristian Barrué	Technical University of Catalonia, Spain
Andrzej Bartoszewicz	Technical University of Lodz, Poland
Bruno Baruque	University of Burgos, Spain
David Becerra Alonso	University of the West of Scotland, UK
Lluís Belanche	Universitat Politecnica de Catalunya, Spain
Sergio Bermejo	Universitat Politecnica de Catalunya, Spain
Francesc Bonin	Universitat de les Illes Balears, Spain
Francisco Bonnín Pascual	Universitat de les Illes Balears, Spain
Julio Brito	University of La Laguna, Spain
Antoni Burguera	Universitat de les Illes Balears, Spain
Joan Cabestany	Universitat Politecnica de Catalunya, Spain
Inma P. Cabrera	University of Malaga, Spain
Tomasa Calvo	University of Alcala, Spain
Jose Luis Calvo Rolle	University of A Coruna, Spain

Rodolfo Garcia Bermudez   Universidad Laica Eloy Alfaro de Manabí, Ecuador
Carlos Garcia Puntonet   Granada University, Spain
Juan M. Garcia-Gomez   UPV
Francisco Garcia-Lagos   Universidad de Malaga, Spain
Jose Garcia-Rodriguez   University of Alicante, Spain
Patricio García Báez   Universidad de La Laguna, Spain
Pablo García Sánchez   University of Granada, Spain
Maribel García-Arenas   Universidad de Granada, Spain
Patrick Garda   Université Pierre et Marie Curie, Paris 6, France
Peter Gloesekoetter   Münster University of Applied Sciences, Germany
Juan Gomez Romero   Universidad Carlos III de Madrid, Spain
Juan Gorriz   University of Granada, Spain
Karl Goser   Technical University Dortmund, Germany
Bernard Gosselin   University of Mons, Belgium
Manuel Grana   University of the Basque Country, Spain
Bertha Guijarro-Berdiñas   University of A Coruña, Spain
Nicolás Guil Mata   University of Málaga, Spain
Alberto Guillen   University of Granada, Spain
Francisco Herrera   University of Granada, Spain
Álvaro Herrero   University of Burgos, Spain
Cesar Hervas   University of Cordoba, Spain
Tom Heskes   Radboud University Nijmegen, The Netherlands
Wei-Chiang Hong   Oriental Institute of Technology, Taiwan
Jose M. Jerez   Universidad de Málaga, Spain
M. Dolores Jimenez-Lopez   Rovira i Virgili University, Spain
Juan Luis Jiménez Laredo   University of Granada, Spain
Gonzalo Joya   University of Malaga, Spain
Vicente Julian   GTI-IA DSIC UPV
Fernando L. Pelayo   University of Castilla-La Mancha, Spain
Alberto Labarga   University of Granada, Spain
Raul Lara Cabrera   University of Malaga, Spain
Nuno Lau   Universidade de Aveiro, Portugal
Amaury Lendasse   The University of Iowa, USA
Miguel Lopez   University of Granada, Spain
Otoniel Lopez Granado   Miguel Hernandez University, Spain
Rafael Marcos Luque Baena   University of Málaga, Spain
Ezequiel López-Rubio   University of Málaga, Spain
Kurosh Madani   LISSI, Université Paris-Est Creteil (UPEC), France
Mario Martin   Universitat Politecnica de Catalunya, Spain
Bonifacio Martin Del Brio   University of Zaragoza, Spain
Jose D. Martin-Guerrero   University of Valencia, Spain
Luis Martí   Universidad Carlos III de Madrid, Spain
Francisco Martínez   ETEA
    Estudillo
José Luis Martínez Martínez   University of Castilla-La Mancha, Spain
José Fco. Martínez-Trinidad   INAOE

Miquel Massot	University of the Balearic Islands, Spain
Francesco Masulli	University of Genoa
Montserrat Mateos	Universidad Pontificia de Salamanca, Spain
Jesús Medina-Moreno	University of Cadiz, Spain
Maria Belen Melian Batista	University of La Laguna, Spain
Mercedes Merayo	Universidad Complutense de Madrid, Spain
Gustavo Meschino	Universidad Nacional de Mar del Plata, Argentina
Margaret Miro	University of the Balearic Islands, Spain
Jose M. Molina	Universidad Carlos III de Madrid, Spain
Augusto Montisci	University of Cagliari, Italy
Antonio Mora	University of Granada, Spain
Angel Mora Bonilla	University Malaga, Spain
Claudio Moraga	European Centre for Soft Computing, Spain
Gines Moreno	University of Castilla-La Mancha, Spain
Jose Andres Moreno	University of La Laguna, Spain
Juan Moreno Garcia	Universidad de Castilla-La Mancha, Spain
J. Marcos Moreno Vega	University of La Laguna, Spain
Susana Muñoz Hernández	Technical University of Madrid, Spain
Pep Lluís Negre Carrasco	University of the Balearic Islands, Spain
Alberto Núñez	Universidad de Castilla La Mancha, Spain
Manuel Ojeda-Aciego	University of Malaga, Spain
Sorin Olaru	Suplec
Iván Olier	The University of Manchester, UK
Madalina Olteanu	SAMM, Université Paris 1, France
Julio Ortega	Universidad de Granada, Spain
Alfonso Ortega de La Puente	Universidad Autonoma de Madrid, Spain
Alberto Ortiz	University of the Balearic Islands, Spain
Emilio Ortiz-García	Universidad de Alcala, Spain
Osvaldo Pacheco	Universidade de Aveiro, Portugal
Esteban José Palomo	University of Málaga, Spain
Diego Pardo	Barcelona Tech, Spain
Miguel Angel Patricio	Universidad Carlos III de Madrid, Spain
Alejandro Pazos Sierra	University of A Coruña, Spain
Jose Manuel Perez Lorenzo	Universidad de Jaen, Spain
Vincenzo Piuri	University of Milan, Italy
Hector Pomares	University of Granada, Spain
Alberto Prieto	Universidad de Granada, Spain
Alexandra Psarrou	University of Westminster, UK
Francisco A. Pujol	University of Alicante, Spain
Pablo Rabanal	Universidad Complutense de Madrid, Spain
Juan Rabuñal	University of A Coruña, Spain
Vladimir Rasvan	Universitatea din Craiova, Romania
Ismael Rodriguez	Universidad Complutense de Madrid, Spain
Juan A. Rodriguez	Universidad de Malaga, Spain
Sara Rodríguez	University of Salamanca, Spain

Fernando Rojas	University of Granada, Spain
Ignacio Rojas	University of Granada, Spain
Samuel Romero-Garcia	University of Granada, Spain
Ricardo Ron-Angevin	University of Málaga, Spain
Eduardo Ros	University of Granada, Spain
Francesc Rossello	University of the Balearic Islands, Spain
Fabrice Rossi	SAMM, Université Paris 1, France
Fernando Rubio	Universidad Complutense de Madrid, Spain
Ulrich Rueckert	University of Paderborn, Germany
Addisson Salazar	Universidad Politecnica Valencia, Spain
Sancho Salcedo-Sanz	Universidad de Alcalá, Spain
Albert Samà	Universitat Politècnica de Catalunya, Spain
Francisco Sandoval	Universidad de Málaga, Spain
Jose Santos	University of A Coruña, Spain
Jose A. Seoane	University of Bristol, UK
Eduardo Serrano	Universidad Autonoma de Madrid, Spain
Luis Silva	University of Aveiro, Portugal
Olli Simula	Helsinki University of Technology, Finland
Jordi Solé-Casals	Universitat de Vic, Spain
Carmen Paz Suárez Araujo	Universidad de las Palmas de Gran Canaria, Spain
Peter Szolgay	Pazmany Peter Catholic University, Hungary
Javier Sánchez-Monedero	University of Cordoba, Spain
Ricardo Tellez	Pal Robotics
Ana Maria Tome	Universidade Aveiro, Portugal
Carme Torras	IRI (CSIC-UPC)
Joan Torrens	University of the Balearic Islands, Spain
Claude Touzet	University of Provence, France
Olga Valenzuela	University of Granada, Spain
Oscar Valero	University of the Balearic Islands, Spain
Miguel Ángel Veganzones	Universidad del País Vasco (UPV/EHU), Spain
Francisco Velasco-Alvarez	Universidad de Málaga, Spain
Sergio Velastin	Kingston University, UK
Marley Vellasco	PUC-Rio
Alfredo Vellido	Universitat Politecnica de Catalunya, Spain
Francisco J. Veredas	Universidad de Málaga, Spain
Michel Verleysen	Université Catholique de Louvain, Belgium
Changjiu Zhou	Singapore Polytechnic, Singapore
Ahmed Zobaa	University of Exeter, UK

Additional Reviewers

Azorín-López, Jorge
Ballesteros Tolosana, Iris
Benítez Caballero, María José
Bermejo, Sergio

Camacho, Carlos
Camacho, David
Castillo, Pedro
Cazorla, Miguel

Cornejo Bueno, Laura
Corona, Francesco
Culita, Janetta
Danciu, Daniela
Dorado-Moreno, Manuel
Durán-Rosal, Antonio Manuel
Fuster-Guillo, Andres
García Sánchez, Pablo
Gomes, João
González Calero, Pedro
Granosik, Grzegorz
Gómez-Martín, Marco A.
Hervas, Cesar
Jimenez Linares, Luis
Lara-Cabrera, Raul
Lobo, David
Luque-Baena, Rafael M.
López-Rubio, Ezequiel

Lu, Trng Hiu
Mohammed, Salem
Nogueira, Mariela
Ortega, Francsico
Orts-Escolano, Sergio
Paderewski, Patricia
Padilla-Zea, Natalia
Palomo, Esteban José
Pucci, Rita
Ramírez Poussa, Eloisa
Riaza Valverde, José Antonio
Rodriguez-Benitez, Luis
Runceanu, Adrian
Souza Junior, Amauri Holanda
Vlad, Cristina
Volosyak, Ivan
Zarychta, Dariusz

Contents – Part II

Machine Learning Applied to Vision and Robotics

Human Activity Recognition for Health and Well-being Applications

Software Testing and Intelligent Systems

Real World applications of BCI Systems

Supervised, Non-supervised, Reinforcement and Statistical Algorithms

Contents – Part I

E-Health and Computational Biology

Human Computer Interaction

Image and Signal Processing

Mathematics for Neural Networks

Self-organizing Networks

Spiking Neurons

Artificial Neural Networks in Industry ANNI'17

Machine Learning for Renewable Energy Applications

Computational Intelligence Tools and Techniques for Biomedical Applications

Prediction of Protein Oxidation Sites

Francisco J. Veredas[1](✉), Francisco R. Cantón[2], and Juan C. Aledo[2]

[1] Dpto. Lenguajes y Ciencias de la Computación, Universidad de Málaga,
29071 Málaga, Spain
franveredas@uma.es
[2] Departamento de Biología Molecular y Bioquímica, Facultad de Ciencias,
Universidad de Málaga, 29071 Málaga, Spain
{frcanton,caledo}@uma.es

Abstract. Although reactive oxygen species are best known as damaging agents linked to aerobic metabolism, it is now clear that they can also function as messengers in cellular signalling processes. Methionine, one of the two sulphur containing amino acids in proteins, is liable to be oxidized by a well-known reactive oxygen species: hydrogen peroxide. The awareness that methionine oxidation may provide a mechanism to the modulation of a wide range of protein functions and cellular processes has recently encouraged proteomic approaches. However, these experimental studies are considerably time-consuming, labor-intensive and expensive, thus making the development of *in silico* methods for predicting methionine oxidation sites highly desirable. In the field of protein phosphorylation, computational prediction of phosphorylation sites has emerged as a popular alternative approach. On the other hand, very few *in-silico* studies for methionine oxidation prediction exist in the literature. In the current study we have addressed this issue by developing predictive models based on machine learning strategies and models— random forests, support vector machines, neural networks and flexible discriminant analysis—, aimed at accurate prediction of methionine oxidation sites.

1 Introduction

Although reactive oxygen species (ROS) are best known as damaging agents involved in aerobic metabolism [1], a more subtle approach has emerged in recent years. It is well-known that some ROS, such as hydrogen peroxide (H_2O_2), can work as effective cellular messengers [2,3] by bringing about post-traslational modifications (PTM) that produce reversible changes in the activity of proteins. Amino acids that often experience PTM are those that have a functional group that can serve as a nucleophile in the modification reaction. To this respect, cysteine and methionine, the two sulphur containing residues in proteins, are liable to be oxidized by H_2O_2. For its part, methionine is oxidized to methionine sulfoxide (MetO) by addition of oxygen to its sulphur atom. This oxidation reaction can be reverted by enzyme-catalyzed reduction reactions [4]: MetO is reduced back to methionine by methionine sulfoxide reductases, which are enzymes that

© Springer International Publishing AG 2017
I. Rojas et al. (Eds.): IWANN 2017, Part II, LNCS 10306, pp. 3–14, 2017.
DOI: 10.1007/978-3-319-59147-6_1

are present in all aerobic cells [5]. However, the role of methionine residues in cellular redox regulation remain thoroughly unexplored [6].

Methionine oxidation is a reversible covalent modification. The addition of an oxygen atom to the sulphur atom of methionine residues is able to produce changes in the physico-chemical properties of the whole protein. This, in turn, can affect the activity and stability of the protein [7]. Thus, methionine oxidation has been shown to both down-regulate [8] and up-regulate [9] protein function, through direct oxidation of specific methionine sites in the protein. Furthermore, methionine oxidation can also affect protein function indirectly by coupling oxidative signals to other sorts of PTMs, such as protein phosphorylation [10].

Different proteomic approaches have recently been inspired by the realisation that methionine oxidation may provide a mechanism to the redox-dependent modulation of protein activity and cellular mechanisms. In this way, proteome-wide studies of methionine oxidation have identified, in both Arabidopsis [11] and human [12], a large number of proteins as potential targets of oxidative signals. Moreover, these proteomic approaches have pointed out the precise sites of oxidation on the target proteins. However, these experimental efforts are considerably labor-intensive, time-consuming and expensive, thus making the development of *in silico* methods for predicting methionine oxidation sites highly desirable. In the field of protein phosphorylation, which can be considered as the most widely studied PTM, computational methods for prediction of phosphorylation sites in proteins have become very popular approaches [13]. Unfortunately, to the best of our knowledge, there are no such methods for methionine oxidation site prediction. Thus, in the current study we have addressed this issue by developing predictive models based on computational intelligence, aimed at accurate prediction of methionine oxidation sites.

2   Materials and Methods

2.1   Datasets

Data regarding methionine peptides that were oxidized in Jurkat cells stressed by H_2O_2 were taken from Table S1 in the supplementary material from reference [12]. This set was further curated to exclude protein entries that have recently been deleted from UniProt (http://www.uniprot.org). The resulting data set was formed by 1646 different proteins accounting for 2616 methionine sulfoxides. A subset of this collection, composed of 774 proteins that exhibit extensive oxidation (degree of oxidation equal or greater than 20%), will be named 'highly H_2O_2-sensitive proteins'.

Using PDB cross-references from the UniProt, we collected a list of PDB identifiers for proteins belonging to the highly H_2O_2-sensitive group. In general, since many proteins were homooligomers, most crystal structures yielded a large number of duplicated observations, which were searched for and eliminated using a R script. Eventually, after removing redundancy and filtering out low quality structures (for instance, those where the target methionine did not appear resolved), we assembled a collection of 127 unique polypeptides of known structure, containing 1118 methyonil residues, 136 of which were oxidation-prone. For each methionine, the distance of the sulfur atom to the geometric center of the aryl moiety of any aromatic residue was computed with the help of an ad doc R script that relies on the package bio3d [14]. Based on a previously established criterion [15], we considered any methionine sulfur atom within 7 Å of the aromatic ring, to be an S-aromatic motif. For each of the 1118 methionines, relevant information such as its redox status, the corresponding PDB identifier, the positions within the structure of both the analysed methionine and their closer aromatic residues, as well as their distances in Ångstroms, is provided.

2.2   Feature Extraction

For each methionine residue being analysed we have extracted several properties, both from the primary and tertiary structure of the protein.

54 independent variables (input):

- 40 protein's primary-structure features:
  - 40 distance variables:
    * NT_X: distance (number of positions in the primary structure) from the analysed methionine to the closest X residue toward the N-terminus.
    * CT_X: distance (number of positions in the primary structure) from the analysed methionine to the closest X residue toward the C-terminus.
- 14 protein's tertiary-structure features:
  - 9 inter-atomic distance variables, with X being either Y (tyrosine), F (phenylalanine), W1 (tryptophan ring #1) or W2 (tryptophan ring #2):
    * Xd: distance in Å between the S and the centre of the nearest ring of the X aromatic residue.
    * Xn: number of X residues at a distance <7 Å.
    * nBonds: number of S-aromatic bonds in which the analysed methionine takes part.
  - 2 accessibility properties:
    * SASA: solvent accessible surface area of the methionine residue.
    * SASA_SD: solvent accessible surface area of the sulphur atom of the methionine residue.
  - 2 entropy variables:
    * H2: Shannon base-2 entropy.
    * H21: Shannon base-21 entropy.
  - 1 frequency variable:
    * fM: relative frequency of methionine at the position of analysed methionine, after multiple sequence alignment (MSA).

Dependent variable (output):

- oxidable: binary variable indicating whether the methionine is oxidized (>20%) or not.

A more detailed explanation from some of the features in the list above needs to be given. Thus, to extract the NT_X-CT_X variables, given an amino acid, for instance alanine, we searched around the methionine of interest for the closest alanine residue toward the N-terminus and for the closest alanine toward the C-terminus. Once these alanine residues were found in the primary structure, we counted the positions away from the methionine being analysed (NT_X and CT_X variables, respectively). This operation was repeated for each of the 20 proteinogenic amino acids, accounting for a 40- dimensional vector. Missing values due to the absence of any particular amino acid either toward the N-terminus or the C-terminus, were imputed by using the protein length.

In a recent work we reported that methionyl residues forming part of an S-aromatic motif are less prone to be oxidized [16]. Therefore, 9 additional features related to this non covalent bond were used. Concretely, Xd was defined as the distance in ångströms between the sulfur atom from the analysed methionine and the nearest X aromatic residue. The variable Xn informs about the number of X aromatic residues at a distance <7 Å from the methionine. The feature nBonds was computed according to:

$$nBonds = \sum_{X \in \{Y, F, W_1, W_2\}} Xn.$$

Four additional features were related to the conservation of the considered methionine during evolution. To assess these features, besides the human sequence, the orthologous proteins from *Pan troglodytes*, *Gorilla gorilla*, *Rattus norvegicus*, *Bos taurus*, *Gallus gallus*, *Xenopus tropicalis* and *Danio rerio* were aligned. These alignments were used to compute the Shannon entropy according to the equation:

$$H2 = -\sum_{i=1}^{2} f_i log_2(f_i),$$

and

$$H21 = -\sum_{i=1}^{21} f_i log_{21}(f_i),$$

where f_i is the relative frequency of the symbol i at the analysed position across the alignment. Thus, for instance, f_M stands for the relative frequency of methionine. The logarithmic base for $H21$ was taken 21 because in addition to the 20 proteinogenic amino acids, the symbol '-' was considered when indels were present. For each analysed methionine, the variables mean.entropy and sd.entropy were computed as the mean and standard deviation, respectively, of the entropy determined at all the positions of the corresponding protein.

For its part, the solvent accessible features (SASA and SASA_SD) of each methionine residue was computed using the POPS program [17]. This software, which is freely available at http://mathbio.nimr.mrc.ac.uk, also provides the accessibility, which is defined as the fraction of the residue surface that is exposed to the solvent.

For those features other than the 40 primary-structure characteristics (i.e. NT_X and CT_X) missing values were imputed by means of a machine learning approach: a bagging tree model for each predictor was fitted as a function of all the others input variables. Though this method has much higher computational cost than other imputation techniques, such as k-nearest neighbour imputation or imputation via medians, among others, it stands out for being accurate [18].

The data set supporting the results of this article can be downloaded from https://github.com/fveredas/PredictionOfMethionineOxidationSites.

2.3   Machine Learning Methods

Random forests (RF) are used in this study to design predictive models of protein oxidation. RFs are ensemble machine learning methods for classification that function by constructing a large pool of decision trees during the training phase, then giving an output that is the mode of the classes given by the individual trees in the pool. The method combines Breiman's 'bagging' idea and the random selection of features (i.e. predictor-set split) in order to construct a collection of decision trees with controlled variation [19].

The quantification of the variable importance is a crucial issue to interpret data and understand underlying phenomena under the methionine oxidation scenario. RFs use two different measures to estimate variable importance: the accuracy importance (AI) and the Gini-index Importance (GI). The AI of a variable is calculated as the average decrease in accuracy on the OOB samples when the values of the respective predictor are randomly permuted. The GI uses the decrease of Gini-index (impurity) after a node split as a measure of variable relevance. The average decrease in Gini-index over all trees in the RF defines the GI.

To account for the potential of RFs as efficient models for protein oxidation prediction, comparisons with other classification models are mandatory. For this purpose, two machine learning approaches, i.e. support vector machines (SVM) [20] and neural networks (NN) [21], as well as a more classical statistical approach, i.e. flexible discriminant analysis (FDA) [22], have been comparatively used.

Model Tuning. For RF model-fitting in our experiments for Met oxidation, the only sensible tuning parameter would be the number of variables (predictors) randomly sampled as candidates at each split (usually known as mtry), but it has been fixed to the optimal recommended value $\lfloor \sqrt{number\ of\ predictors} \rfloor =$ 7 [23]. For its part, the RF parameter number of trees to grow has been fixed to 1000 trees to ensure that every input pattern gets predicted at least a few times [24].

For SVMs, a Gaussian radial basis function (RBF) kernel $k(x, x') = e^{-\sigma \|x - x'\|^2}$ was used (being k a function that calculates the inner product

$\langle \Phi(x), \Phi(x') \rangle$ of two vectors x, x' for a given projection $\Phi : X \to H$). The problem of model selection (parameter tuning) is partially addressed by an empirical observation for the Gaussian RBF kernel, where the optimal values of the hyperparameter σ are known to lie in between the 0.1 and 0.9 quantile of the $||x - x'||$ statistics [25]. A sample of the training set is used to estimate these quantiles, where any value within the quantile interval results in good performance. This way, σ parameter is automatically estimated. Additionally, the optimal hyperparameter *cost*, that represents the cost of constraints violation and stands for the 'C'-constant of the regularisation term in the Lagrange formulation, is tuned as the one of 12 incremental values in $\{2^i\}_{i=-2}^{9}$ that optimises the area under the ROC curve (AUC) of the SVM classifier.

Single-hidden-layer feed-forward NNs are also constructed and trained with different combinations of parameters to search for the best performance rates in the prediction of methionine oxidation. Optimisation of the NNs is done via the quasi-Newton method BFGS (also known as a variable metric algorithm) [26, 27]. The network size (i.e., number of *hidden units*) and *weight* decay are the parameters being tuned, selecting that combination of values giving the highest AUC. All the trained NNs have a number of outputs that is equal to the number of classes (i.e. $n = 2$), and a *softmax* output stage. Weights are randomly initialised, and maximum number of epochs was fixed to 100 [27].

Many classification models, such as ridge regression, the lasso, or adaptive regression splines (MARS) [22], can be extended to create discriminant variables. In particular, MARS can be used to create a set of discriminant functions that are non-linear combinations of the original predictors. This conceptual paradigm is referred to as flexible discriminant analysis (FDA) [18]. In this study we have followed a bagging approach for FDA, which uses MARS basis functions to compute a FDA model for each bootstrap sample. The only parameter to be tuned for model fitting was the maximum number of terms (including intercept) in the pruned model [22] (usually known as nprune), which is used to enforce an upper bound on the model size. The optimal nprune parameter was chosen as that in the range $\{1, \ldots, 25\}$ that gave the highest AUC rate. The maximum degree of interaction (Friedman's mi) was fixed to 1, thus an additive model (i.e., no interaction terms) was used.

Resampling Methods for Model Fitting. The data set has been divided into three independent sets, 70% (96 'positive'; 688 'control') patterns for training, 10% (14 'positive'; 98 'control') patterns for evaluation (this pattern set is used to compute the optimal threshold for the ROC curves) and, finally, 20% (26 'positive'; 196 'control') for testing. In order to preserve the unbalanced nature of the original class distribution within the splits a stratified random sampling strategy was used. To estimate the efficacy of the prediction model across the training set, performance measures (AUC, accuracy, sensitivity and specificity) of the out-of-bag (OOB) samples for 10-fold cross-validation with 5 repetitions (50 re-samplings) were calculated and the mean and standard deviation of those rates are summarised. The entire training set is used to fit a final model and its performance was finally measured on the testing set.

In our study class imbalance is inherent to the procedure being followed for data acquisition (see Sect. 2.1): of the complete set of methionine residues found in the 127 polypeptides analysed, only 136 out of 1118 appeared as oxidised, i.e. a mere 12%. Different approaches to counteracting the negative effects of class imbalance have been proposed in the literature [18], with model tuning (using metrics alternative to accuracy such as ROC, Cohen's Kappa or sensitivity), adjusting of prior probabilities, cost-sensitive training, ROC-curve alternative cutoffs, or sampling methods, among others. In this study a combination of the two latter has given the best results. Prior to model fitting, we have used the synthetic minority over-sampling technique (SMOTE) [28] to get a more balanced training dataset. The general idea of this method is to artificially over-sample the minority class (i.e. 'oxidised' class) patterns by generating new samples using the k-nearest neighbours (KNNs) of these cases ($k = 5$ in our experiments). Furthermore, the majority class (i.e. 'not oxidised' class) cases are also under-sampled.

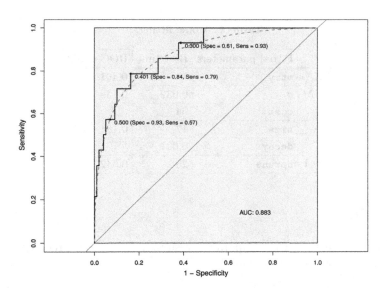

Fig. 1. ROC curve with different thresholds. ROC curve of the RF-Smote classifier on the evaluation data set. Different thresholds have been highlighted in the curve, along with their corresponding specificity and sensibility rates: 0.5 (original), 0.401 (alternative) and 0.3. The theoretical maximal area of reference (i.e. AUC = 1) has been also coloured grey. Dotted grey line represents the smoothed ROC curve.

On the other hand, after model training using the SMOTE samples, we have used the ROC curve to determine alternative cutoffs for the probabilities predicted by the model. Using this ROC curve, an appropriate balance between sensitivity and specificity can be determined. Although several techniques do exist for determining a new cutoff, the more general approach is to find the point on the ROC curve that is closest (i.e., the shortest distance) to the perfect model (with 100% sensitivity and 100% specificity), which is associated with the upper left corner of the plot [5]. To determine this cutoff point without distorting the results obtained from the final testing dataset, an independent evaluation dataset has been used (see above). In Fig. 1 the ROC curve obtained from the RF classifier (trained with the SMOTE samples) on the evaluation dataset is shown together with the computed alternative cutoff (0.401), as well as the original 0.5 and the 0.3 (shown for comparison purposes) cutoffs. As it can be observed in the figure, the alternative cutoff gives the best balance between sensitivity and specificity.

Table 1 shows the list of parameters being fitted. For each predictive model, the best values for the fitted parameters are computed as those giving the highest averaged AUC via 10-fold cross-validation on the training dataset. The ROC cutoffs obtained from the evaluation dataset after model fitting and training are also shown in the table.

Table 1. Model fitting.

	Fitted parameters	Best tune	ROC cutoff
RF	mtry	7	0.401
SVM	σ	0.032	0.138
	cost	64	
NN	size	17	0.181
	decay	0.001	
FDA	nprune	25	0.542

3   Results

In the following sections the comparative results from the four predictive models analysed in this study are presented.

3.1   Predicting Methionine Oxidation in Human Proteins

Table 2 shows the performances rates for the four predictive models analysed when applied on the testing dataset. While accuracy levels are above ∼70% for all the classifiers, it is the RF the model that gives the best results, with the highest accuracy rate as well as the best adjusted balance between sensitivity and specificity.

In Fig. 2 the 30 most important variables as estimated by the RF on the training set are shown along with their averaged decrease in Gini-index (see Sect. 2.3).

Table 2. Performance rates of predictive models.

	AUC	Accuracy	Sensitivity	Specificity
RF	0.8459	81.5315	76.9230	82.1429
SVM	0.7501	69.8198	76.9231	68.8776
NN	0.7388	72.9730	61.5385	74.4898
FDA	0.7329	76.5766	50.0000	80.1020

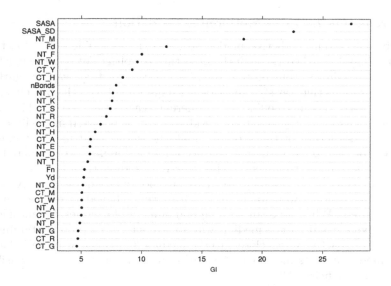

Fig. 2. Variable Importance. The 30 most important variables ordered by GI (averaged decrease in Gini-index) as estimated by the final RF model on patterns in the training dataset.

Predicting oxidation with primary-structure characteristics. Table 3 shows the performance results obtained by training the predictive models with only those protein primary-structure characteristics, i.e. using the NT_X-CT_X distance features (see Sect. 2.2) as the only input variables for the classifiers. The same sampling, tuning and training procedures explained in Sect. 2 are followed again. As can observed in the table, the RF gives again the best performance results on the testing set, followed by far by the SVM. Comparing Tables 2 and 3, i.e. comparing the results obtained when using the complete enriched set of primary and tertiary variables to those that used the primary features only, we can observe a significant decrease of the efficacy rates when the tertiary-structure variables are omitted.

Table 3. Using primary-structure features.

	AUC	Accuracy	Sensitivity	Specificity
RF	0.7805	72.5225	65.3846	73.4694
SVM	0.5389	55.8559	53.8462	56.1225
NN	0.5104	41.8919	57.6923	39.7959
FDA	0.6303	52.7027	57.6923	52.0408

3.2   Predicting *In Vitro* Methionine Oxidation

To double-check our predictive model we tested it on a different dataset. For this purpose, we searched through the literature to collect data on the reactivity of methionyl residues from protein pharmaceuticals. In that way, we gathered data for 8 proteins that satisfy the following requirements: (i) the protein should contain at least two methionine with different reactivities, (ii) the kinetics of oxidation in vitro with H_2O_2 of these residues must be reported in the literature, and (iii) the structure of the protein should be known and its PDB must be available. For each protein the reactivities of its methionines were ordered from lower to higher. Those residues showing reactivities lower than the median were labelled as 'oxidised', all others were treated as 'not oxidised' residues. A total of 35 methionine residues were found, of which 15 were labelled as 'oxidised' and 20 as 'not oxidised'. When we dealt with an odd number of residues, the methionine with reactivity equal to the median was sorted into the group containing the residue whose reactivity was closest to the median. For this new protein dataset, our RF model predicts methionine oxidation with accuracy 74.29%, sensibility 46.67% and specificity 95%.

4   Conclusions

Predictive models of methionine oxidation have been designed and tested in this study. Our results show high accuracy rates, with balanced sensitivity and specificity. The best efficiency results were obtained with random forests, while support vector machines and networks behaved worse, in general.

From the 54 predictors used in the design of the predictive models, some tertiary-structure ones, such as *solvent-accessibility area* have been identified as those with the highest contribution to the predictive power of the random forest model. Moreover, counting on a reliable computational tool to predict methionine oxidation could stimulate further investigation to determine the role of sulphoxidation in cellular oxidative signalling.

Our predictive models also include another important set of characteristics that reinforces their predictive power: *methionine-aromatic motif characteristics*. The *distance in Å between the methionine and the nearest aromatic residue* stands out as one of the most important variables for the predictive mosdels. This result, together with the highest relevance as predictive variable shown by the *solvent accessible surface area of the methionine residue*, emphasises the complex correlation between structural properties and methionine oxidation.

Acknowledgments. This work was partially supported by the Universidad de Málaga and project TIN2014-58516-C2-1-R, MICINN, Plan Nacional de I+D+I.

References

1. Aledo, J.C.: Life-history constraints on the mechanisms that control the rate of ROS production. Curr. Genom. **15**, 217–230 (2014)
2. Collins, Y., Chouchani, E.T., James, A.M., Menger, K.E., Cochemé, H.M., Murphy, M.P.: Mitochondrial redox signalling at a glance. J. Cell Sci. **125**, 801–806 (2012)
3. Veredas, F.J., Cantón, F.R., Aledo, J.C.: Methionine residues around phosphorylation sites are preferentially oxidized in vivo under stress conditions. Sci. Rep. **7**, 40403 (2017)
4. Arnér, E.S., Holmgren, A.: Physiological functions of thioredoxin and thioredoxin reductase. Eur. J. Biochem. **267**, 6102–6109 (2000)
5. Kim, H.Y.: The methionine sulfoxide reduction system: selenium utilization and methionine sulfoxide reductase enzymes and their functions. Antioxid. Redox Sig. **19**, 958–969 (2013)
6. Kim, G., Weiss, S.J., Levine, R.L.: Methionine oxidation and reduction in proteins. BBA-Gen. Subj. **1840**, 901–905 (2014)
7. Jacques, S., Ghesquière, B., Breusegem, F., Gevaert, K.: Plant proteins under oxidative attack. Proteomics **13**, 932–940 (2013)
8. Härndahl, U., Kokke, B.P., Gustavsson, N., Linse, S., Berggren, K., Tjerneld, F., Boelens, W.C., Sundby, C.: The chaperone-like activity of a small heat shock protein is lost after sulfoxidation of conserved methionines in a surface-exposed amphipathic alpha-helix. Biochim. Biophys. Acta **1545**, 227–237 (2001)
9. Drazic, A., Miura, H., Peschek, J., Le, Y., Bach, N.C., Kriehuber, T., Winter, J.: Methionine oxidation activates a transcription factor in response to oxidative stress. Proc. Natl. Acad. Sci. USA **110**, 9493–9498 (2013)
10. Rao, R.S.P., Møller, I.M., Thelen, J.J., Miernyk, J.A.: Convergent signaling pathways—interaction between methionine oxidation and serine/threonine/tyrosine O-phosphorylation. Cell Stress Chaperones **20**, 15–21 (2014)
11. Jacques, S., Ghesquière, B., Bock, P.J., Demol, H., Wahni, K., Willemns, P., Messens, J., Breusegem, F., Gevaert, K.: Protein methionine sulfoxide dynamics in arabidopsis thaliana under oxidative stress. Mol. Cell. Proteomics **14**, 1217–1229 (2015)
12. Ghesquière, B., Jonckheere, V., Colaert, N., Van Durme, J., Timmerman, E., Goethals, M., Schymkowitz, J., Rousseau, F., Vandekerckhove, J., Gevaert, K.: Redox proteomics of protein-bound methionine oxidation. Mol. Cell. Proteomics **10**, M110.006866 (2011)
13. Datta, S., Mukhopadhyay, S.: A grammar inference approach for predicting kinase specific phosphorylation sites. PLoS One **10**, e0122294 (2015)
14. Grant, B.J., Rodrigues, A.P.C., ElSawy, K.M., McCammon, J.A., Caves, L.S.D.: Bio3D: an R package for the comparative analysis of protein structures. Bioinformatics **22**, 2695–2696 (2006)
15. Valley, C.C., Cembran, A., Perlmutter, J.D., Lewis, A.K., Labello, N.P., Gao, J., Sachs, J.N.: The methionine-aromatic motif plays a unique role in stabilizing protein structure. J. Biol. Chem. **287**, 34979–34991 (2012)
16. Aledo, J.C., Cantón, F.R., Veredas, F.J.: Sulphur atoms from methionines interacting with aromatic residues are less prone to oxidation. Sci. Rep. **5** (2015)

17. Cavallo, L.: POPS: a fast algorithm for solvent accessible surface areas at atomic and residue level. Nucleic Acids Res. **31**, 3364–3366 (2003)
18. Kuhn, M., Johnson, K.: Applied Predictive Modeling. Springer, New York (2013)
19. Breiman, L.: Random forests. Mach. Learn. **45**, 5–32 (2001)
20. Vapnik, V.: Statistical Learning Theory. Wiley, New York (1998)
21. Ripley, B.D.: Pattern Recognition and Neural Networks. Cambridge University Press, Cambridge (2007)
22. Friedman, J.H.: Multivariate adaptive regression splines. Ann. Stat. **19**, 1–67 (1991)
23. Díaz-Uriarte, R., Alvarez de Andrés, S.: Gene selection and classification of microarray data using random forest. BMC Bioinform. **7**, 3 (2006)
24. Liaw, A., Wiener, M.: Classification and regression by randomForest. R News **2**, 18–22 (2002)
25. Karatzoglou, A., Smola, A., Hornik, K., Zeileis, A.: kernlab - an {S4} package for kernel methods in {R}. J. Stat. Softw. **11**, 1–20 (2004)
26. Nash, J.C.: Compact Numerical Methods for Computers: Linear Algebra and Function Minimisation, 2nd edn. CRC Press, New York (1990)
27. Venables, W.N., Ripley, B.D.: Modern Applied Statistics with S, 4th edn. Springer, New York (2002)
28. Chawla, N.V., Bowyer, K.W., Hall, L.O., Kegelmeyer, W.P.: SMOTE: synthetic minority over-sampling technique. J. Artif. Intell. Res. **16**, 321–357 (2002)

Neuronal Texture Analysis in Murine Model of Down's Syndrome

Auxiliadora Sarmiento[1]([✉]), Miguel Ángel Fernández-Granero[2], Beatriz Galán[3],
María Luz Montesinos[3], and Irene Fondón[1]

[1] Department of Signal Theory and Communications, University of Seville,
C/Descubrimientos s/n, 41092 Seville, Spain
sarmiento@us.es
[2] Department of Systems and Automatic Engineering,
Electronic Technology and Electronics, University of Cádiz,
Campus Universitario Río San Pedro s/n, 11510 Puerto Real, Cádiz, Spain
[3] Department of Medical Phisiology and Biophysic, University of Seville,
Avda. Dr. Fedriani s/n, 41009 Seville, Spain

Abstract. An alteration of neuronal morphology is present in cognitive neurological diseases where learning or memory abilities are affected. The quantification of this alteration and its evolution by the study of microscopic images is essential. However, the use of advanced and automatic image processing techniques is currently very limited, focusing on the analysis of the morphology of isolated neurons. On this article we present a new methodology, based on texture analysis, to characterize the global distribution of different neural patterns in immunofluorescence images of brain tissue sections, where the neurons can be visualized as they are really distributed. We apply the technique to mice brain tissue section dividing them into two classes: Ts1Cje Down's syndrome model and wild type, free of this neurodegenerative disease. Taking into account CA1 region of the hippocampus, we calculate and compare several state of the art texture descriptors that are subsequently classified using machine learning techniques. Achieving a 95% of accuracy, the assumption that texture characterization is relevant to quantify globally morphological alterations in the neurons, seems to be demonstrated.

Keywords: Texture analysis · Down's syndrome · Pattern recognition · Machine learning

1 Introduction

One of the cognitive and neurodegenerative diseases that has awaken most interest in the scientific field is Down syndrome (DS). DS, originated by a trisomy of the human chromosome 21, is the most frequent cause of intellectual genetic disability. Cognitive neurological diseases, where memory and the learning capability is affected, along with neurodegenerative diseases present a generalized neural morphological alteration. This alteration, if correctly quantified, could

© Springer International Publishing AG 2017
I. Rojas et al. (Eds.): IWANN 2017, Part II, LNCS 10306, pp. 15–25, 2017.
DOI: 10.1007/978-3-319-59147-6_2

provide useful information, specially when assessing the therapeutic potential of drugs aimed at recovery of function and neuronal morphology.

Currently, an objective quantification of the so explained alterations is performed by analyzing the dendritic morphology of isolated neurons. One of the most used techniques to this purpose is the Sholl Analysis [1]. In this method, the dendritic branching pattern that is, the ratio between the number of intersections of the dendrites per unit area and the distance between them and the center of the soma, is calculated. In order to perform this kind of analysis, it is essential to obtain images of isolated neurons. Generally this is achieved by immunofluorescence imaging of low density neuronal cultures, or by studying a kind of tissue staining, Golgi staining for instance, that allows the visualization of just a reduce number of neurons presented in the tissue. This use of neuronal cultures is widespread in the scientific field, although it is an experimental approach to reality. Despite this, in DS study, these staining techniques have been proved to be very useful in the morphogenesis of hippocampal neurons of murine models. As an example, the scientific group led by Dr. Montesinos, director of the Laboratory of Synaptic Local Translation (SLTL) at the University of Seville, has detected morphological differences in dendritic arborization among cultured neurons belonging to Ts1Cje mice, a model of DS, and control mice or wild type (WT), which are free of this neurodegenerative disease [2–6].

On the other hand, staining of tissue sections allows only the visualization of the dendritic structure of certain neurons, usually 5–10% of the total amount of neurons present in the tissue. The actual mechanism that causes the staining of only these neurons remains unknown compromising the results of the studies based on it.

Opposite to these traditional techniques, we propose studying the dendritic pattern in a global way using advanced techniques of image analysis in immunofluorescence images of histological sections of tissues, where all neurons can be visualized as they are distributed in reality. These images are rich on texture information and posses characteristic architectures that could be studied with pattern recognition methods. An example of an immunofluorescence image of a murine hippocampus section is shown in Fig. 1, in which different parts can be differentiated based only on the texture of each of these zones.

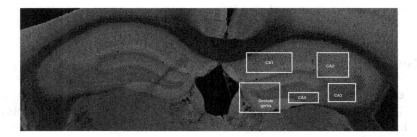

Fig. 1. Confocal image of a coronal section of the murine hippocampus region and its representative areas.

The present research has focused on the study of immunofluorescence images of hippocampal sections to study the neuronal texture of control(wild-type) and trisomic Ts1Cje mice. Due to the morphological difference in cultured isolated neurons, it seems direct to infer that there will also be an overall morphological difference on the sections of neuronal tissues. In the particular case of the CA1 region of the hippocampus, these global morphological differences give rise to patterns that are similar to some well-known patterns present in the traditional Brodatz database, a key image database widely used by the pattern recognition community to validate their algorithms. This similarity can be seen in Fig. 2. However, in most of the images analyzed in this research, the difference between the texture of the CA1 region in WT and DS images is not as clear or appreciable at a glance. In order to study the use of texture descriptors as a way to quantify global morphology, we have designed a series of experiments with the goal of automatically classify images of the hippocampal CA1 region into two categories, WT and DS.

(a)   (b)

(c)   (d)

Fig. 2. Representative images of DS-CA1 region (a), WT-CA1 region (b), Brodatz Straw D15 (c), and Brodatz Grass D9 (d) cases.

2   Related Works

Texture is a fundamental parameter in the description of images, as it provides a measure of properties such as smoothness, roughness and regularity [7]. Unfortunately, the amount of research on texture characterization of biological images is limited and usually focused on magnetic resonance and fluorescence microscopy images.

Within texture image research, Haralick texture descriptors [8] are fundamental. They consist of 14 texture features derived from the so-called Gray Level Co-occurrence Matrix (GLCM) matrix. These descriptors have been used for pattern characterization in fluorescence microscopy images of HELA cells [9] as well as to model the texture of H1-60 cell nuclei [10]. They have also been successfully used in the classification of meniangiomas [11], while in [12] where significant changes in mean temporal lobe texture have been detected in patients with Alzheimer's disease. Recently, other statistical descriptors have been developed improving the performance of GLCM-based descriptors. For instance, in [13] texture characterization is performed on the based of the gray level zonal-size matrix (GLSZM), with an application to the classification of HEp-2 cells.

Other texture descriptor that has been widely accepted in recent years is the so-called Local Binary Patterns (LBP) introduced in [14] for which there are different variants. In the case of biomedical images, the most interesting are Median Binary Pattern [15] and Local Ternary Patterns [16]. An application of these techniques in combination with other texture-based dispersed scatter descriptors can classify images taken with confocal fluorescence microscopy by fibroid lung cancer [17].

Other texture-based methods for image characterization are Gabor filters [18] and Gauss-Markov models [19]. In [20], texture information using the Gabor filter bank is used for automated segmentation of neurons in high-content scanning or High Content Screening (HCS) images.

To our knowledge, the only work in which histological sections of tissue are globally analyzed is [21]. However, in this article, only orientation and anisotropy characteristics are studied by tensor structure analysis, obtaining the same measures that the ones provided by diffusion tensors. On the contrary, our proposal analyses for the first time different texture descriptors, with the objective of obtaining new protocols that allow the objective quantification of the global neuronal morphology.

3   Materials and Methods

This section describes the stages of the proposed method. Initially, we detail the collection and selection of hippocampal images. Next, we describe the preprocessing techniques implemented, ending with the description of the different texture descriptors used in this work. The proposed tool has been developed with MATLAB®R2015a software (The MathWorks Inc., Natick, MA). A block diagram of the system is shown in Fig. 3.

Hippocampal Images Dataset. The images were taken at the Center for Research, Technology and Innovation of the University of Seville (CITIUS) with a ZEISS LSM 7 DUO spectral confocal spectral scanning microscope. Fifteen hippocampal histological preparations were imaged, with six of them having a coronal section. For the histological preparations, the animals were anesthetized and subsequently perfused with paraformaldehyde to fix the brain tissue. The brains were then sectioned using a vibratome. Finally the slices were subjected to labeling of neuronal somatodendritic structures by immunofluorescence, using specific antibodies against the MAP2 protein. A stack of images composed of twelve focal planes was taken to each preparation. The images obtained were

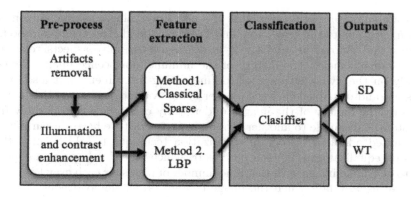

Fig. 3. Block diagram of the proposed method.

in .czi format. These images were converted to .bmp format, a lossless format that facilitates their processing regardless of the operating system. Subsequently, within each hippocampus sample, we selected the focal plane of the stack that had the best contrast and sharpness in the region of interest, and the remaining planes were discarded from the study to avoid correlated data. Also, we discarded images of preparations that were not of sufficient quality.

The final dataset consisted of nine images (one of them with a coronal section), containing a total of ten samples of hippocampus: five of them belonging to the trisomic class, and five others to the wild type. Each initial hippocampus image was converted to gray scale. Then, the CA1 area was manually selected. Finally the selection was rotated in order to visualize the CA1 area horizontally.

3.1   Pre-process

In order to obtain the most relevant texture characteristics, it is necessary to determine a protocol for the pre-processing of the images in order to normalize the structural characteristics of the different parts of the brain to be studied. In the case of the CA1 region, it is necessary first to address the elimination of artifacts, where artifacts are understood as the presence of blood vessels in the images. These blood vessels appear as elliptical areas characterized by a low level of gray. The presence of these vessels is not a differentiating characteristic between DS and WT type, and therefore should not influence the characterization of the texture. The solution adopted in our proposal is to detect the vessels thresholding the image and then restore the detected vessels using the algorithm Fast Image Inpainting [22]. This algorithm, uses as a mask of the undesired points the result obtained by the thresholding procedure. Then, values of the pixels belonging to the mask are replaced propagating the grey level information of their neighbours in a direction from the edges of the vessels to their inner part. The pixels values replacement is performed in a way consistent with human perception to avoid blurred patches in the resulting images. Therefore, the inpainted vessels are substituted in a way that the new gray value and gradient on that location extrapolate the gray value and gradient outside the neighbourhood.

After the removal of artifacts it is necessary to study aspects related to contrast enhancement. The quality of the immunofluorescence images depends not only on the dynamic range of the measuring instrument, but also on the expression and distribution of the marker being evaluated, among other factors. Therefore, the quality of the images under study could vary considerably from one histological sample to another. To improve contrast we have chosen the contrast limited adaptive histogram equalization technique (CLAHE) [23]. Basic histogram equalization (HE) is a process by which pixels' values are mapped in order to obtain an image with the same number of pixels for each gray level. That procedure usually fails when the image content is not homogeneous. To avoid this, improving local contrast and edges definiton, adaptive histogram equalization (AHE) performs a basic HE on neighbourhoods of the pixels instead of considering the whole image at once. The amplification of intensity of the amount of contrast enhancement depends on the slope of the cumulative distribution of grey level function, that is, it depends on the value of the histogram at that pixel value. It causes the amplification of noise in homegeneous regions that tries to be overcomed by CLAHE. This equalization procedure limits de amount of amplification of AHE to certain levels. An example of application of the pre-processing techniques in one of the study images is shown in Fig. 4.

Finally, in order to obtain a larger sample size for both WT and DS, each region of interest was divided into sixteen non-overlapping 50 square blocks. This procedure is generally adopted by texture analysis techniques.

3.2   Feature Extraction Procedure

In this work we have evaluated two groups of texture descriptors. Firstly, we use a set descriptors that combines classic ones that have been proven to be efficient in the past and that are highly validated along to other more recent ones that exploit the sparseness nature of the texture features (method 1) Secondly we use Local Binary Patterns (method 2), descriptors with high impact in state of the art techniques.

Method 1. Classical and Sparse Descriptors. Due to the great robustness and proved performance for image classification, we have decided to use classic texture descriptors along with a new set of features that try to take advantage of the sparsity property of the texture for the characterization of the regions of interest. The resulting feature vector is therefore formed by:

1. First order statistical descriptors: mean, standard deviation, asymmetry or skewness coefficient, kurtosis and entropy.
2. The particularization of the fractal dimension called the Hausdorff dimension [24].
3. Haralick texture descriptors [8] from the co-occurrence matrix.
4. Mean and variance of the sparse texture vector described in [25,26] using a Gaussian mixture of five components.

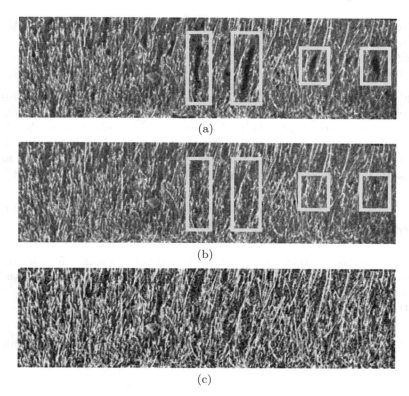

Fig. 4. (a) Original region of interest where Larger blood vessels are highlighted in yellow, (b) image obtained after applying the algorithm of Fast Image Inpainting and (c) image obtained after applying the CLAHE algorithm for contrast enhancement. (Color figure online)

Method 2. Local Binary Patterns. Because of their low computational complexity and their discriminative power, LBP-based texture methods have become very popular in recent years [27]. Among all of the existing LBP variants, a circular and rotation-invariant code [14] was selected, with a neighbourhood of eight pixels and a radius equal to one. For each block, thirty-six LBP codes were obtained. The characteristic vector was formed by the histogram of these thirty-six codes.

3.3 Training and Classification

Once the feature vector is constructed for each block of the region of interest, we proceeded to perform a binary classification step. For each method the following classifiers were used:

1. Support Vector Machine (SVM) [28]: in its versions of linear, quadratic, cubic and Gaussian kernel).

2. K-Nearest Neighbor (KNN) [29]: in their versions of cosine and cubic distance.
3. Complex Tree (a single decision tree).
4. Bagged Tree [30].
5. Random Forests [31].

We have used an external validation method to assess the quality of all tested classifiers. For that, we selected for each type of image, DS and WT, 3 complete CA1 regions, that are in total 48 blocks, for the training stage of the classifiers, and 2 CA1 regions, that is 32 blocks, for the test stage. This is mainly due to the fact that in texture images there may be a high correlation between areas of the same image, so if we do not use complete CA1 regions, we could get false results.

4   Results

The results of classification obtained with the different classifiers for each developed method are shown in Table 1. The metric used to verify the method studied is the accuracy in the DS or WT classification. The "Classification Learner" tool of Matlab 2015 software was used for the simulations.

Table 1. Results of the classification in terms of accuracy (%). The classifier with the best result has been highlighted in bold for each method.

Classifier	Method 1 Accuracy (%)	Method 2 Accuracy (%)
SVM (lineal kernel)	**96.88**	92.18
SVM (quadratic kernel)	93.75	90.62
SVM (cubic kernel)	93.75	89.06
SVM (gaussian kernel)	92.19	90.62
k-NN (cosine distance)	75.00	**95.62**
k-NN (cubic distance)	93.75	89.06
Complex Tree	76.56	76.56
Bagged Tree	85.94	95.31
Random Forest	89.06	93.75

Among all of the possible combinations, the best result, 96.88%, is obtained with method 1 (classical and dispersed texture descriptors) using SVM in its linear kernel version. In the case of method 2, which uses the LBP texture descriptors, the best result, obtained with k-NN cosine, is slightly lower, 95.62%, but still constitutes a high success rate.

5   Discussion

The main objective of this research was to verify the use of texture descriptors to discern neuronal morphological differences globally in sections of brain tissue between murine models of Down syndrome and control mouse subjects. For this, an experiment has been carried out where an image of the region of interest in DS or WT is classified based only on texture descriptors using several recent machine learning techniques. Two sets of different texture descriptors have been used, obtaining high percentages of success with both of the methods studied. From the results obtained it can be inferred that there is a texture pattern that characterizes the CA1 region of the hippocampus in murine models of DS and WT. In addition, the texture of the CA1 hippocampal region of DS mice is altered in relation to the texture of the same region in control subjects. This alteration was foreseeable due to the dendritic morphological differences observed in neurons isolated in culture in previous studies carried out in the Laboratory of Synaptic Local Translation of the University of Seville. However, having been able to identify the type of mouse to which a particular image belongs, based solely on the texture opens a new paradigm of investigation that can be useful for the study of many cognitive and neurodegenerative diseases.

Despite the good results obtained, we believe that it is necessary to improve some aspects of this research. In the first place, it would be necessary to have a larger image base to be able to study the complete CA1 region, without having to divide it into smaller pieces. We are currently working in collaboration with the SLTL of the University of Seville to perform a new imaging in new histological preparations. Secondly, it would be interesting that the selection and trimming of the ROI (the CA1 region) within each image was done completely automatically. In this way, we would eliminate a task that is tedious and usually consumes a lot of time to the specialist.

Finally, in order to develop a more complete investigation, we will analyze other areas of the brain. By quantifying and characterizing the distribution of different neuronal patterns in different areas of the nervous system, it will be possible to extract relevant conclusions at the pre-clinical level about the effect of certain compounds on the recovery of global neuronal morphology, and thus contribute to the progress in knowledge scientist of numerous neurodegenerative diseases.

6   Conclusions

In this research we have proposed a new paradigm of methodology in the investigation of neurodegenerative diseases, through the study of dendritic structures in complete tissues instead of studying isolated neurons. We can consider that the work developed has had more than positive results considering that the approach of the problem was a new line of research totally unknown to date. One of the key points in obtaining these results is the application of a pre-processing stage that addresses the elimination of blood vessels as well as the contrast enhancement. The final solution described is the result of an exhaustive study of different alternatives until finding the most appropriate solution to the immunofluorescence images of hippocampus. All parameters of the algorithms of this stage are fixed and do not need to be adapted to each of the images, which makes the procedure completely automatic. On the other hand, we have made a careful selection of texture parameters that compound the resulting feature vector for each of the two methods studied. In the case of the first method, classic texture descriptors were complemented by dispersed descriptors of more recent creation, and in this way it has been possible to reach excellent results. In the same way, the LBP descriptor chosen is invariant under rotations something relevant in the case of our images due to the fact that rotation of the texture pattern should not be considered as a distinctive feature to distinguish between DS and WT. The results obtained with this set of descriptors have also been very satisfactory. Finally, we have made a complete comparison between different classifiers for the two vectors of characteristics calculated. The best-performing classifiers are those based on Vector Support Machine for the first method and K-NN Cosine and Bagged Tree for the second method. We would also like to highlight the reliability of decision tree-based classifiers such as Random Forest.

References

1. Sholl, D.A.: Dendritic organization in the neurons of the visual and motor cortices of the cat. J. Anat. **87**(4), 387–406 (1953)
2. Troca-Marín, J.A., Alves-Sampaio, A., Montesinos, M.L.: An increase in basal BDNF provokes hyperactivation of the Akt-mammalian target of Rapamycin pathway and deregulation of local dendritic translation in a mouse model of Downs syndrome. J. Neurosci. **31**, 9445–9455 (2011)
3. Alves-Sampaio, A., Troca-Marín, J.A., Montesinos, M.L.: NMDA-mediated regulation of DSCAM dendritic local translation is lost in a mouse model of downs syndrome. J. Neurosci. **30**, 13537–13548 (2010)
4. Troca-Marín, J.A., Alves-Sampaio, A., Montesinos, M.L.: Deregulated mTOR-mediated translation in intellectual disability. Prog. Neurobiol. **96**, 268–282 (2012)
5. Montesinos, M.L.: Roles for DSCAM and DSCAML1 in central nervous system development and disease. In: Berezin, V., Walmod, P.S. (eds.) Cell Adhesion Molecules. AN, vol. 8, pp. 249–270. Springer, New York (2014). doi:10.1007/978-1-4614-8090-7_11
6. Troca-Marín, J.A., Casañas, J.J., Benito, I., Montesinos, M.L.: The Akt-mTOR pathway in down's syndrome: the potential use of rapamycin/rapalogs for treating cognitive deficits. CNS Neurol. Disord. Drug. Targets **13**, 34–40 (2014)
7. González, R., Woods, R.E.: Digital Image Processing, 3rd edn. Prentice Hall, Nueva Jersey (2008)
8. Haralick, R.M., Shanmugam, K., Dinstein, I.: Textural features for image classification. IEEE Trans. Syst. Man Cybern. SMC **3**(6), 610–621 (1973)
9. Boland, M.V., Murphy, R.F.: A neural network classifier capable of recognizing the patterns of all major subcellular structures in fluorescence microscope images of hela cells. Bioinformatics **17**(12), 1213–1223 (2001)
10. Svoboda, D., Kozubek, M., Stejskal, S.: Generation of digital phantoms of cell nuclei and simulation of image formation in 3D image cytometry. Cytometry A **75**(6), 494–509 (2009)

11. Al-Kadi, O.S.: Texture measures combination for improved meningioma classification of histopathological images. Pattern Recogn. **43**, 2043–2053 (2010)
12. Xia, H., Tong, L., Zhou, X., Zhang, J., Zhou, Z., Liu, W.: Texture analysis and volumetry of hippocampus and medial temporal lobe in patients with Alzheimers disease. In: Proceedings of 2012 International Conference on Biomedical Engineering and Biotechnology, pp. 905–908 (2012)
13. Thibault, G., Angulo, J., Meyer, F.: Advanced statistical matrices for texture characterization: application to cell classification. IEEE Trans. Biomed. Eng. **61**(3), 630–637 (2014)
14. Ojala, T., Pietikainen, M., Maenpaa, T.: Multiresolution gray-scale and rotation invariant texture classification with local binary patterns. IEEE Trans. Pattern Anal. Mach. Intell. **24**(7), 971–987 (2002)
15. Hafiane, A., Seetharaman, G., Zavidovique, B.: Median binary pattern for textures classification. In: Kamel, M., Campilho, A. (eds.) ICIAR 2007. LNCS, vol. 4633, pp. 387–398. Springer, Heidelberg (2007). doi:10.1007/978-3-540-74260-9_35
16. Ta, X., Triggs, B.: Enhanced local texture feature sets for face recognition under difficult lighting conditions. IEEE Trans. Image Process. **19**(6), 1635–1650 (2010)
17. Rakotomamonjy, A., Petitjean, C., Salan, M., Thiberville, L.: Scattering features for lung cancer detection in fibered confocal fluorescence microscopy images. Artif. Intell. Med. **61**, 105–118 (2014)
18. Fogel, I., Sagi, D.: Gabor filters as texture discriminator. Biol. Cybern. **61**(2), 103–113 (1989)
19. Rellier, G., Descombes, X., Falzon, F., Zerubia, J.: Texture feature analysis using a Gauss-Markov model in hyperspectral image classification. IEEE Trans. Geosci. Remote Sens. **42**(7), 1543–1551 (2004)
20. Wu, C., Schulte, J.J., Sepp, K.J., Littleton, J.T., Hong, P.: Automatic robust neurite detection and morphological analysis of neuronal cell cultures in high-content screening. Neuroinformatics **8**(3), 83–100 (2010)
21. Budde, M.D., Frank, J.A.: Examining brain microstructure using structure tensor analysis of histological sections. NeuroImage **63**(11), 1–10 (2012)
22. Bornemann, F., März, T.: Fast Image Inpainting Based on Coherence Transport. Springer Science+Business Media, USA (2007)
23. Contrast-Limited Adaptive Histogram Equalization (CLAHE): At MathWorks. http://es.mathworks.com/help/images/ref/adapthisteq.html. Accessed 20 Jan 2017
24. Costa, A.F., Humpire-Mamani, G., Traina, A.J.M.: An efficient algorithm for fractal analysis of textures. In: Proceedings of Graphics, Patterns and Images (SIBGRAPI), Brazil, pp. 1530–1834 (2012)
25. Sarmiento, A., Fondón, I., Velasco, M.M., Qasair, A., Aguilera, P.: Gaussian mixture model generalized for segmentation of melanomas. In: Annual Congress of the Spanish Society of Biomedical Engineering (CASEIB) (2014)
26. Scharfenberger, C.: Statistical textural distinctiveness for salient region detection in natural images. IEEE Conf. Comput. Vis. Pattern Recogn. (CVPR) **24**(1), 457–470 (2013)
27. Local Binary Pattern. http://www.scholarpedia.org/article/Local_Binary_Patterns. Accessed 20 Jan 2017
28. Cortes, C., Vapnik, V.: Support-Vector Networks. AT& T Bell Labs, USA (1995)
29. Altman, N.S.: An introduction to kernel and nearest-neighbor nonparametric regression. Am. Stat. **46**(3), 175–185 (1992)
30. Breiman, L.: Bagging predictors. Mach. Learn. **24**(2), 123–140 (1996)
31. Breiman, L.: Random forests. Mach. Learn. **45**(1), 5–32 (2001)

Architecture for Neurological Coordination Tests Implementation

Michel Velázquez-Mariño[1,2], Miguel Atencia[1(✉)], Rodolfo García-Bermúdez[3,2], Francisco Sandoval[4], and Daniel Pupo-Ricardo[2]

[1] Departamento de Matemática Aplicada, Universidad de Málaga, Málaga, Spain
mvquezm@uma.es, matencia@ctima.uma.es
[2] Grupo de Procesamiento de Datos Biomédicos (GPDB),
Universidad de Holguín, Holguín, Cuba
depupor@uho.edu.cu
[3] Departamento de Informática y Electrónica, Universidad Técnica de Manabí,
Portoviejo, Ecuador
rodgarberm@gmail.com
[4] Departamento de Tecnología Electrónica, Universidad de Málaga, Málaga, Spain
fsandoval@uma.es

Abstract. This paper proposes a generic architecture for devising interactive neurological assessment tests, aimed at being implemented on a touchscreen device. The objective is both to provide a set of software primitives that allow the modular implementation of tests, and to contribute to the standardization of test protocols. Although our original goal was the application of machine learning methods to the analysis of test data, it turned out that the construction of such framework was a pre-requisite to collect enough data with the required levels of accuracy and reproducibility. In the proposed architecture, tests are defined by a set of stimuli, responses, feedback information, and execution control procedures. The presented definition has allowed for the implementation of a particular test, the Finger-Nose-Finger, that will allow the exploitation of data with intelligent techniques.

Keywords: Software architecture · Touchscreen devices · Neurological tests · Machine learning · Finger-Nose-Finger test

1 Introduction

In this paper we propose a framework for the implementation of neurological tests to assess coordination diseases, such as Parkinson's and cerebellar ataxia. The final aim of this research is the assessment of the disease stage, as well as the recommendation of specific test protocols, through the analysis of test results by computational intelligence methods. We have found that the lack of standardized test protocols and software platforms is a severe obstacle for such objective, thus the motivation to undertake the proposed architecture.

© Springer International Publishing AG 2017
I. Rojas et al. (Eds.): IWANN 2017, Part II, LNCS 10306, pp. 26–37, 2017.
DOI: 10.1007/978-3-319-59147-6_3

Nowadays the presence of touchscreen devices is pervasive in health-related applications such as monitoring personal activity [8], at-home assessment of health variables [6], and support to diseases diagnosis [12,17]. Touchscreen devices are also introduced in hospitals for patient monitoring, diagnostic evaluations, and interfacing to larger information systems [9]. The most common touchscreen devices are smartphones and tablets and, in particular, devices with Android Operating System (OS) are very popular. Undoubtedly, the reduced cost, size, and power consumption are key factors in this upsurge. Similar reasons can be given for the inclusion of built-in electronic components into different *mHealth* [5,11,18] applications.

Several tests have been developed for or adapted to touchscreen devices, simplifying the communication between users and computing systems. In the coordination assessment area, which plays an important role in the neurological examination, touchscreen devices have been used to record the movement of upper limbs, in evaluation tasks such as circle tracing [12], Archimedean Spiral [10,13], and handwriting/drawing [4]. The obtained accurate data about the disease progression contribute to the design of objective clinical scales with high sensitivity, the evaluation of therapies, and the exploration of early symptoms.

Although several tests have been proposed and validated for ataxia assessment, they are configured with different application protocols, implemented on heterogeneous devices, and based upon disparate techniques for recording the results [1–3,7,15]. To the best of our knowledge, no module or architecture has been proposed to standardize the implementation of neurological coordination tests using touchscreen devices and external sensors. This lack of standardization is a severe hindrance to the accurate analysis of collected data, as well as to the development of new coordination tests.

Here we present a modular and portable architecture to implement coordination tests, focused on a main touchscreen device (tablet) that can integrate different external devices (sensors, programmable boards, etc.). The control device manages all tests execution, acquires data from its own sensors (touchscreen, accelerometer, etc.) and from external devices, and integrates and saves data. The main elements that such an architecture must manage are the following:

- Data from touchscreen devices and external sensors.
- Stimuli events (visual, sound and haptic) and feedback information.
- Control logic of the implemented neurological tests.
- Data persistence.

The Finger-Nose-Finger (FNF)—often used in neurological examinations and evaluated in previous work [14,16]—is here implemented using the proposed architecture. All along the paper, this test is used as a proof of concept, by presenting a detailed definition of each one of its elements in the architecture.

In Sect. 2 we present the basic definitions that shape the implementation of a generic neurological test, resulting in a key module within the whole framework. The interaction between tests, sensors, and graphical user interface is explained in Sect. 3, by providing a global view on the proposed architecture. Although the framework is intended to be portable, some remarks are pointed out in Sect. 4

regarding implementation requirements. Finally, in Sect. 5 the conclusions and directions for future research are discussed.

2   Coordination Tests

In this section we propose the general scheme of a neurological coordination test. Our goal is to improve the accuracy of data acquisition by minimizing the dependence on specialists' subjectivity. Many of the neurological coordination tests can be automated, thus we construct the necessary infrastructure; in doing so, our final aim will be to approach a standardization of tests that will be both general and flexible to accommodate a wide range of tests.

From the engineering point of view, we grouped the tests in touch-screen-pure tests and touch-screen-non-pure tests. The former only need the touch-screen device for its execution, whereas the latter need more than one device, thus making its design more complex. Even if a test proper does not use the touchscreen to show stimuli or capture responses, this device is used to show information about the test, to control the execution and to manage data captured. Despite differences among tests, there is a set of common characteristics presented in all of them, namely stimuli, responses, feedback information and execution control:

Stimuli are signals provided by the system to the user, intended to generate a definite response. They can consist of visual, audible or haptic events. They are always present in coordination tests, since at least the start stimulus must exist, if the user's expected actuation can be assumed from the context. In general, more than one stimulus occurs. The general execution of the test, the start, the end, and the rhythm of expected tasks are guided by stimuli, which are the reference to assess the timing and accuracy of responses.

Responses comprise the user's actuations driven by stimuli. They are captured using different devices, such as touchscreens, inertial sensors, keyboard buttons, etc. Responses can be continuously polled at a specific sample rate or consist of discrete events depending on subjects' performance velocity, usually guided by either constant or fixed stimuli, respectively. There may also exist other kind of responses, e.g. the occurrence (or not) of a certain action during a time interval, or other more specific accomplishments of subjects. More complex tests can present all of these responses types.

Feedback information acts as an assessment, just after the response, and is usually aimed at both showing a reference and helping to improve response accuracy. Just like stimuli, it can include visual, audible and/or haptic signals. Feedback information allows to explore different neurological behaviours, for instance improvement due to learning or deterioration caused by fatigue. It can be enabled or disabled in the same test depending on the examination goal, e.g. when trials are intended to be independent.

Execution control defines the flow of test actions, by specifying the sequence of stimuli, response measurements, and feedback signals. In addition, this module may pre-process responses and saves all results. The execution control is the kernel of the tests, and it must guarantee that all functionalities occur exactly on time, avoid delays, and accurately perform data acquisition.

Based on these common characteristics, we can represent the neurological test execution in a block diagram (Fig. 1). When a test starts, the execution control block activates a run-loop that owns the precise definitions of the test execution protocol, so as to launch stimuli and feedback information when appropriate. Each neurological test is executed by means of the necessary routines, which can be grouped into five categories: show stimuli, response acquisition, response pre-processing, provide feedback information, and save data. Not always the execution order matches this enumeration, since some complex tests may include repetitions, iterations, and back tracking, so the run-loop is far from linear. Coordination tests start by a stimulus, and finish when some condition is reached. Some common stop conditions include timeout reached, number of responses achieved, number of correct responses achieved, number of wrong responses achieved, and other more specialized conditions for specific tests. All stop conditions are managed by a run-loop controller. The run-loop has a state machine to execute the above mentioned five routines in a coordinated way, as well as to finish the test execution when a stop condition occurs.

In coordination assessment, a session often combines one or several tests into a configuration that we call *evaluation protocol*. For instance, the same test can be configured as a sequence of short repetitive tasks, to evaluate the capacity to learn specific simple movements; or repeated without feedback in long repetitive tasks, in order to train some more complex responses; or a protocol designed to globally evaluate the coordination state could include a combination of different

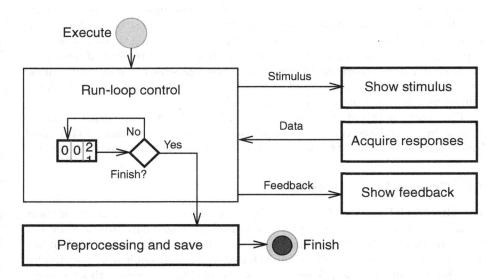

Fig. 1. Execution model of a coordination test.

tests comprising non repetitive tasks. The philosophy of defining protocols is intended to guarantee the extensibility of the proposed architecture, by providing constructive blocks that can serve as modules of sophisticated configurations.

As a proof of concept, we have defined all the elements that are necessary to implement the FNF test. A particular protocol, that is often found in clinical practice, is configured as three slightly modified repetitions of the test. The stimulus consists in a red cross drawn on the screen. Driven by the presentation of such visual stimulus, the subject is expected to touch the red cross, so that the responses include the point actually touched (if any) and the time spent. After the user touches the screen, a yellow circle is drawn on the screen just at the place of the patient's response, providing visual feedback information. The stop condition is defined as ten taps, i.e. each one of the three instances of the test ends when the subject touches the screen ten times. In the first test instance, the stimulus will be placed on the centre of the screen. In the second test occurrence, the stimuli will alternatively be placed at each corner of the screen, moving to the next position after each response, following a clock wise rotation. In the third repetition, the stimulus position will be randomly chosen with a uniform distribution over the screen area. The proposed scheme allows the clinician to easily select a pre-defined protocol, or even to design new protocols by combining and defining different tests.

3   Architecture

In this section we describe the global architecture, which is focused on two critical elements: the execution control and the sensor data acquisition. The execution control manages the stimuli and the feedback information, according to the test definition as described in the previous section. Depending on the kind of stimuli and the feedback information, different low-level routines must be executed, such as simple or complex draw procedures, audio playback, writing to communication ports, and so on. The sensor data acquisition handles the reading of sensors in the main device via *character devices* and of external sensors via communication ports. The punctual execution and coordination of all these tasks, at the instants defined by the test, is crucial for accuracy of results.

The functionality of both execution control and data acquisition is implemented by an architecture that is supported by three levels, namely Graphical User Interface, Protocol Management and Test Execution, as shown in Fig. 2.

3.1   Graphical User Interface

The Graphical User Interface (GUI) is a communication layer between users (medical staff and subjects) and the Protocol Management. Routines in this layer are aware of entities defined by the context, which are involved in the neurological evaluation, such as subject and evaluation clinical protocol. This layer implements functionalities to allow the clinician to select those contextual entities and introduce other important information such as hand used, eyes

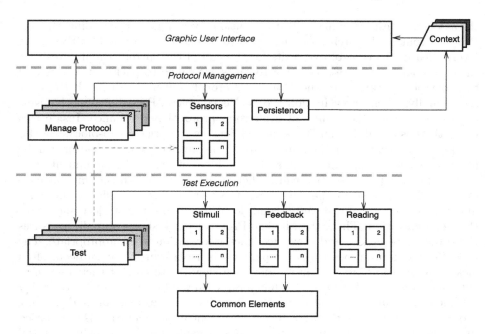

Fig. 2. Architecture diagram.

state (open/close), study type (training/evaluation/rehabilitation), study title or required comments. It is the responsibility of this layer to call the correct protocol type and transfer to Protocol Management all information to perform the evaluation.

Following our implemented example, in the Graphical User Interface implementation for the FNF protocol we present on the screen a widget that allows the user to select basic contextual data of the test, namely:

- The hand that will be used in the execution of tests, left or right.
- Basic personal data of the patient.
- Choice of protocol, as described in the previous section.
- Type of protocol: *evaluation* in this case.

3.2 Protocol Management

The layer Protocol Management implements the mechanisms to control the evaluation protocol execution, manage communication with the main device sensors and external sensors, and save study results. For each evaluation protocol, there exists a block named Manage Protocol$_i$, $i = 1, 2, \ldots, n$, which controls the execution of the protocol by calls to one or more associated tests from the Test Execution layer.

The execution of an evaluation protocol begins with a call from the GUI to the related Manage Protocol block. A test start screen is shown via GUI using the associated test block in the Test Execution layer. The corresponding Manage Protocol routine dispatches the next test to be executed in the Test Execution layer, releases the control and waits for the end of the current running test. At the end of each test, pre-processing routines are called to pack test results and the control is returned to the Manage Protocol routine to dispatch the next test. Finally, when an evaluation protocol is complete, Persistence mechanisms are called to save the results of evaluation protocol performance. Inside this layer, there is an additional Block named Sensors comprising sub-blocks Block$_i$, $i = 1, 2, \ldots, m$ that group the routines to use the involved sensors. Sensors can be heterogeneous with specific commands and different protocol communications.

The process just described can be better understood with the implemented example. In the case of the FNF test, the test start screen shows summary information about the test and a countdown to prepare the subject and foster their concentration. Then, the Manage Protocol block shows stimuli, measures the subject's responses and shows feedback information when appropriate. The need for result pre-processing is apparent when observing that subjects often lack accuracy touching the target, and then they try to correct their responses with fast taps around the target. Therefore, when the test ends, a pre-processing routine must split the correction touches from the valid response ones. Regarding the usage of sensors by the FNF test, the current implementation is only based upon the touchscreen, but in a more advanced version, we plan to include inertial sensors to assess the tremor and other variables related to movement.

3.3   Test Execution

The Test Execution is the layer that owns the mechanisms to actually perform the tests, by implementing the logic of each test. Tests are represented in the diagram by blocks named Test$_i$, $i = 1, 2, \ldots, t$, each one consisting of different stimuli, feedback information and response acquisition mechanisms. They also run with different execution logic, by defining a sequence of tasks, an end condition, and related information about the execution, as described in Sect. 2.

Each Test block interacts with three blocks named Stimuli, Feedback, and Reading, all three in turn comprising the same number t of sub-blocks. Thus, a sub-block Stimuli$_i$ groups the routines that show the stimuli required by the particular i-th test, combining visual, sound and/or haptic signals, either on the touchscreen or through external devices. The Stimuli blocks must account for different screen drawing mechanisms (raster, double-buffer, zone painting, etc.) as well as implement independent threads for read/write operations on communication ports. Stimuli timing is implemented in the Test either by defining them as constant stimuli, or through triggering mechanisms, such as time intervals or subject responses. Each sub-block Feedback$_i$ provides feedback information related to the i-th test, if any is required. When feedback is defined, it is always driven by the subject's responses, thus it can be seen as an stimulus with a fixed trigger mechanism, with the same implementation recommendations. Therefore, the Stimuli and Feedback blocks share a block named Common Elements, which groups basic routines to paint, write information or make calculations used by both blocks. Finally, each sub-block Reading$_i$ contains data acquisition routines, both from the main device or external sensors. Depending on the test type, reading subject's responses can proceed by interruptions or by constant sampling. In order to prevent data loss, reading raw values from the main routine upon every interruption is recommended, whereas in the latter case secondary threads should be implemented to avoid latency in the reading process.

Using the FNF test as example, we can distinguish the following elements. The stimuli consists in painting on the touchscreen a red cross at the position defined by the configuration of the test. For instance, in the first test iteration explained above, the cross position is at the screen centre. The transition to the next stimulus is triggered by the subject response action on the touchscreen. The feedback information is also a draw action, namely to draw a yellow circle at the position of the subject's response, and hide it after two seconds. Finally, data acquisition is accomplished by reading all events on the touchscreen, getting the position of touch or touches, and recording the timestamp of the event.

4   Implementation

The proposed framework is mostly implementation-independent, so the paper does not aim at a detailed description of software coding and used hardware. However, we include in this section some comments on the requirements of both software and hardware for a successful implementation of the architecture.

4.1   Base Classes

Although the proposed architecture was designed to be general and independent of any programming language or technology, we do consider that an object oriented methodology is required to achieve the objectives of modularity and reusability. Consequently, we have coded a number of classes to implement the main functionalities of the architecture. These classes are shown in Fig. 3, following the *Unified Modeling Language*^{TM}, by using generic references of names and data types to make possible any adaptation.

The **ProtocolExcBase** is the base class to implement the specific protocol execution classes. This class has the *runControl* method to control the execution of a set of tests using a state machine implementation. The enumeration **ProtocolState** defines the states of the state machine and the attributes *mProtocol* and *mCurrentTest* provide information about the order of tests execution. The *start* and *end* methods are used to switch between tests, and *saveResults* allows to access the persistence functionalities. We have implemented a protocol for the FNF test, by constructing a new class FNFProtocolExc that inherits from ProtocolExcBase, and then performs the custom implementations.

The class **GenericSensor** is the base class to implement the attributes and hight level methods needed to use sensors that read and write data. Specific sensors class implementations stem from the extension of this class. Methods in this class are invoked from the protocols execution classes, whereas the TestExcBase

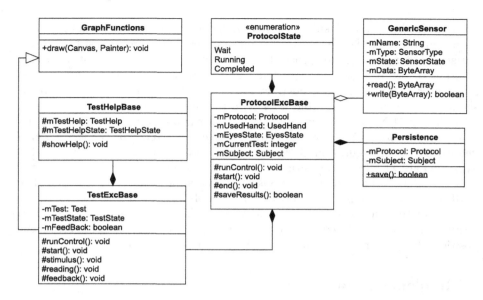

Fig. 3. Base classes diagram.

class is allowed the control of sensors by means of references. The current implementation of our example case, the FNF test, makes no use of specific sensors, but a planned advanced version will use inertial sensors, thus requiring the construction of a class that will inherit from the GenericSensor class and construct its custom functionalities using the read and write methods of the parent class.

The actual execution of tests will be performed by classes that inherit the basic implementation of the class **TestExcBase**, which comprises a set of attributes and methods according to the theoretical definition of a general test, given in Sect. 2. We have also included a **TestHelpBase** class to show some explanations and help of the related test, which can be helpful in some complex tests. All draw operations on the touchscreen of the main device are managed by the class **GraphFunctions**. Drawing techniques are strongly dependent on the programming language and the technology, however this class must provide, at least, high level methods to draw basic shapes (circle, cross, square, text, etc.), whereas other drawing complex tasks should be implemented by the specific TestExc class that requires them. In our example, the FNFTestExc class has been built by inheriting from TestExcBase, and then implementing the custom elements described along the paper, such as execution control routines, stimuli routines, and other test elements.

4.2   Acquisition System

As mentioned above, the core device of the architecture is the touchscreen. Sensors may either be integrated in the main device (the touchscreen itself, accelerometer, gyroscope, light sensor, etc.) or be externally developed on programmable boards such as Arduino, Raspberry pi, cc2640, etc. Communication between the main device and external sensors should be wireless, using Bluetooth, Low Energy Bluetooth, or other wireless technology, in order to ease the subject's performance, minimize wires, and isolate the subject from the main device. Also, each associated sensor must have a trust security code to be connected to the main device. The whole hardware setting is shown in Fig. 4.

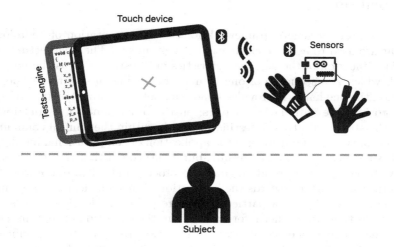

Fig. 4. Architecture environment, showing the main touchscreen device with management acquisition routines, associated sensors and a subject interacting.

In order to achieve the planned functionalities, some requirements are specified on the touchscreen: more than seven screen inches, a capacitive touchscreen with five or more activation points, double core processor, at least one gigabyte of Random Access Memory, and integrated sensors (accelerometer, gyroscope, light sensor, etc.). Additional requirements could be demanded by specific tests. In the main device, the kernel of the OS is responsible for managing interruptions of the integrated sensors. The raw values of sensors are available to the *user space* through *system call interfaces*. For example, when the kernel receives a touchscreen interruption, it writes the raw data (touch position, touch pressure, touch timestamp, etc.) of the event on a *character device*. The routines in the *user space* can then access such character device using *standard libraries*. We propose implementing buffering mechanisms and multi-threads to avoid samples loss. Any OS that includes these mechanisms is, in principle, compatible with the proposed architecture, the most usual being Android, Windows, OS X, iOS and GNU/Linux.

A *beta* version of the end-to-end architecture has been built as a proof of concept, implementing the FNF test. The application was developed in the language C++ using the Qt framework. All described base classes were implemented to provide structures and high level methods to ease the construction of other coordination tests. This implementation is cross-platform and can be ported to Linux, OS X, Windows, and Android OSs. During the actual tests with subjects, only the Android version was used in order to ensure the homogeneity in the experiments. The characteristics of the experiments used to evaluate the implementation and the usefulness of our proposal can be found in [14, 16].

5   Conclusions

We have proposed a generic framework for the implementation of neurological tests that are used to assess coordination of movements. The architecture relies on the identification of the key characteristics of a test, as well as the notion of protocol, which is a sequence of one or more tests. The aim of the proposal is twofold: on the one hand, it provides a modular software infrastructure that can be used to improve automation of tests; on the other hand, it is contribution towards the standardization of testing protocols within the clinical community.

The primary goal of our work is the application of computational intelligence techniques to the analysis of test results, but it turned out that this goal was hindered by the lack of both standard clinical protocols and software libraries. This paper is thus a first step towards the integration of machine learning algorithms within neurological tests. In particular, we are currently developing a layer of cloud services that will gather information from the execution of tests in mobile devices. The massive collected data will be processed to recommend specific tests to patients, according to the progression they show at each performed protocol, or even to improve the configuration of protocols.

Acknowledgement. This work has been partially supported by the Universidad de Málaga, as well as the Universidad de Holguín through the joint project titled *"Mejora del equipamiento para la evaluación de la rehabilitación de enfermedades neurológicas de especial prevalencia en el oriente de Cuba"*.

References

1. Gagnon, C., Mathieu, J., Desrosiers, J.: Standardized finger-nose test validity for coordination assessment in an ataxic disorder. Can. J. Neurol. Sci. Le Journal Canadien Des Sciences Neurologiques **31**(4), 484–489 (2004)
2. Gavriel, C., Thomik, A.A.C., Lourencco, P.R., Nageshwaran, S., Athanasopoulos, S., Sylaidi, A., Festenstein, R., Faisal, A.: Kinematic body sensor networks and behaviourmetrics for objective efficacy measurements in neurodegenerative disease drug trials. In: 2015 IEEE 12th International Conference on Wearable and Implantable Body Sensor Networks (BSN), pp. 1–6 (2015)
3. Harish, K., Rao, M.V., Borgohain, R., Sairam, A., Abhilash, P.: Tremor quantification and its measurements on parkinsonian patients. In: International Conference on Biomedical and Pharmaceutical Engineering, ICBPE 2009, pp. 1–3 (2009)

4. Lefebvre, O., Riba, P., Gagnon-Marchand, J., Fournier, C., Fornes, A., Llados, J., Plamondon, R.: Monitoring neuromotricity on-line: a cloud computing approach. In: Rémi, C., Prévost, L., Anquetil, E. (eds.) 17th Biennial Conference of the International Graphonomics Society, Drawing, Handwriting Processing Analysis: New Advances and Challenges, Pointe-à-Pitre, Guadeloupe (2015)
5. Lennon, T., Bernier, T., Tamayo, D., Goldberg, C., Mankodiya, K.: Multi-sensory system for monitoring dyskinesia in movement disorders. In: 2015 41st Annual Northeast Biomedical Engineering Conference (NEBEC), pp. 1–2. IEEE (2015)
6. Memedi, M., Nyholm, D., Johansson, A., Palhagen, S., Willows, T., Widner, H., Linder, J., Westin, J.: Validity and responsiveness of at-home touch-screen assessments in advanced Parkinson's disease. IEEE J. Biomed. Health Inf. 19(6), 1829–1834 (2015)
7. Notermans, N.C., van Dijk, G.W., van der Graaf, Y., van Gijn, J., Wokke, J.H.: Measuring ataxia: quantification based on the standard neurological examination. J. Neurol. Neurosurg. Psychiatry 57(1), 22–26 (1994)
8. Park, K., Lim, S.: A multipurpose smart activity monitoring system for personalized health services. Inf. Sci. 314, 240–254 (2015)
9. Paschou, M., Sakkopoulos, E., Tsakalidis, A.: easyHealthApps: e-Health apps dynamic generation for smartphones & tablets. J. Med. Syst. 37(3), 9951 (2013)
10. Pullman, S.L.: Spiral analysis: a new technique for measuring tremor with a digitizing tablet. Mov. Disord. 13(Suppl. 3), 85–89 (1998)
11. Salarian, A., Russmann, H., Wider, C., Burkhard, P.R., Vingerhoets, F.J.G., Aminian, K.: Quantification of tremor and bradykinesia in Parkinson's disease using a novel ambulatory monitoring system. IEEE Trans. Bio-med. Eng. 54(2), 313–322 (2007)
12. Say, M.J., Jones, R., Scahill, R.I., Dumas, E.M., Coleman, A., Santos, R.C.D., Justo, D., Campbell, J.C., Queller, S., Shores, E.A., Tabrizi, S.J., Stout, J.C.: TRACK-HD investigators: visuomotor integration deficits precede clinical onset in Huntington's disease. Neuropsychologia 49(2), 264–270 (2011)
13. Surangsrirat, D., Thanawattano, C.: Android application for spiral analysis in Parkinson's Disease. In: 2012 Proceedings of IEEE Southeastcon, pp. 1–6. IEEE (2012)
14. Velázquez-Mariño, M., Atencia, M., García-Bermúdez, R., Pupo-Ricardo, D., Becerra-García, R., Velázquez-Pérez, L.: Contribution analysis of extracted variables from Finger-Nose-Finger test to the classification of SCA2 patients. In: III International Conference on Informatics and Computer Sciences, Havana, Cuba (2016)
15. Velazquez-Perez, L., de la Hoz-Oliveras, J., Perez-Gonzalez, R., Hechavarria, P.R., Herrera-Dominguez, H.: Quantitative evaluation of disorders of coordination in patients with Cuban type 2 spinocerebellar ataxia. Revista De Neurologia 32(7), 601–606 (2001)
16. Velázquez-Mariño, M., Atencia, M., García-Bermúdez, R., Pupo-Ricardo, D., Becerra-García, R., Pérez, L.V., Sandoval, F.: Cluster analysis of Finger-to-Nose test for spinocerebellar ataxia assessment. In: Rojas, I., Joya, G., Catala, A. (eds.) IWANN 2015. LNCS, vol. 9095, pp. 524–535. Springer, Cham (2015). doi:10.1007/978-3-319-19222-2_44
17. Westin, J., Dougherty, M., Nyholm, D., Groth, T.: A home environment test battery for status assessment in patients with advanced Parkinson's disease. Comput. Methods Programs Biomed. 98(1), 27–35 (2010)
18. World Health Organization: mHealth: new horizons for health through mobile technologies. In: Second Global Survey on eHealth, Geneva (2011)

Adaptation of Deep Convolutional Neural Networks for Cancer Grading from Histopathological Images

Stefan Postavaru[1,2], Ruxandra Stoean[3(✉)], Catalin Stoean[3],
and Gonzalo Joya Caparros[4]

[1] Faculty of Mathematics and Computer Science,
University of Bucharest, Bucharest, Romania
[2] Bitdefender, Bucharest, Romania
spostavaru@bitdefender.com
[3] Faculty of Sciences, University of Craiova, Craiova, Romania
{ruxandra.stoean,catalin.stoean}@inf.ucv.ro
[4] School of Telecommunication Engineering, University of Malaga, Málaga, Spain
gjoya@uma.es

Abstract. The paper addresses the medical challenge of interpreting histopathological slides through expert-independent automated learning with implicit feature determination and direct grading establishment. Deep convolutional neural networks model the image collection and are able to give a timely and accurate support for pathologists, who are more than often burdened by large amounts of data to be processed. The paradigm is however known to be problem-dependent in variable setting, therefore automatic parametrization is also considered. Due to the large necessary runtime, this is restricted to kernel size optimization in each convolutional layer. As processing time still remains considerable for five variables, a surrogate model is further constructed. Results support the use of the deep learning methodology for computational assistance in cancer grading from histopathological images.

Keywords: Image processing · Histopathological slides · Classification · Deep convolutional neural networks · Parametrization

1 Introduction

As with all types of cancer, early detection and correct grading (made possible by modern medical imaging technologies) increase the chances of cure by allowing timely counteractive treatment measures. Hence, screening interventions are more and more regular nowadays and the consequent amount of related imaging data is growing vertiginously. With the additional known considerable inter- and intra-observer variability in interpreting the resulting images, computational image evaluation for the automated detection of landmarks and determination of cancer stage is getting a high interest as it can efficiently assist the medical

© Springer International Publishing AG 2017
I. Rojas et al. (Eds.): IWANN 2017, Part II, LNCS 10306, pp. 38–49, 2017.
DOI: 10.1007/978-3-319-59147-6_4

decision-making in terms of precision, invariability and time economy. Among such computational approaches, those that neither need further supervision from the medical experts, in terms of manual delineation or feature specification, nor manual parametrization from the computer scientists are obviously preferred in general, for any diagnosis task in medicine [9–12,21,37].

The current study thus attempts to tackle the medical problem of grading colorectal cancer from available histopathological slides by convolutional neural networks (CNN). The technique is recognized for its high-performing and autonomous learning capability as a result of its special layer types and hierarchical arrangement. It would thus be indeed supportive in diminishing the routine work for the medical experts, since it can configure the important features by itself [36] for the common images and need expert reconfirmation only for the problematic ones. The CNN is nevertheless problem-dependent with respect to parametrization and the paper will consequently furthermore investigate an inexpensive option in design to allow automatic tuning of CNN sensitive variables. A surrogate model is thus constructed to streamline the framework as regards runtime but still maintain accuracy, and an heuristic approach [6,25] is appointed to search for the optimal parameters. The work is not only important from the medical perspective, but also from the CNN side, as the trust in its potential must be also shown on real-world data sets and not only on toy problems [36].

The paper is organized as follows. Section 2 describes the real-world data set containing histopathological colorectal cancer slides with the earlier models for this collection and outlines the state of the art in CNN application for medical image interpretation. The CNN architecture is presented in Sect. 3, together with its augmentation in terms of automatic tuning with an interest in runtime efficiency. Experimental findings are given in Sect. 4, while the conclusions are drawn in Sect. 5.

2   Problem and Prior Medical Computational Diagnosis

The medical data set contains histopathological digitalized slides for colorectal tissues from the Emergency Hospital of Craiova, Romania, with representatives of both benign (denoted by G0) and malignant (G1–G3) classes. The collection[1] contains 357 images (at resolution 800×600), where each is assigned to only one cancer class. There are 62 G0 entries, 96 G1 records, 99 G2 and 100 G3 slides.

2.1   Previous Attempts to Classify the Data

Within the examination of the histopathological slides for the detection of colorectal cancer, the medical expert looks for a uniform pattern of glands and nuclei as a sign of the absence of cancer, while, conversely for a malignant tissue, for a high variation in size, shape and texture. The correct segmentation of glands and nuclei and the appearance of their features is thus important for

---

[1] IMEDIATREAT Project, https://sites.google.com/site/imediatreat/.

the recognition of cancer. Earlier attempts on the current problem consequently started from the segmentation of glands through a watershed algorithm parameterized by evolutionary algorithms (EA) [30,34]. A distance transform algorithm was further employed to additionally find the nuclei. Departing from the discovered landmarks, measurements quantifying the number, area, perimeter, radius for the glands and nuclei, as well as for corresponding Delaunay triangles and Voronoi polygons, resulted into 76 numerical extracted features. Support vector machines (SVM) achieved 79.89% classification accuracy on this numerical collection [33]. Feature selection was further employed to help the SVM and considered a consistency-based filter, a correlation mechanism, principal component analysis and a genetic algorithm (GA). The last led to an accuracy increase of 4 percents and the determination of the more important features [31].

The current work however uses the CNN to allow for direct classification into G0–G3 with its own implicit inner feature detection. This will therefore be an alternative "all-in-one", expert-independent approach to slide interpretation.

2.2   Convolutional Neural Networks for Medical Image Diagnosis

First introduced as "self-organizing" neural networks unaffected by position shifts in [8], CNN have developed over years into powerful learning models, trainable with the efficient method of back-propagation. As presented in [20], convolutional layers present a degree of robustness to translation and distortion, thus being well-suited for problems where the input presents spatial relations. The advancement of computing power in recent years made a range of solutions based on deep-learning with CNN possible, with better results than previous state-of-the-art methods: CIFAR image classification [18], video classification [16], action-recognition [14] and others.

Given their high performance on tasks of artificial vision and even prophesied to be "the most disruptive technology [...] since the advent of digital imaging" [7], CNN were naturally recently applied for medical image classification and diagnosis. [26] implemented a CNN on images obtained with computed tomography to classify the given input into one of five classes. They used an augmented data set with 4298 entries to train and test the CNN model, obtaining an accuracy of 94.1% on the test set. A solution for detecting mitosis in breast cancer histology images with CNN was proposed by [5]. A sliding-window was moved over the full image in order to capture a context for each pixel. These patches were fed into the learned model and labeled accordingly. The IEEE Transactions on Medical Imaging journal even had a dedicated special issue in 2016, i.e. *Deep Learning in Medical Imaging: Overview and Future Promise of an Exciting New Technique*, where CNN was employed for learning from crowds in detecting mitosis in breast cancer images [2], for lung cancer pattern classification from high resolution computed tomography [3] and for brain tumor segmentation from MRI images [24]. Among them there is also the study of [28], which is specifically for the colorectal cancer problem at hand and concerns the detection of nuclei (from images with 20000 annotated nuclei categorized into four classes) by CNN and subsequent, separate classification by a neighboring assemble predictor.

In opposition, the present aim is to use the CNN for direct classification of the histopathological slides into four classes, i.e. normal (G0) and cancer grades G1–G3, with neither intermediate segmentation nor in tandem with other classifiers.

3   Proposed Convolutional Neural Network Methodology

The task of diagnosing histopathological images can be formulated as a supervised learning problem: having a data set of pairs (X_i, c_i), where c_i represents the ground-truth class of the image X_i, the aim is to train a model capable of discriminating between the four possible classes.

3.1   Convolutional Neural Network Architecture

A CNN learning model is proposed to solve this task. It is carefully designed to avoid overfitting, given the small data set size, while being capable to generalize well enough in order to provide a competitive accuracy on the final test suite. The CNN (Fig. 1) receives a volume of $256 \times 256 \times 3$ (image width × image height × three color channels) representing a scaled version of the original images and will output a 4-dimensional vector, interpreted as unnormalized scores for each class. The volume is transformed through 5 convolutional layers, with weight-sharing kernels of sizes (lengths) - KS (7, 5, 3, 3, 3) (i.e. filter sizes of $KS \times KS$), padding (of the volume with zero around the border) chosen such that the first two dimensions of the input are preserved. The first kernel depth - KD (number of filters) is 8, and each consecutive depth is multiplied by 2. Every convolution layer slides each of the $KS \times KS \times KD$ filters over all spatial locations of the image by computing dot products between the shared weight vector within the slice and the image piece. The result of convolving each filter with the image is an activation map of $N \times N$, where there are N unique positions to place the $KS \times KS$ filter. Large values on the activation map are assigned to patters that better stimulate the neurons on that filter.

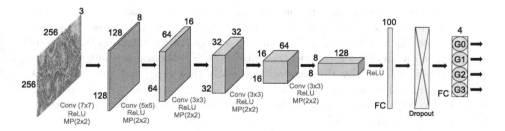

Fig. 1. CNN Architecture: each convolution function (written in green) is followed by ReLU non-linearity (blue) and a Max-pooling (2×2) operation (red). Thus, each consecutive volume doubles in depth and halves its width and height. A dropout layer cuts a subset of activations from the preceding fully-connected one. The last fully connected layer outputs the unnormalized scores for each class. (Color figure online)

The obtained activation maps (whose number is equal to the number of filters) are stacked in a volume $N \times N \times KD$ to enter the next layer [15]. Convolution is followed by a ReLU transfer layer (a Rectified Linear Unit to add nonlinearity) and a Max-Pooling layer (for a downsampling operation of the image following the maximum value in the window) with window size 2×2, effectively reducing the input volume and making the network resistant to spatial shifts.

The learning speed was greatly increased by applying batch-normalization [13] to the last 4 convolutional layers, achieving the same accuracy within 10% of the original number of iterations. The convolutional layers are followed by two fully connected ones. Dropout is also used between the last two fully-connected layers (where neuron activations are randomly blocked within training) [29]. During hyper-parameter tuning, it was found that a dropout rate of 90% manages to greatly reduce overfitting, while maintaining the ability of the network to learn. During training, the KL-Divergence [19] between the ground-truth class distribution and the network outputs interpreted as probabilities is minimized through a softmax function.

3.2 Kernel Size Parametrization

Although acknowledged for autonomous classification and implicit feature learning, parametrization of the network is known to be problem-dependent. To the best of our knowledge, it is only the recent study of [36] that accomplishes the tuning of six hyper-parameters through EA with message passing interface to distribute fitness computation and hence accelerate runtime. The CNN architecture has three convolutional layers and, for each of those, two parameters are heuristically determined: the number of filters and the kernel size. Although known to be flexible, EA are too expensive in runtime and hence the current study alternatively employs Latin hypercube sampling (LHS, [22]) for kernel size automatic generation from the five CNN convolutional layers.

3.3 Surrogate Model Design

In order to reduce runtime, the present paper proposes an accompanying surrogate model to the automatic setting of kernel sizes. Such regression options proved to be efficient before for the same purpose of parametrization [32]. A data set of input parameter values and the obtained CNN accuracy result on validation samples is learnt by the regression models that represent the surrogates. Each such model is next used to estimate the accuracy for new input parameters. A GA [17] is used to search the parameter space for the optimal settings, with the output estimated by the regression models.

4 Experimental Results

The experimental section is split into two parts. While the aim of the first experiment is to reach significantly better classification accuracy results than in the previous attempts on the same data set, the goal of the latter is to better understand the choice for the kernel size parameter. The organization of the two experiments follows the guidelines in [4]. In order to deploy the central architecture, a Python with Tensorflow [1] framework was used.

4.1 Experiment 1: Manual Tuning

Pre-experimental Planning. From the initial experiments it was noticed that the runtime is relatively high, which would have restricted the number of setups to be tried. In search of a balance between accelerating runtime but not decreasing quality, the pictures were resized to a resolution of 256×256 pixels, from the original 800×600. This ensures a smoother training and also allows the usage of a low-volume model, as one of the measures implied in preventing overfitting.

Task. The goal is to test the CNN architecture, with its own implicit inner feature detection, on the colorectal cancer images and achieve a classification accuracy superior to that of the previous multi-stage approaches [31,33].

Experimental Setup. The samples are split into training/validation/test sets with the ratios 0.5, 0.25 and 0.25. This partitioning was chosen for a better control and measurement of the generalization power for the trained model. Given the small data set size, an extra step in ensuring that every class has the same percent of individuals in each set was taken, in order to avoid situations where a class is poorly represented by the training suite.

For fine-tuning the network in search of a successful set of hyper-parameters, it is aimed to maximize the accuracy on the validation set. In order to assign a score to a vector of hyper-parameters, the average accuracy obtained on validation sets of random partitions of the data collection by models trained with the respective hyper-parameters was used, as given by Eq. (1), where $M_{h,X}$ is the model trained with hyper-parameters h and training set X, $Acc_V(M)$ is the accuracy of model M computed on the validation set V as $\frac{\#correct}{|V|}$ and $\mathbf{E}_X$ is the expected value with respect to the distribution of X.

$$score(h) = \mathbf{E}_X[Acc_V(M_{h,X})] \tag{1}$$

After training the model 30 times with random initialization, a mean accuracy of 92.5% is obtained on the validation sets, with learning rate 10^{-3}, batch size 20, kernel sizes (7, 5, 3, 3, 3) and first kernel depth of 16.

Results and Visualization. On the test sets, the model was able to correctly classify, on average, 90.15% of examples. An example confusion matrix is given in Table 1. Figure 2a and b depict the activation intensities of the first and second convolutional layers, given a G0 image.

Table 1. Example of a confusion matrix on one of the test sets.

	G0	G1	G2	G3
G0	0.91	0	0.09	0
G1	0	1	0.	0
G2	0	0.2	0.73	0.07
G3	0	0	0	1

Discussion. Overall, the classification results are significantly better than those obtained in [31,33], as the percentage increase is of over 10%. The confusion matrix in Table 1 is obtained from a single run on the test set and it reveals that samples from G1 and G3 categories are classified with no error and 9% of the normal tissues are classified as grade 2 cancer. This type of errors (from G0 to G2) also occurred in [31], they are probably due to the manner of obtaining the G0 slides, as they are cut from larger images that contained borders between healthy and cancerous tissues and G2 had many instances [33]. The most problematic category is G2, with an accuracy of only 73%, but the vast majority of misclassified examples are labelled as G1, and none are considered normal tissues. Still, the confusion matrix is only computed to have a general overview, as in other runs the distribution might differ.

As observed in Fig. 2a and b, the network learned filters to recognize gland interior, shape and nuclei. They represent sections of the volume that passes through the network. The number of slices on each layer is determined by the depth of the convolution kernels, i.e. 8 and 16. Highlighted areas correspond to image regions producing larger values on that layer, in contrast to darker areas.

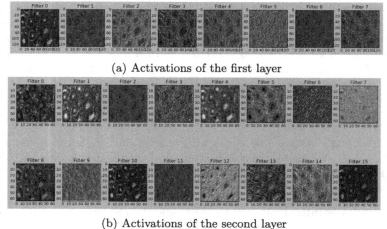

(a) Activations of the first layer

(b) Activations of the second layer

Fig. 2. Activation maps for a histological slice with normal tissue.

4.2   Experiment 2: Inspection of the Kernel Size Parameters Space

Pre-experimental Planning. Despite the image resizing, the runtime of the CNN remains relatively high in order to allow for the investigation of a large variety of parameter settings. The first attempt was to employ a random walk procedure to search for a better set of hyper-parameters. However, the search was still slow, as, in order to reach a relevant score, 10 different models were trained for every solution and the average on the validation sets was taken into account. Next, LHS was considered in order to have a better covering of the search space. All the tried combinations were therefore gathered, as they provided valuable information regarding the proper choice of the values for the kernel sizes. However, an optimization procedure (e.g., a GA) that would automatically search for the best combination of hyper-parameters was still not possible due to the time-consuming evaluations. Each fitness evaluation would assume the run of the CNN using the values provided by the GA candidate solutions. Consequently, the complement solution that was imagined during pre-experimentation was to build surrogate models. These would learn the correspondence between the input parameters and the accuracies obtained on the validation sets and would provide new outputs for novel configurations.

Task. The aim is to investigate the choice of the CNN kernel size hyper-parameters to improve the classification accuracy on the validation and test sets.

Experimental Setup. A random walk algorithm that had a budget of 10 evaluations per solution was firstly considered. Each such evaluation attempts to produce a score for a vector of hyper-parameters, according to Eq. (1). Given the computational constraints, the average validation score of only 10 models is taken. The solution with the highest score is taken to the test phase, where the performance of 30 models with the same hyper-parameters is evaluated.

The total number of tried LHS configurations is 75. The 5 kernel size parameters are chosen from the following intervals: [6, 16], [4, 12], [2, 6], [2, 6] and [2, 6]. As in the case of the random walk procedure, 10 models are considered and the average on the validation set represents the final outcome.

The random walk and the LHS configurations are next gathered and used by 4 different regression models. These are subsequently used to simulate further results in points that have not been previously explored with CNN. In order to search for better parameter settings, a canonical GA is used and the result provided by the surrogate models for each explored configuration is the value of the fitness evaluation. The regression models that are tried in the current experiment are: a linear model, a SVM with radial and linear kernels and regression decision trees. This conducts to 4 versions of the GA, as each one has a different fitness evaluation function. The population size of the GA is of 50, 150 iterations are considered as stop condition, while the crossover and mutation probabilities are of 0.8 and 0.1, respectively. The regression models and the GA are encoded into R implementations [23, 27, 35].

Table 2. The hyper-parameter values as discovered by the GA with the surrogate models. The associated percentages show the regression model estimations beside the actual evaluations of the CNN on the validation and test sets.

Regression model	Kernel size values	Surrogate output	CNN validation	CNN test
Linear model	(6, 11, 2, 5, 5)	93.93%	93.26%	90.91%
SVM radial	(12, 11, 4, 5, 5)	93.87%	93.41%	89.05%
SVM linear	(6, 11, 2, 5, 5)	93.91%	93.26%	90.91%
Regression trees	(10, 6, 3, 2, 5)	94.35%	94.17%	91.44%

Results and Visualization. The best result on the test set as obtained from the random walk is of 91.06% and is achieved from the parameter values (7, 4, 4, 4, 4). Table 2 shows the combinations of kernel size parameters given by the GA with the considered surrogate models and the results obtained for those settings.

Discussion. Generally, Table 2 indicates that there is a good agreement between the surrogate estimated outputs and the CNN actual results on the validation set. As concerns the prediction accuracy on the test set, the only weak result is obtained for the parameters discovered when the GA used the SVM with a radial kernel as the fitness function. Table 2 indicates that there is a wide variety of input values for the kernel sizes that achieve good results. As each surrogate approach models the solution landscape differently, the GA naturally found various sets of parameters, which generally proved to be successful (except for SVM radial) when the actual CNN was tested on them. The most appropriate found solution is the one discovered through the regression decision trees.

5   Conclusions and Future Work

A CNN approach is considered for the automated diagnosis of a set of colorectal cancer histopathological slides. The method provides significantly better results as compared to the previous approaches on the same data set. A drawback of the method is the large training time which does not permit the user to try a wide-range of CNN hyper-parameters. In this sight, by departing from a set of parameters and their results on the validation set, several regression models were employed to explore the regions of a surrogate parameter search landscape. A GA is used to intensify the search and several sets of parameter values are suggested. The approximations on the validation set coincide to a high extent with the actual values obtained by the CNN.

As concerns future work, transfer-learning seems a viable approach to bypass the generalization limits imposed by the relatively small dataset size. By training a set of convolutional layers on a similar dataset and using the resulting weights as a starting point of the main model afterwards, one can possible build a more robust solution. Also, new images are collected in the IMEDIATREAT project, so the data collection will expand, profiting even more from the CNN potential.

Acknowledgments. The second and third authors acknowledge the support of the research grant no. 26/2014, IMEDIATREAT - Intelligent Medical Information System for the Diagnosis and Monitoring of the Treatment of Patients with Colorectal Neoplasm -of the Romanian Ministry of National Education - Research and the Executive Agency for Higher Education Research Development and Innovation Funding.

References

1. Abadi, M., Agarwal, A., Barham, P., Brevdo, E., Chen, Z., Citro, C., Corrado, G.S., Davis, A., Dean, J., Devin, M., Ghemawat, S., Goodfellow, I., Harp, A., Irving, G., Isard, M., Jia, Y., Jozefowicz, R., Kaiser, L., Kudlur, M., Levenberg, J., Mané, D., Monga, R., Moore, S., Murray, D., Olah, C., Schuster, M., Shlens, J., Steiner, B., Sutskever, I., Talwar, K., Tucker, P., Vanhoucke, V., Vasudevan, V., Viégas, F., Vinyals, O., Warden, P., Wattenberg, M., Wicke, M., Yu, Y., Zheng, X.: TensorFlow: Large-scale machine learning on heterogeneous systems (2015). http://tensorflow.org/. Software available from tensorflow.org
2. Albarqouni, S., Baur, C., Achilles, F., Belagiannis, V., Demirci, S., Navab, N.: Aggnet: deep learning from crowds for mitosis detection in breast cancer histology images. IEEE Trans. Med. Imaging **35**(5), 1313–1321 (2016)
3. Anthimopoulos, M., Christodoulidis, S., Ebner, L., Christe, A., Mougiakakou, S.: Lung pattern classification for interstitial lung diseases using a deep convolutional neural network. IEEE Trans. Med. Imaging **35**(5), 1207–1216 (2016)
4. Bartz-Beielstein, T., Preuss, M.: The future of experimental research. In: Bartz-Beielstein, T., Chiarandini, M., Paquete, P., Preuss, M. (eds.) Experimental Methods for the Analysis of Optimization Algorithms, pp. 17–49. Springer, Heidelberg (2010)
5. Cireşan, D.C., Giusti, A., Gambardella, L.M., Schmidhuber, J.: Mitosis detection in breast cancer histology images with deep neural networks. In: Mori, K., Sakuma, I., Sato, Y., Barillot, C., Navab, N. (eds.) MICCAI 2013. LNCS, vol. 8150, pp. 411–418. Springer, Heidelberg (2013). doi:10.1007/978-3-642-40763-5_51
6. Czibula, G., Crişan, G.C., Pintea, C.M., Czibula, I.G.: Soft computing approaches on the bandwidth problem. Informatica **24**(2), 169–180 (2013). http://dl.acm.org/citation.cfm?id=2773202.2773203
7. Daher, N.M.: Deep learning in medical imaging: the not-so-near future. Diagnostic Imaging. http://www.diagnosticimaging.com/pacs-and-informatics/deep-learning-medical-imaging-not-so-near-future
8. Fukushima, K.: Neocognitron: a self-organizing neural network model for a mechanism of pattern recognition unaffected by shift in position. Biol. Cybern. **36**(4), 193–202 (1980)
9. Gorunescu, F., Belciug, S.: Boosting backpropagation algorithm by stimulus-sampling: application in computer-aided medical diagnosis. J. Biomed. Inform. **63**, 74–81 (2016)
10. Gorunescu, F., Belciug, S., Gorunescu, M., Badea, R.: Intelligent decision-making for liver fibrosis stadialization based on tandem feature selection and evolutionary-driven neural network. Expert Syst. Appl. **39**(17), 12824–12832 (2012). http://dx.doi.org/10.1016/j.eswa.2012.05.011
11. Iantovics, B.L.: Agent-based medical diagnosis systems. Comput. Inform. **27**(4), 593–625 (2012). http://www.cai.sk/ojs/index.php/cai/article/view/234

12. Iliescu, D.G., Dragusin, R.C., Cernea, D., Patru, C.L., Florea, M., Tudorache, S.: Intrapartum ultrasound - an integrated approach for best prognosis. Med. Ultrasonography **19**(1), 932 (2017)
13. Ioffe, S., Szegedy, C.: Batch normalization: accelerating deep network training by reducing internal covariate shift. In: Proceedings of the 32nd International Conference on Machine Learning (ICML 2015), pp. 448–456 (2015)
14. Ji, S., Xu, W., Yang, M., Yu, K.: 3D convolutional neural networks for human action recognition. IEEE Trans. Pattern Anal. Mach. Intell. **35**(1), 221–231 (2013)
15. Karpathy, A.: Stanford university cs231n: convolutional neural networks for visual recognition. http://cs231n.github.io/convolutional-networks/
16. Karpathy, A., Toderici, G., Shetty, S., Leung, T., Sukthankar, R., Fei-Fei, L.: Large-scale video classification with convolutional neural networks. In: Proceedings of the 2014 IEEE Conference on Computer Vision and Pattern Recognition, pp. 1725–1732 (2014)
17. Kramer, O.: Genetic Algorithm Essentials. SCI, vol. 679. Springer, Cham (2017). doi:10.1007/978-3-319-52156-5
18. Krizhevsky, A., Sutskever, I., Hinton, G.E.: Imagenet classification with deep convolutional neural networks. In: Advances in Neural Information Processing Systems, pp. 1106–1114 (2012)
19. Kullback, S., Leibler, R.A.: On information and sufficiency. Ann. Math. Stat. **22**(1), 79–86 (1951)
20. Lecun, Y., Bottou, L., Bengio, Y., Haffner, P.: Gradient-based learning applied to document recognition. Proc. IEEE **86**(11), 2278–2324 (1998)
21. Li, R., Emmerich, M.T.M., Eggermont, J., Bovenkamp, E.G.P., Bäck, T., Dijkstra, J., Reiber, J.H.C.: Optimizing a medical image analysis system using mixed-integer evolution strategies. In: Cagnoni, S. (eds.) Evolutionary Image Analysis and Signal Processing. SCI, vol. 213, pp. 91–112. Springer, Heidelberg (2009). doi:10.1007/978-3-642-01636-3_6
22. McKay, M.D., Beckman, R.J., Conover, W.J.: A comparison of three methods for selecting values of input variables in the analysis of output from a computer code. Technometrics **21**(2), 239–245 (1979). http://www.jstor.org/stable/1268522
23. Meyer, D., Dimitriadou, E., Hornik, K., Weingessel, A., Leisch, F.: e1071: Misc Functions of the Department of Statistics, TU Wien (2015)
24. Pereira, S., Pinto, A., Alves, V., Silva, C.A.: Brain tumor segmentation using convolutional neural networks in mri images. IEEE Trans. Med. Imaging **35**(5), 1240–1251 (2016)
25. Riera-Ledesma, J., Salazar-Gonzlez, J.J.: A heuristic approach for the travelling purchaser problem. Eur. J. Oper. Res. **162**(1), 142–152 (2005). http://www.sciencedirect.com/science/article/pii/S037722170300821X. Logistics: From Theory to Application
26. Roth, H.R., Lee, C.T., Shin, H.C., Seff, A., Kim, L., Yao, J., Lu, L., Summers, R.M.: Anatomy-specific classification of medical images using deep convolutional nets. In: 2015 IEEE 12th International Symposium on Biomedical Imaging (ISBI), pp. 101–104 (2015)
27. Scrucca, L.: GA: a package for genetic algorithms in R. J. Stat. Softw. **53**(4), 1–37 (2013)
28. Sirinukunwattana, K., Raza, S.E.A., Tsang, Y.W., Snead, D.R.J., Cree, I.A., Rajpoot, N.M.: Locality sensitive deep learning for detection and classification of nuclei in routine colon cancer histology images. IEEE Trans. Med. Imaging **35**(5), 1196–1206 (2016)

29. Srivastava, N., Hinton, G., Krizhevsky, A., Sutskever, I., Salakhutdinov, R.: Dropout: a simple way to prevent neural networks from overfitting. J. Mach. Learn. Res. **15**(1), 1929–1958 (2014)
30. Stoean, C., Stoean, R., Sandita, A., Mesina, C., Ciobanu, D., Gruia, C.L.: Investigation on parameter effect for semi-automatic contour detection in histopathological image processing. In: 2015 17th International Symposium on Symbolic and Numeric Algorithms for Scientific Computing (SYNASC), pp. 445–451 (2015)
31. Stoean, C.: In search of the optimal set of indicators when classifying histopathological images. In: 2016 18th International Symposium on Symbolic and Numeric Algorithms for Scientific Computing (SYNASC), pp. 449–455 (2016)
32. Stoean, C., Preuss, M., Stoean, R.: EA-based parameter tuning of multimodal optimization performance by means of different surrogate models. In: Genetic and Evolutionary Computation Conference, GECCO 2013, pp. 1063–1070. ACM (2013)
33. Stoean, C., Stoean, R., Sandita, A., Ciobanu, D., Mesina, C., Gruia, C.L.: SVM-based cancer grading from histopathological images using morphological and topological features of glands and nuclei. In: Pietro, G., Gallo, L., Howlett, R.J., Jain, L.C. (eds.) Intelligent Interactive Multimedia Systems and Services 2016. SIST, vol. 55, pp. 145–155. Springer, Cham (2016). doi:10.1007/978-3-319-39345-2_13
34. Stoean, C., Stoean, R., Sandita, A., Mesina, C., Gruia, C.L., Ciobanu, D.: Evolutionary search for an accurate contour segmentation in histopathological images. In: The ACM Genetic and Evolutionary Computation Conference Companion (GECCO 2015), pp. 1491–1492 (2015)
35. Therneau, T., Atkinson, B., Ripley, B.: rpart: Recursive Partitioning and Regression Trees (2015)
36. Young, S.R., Rose, D.C., Karnowski, T.P., Lim, S.H., Patton, R.M.: Optimizing deep learning hyper-parameters through an evolutionary algorithm. In: Proceedings of the Workshop on Machine Learning in High-Performance Computing Environments, pp. 4:1–4:5. ACM (2015)
37. Zaharie, D., Lungeanu, D., Zamfirache, F.: Interactive search of rules in medical data using multiobjective evolutionary algorithms. In: Proceedings of the 10th Annual Conference Companion on Genetic and Evolutionary Computation, GECCO 2008, pp. 2065–2072. ACM, New York (2008). http://doi.acm.org/10.1145/1388969.1389023

Deep Learning to Analyze RNA-Seq Gene Expression Data

D. Urda[1,3]([✉]), J. Montes-Torres[2,3], F. Moreno[2],
L. Franco[2,3], and J.M. Jerez[2,3]

[1] Andalucía Tech, ETSI Informática (España), Universidad de Málaga,
Málaga, Spain
durda@lcc.uma.es
[2] Departamento de Lenguajes y Ciencias de la Computación,
ETSI Informática (España), Universidad de Málaga, Málaga, Spain
[3] Inteligencia Computacional en Biomedicina (España),
Instituto de Investigación Biomédica de Málaga (IBIMA), Málaga, Spain

Abstract. Deep learning models are currently being applied in several
areas with great success. However, their application for the analysis of
high-throughput sequencing data remains a challenge for the research
community due to the fact that this family of models are known to work
very well in big datasets with lots of samples available, just the opposite
scenario typically found in biomedical areas. In this work, a first approximation on the use of deep learning for the analysis of RNA-Seq gene
expression profiles data is provided. Three public cancer-related databases are analyzed using a regularized linear model (standard LASSO)
as baseline model, and two deep learning models that differ on the feature selection technique used prior to the application of a deep neural
net model. The results indicate that a straightforward application of
deep nets implementations available in public scientific tools and under
the conditions described within this work is not enough to outperform
simpler models like LASSO. Therefore, smarter and more complex ways
that incorporate prior biological knowledge into the estimation procedure of deep learning models may be necessary in order to obtain better
results in terms of predictive performance.

Keywords: Deep learning · RNA-Seq · Personalized medicine ·
Machine Learning · Biomarkers discovery

1 Introduction

In the last years, artificial neural networks have raised back interest of the
research community on this family of Machine Learning (ML) models under
the tag "deep learning" [11]. Behind this recent interest, there are well-known
companies, such as Google or Microsoft among other private and public entities, that have made big investments to succeed applying deep neural networks
into several Artificial Intelligence (AI) areas [6,9,10,15]. The implementation of

© Springer International Publishing AG 2017
I. Rojas et al. (Eds.): IWANN 2017, Part II, LNCS 10306, pp. 50–59, 2017.
DOI: 10.1007/978-3-319-59147-6_5

new initialization and training procedures for this ML models [8], supported by the high computing resources available at these entities, has finally allowed to overcome the barrier that artificial neural networks were facing ten years ago.

Deep learning is actively used today in a wide range of fields, including Bioinformatics and Computational Medicine. Its strength working with graphical information has motivated many researches in the last few years to incorporate this ML models in their works. Thus, it has been successfully applied in medical image processing, where deep convolutional neural network have been proven to be robust pixel classifiers [3–5]. Indeed, solving image classification problems is not the only way deep learning can assist biomedical researches. As some recent works show [12,18], deep neural networks are being used for predictive modeling, using RNA-Seq data as input. We may ask ourselves, though, whether the use of deep learning, when it comes to produce predictive models, is as straight forward as it is in other kind of problems.

Despite the increasing amount of papers referencing the use of deep learning models in biomedical related areas, the authors of this work consider that there is still a long way to go in order to achieve a relevant improvement with respect to classical models in certain applications, like predicting the outcome for patients with gene expression datasets. This work aims to provide a first approximation of how to use a multi-layer feed-forward artificial neural network to analyze RNA-Seq gene expression data. For this purpose, three public RNA-Seq dataset are considered in order to predict the vital status of a patient at time t. Two deep learning models, which differ in the feature reduction procedure applied, are compared to a standard linear model with l_1-regularization (LASSO with homogeneous priors). Furthermore, feature selection, models estimation, selection and evaluation are performed using an honest validation scheme.

The rest of the article is organized as follows. Section 2 describes the datasets and ML models considered within the analysis as well as a description of the validation strategy used to compare the performance of the models. Then, Sect. 3 shows the results obtained with each model on the studied datasets. Finally, Sect. 4 provides some conclusions for this work.

2 Materials and Methods

2.1 Datasets

Free-public RNA-Seq gene expression datasets can be easily downloaded from The Cancer Genome Atlas (TCGA) website[1]. In particular, this work analyzes three datasets that have already been pre-processed to take into account batch effects and normalized through the RSEM procedure [13]. The first dataset is linked to Breast Invasive Carcinoma (BRCA) containing 199 cases and 1013 controls. The second database contains 81 cases and 245 controls of Colon Adenocarcinoma (COAD). The last dataset corresponds to a joint cohort of Kidney Chromophobe, Kidney renal clear cell carcinoma and Kidney renal papillary cell carcinoma (KIPAN) with 267 cases and 753 controls. Each sample in

---

[1] https://cancergenome.nih.gov/.

Table 1. Information of the RNA-Seq datasets: number of samples (N), number of genes (P) and class distribution (*control* $= 0$, *cases* $= 1$).

Name	N	P	Controls	Cases
BRCA	1212	20021	1013	199
COAD	326	19467	245	81
KIPAN	1020	20144	753	267

these datasets is finally described by approximately 20000 genes after applying a sanity check procedure where those genes that appeared to be constant across the sample are removed. Additionally, a $log_2(exp + 1)$ transformation of the genes expression levels was performed to make their distribution look as close as possible to a normal distribution. Table 1 shows the overall description of each dataset where the event of interest considered is the vital status of a given patient ($0 =$ "alive" are controls, $1 =$ "dead" are cases).

2.2 Methods

This work uses two different machine learning models to learn a given dataset $D = \{x_i, y_i\}$ of N samples, where $i \in [1, N]$, x_i represents a vector of P genes expression level describing the i-th sample, and y_i is the class label for the i-th sample. On one hand, a linear model is used assuming that the independent variable y_i can be represented as a linear combination of the dependent variables x_i. On the other hand, another model that enables to capture non-linear relationships is also considered as a possible alternative to linear models in order to push forward the predictive performance of this family of models. Next, we describe the models considered in this work:

– **Lasso**: this model is the baseline model in this work and corresponds to a standard LASSO model [16] with homogeneous priors. LASSO is a well-known linear model in the bioinformatics community and it is widely used for several and diverse tasks. LASSO tries to optimize the minimization problem depicted in Eq. 1:

$$\min_{\beta} \sum_{i=1}^{N} (y_i - f(x_i, \beta))^2 + \lambda \sum_{j=1}^{P} |\beta_j| \qquad (1)$$

In contrast to linear or logistic regression, this model includes an l_1-penalty term to set as many features as possible to zero unless the data tells us not to do it. Moreover, this term is controlled by a regularization parameter λ ($\lambda = 0$ would exactly correspond to the objective function in linear or logistic regression). Therefore, LASSO is an embedded method that performs feature selection at the same time that the model is adjusted to data. The R package *glmnet* [7] has been used to estimate a LASSO model due to its

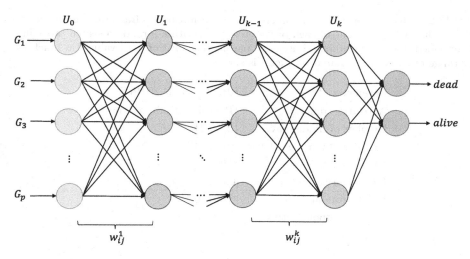

Fig. 1. General architecture of a multi-layer feed-forward artificial neural network to perform binary classification. The input layer composed of P genes expression levels is connected to units of subsequent k-hidden layers through synaptic weights ω_{ij}^k.

easy interface that automatically allows to learn the regularization parameter λ through cross-validation.

- **DeepNets**: among the different deep learning models available, a multi-layer feed-forward artificial neural network was chosen to learn existing non-linearities between the input and output spaces. Figure 1 shows the classical architecture of these artificial neural networks. In concrete, it shows an architecture of $K + 1$ layers (one input layer and K hidden layers) and a vector $U = \{U_0, U_1, ..., U_k\}$ representing the number of units in the 0-th layer (input layer) and the k-th hidden layer, respectively.

  In this type of models, it can be easily inferred that the number of parameters to be learned ω_{ij}^k (synaptic weights), where $k \in [1, K]$, $i \in [1, U_k]$ and $j \in [1, U_{k-1}]$, increases drastically in comparison to the number of parameters β of the standard LASSO. The more layers and the more neurons per layer we add, the more number of parameters ω_{ij}^k will be obtained and needed to be learned. Conversely, this number remains equal to the number of genes in the input space for the LASSO model. Therefore, deep nets require to introduce strategies to avoid overfitting such as regularization or dropout. Regularization aims to impose some constraint in the optimization procedure, being the l_1-penalty (lasso penalty: set as many ω_{ij}^k as possible to zero) or the l_2-penalty (ridge penalty: avoid setting ω_{ij}^k to high values) the most known and used ones. The dropout strategy [14] aims to "disconnect" some of those links between neurons of different layers to decrease the number of parameters that needs to be learned. Both strategies are complementary and applicable jointly. The R package *h2o* [1] has been used in this work to fit a deep net to data using regularization and dropout strategies thus dealing with overfitting issues.

Table 2. Subset of deep nets' parameters included in the *h2o* framework implementation. A list with the name of each parameter tuned within a random search procedure is provided. For each studied dataset and parameter, a list of tried values (in brackets) or range of values (in squared brackets) is shown.

Parameter	BRCA COAD KIPAN
Activation function	{Rectifier, Tanh, Maxout}
Number of hidden layers	{2, 3, 4}
Number of units per layer	[10, 200]
L1 regularization	[0.001, 0.1]
L2 regularization	[0.001, 0.1]
Input dropout ratio	[0.001, 0.1]
Hidden dropout ratios	[0.001, 0.1]

Furthermore, two different feature dimensionality reduction techniques were applied prior to estimating a deep net in order to reduce the input space and, therefore, the number of parameter to be learned:

- *DeepNet$_i$*: this procedure applies a univariate t-test to compare if the difference of genes expression levels in controls and cases are statistically significant. Genes under a *p*-value threshold of 0.001 are retained and then given to a correlation feature reduction procedure that gets rid of highly correlated genes until the number of retained genes is similar to the average number of genes kept in the standard LASSO.
- *DeepNet$_{ii}$*: this procedure uses a standard LASSO model to retrieve the most important genes for the given outcome.

In both cases, the selected genes are given as input to fit a deep net to data, discarding the remaining genes from the analysis. Additionally, a random search was performed to tune some parameters linked to a deep net model in *h2o*, where Table 2 shows the parameters considered to be tuned together with their respective ranges for each dataset analyzed.

2.3   Validation Strategy

A known and valid evaluation strategy is always required in order to estimate generalization error and compare the performance of the models considered in the analysis. In particular, this work implements Z repetitions of k-fold cross-validation, where $Z = 20$ and $k = 10$. For a given repetition, this evaluation strategy divides the complete dataset into k non-overlapping folds of equal sizes and applies an iterative procedure that uses $k - 1$ folds to fit the models and the unseen fold left apart to test the performance (rotating train and test folds on each iteration). The utilization of this validation strategy rather than other

---

Algorithm 1. Pseudocode of our methodological approach

```
 1: dataset ⇐ {"BRCA", "COAD", "KIPAN"} //choose one option
 2: model ⇐ {"lasso", "deepnet"} //choose one option
 3: filtering ⇐ {"ttest − cor", "lasso"} //choose one option
 4: X ⇐ load_design_matrix(dataset)
 5: Y ⇐ load_outcome(dataset)
 6: partitions ⇐ load_partitions(dataset)
 7:
 8: for Z = 1 → 20 do
 9:   folds ⇐ get_folds(partitions, Z)
10:   for k = 1 → 10 do
11:     Xtrain ⇐ get_design_matrix(X, folds, k, "train") //training data
12:     Xtest ⇐ get_design_matrix(X, folds, k, "test") //test data
13:     Ytrain ⇐ get_outcome(Y, folds, k, "train")
14:     Ytest ⇐ get_outcome(Y, folds, k, "test")
15:
16:     if (model=="deepnet") then
17:       retained_genes ⇐ apply_filtering(Xtrain, Ytrain, filtering)
18:       Xtrain ⇐ Xtrain[, retained_genes]
19:     end if
20:
21:     fitted ⇐ fit_model(Xtrain, Ytrain, model) //performs model selection internally
22:     predictions ⇐ predict(fitted, Xtest)
23:     measures[k] ⇐ performance(Ytest, predictions)
24:   end for
25:   results[Z] ⇐ mean(measures)
26: end for
27:
28: print(results)
```

---

well-known strategies such as leave-one-out, bootstrapping, holdout, etc., is motivated on (i) its simplicity and small computational resources needed, and (ii) the proved that there is no universal unbiased estimator of the variance of k-fold cross-validation [2]. Algorithm 1 contains a high-level description of the methodological approach used to carry out the analysis.

The Area Under the Curve (AUC) was computed to compare the performance of each model since the three studied datasets are highly imbalanced (see Table 1). Additionally, both the number of genes retained by each of the feature reduction procedures considered and the total time (in minutes) required to execute the validation strategy described were computed to open a discussion over the results.

3 Results

This work has analyzed three cancer-related RNA-Seq datasets using the models described in Sect. 2.2. The quantitative results are shown in Table 3. In general, it can be seen that the predictive performance in terms of AUC is relatively poor independently of the model used. On two out of three databases, BRCA and COAD, the performance measured by the AUC is not over 0.65, and particularly in the COAD dataset the prediction of the vital status of a patient from RNA-Seq gene expression profiles turned out to be quite difficult (AUC under 0.6, close to random predictions). Conversely, the predictive performance of a simple linear model on the KIPAN dataset seems to be good with AUC values around 0.77.

Table 3. Average AUC results and number of retained genes for 20 repetitions of 10-fold cross-validation over each RNA-Seq dataset using the three models proposed: standard *Lasso*, *DeepNet$_i$* and *DeepNet$_{ii}$*. 95% CI and standard deviation are shown for the AUC and #genes columns respectively. The last column shows the total number of minutes required for the corresponding analysis.

Dataset	Model	AUC	#genes	time (mins.)
BRCA	*Lasso*	0.65 [0.62, 0.67]	285.54 ± 25.83	501.79
	DeepNet$_i$	0.62 [0.58, 0.65]	242.02 ± 8.01	2294.83
	DeepNet$_{ii}$	0.65 [0.63, 0.68]	285.54 ± 25.83	9768.37
COAD	*Lasso*	0.57 [0.52, 0.63]	69.64 ± 11.63	30.89
	DeepNet$_i$	0.58 [0.54, 0.62]	37.29 ± 1.52	2699.84
	DeepNet$_{ii}$	0.57 [0.52, 0.61]	69.64 ± 11.63	2370.15
KIPAN	*Lasso*	0.77 [0.76, 0.78]	268.81 ± 32.54	93.60
	DeepNet$_i$	0.72 [0.68, 0.75]	201.64 ± 3.44	2633.52
	DeepNet$_{ii}$	0.75 [0.73, 0.78]	268.81 ± 32.54	9281.08

Focusing on the models considered in this work (*Lasso*, *DeepNet$_i$*, *DeepNet$_{ii}$*), the results obtained across the three databases confirmed us that the straightforward use of existing implementations of a multi-layer feed-forward artificial neural network (such as the R package *h2o*) will very rarely push the predictive performance further away compared to a simple regularized linear model. In two out of three databases, BRCA and KIPAN, deep learning models do not outperform the baseline model *Lasso*. On the other hand, in the COAD dataset a deep learning model estimated on the retained genes after applying a univariate t-test combined with a correlation filtering procedure turned out to slightly improve the AUC after executing 20 repetitions of 10-fold cross-validation (AUC from 0.57 to 0.58). Nevertheless, this tiny improvement is not statistical significant as indicated by the overlap observed for the 95% confidence intervals, indicating that our baseline model *Lasso* can also achieve similar predictive performance depending on the data used to fit the models.

Regarding the number of genes obtained after filtering reduction to finally estimate the models, two out of the three model (*Lasso* and *DeepNet$_{ii}$*) are using exactly the same average numbers since genes retained by the embedded method *Lasso* are used in both cases (approximately 275 genes in BRCA and KIPAN, or 70 genes in COAD). It turned out that the average numbers of genes retained by *Lasso* is slightly higher across the three databases, although the numbers of genes retained by the filtering procedure used in *DeepNet$_i$* is close to the one in *Lasso* (242 in BRCA, 37 in COAD and 202 in KIPAN), thus making these results comparable. However, it can be highlighted the larger variability in terms of size of the genetic signatures obtained by *Lasso* in contrast to a simple t-test followed by a correlation filtering procedure. Analyzing the robustness of the genetic signatures found is beyond the scope of this work since it would lead to a complete different paper, although that type of analysis will constitute a complement to these results in order to state the (un)suitability of *Lasso* as model for biomarkers discovery [17].

Running time is also an important factor to take into account when using deep learning models. Particularly, the fitting procedure of deep learning models considered within this work required much more time due to the number of parameters that need to be tuned (see Table 2), in contrast to the standard *Lasso* model. $DeepNet_i$ needed minimum five times more minutes (see BRCA dataset in Table 3) to achieve similar predictive performance than *Lasso*, going up to almost 100 times more minutes in the COAD dataset. The case of $DeepNet_{ii}$ is even worse in BRCA and KIPAN, where this model required 3 times more minutes than the $DeepNet_i$. Independently of the quantitative numbers, these results clearly show us how expensive the estimation procedure of deep learning models is in comparison to more simple models. Moreover, in this particular analysis and under the described conditions the use of deep learning models is not suggested since it will take minimum five times more minutes to finally obtain similar predictive performance.

4   Conclusions

This paper has presented a first approximation on the straightforward use of deep learning models existing implementations for the analysis of RNA-Seq gene expression profiles databases. In concrete, it considered a multi-layer feed-forward artificial neural network as deep learning model in combination with two different feature reduction techniques, and a standard LASSO (regularized linear model) as the baseline model to try to outperform. This work has used an honest validation strategy to analyze three public cancer-related databases, where both feature reduction and model estimation were performed in a train dataset and the resulting fitted model was evaluated in an independent test dataset.

In general, the combination of deep learning models with the two considered feature reduction techniques very rarely outperformed a simple standard LASSO in terms of AUC. Furthermore, the estimation of the proposed deep learning models required minimum five times more minutes than LASSO due to the number of parameters that need to be tuned in such models, thus suggesting that using deep learning under the described conditions to predict the vital status of a patient from RNA-Seq data is not suggested. The exploration of this research line lead us to conclude that using a simple feature reduction procedure to reduce the number of genes and subsequently fit a deep learning model will take us much more execution time in order to obtain similar predictive performances.

Despite the discouraging results obtained in this work, there is no need to spread out a negative message in relation to the application of deep learning for RNA-Seq data analysis. Conversely, there is a big hope in pushing predictive performance forward with this type of models. To this end, this work has allowed us to realize that smarter use of deep learning models must be done to be successful in this research line. For instance, deep learning as stack auto-encoders could be used to somehow compress the information of 20000 genes into fewer variables, thus allowing to use any other ML model with these new compressed features as inputs. On the other hand, imposing constrains in the optimization process of deep learning models in such a way that biological knowledge is taken into account may lead to better results in terms of performance. Finally, finding ways of making this type of models interpretable would be desirable from the clinical point of view, and for this purpose using published knowledge of relationships between Single Nucleotide Polymorphisms (SNPs), genes, pathways, proteins, etc., could be a possible way of defining the network architecture.

Acknowledgements. The authors acknowledge support through grants TIN2014-58516-C2-1-R from MICINN-SPAIN which include FEDER funds, and from ICE Andalucía TECH (Spain) through a postdoctoral fellowship.

References

1. Aiello, S., Kraljevic, T., Maj, P., with contributions from the H2O.ai team: h2o: R Interface for H2O (2016). https://CRAN.R-project.org/package=h2o. R package version 3.10.0.8
2. Bengio, Y., Grandvalet, Y.: No unbiased estimator of the variance of K-fold cross-validation. J. Mach. Learn. Res. **5**, 1089–1105 (2004)
3. Cadieu, C., Hong, H., Yamins, D., Pinto, N., Ardila, D., Solomon, E., Majaj, N., DiCarlo, J.: Deep neural networks rival the representation of primate it cortex for core visual object recognition. PLoS Comput. Biol. **10**(12) (2014)
4. Ciompi, F., de Hoop, B., van Riel, S., Chung, K., Scholten, E., Oudkerk, M., de Jong, P., Prokop, M., van Ginneken, B.: Automatic classification of pulmonary peri-fissural nodules in computed tomography using an ensemble of 2D views and a convolutional neural network out-of-the-box. Med. Image Anal. **26**(1), 195–202 (2015)
5. Cireşan, D.C., Giusti, A., Gambardella, L.M., Schmidhuber, J.: Mitosis detection in breast cancer histology images with deep neural networks. In: Mori, K., Sakuma, I., Sato, Y., Barillot, C., Navab, N. (eds.) MICCAI 2013. LNCS, vol. 8150, pp. 411–418. Springer, Heidelberg (2013). doi:10.1007/978-3-642-40763-5_51
6. Deng, L., Li, J., Huang, J.T., Yao, K., Yu, D., Seide, F., Seltzer, M.L., Zweig, G., He, X., Williams, J., Gong, Y., Acero, A.: Recent advances in deep learning for speech research at microsoft. In: ICASSP, pp. 8604–8608. IEEE (2013)
7. Friedman, J., Hastie, T., Tibshirani, R.: Regularization paths for generalized linear models via coordinate descent. J. Stat. Softw. **33**(1), 1–22 (2010)
8. Glorot, X., Bengio, Y.: Understanding the difficulty of training deep feedforward neural networks. In: Proceedings of the International Conference on Artificial Intelligence and Statistics (AISTATS 2010). Society for Artificial Intelligence and Statistics (2010)
9. He, K., Zhang, X., Ren, S., Sun, J.: Deep residual learning for image recognition. In: The IEEE Conference on Computer Vision and Pattern Recognition (CVPR) (2016)
10. Le, Q., Ranzato, M., Monga, R., Devin, M., Chen, K., Corrado, G., Dean, J., Ng, A.: Building high-level features using large scale unsupervised learning. In: International Conference on Machine Learning (2012)
11. LeCun, Y., Bengio, Y., Hinton, G.: Deep learning. Nature **521**(7553), 436–444 (2015)

12. Leung, M., Xiong, H., Lee, L., Frey, B.: Deep learning of the tissue-regulated splicing code. Bioinformatics **30**(12), I121–I129 (2014)
13. Li, B., Dewey, C.N.: RSEM: accurate transcript quantification from rna-seq data with or without a reference genome. BMC Bioinform. **12**(1), 323 (2011)
14. Srivastava, N., Hinton, G., Krizhevsky, A., Sutskever, I., Salakhutdinov, R.: Dropout: a simple way to prevent neural networks from overfitting. J. Mach. Learn. Res. **15**(1), 1929–1958 (2014)
15. Sutskever, I., Vinyals, O., Le, Q.V.: Sequence to sequence learning with neural networks. CoRR abs/1409.3215 (2014)
16. Tibshirani, R.: Regression shrinkage and selection via the lasso: a retrospective. J. Roy. Stat. Soc. Ser. B (Stat. Methodol.) **58**(1), 267–288 (1996)
17. Urda, D., Aragon, F., Veredas, F., Franco, L., Jerez, J.M.: L1-regularization model enriched with biological knowledge. In: Proceedings of the 5th International Work-Conference on Bioinformatics and Biomedical Engineering (IWBBIO 2017) (2017)
18. Wenger, Y., Galliot, B.: Rnaseq versus genome-predicted transcriptomes: a large population of novel transcripts identified in an illumina-454 hydra transcriptome. BMC Genomics **14**(1) (2013)

Assistive Rehabilitation Technology

Designing BENECA m-Health APP, A Mobile Health Application to Monitor Diet and Physical Activity in Cancer Survivors

Mario Lozano-Lozano[1,2], Jose A. Moral-Munoz[3],
Noelia Galiano-Castillo[1,2], Lydia Martín-Martín[1],
Carolina Fernández-Lao[1,2], Manuel Arroyo-Morales[1,2(✉)],
and Irene Cantarero-Villanueva[1,2]

[1] Department Physical Therapy, University of Granada, Granada, Spain
{mlozano,noeliagaliano,lydia,carolinafl,marroyo,
irenecantarero}@ugr.es
[2] Mixed University Sport and Health Institute (IMUDS), Granada, Spain
[3] Department Nursing and Physiotherapy, University of Cadiz, Cadiz, Spain
joseantonio.moral@uca.es

Abstract. This is the abstract of a proposed mobile health application for assessing and monitoring healthy lifestyles (in terms of diet and physical activity levels) in cancer survivors, to be fully exposed at the IWANN 2017. The main goal of this mobile health application is to help cancer patients with energy imbalance, which can increase the risk of recurrence and other associated problems, such as metabolic syndrome and even death, to adhere to the international healthy recommendations in terms of diet and physical activity. The system, called BENECA m-Health app, is still in development, and will be a reliable instrument to assess physical activity and diet in cancer survivors, offering them a real-time energy balance feedback and attempting to overcome specific identified barriers to facilitate the inclusion of exercise and healthy diet into supportive care programs for cancer survivors. This mobile application has been designed to address the new needs for support of breast cancer survivors, reflecting the emerging need to merge new, low cost treatment options. This m-Health System could be a promising approach for dietary and physical assessment, as well as for intervention programs, which can be used whenever and wherever patients want.

Keywords: Cancer survivors · Telemedicine · mHealth · Physical activity · Diet · Energy balance

1 Introduction

"Eat less and move more". The message seems to be simple and is widely known [1], however, the excess in energy intake (diet) and low physical activity remain a major problem in cancer survivors [2]. One of the most important determinant of the risk of recurrence or even death is energy imbalance in most prevalent cancer processes [3]. Clinical experience suggests a high demand and interest among, for example, survivors

I. Rojas et al. (Eds.): IWANN 2017, Part II, LNCS 10306, pp. 63–70, 2017.
DOI: 10.1007/978-3-319-59147-6_6

of breast cancer, based on a healthy lifestyle (increased physical activity, weight loss and healthy diet). Knowledge seems available but there is a clear weakness in getting a social significance for this solution that appears to be simple. Recent research points out that even knowing the benefits of interventions aimed at promoting energy balance among survivors of cancer, it is unrealistic to expect that most of them, with a strong sedentary habit, comply with the current guidelines of good practice [4]. In this sense, it is important to enhance the adherence of cancer survivors to healthy lifestyles through reliable and feasible assessment methods of diet and physical activity and the promotion of tailored cost-effective behavior interventions.

Cancer is one of the most incident diseases worldwide [5]. There are over 14 million of new cancer cases every year but it is estimated that it reaches approximately 22 million by 2030. Breast cancer is the most commonly diagnosed cancer in women, accounting for approximately 16% of all female cancers. An estimated 519,000 women died from breast cancer in 2004 and, although this cancer is considered a disease of the developed world, the majority (69%) of deaths from this cause are recorded in developing countries [6]. Fortunately, the survival rate is very encouraging, due to developments in screening and improved treatments [7].

Currently, accelerometry and direct observation, daily records or 24-hour dietary recall are some assessment instruments, which are known as gold standards for measuring physical activity and diet. However, these procedures are time-consuming for patients and investigators [8].

At present, there is an increase in scientific evidence on the use of new technologies to assess aspects related to different pathologies, including cancer [8–10], not only the e-Health applications (those which use internet or PCs) but also mobile health application. Mobile health (m-Health) applications could be used as an important advance with some advantages, such as ease of use for participants (which can be used whenever and wherever people want), instant self-report data entry, cost reduction and elimination of interviewer-associated bias [11]. In general population there have been developed some m-Health apps to promote healthy lifestyles, or even in some pathologies [12, 13]. However, there is not information about recording energy balance in term of diet and physical activity and giving instant energy balance feedback in cancer survivors. The protocol of the full study has been recent publish [14].

2 Objectives

The BENECA (ENErgy Balance on Cancer) m-Health System is been created with the main goal of helping cancer patients with energy imbalance, which can increase the risk of recurrence and other associated problems, such as metabolic syndrome and even death.

It aims to help cancer survivors complete their nutrition plans and exercise programs, supporting them as they adjust their individual needs according to the International Guidelines [15, 16] for this population.

This system pretends to facilitate adherence of the people who have had cancer to the recommendations on energy balance derived from the American Cancer Society

(American Cancer Society, ACS) and the Global Fund for Research on Cancer (World Cancer Research Fund, WCRF).

The BENECA m-Health System will facilitate the possibility of an individualized checking of balance between the amount and type of food intake and the time spent on physical activity in the context of work or leisure. It will support cancer survivors to complete their nutrition plans and exercise programs, helping them to adjust their individual needs. Moreover, it will serve as a tool that facilitates interaction between patients and health professionals and provide a feedback system regarding the energy balance so necessary to preserve in these patients.

3   System Definition

3.1   Target Behaviors

The BENECA m-Health application is a mobile application with a commercial server and centralized data storage. The BENECA mobile app is a native Android application that was developed using Android Studio. As a database engine, BENECA has a twin engine; on the one hand, in the mobile application, it has a SQLite database as a repository cache and collects patient information to send to the system. Moreover, for integration with the BENECA server (the second database engine), we made requests with the latest technology, RESTful. The BENECA Server is a centralized data management server written in Python 2.7.x with the Django framework. Its engine database is Mysql 5.x, which offers loading and reading information from a RESTful API integration (Fig. 1).

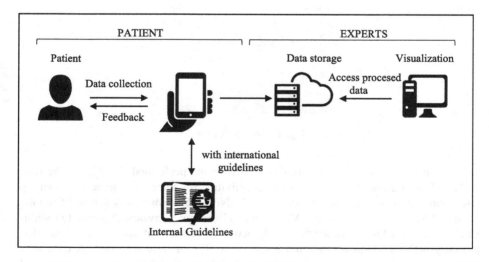

Fig. 1. Top level view of the system

3.2   Description of the Participant's Interaction with the System

Participant. The BENECA m-Health System asks patients to register the food and drinks that they consumed and the different activities that were performed during the previous day. With respect to diet, with an open structure and six time periods, the application takes the form of a diet questionnaire (dietary record) that is structured by intake times. For each time period, participants use the app to report foods and drinks consumed each day (breakfast, including midmorning, lunch, tea, dinner or snack), and they are also asked to record the location of the intake (home, office, etc.). The food and drink items are selected from a predefined list adapted from the Spanish food database BEDCA (©AESAN/BEDCA Base de Datos Española de Composición de Alimentos v1.0 (2010)), and if a food was not available on the list, users were asked to choose the most similar option that was available (Fig. 2).

Fig. 2. Diet collection

Similarly, to report the activities that participants performed throughout the day, BENECA asked about common physical activities for three time periods (morning, afternoon and evening). This section of the BENECA System was developed from the validated Spanish version of the Minnesota Leisure-time Physical Activity Questionnaire [17]. Users had to record their daily activities in terms of duration and intensity, but only activities with a minimum duration of 10 min were recorded. A metabolic equivalent value (MET) was assigned to each activity based on the Compendium of Physical Activities, which was based on the activity count cut-offs that correspond to moderate to vigorous physical activity MET levels ($\geq$ 3.0 METs) [18] (Fig. 3).

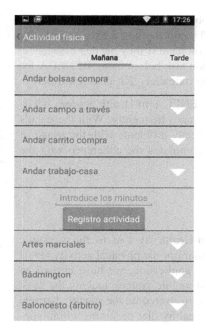

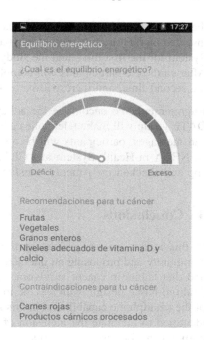

Fig. 3. Physical activity collection

Fig. 4. Energy balance feedback

Users also recorded their weight (kg) and height (cm). At the end of the survey, a summary of the recorded information was displayed.

After entering the information, the system provides the patient with a simple and understood message about the level of their energy balance (Fig. 4), and it detects if there is an energy imbalance (the patient has consumed too much energy relative to the requirements, including basal metabolism and physical activity). Additionally, based on their individual profiles, they receive general physical activity recommendations, which are based on the American College of Sports Medicine reference exercise guidelines in cancer patients [19]. Additionally, they are asked to substitute foods that are considered potentially carcinogenic with others that have protective capacities against cancer. All of these recommendations were developed following the guidelines of the American Cancer Society [20] and the recommendations of the World Cancer Research Fund International (WCRF). Adherence to these guidelines has shown a risk reduction in the overall cancer mortality and, specifically, in colon and breast cancer [21].

Health sciences professionals: a large number of clinicians could benefit due to knowing patient's lifestyles instantly and consistently, better intervention strategies can be developed, following the latest worldwide recommendations.

3.3   Activity Loops

Apprenticeship. Participants will attend to the iMUDS Center (Mixed Institute University Sport and Health) of the University of Granada, and research staff installed the BENECA m-Health System on the patients' mobile phones. Once the app will be installed on their mobile phones, the participants will complete the BENECA m-Health System for the first time, and the study staff will be on hand if they needed assistance. Afterwards, the participants will complete a distraction task before using BENECA for the second time, and there will be an overlap of approximately 2 h between times.

Engagement. To encourage the adherence using BENECA m-Health System, CUIDATE group will make telephone calls and sent messages frequently. With these calls and messages, participants will be also able to solve any problems with the usage of BENECA m-Health System and CUIDATE group stimulated the adherence with the app and checked the patients' satisfaction.

4   Conclusions

In this study, we are presenting the BENECA m-Health System to the scientific community and proposing an interactive method to monitor the physical activity levels and diet habits in cancer survivors. To develop this application, we have chosen an Android-based design, which is the most widespread mobile system, making the most of the smartphone capabilities. In this study, we hope to overcome the specific barriers identified to facilitate the inclusion of exercise and healthy diet programs within supportive care for cancer survivors.

The future results and validation of the BENECA m-Healt System will have many clinical and research implications. The BENECA m-Health System could be suitable for use in cancer units to assess energy balance-related behaviours at one point in time and, potentially, how they change over time, before and after surgical procedures. Moreover, combining the BENECA m-Health System with an objective measurement of physical activity may be a promising approach that is worth investigating in the future.

If this system is reliable, it will highlight the potential advantage and cost reduction provided using a mobile app. This m-Health System could be a promising approach for dietary and physical assessment, as well as for intervention programs, which can be used whenever and wherever patients want. Moreover, the results of this study could garner support for the use of this type of strategy in an increasing number of 17.8 million cancer patients in the European Union.

Acknowledgments. The study is funded by the Spanish Ministry of Economy and Competitiveness (Plan Estatal de I + D + I 2013–2016) and (TIN2016-75850-R), Fondo de Investigación Sanitaria del Instituto de Salud Carlos III (PI14/01627), Fondos Estructurales de la Unión Europea (FEDER) and by the Spanish Ministry of Education (FPU14/01069). This study takes place thanks to the additional funding from the University of Granada, Plan Propio de Investigación 2016, Excellence actions: Units of Excellence; Unit of Excellence on Exercise and Health (UCEES). This is part of a Ph.D. Thesis conducted in the Clinical Medicine and Public Health Doctoral Studies of the University of Granada, Spain.

References

1. Hudisa, C.A., Jones, L.: Promoting exercise after a cancer diagnosis: easier said than done. Br. J. Cancer (2014). doi: 10.1038/bjc.2014.12
2. Ballard-Barbash, R., Friedenreich, C.M., Courneya, K.S., et al.: Physical activity, biomarkers, and disease outcomes in cancer survivors: a systematic review. J. Nat. Cancer Inst. **104**, 815–840 (2012). doi:10.1093/jnci/djs207
3. Fomby, P., Cherlin, A.J.: Interventions to promote energy balance and cancer survivorship: European and North American priorities for research and care. Cancer **119**, 2143–2150 (2013). doi:10.1038/nature13314.A
4. Bourke, L., Homer, K.E., Thaha, M.A., et al.: Interventions to improve exercise behaviour in sedentary people living with and beyond cancer: a systematic review. Br. J. Cancer **110**, 831–841 (2014). doi:10.1038/bjc.2013.750
5. Jemal, A., Vineis, P., Bray, F., et al.: The Cancer Atlas, 2nd edn. American Cancer Society, Atlanta (2014)
6. Soares, E.W.S., Nagai, H.M., Bredt, L.C., et al.: Morbidity after conventional dissection of axillary lymph nodes in breast cancer patients. World J. Surg. Oncol. **12**, 67 (2014). doi:10. 1186/1477-7819-12-67
7. Bray, F., Jemal, A., Torre, L.A., et al.: Long-term realism and cost-effectiveness: primary prevention in combatting cancer and associated inequalities worldwide. J. Nat. Cancer Inst. (2015). doi: 10.1093/jnci/djv273
8. Hillier, F., Batterham, A., Crooks, S., et al.: The development and evaluation of a novel Internet-based computer program to assess previous-day dietary and physical activity behaviours in adults: the synchronised nutrition and activity program for adults (SNAPATM). Br. J. Nutr. **107**, 1221–1231 (2012). doi:10.1017/S0007114511004090
9. Coughlin, S.S., Besenyi, G.M., Bowen, D., De Leo, G.: Development of the physical activity and your nutrition for cancer (PYNC) smartphone app for preventing breast cancer in women. mHealth **3**, 5 (2017). doi:10.21037/mhealth.2017.02.02
10. Robertson, M.C., Tsai, E., Lyons, E.J., et al.: Mobile health physical activity intervention preferences in cancer survivors: a qualitative study. JMIR mHealth uHealth **5**, e3 (2017). doi:10.2196/mhealth.6970
11. Wang, Q., Egelandsdal, B., Amdam, G.V., et al.: Diet and physical activity apps: perceived effectiveness by app users. JMIR mHealth uHealth **4**, e33 (2016). doi:10.2196/mhealth.5114
12. Du, H., Venkatakrishnan, A., Youngblood, G.M., et al.: A group-based mobile application to increase adherence in exercise and nutrition programs: a factorial design feasibility study. JMIR mHealth uHealth **4**, e4 (2016). doi:10.2196/mhealth.4900
13. Bantum, E.O., Albright, C.L., White, K.K., et al.: Surviving and thriving with cancer using a web-based health behavior change intervention: randomized controlled trial. J. Med. Internet Res. **16**, e54 (2014). doi:10.2196/jmir.3020
14. Lozano-Lozano, M., Martín-Martín, L., Galiano-Castillo, N., et al.: Integral strategy to supportive care in breast cancer survivors through occupational therapy and a m-health system: design of a randomized clinical trial. BMC Med. Inform. Decis. Mak. **16**, 1–10 (2016). doi:10.1186/s12911-016-0394-0
15. Denlinger, C.S., Ligibel, J.A., Are, M., et al.: Survivorship: nutrition and weight management, version 2.2014. JNCCN J. Nat. Compr. Cancer Netw. **12**, 1396–1406 (2014)
16. Rock, C.L., Doyle, C., Demark-Wahnefried, W., et al.: Nutrition and physical activity guidelines for cancer survivors. CA Cancer J. Clin. **62**, 243–74. doi:10.3322/caac.21142

17. Ruiz Comellas, A., Pera, G., Baena Díez, J.M., et al.: Validation of a Spanish short version of the minnesota leisure time physical activity questionnaire (VREM). Rev. Esp. Salud. Publica. **86**, 495–508 (2012). doi:10.4321/S1135-57272012000500004

18. Freedson, P.S., Melanson, E., Sirard, J.: Calibration of the computer science and applications inc. accelerometer. Med. Sci. Sports Exerc. **30**, 777–781 (1998). doi:10.1097/00005768-199805000-00021

19. Schmitz, K.H., Courneya, K.S., Matthews, C., et al.: American college of sports medicine roundtable on exercise guidelines for cancer survivors. Med. Sci. Sports Exerc. **42**, 1409–1426 (2010). doi:10.1249/MSS.0b013e3181e0c112

20. Demark-wahnefried, W., Bandera, E.V., Gapstur, S., Patel, A.V.: American cancer society guidelines on nutrition and physical activity for cancer prevention reducing the risk of cancer with healthy food choices and physical activity. CA Cancer J. Clin. **62**, 30–67 (2012). doi:10.3322/caac.20140

21. Thomson, C.A., Mccullough, M.L., Wertheim, B.C., et al.: Nutrition and physical activity cancer risk, and mortality in the women's health initiative. Cancer Prev. Res. **7**, 42–53 (2014). doi:10.1158/1940-6207.CAPR-13-0258

Automatic 2D Motion Capture System
for Joint Angle Measurement

Carlos Bailon[1]($\boxtimes$), Miguel Damas[1], Hector Pomares[1], and Oresti Banos[2]

[1] Department of Computer Architecture and Computer Technology,
CITIC-UGR Research Center, University of Granada, Granada, Spain
cbailon37@correo.ugr.es, {mdamas,hector}@ugr.es
[2] Telemedicine Group, University of Twente, Enschede, Netherlands
o.banoslegran@utwente.nl

Abstract. Joints angles are some of the most common measurements for the evaluation of lower limb injury risk, specially of lower limb joints. The 2D projections of these angles, as the Frontal Plane Projection Angle (FPPA), are widely used as an estimation of the angle value. Traditional procedures to measure 2D angles imply huge time investments, primarily when evaluating multiple subjects. This work presents a novel 2D video analysis system directed to capture the joint angles in a cost-and-time-effective way. It employs Kinect V2 depth sensor to track retro-reflective markers attached to the patient's joints to provide an automatic estimation of the desired angles. The information registered by the sensor is processed and managed by a computer application that expedites the analysis of the results. The reliability of the system has been studied against traditional procedures obtaining excellent results. This system is aimed to be the starting point of an autonomous injury prediction system based on machine learning techniques.

Keywords: Motion capture · 2D analysis · Frontal Plane Projection Angle · Reflective markers · Kinect

1 Introduction

Lower limb injuries are the most common injuries among sport practitioners [1,2]. In particular, injuries in the knee joint complex are very feared due to their complexity and the expensive, long and painful rehabilitation periods involved [3,4]. In the light of that, it appears the opportunity to develop new technologies to support experts during the rehabilitation task of knee injuries, reducing the impact of the process over the wounded subject.

One of the most commonly used measurement to evaluate the knee physical state is its alignment angle, usually known as knee valgus or varus. It gives us information about the knee-complex strength: a low angle value means high knee strength. A quantitative measurement of this angle was introduced by Wilson et al. [5] in 2006, the Frontal Plane Projection Angle (FPPA). It consists of the projection of the knee angle over the frontal body plane (Fig. 1). Subsequent researches have shown a reliable relationship between FPPA and some knee injuries.

© Springer International Publishing AG 2017
I. Rojas et al. (Eds.): IWANN 2017, Part II, LNCS 10306, pp. 71–81, 2017.
DOI: 10.1007/978-3-319-59147-6_7

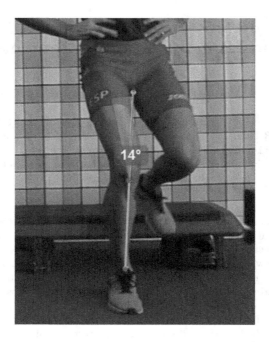

Fig. 1. Frontal Plane Projection Angle (FPPA)

Current FPPA measurement methods involve the offline analysis of previously recorded video frames, placing markers and lines by hand. This process introduces human-made errors due to wrong marker placement. Moreover, the user needs to find exactly the frame with the higher FPPA value, which implies measuring the angle on almost any video frame. In the light of these limitations, this work presents a novel automatic video analysis system intended to support experts in the dynamic measurement of 2D biomechanics angles, including FPPA in a cost-and-time-effective way. The system makes use of an infrared camera to track retro-reflective markers attached to the subject's joints. The measurement of the angles is computed from the marker's coordinates. A computer application with an intuitive user interface has been implemented to support and simplify expert's routine and expedite the analysis of the results.

This system is aimed to be the starting point of a complete injury prediction system that, autonomously, gives experts indications of subject's injury risk and performance status. Some approaches of injury prediction can be found in the literature. In [6], multiple computational intelligence methods such as artificial neural networks and Bayesian networks are compared when performing quantitative risk assessment of injuries. Another research [7] summarizes the predictive factors of sports injuries (which include knee alignment) and proposes a mathematical prediction method based on logistic regression. Hewett et al. [8] also relate the knee valgus load during landing tasks with the anterior cruciate ligament injury. In [9,10], wearable sensors are used to monitor physical performance status.

The paper structures as follows. In Sect. 2, the proposed system is described. The validation procedure followed and the results are shown in Sect. 3. Finally, Sect. 4 summarizes the issues addressed in the paper and presents the conclusions.

2 System Description

The proposed system is composed of three key elements. The first one is the group of retro-reflective markers used to track the exact position of the desired body joints. They are placed over the subject's body. Those markers are made of retro-reflective material in order to avoid problems with the viewpoint of the camera. Any marker shape is allowed provided that the size of the marker is not excessively big. The markers used in this project are depicted in Fig. 2a. They are spherical markers with a diameter of 6.4 mm.

(a)   (b)

Fig. 2. (a) Spherical retro-reflective markers; (b) Kinect V2 sensor

The second key element is the depth sensor. This system has been designed to operate on the depth data provided by the Kinect V2 sensor (Fig. 2b). Kinect is well-known for being a markerless system, but some researches have demonstrated that Kinect's pose estimation algorithm is not accurate enough for precise clinical applications [11,12]. In order to increase the accuracy of the measurement, the aforementioned markers are placed by an expert on the exact anatomical points corresponding to the body joints. The depth sensor tracks the position of the markers and uses Kinect's algorithm to avoid tracking any reflective element in scene that is not a marker. Another purpose of the markers is to allow for tracking points that are not joints, but whose measurement is useful (e.g. the breastbone).

The last element is a computer application that processes data on-the-fly and stores them into a local database. The main feature of the application is the real-time FPPA computing, giving this value to the user with no need of offline analysis. In the following, all the application features are thoroughly described, as well as the process of the data acquisition and processing.

2.1 Data Acquisition and Processing

As it has been explained before, the marker tracking is one of the main tasks of the current work. In order to provide a precise recognition of the markers with no need of external light sources (thereby contributing to the easiness-of-use objective of the system), the depth camera of the Kinect V2 sensor is used.

The Kinect's depth sensor consists of an infrared CMOS camera and three infrared (IR) emitters. Its operation principle is the Time-of-Flight (ToF) [13,14]. It emits a modulated IR pattern which is reflected on the marker and captured by the camera, which captures the intensity of the reflected pattern for each pixel. Each data frame is composed by a 512 × 424 (resolution of the camera) array of 16-bit IR intensity values, one for each pixel. A greyscale image is constructed with these values and shown in the application.

As they are retro-reflective elements, the intensity value of those pixels which belong to a marker is appreciably higher than the rest of values, so it is straight-forward to isolate every pixel that accomplishes this condition through a high-pass filter. Once this filtering has been done, we have an array with the index of those pixels belonging to a reflective element in scene, so the next step is to classify those pixels and determine whether they are a part of a marker or any other reflective element, and which marker they belong to. In this part we apply Kinect's pose estimation algorithm to set a neighborhood of pixels around each estimated joint (Fig. 3). The pixels inside this neighborhood belong to this joint's marker. As the markers are sufficiently far from the rest of them, there is no risk of pixel mismatch. The rejection of unwanted reflective elements is a key part of the process, since it reduces the algorithm workload and improve its robustness.

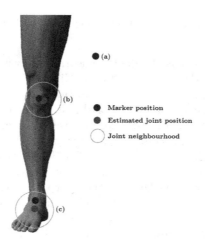

Fig. 3. Reflective pixels: (**a**) reflective element which is not part of a marker, not tracked, (**b**) knee marker pixel, (**c**) ankle marker pixel [15]

After the classification, the centroid of each group of pixels is computed, and its coordinates are considered the marker's coordinated (thus, the joint's coordinates). This consideration increases accuracy and makes the process independent of the subject's distance of the sensor, and the marker size and shape. The aforementioned process is repeated at the arrival of each data frame to perform the motion capture.

These points are used to calculate the desired joint angles. For example, for the FPPA we use the information of hip, knee and ankle position. As the angles are relative and depend only on the markers' position, the measurements are robust against accidental camera rotations. It is worth to notice that the angles are projections over the frontal body plane, so there is some information loss in relation to the 3D angle value.

2.2   Application Description and Implementation

In this section, the computer application used to manage and store the data acquired is described. The aim of this application is to speed up the expert task when performing angle measurements and analyzing those data. The storage of the data and the patients' information relies on a local SQLite database [16] deployed in the user's computer storage disk. The database file is created on the first use of the application. The SQLite database engine has been chosen because it adjusts to the characteristics of this system, since the amount of data managed is able to fit in a single disk file and there will not be concurrent queries to the database. Its good performance on the current task makes it preferable than a client/server engine. In order to provide personal information protection and data anonymity, the information has been decoupled into two data tables. One of them stores patient's personal information and the other one the data registered. Both tables are related by an unambiguous personal ID number, which is assigned to each patient when is registered into the database for the first time. The patient's data table can be modified by the user, who can add, edit or remove profiles. The data table is automatically filled and can only be read.

Once the profiles have been added to the database, the user can perform a great variety of tests to measure the lower limb joint angles, like the FPPA. At this stage one test has been implemented, the Single Leg Landing (SLL) test, which will be explained in Sect. 3. Figure 4 depicts the main screen of the SLL test, which includes the camera image, a control panel and a status bar. The image displays the infrared camera bitmap, which shows the camera image in greyscale, based on the intensity value of each pixel (see Sect. 2.1). For a better angle visualization, a see-through canvas is positioned over the image, where the markers are represented by red circles and connected by yellow lines. Those shapes are attached to the markers' coordinates, so their position is dynamically updated as the subject moves.

The control panel contains multiple buttons to give users full control over the test progress. The buttons allow the user to select a subject from the database and some options about the selected test (e.g. the leg whose angles will be measured or the type of ankle eversion). The start and stop buttons trigger the

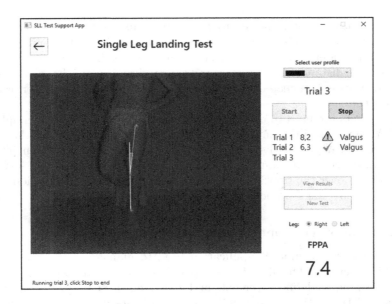

Fig. 4. System application snapshot: SLL test window [15]

respective stages of the test. During the test performance, the angles are shown at runtime and each frame value is stored. If a marker is not properly detected during a frame, it is not displayed and the angle value is not computed. The system gives users the chance to perform up to 3 attempts in each test, in order to repeat those attempts whose performance could be doubtful. Whenever the user stops the test, individual information is offered for each attempt, being possible to visualize a plot of all data captured during the test and the evolution of the angle value (Fig. 5). This feature allows experts to go further than just analysing the value at the maximum flexing point (as it is done with offline video analysis tools). However, an offline analysis of the data is also allowed, since the data is also written on a comma-separated values file (.csv).

The status bar shows both sensor connection state and overall progress test. The control panel is disabled until the sensor becomes available, to avoid possible errors. An historical representation of the data can be also displayed on a chart for each leg and angle type in order to inspect the subject's angle evolution along time.

3 Validation

In order to check the reliability of the developed system it has been validated against a 2D offline video analysis software whose use is very extended, Kinovea software [17]. This software allows the user to analyze previously recorded video frames, running them at slow motion and freezing the video at the desired frame. Once the video has been frozen, the user can place manually lines and markers

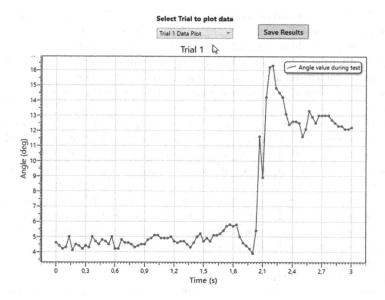

Fig. 5. SLL test data

over the image, and the software returns the angle between the lines. In that way, joint angles can be measured relying on the physical markers placed over the subject's body. 2D video analysis tools have been applied in many studies, and contrasted against 3D techniques [18,19]. Allan G. Munro et al. [20,21] also showed the reliability of 2-D techniques when assessing knee injury risk and established measurement error values for this type of analysis.

There exists a great range of tests on which knee alignment can be measured, regardless of the physical condition of the assessed subject. Most-widely used ones are single leg squat (SLS) [5,18,22], drop vertical jump (DVJ) [8], drop landing (DL) [23] and single leg landing (SLL) [24], as they simulate real movements and interactions produced during sport practice. In this work, the SLL test has been implemented, due to its promptness and simplicity, saving time to experts and making it easier for subjects to understand the procedure. In order to perform the test, the method proposed in [23] is followed and explained in the following paragraph.

In order to simulate the landings encountered during athletic participation, the SLL test requires the subject to perform a unilateral step landing task. This involves stepping off a 30-cm-high bench landing with the opposite leg onto a mark positioned 30 cm from the bench, and holding on the position for at least 2 s. During the test, subjects have to keep the hands on their hips and ensure that the contralateral leg makes no contact with any other surface. The sensor is placed at the subject's knee level, 2 m away from the landing target and aligned perpendicular to the frontal plane. It is recommended to perform various landings with each leg, and the final FPPA value should be measured based on them. Three reflective markers must be strictly attached at the anatomic points proposed in [5]: (i) Anterior Superior Iliac spine, (ii) middle of tibiofemoral joint and (iii) middle of ankle mortise. The markers alignment draws two lines, whose frontal projected angle is recorded as FPPA and is measured at the maximum flexing point. If the knee moves to the subject's sagittal plane, it is known as knee valgus; if it moves outside, it is known as knee varus [25].

To perform the study, ten volunteers, five males and five females were recruited to be evaluated in a Single Leg Landing (SLL) test by a external physical therapist using both Kinovea and the proposed system. Before performing the evaluation, the subjects were informed about the research aims. The execution of the test was recorded simultaneously with the Kinect sensor and with a digital color video camera, one placed above the other with the minimum sensor displacement. The SLL test procedure was explained to the subjects before starting and each of them perform three attempts with a rest time of 30 s. With the proposed system, the FPPA value was automatically obtained, and with the Kinovea software, the recordings had to be analyzed after the experiment by the same physical therapist who directed the test.

The evaluation aims at estimating the inter-rater reliability between both measurement systems. The results of each patient in each attempt are gathered in Table 1. As it can be observed, both results are generally similar. To support this observation, a formal statistical analysis has been performed using the intraclass correlation coefficient (ICC)(ρ) [26]. The results have been extracted using R software v.3.3.2. The one-way random effects ICC(ρ) obtained is **0.996** and its 95% confidence intervals (CI) are **0.984–0.999**. According to [27], values higher than 0.9 reflect excellent reliability.

Table 1. Case study results. Angle values are expressed in degrees ($^\circ$).

Patient ID	1	2	3	4	5	6	7	8	9	10
Age	26	22	21	23	36	18	31	25	32	24
Trial 1 (S)	12.4	13.5	14.9	14.1	18.7	20.2	10.8	9.7	13.4	17.1
Trial 1 (K)	12	14	15	14	18	21	12	9	13	17
Trial 2 (S)	17.6	21.4	12.8	15.1	15.8	16.2	6.7	12.6	15	21
Trial 2 (K)	18	22	14	16	16	17	5	12	15	21
Trial 3 (S)	18.5	16.3	9.6	23.3	18.2	17.4	8.5	8.7	12.4	18.7
Trial 3 (K)	18	16	10	23	18	17	8	9	12	19
Average (S)	16.2	17.1	12.4	17.5	17.6	17.9	8.7	10.3	13.6	18.9
Average (K)	16	17.3	13	17.7	17.3	18.3	8.3	10	13.3	19

(K) Kinovea - (S) Proposed system

Although the results of the study are promising, a study including higher number of participants and different measurements would be required to further confirm this findings.

4   Conclusions

Offline 2-D video analysis is an extended method to quantify knee FPPA. However, this procedure requires large time investments. In this paper we have presented an automatic 2-D video analysis system to support experts during joint angles measurement. The process involves using a depth sensor to track the position of retro-reflective markers attached to the subject's joints. The angles are dynamically measured based on the marker's position.

The comparison between both systems leads us to conclude that the agreement between them is very high, and the reliability of the measurements obtained through the proposed system is excellent. A study including higher number of participants and different measurements and a comparison with 3D motion capture systems will be performed soon and could lead to an extended version of this paper.

Given the promising results of the validation, we are working on the implementation of an autonomous system for injury prediction and performance status evaluation through machine learning, as described in Sect. 1.

Acknowledgements. This work was supported by the University of Granada Research Starting Grant 2015. This work was also partially supported by the Spanish Ministry of Economy and Competitiveness (MINECO) Projects TIN2015-71873-R and TIN2015-67020-P together with the European Fund for Regional Development (FEDER).

References

1. Starkey, C.: Injuries and illnesses in the national basketball association: a 10-year perspective. J. Athl. Train. **35**, 161–167 (2000)
2. Hootman, J.M., Dick, R., Agel, J.: Epidemiology of collegiate injuries for 15 sports: summary and recommendations for injury prevention initiatives. J. Athl. Train. **42**, 311–319 (2007)
3. Feller, J., Webster, K.E.: Return to sport following anterior cruciate ligament reconstruction. Int. Orthop. **37**, 285–290 (2013)
4. Hewett, T.E., Stasi, S.L.D., Myer, G.D.: Current concepts for injury prevention in athletes after anterior cruciate ligament reconstruction. Am. J. Sports Med. **41**(1), 216–224 (2013)
5. Wilson, J.D., Ireland, M.L., Davis, I.: Core strength and lower extremity alignment during single leg squats. Med. Sci. Sports Exerc. **38**, 945–952 (2006)
6. Gregori, D., Berchialla, P., Foltran, F.: Comparing bayesian network, artificial neural networks, classification trees and classical logistic models in quantitative risk assessment: an application to the european registry of foreign body injuries in children. Inj. Prev. **16**, A216 (2010)
7. de la Cruz-Marquez, J.C., de la Cruz-Campos, A., de la Cruz-Campos, J.C., Cueto-Martin, M.B., Garcia-Jimenez, M., Campos-Blasco, M.T.: Prediction of sports injuries by mathematical models. In: An International Prospective on Topics in Sports Medicine and Sports Injury, pp. 333–354. InTech (2012). Number 18

8. Hewett, T.E., Meyer, G.D., Ford, K.R.: Biomechanical measures of neuromuscular control and valgus loading of the knee predict anterior cruciate ligament injury risk in female athletes. Am. J. Sports Med. **33**, 492–501 (2005)

9. Banos, O., Villalonga, C., Damas, M., Gloesekoetter, P., Pomares, H., Rojas, I.: Physiodroid: combining wearable health sensors and mobile devices for a ubiquitous, continuous, and personal monitoring. Sci. World J. **2014**(490824), 1–11 (2014)

10. Banos, O., Moral-Munoz, J.A., Diaz-Reyes, I., Arroyo-Morales, M., Damas, M., Herrera-Viedma, E., Hong, C.S., Lee, S., Pomares, H., Rojas, I., Villalonga, C.: mDurance: a novel mobile health system to support trunk endurance assessment. Sensors **15**(6), 13159–13183 (2015)

11. Bonnechère, B., Sholukha, V., Moiseev, F., Rooze, M., Van Sint, J.S.: From Kinect^{TM} to anatomically-correct motion modelling: preliminary results for human application. In: Schouten, B., Fedtke, S., Bekker, T., Schijven, M., Gekker, A. (eds.) Games for Health, pp. 15–26. Springer, Wiesbaden (2013). doi:10.1007/978-3-658-02897-8_2

12. Wiedmann, L.G., Planinc, R., Nemec, I., Kampel, M.: Performance evaluation of joint angles obtained by the kinect v2. In: IET International Conference on Technlogies for Active and Assisted Living (TechAAL) (2015)

13. Amon, C., Fuhrmann, F.: Evaluation of the spatial resolution accuracy of the face tracking system for Kinect for Windows V1 and V2. In: 6th Congress of Aplh-Adria Acoustics Association (2014)

14. Sarbolandi, H., Lefloch, D., Kolb, A.: Kinect range sensing: structured-light versus time-of-flight Kinect. J. Comput. Vis. Image Underest. **139**, 1–20 (2015)

15. Bailon, C., Damas, M., Pomares, H., Banos, O.: First Approach to Automatic Measurement of Frontal Plane Projection Angle During Single Leg Landing Based on Depth Video, pp. 44–55. Springer International Publishing, Heidelberg (2016)

16. SQLite. http://www.sqlite.org. Accessed 17 April 2016

17. Kinovea Association: Kinovea. http://www.kinovea.org/. Accessed 20 Mar 2016

18. Wilson, J.D., Davis, I.S.: Utility of the frontal plane projection angle in females with patellofemoral pain. J. Orthop. Sports Phys. Ther. **38**(10), 606–615 (2008)

19. McLean, S.G., Walker, K., Ford, K.R., Myer, G.D., Hewett, T.E., van den Bogert, A.J.: Evaluation of a two dimensional analysis method as a screening and evaluation tool for anterior cruciate ligament injury. Br. J. Sport Med. **39**, 355–362 (2005)

20. Munro, A., Herrington, L., Carolan, M.: Reliability of 2-dimensional video assessment of frontal-plane dynamic knee valgus during common athletic screening tasks. J. Sport Rehabil. **21**, 7–11 (2012)

21. Munro, A.G.: The use of two-dimensional motion analysis and functional performance tests for assessment of knee injury risk behaviours in athletes. Ph.D. thesis, School of Health Sciences, University of Salford, Salford, UK (2013)

22. Zeller, B.L., McCrory, J.L., Kibler, W.B., Uhl, T.L.: Differences in kinematics and electromyographic activity between man and women during the single-legged squat. Am. J. Sports Med. **31**(3), 449–456 (2003)

23. Herrington, L., Munro, A.: Drop jump landing knee valgus angle; normative data in a phisically active population. Phys. Ther. Sport **11**, 56–59 (2009)

24. Lawrence, R.K.I., Kernozek, T.W., Miller, E.J., Torry, M.R., Reuteman, P.: Influences of hip external rotation strength on knee mechanics during single-leg drop landings in females. Clin. Biomech. **23**, 806–813 (2008)

25. Kamath, A.F., Israelite, C., Horneff, J., Lotke, P.A.: Editorial: what is varus or valgus knee alignment? A call for a uniform radiographic classification. Clin. Orthop. Relat. Res. **468**(6), 1702–1704 (2010)
26. Bland, J.M., Altman, D.G.: A note on the use of the intraclass correlation coeffcient in the evaluation of agreement between two methods of measurement. Comput. Biol. Med. **20**(5), 337–340 (1990)
27. Prieto, L., Lamarca, R., Casado, A.: La evaluacion de la fiabilidad de las observaciones clnicas: el coeficiente de correlacion intraclase. Medicina Clnica **110**, 142–145 (1998)

Mobile Application for Executing Therapies with Robots

Manuel Martin-Ortiz[1], Min-Gyu Kim[2], and Emilia I. Barakova[1(✉)]

[1] Department of Industrial Design, Eindhoven University of Technology,
Eindhoven, The Netherlands
e.i.barakova@tue.nl
[2] Korea Institute of Robot and Convergence, Pohang, South Korea

Abstract. While robotic technology is being incorporated in therapies, still not enough research has been done to find out how different end-users are willing or able to use robots in their practice. To investigate this issue, a specific study has been designed to determine the preferences of end-users that execute or receive therapies using robots. We applied a participatory design approach which included brainstorming and testing at every stage of the development process. We first determine the preferences of professionals from clinics and schools for children with Autism Spectrum Disorder (ASD). The results indicated that shared (semi-autonomous) control of the robot is preferred in therapies, and mobile devices, like smartphones and tablets, are the preferred interface for the shared robot control. The outcomes of this first stage of research were used as design requirements for the development of a mobile application to be used as an interactive robot control interface. We further developed and tested the application for usability by a broad spectrum of users.

1 Introduction

While robotic technology with some degree of autonomy is being incorporated into therapies, care, and education, there is not yet a standard for intuitive end-user control interfaces. The related topic of development of training content by end-users and creating interface for this purpose has been investigated in several studies already. End-user programming and interaction in technology-supported healthcare applications such as rehabilitation are a promising application domain [1–4]. When using relatively simple technologies for rehabilitation, there are already examples where end-user development can support therapists in the needs of their individual patients. Zubricky et al. [5] have proposed an end-user programming tool for simple robotic devices. In the field of humanoid robotics, the complexity of controlling a robot poses a major challenge for the end-user development [1, 6–8]. A programming environment for real-life dynamic interaction has been provided by Lourens and Barakova [1, 7], however the user interface is not sufficiently intuitive to translate the complexity of all underlying concepts to the level of an end-user. Diprose and colleagues [8] also proposed an end-user development visual language and investigated what abstraction level is appropriate for programming social robot applications and what effects different abstraction levels have on usability. They tested their interface with advanced software

© Springer International Publishing AG 2017
I. Rojas et al. (Eds.): IWANN 2017, Part II, LNCS 10306, pp. 82–92, 2017.
DOI: 10.1007/978-3-319-59147-6_8

engineering students, so their research is not informative for the usability by a layman or professionals from different domains.

To help end-users as therapists in clinics and schools of children with ASD, Kim et al. [21] systematized the creation of therapy dialogue between a robot and a child that results in an easy way to design off-line scenarios including dialogues and motor behaviors for the robot, and for users with technical expertise to also design dynamic scenarios. Gnjatovic [9] proposed a dialogue management system in order to study robots' capacity to autonomously engage in a natural language dialogue in the context of robot-assisted therapy, which is also meant for an off-line use. It is not tested with end-users. Buchina and colleagues [2] have proposed an advanced interface for content creation and real-time interaction with a humanoid robot that combines the use of natural language with graphical visualization. This interface can be used for real-time interaction, and it is designed to support content creation by programming with the use of natural language, restricted to imperative sentences.

All those end-user development interfaces are intended to be used by specialized end-users that create or adapt the therapies. These end-users are usually well educated and can learn to deal with a new program or tool. The control interfaces, however, should be usable by all possible end-users and should be ready for an immediate use with no learning curve needed. The aim of the current study is to develop an interface where end-users can carry on spontaneous interactions within a robot therapy without any help by a robotic expert. We assume that the structure of the robot therapy has already been envisioned and prepared. The interface that is described here accounts for management of dialogues and behaviors that are occurring spontaneously between the human and the robot and are an addition to a pre-designed therapy or part of a spontaneous interaction between the robot and the human.

In the next Sect. 2, we discuss the steps in the participatory design process and the methodology used to receive the user feedback. Then, in Sect. 3 we describe the software architecture of Conversations with NAO, a platform specifically created for the robots NAO and Pepper from Softbank robotics. In Sect. 4 the user tests of the prototype are described and the results commented. Finally, in Sect. 5 we provide discussion and conclusions.

2 Establishing the User Requirements

2.1 Study Design Framework

We used participatory design method that was also applied to similar studies by [5, 10]. Structure-wise the participatory design has a resemblance with the Spiral Model software development methodology [11], however, we performed together with the end-users the ideation, concept validation, and verification phases.

2.2 Materials

The robot NAO was used in the experiment. The available official software for NAO programming, the code-based SDK (Software Development Kit) to the programming suit Choreographe was used. These solutions make possible for different users to easily approach the programming of this robot. However, some background and knowledge in ICT are still needed. We have to consider that therapists, psychologists and other professionals involved in autism therapy are not information workers. They do not rely on ICT for delivering treatment to children with ASD and are usually not familiar with software development and most are not eager to invest time in learning it.

2.3   Experimental Design

We involved the therapists and teachers of children with ASD very early in the design process. Using the robot, first, a live demonstration was performed to show the set of different robot capabilities to the therapists. Afterward, a questionnaire and roundtable discussion took place to gather the qualitative feedback. This experiment has been conducted in two institutions: the Dr. Leo Kannerhuis Foundation(1 http://www. leokannerhuis.nl and the Royal Dutch Kentalis Centre 2 http://www.kentalis.nl. The total of 20 participants, all of them professionals involved in the ASD therapy, took part in the study voluntarily and did not receive any incentives.

In the first phase, the experimenter gave a live demonstration by performing the following interactions with the robot: dialog based on speech recognition, touch-initiated interaction, emotive expressions by the robot composed of movements and synthesized speech. The robot was programmed to speak and behave in an interactive way, so interaction appeared informal and not preprogrammed. The interactive demonstration took around 10 min.

During the second phase of the experiment, we gathered the professionals' feedback via a questionnaire and a round table discussion. A questionnaire was provided to gather information about the therapists' preferences of how the robot should be used. The questionnaire consisted of four different sections as listed below:

- *Therapy practices*: Four questions regarding the current therapist routines in therapy development and administration were asked. The questions were tuned to find out the materials they use, the time they need, and the interaction they have with other therapists.
- *Creating a therapy with the robot*: Questions related to the therapists preferences for robot control or content creation were asked. The term programming has been explicitly avoided along the demonstration and questionnaire to keep the mindset of the therapists as far from technical concepts and jargon as possible. The questions were trying to identify whether they want an assistance from a robotic specialist, and what kind of devices they prefer to use.
- *Using the robot during the therapy*: These questions are about how the therapists want to use the robot in the therapy. They were asked whether the robot should be completely autonomous or just a remotely controlled, and what kind of perceptual skills (to see, to smell, or to hear and understand speech) they want to use.
- *Personal preferences*: Questions about personal impressions. These questions aimed to establish whether the therapists are interested in using this or another robot. If not, why not? The therapists were encouraged with questions to think what the most valuable characteristics of a robot for their work are. Also, a "white space" was provided for other comments.

2.4   Results

The most interesting results are summarized in the following graphs (Figs. 1, 2 and 3). In the context of the most valuable interaction capabilities of the robot, Fig. 1 shows that the speech has been selected as its most desired feature. In the comments after the demonstration, the professionals pointed the importance of speech for the nature of their therapies. This outcome is in line with the current use of the robots in therapies for children with ASD and the therapies based on *Applied behavioral analysis* [12, 13] and *Pivotal response training* [14] which are widely used in autism treatment and are shown to be applicable in robotics [15–18].

Figure 2 shows the professionals preference regarding the robot autonomy. A completely autonomous robot refers to a robot that can interact with the client

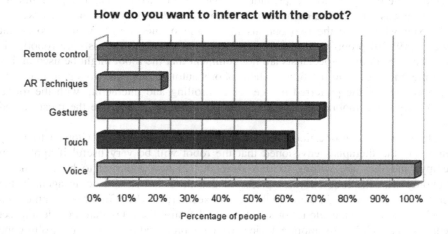

Fig. 1. The preferences of the participants about the robot's interaction modalities that are needed in therapies.

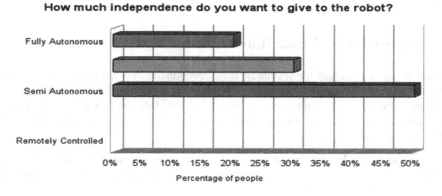

Fig. 2. The preferences of the therapists regarding the interaction autonomy of the robot.

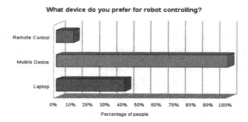

Fig. 3. The preference of the therapists for controlling device.

without the professional intervention while a completely controlled robot refers to a robot that relies entirely on professional intervention. Five point evaluation scale was offered to the therapists with explanations what these options mean. The predominant preference was for shared control between the robot and the therapist, and there was no participant who chose the two options in the teleoperated range. While two recent reviews [19, 20] propose that introduction of more autonomy in assistive robotics is promising, the domain specialists are not confident that the robot might be used on it's on at present or even in the future when robot's autonomous behaviors gain maturity. In the context of the preferred device for controlling and interacting with the robot, Fig. 3 shows that mobile devices, like smartphones and tablets, are the most valued option.

The results of the questionnaire and the round table discussion showed four key trends: (1) the therapists envisioned that the robot will be very useful if applied in therapy, and they are willing to start using it; (2) hey preferred to have a shared autonomy with the robot; (3)they preferred that the robot should be an agent in the therapy, and not the main actor or therapy's executer; (3) they found the synthesized speech as the most valuable robot's feature, "...because the robot can provide a direct feedback to the child"; (4) mobile devices were the preferred devices for controlling the robot.

Considering these findings, we created a custom interface for the therapists: a device and an interface that they could use to control the robot, described in the following Sect. 3.

3   Interface for End-User Robot Control

Taking into consideration the results of the questionnaire and the round table discussion and the supporting or disagreeing literature as a starting point, we designed a software architecture with the capacity to support the different robot features and the user needs. In this sense, we needed to design and develop a software architecture with the following features:

1. The robot has to have a shared control. It needs to be controlled by the user, and at the same time, provide enough independence to the robot to interact with the environment

2. The user controls the robot using a mobile device (like a tablet or mobile phone). Could be interesting if the user can make some basic programming using this device
3. Robot speech is the most valuable robot feature
4. The user should be able to use the robot during therapy, but also create and share therapies

For the second iteration of this research, we used Client-Server architecture. In this case, the server should be running on the robot, receiving commands that are sent from others devices. As controlling device, we choose an Android tablet. The server, running on the robot, should be implemented in a way that any NAO robot can execute it. The server should accept messages from clients and execute the appropriate commands on the robot.

The client part runs over an Android platform. However, the Client-Server architecture enables to any client to send messages to the server. The messages shared between client and server are formed in JSON format and a description of their composition can be found in the correspondence documentation.

Based on the feedback from therapists, we proposed the first solution designed in the Client-Server architecture that utilizes a mobile phone as a client that sends messages with the instructions to a server running on the robot. Due to its suitability to be used for speech interaction at this stage this software was named *Conversations with NAO*.

Every message sent by the client is formed according to the JSON format. JSON is an open standard format that uses human-readable text to transmit data objects consisting of attribute-value pairs. For our application, the attribute command is mandatory in every message and its value indicates the action to perform in the robot. According to the command attribute, other keywords are available to complete the action. The different values for the attribute command are described in Table 1.

The messages are sent by the client running on the mobile device and received by the robot which executes the server role. The server receives the messages, analyses the

Table 1. Action commands and attributes

Command attribute	Attributes	Explanation
Say	sentence attitude	Indicates to the Server that the Robot has to pronounce a sentence. The additional attributes *sentence* and *attitude* indicate the sentence and the robot mood while saying it
Set	parameter value	An attribute used to set different robot parameters and configurations. It allows two more attributes, *parameter* and *value* to complete the information required by the server. The different parameters to modify are *volume, behavior, stiffness* and *language*
Get	parameter	Allows the client to retrieve information from the robot. The purpose of this command is to provide useful information to the user
Stop		Indicates to the server to stop all the behaviors that the robot is executing

message structure and performs the requested action. Then, the server sends a message to the client with the requested information or the result of the operation.

The selected mobile platform for the client has been a 10-inch tablet with Android operating system. However, the software can be run on any Android device. Figure 4 shows the final interface of the developed application. The client-server architecture requires the robot and mobile device to be connected to the same network. We kept the interface, shown in Fig. 4, the most simple and intuitive as possible. The software implemented and tested with users that were not therapists. These tests are described in the next section.

Fig. 4. The interface of the Android mobile application

4   Tests to Establish the Usability of the Interface

We conducted 2 user test to gather information about the usability of the interface. The aim of the first test is to ensure that this interface can be used by the children with ASD. Often the therapies with the robot are performed in a way the robot is used as a mediator of the interaction and one of the interactants is a child with ASD him/herself [10]. The second interactant can be another child [10] or the therapist [14]. Before this test, we piloted the control interface with the teachers and therapists of children with ASD at Orion schools, The Netherlands. Such a pilot is necessary first to get a permission to test with the children. The interface was intuitive to this user group. The second test is aimed to determine usability by an average adult. We did not want to test with therapists only, since the term therapist is used quite freely, and there are different persons involved in the care of the children with ASD – from occupational therapists to volunteers and parents.

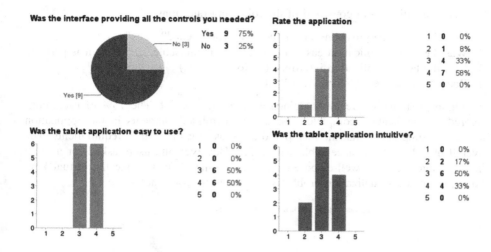

Fig. 5. Usability testing with children with ASD.

4.1 Usability Test I: Playing Game Between Two Children with ASD

Once the client and server applications were designed and programmed, a user test was carried out in order to analyze the usability of the system. Six couples, i.e. in total 12 children (age range 13–15, M 13.7) with ASD played the game as described below. The participants signed a participation consent form allowing us to use the acquired data in this research.

During the user confrontation, the participants played the game called "*Discover what I am seeing*". The game consists of two rounds and in each round, the participants change their roles. In each round, Player A selects a random card from a set. The card has a picture. Player B has to guess the picture asking questions to the robot. Player A uses the tablet in order to answer the questions through the robot. The exact interaction order of each round was as follows:

1. There is a set of cards with pictures
2. The pictures are hidden to the participants
3. Player A picks a random card
4. Player B asks questions to the robot in order to guess the picture
5. Player A uses the tablet in order to answer the questions through the robot
6. After 10 questions, if Player B does not correctly guess the picture, then Player A wins.

The questions that aimed to use some qualitative insights were 'is the tablet application intuitive?', 'is it easy to use?', 'did they find what they were looking for?'. The test had the following protocol: (1) Test introduction, (2) The players can ask about ambiguous items, (3) The players play Round 1 and Round 2 (4) The players fill the evaluation form, and (5) Informal discussion with the experimenter to share impressions.

The evaluation form consisted of the following questions:

1. Was the interface providing all the controls you needed? (Yes/No)
2. Was the tablet application easy to use? (1 No, is unusable ... 5 Yes, it is perfect)
3. Was the tablet application intuitive? (1 No ... 5 Yes)
4. Rate the interface (1 Worst ... 5 Best)

Figure 6 shows the results obtained after the user's test. The 75% of users considered the application complete. The 25% that missed elements in the application asked for components that the therapist did not ask for in the user requirements. The overall usability rate is acceptable with a 3.5. However, the users asked for a more intuitive interface and well-disposed elements. Some elements, like the virtual keyboard, presented visualization problems that the users could not solve.

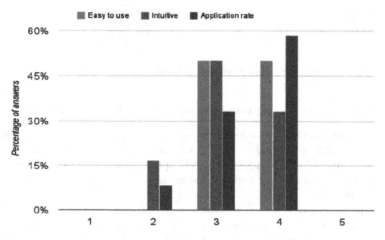

Fig. 6. Evaluation by general users according to the 5 point Likert-scale. The application was rated as fairly easy to use, intuitive and fast.

4.2 Usability Test II: Controlling Robot with the Use of the Interface by General Users

The objective of this test was to gather insights about the ease of use of the application and did the users find in it the elements they were looking for. A total of 12 persons not related to the ASD therapy participated in a 10 min test. In the test, the participant received the tablet with the software installed and a short explanation of the test: they had a card representing an object, their mission was to use the tablet to control the robot and answer the questions made by a third person. This third person was trying to figure out the figure in the card.

After the test, the participant filled in a questionnaire with his/her feedback. The 75% of participants found the interface complete and did not lack any feature for the given task. They were asked to rate different characteristics according to a 5 point Likert Scale. As shown in Fig. 6, the value 1 on this scale represented a poor presence of this characteristic in the interface, while a value of 5 represented a high presence of this characteristic. Also, a final global rate for the application quality was given.

5   Conclusions and Future Work

This study used participatory design to create an interface that was designed to be used in therapies for children with ASD. The interface is designed in a way that can be used in general education as well or for other purposes, however, the study tests only its usability in the scope of therapies of children with ASD. While the participatory design has been used earlier in this setting [5, 10] we have performed end-user evaluations at every stage of the mobile application development. Moreover, we actively involved the therapists in a co-creation process by first demonstrating a set of interactive behaviors that have been found useful in therapies, and some novel interactive behaviors (Figs. 1, 2 and 3).

The demo of all interactive skills that robot can have at this stage of technology development to help to brainstorm with the therapists and caregivers on how the robot could be used, which excluded all the unrealistic suggestions and also the too simple suggestions were ruled out. Based on the outcome of this test and brainstorm session we designed the interface, which was also an iterative process and went through several iterations, not described in this paper. When the interface was ready we piloted it with therapists and tested with the children with ASD, who are one of the interactants that the system consisting of a robot and mobile controlling interface connects. Often the children need to use the interface to interact with the robot, which can be controlled by a therapist with or without the knowledge of the child. In addition we tested with general users with different educational degrees and background, since the term "therapist" is very broadly used, and in clinics and schools for children with ASD the therapies are developed by occupational therapists and psychologists, but the administration or everyday training, and repetitions can be given by less specialized personnel, and in therapies as PRT parents need to take part of the training at home, since these therapies are performed in natural environments of the children and the training takes place also by giving feedback in everyday situations.

Despite the good overall rating of the application, the users provided a lot of useful critical feedback to improve the software before the app can actually be used in therapy. A major drawback is that the interface, being designed to be general, does not address the specific needs of the therapists, and is not designed to support a particular therapy, which was addressed in a follow-up iteration of the interface development.

References

1. Barakova, E.I., Gillesen, J.C.C., Huskens, B.E.B.M., Lourens, T.: End-user programming architecture facilitates the uptake of robots in social therapies. Robot. Auton. Syst. **61**(7), 704–713 (2013)
2. Buchina, N., Kamel, S., Barakova, E.: Design and evaluation of an end-user friendly tool for robot programming. In: *2016* 25th IEEE International Symposium on Robot and Human Interactive Communication (RO-MAN), pp. 185–191 (2016)
3. Zubrycki, I., Granosik, G.: Understanding therapists' needs and attitudes towards robotic support. the roboterapia project. Int. J. Soc. Robot. **8**(4), 553–563 (2016)

4. Tetteroo, D., Seelen, H., Timmermans, A., Markopoulos, P.: Rehabilitation therapists as software creators?: Introducing end-user development in a healthcare setting. Int. J. Sociotechnol. Knowl. Dev. **6**(1), 36–50 (2014)

5. Zubrycki, I., Kolesiński, M., Granosik, G.: A participatory design for enhancing the work environment of therapists of disabled children. In: 2016 25th IEEE International Symposium on Robot and Human Interactive Communication (RO-MAN), pp. 781–786 (2016)

6. Berenz, V., Suzuki, K.: Targets-drives-means: a declarative approach to dynamic behavior specification with higher usability. Robot. Auton. Syst. **62**(4), 545–555 (2014)

7. Lourens, T., Barakova, E.: User-friendly robot environment for creation of social scenarios. In: Ferrández, J.M., Álvarez Sánchez, J.R., de la Paz, F., Toledo, F.J. (eds.) IWINAC 2011. LNCS, vol. 6686, pp. 212–221. Springer, Heidelberg (2011). doi:10.1007/978-3-642-21344-1_23

8. Diprose, J., MacDonald, B., Hosking, J., Plimmer, B.: Designing an API at an appropriate abstraction level for programming social robot applications. J. Vis. Lang. Comput. (2017, in press). https://doi.org/10.1016/j.jvlc.2016.07.005, Accessed 7 Jul 2016

9. Gnjatović, M.: Therapist-centered design of a robot's dialogue behavior. Cogn. Comput. **6**(4), 775–788 (2014)

10. Barakova, E.I., Bajracharya, P., Willemsen, M., Lourens, T., Huskens, B.: Long-term LEGO therapy with humanoid robot for children with ASD. Expert Syst. **32**(6), 698–709 (2015)

11. Boehm, B.W.: A spiral model of software development and enhancement. Computer **21**(5), 61–72 (1988)

12. Harris, S.L., Delmolino, L.: Applied behavior analysis: its application in the treatment of autism and related disorders in young children. Infants Young Child. **14**(3), 11–17 (2002)

13. Schreibman, L., Kaneko, W.M., Koegel, R.L.: Positive affect of parents of autistic children: a comparison across two teaching techniques. Behav. Ther. **22**(4), 479–490 (1991)

14. Koegel, R.L., Koegel, L.K., McNerney, E.K.: Pivotal areas in intervention for autism. J. Clin. Child Adolesc. Psychol. **30**(1), 19–32 (2001)

15. Huskens, B., Verschuur, R., Gillesen, J., Didden, R., Barakova, E.: Promoting question-asking in school-aged children with autism spectrum disorders: effectiveness of a robot intervention compared to a human-trainer intervention. Dev. Neurorehabil. **16**(5), 345–356 (2013)

16. Verschuur, R., Huskens, B., Verhoeven, L., Didden, R.: Increasing opportunities for question-asking in school-aged children with autism spectrum disorder: effectiveness of staff training in pivotal response treatment. J. Autism Dev. Disord. **47**(2), 490–505 (2017)

17. Koegel, R.L., Bradshaw, J.L., Ashbaugh, K., Koegel, L.K.: Improving question-asking initiations in young children with autism using pivotal response treatment. J. Autism Dev. Disord. **44**(4), 816–827 (2014)

18. Kozima, H., Nakagawa, C., Yasuda, Y.: Interactive robots for communication-care: a case-study in autism therapy. In: ROMAN 2005. IEEE International Workshop on Robot and Human Interactive Communication, pp. 341–346 (2005)

19. Thill, S., Pop, C.A., Belpaeme, T., Ziemke, T., Vanderborght, B.: Robot-assisted therapy for autism spectrum disorders with (partially) autonomous control: challenges and outlook. Paladyn **3**(4), 209–217 (2012)

20. Goodrich, M.A., Crandall, J.W., Barakova, E.: Teleoperation and beyond for assistive humanoid robots. Rev. Hum. Factors Ergon. **9**(1), 175–226 (2013)

21. Kim, M.G., Oosterling, I., Lourens, T., Staal, W., Buitelaar, J., Glennon, J., Smeekens, I., Barakova, E.: Designing robot-assisted pivotal response training in game activity for children with autism. In: 2014 IEEE International Conference on Systems, Man, and Cybernetics (SMC), 5–8 October 2014

Computational Intelligence Methods for Time Series

Automated EEG Signals Analysis Using Quantile Graphs

Andriana S.L.O. Campanharo[1(✉)], Erwin Doescher[2], and Fernando M. Ramos[3]

[1] Departamento de Bioestatística, Instituto de Biociências,
Universidade Estadual Paulista, São Paulo, Botucatu, Brazil
andriana@ibb.unesp.br
[2] Departamento de Ciência e Tecnologia, Universidade Federal de São Paulo,
Campus São José dos Campos, São Paulo, Brazil
doescher@unifesp.br
[3] Laboratório de Computação e Matemática Aplicada,
Instituto Nacional de Pesquisas Espaciais, São José dos Campos, São Paulo, Brazil
fernando.ramos@inpe.br

Abstract. Recently, a map from time series to networks has been proposed [7,8], allowing the use of network statistics to characterize time series. In this approach, time series quantiles are naturally mapped into nodes of a graph. Networks generated by this method, called Quantile Graphs (QGs), are able to capture and quantify features such as long-range correlations or randomness present in the underlying dynamics of the original signal. Here we apply the QG method to the problem of detecting the differences between electroencephalographic time series (EEG) of healthy and unhealthy subjects. Our main goal is to illustrate how the differences in dynamics are reflected in the topology of the corresponding QGs. Results show that the QG method cannot only differentiate epileptic from normal data, but also distinguish the different abnormal stages/patterns of a seizure, such as pre-ictal (EEG changes preceding a seizure) and ictal (EEG changes during a seizure).

Keywords: Electroencephalographic time series · Epilepsy · Complex networks · Quantile graphs

1 Introduction

Epilepsy is a neurological disorder characterized by the presence of recurring seizures that affects nearly 1% of the general population [1]. Sudden and abrupt seizures that cause momentarily lapses of consciousness can have significant impact on the daily life of sufferers. Thus, epileptic seizure detection would help these people to have a normal life.

Like many other neurological disorders, epilepsy can be assessed by the electroencephalogram technic. Visual inspection of the EEG data has not yet led to the detection of all characteristic changes preceding seizure onsets. Moreover, detection of seizures by visual inspection of the EEG even by a trained neurologist is challenging for a variety of reasons such as the excessive presence

© Springer International Publishing AG 2017
I. Rojas et al. (Eds.): IWANN 2017, Part II, LNCS 10306, pp. 95–103, 2017.
DOI: 10.1007/978-3-319-59147-6_9

of myogenic artifacts [3]. Hence, several studies have focused on the detection of epilepsy from EEG signals using automatic analysis like methods that detect and quantify non-linear mechanisms and thereby better reflect the characteristics of the EEG signals [3,5]. Therefore, time-domain methods [4,5,16], frequency-domain methods [13,15] and Wavelet-domain methods [12,14] have been recently proposed. However, there is still considerable research toward developing novel methods to quantify or capture additional information in EEG time series in new ways.

In the last two decades, research on complex networks became the focus of widespread attention, with developments and applications spanning different scientific areas, from sociology and biology to physics [2]. One of the reasons behind the growing popularity of complex networks is that almost any discrete structure can be suitably represented as a graph, whose features may be then characterized, analyzed and, eventually, related to its respective dynamics [9]. Recently an approach has been proposed for mapping a time series into a complex network representation, based on the concept of transition probabilities, resulting in a so-called "quantile graph" (QG) [7,8]. This study has shown that distinct features of a time series can be mapped into networks with distinct topological properties, opening the door to the analysis of discrete, time-ordered data sets with mathematical tools usually used in the study of geometric shapes and topological spaces. Here we show that the complex network theory can be effectively used to the problem of detecting differences in EEG signals of patients with and without epilepsy symptoms. It is important to mention that our method does not require assumptions about stationarity, length of the signal, and noise level. The rest of this paper is organized as follows. After this Introduction, in Sect. 2 it is described the QG method for mapping a time series into a network. In Sect. 3 it is described the data set used in this study. Results are presented and discussed in Sect. 4 while an overall conclusion is given in Sect. 5.

2   Methods

Let $\mathcal{M}$ be a map from a time series $X \in \mathcal{T}$ to a network $g \in \mathcal{G}$, with $X = \{x(t) | t \in \mathbb{N}, x(t) \in \mathbb{R}\}$ and $g = \{\mathcal{N}, \mathcal{A}\}$ being a set of nodes $\mathcal{N}$ and arcs $\mathcal{A}$. $\mathcal{M}$ assumes a simple discretization of X that is not sensitive to the distribution of its values. Specifically, once the Q quantiles have been identified, $\mathcal{M}$ assigns each quantile q_i to a node $n_i \in \mathcal{N}$ in the corresponding network. Two nodes n_i and n_j are connected with a weighted arc $n_i, n_j, w_{ij}^k \in \mathcal{A}$ whenever two values $x(t)$ and $x(t + k)$ belong respectively to quantiles q_i and q_j, with $t = 1, 2, \ldots, T$ and the time differences $k = 1, \ldots, k_{max} < T$. For an illustration of the QG method for $k = 1$, see Fig. 1.

Weights w_{ij}^k are simply given by the number of times a value in quantile q_i at time t is followed by a point in quantile q_j at time $t + k$, normalized by the total number of transitions. Repeated transitions through the same arc increase the value of the corresponding weight. With proper normalization, the weighted adjacency matrix becomes a Markov transition matrix $\mathbf{W}_k$, with $\sum_j^Q w_{ij}^k = 1$.

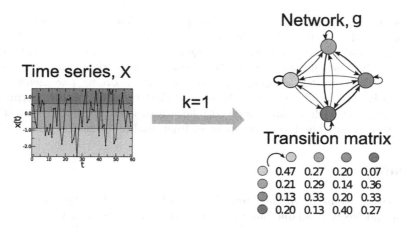

Fig. 1. Illustration of the QG method for $k = 1$. A white Gaussian noise X with $T = 60$ time points is split into $Q = 4$ quantiles (colored shading) and each quantile q_i is assigned to a node $n_i \in \mathcal{N}$ in the corresponding network g. Two nodes n_i and n_j are then connected in the network with a weighted arc $(n_i, n_j, w_{ij}^k) \in \mathcal{A}$ where the weight w_{ij} of the arc is given by the probability that a point in quantile q_i is followed by a point in quantile q_j. Repeated transitions between quantiles results in arcs in the network with larger weights (represented by thicker lines) and therefore higher values in the corresponding transition matrix. (Color figure online)

Since $\mathcal{M}$ is insensitive to the distribution of values of X, the map only requires the specification of the parameter Q, the number of quantiles/nodes, with Q being typically much smaller than T [8].

We previously showed [7] that with $\mathbf{W}_k$ in hand, it is possible to perform a random walk on the quantile graph (QG) g. After $l = L$ jumps of length $\delta_{l,k}(i,j) = |i - j|$, with $i, j = 1, \ldots, Q$ being the node indices as defined by $\mathbf{W}_k$, the mean jump length $\Delta(k)$ can be computed as follows:

$$\Delta(k) = \frac{1}{L} \sum_{l=1}^{L} \delta_{l,k}(i,j). \tag{1}$$

The mean jump length $\Delta(k)$ can be computed directly from the transition matrix $\mathbf{W}_k$ [10], an approach less time-consuming, especially for networks with many nodes. In this case, $\Delta(k)$ given by:

$$\Delta(k) = \frac{1}{Q} tr(\mathbf{P}\mathbf{W}_k^T). \tag{2}$$

where $\mathbf{W}_k^T$ is the transpose of $\mathbf{W}_k$, $\mathbf{P}$ is a $Q \times Q$ matrix with elements $p_{i,j} = |i-j|$, and tr is the trace operation.

Previous works have shown that time series with different properties are mapped into complex networks with different topologies. For example, we have found an association between periodic time series and regular networks, random

time series and random networks, pseudo-periodic time series and small-world
networks [8]. Moreover, we have shown that the bifurcation cascades of two well
known unimodal maps – the Logistic and Quadratic Maps – are mapped through
the QG method into networks whose topological characteristics mimic the main
features of their period-doubling route to chaos as a forcing parameter varies
continuously [6]. Finally, we recently showed that the QG method permits to
quantify features such as long-range correlations or anti-correlations and it can
be used to estimate the Hurst exponent of fractional motions and noises [7].

3   Data

In this study, we use an artifact free EEG database, provided by the University
of Bonn and freely available online [5]. This database has been widely used for
EEG feature extraction and classification in the literature [3]. Our data consists
of five sets (denoted A, B, C, D and E), each containing 100 single-channel EEG
segments, with a duration of 23.6 s, and a sampling rate of 173.61 Hz. Sets A and
B consist of surface EEG recordings from healthy, awake volunteers with eyes
open (A) and closed (B), respectively. Sets C and D correspond to intracranial

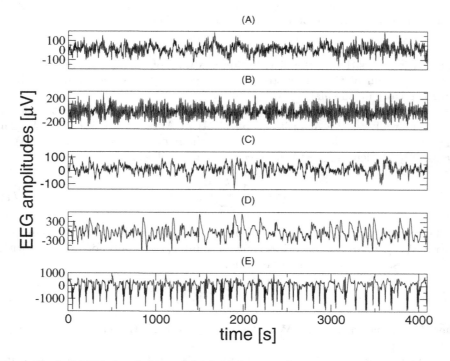

Fig. 2. Typical EEG signals from each of the five sets; from top to bottom: A (healthy,
eyes open), B (healthy, eyes closed), C (epileptic, opposite zone), D (epileptic, epileptogenic zone) and E (seizure).

EEG signals from epileptic patients recorded during seizure-free intervals at the epileptogenic zone (D) and at the hippocampal formation in the opposite hemisphere of the brain (C). Finally, intracranial EEG signals in set E contain seizure activity, measured at sites exhibiting ictal activity.

Figure 2 shows samples from sets A, B, C, D and E. EEG time series recorded extracranially during the relaxed state of healthy subjects with eyes closed (Fig. 2(B)) show a predominant physiological rhythm, the so-called "alpha rhythm" in a frequency range of 8–13 Hz, an activity which is most pronounced at the back of the head. In contrast, broader frequency characteristics are obtained for open eyes (Fig. 2(A)). During a seizure free interval, the EEG recorded from within the epileptogenic zone (Fig. 2(D)) is often characterized by intermittent occurrences of interictal epileptiform activities. Investigation of these steep, sometimes rhythmic high amplitude patterns in EEG recordings contributes to the localization of the epileptogenic zone. Fewer and less pronounced interictal epileptiform activities can be found at recording sites distant from the epileptogenic zone (Fig. 2(C)). As for the EEG recorded during epileptic seizures (Fig. 2(E)), termed ictal activity, it is almost periodic and of high amplitude, resulting from the hypersynchronous activity of large assemblies of neurons [11].

4   Results

We apply the QG algorithm to the problem of differentiating (a) epileptic from normal data, (b) EEG changes preceding a seizure and (c) EEG changes during a seizure. Unless when indicated otherwise, the values $Q = 30$, $T = 4096$ and $k = 1, 2, \ldots, 100$ have been used in the computations. Therefore, we mapped 500 signals into 50,000 quantile graphs and we obtained 50,000 QG transition matrices. After that, for each set and for a given k we took the median over all matrices and we obtained a QG transition matrix of medians. We finally computed $\Delta(k)$ versus k using Eq. (2) for all sets (Fig. 3). Observe that the curves for healthy (A and B) and epileptic (C and D) patients form two distinct clusters with maximum separation at approximately $k = 4$. For $k > 30$, correlations between QG nodes disappear, and all curves almost merge into one.

Figure 4 displays boxplots of $\Delta(k)$, computed over 100 segments, for sets A, B, C, D and E, and $k = 4$. Note how the QG method permits a robust discrimination between healthy and epileptic patients, even more when the EEGs are taken from healthy subjects with the eyes closed. Comparing sets B (healthy, eyes closed) and C (epileptic, opposite zone), there is a statistically significant difference between the two sample means of $\mu_B - \mu_C = 4.5389$, with a 95% confidence interval of $[4.2989, 4.7790]$, $t(df = 195.3079) = 37.2897$ and $p \leq 0.05$. For sets A (healthy, eyes open) and D (epileptic, epileptogenic zone), the less favorable comparison, we still have a statistically significant difference between the two sample means, with $\mu_A - \mu_D = 2.9658$, with a 95% confidence interval of $[2.6689, 3.2625]$, $t(df = 191.8404) = 19.7085$ and $p \leq 0.05$.

Figure 5 displays QG transition matrices of medians for $k = 4$. Observe how well the QG topology mimics the properties of the original time series. For

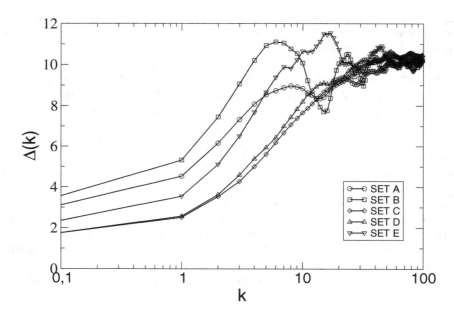

Fig. 3. $\Delta(k)$ versus k for sets A (healthy, eyes open), B (healthy, eyes closed), C (epileptic, opposite zone), D (epileptic, epilectogenic zone) and E (seizure).

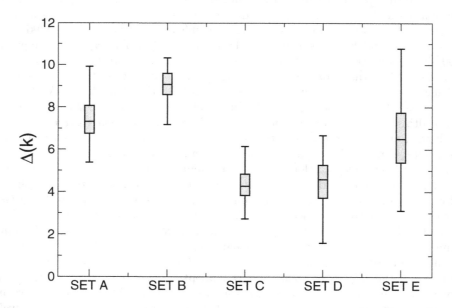

Fig. 4. Boxplots of $\Delta(k)$, computed over 100 segments each, for sets A (healthy, eyes open), B (healthy, eyes closed), C (epileptic, opposite zone), D (epileptic, epilectogenic zone) and E (seizure), and $k = 4$.

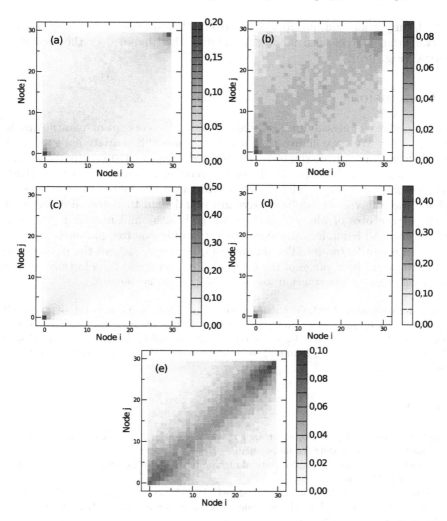

Fig. 5. QG transition matrices for sets A (healthy, eyes open), B (healthy, eyes closed), C (epileptic, opposite zone), D (epileptic, epileptogenic zone) and E (seizure), and $k = 4$.

healthy subjects from set B (Fig. 5(b)), the weights in the corresponding $\mathbf{W}_4$ are more uniformly distributed along its columns and rows when compared with $\mathbf{W}_4$ for healthy subjects from set A (Fig. 5(a)). For unhealthy subjects from set D (Fig. 5(d)), the higher weights in the corresponding $\mathbf{W}_4$ are concentrated in its peripherical quantiles due to the rhythmic high amplitude patterns found in the corresponding time series. Although Fig. 5(c) and (d) are very alike, the fewer and less pronounced interictal epileptiform activities found in set C (when compared to set D) produce a QG transition matrix with heavier weights. Finally, the high amplitude and quasiperiodic patterns found in set E are mapped in a QG transition matrix with its weights mainly distributed over the secondary diagonal (Fig. 5(e)).

5   Conclusions

A novel method for classifying EEG data by using the concept of quantile graphs was presented. We have shown that it can not only differentiate epileptic from normal data, but also distinguish the different abnormal stages/patterns of a seizure, such as pre-ictal (EEG changes preceding a seizure) and ictal (EEG changes during a seizure). These results attest that the QG method is a useful tool for the analysis of nonlinear dynamics and able to detect differences in the data structures of physiological signals of healthy and unhealthy subjects. The QG method is numerically simple and has only one free parameter, Q, the number of quantiles/nodes. The number of quantiles Q defines the partitioning level of the amplitude range of the original time series and its selection involves a trade-off between information loss and computational burden.

Acknowledgments. A.S.L.O. Campanharo acknowledges the support of FAPESP: 2013/19905-3. The support of Conselho Nacional de Desenvolvimento Científico e Tecnológico (CNPq) (Brazil) is acknowledged by F.M. R. All figures were generated with PyGrace (http://pygrace.github.io/) with color schemes from Colorbrewer (http://colorbrewer.org).

References

1. Seizures and epilepsy: hope through research (2004). http://www.ninds.nih.gov/disorders/epilepsy/detail_epilepsy.htm
2. Albert, R., Barabási, A.L.: Statistical mechanics of complex networks. Rev. Modern Phys. **74**, 47 (2002)
3. Alotaiby, T.N., Alshebeili, S.A., Alshawi, T., Ahmad, I., El-samie, F.E.A.: EEG seizure detection and prediction algorithms: a survey. EURASIP J. Adv. Sig. Process. **183**, 1–21 (2014)
4. Andrzejak, R.G., Lehnertz, K., Mormann, F., Rieke, C., David, P., Elger, C.E.: Indications of nonlinear dynamics and finite-dimensional structures in time series of brain electrical activity: dependence on recording region and brain state. Phys. Rev. E **64**, 061907 (2001)
5. Andrzejak, R.G., Schindler, K., Rummel, C.: Nonrandomness, nonlinear dependence, and nonstationarity of electroencephalographic recordings from epilepsy patients. Phys. Rev. E **86**, 046206 (2012)
6. Campanharo, A., Ramos, F.M.: Quantile graphs for the characterization of chaotic dynamics in time series. In: WCCS 2015 – IEEE Third World Conference on Complex Systems. IEEE (2016)
7. Campanharo, A., Ramos, F.M.: Hurst exponent estimation of self-affine time series using quantile graphs. Physica A **444**, 43–48 (2016)
8. Campanharo, A., Sirer, M.I., Malmgren, R.D., Ramos, F.M., Amaral, L.A.N.: Duality between time series and networks. PLoS ONE **6**, e23378 (2011)

9. Costa, L.F., Rodrigues, F.A., Travieso, G., Villas, P.R.: Characterization of complex networks. Adv. Phys. **56**, 167–242 (2007)
10. Doescher, E., Campanharo, A.S.L.O., Ramos, F.M.: Quantile graphs: exact results and applications (2017, in preparation)
11. Güler, I., Übeyli, E.D.: Expert systems for time-varying biomedical signals using eigenvector methods. Expert Syst. Appl. **32**, 1045–1058 (2007)
12. Guo, L., Rivero, D., Pazos, A.: Epileptic seizure detection using multiwavelet transform based approximate entropy and artificial neural networks. J. Neurosci. Meth. **193**, 156–163 (2010)
13. Khamis, H., Mohamed, A., Simpson, S.: Frequency-moment signatures: a method for automated seizure detection from scalp EEG. Clin. Neurophysiol. **124**, 2317–2327 (2013)
14. Liu, Y., Zhou, W., Yuan, Q., Chen, S.: Automatic seizure detection using wavelet transform and svm in long-term intracranial EEG. EEE Trans. Neural Syst. Rehabil. Eng. **20**, 749–755 (2012)
15. Rana, P., Lipor, J., Lee, H., Van Drongelen, W., Kohrman, M.H., Van Veen, B.: Seizure detection using the phase-slope index and multichannel ECoG. IEEE Trans. Biomed. Eng. **59**, 1125–1134 (2012)
16. Vlachos, I., Kugiumtzis, D.: Nonuniform state-space reconstruction and coupling detection. Phys. Rev. E **82**, 016207 (2010)

Hybrid Models for Short-Term Load Forecasting Using Clustering and Time Series

Wael Alkhatib$^{(\boxtimes)}$, Alaa Alhamoud$^{(\boxtimes)}$, Doreen Böhnstedt, and Ralf Steinmetz

Fachgebiet Multimedia Kommunikation, Technische Universität Darmstadt,
S3/20 Rundeturmstr. 10, 64283 Darmstadt, Germany
{wael.alkhatib,alaa.alhamoud,doreen.boehnstedt,
ralf.steinmetz}@kom.tu-darmstadt.de

Abstract. Short-term forecasting models on the micro-grid level help guaranteeing the cost-effective dispatch of available resources and maintaining shortfalls and surpluses to a minimum in the spot market. In this paper, we introduce two time series models for forecasting the day-ahead total power consumption and the fine-granular 24-hour consumption pattern of individual buildings. The proposed model for predicting the consumption pattern outperforms the state-of-the-art algorithm of Pattern Sequence-based Forecasting (PSF). Our analysis reveals that the clustering of individual buildings based on their seasonal, weekly, and daily patterns of power consumption improves the prediction accuracy and increases the time efficiency by reducing the search space.

Keywords: Smart grid · Sequence-based forecasting · Time series models · K-means · Hierarchical clustering

1 Introduction

Until 1998, the electricity market was divided into a set of power supply areas where the grid of each area is owned and supplied by one utility. However, this monopoly ended after the electricity market liberalization in Europe which increased the competition and led to more efficient production and supply of electricity. Power utilities are primarily involved in the trading with electricity suppliers on three different levels, namely: forward, day-ahead and intraday markets [5]. In forward market, utilities and suppliers agree on the deliveries of each year and up to six years with particularly liquid trading for the next three years. In the day-ahead trading, companies agree on the power deliveries for the next day and the deliveries are auctioned during the 12 midday before. However, oscillation of the spot market power consumption can happen due to unexpected events i.e. temperature changes, customers consumption pattern etc. To cope with these issues and to ensure the cost-effective dispatch of the available power generation facilities, the companies can after the day-ahead auction closure, trade on the intraday market level and agree on power deliveries on a very short-term basis from quarter hour to hour blocks which trigger the vital need for forecasting the day-ahead total power consumption and the fine-granular

© Springer International Publishing AG 2017
I. Rojas et al. (Eds.): IWANN 2017, Part II, LNCS 10306, pp. 104–115, 2017.
DOI: 10.1007/978-3-319-59147-6_10

24-hour consumption pattern of individual buildings. Beside the importance of accurate forecasting of power consumption on a very short timescale for utilities, the grid scale hourly power consumption prediction can assist the planning of duty cycles of A/C and ventilators in building management systems to flat the daily power consumption and/or to leverage low price periods based on signals from the power utilities monitoring systems.

The majority of previous research works have concentrated on aggregated energy consumption showing that accurate short-term consumption forecasting at the portfolio of buildings level can be achieved [12,14,16]. However, there is a lack of results related to estimating short-term power consumption of individual customers. In this work, we introduce two time series models, namely the Total Consumption Pattern Matching (TCPM) forecasting model which is used to predict the day-ahead total power consumption, and the Hourly Consumption Pattern Matching (HCPM) model which is used to predict the 24-hours consumption pattern of individual buildings i.e. the detailed consumption values at each hour of the day.

This paper is organized as follows: Sect. 2 gives an overview of related work in the domain of power consumption forecasting. In Sect. 3, we introduce our concept for the long-term and short-term forecasting of power consumption. Section 3 presents in details the short-term prediction models and their comparative analysis. Finally, Sect. 5 summarizes the paper and discusses future work.

2   Related Work

Short-term forecasting of power consumption is considered by the power utilities for economic scheduling [1] and real-time control where accurate and robust short-term forecasting guarantees the cost-effective dispatch of available resources while keeping shortfalls and surpluses to a minimum in the spot market. S. Aman et al. presented in [2] an empirical comparison between several averaging models for short-term forecasting of power consumption. Averaging models are characterized by their simplicity as they make their prediction based on averaging the power consumption of similar points of time horizon such as day, month, and year [8,9].

Time series (TS) models predict future values based on previous historical observations. The commonly used approaches include Moving Average (MA), Auto-Regressive Integrated Moving Average $(ARIMA)$ and the Pattern Sequence-based Forecasting (PSF) [12,14]. Also regression models have been widely adapted to model and forecast power consumption [6,7]. A Semi-Parametric Additive model for short-term (half-hourly) load forecasting model is proposed by Fan et al. [4]. In this work, the authors aim to tackle the non-linearity, volatile consumption pattern and interpret the effect of demand external drivers on power consumption prediction. The study integrates the non-linear and nonparametric driver factors within the regression framework. This is done by proposing semi-parametric additive model to estimate the relation between power consumption and affecting variables. The load demand drivers include the seasonality factors (weekdays, holidays and day of the year). In addition to the previous factors, the historical consumption and temperature data from the previous three hours as well as the same period from the previous six days are used.

Artificial intelligence techniques such as neural networks, support vector machines, and pattern matching techniques are widely applied to predict short-term power consumption [11,13]. An overview of different AI techniques is provided in [10]. Among all AI-based methods, the technique of artificial neural networks (ANNs) has received substantial attention in forecasting power consumption due to its flexibility in learning load series and modeling the non-linearity between power consumption and the exogenous variables influencing it as well as providing fairly acceptable results. S. Wan et al. developed an artificial neural network model for modeling the electricity load of campus buildings in [17]. The input data of the network includes the consumption history of buildings and the time-dependent climate variables such as dew point, rainfall rate, pressure, wind speed, humidity and temperature.

In this work, we improve on previous works by dividing the buildings into homogeneous groups based on extracting features characterizing their consumption pattern over different tie horizons i.e. daily, weekly and monthly. Then we introduce two time series models namely, TCPM which is used to predict the day-ahead total power consumption, and HCPM model which is used to predict the 24-hours consumption pattern of individual buildings i.e. the detailed consumption values at each hour of the day.

3   Hybrid Short-Term Power Consumption Forecasting Model

In this section, we present our design of two time-series models for short-term forecasting of power consumption inspired from the PSF algorithm, TCPM for predicting individual customers day-ahead power consumption and HCPM for day-ahead 24-hours consumption pattern forecasting. Both models consist of two phases. As the first phase of grouping the buildings based on time-derived discriminative features is common for both models, we present it separately in the next section.

3.1   First Phase: Clustering of Buildings Using Time-Derived Discriminative Features

The goal of this phase is to divide the buildings into several homogeneous groups where each group contains only buildings with similar power consumption patterns. As a first step, we extract discriminative features that characterize the power usage pattern of individual buildings by leveraging the seasonal, weekly, and daily patterns in their historical consumption data. The features are extracted with regard to different time horizons, namely day segments, daily, monthly, and yearly. The extracted features are as follows:

– Average power consumption for each week day and each month.
– Percentage of power consumption over six day segments: early morning, morning, early afternoon, afternoon, early night, and late night similar to previous work [15].
– Total power consumption over the trial in kWh.
– Percentage of power consumption over business days, Sunday, and Saturday.

The vital point of using time discriminative features is to identify different classes of buildings without previous knowledge about them and only by relying on the shape of their power consumption during the trial. After the features extraction, the proposed clustering methodology consists of the following steps:

1. Normalizing the extracted features in the range [0,1] and assign different importance levels to them.
2. Clustering buildings using k-means and hierarchical clustering algorithms.
3. Selecting the optimal number of clusters based on validity indexes of the clustering process namely Dunn, Silhouette, and Davies-Bouldin.
4. Selecting the optimal clustering algorithm using the same validity indexes.

Two clustering algorithms, namely k-means and hierarchical clustering are used to identify the different groups of buildings in the designed feature space. In Sect. 4.1, we evaluate our clustering methodology in order to determine the optimal clustering algorithm as well as the optimal number of clusters to which the buildings should be divided.

3.2   Hourly Consumption Pattern Matching (HCPM) Model

As an output of the first phase, we get a set of clusters where each cluster represents buildings with similar consumption pattern. The second phase in HCPM model consists of two independent steps. In the first step, each group from the initial clustering process in the first phase is clustered using hierarchical clustering. Differently from first phase, the clustering in each group is done based on their detailed 24-hour consumption pattern in order to produce a set of different patterns which characterize the different daily consumption patterns of each building. Then a label will be assigned to each resulting cluster. The assigned labels will be used in the next stage for pattern sequence matching. The number of clusters is selected based on majority vote of the previously used clustering indexes. As a result of the labeling process, we will have our 24-hour daily patterns in each group represented as a time series of subsequent labels where each label corresponds to one day. In the second step, HCPM utilizes the labels produced by the clustering phase in the pattern matching process in order to forecast the next day 24-hours consumption pattern. The forecasting process starts by searching for days with same historical power consumption pattern with the window size confirming to the length of the labels sequence.

The algorithm tries to find a matched pattern in the historical consumption data of same building. In case of multiple matches, it takes the closest match in time horizon because the consumption behavior should be more similar in recent past. If the pattern is not found in the historical data of same building, the algorithm searches in the repository of historical consumption data of all buildings. If multiple matches are found, the majority vote over all matched patterns is considered as the predicted pattern. Finally, if no sequence is found in the repository corresponding to window size W, the algorithm searches for the sequences of labels equals to W-1 and thus successively. The searching procedure for a match uses only the historical data of the same day of the week. The full procedure of searching for an equal pattern in HCPM is illustrated in Fig. 1.

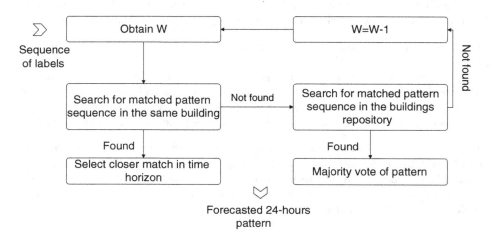

Fig. 1. Workflow of the hourly consumption pattern matching algorithm (HCPM).

3.3 Total Consumption Pattern Matching (TCPM) Model

TCPM forecasting model depends in its second phase on the total power consumption in each day rather than the used labels of each day's 24-hour consumption pattern in HCPM. The percentage error defined in (1) is used as a metric to decide whether two days are considered as a match or not, where $C1$ and $C2$ are the total power consumption in the days to be compared.

$$PR(C1, C2) = |\frac{C1 - C2}{C1}| \tag{1}$$

TCPM follows the same steps as in HCPM. The difference is that the numerical values of the power consumption are the input for the algorithm instead of the sequence of labels in HCPM. Also if multiple matches are found in the repository of all buildings, the average consumption is considered as the final prediction while the majority vote is used in case of HCPM.

4   Evaluation

The used dataset in this work was provided by the Commission of Energy Regulation (CER) in Ireland [3]. Raw data representing the 30-minute power consumption readings in kWh of individual buildings was collected over a period of eighteen months during 2009 and 2010. More than 5000 smart meters were installed in Irish homes and businesses in eight urban areas and three villages. Pre-trial and post-trial surveys were conducted for both residential and business participants. The questions were related to their demographic features, building characteristics, life style and usage patterns of different appliances. For the evaluation of our proposed short-term power consumption forecasting models, we used the data related to small and medium enterprise (SME) buildings as the oscillation in their power consumption have a more regular pattern compared to the residential buildings.

4.1   Buildings Clustering Using Time Derived Discriminative Features

The average daily, monthly, and total power consumption during the trial were given a higher importance level. The different importance levels can be given after normalizing all features values in the range [0,1] by multiplying the range with a constant factor. Based on extensive analysis of different features combinations, the final set of used features consisted of the total consumption as well as the consumption percentages on Sunday, Saturday, and different day segments. We have given the percentage of consumption on Sunday, Saturday and different day segments the same importance level whereas the total consumption was given the highest importance level.

The optimal number of clusters for k-means and hierarchical clustering was selected based on the majority vote of the three validity indexes. Lower value of Davies Bouldin index indicates better clusters quality while higher values for Silhouette and Dunn indexes prove better clustering quality. Figure 2 illustrates the clustering quality for hierarchical and k-means algorithms. Apparently the indices reached their optimal values when the number of clusters K = 8. We proceeded using hierarchical clustering approach based on the relative comparison of the indexes' scores for both algorithms.

Figure 3 illustrates the average daily power consumption in kWh of each cluster generated using hierarchical clustering with an optimal number of clusters K = 8. There is a clear separation between groups 8, 6, 4, and 2. This indicates that the total power consumption is the main factor to distinguish between these groups. However, groups 1, 3, and 7 have close daily average power consumption and the total power consumption is not the main factor to distinguish between these three classes but rather the percentage of power consumed during the weekend in comparison to the whole week as well as the percentage of power consumed over different day segments.

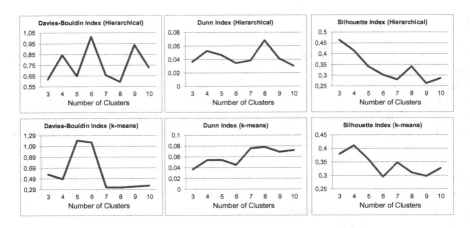

Fig. 2. Derived features clustering indexes.

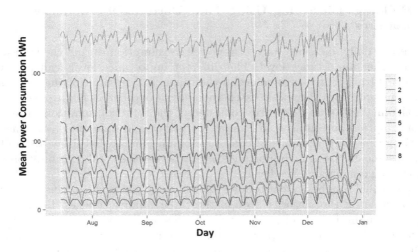

Fig. 3. Daily mean power consumption of each cluster generated using hierarchical clustering algorithm with an optimal number of clusters K = 8.

4.2   HCPM Model

For the evaluation of HCPM model, one group named group 3 was selected with 59 SME buildings. A subset of 5 months of the trial was used as a repository of power consumption historical data that can be used to predict the day-ahead power consumption pattern. Also one month was excluded from the clustering step for out-of-sample evaluation. The input for the hierarchical clustering algorithm is the dataset of detailed 24-hours power consumption for all buildings in the repository of power consumption historical data. Each data-point is a sequence of 24 points representing the consumption during 24 h in kWh. The

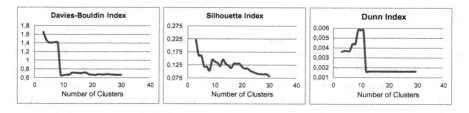

Fig. 4. Selecting the optimal number of clusters for group 3.

optimal number of clusters was selected based on the majority vote of the same validity indexes. Figure 4 demonstrates the quality of clusters generated using hierarchical clustering algorithms.

Based on the majority vote, K = 10 and K = 9 were selected as the optimal number of clusters. Mean Error Relative (MER) is used for analyzing the effect of the obtained window size on the prediction error. Figure 8 demonstrates the relation between the window size and the prediction MER. Mean Error Relative (MER) starts with 24% and falls down to a minimum of 17.1% for a window size equals to 9.

Figure 5 demonstrates the relation between the window size and the percentage of correctly predicted labels. We define the precision as a metric reflecting the percentage of correctly predicted labels divided by the total number of tested samples. It shows that the precision increases significantly from 62.5% for W = 1 to 82.1% for W = 10. Also the number of found matches will decrease with an increasing window size. However, more matches can be found in buildings which are following a consumption pattern close to the targeted building. The results in Fig. 6 are consistent with the MER statistics in Fig. 8.

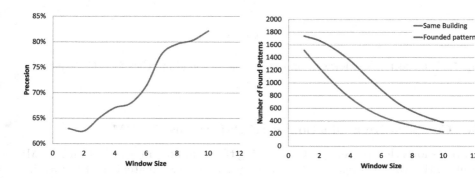

Fig. 5. HCPM precision for different values of window size.

Fig. 6. HCPM number of matches for different values of window size.

4.3 TCPM Model

For the evaluation of the TCPM model, the same group of 59 SME buildings was used. We considered a percentage error of less than 10% to count for a match between a day in the repository of historical data of the buildings and a day in the consumption sequence of previous days of the day in question. Figure 8 depicts the relation between the window size and the prediction MER. The MER is falling gradually from 22.6% and is reaching a minimum value of 9.1% for a window size equal to 8. Then, it slightly increases for a window size W = 9 and again decreases to 9.2% for W = 10. As a conclusion, more knowledge about the power consumption of previous days can help improving the prediction accuracy. However, a larger window might reduce the accuracy as we notice for window size of 9. Moreover, a larger window size will increase the possibility of not finding a match. Figure 7 demonstrates the drop in the number of found patterns based on the increased window size also the relation between the number of found patterns and whether they belong to the consumption history of the same building or not. With a larger window size, the possibility to find a match in the historical data of the same building is higher than finding this pattern in other buildings.

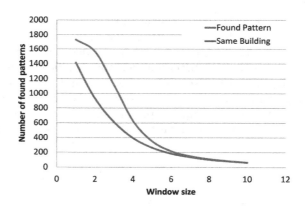

Fig. 7. The effect of window size on the number of found matches for TCPM model.

4.4 Comparative Evaluation of HCPM Against TCPM

Figure 8 shows that pattern matching using the total power consumption TCPM significantly outperforms hourly consumption pattern matching (HCPM) in terms of mean error relative (MER). The reason is that the used Euclidean distance to measure the similarity level between two days based on the 24-hours consumption pattern might assign the same label to two days with a large difference in total power consumption just because both have small distances to the centroid of the cluster they belong to. To sum up, the total power consumption is more appropriate for predicting the day-ahead power consumption.

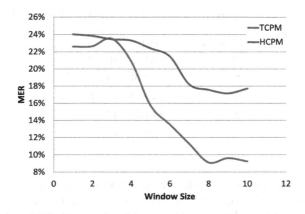

Fig. 8. Relative comparison of MER values.

4.5   Comparative Evaluation of HCPM Against PSF

PSF utilizes the k-means algorithm to realize the clustering of the dataset. The first step was to determine the optimal number of clusters. For this, same indexes were used to validate the clustering quality and select the number of clusters based on the majority vote. The optimal number of clusters has produced clusters with a big gap in the average hourly power consumption as shown in Table 1. Group 4 contains days with an average hourly power consumption around 35.66 kWh. Therefore, we can roughly conclude that it represents retail buildings with high power consumption and not days with different consumption pattern based on day segments.

For the comparative analysis of prediction accuracy, we evaluated the PSF using four different window size values $W = 7, W = 8, W = 9, W = 10$. With $W \geqslant 7$ PSF should be capable of capturing the similarity over a week of consumption data. Table 2 shows the comparative analysis between HCPM and PSF.

Table 1. Average hourly power consumption in kWh of the k-means clusters.

Cluster number	Average hourly power consumption (kWh)
1	6.62
2	10.03
3	4.46
4	35.66
5	1.32
6	18.73
7	15.14

Table 2. Comparative analysis between HCPM and PSF.

Window size	MER	
W	PSF	HCPM
7	46.73%	24.02%
8	45.11%	23.84%
9	43.15%	17.14%
10	39.91%	17.70%

The proposed model outperformed PSF in terms of time efficiency and prediction accuracy by introducing three enhancement on the original PSF approach. Firstly, the proposed model clusters the buildings based on derived features using the seasonal, weekly and daily pattern of the historical power consumption data. This step makes substantial contribution to the overall performance by reducing the search space to include data for buildings belonging to the same cluster only. Moreover, by giving high importance to the total power consumption as one dimension for clustering, we separate buildings belonging to different power consumption classes. In addition, the algorithm tries to find a matched pattern in the same building historical data as a first option. This heuristic reduces the search space and increases the prediction accuracy in case a match is found in same building historical data. With more historical data of each building, the chance for this heuristic to successfully find a match is higher. Finally, searching for a match using only the same day of the week also reduces the search space more. This heuristic will reduce the probability to find a match in the historical data, however, this issue can be overcome by providing more historical consumption data. The experiments prove that the used traditional time series model of PSF cannot forecast the day-ahead power consumption of individual customers in case of inhomogeneous groups. However, additional heuristics can improve the traditional methods ability to predict the individuals short-term consumption.

5   Discussion and Future Work

In this work, two time series models TCPM and HCPM were designed to predict individual customers day-ahead power consumption and the 24-hours consumption pattern respectively. The comparison against PSF showed that the proposed model of HCPM significantly outperformed PSF, also our analysis revealed that, clustering buildings based on their seasonal, weekly and daily patterns can improve the prediction accuracy and increase time efficiency by reducing the search space. In future work, HCPM model will be tested on the post-trial dataset in which the participants were allocated different tariffs, in order to check whether the proposed model is able to handle the changes in customers demand pattern.

References

1. Alfares, H.K., Nazeeruddin, M.: Electric load forecasting: literature survey and classification of methods. Int. J. Syst. Sci. **33**(1), 23–34 (2002)
2. Aman, S., Frincu, M., Chelmis, C., Noor, M.U.: Empirical comparison of prediction methods for electricity consumption forecasting. Department of Computer Science, University of Southern California, Los Angeles, CA, 90089 (2012)
3. Commission for Energy Regulation (CER): Electricity smart metering technology trials findings report. ESB Networks, Belgard Square North, Tallaght, Dublin 24 (2011)
4. Fan, S., Hyndman, R.: Short-term load forecasting based on a semi-parametric additive model. IEEE Trans. Power Syst. **27**(1), 134–141 (2012)
5. Gabriel, S.: An electricity market for germany's energy transition. Federal Ministry for Economic Affairs and Energy (BMWi), PRpetuum GmbH, Munchen (2014)
6. Hong, T., Gui, M., Baran, M., Willis, H.: Modeling and forecasting hourly electric load by multiple linear regression with interactions. In: Power and Energy Society General Meeting 2010, pp. 1–8. IEEE, July 2010
7. Jiang, H., Lee, Y., Liu, F.: Anomaly detection, forecasting and root cause analysis of energy consumption for a portfolio of buildings using multi-step statistical modeling, US Patent App. 13/098,044 (2012)
8. Coughlin, K., Piette, M.A.C.G., Kiliccote, S: Emergency demand response program manual, Sect. 5.2: calculation of customer baseline load (CBL). New York Independent System Operator, Southern California Edison, Technical report (2010)
9. Coughlin, K., Piette, M.A.C.G., Kiliccote, S: 10-day average baseline and day-of adjustment. Southern California Edison, Technical report (2011)
10. Metaxiotis, K., Kagiannas, A.D.A., Psarras, J.: Artificial intelligence in short-term electric load forecasting: a state-of-the-art survey for the researcher. Energy Convers. Manage. **44**, 1525–1534 (2003)
11. Khotanzad, A., Afkhami-Rohani, R., Lu, T.L., Abaye, A., Davis, M., Maratukulam, D.: Annstlf-a neural-network-based electric load forecasting system. IEEE Trans. Neural Netw. **8**(4), 835–846 (1997)
12. Martinez Alvarez, F., Troncoso, A., Riquelme, J., Aguilar Ruiz, J.: Energy time series forecasting based on pattern sequence similarity. IEEE Trans. Knowl. Data Eng. **23**(8), 1230–1243 (2011)
13. Rui, Y., El-Keib, A.: A review of ann-based short-term load forecasting models. In: Proceedings of the Twenty-Seventh Southeastern Symposium on System Theory 1995, pp. 78–82 (1995)
14. Shen, W., Babushkin, V., Aung, Z., Woon, W.: An ensemble model for day-ahead electricity demand time series forecasting. In: Proceedings of the Fourth International Conference on Future Energy Systems, pp. 51–62. ACM, New York (2013)
15. Silipo, R., Winters, P.: Big data, smart energy, and predictive analytics time series prediction of smart energy data (2013). www.KNIME.com
16. Simmhan, Y., Noor, M.: Scalable prediction of energy consumption using incremental time series slustering. In: 2013 IEEE International Conference on Big Data, pp. 29–36, October 2013
17. Wan, S., Yu, X.-H.: Facility power usage modeling and short term prediction with artificial neural networks. In: Zhang, L., Lu, B.-L., Kwok, J. (eds.) ISNN 2010. LNCS, vol. 6064, pp. 548–555. Springer, Heidelberg (2010). doi:10.1007/978-3-642-13318-3_68

Multi-resolution Time Series Discord Discovery

Heider Sanchez and Benjamin Bustos[(✉)]

Department of Computer Science, University of Chile, Santiago, Chile
{hesanche,bebustos}@dcc.uchile.cl

Abstract. Discord Discovery is a recent approach for anomaly detection in time series that has attracted much research because of the wide variety of real-world applications in monitoring systems. However, finding anomalies by different levels of resolution has received little attention in this research line. In this paper, we introduce a multi-resolution representation based on local trends and mean values of the time series. We require the level of resolution as parameter, but it can be automatically computed if we consider the maximum resolution of the time series. In order to provide a useful representation for discord discovery, we propose dissimilarity measures for achieving high effective results, and a symbolic representation based on SAX technique for efficient searches using a multi-resolution indexing scheme. We evaluate our method over a diversity of data domains achieving a better performance compared with some of the best-known classic techniques.

Keywords: Time series · Anomaly detection · Discord discovery · Indexing

1 Introduction

In light of recent advancements in streaming technologies, the anomaly detection in time series has become an important task in different applications of monitoring systems, such as: analysis of video surveillance, multiple sensors in car and aircraft crashes, tracking of objects in riot detection, traffic alert on roads, seismic signals, electrocardiograms (ECG), etc. To address the anomaly detection in time series, we first need to define the type of anomaly that fits the application purpose and the data domain. It is a complex issue, because the anomaly can be associated with outlier points (irregularities, change point), outlier subsequences (unusual patterns, novelty), or anomalous relation between variables. The most common solutions for anomaly detection are built by machine learning methods [2]. Usually, they are supervised learning techniques and need a time series sample of "normal" behavior as training model, and in other cases a set of unusual patterns is required. These sets are provided by a domain expert.

H. Sanchez—Work supported by a research grant from CONICYT-Chile.

B. Bustos—Supported by the Millennium Nucleus Center for Semantic Web Research under Grant NC120004.

© Springer International Publishing AG 2017
I. Rojas et al. (Eds.): IWANN 2017, Part II, LNCS 10306, pp. 116–128, 2017.
DOI: 10.1007/978-3-319-59147-6_11

However, in many real contexts, obtaining this prior knowledge is a difficult task. This is where Unsupervised Learning Techniques are used to try to identify significant patterns, which adjust the knowledge model directly from the input stream.

We focus this research on outlier subsequences, taking as baseline the Discord Discovery Process proposed by Keogh et al. [6]. They previously conducted an important work to identify the most unusual subsequence of long time series designing an unsupervised window-based method. The main characteristic of the subsequences is their level of overlapping; consecutive subsequences are more similar to each other. Accordingly, there is a necessity for optimal structures that consider this property for efficient searching. A series of related works about discord discovery were proposed in the past decade [1,7,8,12]. The main objective of these related works was the detection efficiency, because the brute force algorithm has a quadratic order regarding the total number of subsequences. However, multi-resolution discord discovery has received little attention in this research line.

In this paper, we introduce (1) a new multi-resolution representation based on local trends and mean values of the time series. It becomes a parameter-free technique when we use the maximum level of resolution which will be defined in this work. We also propose (2) a symbolic representation derived from the numeric representation by applying SAX quantization [9] on the trend and value components. It also provides us a lower bounding function for indexing time series collection. The main contribution is (3) a multi-resolution discord discovery technique based on this time series representation. The efficacy and efficiency of our approach is experimentally evaluated over a diversity of data domains [5]. We empirically demonstrate that our method outperforms conventional methods.

2   Background and Related Work

We associate an anomaly into a time series as a subsequence (unusual pattern) which produces a qualitatively significant change in the behavior of data. Unusual patterns are outstanding subsequences that arbitrarily occur and are associated with residual variation of the time series. This is contrary to the frequent patterns that regularly occur and are associated with cyclical or periodic variation of time series, moreover, they are located into the normal observation of the time series. Keogh et al. [6] introduced this new approach to avoid creating a workable definition for "the most unusual subsequence", and furthermore it is an unsupervised method that does not require training data.

Definition 1 (Non-self match). *Given a time series P, containing a subsequence C_i of length w and a matching subsequence C_j of the same length, we say that C_j is a non-self match to C_i if $|i - j| \geq w$, where p and q are their respective starting positions in P.*

Definition 2 (Time Series Discord). *Given a time series P, the subsequence C_j of length w is said to be the discord of P if C_j has the largest distance to its nearest non-self match.*

This problem can be solved by a brute force search using a nested loop. The outer loop takes each subsequence as a possible candidate, and the inner loop is used to search the candidate's nearest non-self match. The candidate that has the greatest such value is the discord. The computational complexity is $O(N^2)$, where N is the number of subsequences. To improve this complexity, Keogh et al. [6] proposed a generic algorithm for efficient detection. This algorithm requires two heuristics that generate two ordered lists of subsequences; one for the outer loop and the other one for the inner loop. The heuristic *Outer* is useful for quickly finding the best candidate, and the heuristic *Inner* is useful for quickly finding the best nearest non-self match. We break out of the inner loop if the distance is less than the best-so-far discord distance. Two main related methods for discord discovery are HOT SAX [7] and HOT iSAX [1], which are based on SAX representation [9]. SAX splits the time series into segments and builds a new symbolic time series quantizing the mean values of each segment. Both techniques build efficient structures to find the time series discord using the discord discovery heuristics.

3   Multi-resolution Trend-Value Approximation

3.1   Why a Trend-Based Representation?

Esmael et al. claim that "using only the value approximation, causes a high possibility to miss some important patterns in some time series data. SAX does not pay enough attention to the shapes of the time subsequences and may produce similar strings for completely different time series" [4]. In this way, several piecewise approximations based on trend and value features have been recently proposed [3,4,11]. We focus on the technique 1d-SAX proposed by Malinowski et al. [11]. This is a compact binary representation to improve the retrieval performance using the same amount of information that SAX. Here, SAX is extended by adding new symbols that represent the slope of each segment. The algorithm uses linear regression to compute the slope. 1d-SAX works with alphabets of different sizes:

$$1\text{d-SAX}(P, m, \alpha^v, \alpha^s) = \{(\hat{v}_1, \hat{s}_1), \cdots, (\hat{v}_i, \hat{s}_i), \cdots, (\hat{v}_m, \hat{s}_m)\},$$

where $\hat{v}_i$ is the average value symbol from an alphabet of size α^v, and $\hat{s}_i$ is the slope symbol from an alphabet of size α^s.

We also compare both methods, SAX and 1d-SAX, using an agglomerative hierarchical clustering to group five time series in three different classes (Fig. 1). The time series is split into four segments. SAX only takes the mean value while that 1d-SAX also considers the slope obtaining a better match between time series 2 and 3 which belong to same class.

In this work, we extend the ability of the local trends to various levels of resolution. While the granularity parameter (number of segments) of the piecewise approximation like SAX and 1d-SAX produces a horizontal segmentation, we propose a hierarchical segmentation induced by the resolution level. This segmentation

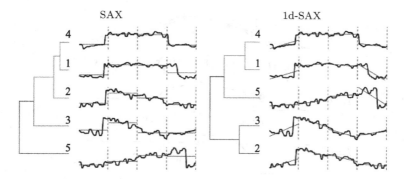

Fig. 1. A comparison of the ability of two time series representations to cluster five members of the CBF dataset using the Euclidean distance.

provides greater advantages in design and optimization that will be discussed. Our time series representation is called Multi-resolution Trend-Value Approximation (MTVA). The basic idea is to generate trend-value pairs on each level of resolution, and then compute the similarity between two MTVA representations using a distance measure. In addition, we design a symbolic representation to build a multi-resolution indexing structure for discord discovery.

3.2   Bottom-Up Construction Algorithm

Given the times series $P = \{p_1, \cdots, p_n\}$ and L as the level of resolution defined by the user, the MTVA representation of P is built following the next steps:

1. We start in the last resolution level L dividing the time series into $M = 2^{L-1}$ segments of size $w = n/M$.
2. Let $Y = \{y_1, \cdots, y_w\}$, be a segment of P in the time segment $X = \{x_1, \cdots, x_w\}$, we compute the linear regression on each segment by the function $lr(x) = ax + b$, where:
   - $a = \frac{\sum_{i=1}^{w}(x_j - \bar{X}) * y_j}{\sum_{i=1}^{w}(x_j - \bar{X})^2}$
   - $b = \bar{Y} - a * \bar{X}$
   - $\bar{X}$ and $\bar{Y}$ are the average value of X and Y, respectively.
   - The trend-value pair (v, s) of the segment Y is defined by:
     - $v = a * \frac{x_1 + x_w}{2} + b$ is the mean value.
     - $s = \arctan(a)$ is the slope,
3. For the next resolution levels $M = 2^{\{L-2, L-3, \cdots, 0\}}$, compute the trend-value pair (v, s) for each segment as follows:
   - $v = \frac{v_i + v_{i+1}}{2}$
   - $s = \arctan\left(\frac{v_{i+1} - v_i}{x_{i+1} - x_i}\right)$.
   - v_i and x_i is the mean value and the average time associated to a segment in the upper level (see Fig. 2).

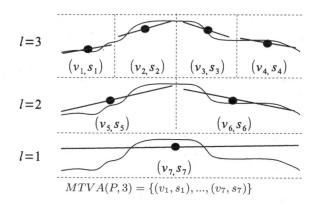

$l=3$

(v_1, s_1) (v_2, s_2) (v_3, s_3) (v_4, s_4)

$l=2$

(v_5, s_5) (v_6, s_6)

$l=1$

(v_7, s_7)

$MTVA(P, 3) = \{(v_1, s_1), ..., (v_7, s_7)\}$

Fig. 2. Construction of the Multi-resolution trend-value approximation.

4. The result is an array of all the trend-value pairs:

$$MTVA(P, L) = \{(v_1, s_1), ..., (v_m, s_m)\}.$$

Figure 2 shows the MTVA representation of the time series P up the third level of resolution ($L = 3$). Parameter L can be automatically computed so that the spatial complexity of the MTVA representation does not exceed the space of the original time series, that is, adjusting the total number of segments $m \leq n/2$. On the other hand, m can be defined in terms of the level of resolution $m = 2^L - 1$. Then solving both equations, we obtain that the maximum level of resolution for P is $L_{max} = \lfloor \log_2(n/2) \rfloor + 1$.

3.3 MTVA Distance

We first need a cost function to measure the distance between trend-value pairs. Given two pairs p_i and q_j, we define the cost function as follows:

$$cost(p_i, q_j) = |v_i^p - v_j^q|^2 + |s_i^p - s_j^q|^2,$$

where both value-domain and slope-domain should have similar range to avoid that the distance is governed by only one of them. The slope ranges are between $-\frac{\pi}{2}$ and $+\frac{\pi}{2}$, we therefore normalize the time series by a standard normalization procedure (e.g. Z-distribution). We then propose a multi-resolution distance $MDist$ to measure the dissimilarity between two MTVA representations executing the cost function on all levels of resolution:

$$MDist(P, Q) = \sum_{l=1}^{L} \sum_{i=2^{(l-1)}}^{2^l - 1} cost(p_i, q_i).$$

The computational time of executing $MDist$ is the sum of the time in each level of resolution:

$$T(L) = \sum_{l=1}^{L} M_l = \sum_{l=1}^{L} 2^{l-1} = 2^L - 1.$$

If we compute the distance in the worst case where L is exactly $\log_2(n/2) + 1$, the computational time is of order $O(n)$, where n is the length of the original time series. Therefore, $MDist$ in the worst case is theoretically as fast as the classic distances that work over the raw representation.

3.4 Symbolic Representation

Discretization techniques are used to transform the numeric representation into a sequence of symbols. This symbolic representation provides us greater ease of interpretation and simplicity to manage time series collections.

Definition 3. *"Breakpoints are a sorted list of numbers $\beta = \{\beta_1, \ldots, \beta_{\alpha-1}\}$, such that the area under a $N(0,1)$ Gaussian curve from β_i to $\beta_{i+1} = 1/\alpha$ (β_0 and β_α area defined as $-\infty$ and $+\infty$, respectively)" [9]. For example, if $\alpha = 4$ then the breakpoints are $\{\beta_1 = -0.67, \beta_2 = 0, \beta_3 = +0.67\}$.*

Gaussian Assumption. To transform the numeric pair $p_i = (v_i, s_i)$ to a symbolic pair $\hat{p}_i = (\hat{v}_i, \hat{s}_i)$, we quantize separately both values using breakpoints that produce equal-size areas under the Gaussian curve $N(\mu, \sigma^2)$ (similar to 1d-SAX). Gaussian discretization is feasible for normalized time series, since statistically the mean value and the slope have a Gaussian distribution [10,11]. As in 1d-SAX, the breakpoints are determined by the curve $N(0,1)$ for the mean value and $N(0, \sigma_L^2)$ for the slope. In this last case, we use the variance σ_L^2 in terms of the level of resolution L because each level of resolution generates different slope distributions (Fig. 3), unlike the 1d-SAX that uses a slope variance in terms of the size of the segment. Additionally, to apply the linear regression between X and Y, we recommend that both variables have similar range. If the time series is normalized in $N(0,1)$, then the temporal component X must fit in this interval size. In this work, we normalize the length of each segment $X = [1, w] \rightarrow X = [-1, 1]$. In this manner, the variance σ_L^2 is defined in terms of the level of resolution independently of the size of the segment.

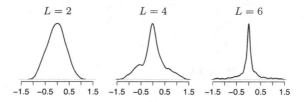

Fig. 3. Density of the slope varying the level of resolution in ECG time series.

Alphabet Size. The alphabet size is delimited by the number of breakpoints (Definition 3) and strongly influences over the compression ratio and the reconstruction error. To quantize the trend-value pair, we need two alphabets with size α_v and α_s for the mean value and the slope respectively. For conformity, we use binary symbols where α is power of two [13]. Thus for example, to compress the numeric MTVA up the level 3 using $\alpha_v = 4$ and $\alpha_s = 4$, we need $(2+2)*(2^3-1)$ bits, it is less than 4 bytes by time series. This symbolic representation will serve us to different applications like indexing and anomaly detection.

3.5 Indexing

To efficiently manage MTVA time series datasets, we use the symbolic representation to build a hash-based index, where each bucket $\hat{P}$ envelops a set of similar MTVA time series. To filter out distances in the similarity search, we design a lower bounding function called MINDIST to measure the distance between the query object Q and a bucket $\hat{P}$, so that it is less than the distance between Q and any object $P \in \hat{P}$. Before defining MINDIST, we first need to define the lower bounding function of the trend-value cost, which is denoted as follows:

$$LB\_cost(\hat{p}_i, q_i) = (\Delta v)^2 + (\Delta s)^2 \le cost(p_i, q_i), \text{where}$$

$$
\Delta v =
\begin{cases}
|v_i^q - \beta_{Ui}| & v_i^q > \beta_{Ui} \\
|\beta_{Li} - v_i^q| & v_i^q < \beta_{Li} \\
0 & \text{else,}
\end{cases}
\qquad
\Delta s =
\begin{cases}
|s_i^q - \beta_{Ui}| & s_i^q > \beta_{Ui} \\
|\beta_{Li} - s_i^q| & s_i^q < \beta_{Li} \\
0 & \text{else,}
\end{cases}
$$

$$: \beta_{Li} \le v_i^p < \beta_{Ui}, \qquad : \beta_{Li} \le s_i^p < \beta_{Ui}.$$

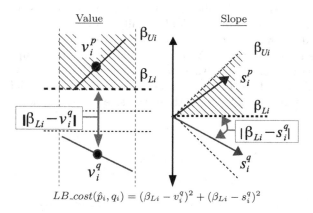

$$LB\_cost(\hat{p}_i, q_i) = (\beta_{Li} - v_i^q)^2 + (\beta_{Li} - s_i^q)^2$$

Fig. 4. Lower bounding trend-value cost. The blue line represents a trend-value pair stored in our database and the green line is the query (Color figure online).

The symbol $\hat{p}_i$ is derived from a trend-value pair p_i that is located between two breakpoints $\beta_{Ui} < p_i \leq \beta_{Li}$, independently for each pair value (Fig. 4). MINDIST is then calculated by following equation:

$$MINDIST(\hat{P}, Q, l, \alpha) = \alpha + \sum_{i=2^{(l-1)}}^{2^l-1} LB\_cost(\hat{p}_i, q_i),$$

where l is the current resolution level and α is the accumulated distance of the previous levels.

4   Multi-resolution Discord Discovery

The main challenge of the discord discovery approach is to face its quadratic complexity. In this sense, our MTVA representation together with the discord discovery heuristics can be used to solve the anomaly detection in time series. We propose a multi-resolution method called HOT MTVA, which increases the level resolution of the index when a hash-bucket is overflowed (Fig. 5). This indexing structure allows the perfect fit to our MTVA representation. Moreover, it is flexible to control the level of resolution of the detected anomaly.

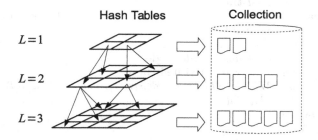

Fig. 5. Multi-resolution index model for the MTVA representation.

4.1   Building Algorithm

Given the times series P of length n, we use an overlapping sliding window of size $w \ll n$ for extracting all the possible subsequences C_i, $i \in \{1, \ldots, (n - w + 1)\}$ from P. The insertion procedure of a MTVA subsequence C_i into the indexing structure $\mathbb{R}$ is described as follows. Unlike HOT iSAX, we apply hierarchical quantization to access to hash tables (where each slot is a node) from the lowest resolution to the maximum resolution. If a terminal node is full, we re-insert all its associated objects into a hash table of higher level to provide additional differentiation, so we create new nodes with the next resolution level of the current node. We use a size threshold th_{max} to control the maximum number of objects in a terminal node (the so-called bucket). As we can see, the indexing level has dynamic behavior, its incremental value depends of the size of dataset and the maximum level of resolution (L_{max}).

4.2   Discord Discovery Heuristics

The discordant subsequence is found applying the optimal discord discovery procedure [7] using the following heuristics:

Outer Loop Heuristic: We first visit all subsequences belonging to the bucket that contains the minimum number of subsequences starting from the lowest resolution level. Afterwards, we visit the rest of buckets in random order. This heuristic ensures that the subsequences that are most isolated, in each resolution level, will be visited at the beginning of the search as potential candidates.

Inner Loop Heuristic: We then use an inner loop to search the best non-self match of each selected candidate C_j. We first visit all subsequences contained

---

Algorithm 1. NNM-Search for the Multi-resolution MTVA Index

Require: (Index $\mathbb{R}$, Query C_j, Window Size w, Threshold Distance th_{dist})
1: $stack.push([\mathbb{R}.getNodeRoot(), 0])$
2: $best\_dist = \infty$
3: $best\_post = -1$
4: **while** $stack \neq \emptyset$ **do**   ▷ inner loop
5:     $[node, min\_d] = stack.pop()$
6:     **if** $min\_d > best\_dist$ **then**
7:         **Break**   ▷ break out of inner loop
8:     **else if** $node$ is internal **then**
9:         $list = \emptyset$
10:         **for** $child\_node \in node.children$ **do**
11:             **if** $child\_node$ was not visited **then**
12:                 $d = MINDIST(child\_node.str, C_j, node.level, min\_d)$
13:                 $list.add([child\_node, d])$
14:             **end if**
15:         **end for**
16:         $sorted\_list = argsort(list)$
17:         $stack.push(sorted\_list)$
18:     **else if** $node$ is terminal **then**
19:         $objects = readBucket(node.str)$
20:         **for** $C_i \in objects$ **do**
21:             **if** $|i - j| \geq w$ **then**   ▷ non-self match?
22:                 $d = MDist(C_i, C_j)$ ▷ Multi-resolution Distance
23:                 **if** $d < best\_dist$ **then**
24:                     $best\_dist = d$
25:                     $best\_post = j$
26:                 **end if**
27:                 **if** $d < th_{dist}$ **then**
28:                     **Break**   ▷ break out of inner loop
29:                 **end if**
30:             **end if**
31:         **end for**
32:     **end if**
33: **end while**
34: **Return** $(best\_dist, best\_pos)$

in the bucket from which C_j is retrieved. Afterwards, we apply the nearest non-self match search algorithm (NNM-Search, Algorithm 1) to visit the rest of the buckets. This heuristic allows to first visit all the subsequences most similar to C_j increasing the probability of early termination of the loop.

The NNM-Search algorithm performs a hierarchical search across the internal nodes using a stack to maintain the nodes ordered in terms of MINDIST (lines 10 – 17). MINDIST measures the minimum distance between the query and the current node. The algorithm also applies two breaking statements to break the inner loop as early as possible: one is associated to MINDIST (line 6) and the other one is associated to the best-so-far discord distance (line 27).

5   Experimental Results

In this section, we evaluate the performance of our approach to address the anomaly detection problem. Effectiveness will be evaluated over a set of 20 real cases of anomalous time series (with different sliding windows) collected by Keogh et al. [5,6]. Efficiency will be evaluated using the same set of long time series described by Sanchez et al. [12].

5.1   Effectiveness of Our MTVA Representation

We first evaluate the accuracy of our trend-value numeric representation over all anomaly cases. The classic techniques use the Euclidean Distance as measure distance over the raw representation of the normalized subsequence. The normalization process is used to homogenize all subsequences applying implicitly two transformations: translation and scaling. This has a problem, though: the presence of local noise is amplified by the scaling transformation. This issue is solved by using a context-dependent parameter $\varepsilon > 0$ for smoothing noisy subsequences [9,12]. An important feature of our MTVA representation is that the slope of the noisy segments trends to zero and thereby the unusualness of noisy subsequences is reduced. Table 1 shows the results obtained by both techniques for six values of ε. We evaluate the MTVA using three different resolution levels. We note that the MTVA representation achieves a higher percentage of true detections when increasing the resolution level to $L = 4$, this is the common maximum level for all the sliding windows. In this way, we assert that our method is more robust to local noise than the classic ED. Furthermore, we can improve this percentage up to 100% of true detections finding the best value for ε in each of the time series. Additionally, we highlight the flexibility of MTVA for dynamically working in different levels of resolution at runtime.

5.2   Efficiency of Our Multi-resolution Index

We accelerate the search using our multi-resolution method HOT MTVA and compare it with two main state-of-the-art techniques: HOT SAX and HOT iSAX (Sect. 2). We set the same quantitative information for each technique: $\alpha_v = 4$

Table 1. Percentage of true detections using our MTVA representation.

ε	ED	MTVA Distance		
		$L = 2$	$L = 3$	$L = 4$
0.025	60%	60%	67%	70%
0.050	77%	70%	83%	87%
0.075	73%	70%	80%	83%
0.100	80%	73%	83%	87%
0.125	83%	77%	83%	87%
0.150	77%	73%	77%	80%
best	100%	–	100%	100%

and $\alpha_s = 4$ for the MTVA index, and $\alpha_v = 8$ for the SAX-based indexes. We set empirically the maximum number of elements in a bucket as $th_{max} = 50$ and the maximum resolution level is restricted to $L = 4$. Additionally, we add a search optimization strategy to the HOT SAX algorithm and call it HOT SAX*, which consists in applying a MINDIST function in the bucket before of visiting their associated subsequences [9]. Figure 6 shows the efficiency of the algorithms in terms of the number of computed distances. We observe that HOT MTVA is much more efficient than the other techniques in terms of computed distances. This efficiency advantage is due to multi-resolution properties of our method, which allows to the outer loop heuristic to find quickly the potential candidates through the resolution levels.

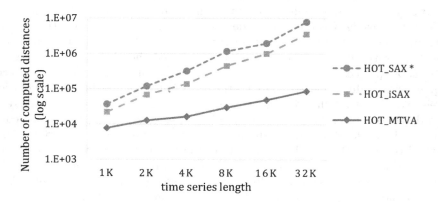

Fig. 6. Efficiency of our multi-resolution method in anomaly detection.

6 Conclusions and Future Work

We proposed a multi-resolution time series representation (MTVA) which is composed of trend-value pairs obtained by applying linear regression linear in each resolution segment. We also provided a distance measure and its lower bounding function to perform efficient searches. We demonstrated the utility of our MTVA representation in Anomaly Detection, where we have highlighted the slope feature for mitigating the false unusualness of noisy subsequences. Furthermore, the efficiency of our multi-resolution discord discovery algorithm outperformed the best existing methods in terms of computed distances. One additional advantage of the MTVA representation is that the level of resolution was more intuitive and easier to fine-tune than the number of segments in piecewise approximations. Nevertheless, one disadvantage of the trend-value approximation is that it requires twice the space per segment. Adding a parameter to represent the trend of the time series, it runs the risk of subtracting simplicity to our concise data model if it is compared with the SAX technique. Our approach may be used as baseline for finding anomalies in different levels of granularity. We plan to focus our method on online anomaly detection for massive streaming data.

References

1. Buu, H.T.Q., Anh, D.T.: Time series discord discovery based on iSAX symbolic representation. In: Proceedings of Third International Conference on Knowledge and Systems Engineering (KSE), pp. 11–18 (2011)
2. Chandola, V., Banerjee, A., Kumar, V.: Anomaly detection: a survey. ACM Comput. Surv. **41**, 1–58 (2009)
3. Dan, J., Shi, W., Dong, F., Hirota, K.: Piecewise trend approximation: a ratio-based time series representation. Abstr. Appl. Anal. **2013**(4) (2013)
4. Esmael, B., Arnaout, A., Fruhwirth, R.K., Thonhauser, G.: Multivariate time series classification by combining trend-based and value-based approximations. In: Murgante, B., Gervasi, O., Misra, S., Nedjah, N., Rocha, A.M.A.C., Taniar, D., Apduhan, B.O. (eds.) ICCSA 2012. LNCS, vol. 7336, pp. 392–403. Springer, Heidelberg (2012). doi:10.1007/978-3-642-31128-4_29
5. Keogh, E., Lin, J., Fu, A.: Univariate Time series discords datasets (2005). http://www.cs.ucr.edu/~eamonn/discords/
6. Keogh, E.J., Lin, J., Fu, A.W.: HOT SAX: efficiently finding the most unusual time series subsequence. In: Fifth IEEE International Conference on Data Mining, pp. 226–233, November 2005
7. Keogh, E.J., Lin, J., Lee, S.H., Herle, H.V.: Finding the most unusual time series subsequence: algorithms and applications. Knowl. Inf. Syst. **11**, 1–27 (2007)
8. Kha, N.H., Anh, D.T.: From cluster-based outlier detection to time series discord discovery. In: Li, X.-L., Cao, T., Lim, E.-P., Zhou, Z.-H., Ho, T.-B., Cheung, D., Motoda, H. (eds.) PAKDD 2015. LNCS, vol. 9441, pp. 16–28. Springer, Cham (2015). doi:10.1007/978-3-319-25660-3_2
9. Lin, J., Keogh, E., Lonardi, S., Chiu, B.: A symbolic representation of time series, with implications for streaming algorithms. In: Proceedings of 8th ACM SIGMOD Workshop on Research Issues in Data Mining and Knowledge Discovery, pp. 2–11 (2003)

10. Lin, J., Keogh, E.J., Wei, L., Lonardi, S.: Experiencing SAX: a novel symbolic representation of time series. Data Min. Knowl. Discov. **15**, 107–144 (2007)
11. Malinowski, S., Guyet, T., Quiniou, R., Tavenard, R.: 1d-SAX: a novel symbolic representation for time series. In: Tucker, A., Höppner, F., Siebes, A., Swift, S. (eds.) IDA 2013. LNCS, vol. 8207, pp. 273–284. Springer, Heidelberg (2013). doi:10.1007/978-3-642-41398-8_24
12. Sanchez, H., Bustos, B.: Anomaly detection in streaming time series based on bounding boxes. In: Traina, A.J.M., Traina, C., Cordeiro, R.L.F. (eds.) SISAP 2014. LNCS, vol. 8821, pp. 201–213. Springer, Cham (2014). doi:10.1007/978-3-319-11988-5_19
13. Shieh, J., Keogh, E.: iSAX: indexing and mining terabyte sized time series. In: Proceedings of 14th ACM SIGKDD International Conference on Knowledge Discovery and Data Mining, pp. 623–631. ACM (2008)

A Pliant Arithmetic-Based Fuzzy Time Series Model

József Dombi[1], Tamás Jónás[2,3]([⊠]), and Zsuzsanna Eszter Tóth[2]

[1] Institute of Informatics, University of Szeged, Árpád tér 2, Szeged 6720, Hungary
[2] Department of Management and Corporate Economics,
Budapest University of Technology and Economics, Magyar tudósok körútja 2,
Budapest 1117, Hungary
jonas@mvt.bme.hu
[3] Research and Development Department, Flextronics International Ltd.,
Hangár u. 5-37, Budapest 1183, Hungary

Dedicated to the memory of Csanád Imreh

Abstract. In this study, a fuzzy arithmetics-based fuzzy time series modeling method is introduced. After input data normalization, the fuzzy c-means clustering algorithm is used for fuzzification and establishment of antecedents of the fuzzy rules. Here, each rule consequent is treated as a fuzzy number composed of a left and a right hand side fuzzy set, each of which is given by a sigmoid membership function. The novelty of the proposed method lies in the application of pliant arithmetics to aggregate separately the left and the right hand sides of the individual fuzzy consequents, taking the activation levels of the corresponding antecedents into account. Here, Dombi's conjunction operator is applied to form the fuzzy output from the aggregates of the left and right hand side sigmoid functions. The introduced defuzzification method does not require any numerical integration and runs in constant time. The output of the pliant arithmetic based fuzzy time series model is obtained by denormalizing the crisp output produced by the fuzzy inference. Lastly, the modeling capability of the introduced methodology was tested on empirical data. Based on these results, our method may be viewed as a viable alternative prediction technique.

Keywords: Fuzzy arithmetics · Fuzzy time series · Pliant

1 Introduction

Following the fuzzy time series model proposed by Song and Chissom [9], many researchers have developed various fuzzy time series modeling techniques to enhance the modeling capability and forecasting performance of previous approaches. These efforts have resulted in a great variety of time series modeling methods with significant achievements that are founded on the application of fuzzy theory (e.g. [8,10–13]).

© Springer International Publishing AG 2017
I. Rojas et al. (Eds.): IWANN 2017, Part II, LNCS 10306, pp. 129–141, 2017.
DOI: 10.1007/978-3-319-59147-6_12

In this study, we introduce a fuzzy time series modeling method which after input data normalization uses the fuzzy c-means clustering algorithm for fuzzification and for the establishment of antecedents of the fuzzy rules. In our fuzzy inference system (FIS), each rule consequent is a fuzzy number composed of a left hand side and a right hand side fuzzy set, each of which is given by a sigmoid membership function. We will then introduce a simple heuristic for building the fuzzy numbers in the rule consequents from sample data. The novelty of our method lies in the application of fuzzy arithmetics to produce the fuzzy output. Namely, pliant arithmetics are used to aggregate separately the left and the right hand sides of individual fuzzy consequents, taking into account the activation levels of the corresponding antecedents [5,6]. Dombi's conjunction operator is applied to form the fuzzy output from the aggregates of the left and right hand side sigmoid functions [4]. Defuzzification is performed by finding the place at which the membership function of the output fuzzy number is maximal. There are two notable advantages of the introduced defuzzification method. On the one hand, it does not require any numerical integration to generate the crisp output, on the other hand, it runs in constant time. The output of our Pliant Arithmetic based Fuzzy Time Series (PAFTS) model is generated by denormalizing the crisp output produced by our Pliant Arithmetic based Fuzzy Inference System (PAFIS).

The practical motivation of our work was to develop a fuzzy arithmetic based time series model that is easy-to-use and can be applied for modeling time series that represent weekly demand for electronic products in electronic manufacturing service businesses.

2   Pliant Arithmetic-Based Fuzzy Inference System

Our goal is to construct a fuzzy inference system that can model the time series $X_1, X_2, \ldots, X_n$. Let $X_j, X_{j+1}, \ldots, X_{j+r-1}$ be an r-period long sub-time series and X_{j+r} its one period long continuation in the time series $X_1, X_2, \ldots, X_n$ ($j = 1, 2, \ldots, n - r$). Here, r is the number of historical periods that we use for generating one-period-ahead forecast, ($r \geq 1$, $r + 1 \leq n$). Here, let $d = n - r$ denote the number of r-period-long sub-time series in $X_1, X_2, \ldots, X_n$.

2.1   Data Preparation

For an r-period long sub-time series $X_j, X_{j+1}, \ldots, X_{j+r-1}$ and its X_{j+r} continuation, let the vector $\mathbf{x}_j = (x_{j,1}, x_{j,2}, \ldots, x_{j,r})$ and the scalar y_j be defined as follows ($j = 1, 2, \ldots, d$).

(1) If $X_j, X_{j+1}, \ldots, X_{j+r-1}$ are not all equal, then

$$x_{j,p} = \frac{X_{j+p-1} - \min_{q=1,\ldots,r}(X_{j+q-1})}{\max_{q=1,\ldots,r}(X_{j+q-1}) - \min_{q=1,\ldots,r}(X_{j+q-1})} \tag{1}$$

$$y_j = \frac{X_{j+r} - \min\limits_{q=1,\ldots,r}(X_{j+q-1})}{\max\limits_{q=1,\ldots,r}(X_{j+q-1}) - \min\limits_{q=1,\ldots,r}(X_{j+q-1})} \tag{2}$$

$p = 1, 2, \ldots, r.$

(2) If $X_j, X_{j+1}, \ldots, X_{j+r-1}$ are all equal and nonzero, say have the value of a ($a > 0$), then $\mathbf{x}_j = (1, 1, \ldots, 1)$, and $y_j = X_{j+r}/a$.

(3) If $X_j, X_{j+1}, \ldots, X_{j+r-1}$ are all zeros, then $\mathbf{x}_j = (0, 0, \ldots, 0)$, and $y_j = X_{j+r}$.

Owing to this transformation, each component of vector $\mathbf{x}_j$ is normalized to the interval $[0, 1]$. The $(\mathbf{x}_j, y_j)$ pairs represent the time series $X_1, X_2, \ldots, X_n$ so that the normalized vector $\mathbf{x}_j$ is followed by y_j, and our goal is to build a FIS that can adequately map the relation between $\mathbf{x}_j$ and y_j ($j = 1, 2, \ldots, d$). So, our aim is to build a fuzzy inference system that has an r-dimensional vector $\mathbf{x}$ as input, and a scalar output y. The system approximates the $y = f(\mathbf{x})$ relation based on the sample $(\mathbf{x}_j, y_j)$, where $\mathbf{x}_j$, y_j are the corresponding observations on $\mathbf{x}$ and y, respectively. In order to identify typical patterns in the time series, the $\mathbf{x}_j$ vectors are clustered using the fuzzy c-means clustering algorithm [2].

2.2   Forming Fuzzy Rules

Let ϕ be the exponent used in the fuzzy c-means clustering algorithm ($\phi \in \mathbb{R}$, $0 < \phi < \infty$) that determines the fuzziness of clusters, and let $\mathbf{C}_1, \mathbf{C}_2, \ldots, \mathbf{C}_m$ be the clusters formed with the cluster centroids $\mathbf{c}_1, \mathbf{c}_2, \ldots, \mathbf{c}_m$, respectively ($1 \leq m \leq d$). (The typical value of ϕ used in practice is 2.)

The $\mathbf{C}_1, \mathbf{C}_2, \ldots, \mathbf{C}_m$ fuzzy clusters represent fuzzy partitions in the input space. From a time series perspective, the r-period-long normalized sub-time series are classified into m clusters, and the cluster centroids represent r-period-long typical normalized patterns. Based on this, the following fuzzy rules can be formed.

$$\begin{array}{ll} \text{Rule 1:} & \text{if } \mathbf{x} \text{ is in } \mathbf{C}_1 \text{ then } \mathbf{B}_1 \\ \text{Rule 2:} & \text{if } \mathbf{x} \text{ is in } \mathbf{C}_2 \text{ then } \mathbf{B}_2 \\ \vdots & \vdots \qquad \vdots \quad \vdots \\ \text{Rule } m: & \text{if } \mathbf{x} \text{ is in } \mathbf{C}_m \text{ then } \mathbf{B}_m, \end{array} \tag{3}$$

where $\mathbf{x} \in \mathbb{R}^r$ is an input vector normalized according to the method described in Sect. 2.1, and $\mathbf{B}_1, \mathbf{B}_2, \ldots, \mathbf{B}_m$ are fuzzy sets defined over set $\mathbf{Y}$ which is the domain of crisp system outputs. For any r-dimensional normalized input vector $\mathbf{x}$, its membership value $\mu_i(\mathbf{x})$ in cluster $\mathbf{C}_i$ can be calculated as follows:

$$\mu_i(\mathbf{x}) = \left(\frac{1}{\|\mathbf{x} - \mathbf{c}_i\|_2^2}\right)^{\frac{1}{\phi-1}} \frac{1}{\sum\limits_{v=1}^{m}\left(\frac{1}{\|\mathbf{x} - \mathbf{c}_v\|_2^2}\right)^{\frac{1}{\phi-1}}}. \tag{4}$$

The value of $\mu_i(\mathbf{x})$ may be viewed as the activation level of Rule i for the input $\mathbf{x}$. As a result of fuzzy c-means clustering, the rule antecedents have been identified.

Each $\mathbf{B}_i$ consequent is treated as a fuzzy number that will be discussed later. In order to make the fuzzy inference system complete, the consequent of each rule needs to be identified; that is, the membership function of each fuzzy number $\mathbf{B}_i$ needs to be supplied $(i = 1, 2, \ldots, m)$.

2.3   Membership Functions of the Fuzzy Rule Consequents

Let the $\mathbf{Y}$ domain of the crisp outputs of our FIS be

$$\mathbf{Y} = [y_l - \Delta, y_h + \Delta], \tag{5}$$

where $y_l = \min_{j=1,\ldots,d}(y_j)$, $y_h = \max_{j=1,\ldots,d}(y_j)$, $\Delta = c(y_h - y_l)$, $c > 0$, $c \in \mathbb{R}$. In our implementation, $c = 0.1$. Sigmoid functions are utilized to compose the membership functions of the fuzzy numbers $\mathbf{B}_1, \mathbf{B}_2, \ldots, \mathbf{B}_m$.

Definition 1. *The sigmoid function $\sigma_a^{(\lambda)}$ with parameters a and λ is given by*

$$\sigma_a^{(\lambda)}(y) = \frac{1}{1 + e^{-\lambda(y-a)}}, \tag{6}$$

where $y, a, \lambda \in \mathbb{R}$ and λ is nonzero.

The main properties of the sigmoid function $\sigma_a^{(\lambda)}(y)$ are as follows. The function is monotone increasing, if λ is positive, and it is monotone decreasing, if λ is negative.

$$\lim_{y \to +\infty} \sigma_a^{(\lambda)}(y) = \begin{cases} 1, & \text{if } \lambda > 0 \\ 0, & \text{if } \lambda < 0 \end{cases}, \tag{7}$$

$$\lim_{y \to -\infty} \sigma_a^{(\lambda)}(y) = \begin{cases} 1, & \text{if } \lambda < 0 \\ 0, & \text{if } \lambda > 0 \end{cases}. \tag{8}$$

$\sigma_a^{(\lambda)}(a) = 0.5$, the slope of $\sigma_a^{(\lambda)}(y)$ in the $(a, 0.5)$ point is determined by λ as

$$\frac{d\sigma_a^{(\lambda)}(a)}{dy} = \frac{\lambda}{4}. \tag{9}$$

Following Dombi's Pliant Inequality Model [5], the upcoming definitions will be used. The indexes l and r stand for "left" and "right", respectively, and will be used to denote left and right hand side components of fuzzy numbers.

Definition 2. *The truth of inequality $a_l < y$ is given by the sigmoid function*

$$\{a_l <_{\lambda_l} y\} = \sigma_{a_l}^{(\lambda_l)}(y) = \frac{1}{1 + e^{-\lambda_l(y-a_l)}}, \tag{10}$$

where $a_l, \lambda_l \in \mathbb{R}$, $\lambda_l > 0$, $y \in \mathbf{Y}$.

Definition 3. *The truth of inequality $a_r > y$ is given by the sigmoid function*

$$\{a_r >_{\lambda_r} y\} = \sigma_{a_r}^{(\lambda_r)}(y) = \frac{1}{1 + e^{-\lambda_r(y-a_r)}}, \tag{11}$$

where $a_r, \lambda_r \in \mathbb{R}$, $\lambda_r < 0$, $y \in \mathbf{Y}$.

Now, Dombi's conjunction operator [4] will be applied to implement intersection of two fuzzy sets.

Definition 4. *(Dombi's t-norm) Let* $\mathbf{A}_1$ *and* $\mathbf{A}_2$ *be fuzzy sets defined over the crisp universe* $\mathbf{Y}$ *given by the membership functions* $\mu_{A_1}(y)$ *and* $\mu_{A_2}(y)$, *respectively. The Dombi intersection of* $\mathbf{A}_1$ *and* $\mathbf{A}_2$ *is given by the membership function* $\mu_{A_1 \cap A_2}(y)$:

$$\mu_{A_1 \cap A_2}(y) = \mu_{A_1}(y) *_{(D)} \mu_{A_2}(y)$$

$$= \frac{1}{1 + \left(\left(\frac{1 - \mu_{A_1}(y)}{\mu_{A_1}(y)} \right)^\alpha + \left(\frac{1 - \mu_{A_2}(y)}{\mu_{A_2}(y)} \right)^\alpha \right)^{1/\alpha}}, \tag{12}$$

where $\alpha \in \mathbb{R}$, $\alpha > 0$, *and* $*_{(D)}$ *denotes the Dombi intersection operator.*

The truth of inequality $a_l < y < a_r$, that is, $\{a_l <_{\lambda_l} y <_{\lambda_r} a_r\}$ is defined as the intersection of fuzzy sets $\{a_l <_{\lambda_l} y\}$ and $\{a_r >_{\lambda_r} y\}$ using Dombi's t-norm in (12), which hereafter is always applied with $\alpha = 1$.

Definition 5. *The truth of inequality* $a_l < y < a_r$ *is given by*

$$\{a_l <_{\lambda_l} y <_{\lambda_r} a_r\} = \sigma_{a_l}^{(\lambda_l)}(y) *_{(D)} \sigma_{a_r}^{(\lambda_r)}(y)$$

$$= \frac{1}{1 + \frac{1 - \sigma_{a_l}^{(\lambda_l)}(y)}{\sigma_{a_l}^{(\lambda_l)}(y)} + \frac{1 - \sigma_{a_r}^{(\lambda_r)}(y)}{\sigma_{a_r}^{(\lambda_r)}(y)}}, \tag{13}$$

where $a_l, \lambda_l, a_r, \lambda_r \in \mathbb{R}$, $a_l < a_r$, $\lambda_l > 0$, $\lambda_r < 0$, $y \in \mathbf{Y}$.

Using (10) and (11),

$$\sigma_{a_l}^{(\lambda_l)}(y) *_{(D)} \sigma_{a_r}^{(\lambda_r)}(y) = \frac{1}{1 + e^{-\lambda_l(y - a_l)} + e^{-\lambda_r(y - a_r)}}. \tag{14}$$

Figure 1 shows the intersection of two fuzzy sets given by an increasing and a decreasing sigmoid function.

In our model, the membership function of the consequent of the ith fuzzy rule (that is, the membership function $\mu_{\mathbf{B}_i}^{(\mathbf{p}_i)}(y)$ of the fuzzy number $\mathbf{B}_i$) has the form of (14):

$$\mu_{\mathbf{B}_i}^{(\mathbf{p}_i)}(y) = \sigma_{a_{l,i}}^{(\lambda_{l,i})}(y) *_{(D)} \sigma_{a_{r,i}}^{(\lambda_{r,i})}(y) = \frac{1}{1 + e^{-\lambda_{l,i}(y - a_{l,i})} + e^{-\lambda_{r,i}(y - a_{r,i})}}, \tag{15}$$

where the parameter vector $\mathbf{p}_i = (a_{l,i}, \lambda_{l,i}, a_{r,i}, \lambda_{r,i})$ contains the parameters of the sigmoid functions $\sigma_{a_{l,i}}^{(\lambda_{l,i})}(y)$ and $\sigma_{a_{r,i}}^{(\lambda_{r,i})}(y)$. $\sigma_{a_{l,i}}^{(\lambda_{l,i})}(y)$ and $\sigma_{a_{r,i}}^{(\lambda_{r,i})}(y)$ are the left hand side and right hand side of $\mathbf{B}_i$, respectively. Here, the fuzzy numbers like $\mathbf{B}_i$ are called Pliant numbers.

Now, we introduce a simple heuristic to identify the parameter vector $\mathbf{p}_i$ of each membership function $\mu_{\mathbf{B}_i}^{(\mathbf{p}_i)}(y)$ $(i = 1, 2, \ldots, m)$. We have the sample of pairs $(\mathbf{x}_j, y_j)$ in which $\mathbf{x}_j$ is the jth input and y_j is the corresponding output

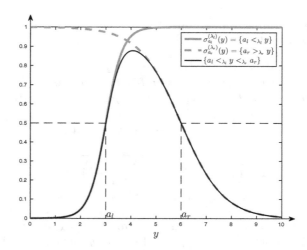

Fig. 1. Dombi intersection of two fuzzy sets given by an increasing and a decreasing sigmoid function

$(j = 1, 2, \ldots, d)$. As mentioned before, the activation level $\mu_i(\mathbf{x}_j)$ of the ith fuzzy rule antecedent for input $\mathbf{x}_j$ may be calculated according to (4); that is, we have the pairs $(y_j, \mu_i(\mathbf{x}_j))$. In order to identify the parameters of $\sigma_{a_{l,i}}^{(\lambda_{l,i})}(y)$ and $\sigma_{a_{r,i}}^{(\lambda_{r,i})}(y)$, the following definitions are used.

Definition 6. *The $y_{0,i}$ separator of the ith rule consequent is given by*

$$y_{0,i} = \frac{\sum\limits_{j=1}^{d} \mu_i(\mathbf{x}_j) y_j}{\sum\limits_{j=1}^{d} \mu_i(\mathbf{x}_j)}. \tag{16}$$

Definition 7. *The left-study-set $\mathbf{S}_{l,i}$ and the right-study-set $\mathbf{S}_{r,i}$ of the ith rule consequent are*

$$\mathbf{S}_{l,i} = \{(y_j, \mu_i(\mathbf{x}_j)) : y_j < y_{0,i}, j = 1, 2, \ldots, d\} \tag{17}$$

$$\mathbf{S}_{r,i} = \{(y_j, \mu_i(\mathbf{x}_j)) : y_j > y_{0,i}, j = 1, 2, \ldots, d\}. \tag{18}$$

Here, $y_{0,i}$ may be viewed as a kind of a sample-based "mean" of the output of rule i. We will group the pairs $(y_j, \mu_i(\mathbf{x}_j))$ into the $\mathbf{S}_{l,i}$ and $\mathbf{S}_{r,i}$ sets and use these two sets to identify the parameters of $\sigma_{a_{l,i}}^{(\lambda_{l,i})}(y)$ and $\sigma_{a_{r,i}}^{(\lambda_{r,i})}(y)$, respectively.

First, the identification of parameters $a_{l,i}$ and $\lambda_{l,i}$ will be described. Let μ_{max} be a positive number, $\mu_{max} \approx 1$, $\mu_{max} < 1$ (e.g. $\mu_{max} = 0.999$). We will set the requirement

$$\sigma_{a_{l,i}}^{(\lambda_{l,i})}(y_{0,i}) = \mu_{max}; \tag{19}$$

that is, we require the point $(y_{0,i}, \mu_{max})$ to be on the curve of $\sigma_{a_{l,i}}^{(\lambda_{l,i})}(y)$. If we have another $(y_{l,i}, \mu_{min})$ point of curve of $\sigma_{a_{l,i}}^{(\lambda_{l,i})}(y)$ given, where $0 < \mu_{min}$, $\mu_{min} \approx 0$ (e.g. $\mu_{min} = 0.001$) and $y_{l,i} \in \mathbf{Y}$, $y_{l,i} < y_{0,i}$, that is,

$$\sigma_{a_{l,i}}^{(\lambda_{l,i})}(y_{l,i}) = \mu_{min}, \tag{20}$$

then solution of the equation system contained in (19) and (20) gives the values of $a_{l,i}$ and $\lambda_{l,i}$:

$$a_{l,i}(y_{l,i}) = y_{0,i} - (y_{0,i} - y_{l,i}) \frac{\ln\left(\frac{\mu_{max}}{1-\mu_{max}}\right)}{\ln\left(\frac{\mu_{max}}{1-\mu_{max}} \frac{1-\mu_{min}}{\mu_{min}}\right)} \tag{21}$$

$$\lambda_{l,i}(y_{l,i}) = \frac{1}{y_{0,i} - y_{l,i}} \ln\left(\frac{\mu_{max}}{1-\mu_{max}} \frac{1-\mu_{min}}{\mu_{min}}\right). \tag{22}$$

The $a_{l,i}(y_{l,i})$ and $\lambda_{l,i}(y_{l,i})$ notations indicate that values of $a_{l,i}$ and $\lambda_{l,i}$ depend on $y_{l,i}$, as μ_{min}, μ_{max} are assumed to be given and $y_{0,i}$ is calculated according to (16). Now, let ε be a small positive number such that $[y_{0,i} - \varepsilon, y_{0,i} + \varepsilon] \subset \mathbf{Y}$. There are two cases here that we need to take into account to determine the unknown value of $y_{l,i}$.

(a) If set $\mathbf{S}_{l,i}$ is nonempty, then $y_{l,i}$ can be obtained by minimizing the

$$d_{l,i} = \sum_{(y_j, \mu_i(\mathbf{x}_j)) \in \mathbf{S}_{l,i}} \left(\sigma_{a_{l,i}(y_{l,i})}^{(\lambda_{l,i}(y_{l,i}))}(y_j) - \mu_i(\mathbf{x}_j)\right)^2 \tag{23}$$

quantity. Here, $d_{l,i}$ can be minimized by using the Interior Point Algorithm [1]. In this algorithm, we initialize $y_{l,i}$ as

$$y_{l,i} = \frac{\min\limits_{y_j \in \mathbf{Y}_{l,i}} (y_j) + y_{0,i}}{2}, \tag{24}$$

where

$$\mathbf{Y}_{l,i} = \{y_j : (y_j, \mu_i(\mathbf{x}_j)) \in \mathbf{S}_{l,i}\}. \tag{25}$$

(b) If set $\mathbf{S}_{l,i}$ is empty, then $y_{l,i} = y_{0,i} - \varepsilon$. In this case, there are no $(y_j, \mu_i(\mathbf{x}_j))$ points on the left hand side of $y_{0,i}$, hence $a_{l,i}(y_{l,i})$ and $\lambda_{l,i}(y_{l,i})$ cannot be determined using curve fitting as in case a). If ε is a small number (in our implementation $\varepsilon = 0.001$), then the $\sigma_{a_{l,i}}^{(\lambda_{l,i})}(y)$ sigmoid function is a quasi unit step function.

The parameters $a_{r,i}$ and $\lambda_{r,i}$ of the right hand side sigmoid function $\sigma_{a_{r,i}}^{(\lambda_{r,i})}(y)$ can be identified by following a method similar to the one we used for identifying the parameters $a_{l,i}$ and $\lambda_{l,i}$ of the left hand side sigmoid function $\sigma_{a_{l,i}}^{(\lambda_{l,i})}(y)$.

Figure 2 shows some examples of the membership functions $\mu_{\mathbf{B}_i}^{(\mathbf{P}_i)}(y)$ that were generated by applying the above described heuristic.

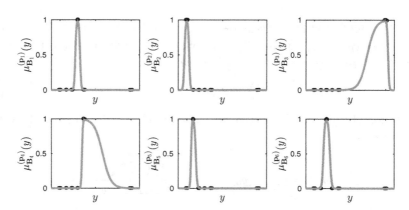

Fig. 2. Examples of membership functions of rule consequents built from two sigmoid functions

2.4 Aggregation of Fuzzy Outputs

As we remarked above, the $\mathbf{B}_i$ consequent of each fuzzy rule is treated as a Pliant number given by two sigmoid functions, namely the left hand side $\sigma_{a_{l,i}}^{(\lambda_{l,i})}(y)$ and the right hand side $\sigma_{a_{r,i}}^{(\lambda_{r,i})}(y)$ $(i = 1, 2, \ldots, m)$. The Pliant numbers are aggregated using the Pliant arithmetics based on the next proposition [5].

Proposition 1. *If* $\mathbf{A}_1, \mathbf{A}_2, \ldots, \mathbf{A}_m$ *are fuzzy sets with the membership functions* $\sigma_{a_1}^{(\lambda_1)}(y),\ \sigma_{a_2}^{(\lambda_2)}(y), \ldots, \sigma_{a_m}^{(\lambda_m)}(y)$, *respectively,*

$$sgn(\lambda_1) = sgn(\lambda_2) = \cdots = sgn(\lambda_m), \tag{26}$$

and the fuzzy set $\mathbf{A}$ *is given by the linear combination*

$$\mathbf{A} = \sum_{i=1}^{m} w_i \mathbf{A}_i, \tag{27}$$

where $\sum_{i=1}^{m} w_i = 1$, *then* $\mathbf{A}$ *is also sigmoid-shaped with the membership function* $\sigma_a^{(\lambda)}(y)$, *where*

$$a = \sum_{i=1}^{m} w_i a_i; \qquad \frac{1}{\lambda} = \sum_{i=1}^{m} \frac{w_i}{\lambda_i}. \tag{28}$$

Proof. See [5].

The $\mu_i(\mathbf{x})$ quantity given by (4) measures how much the input vector $\mathbf{x}$ activates the ith fuzzy rule. In other words, $\mu_i(\mathbf{x})$ may be interpreted as the level of applicability of rule i to input $\mathbf{x}$. Based on this, the normalized weight of rule i, w_i, for input $\mathbf{x}$ is

$$w_i = \frac{\mu_i(\mathbf{x})}{\sum\limits_{k=1}^{m} \mu_k(\mathbf{x})} \quad \text{if} \quad \sum_{k=1}^{m} \mu_k(\mathbf{x}) \neq 0. \tag{29}$$

Notice that nothing is inferred, if $\sum_{k=1}^{m} \mu_k(\mathbf{x}) = 0$. Proposition 1 allows us to aggregate the left and right hand sides of $\mathbf{B}_1, \mathbf{B}_2, \ldots, \mathbf{B}_m$ separately using the $w_i, w_2, \ldots, w_m$ weights and derive the parameters of the left hand side $\sigma_{a_l}^{(\lambda_l)}(y)$ and the right hand side $\sigma_{a_r}^{(\lambda_r)}(y)$ of the fuzzy output $\mathbf{B}$ in the following way:

$$a_l = \sum_{i=1}^{m} w_i a_{l,i}; \qquad a_r = \sum_{i=1}^{m} w_i a_{r,i}; \qquad \frac{1}{\lambda_l} = \sum_{i=1}^{m} \frac{w_i}{\lambda_{l,i}}; \qquad \frac{1}{\lambda_r} = \sum_{i=1}^{m} \frac{w_i}{\lambda_{r,i}}. \tag{30}$$

The membership function $\mu_{\mathbf{B}}(y)$ of the fuzzy output $\mathbf{B}$ is computed according to (14):

$$\mu_{\mathbf{B}}(y) = \frac{1}{1 + e^{-\lambda_l(y-a_l)} + e^{-\lambda_r(y-a_r)}}. \tag{31}$$

It should be emphasized here, that there is no implication applied in our inference method; the left hand side $\sigma_{a_l}^{(\lambda_l)}(y)$ and the right hand side $\sigma_{a_r}^{(\lambda_r)}(y)$, which are the components of the fuzzy output $\mathbf{B}$, are inferred via fuzzy arithmetic operations.

2.5   Defuzzification

As the aim is to use our fuzzy inference system for time series modeling, a defuzzification method, which transforms the fuzzy output $\mathbf{B}$ into a crisp $\hat{y}$, needs to be identified. Here, we will use a maximum defuzzification method; that is, we will represent the fuzzy output $\mathbf{B}$, which is given by $\mu_{\mathbf{B}}(y)$ in (31), by the crisp $\hat{y}$ for which

$$\mu_{\mathbf{B}}(\hat{y}) = \max_{y \in \mathbf{Y}} \mu_{\mathbf{B}}(y). \tag{32}$$

It can be seen from (31) that since $\mu_{\mathbf{B}}(y)$ is maximal when its denominator is minimal, we need to minimize the function

$$g(y) = e^{-\lambda_l(y-a_l)} + e^{-\lambda_r(y-a_r)}. \tag{33}$$

Recall that $\lambda_l > 0$, $\lambda_r < 0$, and so $e^{-\lambda_l(y-a_l)}$ is strictly monotonically decreasing and $e^{-\lambda_r(y-a_r)}$ is strictly monotonically increasing. It can be seen that both $e^{-\lambda_l(y-a_l)}$ and $e^{-\lambda_r(y-a_r)}$ are strictly convex, thus $g(y)$ is strictly convex, too, and there is only one y for which $g(y)$ is minimal and at the same time $\mu_{\mathbf{B}}(y)$ is maximal. The $\hat{y}$ for which $g(y)$ is minimal and $\mu_{\mathbf{B}}(y)$ is maximal can be obtained by solving the equation

$$\frac{dg(y)}{dy} = -\lambda_l e^{-\lambda_l(y-a_l)} - \lambda_r e^{-\lambda_r(y-a_r)} = 0. \tag{34}$$

The $\hat{y}$ solution of (34) is

$$\hat{y} = \frac{\ln\left(-\frac{\lambda_l}{\lambda_r}\right) + \lambda_l a_l - \lambda_r a_r}{\lambda_l - \lambda_r}. \tag{35}$$

Now, let λ_l^* and λ_r^* be defined as $\lambda_l^* = \lambda_l$ and $\lambda_r^* = -\lambda_r$. As $\lambda_l > 0$ and $\lambda_r < 0$, both λ_l^* and λ_r^* are positive. By applying these substitutions, $\hat{y}$ may be written in the following form.

$$\hat{y} = \frac{\lambda_l^*}{\lambda_l^* + \lambda_r^*} a_l + \frac{\lambda_r^*}{\lambda_l^* + \lambda_r^*} a_r + \frac{1}{\lambda_l^* + \lambda_r^*} \left(\ln(\lambda_l^*) - \ln(\lambda_r^*) \right) \tag{36}$$

In this form, $\hat{y}$ may be interpreted as the weighted average of a_l and a_r corrected with the third term in (36). If $\lambda_l^* = \lambda_r^*$, namely, if $\lambda_l = -\lambda_r$, then the curve of $\mu_{\mathbf{B}}(y)$ is symmetric and (36) becomes

$$\hat{y} = \frac{a_l + a_r}{2}, \tag{37}$$

which is in line with the expectation that if the curve of $\mu_{\mathbf{B}}(y)$ is symmetric, then the place of its maximum is the midpoint of the (a_l, a_r) interval.

The introduced defuzzification method has two notable advantages. On the one hand, it does not require any numerical integration to generate the crisp output, on the other hand, it runs in a constant time as its speed does not depend on the number of fuzzy rules.

3   Pliant Arithmetic-Based Fuzzy Time Series Modeling

Let $X_1, X_2, \ldots, X_r$ be an r-period-long known time series and X_{r+1} be its one-period-long unknown continuation, $r \geq 1$. Furthermore, let $\hat{X}_{r+1}$ denote the forecast for X_{r+1} generated based on the $X_1, X_2, \ldots, X_r$ values using a fuzzy inference system constructed based on the method that was introduced in Sect. 2. Here, $\hat{X}_{r+1}$ can be generated as follows. Depending on the $X_1, X_2, \ldots, X_r$ values, the normalized input vector $\mathbf{x} = (x_1, x_2, \ldots, x_r)$ to the PAFIS is created by applying one of the following cases.

(1) If $X_1, X_2, \ldots, X_r$ are not all equal, then

$$x_p = \frac{X_p - \min_{q=1,\ldots,r}(X_q)}{\max_{q=1,\ldots,r}(X_q) - \min_{q=1,\ldots,r}(X_q)} \tag{38}$$

$p = 1, 2, \ldots, r$.

(2) If $X_1, X_2, \ldots, X_r$ are equal and nonzero, say have the value of a $(a > 0)$, then $\mathbf{x}_j = (1, 1, \ldots, 1)$.

(3) If $X_1, X_2, \ldots, X_r$ are all zeros, then $\mathbf{x} = (0, 0, \ldots, 0)$.

Let $\hat{y}$ be the PAFIS output for the input vector $\mathbf{x}$. Depending on how vector $\mathbf{x}$ was created from the time series $X_1, X_2, \ldots, X_r$ based on the cases above, the $\hat{X}_{r+1}$ forecast is computed according to one of the following denormalizations.

(1) If $x_1, x_2, \ldots, x_r$ are not all equal, then

$$\hat{X}_{r+1} = \hat{y} \left(\max_{q=1,\ldots,r}(X_q) - \min_{q=1,\ldots,r}(X_q) \right) + \min_{q=1,\ldots,r}(X_q). \tag{39}$$

(2) If $\mathbf{x} = (1, 1, \ldots, 1)$, then $\hat{X}_{r+1} = ay$.
(3) If $\mathbf{x} = (0, 0, \ldots, 0)$, then $\hat{X}_{r+1} = y$.

If we apply this forecasting method for X_{j+r} continuation of each r-period-long sub-time series $X_j, X_{j+1}, \ldots, X_{j+r-1}$ in the time series $X_1, X_2, \ldots, X_n$, then $\hat{X}_{r+1}, \hat{X}_{r+2}, \ldots, \hat{X}_n$ are the predicted (simulated) values of $X_{r+1}, X_{r+2}, \ldots, X_n$, respectively. Namely, the $\hat{X}_{r+1}, \hat{X}_{r+2}, \ldots, \hat{X}_n$ values model the $X_{r+1}, X_{r+2}, \ldots, X_n$ ones $(j = 1, 2, \ldots, n - r)$.

4   Empirical Results

The performance of our Pliant Arithmetic-based Fuzzy Time Series Model was tested on 12 time series. Each of these time series represents the weekly customer demand for an electronic product with a typical size between 120 and 180 data. For each time series, the number of historical periods r in our PAFTS model was initialized with the lag value of the best auto-correlation in the time series. 90% of the data in each time series were used for training purposes, while rest of the data were utilized to test our PAFTS. The mean square error (MSE) values calculated for each simulation were compared to those of the Autoregressive Integrated Moving Average (ARIMA), Adaptive Neuro-Fuzzy Inference System (ANFIS) and Neural Autoregressive (NAR) methods. All these three methods were implemented in MATLAB R2016b by using its built-in tools. For each time series, the length of the input sub-time series for the ANFIS and NAR methods was set to the same r value as for the PAFTS model. We applied the Hyndman-Khandakar algorithm [7] to obtain the best ARIMA model. The Calinski-Harabasz [3] index was used to find the quasi optimal number of clusters (number of fuzzy rules) in the PAFTS method.

Table 1. The MSE results of simulations

#	ARIMA		ANFIS		NAR		PAFTS		PAFTS vs. Best (%)	
	Training	Test	Training	Test	Training	Test	Training	Test	Training	Test
1	133.6	156.9	141.3	166.2	118.8	142.2	129.4	149.8	8.92%	5.34%
2	192.6	221.4	188.8	217.5	181.7	223.8	186.7	230.7	2.75%	6.07%
3	202.9	245.7	198.3	211.5	204.8	228.9	196.8	229.5	0.00%	8.51%
4	108.2	155.3	105.2	147.9	109.8	139.6	112.8	137.8	7.22%	0.00%
5	97.2	125.7	101.2	132.6	95.8	112.7	102.8	120.1	7.31%	6.57%
6	234.7	277.8	222.1	288.2	225.8	251.1	219.8	261.7	0.00%	4.22%
7	134.7	171.2	129.7	166.3	127.6	153.8	131.8	161.6	3.29%	5.07%
8	118.8	153.2	121.8	162.7	110.2	145.8	119.7	141.8	8.62%	0.00%
9	87.1	104.6	88.3	109.7	80.1	110.2	84.3	113.7	5.24%	8.70%
10	117.1	134.7	123.2	140.2	115.6	138.7	110.3	141.2	0.00%	4.83%
11	303.2	340.2	288.7	327.7	291.2	351.1	291.7	345.7	1.04%	5.49%
12	67.8	99.3	72.2	87.6	65.1	91.2	68.3	84.3	4.92%	0.00%

Table 1 summarizes the MSE results of the compared models. We can see from this table that both on the training and test data sets, the four evaluated methods give similar results. The column PAFTS vs Best (%) shows the relative difference between the MSE of the best performing method and the MSE of the PAFTS model. Based on these empirical results, our PAFTS model may be viewed as a novel alternative of the ARIMA, ANFIS and NAR methods.

5 Conclusions

In this article, a novel time series model, called Pliant Arithmetic-based Fuzzy Time Series model was introduced. With this method, the fuzzification and establishment of fuzzy rule antecedents are carried out using the fuzzy c-means clustering algorithm, rule consequents are treated as fuzzy numbers built from sample data, and fuzzy inference is obtained by applying pliant arithmetics. Based on the introduced method, fuzzy inference is a fuzzy number which is converted into a crisp output by finding the place at which the membership function of the output fuzzy number is maximal. This defuzzification method does not require any numerical integration to generate the crisp output and it runs in constant time.

Based on the results from the presented real-life application, our method may be viewed as a viable alternative time series modeling technique. In addition to this, PAFTS model is more convenient from computational complexity point of view than the other three evaluated methods.

References

1. Bazaraa, M.S., Sherali, H.D., Shetty, C.M.: Nonlinear Programming: Theory and Algorithms, 3rd edn. Wiley, New Jersey (2006)
2. Bezdek, J.C.: Pattern Recognition with Fuzzy Objective Function Algorithms. Plenum Press, New York (1981)
3. Caliński, T., Harabasz, J.: A dendrite method for cluster analysis. Commun. Stat. Simul. Comput. **3**(1), 1–27 (1974)
4. Dombi, J.: Towards a general class of operators for fuzzy systems. IEEE Trans. Fuzzy Syst. **16**(2), 477–484 (2008)
5. Dombi, J.: Pliant arithmetics and pliant arithmetic operations. Acta Polytech. Hung. **6**(5), 19–49 (2009)
6. Dombi, J., Szépe, T.: Pliant control system: implementation. In: IEEE 8th International Symposium on Intelligent Systems and Informatics, pp. 225–230 (2010)
7. Hyndman, R.J., Khandakar, Y.: Automatic time series forecasting: the forecast package for R. J. Stat. Softw. **27**(3), 1–22 (2008)
8. Li, S.T., Cheng, Y.C., Lin, S.Y.: A FCM-based deterministic forecasting model for fuzzy time series. Comput. Math. Appl. **54**(12), 3052–3063 (2008)
9. Song, Q., Chissom, B.: Fuzzy time series and its models. Fuzzy Sets Syst. **54**, 269–277 (1993)
10. Tanii, H., Nakajima, H., Tsuchiya, N., Kuramoto, K., Kobashi, S., Hata, Y.: A fuzzy-AR model to predict human body weights. In: 2012 IEEE International Conference on Fuzzy Systems, pp. 1–6 (2012)

11. Valenzuela, O., Rojas, I., Rojas, F., Pomares, H., Herrera, L., Guillen, A., Marquez, L., Pasadas, M.: Hybridization of intelligent techniques and ARIMA models for time series prediction. Fuzzy Sets Syst. **159**, 821–845 (2008)
12. Štěpnička, M., Dvořák, A., Pavliska, V., Vavříčková, L.: A linguistic approach to time series modeling with the help of F-transform. Fuzzy Sets Syst. **180**(1), 164–184 (2011)
13. Yu, H.K.: Weighted fuzzy time series models for TAIEX forecasting. Phys. A: Stat. Mech. Appl. **349**(3–4), 609–624 (2005)

Robust Clustering for Time Series Using Spectral Densities and Functional Data Analysis

Diego Rivera-García[1], Luis Angel García-Escudero[2], Agustín Mayo-Iscar[2], and Joaquín Ortega[1(✉)]

[1] Centro de Investigación en Matemáticas, Guanajuato, Mexico
{driver,jortega}@cimat.mx
[2] Universidad de Valladolid, Valladolid, Spain
{lagarcia,agustinm}@eio.uva.es

Abstract. In this work a robust clustering algorithm for stationary time series is proposed. The algorithm is based on the use of estimated spectral densities, which are considered as functional data, as the basic characteristic of stationary time series for clustering purposes. A robust algorithm for functional data is then applied to the set of spectral densities. Trimming techniques and restrictions on the scatter within groups reduce the effect of noise in the data and help to prevent the identification of spurious clusters. The procedure is tested in a simulation study, and is also applied to a real data set.

Keywords: Time series clustering · Functional Data Analysis · Clustering · Robustness · Trimming · Functional principal components analysis

1 Introduction

Time series clustering has become a very active research area, with applications in many different fields. However, most methods developed so far do not take into account the possible presence of contamination by outliers or spurious information. In this work, we propose a clustering algorithm for stationary time series that is based on considering the estimated spectral density functions as functional data. This procedure has robust features that mitigate the effect of noise in the data and help to prevent the identification of spurious clusters.

Comprehensive revisions of the area can be found in [1,8,19]. [21] present a package in R for time series clustering with a wide range of alternative methods. According to Liao [19], there are three approaches to clustering of time series: methods that depend on the comparison of the raw data, methods based on models fitted to the data and, methods based on features derived from the time series. Our proposal falls within the third approach and the spectral density is the characteristic used to gauge the similarity between time series in the sample.

Spectral characteristics have been previously considered as the main tool for time series clustering. Caiado et al. [6,7] use the periodogram and normalized periodogram ordinates for clustering time series. Maharaj and D'Urso [20] propose a fuzzy clustering algorithm based on the estimated cepstrum, which is the

© Springer International Publishing AG 2017
I. Rojas et al. (Eds.): IWANN 2017, Part II, LNCS 10306, pp. 142–153, 2017.
DOI: 10.1007/978-3-319-59147-6_13

spectrum of the logarithm of the spectral density. Alvarez-Esteban et al. [3] and Euán et al. [13] consider the use of the total variation distance on normalized estimates of the spectral density as a similarity measure for clustering. A brief description of the last two algorithms will be given in Sect. 2.

Other works have focused on developing robust clustering algorithms for time series. D'Urso et al. [11] use a fuzzy approach to propose a robust clustering model based on autoregressive models. D'Urso et al. [12] present robust fuzzy clustering schemes for heteroskedastic time series based on parametric models, Bahadori et al. [4] propose a clustering framework for functional data, which can be applied to time series with warped alignments.

Our proposal is based on the use of spectral densities, considered as functional data, and the application of the clustering algorithm recently developed in [24], which will described in Sect. 3. Several clustering methods for functional data have been proposed in the literature as, for instance, [5,17,18] but these methods are not aimed at dealing with outlying curves. Trimming techniques for robust clustering have been applied in [9,15].

The rest of the paper is organized as follows: Sect. 2 describes the idea behind our proposal for time series clustering. Section 3 gives a brief description of the robust clustering procedure for functional data that supports the time series clustering algorithm. Section 4 presents a simulation study that compares the performance of the algorithm with existing alternatives and Sect. 5 gives an application to a real data set. The paper ends with a discussion of the results.

2 Time Series Clustering

Consider a collection of n stationary time series $X_{1,t}, X_{2,t}, \ldots, X_{n,t}$ with $1 \leq t \leq T$. For ease of notation we take all series to have the same length, but this is not a requirement of the procedure. The spectral density of each time series is estimated by one of the many procedures available. Previous clustering methods based on the use of spectral densities relied on similarity measures for discriminating between them. In this work, the spectra are considered as functional data to which the robust clustering procedure developed in [24] is applied. The resulting clusters correspond to time series whose spectral densities have similar shapes, and hence similar oscillatory behavior. The procedure is able to detect outliers among the spectral densities, which correspond to time series having atypical oscillatory characteristics.

Two methods based on estimated spectral densities are presented in [3,13]. We describe them in more detail since they will be used later for comparison purposes. We refer to them as "TVDClust" and "HSMClust", respectively. In both cases the total variation distance (TVD) is used to measure similarity between spectral densities. TVD is a frequently-used distance between probability measures that, in the case of probability distributions having a density, measures the complement of the common area below the density curves. Thus, the more alike the densities are, the larger this common area and the smaller the TV distance. To use this distance to compare spectral densities, they need to be normalized so that the total area below the curve is equal to 1, which is equivalent to normalizing the original time series so that it has unit variance. Thus, it is the oscillatory behavior of the series, and not the magnitude of the oscillations that is taken into account in these clustering algorithms.

For "TDVClust", a dissimilarity matrix is built up by measuring the TVD distance between all pairs of normalized estimated spectral densities. This matrix is then fed to a hierarchical agglomerative algorithm with the complete or average linkage functions. The result is a dendrogram which can be cut to obtain the desired number of groups. To decide on the number of clusters an external criteria such as the Silhouette or Dunn's index is used. More details can be found in [3].

The second method, "HSMClust", is a modification in which every time two clusters are joined together, all the information in them is used to obtain a representative spectrum for the new cluster. There are two ways to do this, either all the spectral densities are averaged, which is the *average* option in the algorithm, or else all the time series in the two groups are concatenated and a new spectral density is estimated, which corresponds to the *single* option. Under the assumption that the series in the same cluster have common oscillatory characteristics, either of this procedures will give a more accurate estimation of the common spectral density for the whole group. This algorithm is known as the Hierarchical Spectral Merger (HSM) algorithm, and its implementation in R is available at http://ucispacetime.wix.com/spacetime#!project-a/cxl2.

Every time two clusters are merged, the dissimilarity matrix reduces its size. In "TVDClust", this matrix remains the same throughout the procedure and the distances between clusters are calculated using linear combinations of the distances of the individual points in each cluster. The linear combination used is determined by the linkage function employed. More details can be found in [3].

3   Robust Clustering for Functional Data

We now give a brief description of the algorithm proposed in [24], where more details can be found. Let X be a random variable taking values in the Hilbert space $L^2([0,T])$ of square integrable functions defined in the interval $[0,T]$, with inner product given by $\langle f, g \rangle = \int f(t)g(t)\, dt$. If $\mu(t) = E\{X(t)\}$ and $\Gamma(s,t) = \text{cov}\{X(s), X(t)\}$, then it is usual to represent X through its Karhunen-Loève expansion $X(t) = \mu(t) + \sum_{j=1}^{\infty} C_j(X)\psi_j(t)$. In that expansion, the ψ_j are an orthonormal system of functions obtained as eigenfunctions of the covariance operator Γ, i.e. $\langle \Gamma(\cdot, t), \psi_j \rangle = \lambda_j \psi_j(t)$, and the eigenvalues λ_j are taken in decreasing order and assumed to satisfy $\sum_{j=1}^{\infty} \lambda_j < \infty$. The principal component scores $C_j(X) = \langle X - \mu, \psi_j \rangle$ are uncorrelated univariate random variables with zero mean and variance equal to λ_j. Delaigle and Hall [10] show that $\log P(\|X - x\| \leq h)$ can be approximated by $\sum_{j=1}^{p} \log f_{C_j}(c_j(x))$, for any $x \in L_2([0,T])$ and small h, where f_{C_j} corresponds to the probability density function of $C_j(X)$ and $c_j(x) = \langle x, \psi_j \rangle$. This approximation entails a kind of "small-ball pseudo-density" approach for Functional Data Analysis by taking into account that probability density functions in the finite dimensional case can be seen as the limit of $P(||X - x|| \leq h)/h$ when $h \to 0$. In the particular case of X being a Gaussian process, the $C_j(X)$ are independent normally distributed random variables with mean equal to 0 and variance equal to λ_j.

With these ideas in mind, Jacques and Preda [17] proposed a "model-based" approach for clustering of functional data, where a finite number of independent normally distributed principal component scores are assumed and different variances are also allowed for each cluster. Previously, Bouveyron and Jacques [5] had already considered a different approach, where a certain fraction of the smallest variances are constrained to be equal for each cluster.

In Rivera-García et al. [24], starting from Bouveyron and Jacques [5] and Jaques and Preda [17], a robust functional clustering procedure is proposed where a proportion α of curves are allowed to be trimmed and constraints on the variances are considered. If $\{x_1, ..., x_n\}$ is a set of curves in $L^2([0, T])$, we consider the maximization of a trimmed mixture-loglikelihood defined as

$$\sum_{i=1}^{n} \eta(x_i) \log \left(\sum_{g=1}^{K} \pi_g \left[\prod_{j=1}^{q_g} \frac{1}{\sqrt{2\pi a_{jg}}} \exp \left(\frac{-c_{ijg}^2}{2a_{jg}} \right) \prod_{j=q_g+1}^{p} \frac{1}{\sqrt{2\pi b_g}} \exp \left(\frac{-c_{ijg}^2}{2b_g} \right) \right] \right)$$

(1)

where $c_{ijg} = c_{jg}(x_i)$ is the j-th principal component score of curve x_i in group g, $g = 1, ..., K$, and, $\eta(\cdot)$ is an indicator function with $\eta(x_i) = 0$ if the x_i curve is trimmed and 1 if it is not and $\pi_g, g = 1, ..., K$ are mixing weights that add up to 1. A proportion α of curves is trimmed, so that $\sum_{i=1}^{n} \eta(x_i) = [n(1 - \alpha)]$. The main variance contributions in the g-th cluster are assumed to be $a_{1g}, ...,$ $a_{q_g g}$, for the first q_g components, while we assume that each of the remaining $p - q_g$ components contribute with the same b_g variance. Notice that we take an equal number of principal components p in every cluster but the number of main components q_g may vary across clusters. Finally, to prevent the detection of spurious clusters, two constants $d_1 \geq 1$ and $d_2 \geq 1$ were fixed such that the maximization of (1) is done under the constraints:

$$\frac{\max_{g=1,...,K; j=1,...,q_j} a_{jg}}{\min_{g=1,...,K; j=1,...,q_j} a_{jg}} \leq d_1 \quad \text{and} \quad \frac{\max_{g=1,...,K} b_g}{\min_{g=1,...,K} b_g} \leq d_2.$$

(2)

A feasible algorithm for performing the constrained maximization, detailed in [24], is a modification of the traditional EM algorithm used in model-based clustering where a "trimming" step (T-step) is added. In the T-step, those curves with smallest contributions to the trimmed likelihood are temporarily not taken into account in each iteration of the algorithm. The trimming step is similar to that applied in the "concentration" steps applied when performing the fast-MCD algorithm [25]. To enforce the required constraints on the variances, optimally truncated variances as done in [14] are adopted if needed. For the estimation of the dimension q_g in each cluster, a Bayesian Information Criterion (BIC) approach was proposed in [24].

4 Simulation Study

To evaluate the performance of the proposed Robust Functional Clustering (RFC) methodology, a simulation study was carried out. We now describe the different scenarios and contamination types. As in [13], the simulations are based on combinations of autoregressive processes of order 2, AR(2), which are defined as $X_t = u_1 X_{t-1} + u_2 X_{t-2} + \epsilon_t$ where ϵ_t is a white noise process. The associated characteristic polynomial is $h(y) = 1 - u_1 y - u_2 y^2$ and its roots, denoted by y_1 and y_2, are related to the spectrum of the time series. If the roots are complex-valued, they must be conjugate, i.e. $y_1 = \overline{y_2}$ and their polar representation is $|y_1| = |y_2| = M$ and $\arg(y_i) = 2\pi\nu/w_s$ where w_s is the sampling frequency in Hertz; M is the magnitude of the root ($M > 1$ for causality) and ν the frequency index, $\nu \in (0, w_s/2)$. The spectrum will have modal frequency in ν, which will be broader as $M \to \infty$ and narrower as $M \to 1^+$. Then, given (ν, M, w_s) and with $\omega_0 = \frac{2\pi\nu}{w_s}$ we have $u_1 = 2M^{-1}\cos\omega_0$ and $u_2 = -M^{-2}$.

Two groups of 50 time series each were simulated, with parameters $\nu_1 = 0.21$, $\nu_2 = 0.22$, $M_1 = M_2 = 1.15$, $w_s = 1$ and length $T = 1000$. From the simulated time series, the spectral densities were estimated using a smoothed lag-window estimator with a Parzen window and bandwidth $100/T$. The estimated spectral densities are shown in Fig. 1(a). The functional form of the estimated spectral densities was recovered using a B-Spline basis of degree 3 with 14 equispaced nodes and smoothing parameter $\lambda = 0.000003$ (see e.g. [22], Chap. 3) We want to test the performance of the different algorithms in recovering these two groups, even in the presence of contaminating data. In the absence of contamination we have 100 observations divided into two groups.

We introduce the mixtures of AR(2) processes that will be used in the contamination schemes. Let $Y_t^i, i = 1, 2$ be two AR(2) processes with parameters M_i and ν_i, $i = 1, 2$. Their mixture is given by $X_t = a_1 Y_t^1 + a_2 Y_t^2 + \epsilon_t$ where $a_i, i = 1, 2$ are the weights and ϵ_t is a white noise process. This mixture creates a signal that combines the oscillatory behavior of the processes $Y_t^i, i = 1, 2$.

Starting from the two groups of 50 AR(2) time series described previously, which are considered as the clean data, we added another 11 time series (around 10% contamination level), generated according to the following schemes:

(i) AR(2) processes with parameters ν_i chosen randomly with uniform distribution in the interval $(.20, .25)$, denoted $U(.20, .25)$, $M = 1.2$ and $w_s = 1$. The contaminating series have smaller variance than the series in the clusters. See Fig. 1(b).

(ii) A mixture of two AR(2) processes having parameters $\nu_i = .20$ and $.25$; $M_i = 1.05, 1.1$, $i = 1, 2$ y $w_s = 1$. See Fig. 1(c).

(iiii) A mixture of two $AR(2)$ processes with random parameters $\nu_1 = U(.19, .22)$ y $\nu_2 = U(.24, .26)$; $M_i = 1.05, 1.1$, $i = 1, 2$ and $w_s = 1$, See Fig. 1(d).

Figure 1(b), (c) and (d) show the spectral densities for the simulated time series with the three contamination schemes described.

In order to test the performance of the RFC methodology, the simulated process and their estimated spectral densities were used to compare with the

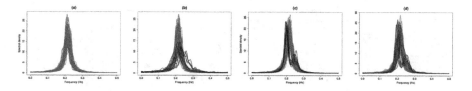

Fig. 1. Spectral density of the simulated time series: (a) No contamination, (b) Contamination type (i), (c) Contamination type (ii) and (d) Contamination type (iii)

results obtained when using the "Funclust" algorithm [17] and hierarchical methods using the total variation distance: "HSMClust" [13] and "TVDClust" [2,3].

It is important to recall that we assume the q_g dimensions in the RFC procedure to be unknown parameters and that the BIC criterion is used to estimate them when applying this algorithm. The results in [24] already show the importance of trimming. Trimming levels $\alpha = 0$ and $\alpha = 0.1$ are used. As regards the constraints, we are assuming $d_1 = d_2 = d$ to simplify the simulation study. Values of $d = 3$, 10 and 10^{10} (i.e., almost unconstrained in this last case) were used. We always return the best solution in terms of the highest BIC value for each combination of all those fixed values of trimming level and constraints. We use 100 random initializations with 20 iterations.

For the "Funclust" method we used the library **Funclustering** [26] in R where the EM algorithm has been initialized with the best solutions out of 20 "short" EM algorithms with only 20 iterations and threshold values of $\varepsilon = 0.001, 0.05, 0.1$ in the Cattell test. For the agglomerative methods we use the library **HSMClust** in R for "HSMClust" and "TVDClust" by means of the algorithm described in [2,3].

Figure 2 shows the results for the simulation study. It is composed of a matrix of graphs, where the rows correspond to the different contamination schemes (uncontaminated in the first row) while the columns correspond to the methodologies tested. The first column corresponds to "Funclust", the second to "HSMClust", the third shows the results for the RFC procedure with trimming levels $\alpha = 0$ (untrimmed) and $\alpha = 0.1$ and three constraint levels $d = 3$, 10 and 10^{10} (i.e., almost unconstrained in this last case). The fourth column shows the results corresponding to "TVDClust". The x-axis corresponds to the threshold applied in the Cattell test for "Funclust", the procedure in "HSMClust", the constraint level for RFC and the linkage function for the agglomerative method "TVDClust", while the y-axis corresponds to the correct classification rate (CCR).

Results show that the hierarchical methods, "HSMClust" and "TVDClust" are better in the absence of contamination, giving very consistent results. However, their performance degrades sharply in the presence of noise. This is not surprising since these procedures were not designed to handle contamination in the sample. The joint use of trimming and constraints in RFC improve the results (CCR) substantially. Results are very good for moderate ($d = 10$) and small ($d = 3$) values of the constraint constants, while for high values the results are

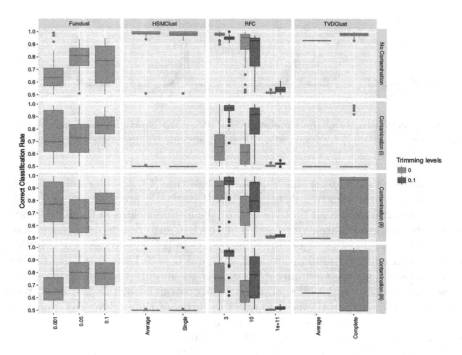

Fig. 2. classification rate (CCR) for the four methods considered, represented in different columns. Rows correspond to the different contamination schemes, starting with no contamination in the first row and following with schemes (i), (ii) and (iii) described in the text. Constraint levels $d_1 = d_2 = 3$, 10 and 10^{10}, trimming levels $\alpha = 0$ and 0.1 were used for the RFC method. Threshold values $\varepsilon = 0.001, 0.05$ and 0.1 were used for the "Cattell" procedure in "Funclust". Single and average versions were used for "HSMClust" while average and complete linkage functions were used for "TVDClust".

poor. Very high values for these constants are equivalent to having unconstrained parameters. Trimming turns out to be very useful in all the contaminated cases while the results are not affected by trimming in the uncontaminated case.

In the presence of contamination, the results for "Funclust", "HSMClust" and "TVDClust" fall below those of RFC when applying the $\alpha = 0.1$ trimming and small/moderate values d_1 and d_2 for the variance parameters.

5   Analysis of Real Data

We now consider wave-height data measured by a buoy located in Waimea Bay, Hawaii, at a water depth of approximately 200 m. This buoy is identified as number 106 (51201 for the National Data Buoy Centre). The data, which corresponds to 72.5 h divided into 30-minute intervals, was collected in June 2004 and has previously been analyzed by [2] where more details can be found.

In [2] the spectrum for each 30-minute interval was estimated and normalized The TV distance between all spectral densities was used to build a dissimilarity matrix, that was fed into a hierarchical agglomerative clustering algorithm. For more details see [3]. The 145 normalized densities are shown in Fig. 3(a).

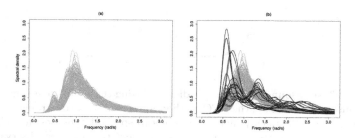

Fig. 3. Spectral densities for the sea wave data after normalization. (a) Original data. (b) Original data plus 22 additional densities in black, considered as noise.

The RFC method was applied to this data set in order to obtain an alternative clustering. The functional form of the data was recovered using B-splines of order 3 with 31 equispaced nodes. We use 100 initializations with 20 iterations each. The constraint level considered was $d_1 = d_2 = 3$, and the trimming level $\alpha = 0.13$. In [2] two different clusterings were obtained, depending on the linkage function used: 4 clusters for the complete linkage and 3 for average. We will only consider the clustering into 4 groups for comparison purposes in what follows.

To compare the two results, the Adjusted Rand Index (ARI) [16] was used. This is an improvement of the original Rand Index [23] and measures the similarity between two partitions of the same set, having value 1 when there is exact coincidence and close to 0 when considering a completely "random" clustering of the data.

One can see that the effect of trimming and constraints is not harmful, even in the absence of contamination. For instance, we can see that the ARI of RFC with $d_1 = d_2 = 3$ and $\alpha = 0.13$ is equal to 0.513 with respect to the "reference" partition, which is obtained when applying [2] with 4 groups. To compute this ARI index we assign all the time series (trimmed and non-trimmed) to clusters by using posterior probabilities from the fitted mixture model that was described in Sect. 3. The two rows in Fig. 4 show the clusters found when using the TVD and RFC, respectively. Even though the groups have differences in membership and size, it is possible to see from the figures that the shape of the functions in the corresponding clusters are very similar and the mean functions are close. The variations are probably due to the different clustering techniques employed, but the similarity in the groups obtained point to consistent results for both methods. Observe that the trimmed curves for the RFC method are different from the rest of the functions in their cluster. For "HMSClust" both versions gave a value of 0.723 for the ARI, higher than that obtained with RFC, while

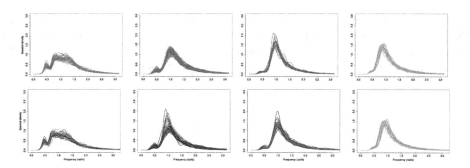

Fig. 4. (Top) Clusters found using the TV distance using the complete linkage function. (Bottom) Clusters found using the RFC method for $K = 4$ with constrains $d_1 = d_2 = 3$ and trimming $\alpha = 0.13$. Each panel corresponds to spectral densities in each cluster, grey lines represent the means and black lines represent the trimmed observations.

for "Funclust", values were lower, with a maximum of 0.315 with a threshold value of 0.01 or 0.1 in the "Cattell" test.

In order to test the performance of the different methods with real data and in the presence of contamination, 22 time series were added to the sample. These measurements were recorded at the same location and during the same month, but during different days. The corresponding estimated spectral densities are shown in black in Fig. 3(b). Some of these densities are bimodal while others are unimodal but have lower modal frequency than those in the original sample.

The four clustering procedures considered were applied to this contaminated sample and the results were compared using again as "reference" the clustering obtained in [2] with 4 groups applied to the clean data (i.e., before adding the contaminating curves). The ARI was computed by taking only into account the classification of the original (non-contaminating) densities. In the case of the RFC methodology, the assignments based on "posterior" probabilities were considered for the wrongly trimmed observations.

The ARI for the RFC method with $K = 4$ and $d_1 = d_2 = 3$ are equal to 0.167, 0.723 and 0.599 when trimming levels $\alpha = 0$, $\alpha = 0.13$ and $\alpha = 0.2$, respectively, are used. The associated ARI when using Funclust are always below 0.21 for all three Cattell thresholds tested (0.002, 0.05 and 0.1) as this method is not designed to cope with outlying curves. The other methods tested, "average TVD" and "average HMSClust", have even worse results in this contaminated case reaching ARI values equal to 0 in both cases. Therefore, the best results overall were obtained using RFC with a $\alpha = 0.13$ while the other methods show poor results in the presence of contaminating data.

To reinforce previous claims, Fig. 5 shows in the first row, the partition obtained in [2] with four clusters before adding the contaminating time series, in the second, the results when using RFC with four clusters, $d_1 = d_2 = 3$ and trimming level $\alpha = 13\%$ to the "contaminated" data set. In the third row the results obtained with "TDVClust", then "HMSClust" and "Funclust", also in case that

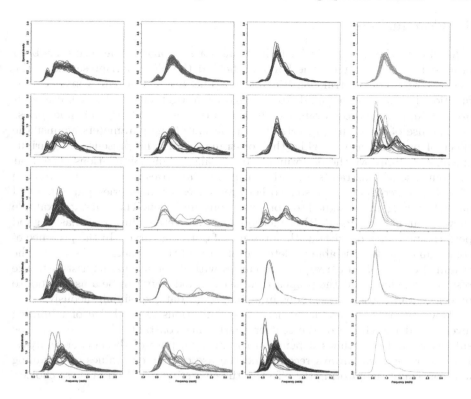

Fig. 5. Clusters found with the different procedures when $K = 4$. Each panel corresponds to a cluster of spectral densities. Gray lines represent the means and black lines represent the trimmed observations. *First row:* Original clusters for the TV distance using the complete link before adding contaminating time series. *Second row:* Clusters for the RFC method for $K = 4$ with constrains $d_1 = d_2 = 3$ and trimming $\alpha = 0.13$. *Third row:* Clusters for the "TVDClust" method with complete linkage. *Fourth row:* Clusters for "HSMClust" method. *Fifth row:* Clusters for "Funclust"

the contaminating time series were added. Once again, the clusters obtained with RFC differ slightly from those obtained in [2] but, in spite of the presence of contamination, the shape of the spectral densities in the corresponding clusters are very similar and the average densities are very close. The trimmed functions when using level $\alpha = 13\%$ are shown in black in the second row. The last three rows show the poor results obtained with the other three methods. For instance, in the third row, corresponding to "TVDClust", the original sample is clustered together in a single group in the leftmost panel, while the other three groups only contain contaminating functions that were added as noise. Since $\alpha = 0.13$, 19 curves were trimmed with the RFC procedure, most of which come from the contaminating series that were added. Finally, it is also important to point out that trimming and clustering are performed simultaneously in the RFC approach.

6   Conclusions

A feasible methodology of robust clustering for stationary time series has been proposed and illustrated. The key idea behind the algorithm presented is the use of estimated spectral densities of the time series, that are considered as functional data. A robust model-based algorithm together with the simultaneous use of trimming and constraints is then used to cluster the original time series.

The use of trimming protects the estimation of the parameters against the effect of outlying curves, while the constraints avoid the presence of spurious clusters and improve the performance of the algorithms. Simulations show that the joint use of constraints and trimming tools improves results in the presence of outliers, in comparison to some other procedures for time series and functional data clustering, not designed to work with contamination. The real data example shows that the proposed RFC method for time series clustering has a good performance, with or without the presence of outlying curves. In the presence of contamination, RFC is able to detected almost all the outliers in the data. The trimmed curves often correspond to curves with different characteristics to the rest. We conclude that the proposed robust methodology can be a useful tool to detect contamination and groups in a time series data set simultaneously.

However, this methodology has some limitations. The choice of trimming level α and the choice of the scatter constraints constants d_1 and d_2, can be subjective and sometimes depend on the final purpose of the cluster analysis. For this reason, we always recommend the use of different values of trimming and constraint, monitoring the effect in the clustering partition of these choices.

Acknowledgements. Data for station 160 were furnished by the Coastal Data Information Program (CDIP), Integrative Oceanographic Division, operated by the Scripps Institution of Oceanography (http://cdip.ucsd.edu/). Research by DRG and JO was partially supported by Conacyt, Mexico Proyecto 169175 Análisis Estadístico de Olas Marinas, Fase II. Research by LA G-E and A M-I was partially supported by the Spanish Ministerio de Economía y Competitividad y fondos FEDER, grant MTM2014-56235-C2-1-P, and by Consejería de Educación de la Junta de Castilla y León, grant VA212U13.

References

1. Aghabozorgi, S., Shirkhorshidi, A.S., Wah, T.Y.: Time-series clustering - a decade review. Inf. Syst. **53**, 16–38 (2015)
2. Alvarez-Esteban, P.C., Euán, C., Ortega, J.: Statistical analysis of stationary intervals for random waves. In: Proceedings of International Society of Offshore and Polar Engineering Conference, vol. 3, pp. 305–311 (2016)
3. Alvarez-Esteban, P.C., Euán, C., Ortega, J.: Time series clustering using the total variation distance with applications in oceanography. Environmetrics **27**(6), 355–369 (2016)
4. Bahadori, M.T., Kale, D.C., Fan, Y., Liu, Y.: Functional subspace clustering with application to time series. In: Proceedings of 32nd International Conference on Machine Learning, pp. 228–237 (2015)

5. Bouveyron, C., Jacques, J.: Model-based clustering of time series in group-specific functional subspaces. Adv. Data Anal. Classif. **5**(4), 281–300 (2011)
6. Caiado, J., Crato, N., Peña, D.: A periodogram-based metric for time series classification. Comput. Stat. Data Anal. **50**(10), 2668–2684 (2006)
7. Caiado, J., Crato, N., Peña, N.: Comparison of times series with unequal length in the frequency domain. Commun. Stat. Simul. Comput. **38**(3), 527–540 (2009)
8. Caiado, J., Maharaj, E.A., D'Urso, P.: Time Series Clustering. Chapman & Hall/CRC Handbooks of Modern Statistical Methods. Taylor & Francis, London (2015). Chap. 12
9. Cuesta-Albertos, J.A., Fraiman, R.: Impartial trimmed k-means for functional data. Comput. Stat. Data Anal. **51**(10), 4864–4877 (2007)
10. Delaigle, A., Hall, P.: Defining probability density for a distribution of random functions. Ann. Stat. **38**(2), 1171–1193 (2010)
11. D'Urso, P., De Giovanni, L., Massari, R.: Time series clustering by a robust autoregressive metric with application to air pollution. Chemometr. Intell. Lab. Syst. **141**, 107–124 (2015)
12. D'Urso, P., De Giovanni, L., Massari, R.: GARCH-based robust clustering of time series. Fuzzy Sets Syst. **305**, 1–28 (2016)
13. Euán, C., Ombao, H., Ortega, J.: The hierarchical spectral merger algorithm: a new time series clustering procedure. J. Classif. (2017, accepted)
14. Fritz, H., García-Escudero, L.A., Mayo-Iscar, A.: A fast algorithm for robust constrained clustering. Comput. Stat. Data Anal. **61**, 124–136 (2013)
15. García-Escudero, L.A., Gordaliza, A.: A proposal for robust curve clustering. J. Classif. **22**(2), 185–201 (2005)
16. Hubert, L., Arabie, P.: Comparing partitions. J. Classif. **2**, 193–218 (1985)
17. Jacques, J., Preda, C.: Funclust: a curves clustering method using functional random variables density approximation. Neurocomputing **112**, 164–171 (2013)
18. James, G.M., Sugar, C.A.: Clustering for sparsely sampled functional data. J. Am. Stat. Assoc. **98**(462), 397–408 (2003)
19. Liao, T.W.: Clustering of time series data - a survey. Pattern Recogn. **38**, 1857–1874 (2005)
20. Maharaj, E.A., D'Urso, P.: Fuzzy clustering of time series in the frequency domain. Inf. Sci. **181**(7), 1187–1211 (2011)
21. Montero, P., Vilar, J.: TSclust: an R package for time series clustering. J. Stat. Softw. **62**(1), 43 (2014)
22. Ramsay, J.O., Silverman, B.W.: Functional Data Analysis. Springer Series in Statistics, 2nd edn. Springer, New York (2005)
23. Rand, W.M.: Objective criteria for the evaluation of clustering methods. J. Am. Stat. Assoc. **66**, 846–850 (1971)
24. Rivera-García, D., García-Escudero, L.A., Mayo-Iscar, A., Ortega, J.: Robust clustering for functional data based on trimming and constraints. arXiv:1701.03267 (2017)
25. Rousseeuw, P.J., Van Driessen, K.: A fast algorithm for the minimum covariance determinant estimator. Technometrics **41**, 212–223 (1999)
26. Soueidatt, M.: Funclustering: A package for functional data clustering. R package version 1.0.1 (2014)

Introducing a Fuzzy-Pattern Operator in Fuzzy Time Series

Abel Rubio, Enriqueta Vercher, and José D. Bermúdez$^{(\boxtimes)}$

Department of Statistics and Operational Research,
University of Valencia, Burjassot, Spain
{abel.rubio,enriqueta.vercher,jose.d.bermudez}@uv.es

Abstract. In this paper we introduce a fuzzy pattern operator and propose a new weighting fuzzy time series strategy for generating accurate ex-post forecasts. A decision support system is built for managing the weights of the information provided by the historical data, under a fuzzy time series framework. Our procedure analyzes the historical performance of the time series using different experiments, and it classifies the characteristics of the series through a fuzzy operator, providing a trapezoidal fuzzy number as one-step ahead forecast. We also present some numerical results related to the predictive performance of our procedure with time series of financial data sets.

Keywords: Fuzzy time series · Fuzzy pattern · Pattern analysis · Forecasting · Fuzzy numbers · Decision support system

1 Introduction

A large number of experts on financial economics have analyzed the behavior of stock market indices from different perspectives, trying to acquire information and knowledge about the future performance of stock market indices. The main difficulty of this analysis is due to the ambiguity of the financial data and their asymmetric volatility over time. We propose to analyze the time series of stock market indices from a fuzzy point of view. Fuzzy time series (FTS) is applied when the historical data are expressed through linguistic variables.

In our approach, we use the basic methodology of the weighting fuzzy time series (FTS), where the weights are built using the information extracted from the fuzzy relationships between the linguistic variables and the fuzzy relations of groups [1,9,17,20]. This fuzzy framework allows us to establish a new approach for the treatment of non-linear time series, giving fuzzy estimates of the future values. In addition, in this paper we analyzed and forecast the daily quotation of two benchmark stock indices (DAX and TAIEX) in order to learn about their behavior using past observed values, analyzed in an independent way, for building one-step ahead forecasts of their future behavior, which is considered as the ex-post forecast.

The seminal paper on FTS, authored by Song and Chissom [17], introduces the fundamental concepts of this topics. Although, the FTS methodologies were

© Springer International Publishing AG 2017
I. Rojas et al. (Eds.): IWANN 2017, Part II, LNCS 10306, pp. 154–164, 2017.
DOI: 10.1007/978-3-319-59147-6_14

suitably established and developed by Chen [1,2]. The main steps of Chen's FTS model are:

1. definition and partitioning of the universe of discourse U,
2. define the linguistic values that be represented by fuzzy sets and classify the data,
3. define the fuzzy relationships between sets and the fuzzy relations of groups, and
4. calculate the crisp value of the fuzzy forecast, based on previously established prediction rules.

Concerning the first step, Huarng [8] proves that the partitioning of the universe U affects the forecasts; then, he proposes two algorithms (distribution-based length and average-based length) to determine the interval length of the initial partitioning of the universe of discourse. Although his FTS models follow the basic scheme proposed by Chen, these partitioning proposals allow a notable diminution of the fitting errors. Recently, other partitions have been proposed with intervals of different length based on granular information procedures and clustering techniques [6,11,18]. Some researchers have also proposed to subdivide and develop the main steps of FTS methods in multiple sub-steps, in order to increase the accuracy of the forecasts [11,19]. For taking into account recurrences in the fuzzy relations of groups (Step 3), Yu [20] introduces weighted fuzzy time series, and other authors propose several weighting ways that improve forecasting accuracy (see, for instance, [3,12,14,16] and references therein). Recently, the introduction of trend in the FTS methodology has been also analyzed using different strategies [4,15].

Concerning the forecasting rules, classical FTS models usually focus on improving fitting errors rather than improving ex-post forecasting accuracy, since they are designed for improving the forecast of more segmented or linguistic series. But, financial time series presents disturbances, fluctuations and changes of trend which need models that capture this erratic performance and efficiently explain their current state analyzing global properties of the time series. In previous works, we have propose several weighting methods in order to include the effect of recurrence and trend in financial time series data sets [12,13]. Now, we propose a new fuzzy pattern operator which has been incorporated to a weighted FTS method, allowing to forecast the future value of some stock market indices while improving the accuracy and robustness of the fuzzy forecast by means of the pattern analysis of *runs* of the time series.

The remaining content of this paper is organized as follows. In Sect. 2 we introduce the fuzzy pattern operator and present a weighted FTS model. The decision support system for different proposals of weighting FTS methods is explained in Sect. 3. Finally, in Sect. 4 we have included numerical experiments performed with daily quotes of two stock market indices: the German Stock DAX and the Taiwanese TAIEX Index.

2   A Weighted Fuzzy Time Series Model

In previous papers we have proposed several modifications of the main steps of Chen's FTS model. In Rubio et al. [12], we postulated that the interval length of the partitioning of the universe of discourse, U, should depend on the variability of the time series. So, here we propose to use equal length intervals, given by $len = \sigma/\sqrt{(N)}$, being σ the standard deviation of the observed time series and N the number of observations. In this way, we obtain m consecutive intervals $\{u_1, \ldots, u_m\}$, which constitutes a partition of one interval in $\Re$ that includes the universe of discourse U; those intervals define the m linguistic values $\{A_1, \ldots, A_m\}$ used for fuzzifying the time series. In that paper we also proposed to represent those linguistic values by trapezoidal fuzzy numbers, previously also suggested by Liu [10].

Definition 1. *Trapezoidal Fuzzy Number*
   A fuzzy number A is said to be a trapezoidal fuzzy number, $A = (a_1, a_2, a_3, a_4)$, if its membership function has the following form:

$$\mu_A(y) = \begin{cases} \frac{y-a_1}{a_2-a_1} & \text{if } a_1 \leq y \leq a_2 \\ 1 & \text{if } a_2 \leq y \leq a_3 \\ \frac{a_4-y}{a_4-a_3} & \text{if } a_3 \leq y \leq a_4 \\ 0 & \text{otherwise} \end{cases}$$

where $[a_2, a_3]$ is the core of A, and the support of A, $supp(A) = \{y : \mu_A(y) \geq 0\}$, is $[a_1, a_4]$.

   In [12], we also introduced a *weighted chronological operator* to be used as a prediction rule. Let $Y(t) = x_t$, for $t = 1, \ldots, N$ the observed time series, and let $F(t)$, $t = 1, \ldots, N$, the fuzzifyed series, so each $F(t)$ will be an element of the set $\{A_1, \ldots, A_m\}$. For each $t_0 = 1, \ldots, N-1$, and each $i, j = 1, \ldots, m$ let $\beta_{i,j}^{t_0} = t_0$ if $F(t_0) = A_i$ and $F(t_0 + 1) = A_j$, being zero otherwise. Then we define the matrix of weights W with generic element $w_{i,j} = \sum_t \beta_{i,j}^t$, for $i = 1, \ldots, m$ and $j = 1, \ldots, m$; finally, we normalize the matrix W in such a way that each row add up to one, obtaining the normalized matrix W'. Our chronological operator use as input the matrix of weights W' and the last observed linguistic value $F(N)$ of the time series; if $F(N) = A_p$, then the output is a weighted mean of the m trapezoidal fuzzy numbers $\{A_1, \ldots, A_m\}$ with weights given by the p row of matrix W'.
   The chronological operator try to incorporate information from identical patterns observed in the FTS, so they could be useful when the time series does not show a clear trend. On the contrary, identical patterns will be rare if there is a tendency, so this operator could not be useful. In [13] we introduce a *weighted jump operator* based on the concept of size of a jump in a time series.

Definition 2. *Size of a jump*
   If $F(t-1) = A_i$ and $F(t) = A_j$ let $J(t) = j - i$ be the size of the jump associated to time $t = 2, \ldots, N$. The size of the jump a fulfills that $-m + 1 \leq J(t) \leq m - 1$ for $t = 2, \ldots, N$.

Let k be a jump size, for $-m + 1 \leq k \leq m - 1$, and let $s_k = \sum_t \sum_p \beta_{p,p+k}^{t-1}$ the chronological weight associated to a jump of size k. Let s_k' be the normalized weights, in such a way that they add up to one. The jump operator use as input both the vector of normalized weights and the last observed linguistic value $F(N)$ of the time series. If $F(N) = A_p$, then the output is a weighted mean of the $2m - 1$ trapezoidal fuzzy numbers $A_{p-m+1}, \ldots, A_{p+m-1}$ with weights given by s_k', $-m + 1 \leq k \leq m - 1$. To do that, the universe of discourse has to be extended appropriately [13]. The weighted jump operator try to learn from affine patterns -specifically, identical patterns after a translation on the scale of the universe of discourse-, and they could be very useful when the time series has a clear trend because, in that case, the fuzzy time series presents few identical patterns but a lot of affine patterns.

2.1   Introducing a New Pattern Operator

The aforementioned operators are based on two consecutive outcomes in the fuzzy time series. In this paper we introduce a *weighted pattern operator* based on patterns of more than two consecutive outcomes. It is based on the concept of *run*, which have been extensively used in other contexts.

Definition 3. *Run*

A run or a clump of length k, with $k > 1$ is a sequence of k consecutive outcomes in which repetitions are allowed.

The weights used by the pattern operator are computed from the frequency, chronologically penalized, of runs of length k in the series of sizes of jumps. Let us explain how they are calculated. For example, if it is decided to use runs of length $k = 2$, we compute the weight associated to the run (k, l) as $\alpha_{k,l} = \sum_t \sum_p \sum_q \beta_{p,p+k}^{t-2} \beta_{q,q+l}^{t-1} / t$; then, we normalize the matrix of weights with the generic element (k, l) in such a way that each row add up to one, obtaining the matrix W^*. The pattern operator based on runs of length 2 use as input the matrix W^*, the last observed jump value $J(N)$ and the last observed linguistic value $F(N)$; if $J(N) = k$ and $F(N) = A_m$, then the output is a weighted mean of the $2m - 1$ trapezoidal fuzzy numbers $A_{p-m+1}, \ldots, A_{p+m-1}$ with weights given by the k row of W^*. This operator could easily be extended for runs of length greater than 2.

This new pattern operator is a generalization of the jump operator, previously introduced in [13], and it will be useful when there is a trend, not necessarily linear. The weakness of this operator is that the set of possible runs could be very large, so it is only advisable when N is large.

3   Decision Support System

The three prediction operators introduced in the previous section allow us to propose a general prediction rule based on all of them. Let O_C, O_M and O_T be the chronological, jump and pattern operators, respectively. Given an observed time series, and once the associated linguistic time series was obtained, we propose to use the following predictor:

$$\hat{F}(N+1) = \gamma_1 O_C(F(1),\ldots,F(N)) + \gamma_2 O_M(F(1),\ldots,F(N)) + \gamma_3 O_T(F(1),\ldots,F(N))$$

where $\{\gamma_1,\gamma_2,\gamma_3\}$ are non-negative coefficients such that $\gamma_1 + \gamma_2 + \gamma_3 = 1$, and $\hat{F}(N+1)$ is the fuzzy predictor, a trapezoidal fuzzy number from which is also possible to obtain a point-wise forecast, as well as a measure of the uncertainty still remaining in the prediction. Specifically, that trapezoidal fuzzy number allows us to calculate the value and ambiguity of the prediction, as well as the possibilistic moments from the concept of interval-valued mean [5,7].

We have previously work with the first two operators (O_C and O_M) and a linear combination of them, whose weights were evaluated using the fitting errors attained with the training data set [12,13]. Now, we are interested in the performance of the pattern operator and the accuracy of the predictions obtained working with it; however, this operator needs a large number of observations in order to learn and built suitable pattern structures. This is the main constraint of the applicability of the new operator, although the size of the data set also plays an important role when using the other operators.

Based on the results provided by numerical experiments developed using time series of different sizes, we propose to take into account the size N of the series into the DSS scheme, in such a way that:

- If $N < 30$, use the chronological operator ($\gamma_1 = 1$)
- If $30 \leq N < 250$, add the jump operator ($\gamma_1 + \gamma_2 = 1$)
- If $250 \leq N < 500$, use the pattern operator with runs of length 2 ($\gamma_1 + \gamma_2 + \gamma_3 = 1$)
- If $N \geq 500$, use the pattern operator with runs of length 3 ($\gamma_1 + \gamma_2 + \gamma_3 = 1$)

Other runs of greater length could be used if N is larger than 500.

4   Numerical Experiments

To analyze the predictive performance of the decision support system presented in Sect. 3, we use two reference indices, the German Stock (DAX) and the Taiwanese (TAIEX) Index. All information about the daily quotes of these indexes have been obtained from www.finance.yahoo.com.

The performance of our DSS procedure is compared with Yu's [20] and Cheng et al.'s [3] weighted methods and Wang et al. proposal [18]. It must be noted that the aforementioned FTS methods use different strategies for partitioning the universe of discourse U, at Step 1. Additionally, we have considered the fuzzy-trend time series method [13], which only uses a linear combination of the chronological (O_C) and jump (O_M) operators. Here, in the numerical experiments, we are going to use $\gamma_1 = \gamma_2 = 0$ and $\gamma_3 = 1$, focussing on the pattern operator model; in such a way that the performance of the new pattern operator is clearly analyzed and compared.

The basic algorithm with the DSS strategies and the different weighting methods have been implemented on a personal computer using the R programming environment (www.r-project.org).

The data sets of daily quotes have been dynamically divided, in a rolling horizon scheme, into two parts: the training data set (for the ex-ante forecasts) and the test data set (for the ex-post forecasts). We use the observations of the training data set to compute the weights of our prediction methods, which will be applied to generate the one-step ahead forecast. The forecasting accuracy of all FTS methods is measured by means of the root mean square error (RMSE), the mean absolute deviation (MAD) and the bias of the forecasting errors.

4.1 Forecasting of the DAX Index

For the first experiment we deal with daily quotes from the German Stock DAX Index from January 2011 to December 2013. There, the first 700 quotes will serve as training data set, to establish the weights of a run of five 5 predictions (as representative of one standard five-day week). Our procedure build a one-step ahead forecast for the first day of the run, and posteriorly incorporates the observed daily quote in order to forecast the quote of the second day, maintaining the weights provided in the training period. The procedure works in a rolling horizon scheme, in such a way that these 5 observations are added to the training set, while the first 5 observations of January 2011 are eliminated. For this new training data set the procedure evaluates new weights, and applies the same strategy for the next 5 ex-post forecasts. This weekly prediction scheme has been repeated 5 times, obtaining a total amount of 25 daily forecasts.

Figure 1 shows at the left-hand side the first 700 quotation, which corresponds to the data training set, and at the right-hand side both the trace of the time series and the one-step ahead forecasts provided by the DSS procedure using the above explained weekly rolling horizon scheme.

Note that our approach assumes that every observed value belongs to a fuzzy set, and every one-step ahead forecast is a trapezoidal fuzzy number, for which its mean value is considered as the crisp forecast. These facts are showed in Fig. 2, where a partial representation of the fuzzy time series of 20 quotations of the data training set and the first fuzzy forecast are included.

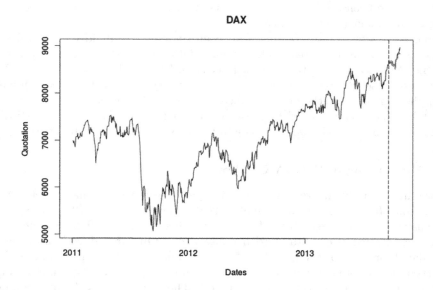

Fig. 1. Trace of the time series of daily quotes of German Stock Dax Index, from January 2011 to December 2013, and one-step ahead forecasts for the last 25 quotes.

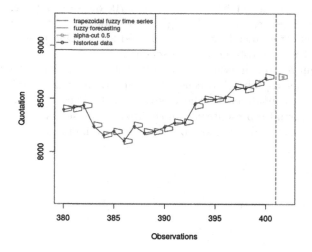

Fig. 2. Partial representation of the fuzzy time series of 20 daily quotes of German DAX Index and the first one-step ahead fuzzy forecast.

For this experiment, the selected prediction method using the DSS scheme is a pattern operator (using $\gamma_3 = 1$) with runs of length 3. Note that the procedure manages a large training set, which has allowed us to use *runs* of longer length. The accuracy of our forecasting DSS proposal is compared with those obtained by applying Yu's, Cheng et al., Wang et al. and Rubio el al. (using $\gamma_3 = 0$) FTS models. Table 1 shows the obtained results; our proposal getting higher forecasting accuracy.

Table 1. Mean forecast errors for different FTS models, with DAX data set.

	Yu [20]	Cheng et al. [3]	Wang et al. [18]	Rubio et al. [13]	Pattern Op.
	Model	Model	Model	Model	$\gamma_3 = 1$
RMSE	77.63	107.11	270.35	72.78	57.33
MAD	60.48	78.59	266.34	63.60	48.49
BIAS	−24.08	−46.76	−266.34	−33.82	−23.03

These results are only statistically significant (using pairwise comparison of the one-step ahead forecast errors) with respect to the forecasts obtained with the Wang et al. model, which has the particularity of using a fixed number of intervals in the partitioning of U (7 intervals of different length).

4.2   TAIEX Predictions

The second experiment deals with TAIEX daily quotation from January 2013 to August 2014. We have apply the same weekly rolling horizon scheme than in the first experiment, although now the training data set only contains 400 quotes. This scheme preserves the size of the training set, and the last 5 observed quotes are entered into the training data set for recalculating all the parameters of the weighted FTS model, before generating the next 5 one-step ahead predictions. Now, the process is repeated for 20 times, obtaining a total of 100 daily forecasts.

Figure 3 shows at the left-hand side the first 400 quotation, which corresponds to the initial training set, and at the right-hand side both the trace of the time series and the forecasts provided by the DSS procedure using the weekly rolling horizon scheme.

Following the DSS scheme for this data set, the pattern operator (using $\gamma_3 = 1$) works with runs of length 2. Finally, we have calculated the forecasting errors for the day of each week, the mean of their absolute values and the root of the mean of the squared errors and the bias of the errors. The procedure evaluates these means weekly for the 20 weeks (100 daily quotes), and finally the averaged RMSE, MAD and Bias are obtained, which are shown in Table 2.

Note that the differences among the weighted FTS procedures are not so important for this time series, that present a behavior with less fluctuations than the previous one.

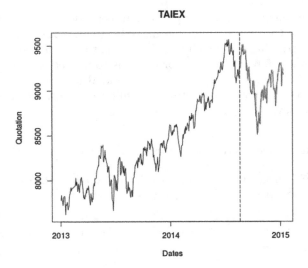

Fig. 3. Trace of the time series of daily quotes of Taiwanese TAIEX Index, from January 2013 to August 2014, and forecasts for the last 100 quotes.

Table 2. Mean forecast errors for different FTS models, with TAIEX data set.

	Yu [20] Model	Cheng et al. [3] Model	Wang et al. [18] Model	Rubio et al. [13] Model	Pattern Op. $\gamma_3 = 1$
RMSE	81.43	75.22	160.49	74.10	71.79
MAD	67.31	61.86	143.07	61.92	59.91
BIAS	2.94	1.40	6.85	5.05	7.95

5   Conclusions

In this paper we introduce a new fuzzy pattern operator and a DSS for the analysis of fuzzy time series. The strategies included in the decision tree are mainly based on the concept of jump between fuzzy sets in a FTS framework, allowing the definition of weighting operators that analyze the importance of the patterns of runs in the jump series.

The one-step ahead forecasts provided by our procedure are trapezoidal fuzzy numbers, and its application on series of daily quotes of stock market indices has provided promising results.

Future research could focus on the incorporation of more sophisticated rules for selecting the different operators that work in the DSS, which must be applied for providing more accurate predictions for several types of time series.

Acknowledgments. Research partially supported by the Ministerio de Economía y Competitividad, España (project MTM2014-56233-P, co-financed by FEDER funds).

References

1. Chen, S.: Forecasting enrollments based on fuzzy time series. Fuzzy Sets Syst. **81**(3), 311–319 (1996). doi:10.1016/0165-0114(95)00220-0
2. Chen, S., Hsu, C.: A new method to forecast enrollments using fuzzy time series. Int. J. Appl. Sci. Eng. **3**(2), 234–244 (2004)
3. Cheng, C.H., Chen, T.L., Teoh, H.J., Chiang, C.H.: Fuzzy time-series based on adaptive expectation model for taiex forecasting. Expert Syst. Appl. **34**(2), 1126–1132 (2008). doi:10.1016/j.eswa.2006.12.021
4. Cheng, S.H., Chen, S.M., Jian, W.S.: Fuzzy time series forecasting based on fuzzy logical relationships and similarity measures. Inf. Sci. **327**, 272–287 (2016). doi:10.1016/j.ins.2015.08.024
5. Dubois, D., Prade, H.: The mean value of a fuzzy number. Fuzzy Sets Syst. **24**, 279–300 (1987). doi:10.1016/0165-0114(87)90028-5
6. Duru, O., Bulut, E.: A non-linear clustering method for fuzzy time series: histogram damping partition under the optimized cluster paradox. Appl. Soft Comput. **24**, 742–748 (2014). doi:10.1016/j.asoc.2014.08.038
7. Fuller, R., Majlender, P.: On weighted possibilistic mean and variance of fuzzy numbers. Fuzzy Sets Syst. **136**, 363–374 (2003). doi:10.1016/S0165-0114(02)00216-6
8. Huarng, K.: Effective lengths of intervals to improve forecasting in fuzzy time series. Fuzzy Sets Syst. **123**(3), 387–394 (2001). doi:10.1016/S0165-0114(00)00057-9
9. Huarng, K.: Heuristic models of fuzzy time series for forecasting. Fuzzy Sets Syst. **123**, 369–386 (2001). doi:10.1016/S0165-0114(00)00093-2
10. Liu, H.T.: An improved fuzzy time series forecasting method using trapezoidal fuzzy numbers. Fuzzy Optim. Decis. Mak. **6**(1), 63–80 (2007). doi:10.1007/s10700-006-0025-9
11. Lu, W., Chen, X., Pedrycz, W., Liu, X., Yang, J.: Using interval information granules to improve forecasting in fuzzy time series. Int. J. Approx. Reason. **57**, 1–18 (2015). doi:10.1016/j.ijar.2014.11.002
12. Rubio, A., Bermúdez, J.D., Vercher, E.: Forecasting portfolio returns using weighted fuzzy time series methods. Int. J. Approx. Reason. **75**, 1–12 (2016). doi:10.1016/j.ijar.2016.03.007
13. Rubio, A., Bermúdez, J.D., Vercher, E.: Improving stock index forecasts by using a new weighted fuzzy-trend time series method. Expert Syst. Appl. **76**, 12–20 (2017). doi:10.1016/j.eswa.2017.01.049
14. Sadaei, H.J., Enayatifar, R., Lee, M.H., Mahmud, M.: A hybrid model based on differential fuzzy logic relationships and imperialist competitive algorithm for stock market forecasting. Appl. Soft Comput. **40**, 132–149 (2016). doi:10.1016/j.asoc.2015.11.026
15. Chen, S.-M., Manalu, G.M., Pan, J.-S., Liu, H.-C.: Fuzzy forecasting based on two-factors second-order fuzzy-trend logical relationship groups and particle swarm optimization techniques. IEEE Trans. Cybern. **43**(3), 1102–1117 (2015). doi:10.1109/TSMCB.2012.2223815
16. Singh, P., Borah, B.: An efficient time series forecasting model based on fuzzy time series. Eng. Appl. Artif. Intell. **26**(10), 2443–2457 (2013). doi:10.1016/j.engappai.2013.07.012

17. Song, Q., Chissom, B.S.: Fuzzy time series and its models. Fuzzy Sets Syst. **54**(3), 269–277 (1993). doi:10.1016/0165-0114(93)90372-O
18. Wang, L., Liu, X., Pedrycz, W.: Effective intervals determined by information granules to improve forecasting in fuzzy time series. Expert Syst. Appl. **40**(14), 5673–5679 (2013). doi:10.1016/j.eswa.2013.04.026
19. Ye, F., Zhang, L., Zhang, D., Fujita, H., Gong, Z.: A novel forecasting method based on multi-order fuzzy time series and technical analysis. Inf. Sci. **367–368**, 41–57 (2016). doi:10.1016/j.ins.2016.05.038
20. Yu, H.K.: Weighted fuzzy time series models for TAIEX forecasting. Physica A: Stat. Mech. Appl. **349**(3–4), 609–624 (2005). doi:10.1016/j.physa.2004.11.006

Scalable Forecasting Techniques Applied to Big Electricity Time Series

Antonio Galicia, José F. Torres, Francisco Martínez-Álvarez,
and Alicia Troncoso[✉]

Division of Computer Science, Universidad Pablo de Olavide, 41013 Seville, Spain
{agalde,jftormal}@alu.upo.es, {fmaralv,ali}@upo.es

Abstract. This paper presents different scalable methods to predict
time series of very long length such as time series with a high sampling
frequency. The Apache Spark framework for distributed computing is
proposed in order to achieve the scalability of the methods. Namely, the
existing MLlib machine learning library from Spark has been used. Since
MLlib does not support multivariate regression, the forecasting problem
has been split into h forecasting subproblems, where h is the number
of future values to predict. Then, representative forecasting methods of
different nature have been chosen such as models based on trees, two
ensembles techniques (gradient-boosted trees and random forests), and
a linear regression as a reference method. Finally, the methodology has
been tested on a real-world dataset from the Spanish electricity load data
with a ten-minute frequency.

Keywords: Big data · Scalable · Electricity time series · Forecasting

1 Introduction

It is known that advances in technology have meant that the amount of data
being generated and stored is increasing to the point that 90% of the data in
the world have been generated in the last years. The need to process this huge
amount of data has become essential for the evolution of the data mining tools
giving rise to the term big data. On the other hand, an essential component in
the nature of the big data is that they are commonly indexed over time, called
here big time series, and its prediction in future time periods can be extremely
important in diverse areas such as energy, traffic, pollution and so forth.

Nowadays, the main existing frameworks for processing big time series have
been developed by over the top tech companies like Google or Yahoo. Google developed the MapReduce technology [5], which divides input data for processing in
blocks and then integrates the output information of each block in a single solution.
Later, Yahoo developed Hadoop technology [22], an open code implementation
of the MapReduce paradigm, currently integrated with the Apache foundation.
The limitations of MapReduce in the implementation of algorithms, which iterate

© Springer International Publishing AG 2017
I. Rojas et al. (Eds.): IWANN 2017, Part II, LNCS 10306, pp. 165–175, 2017.
DOI: 10.1007/978-3-319-59147-6_15

over the data, have required the creation of new tools, such as Spark [9], developed by the University of Berkeley and also today in the Apache Foundation. Spark installed on a Hadoop distributed file system (HDFS) allows in-memory parallel data processing, achieving a much higher processing speed than Hadoop. Apache Spark is also an open source software project that allows the multi-pass computations, provides high-level operators, uses diverse languages (Java, Python, R) in addition to its own language called Scala, and finally, offers the machine learning library MLlib [8].

In this work, a collection of scalable algorithms are proposed in order to forecast big data time series. In particular, representative prediction methods of different nature have been chosen such as models based on trees, linear regression and two ensembles techniques (gradient-boosted trees and random forests). The algorithms have been developed in the framework Apache Spark under the Scala programming language by using the library MLlib. All the methods have been tested on a real-world big time series related to energy consumption.

The rest of the paper is structured as follows. Section 2 reviews of the existing literature related to the machine learning algorithms for big data. In Sect. 3 the proposed methodology to forecast big data time series is introduced. Section 4 presents the experimental results corresponding to the prediction of the energy consumption. Finally, Sect. 5 closes the paper giving some final conclusions.

2 Related Work

The prediction of future events has always fascinated humankind. Not in vain, many of these efforts can be seen in everyday activities, such as weather forecasting, the prediction of exchange rate fluctuations or of pollution.

The methods for time series forecasting can be roughly classified as follows: classical Box and Jenkins-based methods such as ARMA, ARIMA, ARCH or GARCH [1] and data mining techniques (the reader is referred to [12] for a taxonomy of these techniques applied to energy time series forecasting). However, the majority of the data mining techniques cannot be applied when big data have to be processed due to the high computational cost. Therefore, big data mining techniques [21,24] are being developed for distributed computing in order to solve typical tasks as clustering, classification or regression. A brief description of the main advances is made below.

Increased attention has been paid to big data clustering in recent years [11,15]. A survey on this topic can be found in [7]. Specifically, several approaches have been recently proposed to apply clustering to big data time series. Namely, in [6] the authors propose a new clustering algorithm based on a previous clustering of a sample of the input data. The dynamic time warping was tested to measure the similarity between big time series in [16]. In [23] a data processing based on MapReduce was used to obtain clusters. A distributed method for the initialization of the k-means is proposed in [3].

Regarding classification tasks, several MapReduce-based approaches in big data scenarios have been recently provided. A MapReduce-based framework focused on several instance reduction methods is proposed in [20] to reduce the computational cost and storage requirements of the k Nearest Neighbors (kNN) classification algorithm. Also, several parallel implementations of the kNN algorithm based on Spark have been proposed in the literature [17,19]. Support vector machines (SVM) were recently adapted to the field of high performance computing giving rise to parallel SVMs [4].

In the regression field, there is still much research to be conducted, especially considering that very few works have been published. For instance, the ensemble techniques based on trees have been the most studied topic in the literature due to its easy adaptation to a distributed computing framework. Random forests have been applied to some particular problems showing a good performance for high-dimensional data [10]. On the other hand, regression trees have been built by parallel learning based on MapReduce on computer clusters in [14]. However, these methods based on a distributed computing have not used for big time series forecasting in to the best of authors' knowledge, and therefore, this work aims at filling this gap.

3 Methodology

This section describes the methodology proposed in order to forecast big data time series by using the MLlib library.

Given a time series recorded in the past up to the time t, $[x_1,...,x_t]$, the problem consists in predicting the h next values for the time series from a historical windows composed of w-values (h is known as the prediction horizon). This can be formulated as:

$$[x_{t+1}, x_{t+2}, \ldots, x_{t+h}] = f(x_t, x_{t-1}, \ldots, x_{t-(w-1)}) \tag{1}$$

where f is the model to be found by the forecasting method in the training phase.

Nevertheless, the existing regression techniques in MLlib do not support the multivariate regression, that is, the multi-step forecasting. Therefore, the first stage splits the problem into h forecasting subproblems as follows:

$$x_{t+1} = f_1(x_t, x_{t-1}, \ldots, x_{t-(w-1)})$$
$$x_{t+2} = f_2(x_t, x_{t-1}, \ldots, x_{t-(w-1)})$$
$$\ldots$$
$$x_{t+h} = f_h(x_t, x_{t-1}, \ldots, x_{t-(w-1)}) \tag{2}$$

The existing possible relations between the h consecutive values $x_{t+1}, ..., x_{t+h}$ are missed with this formulation. However, if the prediction of previous values is used to predict the next values a greater error is obtained, as the errors are accumulated in the last time stamps of the prediction horizon. Additionally, to obtain h models $f_1, ..., f_h$ to predict h values has a greater computational cost than the building of a just model f to predict all the values.

The next stage consists in solving each forecasting subproblem in the Spark distributed computing framework by using the regression methods of the MLlib library. The main variable in Apache Spark is the Resilient Distributed Dataset (RDD), which is an immutable and partitioned collection of elements that can be operated in a distributed way. Thus, every RDD created is split in blocks of the same size approximately across the nodes that integrate the cluster, as it is shown in Fig. 1.

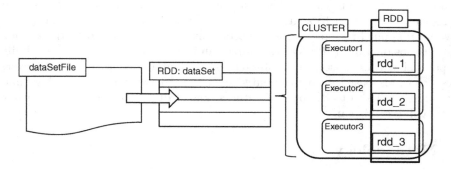

Fig. 1. A RDD variable in a spark cluster.

Once the dataset has been distributed, the MLlib algorithms firstly obtain a model from each worker node, and later, aggregate the predictions obtained for each model in a stage called reducer. It is important to highlight that RDD variables do not preserve the order, and therefore, all instances have to be indexed to deal with time series by using MLlib. An illustration of the methodology is presented in Fig. 2.

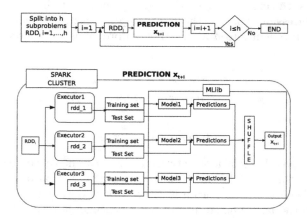

Fig. 2. Illustration of the proposed methodology.

Regression methods from MLlib have been selected according to cover different paradigms such as linear models, models based on trees and, finally, techniques ensembles.

The models based on trees have been mainly proposed because interpretable results are always desirable for the end-user. Furthermore, the ensemble techniques usually improve the results obtained by a single regressor in addition to obtain very good results for many real applications. Finally, a linear model has been selected as a state-of-the-art reference method. A brief description of the methods used for each paradigm is made below.

Within the models based on trees, a greedy algorithm [18] that performs a recursive binary partitioning of the feature space in order to build a decision tree has been used. The tree predicts the same value for all instances that reach the same leaf node. The root nodes are selected from a set of possible splits, but no from all attributes, by maximizing the information gain. In this approach, the possible split candidates are a quantile over the block of the data, which is being processed by a certain worker machine in the cluster. Moreover, once the splits are ordered, a maximum number of bins is allowed.

Two ensemble of decision trees have been considered: random forests [2] and the gradient-boosted trees (GBTs) [13]. Both algorithms learn ensembles of trees, but the training processes are very different. GBTs train one tree at a time, being the longer training than random forests, which can train multiple trees in parallel. Random forests improves the performance when the number of trees increases, however, GBTs can present overfitting if the number of trees grows too large.

Random forests is an ensemble of decision trees trained separately in the same way as detailed above for individual decision trees. The trees generated are different because of different training sets from a bootstrap subsampling and different random subsets of features to split on at each tree node are used. To make a prediction on a new instance, a random forest makes the average of the predictions from its set of decision trees.

GBTs iteratively train a sequence of decision trees. On each iteration, the algorithm uses the current ensemble to predict the label of each training instance and then compares the prediction with the true label by computing the mean square error. The training instances with poor predictions are re-labeled, and therefore, in the next iteration, the decision tree will help correct for previous mistakes.

Finally, a linear regression has been selected as linear model. The well-known stochastic gradient descent method has been used to minimize the mean square error for the training set in order to obtain the model.

4   Results

This section presents the results obtained from the application of the proposed methodology to electricity consumption big data time series to predict the 24 next values, that is, the forecast horizon set to $h = 24$ (4 h). Hence, Sect. 4.1

describes the used dataset. The experimental setup carried out is detailed in Sect. 4.2. Finally, the results are discussed in Sect. 4.3.

4.1   Datasets Description

The time series used is related to the electrical energy consumption, which ranges from January 1st 2007 at 00:00 am to June 21st 2016 at 23:40 am. The consumption is measured every ten minutes during this period. This makes a time series with a total length of 497832 measurements, which have been split into 298608 samples for the training set corresponding to the period from January 1st, 2007 at 00:00 am to September 8th 2012 at 10:30 am and 199080 samples for the test set corresponding to the period from September 8th 2012 at 10:40 am to June 21st 2016 at 11:40 pm.

4.2   Design of Experiments

The experimental setting of the algorithms is as follows:

1. The number of past values used to predict the 24 next values has been set to 144 (window $w = 144$), which represents all the values for a whole day.
2. In the linear regression, the stochastic gradient descent method requires an adequate number of iterations and rate of learning in order to guarantee the convergence of the optimization technique. In this work, values of $1.0E - 10$ for the rate and 100 for the iterations have shown to be suitable.
3. The number of trees and the maximum depth are the main inputs for random forests and GBTs. Different depth levels have been tested for both ensembles, namely, four and eight. A number of five trees has been set for GBTs and values of 50, 75, 100, 125 and 150 trees for random forests.

The experimentation has been launched on a cluster, which is composed of three nodes: the master and two slaves nodes. Each node has two Intel Xeon E7-5820K processors at 3.3 GHz, 15 MB cache, 6 cores per processor and 16 GB of main memory working under Linux Ubuntu. The cluster works with Apache Spark 2.0.2 and Hadoop 2.6.

Finally, the well-known mean relative error (MRE) measure has been selected to assess the accuracy of the predictions. Its formula is:

$$MRE = \frac{1}{N} \sum_{i=1}^{N} \frac{|\widehat{x}_i - x_i|}{x_i} \tag{3}$$

where $\widehat{x}_i$ stands for the predicted values and x_i for the actual consumption values.

4.3   Electricity Consumption Big Data Time Series Forecasting

Table 1 summarizes the MRE obtained by all methods based on trees when predicting the test set. A study of how the number of trees has an influence on the error is made for the random forests ensemble. In addition, the depth of the trees used for all methods has been analyzed. It can be seen that a greater accuracy is provided when the depth of the trees increases due to trees more specific are obtained. By contrast, it seems that the number of trees to be used by the random forest has not a high impact over the error, and therefore, fifty trees was a sufficient number to obtain a good performance of the method.

Table 1. MRE for different depth levels and number of trees.

	Decision tree	Random forests					GBTs
Number of trees	1	50	75	100	125	150	5
Depth 4	5.1516	4.2823	4.2583	4.2415	4.2415	4.2427	4.3402
Depth 8	2.8783	2.2005	2.1853	2.1842	2.1810	2.1773	2.7190

Table 2 shows the MRE for the methods based on trees when a depth of 8 and a number of 50 trees for random forests has been used. Additionally, it shows the MRE obtained by means of a linear regression as baseline method to establish a benchmarking. All non linear methods based on trees achieved better errors than the linear regression, namely a difference of 5% approximately. Although the best results are obtained by the random forests ensemble technique, it can be concluded that the decision tree is the more adequate method in terms of accuracy and CPU time to predict big data time series.

Table 2. MRE for the test set and CPU time for training.

	MRE (%)	Time (seconds)
Linear regression	7.3395	553
Decision tree	2.8783	81
Random forests	2.2005	277
GBTs	2.7190	417

Figures 3 and 4 present the predicted values along with the actual values for the random forest algorithm for the two days from the test set leading to the largest and smallest errors, respectively. The worst prediction corresponds to an error of 9.12% associated to the period from December 24th 2013 at 10:50 am to December 25th 2013 at 10:40 am and the error of the best prediction is 0.67% corresponding to the day from September 20th 2012 at 10:40 am to September 21st 2012 at 10:30 am. It can be noted that the worst day is a special day, namely, Christmas Eve.

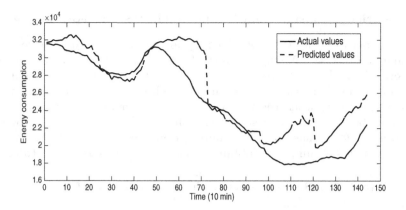

Fig. 3. The day corresponding to the worst prediction when using random forests.

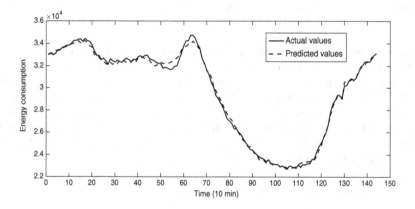

Fig. 4. The day corresponding to the best prediction when using random forests.

Finally, the training time versus the length of the time series for all algorithms proposed here are shown in the Fig. 5. The execution time has been obtained with time series of two, four, eight, sixteen and thirty and two times the length of the original time series. It is necessary to highlight the building of the dataset from the time series for each subproblem is not included in the training time as that is not made in a distributed way, but in an iterative way. From this figure, it can be observed that the most scalable method is the decision tree.

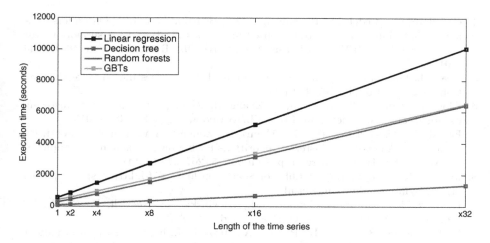

Fig. 5. Runtime and scalability for all algorithms.

5   Conclusions

In this work, a new formulation has been proposed for multi-step forecasting problems in order to be able to use the MLlib library from Apache Spark framework. The use of this library guarantees that the methods applied to predict the energy consumption for the next twenty four values are scalable, and therefore, they can be used for big data time series. A pool of linear and non linear methods have been selected, e.g., methods based on trees, ensemble techniques based on trees and a linear regression. Results for the Spanish electricity demand time series have been reported, showing the good performance of the methods proposed here and the grade of scalability for each of them.

Future work is directed towards solving the forecasting subproblems in a distributed way by using technology based on multithreads.

Acknowledgments. The authors would like to thank the Spanish Ministry of Economy and Competitiveness and Junta de Andalucía for the support under projects TIN2014-55894-C2-R and P12-TIC-1728, respectively.

References

1. Box, G., Jenkins, G.: Time Series Analysis: Forecasting and Control. Wiley, New York (2008)
2. Breiman, L.: Random forests. Mach. Learn. **45**(1), 5–32 (2001)
3. Capó, M., Pérez, A., Lozano, J.A.: A Recursive k-means initialization algorithm for massive data. In: Proceedings of the Spanish Association for Artificial Intelligence, pp. 929–938 (2015)

4. Cavallaro, G., Riedel, M., Richerzhagen, M., Benediktsson, J.A.: On understanding big data impacts in remotely sensed image classification using support vector machine methods. IEEE J. Sel. Top. Appl. Earth Obs. Remote Sens. **8**, 4634–4646 (2015)
5. Dean, J., Ghemawat, S.: Mapreduce: simplified data processing on large clusters. Commun. ACM **51**(1), 107–113 (2008)
6. Ding, R., Wang, Q., Dan, Y., Fu, Q., Zhang, H., Zhang, D.: Yading: fast clustering of large-scale time series data. Proc. VLDB Endow. **8**(5), 473–484 (2015)
7. Fahad, A., Alshatri, N., Tari, Z., Alamri, A., Zomaya, A.Y., Khalil, I., Sebti, F., Bouras, A.: A survey of clustering algorithms for big data: taxonomy & empirical analysis. IEEE Trans. Emerg. Top. Comput. **5**, 267–279 (2014)
8. Machine Learning Library (MLlib) for Spark. On-line (2016). http://spark.apache.org/docs/latest/mllib-guide.html
9. Hamstra, M., Karau, H., Zaharia, M., Knwinski, A., Wendell, P., Spark, L.: Lightning-Fast Big Analytics. O' Really Media, USA (2015)
10. Li, L., Bagheri, S., Goote, H., Hassan, A., Hazard, G., Risk adjustment of patient expenditures: a big data analytics approach. In: Proceedings of the IEEE International Conference on Big Data, pp. 12–14 (2013)
11. Luna-Romera, J.M., Martínez-Ballesteros, M., García-Gutiérrez, J., Riquelme-Santos, J.C.: An approach to Silhouette and Dunn clustering indices applied to big data in spark. In: Luaces, O., Gámez, J.A., Barrenechea, E., Troncoso, A., Galar, M., Quintián, H., Corchado, E. (eds.) CAEPIA 2016. LNCS, vol. 9868, pp. 160–169. Springer, Cham (2016). doi:10.1007/978-3-319-44636-3_15
12. Martínez-Álvarez, F., Troncoso, A., Asencio-Cortés, G., Riquelme, J.C.: A survey on data mining techniques applied to electricity-related time series forecasting. Energies **8**(11), 13162–13193 (2015)
13. Mason, L., Baxter, J., Bartlett, P., Frean, M.: Boosting algorithms as gradient descent. In: Proceedings of the Neural Information Processing Systems Conference, NIPS, pp. 512–518 (1999)
14. Panda, B., Herbach, J.S., Basu, S., Bayardo, R.J.: PLANET: massively parallel learning of tree ensembles with mapreduce. In: Proceedings of the Very Large Databases, pp. 1426–1437 (2009)
15. Perez-Chacon, R., Talavera-Llames, R.L., Martinez-Alvarez, F., Troncoso, A.: Finding electric energy consumption patterns in big time series data. In: Omatu, S. (ed.) Proceedings of the International Conference on Distributed Computing and Artificial Intelligence. Advances in Intelligent Systems and Computing, vol. 474. Springer, Cham (1991)
16. Rakthanmanon, T., Campana, B., Mueen, A., Batista, G., Westover, B., Zhu, Q., Zakaria, J., Keogh, E.: Addressing big data time series: mining trillions of time series subsequences under dynamic time warping. ACM Trans. Knowl. Discov. Data **7**(3), 267–279 (2014)
17. Reyes-Ortiz, J.L., Oneto, L., Anguita, D.: Big data analytics in the cloud: spark on Hadoop vs MPI/OpenMP on Beowulf. Procedia Comput. Sci. **53**, 121–130 (2015)
18. Rokach, L., Maimon, O.: Top-down induction of decision trees classifiers - a survey. IEEE Trans. Syst. Man Cybern. Part C **35**(4), 476–487 (2005)
19. Talavera-Llames, R.L., Pérez-Chacón, R., Martínez-Ballesteros, M., Troncoso, A., Martínez-Álvarez, F.: A nearest neighbours-based algorithm for big time series data forecasting. In: Martínez-Álvarez, F., Troncoso, A., Quintián, H., Corchado, E. (eds.) HAIS 2016. LNCS, vol. 9648, pp. 174–185. Springer, Cham (2016). doi:10.1007/978-3-319-32034-2_15

20. Triguero, I., Peralta, D., Bacardit, J., García, S., Herrera, F.: MRPR: a mapreduce solution for prototype reduction in big data classification. Neurocomputing **150**, 331–345 (2015)
21. Tsai, C.-W., Lai, C.-F., Chao, H.-C., Vasilakos, A.: Big data analytics: a survey. J. Big Data **2**(1), 21 (2015)
22. White, T.: Hadoop, The Definitive Guide. O' Really Media, USA (2012)
23. Zhao, W., Ma, H., He, Q.: Parallel k-means clustering based on mapreduce. In: Jaatun, M.G., Zhao, G., Rong, C. (eds.) Cloud Computing. LNCS, vol. 5391, pp. 674–679. Springer, Heidelberg (2009). doi:10.1007/978-3-540-95885-7_24
24. Zhou, L., Pan, S., Wang, J., Vasilakos, A.V.: Machine learning on big data: opportunities and challenges. Neurocomputing **237**, 350–361 (2017)

Forecasting Financial Time Series with Multiple Kernel Learning

Luis Fábregues, Argimiro Arratia, and Lluís A. Belanche$^{(\boxtimes)}$

Department of Computer Science, Universitat Politècnica de Catalunya,
Jordi Girona, 1-3, 08034 Barcelona, Spain
luis.fabregues@est.fib.upc.edu, {argimiro,belanche}@cs.upc.edu

Abstract. This paper introduces a forecasting procedure based on multivariate dynamic kernels to re-examine –under a non linear framework– the experimental tests reported by Welch and Goyal showing that several variables proposed in the academic literature are of no use to predict the equity premium under linear regressions. For this approach kernel functions for time series are used with multiple kernel learning in order to represent the relative importance of each of these variables.

Keywords: Forecasting · Support vector classification · Financial time series · Multiple Kernel Learning · Time series kernels

1 Introduction

There is a long history of attempts to predict stock market returns by specifying some regression on lagged predictor variables independent of the stock market returns. Shiller [20], Campbell and Shiller [5], Cochrane [9], among others, have studied the forecasting of future excess returns using the dividend price ratio as predictor. Other popular predictor variables explored in the literature are the dividend yield, earnings price ratio, dividend-to-earnings ratio, volatility, interest rates, exchange rates, consumption indices and inflation rates (see, e.g., [11,15–18] for a general discussion). The list of valuation ratios sought of as forecasters of expected excess returns is much longer and show *"... a pervasive pattern of predictability across markets wherein the cashflow or price change one may have expected is not what is forecast."* [11]. In view of this and further evidence showing the spurious nature of predictor models (mostly linear regressions on the aforementioned valuation ratios), several authors have conducted extensive studies on the forecasting performance of various economic variables and different models (to mention a few, e.g., [2,6,10,21]). The work by Welch and Goyal [21] is of particular interest since the authors do a comprehensive revision of the empirical performance of the most widely accepted variables as predictors of equity premium, under *linear* regression models, and conclude that these models have poor predictive capacity both in-sample and out-of-sample.

Supported by MINECO project APCOM (TIN2014-57226-P) and *Generalitat de Catalunya* 2014 SGR 890 (MACDA).

© Springer International Publishing AG 2017
I. Rojas et al. (Eds.): IWANN 2017, Part II, LNCS 10306, pp. 176–187, 2017.
DOI: 10.1007/978-3-319-59147-6_16

In this work we extend the stock return predictability test of Goyal and Welch to non-linear (and semi-parametric) classification models on the different valuation ratios. We are thus considering the possible non-linear relationship between the stock returns and the predictor variables and expanding the framework of predictability from linear to more complex models. The models we consider come from statistical learning adapted to the objective of time series forecasting. We use ν-Support Vector Machines for classification [7] with dynamic kernel functions able to analyze multivariate temporal structures, and Multiple Kernel Learning [3] to integrate different financial information and different kernels.

2   Multivariate Dynamic Kernels for Time Series

Kernels are two-place symmetric functions that evaluate an inner product of the arguments in some feature space, thereby inducing an implicit mapping that creates an image of the input into the desired feature space. It is then possible to compare input data in a higher-dimensional space without the need of calculating the exact coordinates of the transformation. Kernel functions can be defined as $k(\boldsymbol{x}, \boldsymbol{z}) = \langle \phi(\boldsymbol{x}), \phi(\boldsymbol{z}) \rangle$, where $\boldsymbol{x}, \boldsymbol{z} \in X$ are input vectors. ϕ represents the mapping from the original feature space X into a new feature space F as $\phi : \boldsymbol{x} \to \phi(\boldsymbol{x}) \in F$. It is known that a function is a valid kernel function if and only if it induces positive semi-definite (p.s.d.) matrices $K = [k(\boldsymbol{x}_i, \boldsymbol{x}_j)]$. This property can be defined as $\boldsymbol{c}^\top K \boldsymbol{c} \geq 0$, for all $N \in \mathbb{N}, \boldsymbol{x}_1, ..., \boldsymbol{x}_N \in X$ and $\boldsymbol{c} \in \mathbb{R}^N$.

Kernels for time series can be constructed using two approaches: structural similarity and model similarity. Structural similarity employs methods to find an alignment of the data that makes possible the comparison between series. Model similarity changes the structure of the data by constructing a higher level representation thereof and the comparison is performed using this new representation.

2.1   Vector Auto-Regression Kernel

The Vector Auto-Regression (VAR) model relates the data at the observation at point $x(t)$ with a linear combination of lagged values of the observation. In order to fit a model, a lag parameter is provided, which defines how many time steps the function will be looking at in the past to assess the linear combination parameters. The formulation is as follows: $x(t) = \sum_{l=1}^{L} A_l x(t-l) + b + \varepsilon_t$, where $x(t)$ is the sample at time t, L is the number of lags of the model, A is the transition matrix (a square matrix with as many dimensions as features the data has), b is the intercept (a vector of the dimension equal to the number of features) and ε_t is the Gaussian noise at time t.

The VAR function can be used to build a model similarity kernel. In order to compare VAR models one considers the binding of transition matrices and the intercept vector as an additional column. This results in $\hat{B} = \left[A_1 | A_2 | ... | A_L | [b] \right]$.

The distance between series s_1 and s_2 is the Frobenius norm of the difference between $\hat{B}_{s_1}$ and $\hat{B}_{s_2}$:

$$\text{FD}(s_1, s_2) = \sqrt{Trace\left\{(\hat{B}_{s_1} - \hat{B}_{s_2})(\hat{B}_{s_1} - \hat{B}_{s_2})^\top\right\}}. \tag{1}$$

Once this Frobenius distance is calculated, the distance can be transformed into a valid kernel using the Radial Basis Function (RBF) transform:

$$k_{\text{VAR}} = \exp\left\{\frac{-FD(s_1, s_2)}{2\sigma}\right\} \tag{2}$$

The parameters of this kernel methodology are the number of lags L and σ. The value of L will be fixed to 5, whereas σ will be set to the median Frobenius distance between the time series being compared. Both parameters are set following [13] and our own previous experience [19].

2.2   Global Alignment Kernel

Global Alignment(GA) [12] is a generalization of a family of distance and similarities called Dynamic Time Warping (DTW), the goal of which is to measure the distance between two series. In order to do so, both series should be aligned. The core of the problem is how to determine the best alignment between the two series and, using that alignment, to measure their similarity.

An *alignment* in DTW is represented by a set of relationships between a point in the series and another point of the same series (or the other one). Considering s_1 and s_2 as two time series, those relationships are the following: $s_1(t)$ with $s_2(t)$ denoted by $\rightarrow$, $s_1(t)$ with $s_1(t+1)$ denoted by $\uparrow$ and $s_1(t)$ with $s_2(t+1)$ denoted by $\nearrow$. The relationships are represented as two integer vectors π_1, π_2 of the same length with binary increases. The length of these vectors is always equal or less to the length of the smallest series. Each relationship can be represented on those vectors as follows: $(0, 1)$ for $\rightarrow$, $(1, 0)$ for $\uparrow$ and $(1, 1)$ for $\nearrow$. Intuitively, each vector $\pi_1(t)$ indicates an element of s_1 that forms a relationship with the element $\pi_2(t)$ of s_2. For the sake of simplicity the two vectors that represent the alignment will be denoted as π. Those alignments, by definition, only consider values of zero or one lag in both series.

After obtaining a satisfying alignment, the distance between the series can be obtained as: $D_\pi = \sum_{i=1}^{|\pi|} d(x_{\pi_1(i)}, y_{\pi_2(i)})$, where the distance function d can be any metric, most often the Euclidean distance. The presented algorithm is capable of finding more than one alignment. The selected alignment will be the one that minimizes the distance between the series. The formulation of the final MDTW distance is:

$$\text{MDTW}(s_1, s_2) = \frac{1}{|\pi^*|} \min_{\pi \in A(s_1, s_2)} D_\pi(s_1, s_2), \tag{3}$$

where $|\pi^*|$ is the length of the alignment with less distance and A is the set of all possible alignments.

This distance measurement does not fulfill the p.s.d.-ness requirement to form a kernel, even after applying the RBF transformation. For this reason the GA generalization is applied, which delivers correct kernels and enables the creation of a structural similarity kernel. This follows the same computational steps as DTW; however, instead of selecting the alignment with minimum distance, it considers all the alignments. This makes kernels defined by this metric p.s.d. under mild assumptions, based in the notion that all alignments provide information about the similarities between both series.

The formulation of this kernel function can be expressed using several distance metrics. The following formula is the one used in the context of this work:

$$k_{\mathrm{GA}}(s_1, s_2) = \sum_{\pi \in A(s_1, s_2)} \exp(-D_\pi(s_1, s_2)) \tag{4}$$

An improvement over GA was introduced under the name of Fast Global Alignment or Triangular Global Alignment [14]. This improved version aims at reducing the computational time of the procedure; this is accomplished by using an extra parameter T that restricts the number of alignments taken into account during the final calculation of k_{GA}. In particular, lower values of T make the kernel function use alignments close to the diagonal. Increasing the value of T increases the range of alignments that are taken into account.

2.3   Multivariate Dynamic Euclidean Distance Kernel

In the same lines as Global Alignment Kernel, Multivariate Dynamic Euclidean Distance Kernel (MDED) is a structural similarity model that creates an alignment of data between two series of different size in order to be able to compute the distance measure. MDED opts for a much simpler approach, as it removes the first elements of the longest series until it matches the size of the shortest time series. Even if this alignment is potentially worse in most of the cases with respect to MDTW, MDED is computationally less expensive. The approach is also backed by financial theory: observations generated in later time stamps contains information from older ones. Having that $L_1 \leqslant L_2$ where L_1 and L_2 are the lengths of vectors s_1 and s_2 respectively, we define this alignment in the notation of DTW as $\pi_1 = [0, 1, 2, ..., L_1 - 1, L_1]$ and $\pi_2 = [L_2 - (L_1 - 1), L_2 - (L_1 - 2), ..., L_2 - 1, L_2]$. Using this alignment, the distance between the series is computed as in Eq. (5). Again, the metric employed is the Euclidean distance. In a similar line to the k_{VAR} calculation, it is necessary to calculate the RBF kernel using the defined dissimilarity measure in order to obtain a p.s.d. kernel:

$$k_{\mathrm{MDED}} = \exp\left\{ \frac{-D_\pi(s_1, s_2)}{2\sigma} \right\} \tag{5}$$

The σ parameter of this kernel function is estimated following the author's past experience [19], using the median of all D_π of the available data.

2.4 Multivariate Dynamic Arc-Cosine Kernel

Arc-Cosine kernels have several interesting properties, because their construction is related to neural networks with an infinite hidden layer [8]. A structural similarity model can be also built using this formulation. The kernel as used here depends on the angle between the series, given by $\theta = \cos^{-1}\left(\dfrac{s_1^\top s_2}{\|s_1\|\|s_2\|}\right)$.

The full formulation of this kernel function introduces a "degree" n, which regulates its complexity, as follows:

$$k_n(s_1, s_2) = \frac{1}{\pi}\|s_1\|^n\|s_2\|^n J_n(\theta) \tag{6}$$

where J_n is a family con the chosen degree and the current angle; in particular, $J_0(\theta) = \pi - \theta$ and $J_1(\theta) = \sin(\theta) + (\pi - \theta)\cos(\theta)$.

2.5 Multiple Kernel Learning

Multiple Kernel Learning (MKL) [3] aims at finding the best combination of kernels to solve a task. It is possible for a problem to have several kernel functions that cover different characteristics of the data, or different representations of the same data. Those procedures create different kernel matrices that can be used to train a predictive model. Although it is possible to combine the information of those matrices into a single combined kernel matrix, MKL makes possible to combine in the same predictive model information obtained using different techniques. The mathematical formulation of this process is the following:

$$k_\eta(\boldsymbol{x}_i, \boldsymbol{x}_j) = f\left(\left\{k_m(x_i^m, x_j^m)\right\}_{m=1}^{P}; \eta\right), \tag{7}$$

where k_η represents the combined kernel, f_η is the (linear or non-linear) combination function, k_m represents the m-th kernel function and η parametrizes the combination. In this paper, the MKL algorithm of choice is EasyMKL [1], which obtains the parameters η of the combination function using an optimization approach. Specifically, a max-min problem is solved involving the η parameters and the probability distribution γ of each class. After the η weights are obtained they are combined convexly (the optimization restrictions ensuring that the weights are positive and sum to one). The L_1 norm is used as a structural risk function to guide the process. As the base learner EasyMKL uses KOMD, a kernel classifier that performs direct optimization of the margin distribution.

3 A Non-linear Predictability Framework

As stated in the introduction, the objective of this work is to replicate the experiments performed by Welch and Goyal in [21] with kernel functions and using multiple kernel learning to weight each feature in order to determine its relative influence in the prediction of the equity premium. One notable difference is that this experiment will consider as output variable not the equity premium but its sign, or direction. A positive value indicates a good precondition to hold a share and a negative value indicates a proper time to sell the stock shares. This paradigm shift implies that the problem is no longer a regression one but one of classification.

The experiment uses the same data as in [21] for a partial set of the variables considered by the authors. It is a data set that comprises several financial variables measured monthly, quarterly and yearly in the range of years between 1871 and 2005, and compiled from several sources[1]. The considered variables are described below. The experimental process follows two phases: the evaluation of each combination of kernels using only endogenous variables and the comparison between exogenous variables using multiple kernel learning.

3.1   Considered Variables

Equity Premium. A representation of the stock market returns. It is calculated combining several variables from the original data $ep_t = \log((Index_t + D12_t)/Index_{t-1}) - \log(Rfree_t + 1)$ where $Rfree$ is the risk-free rate. This variable is theoretically the return rate of an investment with zero risk, however in practice it is obtained from the interest rate of a three month U.S. Treasury bill. The variables associated with $Index$ and $D12$ are the S&P 500 index and the dividends, respectively, and will be discussed below.

Stock Returns. The original problem uses S&P 500 index returns (denoted as SPX or simply *Index*), obtained from Center for Research in Security Press (CRSP) and the website of R. Shiller. This variable encapsulates the largest market capitalizations of public companies serving as an indicator of the U.S. economy and will be considered an endogenous variable.

Dividend Price Ratio. Both this variable and the next are dependent on the dividends ($D12$), which are a moving sum with a window of 12 months of the dividends paid on the S&P 500 index. The dividends data is obtained from the Shiller's website and the S&P Corporation. The formula for Dividend Price Ratio is $dp_t = \log(D12_t) - \log(Index_t)$.

Dividend Yield. Very similar to Dividend Price, but Dividend Yield considers past values of SPX $dy_t = \log(D12_t) - \log(Index_{t-1})$.

Earning Price. In a similar line to the variables created using dividends, earning price($E12$) is the moving sum of earnings from S&P 500 index in a window of 12 months. Part of this data is extracted from Shiller's website and the other part is the result from an interpolation process by the authors. The Earning Price is formulated as $ep = \log(E12_t) - \log(Index_t)$.

---

[1] Publicly available from http://www.hec.unil.ch/agoyal/.

Stock Variance. The sum of squared daily returns of SPX. This data was obtained by Welch and Goyal with the help of G. William Schwert and CRSP.

3.2 Experimental Methodology

The data has to be grouped in blocks of a given size as the kernel functions of VAR and GA need a sequence of events to extract their similarity. A simple approach is to group monthly data into yearly blocks, creating data structures that contain 12 months.

The experimental approach will be a moving window of a certain number of years. A predictive model will be validated and built with the size of the window. The output variable is the sign of the equity premium in the first month of the next year, so it becomes a classification problem. The input variables are calculated as stated in the previous section. It is included in the data frames lagged values of each feature in order to increase the information of each data entry and help the models form better temporal dependencies.

Our approach is then divided in two steps. The first is to determine which is the best kernel function or combination of them to perform the described predictive task. The second is to use the selected method to determine the relative importance of each exogenous variable using the weights of the trained Multiple Kernel Learning (MKL) model.

The first step is in accordance with the classical methodology to evaluate several models: perform a predictive task using the same data and compare the results using a predefined metric. Being a classification problem, the easiest way to compare models is to measure the accuracy of training, validation and testing. In this step, the different kernel functions considered will be used both individually and in conjunction using MKL. Each of these models will have its parameters ν, σ and λ individually tuned.

The implementation is able to use two types of validation: Out Of Sample (OOS) and Cross-Validation (CV). The inclusion of two validation procedures is due to the somewhat controversial use of CV for time series. By the nature of this approach the folds do not respect any particular order (*e.g.*, the validation part could be the oldest, using a model trained with future data). This fact often prevents the use of CV in time-dependent datasets, as it is considered 'unrealistic'. However, as it has been recently shown, CV *can* be used with stationary time series in which the predictor values are lagged versions of the response value [4].

In the case of the single kernel functions, the creation and evaluation of the model is straightforward: a support vector machine is fitted using the training data and tested or validated with the rest of the data. In the case of Multiple Kernel Learning, all kernel functions introduced in Sect. 2 will create a kernel matrix that will be used in the MKL procedure. However, the implementations of EasyMKL do not admit any parameter, reducing the adaptability of the model. To correct this, the training procedure will fit an EasyMKL model, obtain the weights, build the combined matrix of kernels and then train a standard ν-SVM.

The second step tries to determine the relative importance of each exogenous variable by using the MKL weights. To this end, one data frame for each exogenous variable is created, each containing the exogenous variable and SPX, with different number of lags. In the list of data frames for the MKL are also included the matrix with the endogenous variable, SPX, and a data frame with all the variables. Each data frame will serve as training and validation data for the methodology selected in the first step, creating a kernel matrix that will be included in the MKL list. At the end, the performance of the models can be measured using their accuracy and the distribution of MKL weights.

Using this representation, the input to each experiment will be a list containing six matrices encapsulating the following features sets:

1. S&P 500 Index;
2. S&P 500 Index and Dividend-Price Ratio;
3. S&P 500 Index and Earning Price;
4. S&P 500 Index and Dividend Yield;
5. S&P 500 Index and Stock Variance;
6. All the features

The experiments will be considered including lagged versions of these features as additional information for the modeling process.

3.3   Performance Metrics

The performance metric that was selected for this task is the accuracy, defined as the number of correctly classified samples over the total number of samples. Three different accuracies will be taken into account in the results: train accuracy, validation accuracy and test accuracy. A second metric is given by the weights of the multiple kernel learning model. As stated in the definition of the work, the weights of a multiple kernel learning can be helpful to determine the relative importance of each kernel. This metric will help to determine which kernel function creates the best models from the data and which kernel matrix, constructed with the different variables, has a higher impact on the prediction of the results.

3.4   Empirical Results

Table 1 shows the evaluation of the different kernel methods and multiple kernel learning using EasyMKL. These results are obtained using the same data for each method but adjusting the parameters individually. All the methods are also compared using out of sample validation and cross-validation.

The results of this table show several interesting points. All the test accuracies fall within the range from 55% to 70%, which indicates the general capacity of Multivariate Dynamic kernels for this task. Individual kernels share similar test accuracies, around the 64%, including very simple kernels like k_{MDED} and k_{MDARC0}. It is also worth to mention that the kernels based on the arc-cosine

184   L. Fábregues et al.

Table 1. Results of kernel combinations.

	Out of sample validation						
	k_{VAR}	k_{GA}	k_{MDED}	k_{MDARC0}	k_{MDARC1}	MKL	MKL norm
Train Acc.	0.588	0.739	0.722	0.674	0.734	0.982	0.777
Validation Acc.	0.872	0.859	0.74	0.491	0.452	0.529	0.763
Test Acc.	0.631	0.64	0.64	0.64	0.658	0.64	0.694
	Cross-validation						
	k_{VAR}	k_{GA}	k_{MDED}	k_{MDARC0}	k_{MDARC1}	MKL	MKL norm
Train Acc.	0.61	0.762	0.742	0.675	0.729	1.000	0.898
Validation Acc.	0.751	0.691	0.674	0.559	0.445	0.417	0.568
Test Acc.	0.568	0.649	0.631	0.640	0.640	0.667	0.658

kernel perform as well if not better than other more common kernel functions. In particular, k_{MDARC1} is the best performing individual kernel, surpassing Global Alignment in the version that uses out-of-sample validation.

By a considerable margin, the best performing method is the combination of kernels created with EasyMKL, surpassing all the individual kernels. This technique tends to over-fit, as it can be observed in the difference of accuracy between training, validation and testing. It is most prominent in the case of cross-validation, were training accuracy is much higher than test and validation accuracies. The normalization also has interesting repercussions on the results: it reduces training accuracy and increases validation accuracy, reducing over-fitting. The results of the normalization technique seem to differ depending of the validation technique employed.

Finally, it is also worth to comment the application of the different validation techniques on the results. There is no clear pattern to determine if cross-validation is better or worse than out-of-sample since all kernels react different. In the cases of k_{GA} and not-normalized MKL the results improve, but the rest of kernel functions have the same or worst tested accuracy. The results clearly differ with [4], specially in the case of k_{VAR}.

What follows is a table containing the resulting weights of the process of EasyMKL. It can be an interesting source of information to determine how the algorithm determines which kernels are relatively more important.

From the results it can be observed that k_{MDARC1} usually is the kernel with most weight. This further supports the fact that this kernel function is possibly the best performing one for the problem. However, in the case of MKL with out-of-sample validation and normalization, k_{GA} is the kernel with highest weight and this combination is also the one with higher test accuracy. k_{MDARC0} also receives weights higher than the mean, signaling that it is also important in the construction of the model and adds additional information. Finally, it is worth to comment the impact of the normalization in the weights: in all the cases it decreases the weights of k_{VAR} and k_{MDED}, the worst performing kernel functions (Table 2).

Table 2. Weights of the different MKL procedures.

Method	Kernel weight				
	k_{VAR}	k_{GA}	k_{MDED}	k_{MDARC0}	k_{MDARC1}
MKL OOS	0.135	0.182	0.121	0.134	0.429
MKL OOS norm	0.128	0.339	0.110	0.159	0.264
MKL CV	0.125	0.163	0.107	0.158	0.449
MKL CV norm	0.095	0.246	0.093	0.200	0.366

The following table contains the experiments performed with exogenous variables and their results. All the experiments are executed using the best performing kernel method for the data, normalized multiple kernel learning using out-of-sample validation technique. Each variable is contained in a data frame that will be transformed into a kernel matrix using the MKL procedure. This creates four data frames (one for each exogenous variable) plus one containing only the endogenous variable (the S&P 500 index) and an additional data frame containing all the variables. In order to extract more information from these variables, four different methods of building the data frames were tested. DF1 only considers the exogenous variables without lags and SPX with four lags is included in each data frame. DF2 adds to the information of DF1 each exogenous variable with a lag of 3; this is motivated by the fact that the resulting data frame will contain more information but without adding too much redundancy. DF3 includes four lags of each exogenous variable. DF4 contains the same information as DF3 for the exogenous variables, but only includes the SPX without lags (Table 3).

Table 3. Tests with exogenous variables.

Accuracy				Weights					
	Train Acc.	Validation Acc.	Test Acc.		SPX	DP	EP	DY	SV	All
DF1	0.767	0.757	0.660	DF1	0.170	0.167	0.167	0.167	0.166	0.162
DF2	0.768	0.757	0.660	DF2	0.171	0.166	0.167	0.166	0.168	0.162
DF3	0.759	0.744	0.680	DF3	0.197	0.140	0.141	0.140	0.194	0.188
DF4	0.744	0.787	0.649	DF4	0.141	0.176	0.181	0.176	0.190	0.136

The best performing method is DF3 in terms of test accuracy. DF1 and DF2 are fairly similar, indicating that the inclusion of the third lag does not affect too much the model. DF4 is the worst performing one, which can mean that the model heavily rely on the lags of SPX to predict the output. The weights of DF1 and DF2 are near the mean, with slightly higher weight for the endogenous variable. This result does not mean that the rest of variables are not important to predict the result (that would be represented by weights near zero) but that they are mostly equally important. DF3 shows a shift in the weights, as SV (Stock Variance) is near SPX. The data frame containing all the variables also increases in weight, indicating a better predictive capability of all the variables by including all their lags until 4. Further experimental research shows that the fourth lag of the stock variance provides relevant information that helps the predictive process. Finally, DF4 has its weights shifted towards the exogenous variables. As those data frames lack of lags of SPX it is possible that the information contained in the exogenous variables takes a more active role in the prediction procedure, however the model performs worse in comparison to the rest. It is also worth noting that stock variance is still one of the most relevant variables in this case.

4   Conclusions

In this article, a non-linear approach was applied to the Welch and Goyal experiments [21] to measure the influence of several economic indicators in the prediction of the equity premium. The experiment is designed around several kernel functions for time series that aimed to extract relevant information from these variables and create a predictive model. The experimental procedure was based on selecting the best kernel or combination thereof for this problem, applying the selected model to each of the variables creating several kernel matrices and then obtaining the multiple kernel learning weights of each matrix. These weights indicate the relative importance of each variable.

The results indicate that, in this experimental procedure, the exogenous variables have a relative importance comparable with the S&P 500 index. However, the predictive capabilities of the model are not improved upon the introduction of these values and the weights are not consistent across the different experiments. This instability was previously reported in Welch and Goyal's work and it is now confirmed in our work.

As future works, the influence of each individual variable and the number of lags included could be further explored to find deeper relationships. More and different kernels can also be introduced to enrich the descriptive capabilities of the multiple kernel learning models. In this sense, a finer tuning of the parameters of each kernel should be performed to extract the most of the approach.

References

1. Aiolli, F., Donini, M.: EasyMKL: a scalable multiple kernel learning algorithm. Neuro Comput. **169**, 215–224 (2015)
2. Ang, A., Bekaert, G.: Stock return predictability: is it there? Rev. Financ. Stud. **20**(3), 651–707 (2007)
3. Bach, F.R., Lanckriet, G.R., Jordan, M.I.: Multiple kernel learning, conic duality, and the SMO algorithm. In: Proceedings of the Twenty-First International Conference on Machine learning, p. 6. ACM (2004)
4. Bergmeir, C., Hyndman, R., Koo, B.: A note on the validity of cross-validation for evaluating time series prediction. Department of Econometrics and Business Statistics, Working Paper (2015). ISSN 1440–771X

5. Campbell, J.Y., Shiller, R.J.: The dividend-price ratio and expectations of future dividends and discount factors. Rev. Financ. Stud. **1**, 195–228 (1988)
6. Campbell, J.Y., Thompson, S.B.: Predicting excess stock returns out of sample: can anything beat the historical average? Rev. Financ. Stud. **21**(4), 1509–1531 (2008)
7. Chang, C., Lin, C.: Training ν-support vector classifiers: theory and algorithms. Neural Comput. **13**(9), 2119–2147 (2001)
8. Cho, Y., Saul, L.: Kernel methods for deep learning. Adv. Neural Inf. Process. Syst. **22**, 342–350 (2009)
9. Cochrane, J.H.: Explaining the variance of price-dividend ratios. Rev. Financ. Stud. **5**, 243–280 (1992)
10. Cochrane, J.H.: The dog that did not bark: a defense of return predictability. Rev. Financ. Stud. **21**, 1533–1575 (2006)
11. Cochrane, J.H.: Presidential address: discount rates. J. Financ. **56**(4), 1047–1108 (2011)
12. Cuturi, M., Vert, J.-P., Birkenes, Ø., Matsui, T.: A kernel for time series based on global alignments. In: IEEE International Conference on ICASSP 2007, p. II-413. IEEE (2007)
13. Cuturi, M., Doucet, A.: Autoregressive kernels for time series. Technical Report (2011). arXiv:1101.0673
14. Cuturi, M.: Fast global alignment kernels. In: Proceedings of the 28th International Conference on Machine Learning (ICML-11), pp. 929–936 (2011)
15. Fama, E.F., French, K.R.: Dividend yields and expected stock returns. J. Financ. Econ. **22**, 3–25 (1988)
16. Hansen, L.P., Hodrick, R.J.: Forward exchange rates as optimal predictors of future spot rates: an econometric analysis. J. Polit. Econ. **88**, 829–853 (1980)
17. Kothari, S.P., Shanken, J.: Book-to-market, dividend yield, and expected market returns: a time-series analysis. J. Financ. Econ. **44**(2), 169–203 (1997)
18. Lettau, M., Ludvigson, S.: Consumption, aggregate wealth, and expected stock returns. J. Financ. **56**(3), 815–849 (2001)
19. Peña, M., Arratia, A., Belanche, L.A.: Multivariate dynamic kernels for financial time series forecasting. In: Villa, A.E.P., Masulli, P., Pons Rivero, A.J. (eds.) ICANN 2016. LNCS, vol. 9887, pp. 336–344. Springer, Cham (2016). doi:10.1007/978-3-319-44781-0_40
20. Shiller, R.J.: Do stock prices move too much to be justified by subsequent changes in dividends? Am. Econ. Rev. **71**, 421–436 (1981)
21. Welch, I., Goyal, A.: A comprehensive look at the empirical performance of equity premium prediction. Rev. Financ. Stud. **21**(4), 1455–1508 (2008)

Spatial-Temporal Analysis for Noise Reduction in NDVI Time Series

Fernanda Carneiro Rola Servián[(✉)] and Julio Cesar de Oliveira

Universidade Federal de Viçosa, Viçosa, Brazil
{fernanda.servian,oliveirajc}@ufv.br

Abstract. MODerate resolution Imaging Spectroradiometer (MODIS) data are largely used in multitemporal analysis of various Earth-related phenomena, such as mapping patterns of vegetation phenology and detecting land use/land cover change. NDVI time series are composite mosaics of the best quality pixels over a period of sixteen days. However, it is common to find low quality pixels in the composition that affect the time series analysis due to errors in the atmosphere conditions and in data acquisition. We present a filtering methodology that considers the pixel position (location in space) and time (position in the temporal data series) to define a new value for the low quality pixel. This methodology estimates the value of the point of interest, based first on a linear regression excluding pixels with low coefficient of determination R^2 and second on excluding outliers according to a boxplot analysis. Thus, from the remaining group of pixels, a Smooth Spline is generated in order to reconstruct the time series. The accuracies of estimated NDVI values using Spline were higher than the Savitzky–Golay method.

Keywords: MOD13Q1 · Filtering · Noises · Spatial-Temporal Analysis

1 Introduction

Remote sensing instruments aboard multiple satellite platforms have constantly provided observations of the Earth's surface. Data from image series with high temporal resolution and low spatial resolution can help monitoring and continuously mapping terrestrial coverage changes, as they contain key information in almost real time of large areas [1].

One of the products of the MODerate resolution Imaging Spectroradiometer (MODIS) sensor is the Normalized Difference Vegetation Index (NDVI). The NDVI time series have been widely used in several applications such as: monitoring vegetation phenology to quantify the effects of climate change on terrestrial ecosystems [2]; mapping the vegetation classes and assessing the vegetation seasonal dynamic [3]; classifying the diverse agricultural land-use practices and soil coverage in regional scale [4]; estimating post-fire vegetation greenness [5]; and may be used in vegetation patterns studies, as the NDVI variation is correlated with precipitation and relative humidity [6].

However, time series may have their applications hampered by noises due to variations of atmospheric conditions, effects from data acquisition geometry

© Springer International Publishing AG 2017
I. Rojas et al. (Eds.): IWANN 2017, Part II, LNCS 10306, pp. 188–197, 2017.
DOI: 10.1007/978-3-319-59147-6_17

(source-target-sensor position), instrument failures and data processing. Thus, with the increasing use of NDVI time series, it is of great importance to have a deeper knowledge regarding the reduction of data existent noises [7].

Existing methods for noise filtering in time series of vegetation index usually work only in one domain - frequency or time, i.e., analyzing the pixel behavior in only one dimension of the information [8]. In this perspective, in order to minimize the noises in NDVI time series using the two dimension of the information, we propose a method that estimates NDVI values based on temporal-spatial neighborhood analysis.

2   Materials and Methods

2.1   Database

The data used in the present article comes from an agricultural area in the State of São Paulo, Brazil, where a field reference map was available in case there was the need to prove a crop type [9]. It consists of the NDVI temporal series obtained between the years 2003 and 2004 (44 images – Table 1) coming from the product MOD13Q1 (NDVI with spatial resolution of 250 m and compositions of 16 days).

Table 1. Dates from the temporal series used and corresponding day of year (DOY).

2003				2004			
Date	DOY	Date	DOY	Date	DOY	Date	DOY
1	33	13	225	22	1	34	193
2	49	14	241	23	17	35	209
3	65	15	257	24	33	36	225
4	81	16	273	25	49	37	241
5	97	17	289	26	65	38	257
6	113	18	305	27	81	39	273
7	129	19	321	28	97	40	289
8	145	20	337	29	113	41	305
9	161	21	353	30	129	42	321
10	177			31	145	43	337
11	193			32	161	44	353
12	209			33	177		

For the study area, a subarea of 8 × 8 (row, column) pixels (64 pixels) was selected (Fig. 1). The chosen subarea contains good quality pixels according to the Quality Assessment (QA), which is an evaluation pixel by pixel of the scientific quality of the product with respect to the performance expected from it and it is provided by MODIS [10].

From study area, going from image to image, knowing that it is an agricultural site, the vegetation phenology can be analyzed. Looking at Date "X2", an area with the same or similar crop type is seen. Looking at Date "X3", half of the subarea was

Original NDVI series

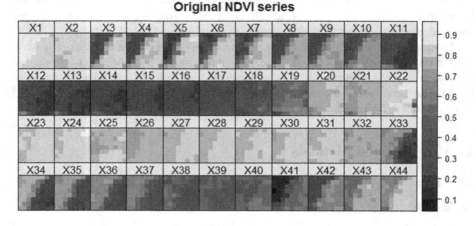

Fig. 1. Original NDVI time series of study area.

cropped and the other half continued to grow until Date "X11" where all the vegetation was cropped. From Date "X11" to "X18" the land was resting and preparing to grow again, which began in "X19", following the same cycle.

The spectral response of vegetation measured by the Normalized Difference Vegetation Index (NDVI) is a result of the 16-day composite period (product MOD13Q1), and noise can change NDVI values over time and space. Therefore, the present method is to make the NDVI time series more homogeneous and to predict NDVI values.

2.2   Sampling Data for Analysis

In order to apply the method and analyze the results, noises were introduced in the database to simulate low quality NDVI time series. For noise simulation, two samplings were conducted: (I) one pixel randomly selected from random dates; (II) a group of pixels selected from random dates (Fig. 2).

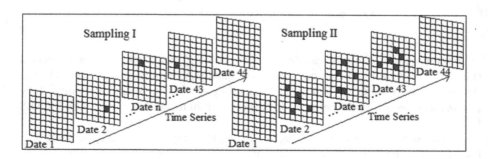

Fig. 2. Examples of the two samplings conducted.

The two samplings were conducted in order to analyze pixels and dates throughout the temporal series. The entire process that involves the steps of sampling, the method application to reduce de noise and results evaluation was repeated a thousand times with random levels of noise. This procedure was carried out in order to achieve consistent results.

Sampling "I" is a temporal sampling and caused a temporal degradation in the NDVI values. It consisted of a random selection of 30% of the dates and only 42 dates were considered, Date 1 (first date of the time series) and Date 44 (last date of the time series) were excluded from the noise process. One pixel was randomly chosen for each of the 12 selected dates. Then, for each of the 12 pixels selected, a noise varying on the levels of 10%, 30% or 50% (subtractive and additive) of their NDVI value was introduced. The noise for each pixel was also chosen randomly. Figure 3 shows the noises introduced to some of the selected dates for iteration #1000.

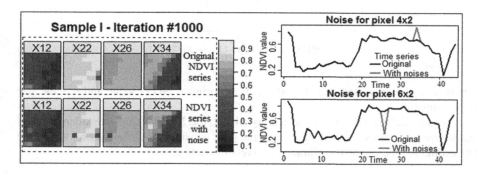

Fig. 3. Images and pixels' time series with random noise levels for iteration #1000 of Sample I.

Sampling "II" simulates noise in time and space. For each date of the selected dates, noise was applied for more than one pixel (for this sampling, 5 pixels were selected) of the image. This degradation was performed with three intensities equivalent to 10%, 30%, or 50% (subtractive or additive) for each pixel selected. Figure 4 shows the noises introduced to some of the 12 dates for iteration #1.

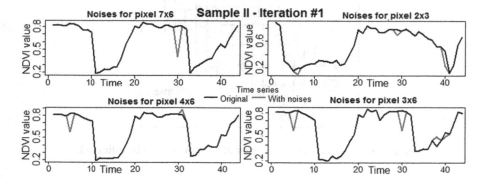

Fig. 4. Pixels' time series with random level of noises for iteration #1 of Sample II.

2.3   Technique Evaluation

The evaluation of the method is focused on the accuracy between the estimated and the original NDVI values of the pixels selected with noise. The Savitzky-Golay (SGolay) method [11], restructured recently [12], was chosen for results comparison. It is a commonly used method based on temporal estimation and on a comparison between few methods [7], it obtained a reliable performance and showed the ability to minimize overall noise in NDVI time series.

The root mean square error (RMSE) and the mean absolute percent error (MAPE) are going to be used to assess the accuracy of the estimated pixels. Low values of RMSE and MAPE indicate high accuracy of the estimated NDVI value. The MAPE and RMSE were calculated using Eqs. (1) and (2):

$$MAPE = \frac{1}{n}\sum_{i=1}^{n}\left|\frac{X_e-X_i}{X_i}\right| \times 100 \tag{1}$$

$$RMSE = \sqrt{\frac{1}{n}\sum_{i=1}^{n}(X_e - X_i)^2} \tag{2}$$

where X_e is the estimated NDVI value, X_i is the original NDVI value and n is the size of the sample.

As the database contains good quality pixels, the NDVI values considered "true" values are the pixel's original values and the ones considered as observation values are the estimated values for the eliminated pixels of each sample.

2.4   Method

The method titled in this article, the "Spline", consists in the creation of a mask that goes through the image estimating new values for its central pixel (Fig. 5).

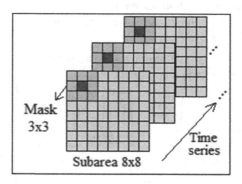

Fig. 5. Mask 3 × 3 over the image with central pixel to be estimated.

To apply the mask, the following procedure is conducted:

1. A linear regression of each neighboring pixel with the central pixel of the mask is performed in order to identify the pixels that result in a low spatial correlation over time (low coefficient of determination R^2). The neighboring pixel that has a low linear correlation with the central pixel will receive the value "no data available" in its entire time series and will not be used in the time series reconstruction process.
2. A boxplot analysis is performed, which statistically analyzes in a database what values are considered outliers (point considered atypical or that has a distinct behavior when compared to the data set present in a statistical series). Thus, the pixels considered as outliers are identified and they are also given the value of "no data available" so that they will not be used for the time series reconstruction. This boxplot analysis will only analyze the data for the same date, i.e. it is verified if within the remaining pixels values selected by the 3×3 mask there is an outlier.
3. The pixels considered as able for analysis by the previous steps are used to get a Smooth Spline curve that best fits the database. At the end of this interpolation it is expected to reconstruct the NDVI time series for the central pixel of the 3×3 mask.

The value calculated from a mask will be used for the calculation of the next mask. This means that to reconstruct the entire time series, the mask goes around the image pixel by pixel. But the borders pixels are not analyzed because the mask only considers the central pixel to be estimated.

As the study area is 8×8, there is an area of 6×6 that the mask will go through. So at the end of the process, there will be 36 masks.

It is important to mention that the whole process acts as a filtering and making the NDVI values across the time series more homogenous because of the Smooth Spline curve.

3   Results and Discussion

3.1   Sample I

Table 2 shows a few statistical measures of MAPE and RMSE for both of the methods "Spline", method proposed in this article, and Savitzky-Golay (SGolay) from the thousand iterations executed for Sample I.

Table 2. Analysis of MAPE and RMSE values for the 1000 iterations of Sample I.

Measures	Methods			
	MAPE (%)		RMSE	
	Spline	SGolay	Spline	SGolay
Min.	2.01	7.34	0.01	0.05
Median	7.46	16.94	0.04	0.11
Mean	8.35	17.39	0.05	0.11
Max.	24.55	33.53	0.11	0.17
Range	22.54	26.19	0.09	0.11

Table 2 shows that the accuracies of the estimated NDVIs for Sample I using the Spline method were clearly higher than those of the SGolay method. Both of them had a similar MAPE range, but the Spline ranged from 2.01% to 24.55%, while the SGolay ranged from 7.34% to 33.53%. This means that even though the range was similar (difference of 4%), the NDVI estimated values obtained with the Spline method had a better performance and varied less when compared to the original value than the ones obtained with the SGolay method. Overall, the NDVI values after the reconstruction have a mean error of about 8%, which need to be analyzed regarding the application that will be based on this data set.

From the RMSE values, the results obtained with the Spline presented a better fit to the original NDVI values than the results obtained with the SGolay. Figure 6 presents an example of the reconstructed NDVI time series for the pixel 6 × 2 at iteration #1000 with the Spline and SGolay, and it also shows the original time series and the time series with noise. Looking at the graph through the entire time series, the Spline had a better fit to the data than SGolay. It seems to be filtering the data and making it smoother, and it also had a better estimation of the date with noise.

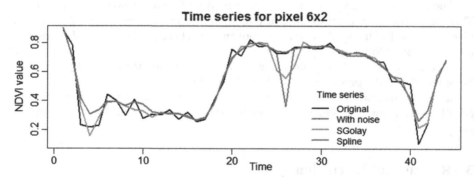

Fig. 6. Raw, degraded and reconstructed (SGolay and Spline) time series of pixel 6 × 2 (iteration #1000).

3.2   Sample II

Table 3 shows a few statistical measures of MAPE and RMSE for both of the methods Spline and SGolay from the thousand iterations executed for Sample II.

From Table 3, it is clear that the accuracies for Sample II were higher using the Spline method than using the SGolay method. Even though the Spline had a bigger MAPE range (13.54%), its MAPE's values ranged from 3.76% to 17.30%, while the SGolay ranged from 12.90% to 24.74%, meaning that the maximum error the estimated value will present with Spline is 17.30%, while with SGolay is 24.74%.

The RMSE Spline values were a lot better than the SGolay values. This means that the reconstructed NDVI time series using the Spline presented a much better fit to the data than the one reconstructed using the SGolay. As the SGolay only works in the frequency of time (it does not depend on the neighboring pixels), it was expected a

Table 3. Analysis of MAPE and RMSE values for the 1000 iterarions of Sample II.

Measures	Methods			
	MAPE (%)		RMSE	
	Spline	SGolay	Spline	SGolay
Min.	3.76	12.90	0.03	0.09
Median	8.73	17.54	0.05	0.12
Mean	8.94	17.66	0.05	0.12
Max.	17.30	24.74	0.09	0.15
Range	13.54	11.84	0.06	0.07

better result from it. In this sample, several pixels from one image had noise introduced to their NDVI values, and even this way, the Spline had a better result using neighboring pixels with noise.

Figure 7 presents two examples of the reconstructed NDVI time series from Sample II for the pixels 3×2 and 2×3 at iteration #1000 with the Spline and SGolay, it also shows the original time series and the one with noise. Looking at the images, specifically at the dates with noise, the Spline has a better response than the SGolay.

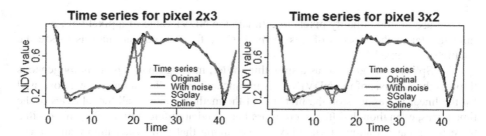

Fig. 7. Raw, degraded and reconstructed (SGolay and Spline) time series of pixels 3×2 and 2×3 (iteration #1000)

In Fig. 8, looking at the MAPE boxplot analysis, it is clear that the Spline method obtained better results than the SGolay method for both of the samplings.

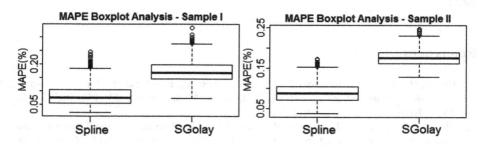

Fig. 8. Boxplot analysis of the Spline and SGolay methods' results for two samplings.

4   Conclusion

This article proposed a method to estimate NDVI values of MODIS13Q1 time series with existent noises using the correlation between time and space of neighboring pixels. It consists of a 3 × 3 mask that goes pixel by pixel estimating new NDVI values for its central pixel's time series. First, from a linear regression of the mask's central pixel and each neighboring pixel, the pixel with the lowest coefficient of determination (R^2) was excluded from the process. Then, a boxplot analysis is done on each date and pixels considered as outliers were eliminated. From the remaining pixels, a Smooth Spline was generated in order to fit the data set, estimating new NDVI values for the mask's central pixel's time series, reconstructing the NDVI time series. Once those steps were followed, the mask went to the next pixel, beginning the process again until it reached the last pixel of the image.

The results indicated that the accuracy of the estimated NDVI values obtained with the Spline method were higher than those estimated with the Savitzky–Golay method. The accuracies from both of the methods were better in the data that had one noise per date in 30% of the dates of the database, when compared with the accuracies from data with several noises per date. In the first case, the Spline's MAPE ranged from 2.01% to 24.55%, while the SGolay's ranged from 7.34% to 33.53%. And in the second case, the Spline's ranged from 3.76% to 17.30%, while the SGolay's ranged from 12.90% to 24.74%.

The use of good quality pixels to estimate the NDVI value of a pixel with noise is fundamental to the process of this method, since there is not an analysis to see the quality of the pixel.

The Spline method acted as a smoothing filter, making images more homogeneous through the analysis of the NDVI values in time and space. The temporal window used for the linear regressions on the first step of the method contains all the 44 dates of the time series, even though different crops are harvested at different times. This means that there is a great variation of the NDVI values along that temporal window and values considered as outliers can be wrongly eliminated. The temporal windows could be smaller, containing fewer dates and trying to follow the crop phenology, so that the linear regression curves can better fit the data.

Acknowledgments. This study was supported by the Institutional Scholarship Program for Scientific Initiation funded by the Foundation for Research Support of the State of Minas Gerais – Brazil.

References

1. Eerens, H., Haesen, D., Rembold, F., Urbano, F., Tote, C., Bydekerke, L.: Image time series processing for agriculture monitoring. J. Environ. Model. Softw. **53**, 154–162 (2014). doi:10.1016/j.envsoft.2013.10.021
2. Zhang, G.L., Zhang, Y.J., Dong, J.W., Xiao, X.M.: Green-up dates in the Tibetan Plateau have continuously advanced from 1982 to 2011. J. PNAS **110**, 4309–4314 (2013). doi:10.1073/pnas.1210423110

3. Silveira, E.M.D., Carvalho, L.M.T., Acerbi-Junior, F.W., Mello, J.M.: The assessment of vegetation seasonal dynamics using multitemporal NDVI and EVI images derived from MODIS. J. Cerne **14**, 177–184 (2008)
4. Brown, J.C., Kastens, J.H., Coutinho, A.C., Victoria, D.D., Bishop, C.R.: Classifying multiyear agricultural land use data from Mato Grosso using time-series MODIS vegetation index data. J. Remote Sens. Environ. **130**, 39–50 (2013). doi:10.1016/j.rse.2012.11.009
5. Villareal, M.L., Norman, L.M., Buckley, S., Wallace, C.S.A., Coe, M.A.: Multi-index time series monitoring of drought and fire effects on desert grasslands. J. Environ. Earth Sci. **183**, 186–197 (2016). doi:10.1016/j.rse.2016.05.026
6. Jin, X.M., Guo, R.H., Zhang, Q., Zhou, Y.X., Zhang, D.R., Yang, Z.: Response of vegetation pattern to different landform and water-table depth in Hailiutu River basin, Northwestern China. J. Environ. Earth Sci. **71**, 4889–4898 (2014). doi:10.1007/s12665-013-2882-1
7. Hird, J.N., McDermid, G.J.: Noise reduction of NDVI time series: an empirical comparison of selected techniques. J. Remote Sens. Environ. **113**, 248–258 (2009). doi:10.1016/j.rse.2008.09.003
8. Oliveira, J.C.: Janela de regressão: uma análise espacial e temporal para estimar valores de NDVI classificados com baixa qualidade em séries temporais MODIS. INPE, São José dos Campos (2014)
9. Sanches, I.D.: Sensoriamento remoto para o levantamento espectro-temporal e estimativa de área de culturas agrícolas. INPE, São José dos Campos (2004)
10. Roy, D.P., Borak, J.S., Devadiga, S., Wolfe, R.E., Zheng, M., Descloitres, J.: The MODIS land product quality assessment approach. Remote Sens. Environ. **83**, 62–76 (2002). doi:10.1016/S0034-4257(02)00087-1
11. Savizky, A., Golay, M.J.E.: Smoothing and differentiation of data by simplified least squares procedures. Anal. Chem. **36**, 1627–1639 (1964). doi:10.1021/ac60214a047
12. Chen, J., Jönsson, P., Tamura, M., Gu, Z., Matsushita, B., Eklundh, L.: A simple method for reconstructing a high-quality NDVI time-series data set based on the Savitzky-Golay filter. Remote Sens. Environ. **91**, 332–344 (2004). doi:10.1016/j.rse.2004.03.014

Hidden-Markov Models for Time Series of Continuous Proportions with Excess Zeros

Julien Alerini[1], Marie Cottrell[2]($\boxtimes$), and Madalina Olteanu[2]($\boxtimes$)

[1] IHMC-PIREH, UMR 8589, Université Paris 1 Panthéon-Sorbonne, Paris, France
`julien.alerini@univ-paris1.fr`
[2] SAMM, EA 4543, Université Paris 1 Panthéon-Sorbonne, Paris, France
{`marie.cottrell,madalina.olteanu`}`@univ-paris1.fr`

Abstract. Bounded time series and time series of continuous proportions are often encountered in statistical modeling. Usually, they are addressed either by a logistic transformation of the data, or by specific probability distributions, such as Beta distribution. Nevertheless, these approaches may become quite tricky when the data show an over-dispersion in 0 and/or 1. In these cases, the zero-and/or-one Beta-inflated distributions, $\mathcal{ZOIB}$, are preferred. This manuscript combines $\mathcal{ZOIB}$ distributions with hidden-Markov models and proposes a flexible model, able to capture several regimes controlling the behavior of a time series of continuous proportions. For illustrating the practical interest of the proposed model, several examples on simulated data are given, as well as a case study on historical data, involving the military logistics of the Duchy of Savoy during the XVIth and the XVIIth centuries.

1 Introduction

Time series of continuous proportions or percentages are often encountered in various research fields such as economy, biology or history. For instance, one may be interested in modeling the fraction of income a family devotes to lodging or taxes; or in modeling the proportion of a population exposed to fine particles pollution or subject to a certain type of disease. In our case, as it will be shown later, we are interested in understanding the rhythms of the Sabaudian State during the XVIth and the XVIIth centuries, and more particularly the evolution of the ratio of legislative texts issued by the Duchy and related to military logistics, among the entire production of law.

In statistical modeling, the two common approaches for dealing with continuous proportions are, on the one hand, a logistic transformation of the data [1], and, on the other hand, the use of specific probability distributions such as Beta or Dirichlet, [2]. However, both of these approaches have a major drawback, since they do not take into account the possibility of an over-dispersion in the limit values, 0 and/or 1. During the last few years, this issue has been addressed by several authors, who proposed either further transforming the data [3], or introducing specific probability masses in 0 and/or 1, hence using zero-and/or-one Beta Inflated distributions. The latter approach has been intensively studied during the last five years, mainly in a regression context [4,5].

© Springer International Publishing AG 2017
I. Rojas et al. (Eds.): IWANN 2017, Part II, LNCS 10306, pp. 198–209, 2017.
DOI: 10.1007/978-3-319-59147-6_18

In this manuscript, we aim at using the Beta inflated distributions in a framework different from that of regression, since our main interest is to uncover and to highlight the possible existence of several regimes in a time series of proportions. With this in mind, we introduce a hidden-Markov model, having as emission distribution a Beta inflated distribution, $\mathcal{ZOIB}$-HMM in abbreviated form. Originally introduced for speech recognition [6], hidden Markov models (HMM hereafter) are especially interesting in the context of the presumed existence of several regimes controlling the parameters of the model, or the parameters of the emission distribution.

The next sections are organized as follows: in Sect. 2, we recall the definition and the properties of the zero-and-one Beta-inflated distribution and then, in Sect. 3, we introduce the $\mathcal{ZOIB}$-HMM model and describe the estimation procedure. Section 4 contains several experimental results, with a discussion on the convergence properties, while Sect. 5 is devoted to presenting the results on a real data set, coming from medieval history. Finally, a conclusion follows in Sect. 6.

2 Zero-and-One Beta-Inflated Distributions

As mentioned in the introduction, statistical models based on Beta distributions assume the data to be valued in the open interval $]0,1[$. In practical applications, this is rarely the case, and the situation of an over-dispersion in 0 and/or 1 appears quite often. The solution for dealing with this is to mix the Beta distribution either with a Dirac mass (in 0 for data valued in $[0,1[$, in 1 for data valued in $]0,1]$), or with a Bernoulli distribution (for data valued in $[0,1]$). Only the latter case will be addressed here, but the reader may refer to [7] for a complete review of Beta inflated distributions.

The probability density function of a Beta distribution with parameters $\alpha, \beta > 0$, denoted $\mathcal{B}e(\alpha, \beta)$, is

$$f_B\left(x; \alpha, \beta\right) = \frac{\Gamma(\alpha + \beta)}{\Gamma(\alpha)\Gamma(\beta)} x^{\alpha-1}(1-x)^{\beta-1}\mathbb{1}_{x\in]0,1[}, \tag{1}$$

where Γ is the Gamma function. Also, if $X \sim \mathcal{B}(\alpha, \beta)$ and $\mu = \frac{\alpha}{\alpha+\beta}$, $\phi = \alpha + \beta$, the expectation and the variance of X may be expressed as:

$$\mathbb{E}(X) = \mu \ ; \ \mathbb{V}(X) = \frac{\mu(1-\mu)}{\phi + 1}. \tag{2}$$

Besides its support reduced to the interval $]0,1[$, the interest of using a Beta distribution for statistical modeling also resides in the large variety of shapes for its density, which makes it quite appealing for applications (see Fig. 1).

Let us now define the zero-and-one Beta-inflated as a mixture between a Bernoulli and a Beta distribution, using a latent variable Y. If $Y \sim \mathcal{B}(\eta)$, define X conditionally to Y such that $X|(Y=1) \sim \mathcal{B}(\gamma)$ and $X|(Y=0) \sim \mathcal{B}e(\alpha, \beta)$. The marginal density[1] of X is

$$f(x; \eta, \gamma, \alpha, \beta) = (\eta\gamma)^{\mathbb{1}_{x=1}}(\eta(1-\gamma))^{\mathbb{1}_{x=0}}((1-\eta)f_B(x; \alpha, \beta))^{\mathbb{1}_{x\in]0,1[}}, \tag{3}$$

---

[1] The density is taken with respect to the probability measure $\lambda + \delta_0 + \delta_1$, where λ is the Lebesgue measure on $[0,1]$, and δ_0 and δ_1 are Dirac masses in 0 and 1.

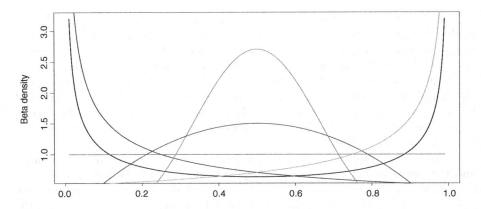

Fig. 1. Examples of Beta distributions for various parameters α and β.

where $\eta \in {]}0,1[$ is the mixture parameter, $\gamma \in {]}0,1[$ is the Bernoulli-distribution parameter and $\alpha, \beta > 0$ are the Beta-distribution parameters. Throughout the rest of the paper, let $\xi = (\eta, \gamma, \alpha, \beta)$ be the four-dimensional parameter of a zero-and-one Beta-inflated distribution, $\mathcal{ZOIB}(\xi)$.

Consider now $X_1^T = (X_1, ..., X_T)$ an i.i.d. T-sample of $\mathcal{ZOIB}(\xi)$. The likelihood may be written as :

$$\mathcal{L}(X_1^T; \xi) = \prod_{t=1}^{T} \left[\eta^{\mathbb{1}_{X_t \in \{0,1\}}} (1-\eta)^{\mathbb{1}_{X_t \in]0,1[}}\right] \times \prod_{t=1}^{T} \left[\gamma^{\mathbb{1}_{X_t=1}} (1-\gamma)^{\mathbb{1}_{X_t=0}}\right]$$

$$\times \prod_{t=1}^{T} f_B(X_t; \alpha, \beta)^{\mathbb{1}_{X_t \in]0,1[}} = \mathcal{L}_1(X_1^T; \eta) \mathcal{L}_2(X_1^T; \gamma) \mathcal{L}_3(X_1^T; \alpha, \beta). \quad (4)$$

Maximizing the likelihood consists in maximizing each of the three terms in the product, which are independent in the parameter components. For the mixture parameter η, the maximum likelihood estimate (MLE) is computed by maximizing

$$\ln \mathcal{L}_1(X_1^T; \eta) = \ln \eta \sum_{t=1}^{T} \mathbb{1}_{X_t \in \{0,1\}} + \ln(1-\eta) \sum_{t=1}^{T} \mathbb{1}_{X_t \in]0,1[}, \quad (5)$$

which yields

$$\hat{\eta} = \frac{\sum_{t=1}^{T} \mathbb{1}_{X_t \in \{0,1\}}}{\sum_{t=1}^{T} \mathbb{1}_{X_t \in \{0,1\}} + \sum_{t=1}^{T} \mathbb{1}_{X_t \in]0,1[}} = \frac{T_1}{T}, \quad (6)$$

where $T_1 = \sum_{t=1}^{T} \mathbb{1}_{X_t \in \{0,1\}}$. For the Bernoulli parameter γ, the MLE is computed by maximizing

$$\ln \mathcal{L}_2(X_1^T; \gamma) = \ln \gamma \sum_{t=1}^{T} \mathbb{1}_{X_t=1} + \ln(1-\gamma) \sum_{t=1}^{T} \mathbb{1}_{X_t=0}, \quad (7)$$

which yields

$$\hat{\gamma} = \frac{\sum_{t=1}^{T} \mathbb{1}_{X_t=1}}{\sum_{t=1}^{T} \mathbb{1}_{X_t=1} + \sum_{t=1}^{T} \mathbb{1}_{X_t=0}} = \frac{T_2}{T_1}, \qquad (8)$$

where $T_2 = \sum_{t=1}^{T} \mathbb{1}_{X_t=1}$. Finally, for the Beta parameters, α and β, one has to maximize

$$\ln \mathcal{L}_3(X_1^T; \alpha, \beta) = \sum_{X_t \in]0,1[} \ln f_B(X_t; \alpha, \beta). \qquad (9)$$

In this case, since there is no analytical form for the MLE of a Beta distribution, the solution may be found using numerical optimization. However, in order to avoid numerical issues linked to the initial values of the gradient-descent based algorithms, an approximation of the MLE with the moment estimates is preferred. Following Eq. 2, the moment estimates of α and β are:

$$\tilde{\alpha} = \tilde{\mu}\tilde{\phi}, \quad \tilde{\beta} = (1 - \tilde{\mu})\tilde{\phi}, \qquad (10)$$

where

$$\tilde{\mu} = \frac{1}{n - T_1} \sum_{X_t \in]0,1[} X_t, \quad \tilde{\phi} = \frac{\tilde{\mu}(1 - \tilde{\mu})}{s^2} - 1,$$

$$\text{and } s^2 = \frac{1}{n - T_1} \sum_{X_t \in]0,1[} (X_t - \tilde{\mu})^2. \qquad (11)$$

3 Estimation Procedure for the $\mathcal{ZOIB}$-HMM Model

In this section, the parameters of a zero-and-one Beta-inflated distribution are supposed to be controlled by a hidden Markov chain with a finite number of states.

3.1 The Model

Let $(X_t)_{t \in \mathbb{Z}}$ be the observed time series, valued in $[0,1]$, and let $(S_t)_{t \in \mathbb{N}}$ be the unobserved process, controlling the parameters of the distribution of X_t. Throughout the rest of the paper, S_t is supposed to be a homogeneous Markov chain, irreducible, recurrent and aperiodic, valued in a finite state-space $E = \{e_1, ..., e_q\}$ and defined by its transition matrix $\Pi = (\pi_{ij})_{i,j=1,...,q}$, $\pi_{ij} = \mathbb{P}(S_t = e_j | S_{t-1} = e_i)$, with $\pi_{ij} > 0$, $\sum_{j=1}^{q} \pi_{ij} = 1$, and by its initial probability distribution π^0, $\pi_i^0 = \mathbb{P}(S_1 = e_i)$, $\forall i = 1, ..., q$.

Furthermore, let us suppose that X_t are independent conditionally to S_t, and that X_t conditionally to S_t is distributed according to a zero-and-one Beta-inflated distribution, $\mathcal{ZOIB}(\xi_i)$, with $\xi_i = (\eta_i, \gamma_i, \alpha_i, \beta_i) \in]0,1[^2 \times]0, +\infty[^2$. For a fixed number of states q in the hidden Markov chain, the set of possible values for the parameters may be written as:

$$\Theta = \left\{ \theta = ((\xi_i)_{i=1,\dots,q}, \Pi) \in (]0,1[^2 \times]0,+\infty[^2)^q \times]0,1[^{q^2}, \right.$$

$$\left. \text{and } \forall i \in \{1 \cdots q\}, \sum_{j=1}^{q} \pi_{ij} = 1 \right\}. \tag{12}$$

3.2   The EM Algorithm for $\mathcal{ZOIB}$-HMM

Since the above model involves a hidden Markov chain, the estimation procedure is carried out using the EM algorithm [6,8]. With the previous assumptions, and with the notations $X_1^T = (X_1, ..., X_T)$, $S_1^T = (S_1, ..., S_T)$, the complete likelihood is given by:

$$\mathcal{L}(X_1^T, S_1^T; \theta) = \prod_{t=1}^{T} \prod_{i=1}^{q} f(X_t|S_t = e_i; \xi_i)^{\mathbb{1}\{S_t=e_i\}} \prod_{t=2}^{T} \prod_{i,j=1}^{q} \pi_{ij}^{\mathbb{1}\{S_{t-1}=e_i, S_t=e_j\}} \times C, \tag{13}$$

where $f(X_t|S_t = e_i; \xi_i)$ is the Beta-inflated density, conditionally to the hidden state $S_t = e_i$ and defined in Eq. 3, and $C = \prod_{i=1}^{q} (\pi_i^0)^{\mathbb{1}\{S_t=e_i\}}$ is the likelihood of the initial state of the Markov chain.

When expressing the Beta-inflated density in its analytical form, the complete likelihood may further be written as:

$$\mathcal{L}(X_1^T, S_1^T; \theta) = \prod_{t=1}^{T} \prod_{i=1}^{q} \left(\eta_i^{\mathbb{1}X_t \in \{0,1\}} (1 - \eta_i)^{\mathbb{1}X_t \in]0,1[} \right)^{\mathbb{1}\{S_t=e_i\}} \times$$

$$\prod_{t=1}^{T} \prod_{i=1}^{q} \left(\gamma_i^{\mathbb{1}X_t=1} (1 - \gamma_i)^{\mathbb{1}X_t=0} \right)^{\mathbb{1}\{S_t=e_i\}} \times$$

$$\prod_{t=1}^{T} \prod_{i=1}^{q} (f_B(X_t, \alpha_i, \beta_i))^{\mathbb{1}\{X_t \in]0,1[, S_t=e_i\}} \times \prod_{t=2}^{T} \prod_{i,j=1}^{q} \pi_{ij}^{\mathbb{1}\{S_{t-1}=e_i, S_t=e_j\}} \times C$$

$$= \mathcal{L}_1(X_1^T, S_1^T; \eta) \mathcal{L}_2(X_1^T, S_1^T; \gamma) \mathcal{L}_3(X_1^T, S_1^T; \alpha, \beta) \mathcal{L}_4(X_1^T, S_1^T; \Pi), \tag{14}$$

where $\eta = (\eta_1, ..., \eta_q)$, $\gamma = (\gamma_1, ..., \gamma_q)$, $\alpha = (\alpha_1, ..., \alpha_q)$ and $\beta = (\beta_1, ..., \beta_q)$. The EM algorithm consists in iteratively maximizing the expected value of the complete likelihood with respect to θ and conditionally to the observed data set X_1^T and a fixed value of the parameter $\theta^\star$, and then updating $\theta^\star$ at each step.

E-Step. The E-step is given by the computation of the expected value of the complete likelihood, conditionally to the observed data,

$$Q(\theta|\theta^\star) = \mathbb{E}_{\theta^\star} \left[\ln \mathcal{L}(X_1^T, S_1^T; \theta) | X_1^T \right]. \tag{15}$$

According to Eq. 14, $Q(\theta|\theta^\star)$ can be split into:

$$Q(\theta|\theta^\star) = Q_1(\eta|\theta^\star) + Q_2(\gamma|\theta^\star) + Q_3(\alpha, \beta|\theta^\star) + Q_4(\Pi|\theta^\star), \tag{16}$$

where

$$Q_1(\boldsymbol{\eta}|\theta^\star) = \mathbb{E}_{\theta^\star}\left[\ln \mathcal{L}_1(X_1^T, S_1^T; \boldsymbol{\eta})|X_1^T\right]$$

$$= \sum_{i=1}^q \left[\sum_{X_t \in \{0,1\}} \omega_t(e_i) \ln \eta_i + \sum_{X_t \in]0,1[} \omega_t(e_i) \ln(1 - \eta_i)\right], \quad (17)$$

with $\omega_t(e_i) = \mathbb{P}_{\theta^\star}(S_t = e_i|X_1^T)$;

$$Q_2(\boldsymbol{\gamma}|\theta^\star) = \mathbb{E}_{\theta^\star}\left[\ln \mathcal{L}_2(X_1^T, S_1^T; \boldsymbol{\gamma})|X_1^T\right]$$

$$= \sum_{i=1}^q \left[\sum_{X_t = 1} \omega_t(e_i) \ln \gamma_i + \sum_{X_t = 0} \omega_t(e_i) \ln(1 - \gamma_i)\right]; \quad (18)$$

$$Q_3(\boldsymbol{\alpha}, \boldsymbol{\beta}|\theta^\star) = \mathbb{E}_{\theta^\star}\left[\ln \mathcal{L}_3(X_1^T, S_1^T; \boldsymbol{\alpha}, \boldsymbol{\beta})|X_1^T\right]$$

$$= \sum_{i=1}^q \sum_{X_t \in]0,1[} \omega_t(e_i) \ln\left(f_B\left(X_t, \alpha_i, \beta_i\right)\right); \quad (19)$$

$$Q_4(\Pi|\theta^\star) = \mathbb{E}_{\theta^\star}\left[\ln \mathcal{L}_4(X_1^T, S_1^T; \Pi)|X_1^T\right] = \sum_{i,j=1}^q \sum_{t=2}^T \omega_t(e_i, e_j) \ln \pi_{ij}, \quad (20)$$

with $\omega_t(e_i, e_j) = \mathbb{P}_{\theta^\star}(S_{t-1} = e_i, S_t = e_j|X_1^T)$. The probabilities $\omega_t(e_i)$ and $\omega_t(e_i, e_j)$ may be easily computed using the forward-backward procedure, typical for the EM algorithm [9].

M-Step. Thanks to the factorization of the complete likelihood, the optimization step may be performed by independently maximizing each term of $Q(\theta|\theta^\star)$. For η_i, γ_i and π_{ij}, the following analytical expressions are straightforward:

$$\hat{\eta}_i = \frac{\sum_{X_t \in \{0,1\}} \omega_t(e_i)}{\sum_{t=1}^T \omega_t(e_i)} , \quad \hat{\gamma}_i = \frac{\sum_{X_t = 1} \omega_t(e_i)}{\sum_{X_t \in \{0,1\}} \omega_t(e_i)} ,$$

$$\text{and } \hat{\pi}_{ij} = \frac{\sum_{t=2}^T \omega_t(e_i, e_j)}{\sum_{t=1}^T \omega_t(e_i)} . \quad (21)$$

As for α_i and β_i, there are no analytical expressions of the estimates, directly tractable from $Q_3(\boldsymbol{\alpha}, \boldsymbol{\beta}|\theta^\star)$. Rather than numerically optimizing this function, operation which would slow down the algorithm and possibly introduce numerical instability, we prefer the use of moment estimates, which appear as good substitues for the MLE:

$$\tilde{\alpha}_i = \tilde{\mu}_i \tilde{\phi}_i , \quad \tilde{\beta}_i = (1 - \tilde{\mu}_i)\tilde{\phi}_i, \quad (22)$$

where

$$\tilde{\mu}_i = \frac{\sum_{X_t \in]0,1[} \omega_t(e_i) X_t}{\sum_{X_t \in]0,1[} \omega_t(e_i)} \; , \; \tilde{\phi}_i = \frac{\tilde{\mu}_i(1 - \tilde{\mu}_i)}{s_i^2} - 1,$$

$$s_i^2 = \frac{\sum_{X_t \in]0,1[} \omega_t(e_i) \left(X_t - \tilde{\mu}_i\right)^2}{\sum_{X_t \in]0,1[} \omega_t(e_i)}. \tag{23}$$

As one may easily notice, Eqs. 22 and 23 are very similar to Eqs. 10 and 11, except for the weights $\omega_t(e_i)$, which are introduced by the hidden Markov chain and represent the conditional probabilities of being in state e_i at time t, given the observed data, X_1^T.

4 Experimental Results

In order to test the quality of the estimates and its convergence rate, the algorithm was trained on several simulated examples. For each of the following scenarios and for sample sizes ranging from 500 to 1 000, 100 different trajectories of a two-state ($q = 2$) $\mathcal{ZOIB}$-HMM are simulated. The values of the parameters used for the simulations are the following:

$$\Pi = \begin{pmatrix} 0.9 & 0.1 \\ 0.1 & 0.9 \end{pmatrix}, (\alpha_1, \alpha_2) = (1; 0.5), (\beta_1, \beta_2) = (1; 2), (\gamma_1, \gamma_2) = (0.5; 0.9),$$

and $(\eta_1, \eta_2) = (\eta_1, 0.8)$, where $\eta_1 \in \{0.1, 0.3, 0.5, 0.7\}$. The results are detailed in Tables 1, 2, 3 and 4 below. In each case are reported the mean values of the estimates, as well as their standard errors and medians. We also provide the squared bias and the ratio of errors in the a posteriori identification of the hidden regimes (mean-values, standard errors and medians).

According to these first results on synthetic data, most of the parameters (Π, the η's and the γ's) are generally correctly estimated, even for short time series. However, the quality of the estimated transition matrix diminishes when η_1 has larger values (the ratio of zeros and ones is overriding the ratio of values in $]0, 1[$). The α's and the β's are correctly estimated for small values of η_1 and sufficiently large time series, with a length at least equal to 1 000. However, the algorithm fails in fairly approaching them when η_1 is greater than 0.5 or for time series shorter than 1 000 observations.

Furthermore, when comparing the mean values of the estimates with their medians, one may easily see that, if considering the medians, the performances of the algorithm are eventually not bad even in this limit cases. When looking into details, some of the incoherences come from atypical time series in the simulations which rise identifiability issues. In all cases, the experimental section should be further improved with more examples, involving more diverse scenarios for the parameters and longer time series.

Table 1. Simulation results for $\eta_1 = 0.1$ and 100 time-series of length T. Mean, standard error (italics) and median (bold) of the estimates.

	$T = 500$		$T = 1000$	
$\hat{\Pi}$	0.82(0.18)	0.18(0.18)	0.84(0.16)	0.16(0.16)
	0.89	**0.11**	**0.89**	**0.11**
	0.19(0.20)	0.81(0.20)	0.16(0.16)	0.84(0.16)
	0.11	**0.89**	**0.11**	**0.89**
$\hat{\alpha}_1, \hat{\alpha}_2$	0.99(0.27)	0.74(0.66)	0.99(0.14)	0.54(0.15)
	0.96	**0.60**	**1.00**	**0.52**
$\hat{\beta}_1, \hat{\beta}_2$	0.98(0.11)	2.70(1.83)	1.00(0.11)	2.11(0.75)
	0.99	**2.35**	**1.00**	**2.09**
$\hat{\gamma}_1, \hat{\gamma}_2$	0.55(0.22)	0.88(0.09)	0.52(0.18)	0.89(0.07)
	0.53	**0.90**	**0.51**	**0.90**
$\hat{\eta}_1, \hat{\eta}_2$	0.15(0.13)	0.76(0.15)	0.15(0.14)	0.77(0.13)
	0.10	**0.80**	**0.10**	**0.80**
$Bias(\theta)^2$	1.39(1.65)		0.71(0.57)	
	0.89		**0.53**	
$\%ERR$	14.6(18.4)		12.1(14.8)	
	6.9		**6.7**	

Table 2. Simulation results for $\eta_1 = 0.3$ and 100 time-series of length T. Mean, standard error (italics) and median (bold) of the estimates.

	$T = 500$		$T = 1000$	
$\hat{\Pi}$	0.78(0.19)	0.22(0.19)	0.84(0.15)	0.16(0.15)
	0.86	**0.14**	**0.90**	**0.10**
	0.24(0.24)	0.76(0.24)	0.16(0.15)	0.84(0.15)
	0.11	**0.89**	**0.11**	**0.89**
$\hat{\alpha}_1, \hat{\alpha}_2$	1.06(0.78)	1.12(3.49)	1.01(0.34)	0.55(0.14)
	0.97	**0.56**	**1.00**	**0.54**
$\hat{\beta}_1, \hat{\beta}_2$	1.01(0.18)	2.91(3.30)	0.99(0.11)	2.08(0.81)
	0.99	**1.88**	**0.98**	**1.94**
$\hat{\gamma}_1, \hat{\gamma}_2$	0.51(0.21)	0.86(0.15)	0.50(0.15)	0.89(0.06)
	0.48	**0.89**	**0.49**	**0.90**
$\hat{\eta}_1, \hat{\eta}_2$	0.34(0.14)	0.75(0.16)	0.31(0.12)	0.78(0.11)
	0.29	**0.80**	**0.30**	**0.79**
$Bias(\theta)^2$	2.23(4.51)		0.76(0.62)	
	0.94		**0.56**	
$\%ERR$	21.2(16.5)		15.1(10.6)	
	12.2		**11.6**	

Table 3. Simulation results for $\eta_1 = 0.5$ and 100 time-series of length T. Mean, standard error (italics) and median (bold) of the estimates.

		$T = 500$		$T = 1000$	
$\hat{\Pi}$		0.69(0.23)	0.31(0.23)	0.77(0.19)	0.23(0.19)
		0.76	**0.24**	**0.88**	**0.12**
		0.24(0.19)	0.76(0.19)	0.24(0.22)	0.76(0.22)
		0.18	**0.82**	**0.13**	**0.87**
$\hat{\alpha}_1, \hat{\alpha}_2$		13.08(80.91)	4.01(27.24)	0.94(0.19)	0.58(0.30)
		0.93	**0.60**	**0.97**	**0.54**
$\hat{\beta}_1, \hat{\beta}_2$		4.30(20.91)	7.58(45.68)	0.98(0.12)	2.27(2.59)
		1.00	**1.74**	**0.98**	**1.71**
$\hat{\gamma}_1, \hat{\gamma}_2$		0.50(0.24)	0.81(0.20)	0.54(0.21)	0.83(0.16)
		0.48	**0.88**	**0.51**	**0.89**
$\hat{\eta}_1, \hat{\eta}_2$		0.51(0.22)	0.75(0.18)	0.50(0.13)	0.76(0.15)
		0.49	**0.79**	**0.50**	**0.79**
$Bias(\theta)^2$		20.31(98.05)		1.25(2.39)	
		1.34		**0.79**	
$\%ERR$		30.6(18.7)		26.4(16.2)	
		23.2		**18.2**	

Table 4. Simulation results for $\eta_1 = 0.7$ and 100 time-series of length T. Mean, standard error (italics) and median (bold) of the estimates.

		$T = 500$		$T = 1000$	
$\hat{\Pi}$		0.59(0.25)	0.41(0.25)	0.61(0.24)	0.39(0.24)
		0.60	**0.40**	**0.61**	**0.39**
		0.33(0.21)	0.67(0.21)	0.30(0.19)	0.70(0.19)
		0.26	**0.74**	**0.27**	**0.73**
$\hat{\alpha}_1, \hat{\alpha}_2$		4.15(31.39)	10.58(53.98)	0.89(0.37)	5.96(51.40)
		0.81	**0.56**	**0.86**	**0.54**
$\hat{\beta}_1, \hat{\beta}_2$		2.15(11.52)	80.26(664.83)	0.97(0.20)	5.77(36.67)
		0.99	**1.53**	**0.99**	**1.56**
$\hat{\gamma}_1, \hat{\gamma}_2$		0.51(0.26)	0.80(0.19)	0.53(0.24)	0.81(0.16)
		0.53	**0.83**	**0.50**	**0.87**
$\hat{\eta}_1, \hat{\eta}_2$		0.64(0.23)	0.77(0.19)	0.64(0.19)	0.78(0.17)
		0.69	**0.80**	**0.68**	**0.80**
$Bias(\theta)^2$		85.93(666.99)		7.86(63.01)	
		1.46		**1.24**	
$\%ERR$		37.8(15.2)		30.0(15.0)	
		35.8		**31.0**	

5   A Case Study on Historical Data

The aim of this section is to apply the proposed model in studying the rhythms of the Duchy of Savoy, during the XVIth and the XVIIth centuries. These two centuries were marked by deep political changes and by several long and intense wars. It was a period during which the Duchy changed and shaped its structure and its functioning as a state. More specifically, we are interested in the production of legislative texts related to military logistics, as compared to the entire production of law. The corpus of data comes from the massive work of F-A. Duboin [10,11]. According to [11], this edition would be exhaustive, and few texts would be missing. Between 1559 and 1661, the State as a whole issued 55.5 law texts per year (in average), of which 4.8 in connection with military logistics. The ratio between the texts on military logistics and the total number of texts of law varies between 0 and 1, and, as one may see in Fig. 2, the importance of military logistics is far from negligible.

In order to evaluate the temporality of the State, the selected corpus of documents is thus represented as a time series. After having considered several representations (yearly, quarterly, monthly), the monthly approach was selected, as being sufficiently fine for one to observe the closeness between making the decision and issuing the associated text of law. A full description of the corpus and of its construction is available in [12]. In a previous work [13], the series of texts related to military logistics only was studied, using hidden Markov models for count data. The results appeared as very encouraging and they allowed to point out several specificities of the historical period of interest such as short-term events, but also a long transition between two normative systems of the Duchy. However, in [13] the relation between the texts on military logistics and the entire production of law was left out. In this manuscript, we specifically focus on this point and study the ratio computed between the number of texts on military logistics and the complete production of law, recorded each month. The resulting series is of length 1 236 and has an important over-dispersion in 0.

After having trained a two-state hidden Markov model with Beta-inflated distributions, the estimated parameters are the following:

$$\hat{\Pi} = \begin{pmatrix} 0.83 & 0.17 \\ 0.26 & 0.74 \end{pmatrix} \ , \ (\hat{\alpha}_1, \hat{\alpha}_2) = (5.92; 4.11) \ , \ (\hat{\beta}_1, \hat{\beta}_2) = (8.07; 15.83),$$

$(\hat{\gamma}_1, \hat{\gamma}_2) = (0.01; 0.02)$ and $(\hat{\eta}_1, \hat{\eta}_2) = (0.87, 0.45)$.

The Viterbi algorithm allows to track the a posteriori probabilities of the hidden states and to represent the estimated trajectory of the hidden Markov chain, as shown in Fig. 3. The results are globally consistent with our previous findings in [13]. The second regime is more persistent during the XVIIth century, which mainly corresponds to the end of the long transition period found in our previous work. However, the two regimes do not appear as being stable in time and the Markov chain is changing often from one state to another. From this point of view, the results on the ratio data appear as less convincing than the results on logistic texts only. Some adjustments should thus be made in order

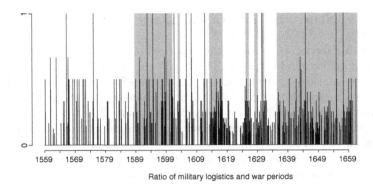

Fig. 2. Ratio of texts on military logistics among the entire production of law. The periods of war for the Duchy are in grey.

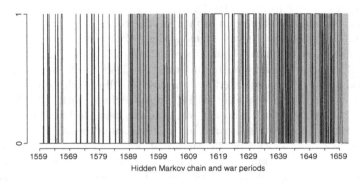

Fig. 3. A posteriori estimated hidden states of the model. The periods of war for the Duchy are in grey.

to improve our approach and one way to tackle this would be to restrict the Beta-inflated distributions in the HMM either to 0 or to 1 additional probability masses. Indeed, this assumption would favor the hypothesis of a regime with an intense production of texts on military logistics versus a regime with a low production.

6   Conclusion

The present manuscript introduces the Beta-inflated distributions in the framework of hidden-Markov models. The results on simulated examples show that the EM algorithm usually manages quite well in estimating the parameters of the model, provided the time series is sufficiently long and provided the parameters of the regimes are sufficiently different. We aim at further improving this section, by adding more, and more various, examples. Finally, the results on the real dataset are consistent with previous findings, although less convincing in terms of stability of the regimes. We are currently studying the possibility of restricting the Beta-inflated distribution in the model to 0 or to 1 additional probability masses.

References

1. Wallis, K.F.: Time series analysis of bounded economic variables. J. Time Ser. Anal. **8**(1), 115–123 (1987)
2. Ferrari, S., Cribari-Neto, F.: Beta regression for modelling rates and proportions. J. Appl. Stat. **31**(7), 799–815 (2004)
3. Smithson, M., Verkuilen, J.: A better lemon squeezer? maximum-likelihood regression with beta-distributed dependent variables. Psychol. Methods **11**(1), 54 (2006)
4. Simas, A.B., Barreto-Souza, W., Rocha, A.V.: Improved estimators for a general class of beta regression models. Comput. Stat. Data Anal. **54**(2), 348–366 (2010)
5. Ospina, R., Ferrari, S.L.: A general class of zero-or-one inflated beta regression models. Comput. Stat. Data Anal. **56**(6), 1609–1623 (2012)
6. Baum, L., Petrie, T.: Statistical inference for probabilistic functions of finite state markov chains. Ann. Math. Stat. **37**(6), 1554–1563 (1966)
7. Ospina, R., Ferrari, S.L.P.: Inflated beta distributions. Stat. Pap. **51**(1), 111 (2008)
8. Dempster, A., Laird, N., Rubin, D.: Maximum likelihood from incomplete data via the em algorithm. J. Roy. Stat. Soc. (B) **39**(1), 1–38 (1977)
9. Rabiner, L.R.: A tutorial on hidden markov models and selected applications in speech recognition. Proc. IEEE **77**(2), 257–286 (1989)
10. Duboin, F.A.: Raccoolta per ordine di materie delle leggi cio editti, manifesti, ecc., pubblicati negli stati della Real Casa di Savoia fino all'8 dicembre 1798, Torino (1818–1869)
11. Couzin, T.: Contribution piémontaise à la genèse de l'État italien. L'historicité de la Raccolta per ordine di materie delle leggi (1818–1868). Bolettino Storico-Bibliografico Subalpino **CVI**, 101–120 (2008)
12. Alerini, J.: La Savoie et le "Chemin espagnol", les communautés alpines à l'épreuve de la logistique militaire (1560–1659). Ph.D thesis, Université Paris 1 Panthéon-Sorbonne (2012)
13. Alerini, J., Olteanu, M., Ridgway, J.: Markov and the Duchy of Savoy: segmenting a century with regime-switching models. https://hal.archives-ouvertes.fr/hal-01442314 (January 2017, to appear in Journal de la SFdS)

Forecasting Univariate Time Series by Input Transformation and Selection of the Suitable Model

German Gutierrez[✉], M. Paz Sesmero, and Araceli Sanchis

Computer Science Department,
Carlos III University of Madrid, Leganes, Madrid, Spain
{ggutierr, msesmero, masm}@inf.uc3m.es

Abstract. Several tasks in science, engineering, or financial are related with sequences of values throughout the time (time series). This paper is focused in univariate time series, so unknown future values are obtained from k previous (and known) values. To fit a model between independent variables (present and past values) and dependent variables (future values), Artificial Neural Networks, which are data driven, can get good results in its performance results. In this work, we present a method to find some alternatives to the ANN trained with the raw data. This method is based on transforming the original time series into the time series of differences between two consecutive values and the time series of increment (-1, 0, $+1$) between two consecutive values. The three ANN obtained can be applied in an individual way or combine to get a fourth alternative which result from the combination of the other. The method evaluates the performance of all alternatives and take the decision, on validation subset, which of the alternatives could improve the performance, on test subset of the ANN trained with raw data.

Keywords: Artificial neural networks · Computational intelligence · Time series forecasting · Taguchi's method · Design of experiments

1 Introduction

Data in science, engineering, and financial markets, are found to be a temporal sequence forming the well-known time series. This kind of temporal relation of data is found in different fields, from nature (i.e. tides or the Niño Phenomenon) to industry or business (load and demand of electricity, stock markets) among others.

A sequence of values which are usually regularly sampled in time $(x_1, x_1, \ldots, x_{t-1}, x_t)$ is a time series. Depending of the relation of the values, two kinds of series can be found: univariate and multivariate time series. Univariate time series refers to a sequence of one scalar measure (e.g. *NOx* concentrations at main cities taken hourly or twice a day); in this case, the forecast models try to give an estimation for the relation between past and future values in the time series. On the other hand, multivariate time series show more than one different measures (or time series variables), and the aim of the studies related with them is "*to model and explain the interactions and co-movements among a group of time series variables*".

© Springer International Publishing AG 2017
I. Rojas et al. (Eds.): IWANN 2017, Part II, LNCS 10306, pp. 210–221, 2017.
DOI: 10.1007/978-3-319-59147-6_19

When dealing with time series data, two main tasks can be considered: analysis, and forecasting. Analysis is related with calculating the statistical characteristics of the data, and their internal structure. Forecasting deals with computing the new and unknown future values. The unknown values are obtained applying a model as a function f that relates h values ahead (future values to be forecasted) and k target values (previously known), Eq. (1). This work is focused on univariate time series, and forecast one step ahead ($h = 1$).

$$x(t+h), x(t+h-1), \ldots, x(t+1) = f(x(t), x(t-1), \ldots x(t-k+1)) \qquad (1)$$

In the literature, different techniques can be found to solve the time series fore-casting problem. Some of them are statistical approaches that result in linear models as ARIMA (Autoregressive Integrated Moving Average), which extract the autocorrela-tion(s) of the values considering the time gap [1]. Nevertheless, there are other approaches related with Machine Learning techniques: Artificial Neural Networks (ANN) [2, 3], Fuzzy techniques [4] and Support Vector Machines [5, 6]. This work is focused on applying ANN as a part of a more complex system to forecast univariate time series.

Additionally, in Machine Learning there are sophisticated techniques that improve the performance of the models obtained individually for different kind of tasks (clas-sification or regression). Some of these techniques combine the different answers given by several models following the hypothesis that the answer obtained from a group of experts is better than the decision given by only one of them.

1.1 Objectives

In this work, we propose a method to decide if an alternative to a single ANN model for the raw (original) time series data can improve its forecasting performance. The alternative system is based on the combination of three ANN models. The original input data (i.e. the raw time series data) is transformed into two additional time series: differencing (*dif*) time series ($d_t = x_t - x_{t-1}$) and the increment (*inc*) time series ($inc_t = \{+1, 0, -1\}$)

Then a different ANN for each of these three time series (*raw*, *dif* and *inc*) will be obtained. To get this three different ANN, and as hyper-parameter setting method, the Taguchi's method, a technique from Design of Experiments (DoE), will be applied. In fact, this technique (Taguchis's method and DoE) needs additional studies and further experimentation to show its full potential within machine learning. Finally, as a first approach, a simple rule based on the performance of the three individual ANNs (*raw*, *dif* and *inc*) and the combination of the output of these three ANNs, evaluated on the validation subset, will decide if the alternative (i.e. the combination of the three models for *raw*, *dif* and *inc*) to the single ANN for raw data could get a better result for test subset.

The rest of the paper is structured as follows. In Sect. 2 review briefly the related work. In Sect. 3 we explain our alternative method, Sect. 4 is for the experimental results, and finally the Conclusions and future works.

2   Related Work

In time series domain, there are different problems to be accomplished as: time series classification, outlier identification, detecting unusual patterns [7], but also time series forecasting.

The time series forecasting task lies in computing the unknown future values by the application of a model (f). The inputs (independent variables) of the model are the known previous values, and it will obtain one or several ahead values

$$x_{t+h}, x_{t+h-1}, \ldots, x_{t+1} = f(x_t, x_{t-1}, \ldots x_{t-k+1}) \qquad (2)$$

There are several techniques to solve the time series forecasting computational task. Some of these techniques are based on statistical techniques as autoregressive integrated moving average (ARIMA), which extract the autocorrelation(s) of the values considered the time [1], but other on Computational Intelligence techniques as Artificial Neural Networks [2, 3], Fuzzy techniques [4] and Support Vector Machines [5, 6].

In machine learning, learning algorithms find a hypothesis for the underlying relation between the label (classification) or an output value (regression), and the attributes of an input pattern. Additionally, the combination of the answer given by a set of (different) models is considered an alternative for obtaining an improvement in forecasting accuracy. This same idea is behind the autoregressive integrated moving average (ARIMA), based on statistical techniques. Its goal is to combine the information given by the previous values, the error terms, and the difference among the values through the time. Some of the main computational intelligence techniques to solve the time series forecasting task are artificial neural networks, fuzzy systems, and support vector machines.

2.1   Artificial Neural Networks for Time Series Forecasting

Artificial Neural Networks (ANN) are applied to several different computational tasks as classification or ordinal regression problems, and regression. Therefore, they (ANN) are also applied to the specific domain of time series, where there are different issues to be accomplished as: time series classification, outlier identification or detecting unusual patterns, but also time series forecasting. All different kinds of ANN which can be applied to regression task, as multilayer perceptron, radial basis function networks, and recurrent neural networks can also be applied to time series forecasting (references).

Designing the Artificial Neural Network

The performance of an ANN is established by their computational elements (the artificial neurons), their pattern of links (how are they interconnected) which carry the values throughout the neurons, and the weights (parameters) of those links. The final ANN weights are stablished by a learning algorithm that, starting from an initial set of random weights and using a set of learning patterns, adapt them to decrease the error of the output (the difference between the output of the net and the target values). However, the topology of the net (the nodes, and how are they linked), but also other parameters (hyper-parameters) as the specific learning algorithm applied, etc., are obtained by trial and error methods.

The problem of setting the hyper-parameters of an ANN which outperform its error, i.e. a minimization problem, can be seen as a search problem. Of course, the full factory design: compute the ANN performance for any available combination of all the values (a.k.a. levels) for all the hyper-parameters (a.k.a. factors), ensures to find the best combination. However, the computational complexity of the full factory design is exponential with the number of parameters. Nonetheless, there are in the literature different alternatives to solve this issue as a *"stepwise selection"* method [3], or metaheuristics as evolutionary computation [8].

Design of Experiments
The hyper-parameter optimization with limited resources is also afforded in other areas, as statistics or industrial engineering, applying a set of techniques grouped under the name of Design of Experiments (DoE). These include (among others) full factorial design, fractional factorial designs, or Taguchi's designs [10]. In fact, Taguchis's design is a fractional factorial design based on orthogonal arrays, and one of its main contributions is the hyper-parameter design. Some works [11, 12] have applied these techniques within machine learning community. Nevertheless, there are few examples (and in very specific tasks) for the claim of DoE as an application of statistics methodology to the network topology and hyper-parameter optimization. So, to check the full potential of this technique requires further studies and more extensive experimentation.

3   Approach Description

In this section, we explain the approach: (i) how to transform the time series into two additional time series; (ii) how to get the hyper-parameters which optimize the performance of the Model for the three versions of the time series (*raw*, *dif* and *inc*); (iii) how to set up the combination of the three (*raw*, *dif*, and *inc*) ANN models, and (iv) how to decide if one of the alternatives, *dif*, *inc* or the combination of the three, to the single ANN for the raw data could get a better result in forecasting task.

This approach is made up of following steps:

i. turn the initial time series forecasting task (called as *raw* forecasting task) into two additional time series, *differential* (*dif*) and *increment* (*inc*).
   (1) Differential time series. To get this transformed data, the difference between two consecutive raw data is assign to the value at time t

$$dif_t = x_t - x_{t-1} \qquad (3)$$

(2) Incremental time series. The raw time series data is transformed into a sequence of values in the set $\{-1, 0, +1\}$ which indicates at time t if the raw time series data decreases, stays at the same value or increases,

$$inc_t = \begin{cases} +1, x_t > x_{t-1} \\ -1, x_t < x_{t-1} \\ 0, x_t = x_{t-1} \end{cases} \tag{4}$$

ii. Get a different model (ANN), to accomplish its own forecasting task, for each of the three time series, *raw* (the original data), *dif*, and, *inc*. To get each model, we apply Taguchi's orthogonal arrays [10] on each specific forecasting task. The details of this step are explained in Subsect. 3.1

iii. Combine the three models obtained in step (ii), getting a new fourth model. Once we have four alternatives (*raw*, *dif*, *inc*, and its combination) to model the time series forecasting task, they are evaluated on the validation (known data values) subset. The option selected for the test subset (i.e. the future unknown values to be predicted at a competition or to be provided when the system is production phase) will be the one that has a better performance for the validation set.

In the Sect. 4, Experimental Results we will check if when the model selected in step ii is different than the single model for raw data (or baseline), it gets a better result for the test subset.

3.1   ANN Hyper-parameter Setting

To apply an ANN to forecast time series, it is necessary to pre-process the time series data generating the pattern set. For each $t \in [1, tslenght]$, the attributes (inputs) of pattern t are obtained from the k past values $x_t, x_{t-1}, \ldots, x_{t-k+1}$, the h values to forecast $x_{t+h}, \ldots, x_{t+1}$ will be the targets for that pattern. The time series is split in two parts, the first part for learning (training and validation) and test.

In this work the forecast is carried out one step ahead, so the model gives just the next value in time. Then, to forecast several future values with that model, each of them, x_{t+1}^f, x_{t+2}^f, etc. will be computed one by one, Eq. (5). This process continues till value at $t+h$, x_{t+h}^f, is computed.

$$\begin{aligned} x_{t+1}^f &= f_{ANN}(x_t, x_{t-1}, \ldots, x_{t-k}), \\ x_{t+2}^f &= f_{ANN}(x_{t+1}, x_t, x_{t-1}, \ldots, x_{t-k+1}) \end{aligned} \tag{5}$$

For each of the three versions of the time series (*raw*, *dif*, and *inc*), a different ANN is obtained. The parameters (the weights of the net) are obtained by executing a learning algorithm. However, and before the training process, other hyper-parameters of the net (a multilayer perceptron with a single hidden layer) must be also settled. In this work, the hyper-parameters are the following:

- the *number of inputs nodes*, which are related with how many backward steps, k in Eqs. 2 and 5, are considering to compute the one step ahead value.
- the *number of neurons* in the single hidden layer, *hidden nodes*
- the *learning algorithm* to compute set the weights of the net
- the *learning rate* (α) of the learning, and
- the *training cycles* of the learning algorithm.

The method to fix the values of the hyper-parameters that maximize the performance of each ANN for *raw* (x_t), *dif* (dif_t) and *inc* (inc_t) time series, is an experimental design (*"or a method to ensure good performance at design stage of the ANN"* [10])[1]. Therefore, we perform in this work the Taguchi's orthogonal arrays method from Design of Experiments (DoE) to obtain a set of values of the hyper-parameters to optimize the performance of each ANN trained within the system. This method is an alternative to execute the full factorial combinations of all the values (*levels*) established by an expert for all the hyper-parameters (*factors*) indicated above. The Taguchi's method [12] set a table for a list of specific combinations of factors and levels. This table is based in orthogonal arrays (Latin squares), and the number of combinations of K factors (in this work, $K = 5$) and L levels given by the table is much smaller (about 1.5 order of magnitude lower) than the number of combinations given by the full factory. In this work, the list of factors and their values (levels) are shown in Table 1.

Table 1. Factors (hyper-parameters) and Levels (values) to establish a better model

Factor	Levels
Number of inputs	5, 10, 15, 20
Hidden Nodes (HN)	5, 10, 15, 20, 25, 30, 35, 40
Learning algorithm	Scaled conjugate gradient [13]. Resilient back-propagation [14]
Learning rate	0.2, 0.1, 0.0.5, 0.01
Training cycles	5000, 10000, 20000, 50000

In this work, there are one factor with 2 levels, three factors with 4 levels, and one factor with 8 levels. So, we approximate the Hyper-Graeco-Latin Square Designs for 4-and-5-Level Factors 1 with 16 combinations. In this approximation, first we "translate" the 2 levels factor (learning algorithm) to a 4-level factor, where 2 of the levels means the algorithm, and second double the table, the first part with 4 levels (hn: 5, 10, 15, 20) up to 8 for the factor of hidden nodes, and then double the table for the other 4 values (hn: 25, 30, 35, 40). The final sequence of 32 combinations is shown in Table 2, instead of the 1024 ($4 \times 8 \times 2 \times 4 \times 4$) combinations for full factorial design.

Each run (i.e. combination) will be executed for training each single ANN for *raw*, *dif*, and *inc* time series data. But also, in order, to tackle the variability of the ANN obtained because the random set if initial weights, the 32 run will be executed for 10 different initializations.

---

[1] http://www.itl.nist.gov/div898/handbook/pri/section3/pri3323.htm.

Table 2. List of factors-levels combinations, to get the hyper-parameters

	in	hn	la	α	trc	run	in	hn	la	α	trc	run	in	hn	la	α	trc
	in: inputs; hn: hidden nodes; la: learning algorithm; α: learning rate; trc: training cycles.																
run	in	hn	la	α	trc	run	in	hn	la	α	trc	run	in	hn	la	α	trc
01	05	05	05	0,20	05	12	15	20	05	0,10	05	23	10	35	06	0,01	05
02	05	10	06	0,10	10	13	20	05	05	0,01	10	24	10	40	05	0,05	10
03	05	15	05	0,05	20	14	20	10	06	0,05	05	25	15	25	06	0,05	50
04	05	20	06	0,01	50	15	20	15	05	0,10	50	26	15	30	05	0,01	20
05	10	05	06	0,10	20	16	20	20	06	0,20	20	27	15	35	06	0,20	10
06	10	10	05	0,20	50	17	05	25	05	0,20	05	28	15	40	05	0,10	05
07	10	15	06	0,01	05	18	05	30	06	0,10	10	29	20	25	05	0,01	10
08	10	20	05	0,05	10	19	05	35	05	0,05	20	30	20	30	06	0,05	05
09	15	05	06	0,05	50	20	05	40	06	0,01	50	31	20	35	05	0,10	50
10	15	10	05	0,01	20	21	10	25	06	0,10	20	32	20	40	06	0,20	20
11	15	15	06	0,20	10	22	10	30	05	0,20	50	-	-	-	-	-	-

3.2   Combination of the Models

The aim of this module is to get three different alternatives to forecast the original (*raw*) time series data on the test set, to be forecasted at the final application of the system. First, all the error measures for the forecasting performance (e.g. root mean square error, etc.) are affected by the range of values to be forecasted, so all the models and their combination are evaluated on the same validation subset for raw data. Of course, the ANN which is focused on *dif* and *inc* data needs the last raw value of train subset to accomplish the forecast, Eq. (6). Besides, the ANN focused on *inc* data also needs the absolute value from the output given by the ANN for *dif* data, Eq. (7). And, the output of the ANN focused on *inc* version can or cannot round that output to one of the values of the input $\{-1, 0, +1\}$; this option turns into a parameter of this module, and Eq. (8).

Now for each of the 10 different initializations, we get the best of the 32 factor-level combination, and then from that 10 best, we could get the best of them (i.e. the best of the 32×10 ANN obtained) or the one closest to the median of the performance. So, we have a second parameter in this module: best of the bests or median of the bests, $bBmB \in \{0, 1\}$ respectively.

Answer from *ANNdif*:

$$x_{t+1}^{dif} = d_{t+1}^{dif} + x_t^{raw}. \tag{6}$$

Answer from *ANNinc*:

$$x_{t+1}^{inc} = inc_{t+1} * abs\left(x_{t+1}^{dif}\right) + x_t^{raw} \tag{7}$$

$$x_{t+1}^{raw} = round\left(x_{t+1}^{inc}\right) * abs\left(x_{t+1}^{dif}\right) + x_t^{raw} \tag{8}$$

Finally the output of the three ANN focused on *raw* (x_{t+1}^{raw}), *dif* (x_{t+1}^{dif}) and *inc* (x_{t+1}^{inc}) time series data are combined (x_{t+1}^{comb}) following one of the three options as indicated in Table 3. The combination of the output of each net is weighted, that weight are related with the error of the elements of the combination, so that if the error of one net is lower than the others, its weight in the combination will be higher.

Table 3. Three options for the combination of the outputs of x^{raw}, x^{dif}, x^{inc}, and its weights.

$x_{t+1}^{comb} = \left(w_{raw} \cdot x_{t+1}^{raw}\right) + \left(w_{dif} \cdot x_{t+1}^{raw}\right),$	$w_{raw} = \dfrac{error_{valid}^{dif}}{error_{valid}^{raw} + error_{valid}^{dif}};$ $w_{dif} = \dfrac{error_{valid}^{raw}}{error_{valid}^{raw} + error_{valid}^{dif}};$
$x_{t+1}^{comb} = \left(w_{raw} \cdot x_{t+1}^{raw}\right) + \left(w_{dif} \cdot x_{t+1}^{raw}\right)$ $+ \left(w_{inc} \cdot x_{t}^{raw}\right)$	$w_{raw} = \dfrac{error_{valid}^{dif} + error_{valid}^{inc}}{error_{valid}^{raw} + error_{valid}^{dif} + error_{valid}^{inc}}$ $w_{dif} = \dfrac{error_{valid}^{dif} + error_{valid}^{inc}}{error_{valid}^{raw} + error_{valid}^{dif} + error_{valid}^{inc}}$ $w_{inc} = \dfrac{error_{valid}^{dif} + error_{valid}^{inc}}{error_{valid}^{raw} + error_{valid}^{dif} + error_{valid}^{inc}}$
$x_{t+1}^{comb} = \left(w_{raw} \cdot x_{t+1}^{raw}\right) + \left(w_{inc} \cdot x_{t+1}^{inc}\right)$	$w_{raw} = \dfrac{error_{valid}^{raw}}{error_{valid}^{raw} + error_{valid}^{dif}};$ $w_{inc} = \dfrac{error_{valid}^{inc}}{error_{valid}^{raw} + error_{valid}^{inc}};$

So, finally our approach will have four models to map the relation between past values and future (and forecasted) values: *(i)* the ANN that model the raw time series values (x_t); *(ii)* the ANN that model the differencing values $(d_t = x_t - x_{t-1})$; *(iii)* the ANN that model the increment values (inc_t); *(iv)* a combination of two or three of the first three models. Each of these four model is evaluated on validation data set, and for a specific time series at hand (e.g. Mackey Glass) the best of them for validation subset (minimum error measure) will be selected to be the one which provides the output for the test subset.

With the experimental results, we will show if when the model selected from the four options is different than ANN_{raw} (the ANN that learns with the raw time series data), will this alternative to be applied on the test subset get a better result than ANN_{raw} on the test subset?

4   Experimental Evaluation

In this section, first we are going to explain the experimental setup, and then show the experimental results and what is the answer (yes or no) to the question already indicated above.

4.1   Experimental Setup

The approach shown in this work is evaluated in four time series widely applied in other works in the literature: Mackey-Glass, Dow-Jones, Quebec, and Temperature.

The last three time series has been collected from Hyndman's Time Series Data Library [15]. For each time series, the whole data is split in three subsets: train, validation, and test. Train subset is used by the learning algorithm (back-Propagation) to modify and fixed the ANN weights. Validation subset allows to state the generalization ability of the ANN result from learning process. Test subset is never seen by any ANN in the learning process, so it is incorporated to test the generalization ability of the system.

In Table 4 are shown, for each time series, the number of periods of the whole time series, and the number of times for train (60%), validation (20%) and test (forecast) (20%).

Table 4. Description of the datasets (time series)

Time series	Training (60%)	Validation (20%)	Test (20%)
Dow-Jones	441	146	146
Mackey-Glass	97	30	30
Quebec 584	438	146	146
Temperature	144	48	48

Different accuracy measures can be found in literature: mean square error, root mean square error (RMSE), mean absolute error, or symmetric mean absolute percentage error. In this work if there is a choice to be made, it will be made based on RMSE.

Table 5. Results for Dow-Jones (D) and Mackey-Glass (M)

ts	rO	wO	bO	VAL SUBSET (20%) - RMSE				TEST SUBSET (20%) -RMSE				Be
				raw	dif	inc	comb	raw	dif	inc	comb	
D	0	1	1	22,510	21,819	24,703	**19,961**	83,256	30,619	**27,198**	53,319	+
D	0	2	1	22,510	21,819	24,703	**20,659**	83,256	30,619	**27,198**	43,369	+
D	0	3	1	22,510	21,819	24,703	**20,828**	83,256	30,619	**27,198**	53,041	+
D	0	1	2	55,024	**21,819**	27,440	26,824	100,593	30,619	**24,828**	44,345	+
D	0	2	2	55,024	**21,819**	27,440	27,002	100,593	30,619	**24,828**	38,313	+
D	0	3	2	55,024	**21,819**	27,440	30,831	100,593	30,619	**24,828**	43,032	+
D	1	1	1	22,510	21,819	25,778	**19,961**	83,256	30,619	**26,779**	53,319	+
D	1	2	1	22,510	21,819	25,778	**20,878**	83,256	30,619	**26,779**	43,438	+
D	1	3	1	22,510	21,819	25,778	**21,089**	83,256	30,619	**26,779**	53,686	+
D	1	1	2	55,024	**21,819**	26,668	26,824	100,593	30,619	**25,877**	44,345	+
D	1	2	2	55,024	**21,819**	26,668	26,730	100,593	30,619	**25,877**	38,871	+
D	1	3	2	55,024	**21,819**	26,668	30,417	100,593	30,619	**25,877**	43,707	+
M	0	1	1	0,007	0,010	0,049	**0,005**	0,006	0,014	0,073	0,007	x
M	0	2	1	**0,007**	0,010	0,049	0,021	0,006	0,014	0,073	**0,031**	x
M	0	3	1	**0,007**	0,010	0,049	0,009	0,006	0,014	0,073	0,011	-
M	0	1	2	0,007	0,012	0,066	**0,006**	0,008	0,013	0,070	**0,006**	+
M	0	2	2	**0,007**	0,012	0,066	0,029	0,008	0,013	0,070	0,030	-
M	0	3	2	**0,007**	0,012	0,066	0,009	0,008	0,013	0,070	0,010	-
M	1	1	1	0,007	0,010	0,041	**0,005**	0,006	0,014	0,077	0,007	x
M	1	2	1	**0,007**	0,010	0,041	0,018	0,006	0,014	0,077	0,032	-
M	1	3	1	**0,007**	0,010	0,041	0,008	0,006	0,014	0,077	0,012	-
M	1	1	2	0,007	0,012	0,079	**0,006**	0,008	0,013	0,080	**0,006**	+
M	1	2	2	**0,007**	0,012	0,079	0,035	0,008	0,013	0,080	0,035	-
M	1	3	2	**0,007**	0,012	0,079	0,009	0,008	0,013	0,080	0,009	

4.2   Results

In Tables 5 and 6 it is shown the performance of the system for 4 time series through all the combinations of parameters explained in Sect. 3.2. In Table 5 are show the results for Dowjones and MackeyGlass. In Table 6 the results for Quebec and Temperature are shown.

Table 6. Results for Quebec (Q) and Temperature (T)

ts	rO	wO	bO	VAL SUBSET (20%) - rmse				TEST SUBSET (20%) -RMSE				Be
				raw	dif	inc	comb	raw	dif	inc	comb	
Q	0	1	1	20,273	20,593	25,752	**19,857**	23,222	22,515	28,147	**22,121**	+
Q	0	2	1	**20,273**	20,593	25,752	20,542	23,222	22,515	28,147	22,518	-
Q	0	3	1	**20,273**	20,593	25,752	21,056	23,222	22,515	28,147	23,597	-
Q	0	1	2	20,477	22,068	27,602	**20,337**	22,652	24,848	31,251	**22,454**	+
Q	0	2	2	**20,477**	22,068	27,602	21,290	22,652	24,848	31,251	23,795	-
Q	0	3	2	**20,477**	22,068	27,602	21,581	22,652	24,848	31,251	24,113	-
Q	1	1	1	20,273	20,593	26,887	**19,857**	23,222	22,515	30,644	**22,121**	+
Q	1	2	1	**20,273**	20,593	26,887	20,933	23,222	22,515	30,644	22,803	-
Q	1	3	1	**20,273**	20,593	26,887	21,381	23,222	22,515	30,644	24,032	-
Q	1	1	2	20,477	22,068	29,543	**20,337**	22,652	24,848	33,569	**22,454**	+
Q	1	2	2	**20,477**	22,068	29,543	21,802	22,652	24,848	33,569	24,218	-
Q	1	3	2	**20,477**	22,068	29,543	22,044	22,652	24,848	33,569	24,544	-
T	0	1	1	2,259	2,216	2,455	**2,034**	2,588	2,743	3,103	**2,460**	+
T	0	2	1	2,259	2,216	2,455	**2,098**	2,588	2,743	3,103	**2,487**	+
T	0	3	1	2,259	2,216	2,455	**2,133**	2,588	2,743	3,103	**2,522**	+
T	0	1	2	2,259	2,846	2,721	**2,220**	2,588	3,065	3,982	**2,589**	+
T	0	2	2	**2,259**	2,846	2,721	2,307	2,588	3,065	3,982	2,680	-
T	0	3	2	2,259	2,846	2,721	**2,153**	2,588	3,065	3,982	2,700	x
T	1	1	1	2,259	2,216	2,458	**2,034**	2,588	2,743	3,025	**2,460**	+
T	1	2	1	2,259	2,216	2,458	**2,109**	2,588	2,743	3,025	**2,464**	+
T	1	3	1	2,259	2,216	2,458	**2,157**	2,588	2,743	3,025	**2,494**	+
T	1	1	2	2,259	2,846	2,707	**2,220**	2,588	3,065	3,992	**2,589**	+
T	1	2	2	2,259	2,846	2,707	**2,303**	2,588	3,065	3,992	2,686	x
T	1	3	2	2,259	2,846	2,707	**2,148**	2,588	3,065	3,992	2,716	x

In each table, the 3 columns after time series name are the three parameters explained in Sect. 3.2: rO, round (1) or not (0) the output of the *ANNinc*; bO indicates if we take the best of the bests (1), or take the median from the 10 best results (2); wO indicates which combination of the three models is considered (1, *raw* and *dif*; 2, *raw*, *diff* and *inc*). Then the RMSE error for each ANN: *raw*, *dif*, *inc* and its combination, is shown for validation subset, and then test subset. The cell in grey are those which are better than the other in validation subset, and test subset. The last column (Be) indicates if best alternative in validation subset is the raw data (−), or a different one. If the option selected to be applied on the test subset is not the model for raw data, it is check if this option has got a better (+) result or not (x).

So the total number of combinations are 12 for each time series, for a total of 48 experiments. In 17 tries (rows) from 48 experiments, the model selected is the ANN which is learned with the raw data, and for 31 experiments the method selected at validation subset is different. For this 31 experiments it is checked in the table (column Be) if the performance of the model selected in validation gets a better result in test subset than the ANN based in data raw. Checking the table, we can observe that for 26 of 31 (83.8%) experiments if there is a better alternative to learn only from raw data in validation subset, that alternative get also better results in 83.8% of the cases.

5   Conclusions and Future Works

In this work, we initially define an approach to propose an alternative to the ANN (or model) that learns only from the *raw* time series data. This alternative can be the model which learns with the differential (*dif*) data and the incremental (*inc*) data. But also, we define an additional and alternative model which is constructed with a weighted combination of the three (*raw, dif, inc*) models.

To get the four model, instead of a full factorial or trial and error, it is established a set of hyper-parameter combinations based on Taguchi's orthogonal vectors to optimize the performance of the nets.

Finally, the alternative selected, with a very single rule (the one with better performance on validation subset) is the one to be applied for the test (unknown and future value) never seen before for the models. If our approach finds an alternative to ANN which learns from raw data, an 83 of the times, this alternative model is going to outperform the *ANNraw* model.

Future Works

Among the future works we must extend the experimentation to an additional number of time series to endorse the result from this works. Also, the same framework could be applied but in this case with different computational intelligence techniques, for instance Support Vector Machine. And additionally, we could apply an additional machine learning (artificial neural networks, or support vector machines) to learn how to combine the model (stacking).

Acknowledgement. This work has been supported by the Spanish MICINN under projects: TRA2015–63708-R, TRA2013–48314-C3-1-R, and TRA2016-78886-C3-1-R.

References

1. Box, G.E.P., Jenkins, G.: Time Series Analysis, Forecasting and Control, 4th edn. Wiley, Hoboken (2007)
2. Haykin, S.: Neural Networks: A Comprehensive Foundation, 3rd edn. Prentice-Hall Inc., Upper Saddle River (2007)
3. Crone, S.: Stepwise selection of artificial neural network models for time series prediction. J. Intell. Syst. **14**, 99–122 (2005)

4. Stepnicka, M., Dvorak, A., Pavliska, V., Vavrickova, L.: A linguistic approach to time series modeling with the help of F-transform. Fuzzy Sets Syst. **180**(1), 164–184 (2011). Fuzzy Transform as a New Paradigm in Fuzzy Modeling. http://www.sciencedirect.com/science/article/pii/S0165011411000996
5. Cao, L.J., Tay, F.E.H.: Support vector machine with adaptive parameters in financial time series forecasting. IEEE Trans. Neural Netw. **14**(6), 1506–1518 (2003)
6. Iglesias, A., Gutierrez, G., Ledezma, A., Sanchis, A.: Time series forecasting using artificial neural networks vs. evolving models. In: 2014 IEEE Conference on Evolving and Adaptive Intelligent Systems (EAIS), pp. 1–7, June 2014
7. Kourentzes, N., Crone, S.F.: Semi-supervised monitoring of electric load time series for unusual patterns. Int. Joint Conf. Neural Netw. **2011**, 2852–2859 (2011)
8. Ojha, V.K., Abraham, A., Snášel, V.: Metaheuristic design of feedforward neural networks: a review of two decades of research. Eng. Appl. Artif. Intell. **60**, 97–116 (2017)
9. Gatti, C.: Design of Experiments for Reinforcement Learning. Springer Theses, Springer (2014)
10. NIST/SEMATECH e-Handbook of Statistical Methods (2013). http://www.itl.nist.gov/div898/handbook/pri/pri.htm. Accessed Feb 2017
11. Balestrassi, P.P., Popova, E., Paiva, A.P., Marangon, J.W.: Design of experiments on neural network's training for nonlinear time series. Forecast. Neurocomput. **72**(4–6), 1160–1178 (2009)
12. Roy, R.K.: A Primer on the Taguchi Method, p. xiii, 247 p. Van Nostrand Reinhold, New York (1990). ISBN 0442237294
13. Braun, H., Riedmiller, M.: Rprop: a fast and robust backpropagation learning strategy. In: Proceedings of the ACNN (1993)
14. Riedmiller, M.: Rprop – description and implementation details. Technical report (1994)
15. Hyndman, R.: Time series data library (2016). http://data.is/TSDLdemo. Accessed Apr 2016

Machine Learning Applied to Vision and Robotics

Vehicle Classification in Traffic Environments Using the Growing Neural Gas

Miguel A. Molina-Cabello$^{(\boxtimes)}$, Rafael Marcos Luque-Baena,
Ezequiel López-Rubio, Juan Miguel Ortiz-de-Lazcano-Lobato,
Enrique Domínguez, and José Muñoz Pérez

Department of Computer Languages and Computer Science,
University of Málaga, Bulevar Louis Pasteur, 35, 29071 Málaga, Spain
{miguelangel,rmluque,ezeqlr,jmortiz,enriqued,munozp}@lcc.uma.es
http://www.lcc.uma.es/~ezeqlr/index-en.html

Abstract. Traffic monitoring is one of the most popular applications of automated video surveillance. Classification of the vehicles into types is important in order to provide the human traffic controllers with updated information about the characteristics of the traffic flow, which facilitates their decision making process. In this work, a video surveillance system is proposed to carry out such classification. First of all, a feature extraction process is carried out to obtain the most significant features of the detected vehicles. After that, a set of Growing Neural Gas neural networks is employed to determine their types. A qualitative and quantitative assessment of the proposal is carried out on a set of benchmark traffic video sequences, with favorable results.

Keywords: Foreground detection · Background modeling · Probabilistic self-organising maps · Background features

1   Introduction

The field of traffic monitoring has generated great excitement in recent years within the intelligent transport systems community due to the increase of hardware development, the low cost sensor technologies and the improvement in the development and optimization of data processing algorithms. Specifically, the video detection and monitoring solutions for traffic applications can help to improve the performance in traffic management [3,10,13]. Thus, for example, if a high frequency of heavy vehicles is detected in one of the analyzed road sections, it is possible to redirect the traffic in a previous point with the aim of avoiding traffic congestion.

Automatic video surveillance systems can be divided into several phases [1,2]. A first step involves the detection of moving objects within the scene; a second stage performs monitoring tasks to associate the same vehicle detected in all frames of the sequence in which it appears; and finally a feature detection phase to extract relevant knowledge of the movement of these objects, their behavior and appearance.

© Springer International Publishing AG 2017
I. Rojas et al. (Eds.): IWANN 2017, Part II, LNCS 10306, pp. 225–234, 2017.
DOI: 10.1007/978-3-319-59147-6_20

Each stage builds on the previous one, which implies that it is needed to have implemented the steps of detecting and tracking objects if it is intended to conduct an analysis of the detected vehicle. In this paper, some methods previously developed by our research team for object detection and tracking are combined with other techniques to yield a vehicle classification system. Specifically, a self-organizing neural network is applied to cluster the pixels in background and foreground layers in order to detect which pixels are in motion inside the scene [12]. Subsequently, a Kalman model for multiple objects is applied to determine the trajectory of each vehicle which appears in the scene [16].

Therefore, the aim of this work is to classify the detected vehicles in four categories: car, motorcycle, truck and van. A feature extraction process is required in order to obtain robust and discriminant characteristics which can differentiate correctly among the groups of vehicles. This analysis would help to manage and distribute the traffic more efficiently in the analyzed area. Other works in the literature have the same aim, although some of them apply different classification or clustering methods [4,8] or start from other methodologies associated to video surveillance systems [11]. In this case, the Growing Neural Gas model (GNG) is considered, since it has been used successfully in different classification problems, from novelty detection [5] to text classification [17], or even issues related to medicine or biology such as osteoporosis detection [15].

The rest of the paper is organized as follows: Sect. 2 presents our vehicle classification model, divided in an explanation of the base method (Subsect. 2.1) and its application to our classification problem (Subsect. 2.2). Section 3 shows several experimental results over several well-known public traffic surveillance sequences. Finally, Sect. 4 outlines the conclusions of the paper.

2   Model

This section describes a multi-class classification system consisting of a combination of several self-organizing models based on the growing neural gas approach.

2.1   Growing Neural Gas

The growing neural gas (GNG), [6], is an unsupervised neural network suited to vector quantization and clustering due to its capability of finding a topological structure which reflects the topology of the input distribution.

Let $X = \{x \in \mathbb{R}^d\}$ be a set of M training patterns in a d-dimensional space. Let GNG be a growing neural gas model composed of N neurons. Initially, GNG starts placing $N = 2$ neurons randomly in the input space, which are supposed to be neighbors in the topological structure and, thus, connected by an edge.

Throughout the training phase, the topology is modified in two ways. First, the centroids of the clusters which the neurons represent w_i change, in order to adapt the self-organizing map that the GNG is creating to the input data distribution. For that purpose, the competitive learning rule is used. The competitive rule states that in a training iteration t only the neuron that best represents the currently presented training pattern x_t, i.e. the neuron whose centroid is closest to it, can modify its own internal data, which in case of a GNG neural network are the neuron centroid w_{win} and a variable $error_{win}$ containing the accumulated error that would be obtained if a vector quantization was carried out.

$$win(t) = argmin_{1 \le j \le N}\{\|x_t - w_j(t-1)\|^2\} \tag{1}$$

$$w_i(t) = \begin{cases} w_i(t-1) + \eta_{win}(t)\ (x_t - w_i(t-1)) & \text{if } i = win(t) \\ w_i(t-1) & \text{otherwise} \end{cases} \tag{2}$$

$$error_{win}(t) = error_{win}(t-1) + \|w_{win}(t-1) - x_t\|^2 \tag{3}$$

where $\|\cdot\|$ is the Euclidean norm.

In order to allow the map to self-organize and maintain the shape of the input distribution that it has captured, the neurons that are neighbors of the winner neuron are also able to slightly update their centroid position. In this case, a much smaller learning rate η_{neigh} is used for those neurons adjacent to the winner. Therefore, that new case is added to Eq. 2 and the final equation to model the centroid position update is

$$w_i(t) = w_i(t-1) + \begin{cases} \eta_{win}(t)\ (x_t - w_i(t-1)) & \text{if } i = win(t) \\ \eta_{neigh}(t)\ (x_t - w_i(t-1)) & \text{if } i \in Neighbors(win(t)) \\ 0 & \text{otherwise} \end{cases} \tag{4}$$

where $\eta_{win} : \mathbb{N} \longrightarrow [0,1]$ and $\eta_{neigh} : \mathbb{N} \longrightarrow [0,1]$ are two constant or monotone decreasing functions which satisfy $\forall t \in \mathbb{N} \quad \eta_{win}(t) > \eta_{neigh}(t)$.

Second, during the training phase new neurons are created periodically and useless neurons, also known as dead ones, are removed from the self-organizing-map. This is one of the advantages of growing neural networks such as the GNG, the number of neurons which are part of the model does not have to be fixed a priori by the user. However, a maximum number of neurons may be defined and it prevents GNG from creating too many neurons.

Every λ iterations a new neuron is inserted in the self-organizing map. The selection of the place to insert the neuron is based on the performance of the existing neurons. Those neurons with a high accumulated error perform poorly because the cluster which they represent are heterogeneous or contain many more elements than other clusters. Therefore, the new neuron centroid will be placed in the middle point of the edge that connect the neuron u with the largest error and the neuron v which is the neighbor of u accumulating the largest error.

$$w_{new} = \frac{w_u + w_v}{2} \tag{5}$$

As *new* is between u and v the edge connecting both of them is replaced by two new edges: one from u to *new* and other one from *new* to v. The new neuron is assumed to represent some of the patterns which previously belonged to the clusters corresponding to u and v, thus the accumulated error of the three neurons are updated accordingly.

$$error_u(t) = \alpha \cdot error_u(t) \tag{6}$$

$$error_v(t) = \alpha \cdot error_v(t) \tag{7}$$

$$error_{new}(t) = error_u(t) \tag{8}$$

where $\alpha \in [0, 1]$ is considered the estimated fraction of accumulated error that is reduced after inserting the new neuron.

On the other hand, the edges between neurons have an associated variable *age* which is incremented as the training advances. Age of the edges connecting neurons which keep on winning the competition, i.e. those that are not dead, are refreshed and set to 0 (see Subsect. 2.1). Those edges whose age is greater than a given threshold a_{max} are removed, since the neurons they connect did not win the competition recently. If some neurons become isolated after removing edges, they are considered dead and are also removed from the map.

GNG Training Algorithm

1. The network is initialized by creating two randomly positioned nodes which are connected by an edge. Their accumulated errors are set to 0.
2. In time instant t, select randomly a vector $\boldsymbol{x}_t$ which has not been presented to the network previously, if possible.
3. Determine the winning neuron win using Eq. 1 and also the runner-up neuron rup with a reference vector $\boldsymbol{w}_{rup}$ such that $\|\boldsymbol{x}_t - \boldsymbol{w}_{rup}(t-1)\|^2$ is the second smallest, for all neurons N.
4. Update the accumulated error of the winning neuron by means of Eq. 3. Update the network topology using Eq. 4.
5. Age of all edges connecting win to its topological neighbors are incremented by one.
   If win and rup are connected by an edge, the age of that edge is set to 0. Otherwise, a new edge is created between them.
6. Edges with age larger than a_{max} are removed. If that results in some neurons without connections, those isolated neurons are also removed.
7. In case the maximum number of nodes has not been reached and $\{\exists j \in \mathbb{N} - \{0\} \mid t = j \cdot \lambda\}$ then a new neuron is created.
   (a) The neuron u with the largest error is determined as well as its neighbor neuron v with the largest error. The new neuron is inserted between them (see Eq. 5)
   (b) The edge between u and v is removed and two new edges are added, from new to u, and from new to v.
   (c) The accumulated errors for neurons u, v and new are updated following Eqs. 6, 7 and 8.
8. The accumulated errors are decreased by a factor β

$$\forall i \in [1..N] \quad error_i(t) = error_i(t) - \beta \cdot error_i(t) \tag{9}$$

9. If the finish requirements have not been meet, i.e. the maximum number of training steps has not been reached, go to step 2.

2.2 Classification System

The proposed multi-class classification system is based on a one-vs.-all approach, which involves training a GNG neural network for each one of the C classes to which the patterns can belong.

$$Classifier = \{GNG_i, \quad 1 \leq i \leq C\} \tag{10}$$

A given pattern $\boldsymbol{x} \in \mathbb{R}^d$ is assigned to the class i corresponding to the GNG which has the closest neuron to that pattern in the input space.

$$class(\boldsymbol{x}) = argmin_{1 \leq i \leq C}\{\|\boldsymbol{x} - \boldsymbol{w}^i_{winner}\|^2\} \tag{11}$$

where $\boldsymbol{w}^i_{winner}$ is the winner neuron (see Eq. 1) of GNG neural network GNG_i, which is trained to detect patterns that belong to class i.

The maximum number of neurons which form the different self-organizing maps has been set to the same value, $N = N_{max}$. That way none of the particular networks GNG_i is given an unfair advantage during the training phase.

3 Experimental Results

In this section we present the obtained results from our tests. The selected sequence to test our approach is the traffic video named *US-101 Highway* which is available in the dataset of Next Generation Simulation (NGSIM) program, provided by the Federal Highway Administration (FHWA). This sequence presents several difficulties that should be dealt with, such as a perspective view from the outdoor scene, overlapping objects or occlusions.

The dataset, which contains information about the trajectories of several vehicles detected in a video sequence, presents small quantity of labeled vehicles and a large amount of them without any tag. Therefore, our proposal attempts to classify the vehicles that appear in the sequence into 4 possible classes: motorcycle, car, van and truck. An example of this different classes can be observed in Fig. 1, which shows 4 vehicles and its corresponding trajectories, one vehicle of each class.

Among the distinct features that can be extracted after segmenting an image, the selected ones for the tests have been the area, the perimeter, the width and the height of each object. Features like the object position in the scene are irrelevant when trying to identify its type and are not taken into consideration. Thus, for each segmented object O_i a set $\{\boldsymbol{x}^f_i \in \mathbb{R}^4 \mid f \in [1..MaxFrame]\}$ defines its trajectory during the video sequence. In order for the proposed classifier to work properly, a representative for each object is necessary. The chosen object description $\boldsymbol{x}^f_i$ is the one whose area coincides with the median of the area values in the members of O_i. Due to the robustness of the median statistic, those object descriptions corresponding to frames in which the segmentation failed or several objects overlap are never selected because they are far from the median value. Furthermore, only trajectories O_i with a minimum cardinality have been considered, in an attempt to avoid problems involving overlapping objects.

Fig. 1. Different vehicles and its trajectories detected by our approach. From top to bottom: vehicle 422 corresponds to the moto class, the trajectory 408 is a van, a truck can be observed with the vehicle 2776 and the vehicle 426 is a car.

The SOM methodology we have used is composed by 4 neural networks, one per class, that represents each considered vehicle type. Each network consists of 4 neurons. This number of neurons is low because the approach is based on an online learning strategy so that a desired real-time operation is attained.

To test the goodness of our proposal we have separated the labeled data in two random groups: the first one with the 90% of the data is used to train the model, and the second one with the remaining 10% of the data is managed to compare their vehicle type label with the proposed vehicle type result by the approach. The division of the groups has been carried out applying a stratified random selection. So each group has the same class object proportion.

Thus, the process to obtain a classification of the objects is as follows. First of all, the labeled data is divided into 2 groups: training data and test data. Then the model is created and trained with the training data. Afterwards the model is proved with the test data. Finally, the unlabeled data are classified. This process has been carried out 10 times.

The distribution of one model with its neurons of each class model can be observed in Fig. 2: (a) shows the distribution organized by area and perimeter, and (b) exhibits the same distribution organized by width and height.

From a qualitative point of view, the produced classification results by the implemented model are shown in Fig. 3. The results organized by the area and the perimeter of each vehicle are presented in (a). In addition the same classification organized by width and height is visible in (b).

On the other hand, the quantitative results of the performance of the approach can be observed in Table 1. It shows some different well-known measures like the Accuracy or the Mean Square Error. Accuracy (Acc) is a value between 0 and 1, where higher is better. On the other hand, Mean Square Error (MSE) is a positive real number, where lower is better.

Let k be the observed object of the K existing objects, let $\mathbf{x}_k$ and $\mathbf{w}_k$ be the class of the object and the class suggested by the approach, respectively, where

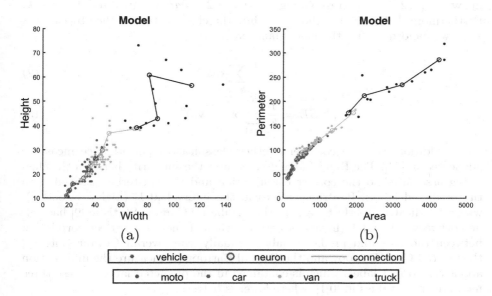

Fig. 2. Training data and the modelled neural network using 4 neurons per class. First column (a) shows data ordered by area and perimeter. Second column (b) presents the data sorted by width and height.

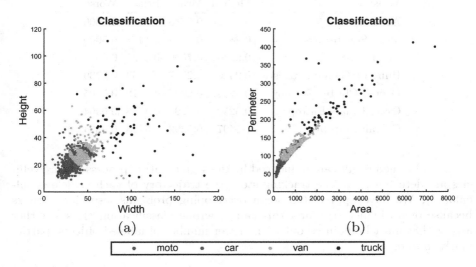

Fig. 3. Classification produced by the model of the Fig. 2. First column (a) shows data ordered by area and perimeter. Second column (b) presents the data sorted by width and height.

$\mathbf{x_k}, \mathbf{w_k} \in \{1, 2, 3, 4\}$, corresponding $1 = moto, 2 = car, 3 = van$ and $4 = truck$. Furthermore, Let $\mathbf{q}_k = 1$ if the model hits the classification of the object k (so $\mathbf{x_k} = \mathbf{w_k}$) and $\mathbf{q}_k = 0$ if the model fails ($\mathbf{x_k} \neq \mathbf{w_k}$):

$$Acc = \frac{1}{K} \sum_{k=1}^{K} \mathbf{q}_k \tag{12}$$

$$MSE = \frac{1}{K} \sum_{k=1}^{K} (\mathbf{x}_k - \mathbf{w}_k)^2 \tag{13}$$

In addition we have considered other classification performance measures advocated in [14]. The Rand Index [9] measures the similarity between the clustering associated to the correct classification and the clustering associated to the predicted labels. It attains values in the interval $[0, 1]$ (higher is better), where 1 indicates a perfect classification. Hubert's Gamma Statistic [9] has values between -1 and 1 (higher is better), where 1 means a perfect correlation between the true and predicted labels. Finally, the Overall Cluster Entropy, the Overall Class Entropy and the Overall Entropy [7] measure the information about the correct labels which is contained in the predicted labels. These three measures have values in $[0,1]$, where lower is better.

Table 1. Quantitative measures of the results. Each row is a measure and each column presents the median, the mean, the standard deviation, the best and the worst result for these measure, respectively.

Measure	Median	Mean	Best	Worst
Accuracy	0.7142	0.6785	0.8571	0.5000
Mean Square Error	0.2857	0.3214	0.1428	0.5000
Rand Index	0.6978	0.7000	0.8021	0.6043
Hubert's Gamma Statistic	0.2470	0.2737	0.5153	0.0739
Overall Cluster Entropy	0.5228	0.5290	0.3718	0.7305
Overall Class Entropy	0.5252	0.5294	0.3718	0.6364
Overall Entropy	0.5407	0.5292	0.3718	0.6834

The obtained results are influenced by the segmentation process (where vehicles are detected), and the tracking step (the trajectory of each vehicle is calculated). We have appreciated some overlapping problems, especially trucks because of its higher size, and this causes wrong classification classes. Other aspect that must be pointed out is the lower number of labeled objects, particularly motorcycles.

4   Conclusion

This paper proposes a novel approach based on the GNG (Growing Neural Gas) to determine the types of the vehicles appearing in traffic scenes. The detected vehicles are classified into four categories (cars, motorcycles, trucks and vans) by the system, based on a feature extraction process which provides the input data to carry out the classification.

The proposed neural architecture is composed by four GNGs, so that each GNG represents one of the considered vehicles types. Each vehicle is classified into the class associated to the GNG which best represents the features of the vehicle. Therefore, the proposed multiclass classification system is based on a one-vs-all approach, which involves the four GNGs. Due to the real-time operation requirement, an online learning strategy was used for the training of the four GNGs, whose number of neurons was also limited for the same reason.

The reported results of the simulation experiments in different video streams showed that the proposed approach achieved satisfactory classification results. The performance of the proposed approach depends on both the segmentation process and the tracking algorithm. Qualitative and quantitative evaluations of these results showed that the proposed approach performed at a high degree of vehicle detection rate, while the accuracy rate was satisfactory in the classification process.

Acknowledgments. This work is partially supported by the Ministry of Economy and Competitiveness of Spain under grant TIN2014-53465-R, project name Video surveillance by active search of anomalous events. It is also partially supported by the Autonomous Government of Andalusia (Spain) under projects TIC-6213, project name Development of Self-Organizing Neural Networks for Information Technologies; and TIC-657, project name Self-organizing systems and robust estimators for video surveillance. Finally, it is partially supported by the Autonomous Government of Extremadura (Spain) under the project IB13113. All of them include funds from the European Regional Development Fund (ERDF). The authors thankfully acknowledge the computer resources, technical expertise and assistance provided by the SCBI (Supercomputing and Bioinformatics) center of the University of Málaga. They also gratefully acknowledge the support of NVIDIA Corporation with the donation of the Titan X GPU used for this research.

References

1. Baumann, A., Boltz, M., Ebling, J., Koenig, M., Loos, H., Merkel, M., Niem, W., Warzelhan, J., Yu, J.: A review and comparison of measures for automatic video surveillance systems. EURASIP J. Image Video Process. **2008**, 1–30 (2008)
2. Buch, N., Velastin, S., Orwell, J.: A review of computer vision techniques for the analysis of urban traffic. IEEE Trans. Intell. Transp. Syst. **12**(3), 920–939 (2011)
3. Cheng, H.Y., Hsu, S.H.: Intelligent highway traffic surveillance with self-diagnosis abilities. IEEE Trans. Intell. Transp. Syst. **12**(4), 1462–1472 (2011)
4. Crouzil, A., Khoudour, L., Valiere, P., Truong Cong, D.: Automatic vehicle counting system for traffic monitoring. J. Electron. Imag. **25**(5), 051207 (2016)
5. Fink, O., Zio, E., Weidmann, U.: Novelty detection by multivariate kernel density estimation and growing neural gas algorithm. Mech. Syst. Sig. Process. **5051**, 427–436 (2015)
6. Fritzke, B.: A growing neural gas network learns topologies. In: Advances in Neural Information Processing Systems 7, pp. 625–632. MIT Press, Cambridge (1995)

7. He, J., Tan, A.H., Tan, C.L., Sung, S.Y.: On quantitative evaluation of clustering systems. In: Wu, W., Xiong, H., Shekhar, S. (eds.) Clustering and Information Retrieval. Network Theory and Applications, vol. 11, pp. 105–133. Springer, Heidelberg (2004)

8. Huang, D.Y., Chen, C.H., Chen, T.Y., Hu, W.C., Lin, Y.L.: A vehicle flow counting system in rainy environment based on vehicle feature analysis. J. Inf. Hiding Multimedia Sig. Process. **7**(1), 101–114 (2016)

9. Jain, A.K., Dubes, R.C.: Algorithms for Clustering Data. Prentice-Hall Inc., Upper Saddle River (1988)

10. Kamijo, S., Matsushita, Y., Ikeuchi, K., Sakauchi, M.: Traffic monitoring and accident detection at intersections. IEEE Trans. Intell. Transp. Syst. **1**(2), 108–117 (2000)

11. Liang, M., Huang, X., Chen, C.H., Chen, X., Tokuta, A.: Counting and classification of highway vehicles by regression analysis. IEEE Trans. Intell. Transp. Syst. **16**(5), 2878–2888 (2015)

12. López-Rubio, E., Luque-Baena, R.M.: Stochastic approximation for background modelling. Comput. Vis. Image Underst. **115**(6), 735–749 (2011)

13. Luque-Baena, R.M., López-Rubio, E., Domínguez, E., Palomo, E.J., Jerez, J.M.: A self-organizing map to improve vehicle detection in flow monitoring systems. Soft Comput. **19**(9), 2499–2509 (2015)

14. Moschou, V., Ververidis, D., Kotropoulos, C.: Assessment of self-organizing map variants for clustering with application to redistribution of emotional speech patterns. Neurocomputing **71**(1–3), 147–156 (2007)

15. Podolak, I., Jastrzebski, S.: Density invariant detection of osteoporosis using growing neural gas. Adv. Intell. Syst. Comput. **226**, 629–638 (2013)

16. Rad, R., Jamzad, M.: Real time classification and tracking of multiple vehicles in highways. Pattern Recogn. Lett. **26**(10), 1597–1607 (2005)

17. Wang, X.J., Shen, H.: Improved growing learning vector quantification for text classification. Jisuanji Xuebao/Chin. J. Comput. **30**(8), 1277–1285 (2007)

Recognizing Pedestrian Direction Using Convolutional Neural Networks

Alex Dominguez-Sanchez, Sergio Orts-Escolano, and Miguel Cazorla[✉]

Instituto de Investigación en Informática, University of Alicante, Alicante, Spain
alexdominguez09@yahoo.co.uk, {sorts,miguel}@dccia.ua.es

Abstract. Pedestrian movement direction recognition is an important factor in autonomous driver assistance and security surveillance systems. Pedestrians are the most crucial and fragile moving objects in streets, roads and events where thousands of people may gather on a regular basis. People flow analysis on zebra crossings and in commercial centres or events such as demonstrations, are a key element to improve safety and to enable autonomous cars to drive in real life environments. This paper focuses on deep learning techniques such as Convolutional Neural Networks (CNN) to achieve a good and reliable detection of pedestrians moving in a particular direction. We present a novel input representation that leverages current pedestrian detection techniques to generate a sum of subtracted frames, which are used as an input for the proposed CNN. Moreover, we have also created a new dataset for this purpose.

Keywords: Pedestrian detection · Advance driver assistance system · Convolutional neural networks · Pedestrian intention recognition

1 Introduction

Traffic control, risk detection and autonomous driver assistance systems (ADAS) are key elements for the development of future intelligent transportation systems. Furthermore, dynamic pedestrian movement in traffic environments makes it necessary to develop people flow analysis and movement intention recognition systems. In the last years, Convolutional Neural Networks (CNN) and other deep learning techniques have demonstrated impressive performance in many computer vision problems and therefore we believe they could be the perfect approach for the aforementioned problems. Moreover, computer vision and machine learning techniques have been transformed due to the rapid evolution and remarkable performance of Graphics Processing Units (GPUs), which has enabled the development of deep learning-based systems. In this work, our objective is the detection and recognition of pedestrian intention on streets, zebra crossings or road junctions, so as to be able to alert drivers or monitoring systems about possible risk situations.

© Springer International Publishing AG 2017
I. Rojas et al. (Eds.): IWANN 2017, Part II, LNCS 10306, pp. 235–245, 2017.
DOI: 10.1007/978-3-319-59147-6_21

1.1   Areas of Interests

What is needed to classify the images from a video to be able to distinguish if the pedestrian is moving towards the left or the right? Image classification and, consequently, convolutional neural networks in computer vision, are becoming a popular subject of research. In this work, we will focus on convolutional neural networks (CNN) as a tool for our recognition problem together with Histograms of Oriented Gradients (HOG) and other pixel-based techniques for dataset creation.

1.2   Related Works

Until 2012, most recognition, segmentation and classification image problems were approached by extracting hand-designed features and applying specific algorithms for those particular features. For example, if a number plate on a car needed to be detected, we segmented the image by looking for straight lines, then corners and finally reducing until we had an area similar to the geometry of a number plate. Basically, we looked for the particular features that could solve an specific problem.

A common hand-crafted feature used for pedestrian detection is the Histogram of Oriented Gradients (HOG) [1]. The main idea behind this descriptor is that local object appearance and shape within an image can be described by the intensity distribution of gradients or edge directions. The image is divided into small connected areas, and for the pixels within each area, a histogram of gradient directions is generated. Finally, the descriptor is the concatenation of these histograms.

Recent work in this area added a local sub-descriptor called Colour Self Similarity (CSS) [2] where colour histograms are compared within a HOG detected window, and for example, colour histograms from the two arms have a high similarity.

In addition, extensive research has been done on pedestrian detection [3,4], where more than sixteen different detectors were benchmarked [5] against several public datasets. Most of these hand-designed features were studied in [5]. The features were mostly based on window-sliding techniques and the detection was performed using support vector machines (SVM) for classification. Moreover, other approaches based on the Adaboost work of Viola and Jones [6], and many others based on HOG and variants of the same were extensively evaluated in [5].

Since 2012, new approaches for pedestrian detection and related problems emerged with the advent of deep learning techniques. Deep learning is a new way of applying machine learning algorithms, where neural networks are made deeper and deeper (more layers); not just two or three layers, but tens, or even hundreds. Specifically, in computer vision, a lot of work was done before 2012 in this regard, using multi-layer neural networks and obtaining poor results. Recognition of characters was conducted using a CNN [7] with a deeper layer structure. However, it was after 2012, with the proposal of Alex Krizhevsky CNN, AlexNet, when the real capabilities of CNNs became clear. These methodologies were first used at the Imagenet Competition [8] where the novel techniques, of deep multi-layer neural networks, were accelerated using GPUs. Since then, new and better hardware has appeared. This increases the possibility of bigger and deeper CNNs, providing better classification accuracy and making the training of existing deep networks an affordable scientific tool in terms of training time.

Computer vision research groups focused on pedestrian detection have also benefited from the rise of CNNs, and recent analyses have proved that better and more reliable results can be achieved [9]. However, our work focuses, not just on pedestrian detection, but also the pedestrian's movement direction recognition. Analyzing for example if the pedestrian moved to the left, right or to the front of the scene. There are few studies in this area. Enzweiler and Gavrila [10] and Gandhi et al. [11] focused on that aspect using the HOG descriptor and SVM as a classifier while Mogelmose et al. [12] used pedestrian tracking techniques and trajectory analysis for estimating pedestrian direction.

In general, the estimation of pedestrians' trajectories have been traditionally addressed using naive movement models based on human gait estimation and analysis of simple heuristics based on that information [13,14]. Other traditional approaches focused on the use of Kalman Filters (KF) to estimate pedestrian trajectories, most of these existing techniques produced poor results due to the impossibility to properly handle and adapt to changes on pedestrians' movements [15]. More recently, a more complex method based on Artificial Neural Networks (ANN) has been proposed for pedestrian trajectory estimation and intention recognition [14]. This work is able to estimate pedestrian trajectory based on pedestrian head detection and the use of its position for tracking along the sequence. Other existing works in the literature make use of features extracted from a dense optical which is compensated with ego-motion techniques (car movement) [16]. Using this approach, they are not only able to estimate pedestrian's path but also to roughly estimate pedestrian intentions towards specific situations like crossing at intersections [17].

Finally, it is worth to mention that there exist related works that address this problem from a different point of view. Most of these works are based on the information gathered by inertial measurement units (IMUs) and similar technologies (accelerometers, gyroscopes, etcetera). These kind of approaches are very intrusive from the pedestrian viewpoint and do not provide enough information to distinguish between different pedestrians actions.

After reviewing state-of-the-art techniques we can conclude that even though in the last years a lot of progress has been made towards pedestrians recognition systems, it is still required more research on systems and new techniques that can provide better classification accuracy, improved performance and ease of integration in current ADAS and security surveillance systems.

In this work, we contribute to pedestrian movement direction recognition by proposing a modified CNN based on Alex Krizhevsky [8] network, *AlexNet*, which we trained with a novel dataset that was recorded in different scenarios using a static camera. Pedestrians were video recorded and the convnet was fed with output images produced as a result of several image operations at pixel level

from that input video. The main purpose of this additional image processing was
to visually highlight image characteristics that may be relevant for pedestrian
trajectory recognition. Some videos showing the processing pipeline and the
proposed dataset are available on our project website[1].

From the best of our knowledge, no work has been done in the classification of
pedestrians according to their motion direction using deep learning techniques.
The rest of the paper is organized as follows: The proposed dataset is explained
in Sect. 2. In Sect. 3, we describe a CNN based on AlexNet [8] and how we fine-
tuned this network to achieve the best results for pedestrian movement direction
estimation. Section 4 explains how we have used our CNN in several experiments,
including hyper-parameter search to find optimum parameters for our network.
In Sects. 5 and 6 we present and discuss the results obtained. Finally, we draw
conclusions and indicate a few lines of future research work.

2   Dataset

To test the proposed CNN-based system we needed a specific dataset, designed
to feed our network with images of pedestrians moving in different directions and
in different scenarios, boardwalks, zebra crossings, sidewalks, etcetera. Video was
recorded with a colour static camera capturing at 30fps, 640 × 480 resolution.

Once the video was acquired, we fed the video stream through an image
preprocessing pipeline for image filtering, obtaining our final added video frames
(Fig. 1), which were used as input data for the proposed CNN architecture. Only
one of every six frames was used, as we saw that consecutive frames at 30 frames
per second contributed no new relevant features to our network.

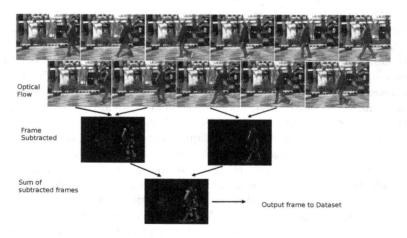

Fig. 1. Full process for video frame extraction of a pedestrian in motion

---

[1] http://www.rovit.ua.es/dataset/pedirecog/.

First, we computed a dense optical flow using the algorithm proposed in [18], allowing us to detect changes in the scene or camera movements. Even though some works directly used optical flow and some simple heuristics to predict pedestrians' direction, we observed that the global optical flow does not provide proper direction data due to movement of other objects or camera movements or a mix of both in the same scene, resulting in poor classification results.

In our pipeline, once we detect movement in a frame, we passed it to the next stage, detecting pedestrians in the scene using an existing technique based on HOG features and a linear SVM [1]. If this detection is positive, then we jump ahead six frames and carry out the same image processing steps, obtaining two frames with a pedestrian in movement. Next, we subtract the second frame from the first one, obtaining a black background image, in the cases where the camera is static, with the contour of the pedestrian slightly shifted towards a particular direction, as seen in the lower part of the Fig. 1. We repeated the same process after six frames, subtracting the third frame from the second one, obtaining a similar black background frame, and then we added those two resulting frames at pixel level. This output image is used as input data for the proposed CNN architecture. Several hours of video were recorded in five different locations, acquiring different datasets of background-removed-summed-frames within different environments. The images were then resized (downscaled) to 216×160 pixels and coded into JPG files converted to gray.

This dataset was classified manually, creating a ground truth split for training the proposed CNN architecture. The dataset comprised 7416 images for training and 1752 for validation. Images were classified according to three different categories, right, left and front: 2907 images were assigned to *Right*, 3099 images to *Left* and 1410 to: *Front*.

Once the previously discussed dataset was cleaned and classified, we trained a CNN using a modified version of AlexNet [8] network, which was modified for the recognition of the proposed classes and fine tuned for our specific problem. The proposed dataset is available on our project website http://www.rovit.ua.es/dataset/pedirecog/.

3   Proposed Method

Our method is based on a CNN network that was trained with our own dataset and takes as an input preprocessed data from a video feed as input.

3.1   Methodology

A Convolutional Neural Network is very similar to a standard Neural Network. This kind of network is made of neurons that have learnable weights. Each neuron receives some inputs from the input data, and performs a dot product with some initial weights that will be modified during the training process (backpropagation). The network expresses a single differentiable score function, the gradient, from the raw image pixels, and outputs a classification class as a result. Additionally, they have a loss function that is used to minimize the score.

A CNN takes advantage of having the input from images and builds the neural network model in a more intelligent way than conventional neural networks. In particular, unlike a normal neural network, the layers of a CNN have neurons organized in three dimensions: x, y, z, instead of 2D vectors.

In particular, we use a modified version of AlexNet [8] which has five convolutional layers and three fully connected layers. The last layer only contains 3 neurons with a Softmax function, being able to recognize the three classes of pedestrians trajectories.

All layers have their neurons activated using the ReLU (Rectified Linear Unit) activation function, where given an input value z, the ReLU layer computes the output as z if $z > 0$, with no negative slope.

Once the CNN had been defined, and the dataset was annotated with ground truth information, we started to feed the network with training data from the recorded dataset.

3.2   Technique

We trained the CNN using Stochastic Gradient Descent (SGD) as solver type, with a batch size of 100 images. For the momentum parameter we tried values ranging from 0.8 to 0.95, and a weight decay strategy ranging from 0.0001 to 0.01. In most of the layers, the weights were initialized using a Gaussian distribution with a standard deviation of 0.01. Then, the sixth and the seventh layers were initialized with a standard deviation of 0.05. The learning rate started as 0.01; we used an automatic reduction of 0.1 every 372 iterations, with a learning rate value of 0.0001 in the last iterations.

The layers structure is similar to the AlexNet [8] network. However, in our case, the input layer is a single image of 216×160 pixels (the image is converted to greyscale) and the last fully connected layer has only 3 neurons instead of 1000. No pre-trained model was used in the main experiment.

3.3   Description

The methodology followed to train the model has been separated in two phases:

- The first one consists of training the previously described CNN architecture several times, using a split containing 80% images from the dataset; we then used the remaining 20% images as validation data. We obtained a classification accuracy of 82% using a Gaussian distribution for weights initialization and the SGD solver.
- In the second stage, we used the defined model and then we ran a hyperparameter search, in order to find the best values for the momentum and weight decay parameters. After this process, we improved classification accuracy for the proposed architecture, finding optimum parameter values for our CNN and obtaining a classification accuracy of 83.8%

4   Experiments

We carried out different experiments to test different aspects of the proposed CNN architecture, such as layers configuration, solver values and train/test dataset splits.

4.1   Key Points

One of the key elements for this work was the proposal of a novel input representation for the convolutional neural network. An example can be seen in Fig. 2, where the output frame produced as a result of the pre-processing steps highlights pedestrian motion from the rest of the scene. In addition, background removal helped during the training process by avoiding learning irrelevant features.

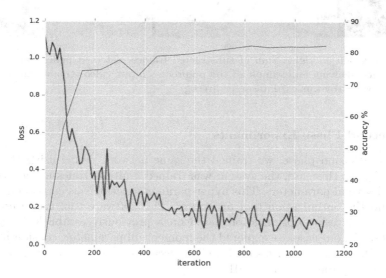

Fig. 2. Results obtained after the first training process using initial training/test dataset splits and without any further optimization. 82% classification accuracy. Red line: loss. Blue line: accuracy. (Color figure online)

4.2   First Phase Experiments

The solver strategy we followed for training the proposed CNN using *Caffe* during the first phase took 1125 iterations, starting with a learning rate of 0.01, and reducing it by 0.1 every 372 iterations. In all cases, we applied a SGD solver, with a momentum of 0.9 and a weight decay of 0.0005.

Finally, two fully connected layers were configured with a 50% dropout to avoid over-fitting as it was demonstrated in Hinton et al. [19]. This configuration was found to work very well with three different splits of our dataset, so we joined them together to achieve a bigger dataset of 7416 images for training and 1752 for validation.

We defined three different classes or categories: pedestrian moving to the left, to the right and to the front, considering each direction of pedestrians motion. An example can be seen in Fig. 2. There are 3099 images assigned to the left class, 2907 to the right one, and 1410 to the front one. After fully classifying the dataset manually, we achieved an 82% classification accuracy, using 1752 validation images as reflected in Fig. 3.

Fig. 3. From left to right: Sum of substracted frames, detected pedestrian, Optical flow, a video showing an example of the preprocess input video stream can be found at: http://www.rovit.ua.es/dataset/pedirecog/

4.3   Second Phase Experiments

During the second phase, we trained the same network model using the same dataset, but in this case, the system was trained multiple times using a combination of several parameters. This hyper-parameterization was carried out using *Caffe* and the optimization tool *Spearmint*. This software tool uses *Caffe* and performs a Bayesian optimization based on a previously established range of parameters. In our case, we explored the momentum parameter, considering values from 0.8 to 0.95, and also the weight decay parameter, exploring a set of values ranging from 0.0001 to 0.01.

4.4   Additional Experiments

In order to increase further the average 82% classification accuracy, we conducted a series of tests experimenting with several aspects of the whole training process, such as using Nesterov's Accelerated Gradient Descent [20] and a pre-trained CNN (fine tuning). Moreover, in order to provide our deep learning-based system with more ground truth information, we increased the dataset using data augmentation techniques. Finally, different batch sizes were evaluated.

Nesterov and Pre-trained Model Experiment. We ran a few training steps using Nesterov's accelerated gradient solver and also using the same dataset that we had previously used. In all executions a 50% dropout was used, with no sample cropping and a batch size of 100 samples. We used a pre-trained model for the AlexNet network and for the first experiment we initialised our own weights for the first convolutional layers, and the last fully-connected layers (6th and 8th layers). This change improved the accuracy of the network obtaining an 83.8% classification accuracy. For the second round of experiments we initialised the weights using Xavier's [21] method instead of the Gaussian one, and in the third one performed a similar weight initialization but only applying the changes to the sixth and final fully connected layers.

Data Augmentation Experiment: Sharp, Blur and Translate. Finally, an additional test was carried out using a sharpened version of the recorded dataset. The whole dataset was filtered using a sharpen filter. The sharpen filter accentuates the edges of the input images. This proved to be produce better results in terms of the test classification accuracy, but the loss in the validation data yielded worse results. Additionally, reducing the batch size without increasing the learning rate proved to be totally wrong. According to Krizhevsky et al. [8], reducing the batch size by x, should be followed by increasing the learning rate by x. Some experiments tend to increase using the $\sqrt{x}$.

5   Results

The accuracy of the proposed CNN was quantified using the validation data obtained during the training process. The higher the accuracy in classifying the validation images, the better the network performed. To double check the accuracy, it was measured with a manual test consisting of a batch of 120 images that were not used during the training or the validation process. Therefore, these were absolutely unknown images for the trained model. It also proved that the way we trained our system was not overfitting to the training dataset split.

After many different executions using hyper-parameter searching, we obtained several results ranging from a 74.8% to 83.8% classification accuracy. Best results were obtained for a value of 0.95 for the momentum parameter, and a value of 0.008127 for weight decay parameter. It can be noticed, that the loss function in all the cases also tends to a minimum between 0.1 and 0.3 (Table 1).

Table 1. Experiment results using Nesterov's Accelerated Gradient Descent solver

Learning rate	Layers modified	Xavier init	Accuracy	Loss	120 test
0,01	no	no	80%	0,12	84%
0,01	Conv1, fc6, fc8	no	**82%**	0,14	85%
0,01	Conv1, fc6, fc8	Conv1, fc6, fc8	80%	0,09	86%
0,01	Conv1, fc6, fc8	Fc6, fc8	80%	0,04	85%

Table 2. Results obtained using the sharpened version of the dataset

Batch size	Layers modified	Xavier init	Accuracy	Loss	120 test
50	Conv1, fc6 fc8	Conv1, fc6 fc8	84%	0.05	64%
100	Conv1, fc6 fc8	Conv1, fc6 fc8	82%	0.47	83%
100	Conv1, fc6 fc8	Conv1, fc6 fc8	**83%**	0.53	84%

Testing Nesterov's solver, we noticed that just initializing our own weights during the first, sixth and final layers, it produced better results than using just their pre-trained weights (Table 2).

6 Conclusions and Future Work

We have presented a method to differentiate the motion of pedestrians in real life environments. By building a novel input-filtered image based on the post-processing of static recorded video frames, we have managed to successfully distinguish three different pedestrian movement directions. Additionally, it has been proved how a reasonably small CNN can impressively perform in such a task by training it with a specialised dataset. Moreover, we have demonstrated how the results can be pushed even further by searching for the best hyperparameters once the CNN has been fine-tuned for our specific problem, in this case tuning the momentum and weight decay CNN parameters.

As future directions, we are working on a better and more robust use of data augmentation, that should provide a more robust model. Online learning or incremental learning could also be beneficial, which would entail training the pre-trained CNN with new classified images from the same network.

Acknowledgment. This work has been funded by the Spanish Government TIN2016-76515-R grant for the COMBAHO project, supported with Feder funds. Experiments were made possible by a generous hardware donation from NVIDIA.

References

1. Dalal, N., Triggs, B.: Histograms of oriented gradients for human detection. In: Proceedings of the IEEE Computer Society Conference on Computer Vision and Pattern Recognition, Washington, D.C., USA, pp. 886–893 (2005)
2. Walk, S., Majer, N., Schindler, K., Schiele, B.: New features and insights for pedestrian detection. In: CVPR, pp. 1030–1037 (2010)
3. Enzweiler, M., Gavrila, D.M.: Monocular pedestrian detection: survey and experiments. IEEE Trans. Pattern Anal. Mach. Intell. **31**, 2179–2195 (2009)
4. Benenson, R., Omran, M., Hosang, J., Schiele, B.: Ten years of pedestrian detection, what have we learned? In: Agapito, L., Bronstein, M.M., Rother, C. (eds.) ECCV 2014. LNCS, vol. 8926, pp. 613–627. Springer, Cham (2015). doi:10.1007/978-3-319-16181-5_47

5. Dollar, P., Wojek, C., Schiele, B., Perona, P.: Pedestrian detection: an evaluation of the state of the art. IEEE Trans. Pattern Anal. Mach. Intell. **34**(4), 743–761 (2012)
6. Viola, P., Jones, M.J.: Robust real-time face detection. Int. J. Comput. Vis. **57**(2), 137–154 (2004)
7. Lecun, Y., Bottou, L., Bengio, Y., Haffner, P.: Gradient-based learning applied to document recognition. Proc. IEEE **86**, 2278–2324 (1998)
8. Krizhevsky, A., Sutskever, I., Hinton, G.E.: Imagenet classification with deep convolutional neural networks. In: Pereira, F., Burges, C.J.C., Bottou, L., Weinberger, K.Q. (eds.) Advances in Neural Information Processing Systems 25, pp. 1097–1105. Curran Associates Inc., New York (2012)
9. Zhang, S., Benenson, R., Omran, M., Hosang, J., Schiele, B.: How far are we from solving pedestrian detection? In: The IEEE Conference on Computer Vision and Pattern Recognition (CVPR), June 2016
10. Enzweiler, M., Gavrila, D.M.: Integrated pedestrian classification and orientation estimation. In: CVPR, pp. 982–989. IEEE Computer Society (2010)
11. Gandhi, T., Trivedi, M.M.: Image based estimation of pedestrian orientation for improving path prediction. In: IEEE Intelligent Vehicles Symposium (2008)
12. Mogelmose, A., Trivedi, M.M., Moeslund, T.B. Trajectory analysis, prediction for improved pedestrian safety: integrated framework and evaluations. In: Intelligent Vehicles Symposium, pp. 330–335. IEEE (2015)
13. Fugger, T., Randles, B., Stein, A., Whiting, W., Gallagher, B.: Analysis of pedestrian gait and perception-reaction at signal-controlled crosswalk intersections. Transportation Research Record **1705**, 20–25 (2000)
14. Goldhammer, M., Hubert, A., Köhler, S., Zindler, K., Brunsmann, U., Doll, K., Sick, B.: Analysis on termination of pedestrians' gait at urban intersections. In: IEEE 17th International Conference on Intelligent Transportation Systems, pp. 1758–1763 (2014)
15. Schneider, N., Gavrila, D.M.: Pedestrian path prediction with recursive bayesian filters: a comparative study. In: Weickert, J., Hein, M., Schiele, B. (eds.) GCPR 2013. LNCS, vol. 8142, pp. 174–183. Springer, Heidelberg (2013). doi:10.1007/978-3-642-40602-7_18
16. Keller, C.G., Gavrila, D.M.: Will the pedestrian cross? A study on pedestrian path prediction. IEEE Trans. Intell. Transp. Syst. **15**(2), 494–506 (2014)
17. Koehler, S., Goldhammer, M., Bauer, S., Zecha, S., Doll, K., Brunsmann, U., Dietmayer, K.: Stationary detection of the pedestrian's intention at intersections. IEEE Intell. Transp. Syst. Mag. **5**(4), 87–99 (2013)
18. Farnebäck, G.: Two-frame motion estimation based on polynomial expansion. In: Bigun, J., Gustavsson, T. (eds.) SCIA 2003. LNCS, vol. 2749, pp. 363–370. Springer, Heidelberg (2003). doi:10.1007/3-540-45103-X_50
19. Hinton, G.E., Srivastava, N., Krizhevsky, A., Sutskever, I., Salakhutdinov, R.: Improving neural networks by preventing co-adaptation of feature detectors. CoRR abs/1207.0580 (2012)
20. Nesterov, Y.: A method for solving a convex programming problem with rate of convergence $O(1/k^2)$. Sov. Math. Dokl. **269**(3), 543–547 (1983)
21. Glorot, X., Bengio, Y.: Understanding the difficulty of training deep feedforward neural networks. In: Proceedings of the International Conference on Artificial Intelligence and Statistics (2010)

XRAY Algorithm for Separable Nonnegative Tensor Factorization

Rafał Zdunek$^{(\boxtimes)}$ and Tomasz Sadowski

Faculty of Electronics, Wroclaw University of Science and Technology,
Wybrzeze Wyspianskiego 27, 50-370 Wroclaw, Poland
rafal.zdunek@pwr.edu.pl

Abstract. Many computational problems in machine learning can be represented by separable matrix factorization models. In a geometric approach, linear separability means that the whole set of data points can be modeled by a convex combination of a few data points, referred to as the extreme rays. The aim of the XRAY algorithm is to find the extreme rays of the conic hull, generated by observed nonnegative vectors. In this paper, we extend the concept of this algorithm to a multi-linear data representation. Instead of searching into a vector space, we attempt to find the equivalent extreme rays in a space of tensors, under the linear separability assumption of subtensors, ordered along the selected mode. The proposed multi-way XRAY algorithm has been applied to Blind Source Separation (BSS) of natural images. The experiments demonstrate that if multi-way observations are at least one-mode linearly separable, the proposed algorithms can estimate the latent factors with high Signal-to-Interference (SIR) performance. The discussed methods may also be useful for analyzing video sequences.

Keywords: Nonnegative Tensor Factorization · Nonnegative Matrix Factorization · Separable factorization model · XRAY algorithm · Blind Source Separation

1 Introduction

Representing a large-scale dataset by a product of lower-rank factors is a well-known strategy in numerous research areas. When the data is nonnegatively constrained and stored in the form of a 2D algebraic object (matrix), probably the most known approach is to represent it by Nonnegative Matrix Factorization (NMF) [1,2]. The fundamental model assumes a decomposition of the nonnegative input matrix $\boldsymbol{Y} \in \mathbb{R}_+^{I \times T}$ into two lower-rank nonnegative factors (matrices) $\boldsymbol{A} \in \mathbb{R}_+^{I \times J}$ and $\boldsymbol{X} \in \mathbb{R}_+^{J \times T}$, such that $\boldsymbol{Y} \cong \boldsymbol{AX}$. The latent factors usually have some physical sense or easy interpretation.

From a geometric viewpoint, the column vectors in $\boldsymbol{Y}$ form a cloud of points in the nonnegative orthant $\mathbb{R}_+^I$. Assuming an exact generative model $\boldsymbol{Y} = \boldsymbol{AX}$ with rank$(\boldsymbol{A}) = J$, the points reside inside the cone $\mathcal{C}(\boldsymbol{Y})$, generated by the

© Springer International Publishing AG 2017
I. Rojas et al. (Eds.): IWANN 2017, Part II, LNCS 10306, pp. 246–256, 2017.
DOI: 10.1007/978-3-319-59147-6_22

extreme rays that are determined by all column vectors of $\boldsymbol{A}$. Hence, $\forall t : \boldsymbol{y}_t = \boldsymbol{A}\boldsymbol{x}_t$, where $\boldsymbol{y}_t$ is a conic combination of J vectors from $\boldsymbol{A}$, and $\boldsymbol{x}_t \in \mathbb{R}_+^J$ contains nonnegative coefficients of this combination. Regarding the above exact model, the aim of finding the factors $\boldsymbol{A}$ and $\boldsymbol{X}$ is equivalent to the problem of finding the extreme rays in $\mathcal{C}(\boldsymbol{Y})$. As reported by Vavasis [3], it belongs to a class of NP-hard problems. Moreover, the exact factorization given by the above-model is rarely encountered in practical applications of NMF because the observed data in $\boldsymbol{Y}$ are usually perturbed with disturbances. Even weak noise considerably affects positions of the extreme rays of $\mathcal{C}(\boldsymbol{Y})$, which makes the task of finding the true rays very difficult [4]. Hence, in practice, the factors in the basic model of NMF are usually estimated by alternating minimization of a given objective function, usually with additional constraints. A survey of such strategies can be found in [2,5].

However, a geometrical approach to NMF is still important for practical reasons if a different model of factorization is used. One may assume that the extreme rays are determined by a convex combination of the selected observed vectors $\{\boldsymbol{y}_t\}$. It leads to the model of convex NMF (CNMF), i.e.

$$Y = YWX, \tag{1}$$

where $\boldsymbol{W} = [\boldsymbol{w}_1, \ldots, \boldsymbol{w}_J] \in \mathbb{R}_+^{T \times J}$, $\forall j : ||\boldsymbol{w}_j||_1 = 1$. Due to the l_1-norm and non-negativity constrains on $\boldsymbol{W}$, the factor $\boldsymbol{A} = \boldsymbol{Y}\boldsymbol{W}$ contains convex combinations of the observed vectors. The column vectors in $\boldsymbol{A}$ lie in the column space spanned by $\{\boldsymbol{y}_t\}$, and may not be constrained to nonnegative values as in the standard NMF model. This model was first proposed by Ding *et al.* [6] for clustering of unsigned data, and it is conceptually closely related to the k-means, however the experiments carried out in [6] demonstrated its superiority over the k-means with respect to clustering accuracy. Then, CNMF was further developed and improved [7,8].

A special case of the model in (1) takes place if $\boldsymbol{W}$ is expressed by a binary matrix in such a way that $\boldsymbol{A} = \boldsymbol{Y}\boldsymbol{W} = \boldsymbol{Y}(:,\mathcal{K})$, where $\mathcal{K}$ is a subset of indices of the column vectors in $\boldsymbol{Y}$ with the cardinality J. This case is equivalent to the form:

$$Y = Y(:,\mathcal{K})X = A \begin{bmatrix} I_J & \tilde{X} \end{bmatrix} \Pi, \tag{2}$$

where $\boldsymbol{I}_J \in \mathbb{R}_+^{J \times J}$ is an identity matrix of the order J, and $\boldsymbol{\Pi} \in \mathbb{R}_+^{T \times T}$ is a permutation matrix. If the model in (2) is satisfied, then the matrix $\boldsymbol{Y}$ is *separable* [9,10]. In other words, all the extreme rays of $\mathcal{C}(\boldsymbol{Y})$ can be identified among the columns of $\boldsymbol{Y}$.

Separable NMF has already found important applications, especially in hyperspectral imaging and text mining. In the former, assuming existence of so-called "pure pixels", i.e. the pixels in abundance maps that correspond to only one endmember, the extreme rays determine the unknown pure spectra (end-members). This approach has been analyzed in many research papers [11–13]. In the latter, the separable NMF model is used for textual document classification or clustering [10,14,15], if a collection of the analyzed documents contains such documents that uniquely represent the topics or there exist words that occur only in the documents of a given topic.

Many computational strategies exist to estimate extreme rays of a convex cone, e.g. see [9,16–20]. One of the simplest and probably best known is the Successive Projection Algorithm (SPA) [21,22] that is often applied in hyperspectral imaging, It is a fast algorithm, very robust for noise-free data but does not work well if A is a rank-deficient matrix. To relax this drawback, Gillis proposed the modified version, referred to as the Successive Nonnegative Projection Algorithm (SNPA) [13]. An alternative algorithm to SNPA is the XRAY that was proposed by Kumar *et al.* [10] for solving near-separable NMF problems. It consists of two stages: projection and selection. In the first stage, each data point is projected onto the cone generated by the extreme rays, already selected in previous recursive steps. In the selection stage, the projection results are analyzed to select a candidate for the next extreme ray. The selection in XRAY can be implemented in various ways. Both steps can be performed with the computational complexity $O(IJT)$.

Motivated by the success of XRAY, we extend this concept to multi-way arrays that are also known in machine learning as tensors. Many datasets can be represented by such objects, e.g. a set of images ordered along various modes. Extraction of the representative images along one mode might be an important task in their analysis, including clustering and classification. In this paper, we attempt to blindly separate images, mixed along one mode, under the separability assumption. Each observed image can be modeled by a conic combination of source images, and the mixed images are collected in the form of a higher-order tensor. The source images can therefore be represented by the extreme rays of a cone in the space of images, or generally, in the space of tensors. Instead of searching for vectors (as in XRAY), we propose to search for multi-way arrays. Due to this assumption, there is no need to vectorize images, which relaxes the problem of losing 2D interactions between pixels. Moreover, this approach is computationally more efficient than the standard CP decomposition, obviously under the separability assumption along one mode of an observed tensor.

The paper is organized as follows: Sect. 2 discusses the XRAY algorithm for separable NMF problems. The XRAY algorithm for a separable tensor decomposition is presented in Sect. 3. The experiments carried out for the blind image separation problem are described in Sect. 4. Finally, the conclusions are drawn in Sect. 5.

2   XRAY Algorithm

The XRAY is a recursive algorithm that in each step finds one anchor, i.e. the column from Y that determines an extreme ray. Each step consists of two stages: selection and projection.

The first stage is based on the observation [10] that for any y_t projected onto the cone $\mathcal{C}(Y_A)$, we have $r_t^T Y_A \leq 0$ and $r_t^T y_t > 0$ if y_t is exterior to

$C(\boldsymbol{Y}_A)$. The matrix $\boldsymbol{Y}_A = \boldsymbol{Y}(:,\mathcal{K}) \in \mathbb{R}_+^{I \times |\mathcal{K}|}$ contains the anchors found so far, and $\boldsymbol{r}_t = \boldsymbol{y}_t - \boldsymbol{Y}_A \boldsymbol{x}_t$ is the residual vector obtained after projecting $\boldsymbol{y}_t$ onto $C(\boldsymbol{Y}_A)$. Assuming $\boldsymbol{y}_t = 0$ are removed from $\boldsymbol{Y}$, and $\forall i,t : y_{it} \geq 0$, then the following inequality $\boldsymbol{r}_t^T \tilde{\boldsymbol{y}}_t > 0$ holds for any $\boldsymbol{y}_t \notin C(\boldsymbol{Y}_A)$, where $\tilde{\boldsymbol{y}}_t = \frac{\boldsymbol{y}_t}{||\boldsymbol{y}_t||_1}$. It is also intuitively justified that the exterior point located the furthest from $C(\boldsymbol{Y}_A)$ has the largest length of the residual vector, obtained after projecting it onto the current cone. Each residual vector is also orthogonal to one of the faces of the current $C(\boldsymbol{Y}_A)$. Hence, the point $\boldsymbol{y}_t$ that maximizes the inner product $\boldsymbol{r}_k^T \tilde{\boldsymbol{y}}_t$ for any k for which $||\boldsymbol{r}_k||_2 > 0$ seems to be a good candidate for an anchor. The selection stage is therefore defined by the rule:

$$j = \arg\max_t \frac{\boldsymbol{r}_k^T \boldsymbol{y}_t}{||\boldsymbol{y}_t||_1}, \quad \text{for any } k : ||\boldsymbol{r}_k||_2 > 0. \tag{3}$$

This stage can be implemented in various ways. Kumar *et al.* defined various rules for setting k:

- random: i.e. any random k for which $||\boldsymbol{r}_k||_2 > 0$;
- maximum residual:

$$k = \arg\max_v ||\boldsymbol{r}_v||_2; \tag{4}$$

- maximum distance:

$$k = \arg\max_v ||\left[\boldsymbol{r}_v^T \boldsymbol{Y}\right]_+||_2. \tag{5}$$

Another option is the use of the greedy rule:

$$j = \arg\max_t \frac{||[\boldsymbol{R}^T \boldsymbol{y}_t]_+||_2}{||\boldsymbol{y}_t||_2}, \tag{6}$$

where $[\xi]_+ = \max\{0, \xi\}$. The index j in (3) or (6) determines the new anchor, hence $\mathcal{K} \leftarrow \mathcal{K} \cup \{j\}$.

The projection stage is defined by: $\boldsymbol{x}_t^* = \arg\min_{\boldsymbol{z} \in C(\boldsymbol{Y}_A)} ||\boldsymbol{y}_t - \boldsymbol{z}||_2^2$, which is equivalent to the nonnegative least squares problem:

$$\boldsymbol{X}^* = \arg\min_{\boldsymbol{X} \geq 0} ||\boldsymbol{Y} - \boldsymbol{Y}_A \boldsymbol{X}||_2^2. \tag{7}$$

The residual matrix is computed by $\boldsymbol{R} = \boldsymbol{Y} - \boldsymbol{Y}_A \boldsymbol{X}^*$. The recursive process should be repeated until a given number of extreme rays is found.

3   Separable NTF

The basic model of NMF can be presented in various equivalent forms:

$$\boldsymbol{Y} = \boldsymbol{A}\boldsymbol{X} = \sum_{j=1}^J a_j \underline{\boldsymbol{x}}_j = \sum_{j=1}^J \boldsymbol{u}_j^{(1)} \circ \boldsymbol{u}_j^{(2)}, \tag{8}$$

where $\underline{\boldsymbol{x}}_j \in \mathbb{R}_+^{1 \times T}$ is the j-th row vector of $\boldsymbol{X}$, $\boldsymbol{u}_j^{(1)} \in \mathbb{R}_+^I$ and $\boldsymbol{u}_j^{(2)} \in \mathbb{R}_+^T$ are the j-th column vectors of the matrices $\boldsymbol{U}^{(1)} = \boldsymbol{A}$ and $\boldsymbol{U}^{(1)} = \boldsymbol{X}^T$, respectively. The symbol $\circ$ stands for the outer product of vectors.

For the N-th order tensor $\mathcal{Y} \in \mathbb{R}^{I_1 \times \cdots \times I_N}$, the model (8) can be intuitively extended to the form:

$$\mathcal{Y} = \sum_{j=1}^{J} \boldsymbol{u}_j^{(1)} \circ \ldots \circ \boldsymbol{u}_j^{(N)} = \mathcal{I} \times_1 \boldsymbol{U}^{(1)} \times_2 \ldots \times_N \boldsymbol{U}^{(N)}, \tag{9}$$

where $\mathcal{I} \in \mathbb{R}^{J \times \cdots \times J}$ is the N-th order identity tensor, and $\times_n$ denotes the tensor-matrix multiplication along the n-th mode of a tensor. The model (9) expresses the well-known CANDECOMP/PARAFAC decomposition [2] of the tensor $\mathcal{Y}$.

Assuming $\mathcal{A} = \mathcal{I} \times_1 \boldsymbol{U}^{(1)} \times_2 \ldots \times_{N-1} \boldsymbol{U}^{(N-1)} \in \mathbb{R}^{I_1 \times \cdots \times I_{N-1} \times J}$ and $\boldsymbol{C} = \boldsymbol{U}^{(N)} \in \mathbb{R}^{I_N \times J}$, the model (9) has the form:

$$\mathcal{Y} = \mathcal{A} \times_N \boldsymbol{C}. \tag{10}$$

The model (10) is closely related to the Tucker-1 model [2], but it is not restricted to a 3-way tensor. Imposing the nonnegativity constraints onto $\mathcal{Y}$, $\mathcal{A}$ and $\boldsymbol{C}$, and unfolding it along the N-th mode, we obtain the basic form of NMF: $\boldsymbol{Y}_{(N)} = \boldsymbol{C} \boldsymbol{A}_{(N)}$, where $\boldsymbol{Y}_{(N)} \in \mathbb{R}^{I_N \times \prod_{n=1}^{N-1} I_n}$ and $\boldsymbol{A}_{(N)} \in \mathbb{R}^{J \times \prod_{n=1}^{N-1} I_n}$. Despite the models are equivalent, they may lead to different results if the factors in (10) are computed in a space of tensors, i.e. by omitting the unfolding. With the nonnegativity constraints, the model (10) can be also regarded as a special case of Nonnegative Tensor Factorization (NTF).

Definition 1. *The model (10) with the nonnegativity constraints is separable along its N-th mode, if $\boldsymbol{C}$ can be expressed by:*

$$\boldsymbol{C} = \boldsymbol{\Pi} \begin{bmatrix} \boldsymbol{D} \\ \tilde{\boldsymbol{C}} \end{bmatrix}, \tag{11}$$

where $\boldsymbol{D} \in \mathbb{R}_+^{J \times J}$ is a diagonal matrix, $\tilde{\boldsymbol{C}} \in \mathbb{R}^{(I_N - J)_+ \times J}$, and $\boldsymbol{\Pi} \in \mathbb{R}^{I_N \times I_N}$ is a permutation matrix.

If the model (10) is separable by Definition 1, then following (2), we have:

$$\mathcal{Y} = \mathcal{Y}(:, \ldots, :, \mathcal{K}) \times_N \boldsymbol{C} = \mathcal{Y}_A \times_N \boldsymbol{C}, \tag{12}$$

where $\mathcal{Y}_A \in \mathbb{R}_+^{I_1 \times \cdots \times I_{N-1} \times |\mathcal{K}|}$ is the part of $\mathcal{Y}$, obtained by selecting the subtensors $\{\mathcal{Y}_j\}_{j \in \mathcal{K}}$ along the N-th mode of $\mathcal{Y}$ that are indicated by the set $\mathcal{K}$. Each subtensor $\mathcal{Y}_t \in \mathbb{R}^{I_1 \times \cdots \times I_{N-1}}$ can therefore be expressed by a conic combination of the subtensors $\mathcal{Y}_j$ with the coefficients $c_{tj} \geq 0$. Thus:

$$\mathcal{Y}_t = \sum_{j=1}^{J} \mathcal{Y}_j c_{tj} = \mathcal{Y}_A \times_N \underline{\boldsymbol{c}}_t, \tag{13}$$

where $\underline{c}_t \in \mathbb{R}_+^{1 \times J}$ is the t-th row vector of C. The subtensors $\{\mathcal{Y}_j\}$ can be treated as the anchors that determine the extreme rays in the space of tensors $\mathbb{R}_+^{I_1 \times \ldots \times I_{N-1}}$.

Considering (13), the residual tensor for the t-th sample is given by: $\mathcal{R}_t = \mathcal{Y}_t - \mathcal{Y}_A \times_N \underline{c}_t$. Let $\mathcal{C}(\mathcal{Y}_A)$ be the cone generated by the extreme rays in $\mathcal{Y}_A$. Similarly to the vector case, if any $\mathcal{Y}_t$ is exterior to $\mathcal{C}(\mathcal{Y}_A)$, then $\langle \mathcal{R}_t, \mathcal{Y}_t \rangle > 0$, and $\langle \mathcal{R}_t, \mathcal{Y}_j \rangle \leq 0$ for $j \in \mathcal{K}$, where $< \cdot, \cdot >$ denotes the inner product.[1]

Applying the concept of the XRAY algorithm to NTF given by the model (10), the rule for selection of a new anchor is given by:

$$j = \arg \max_t \langle \mathcal{R}_k, \tilde{\mathcal{Y}}_t \rangle, \quad \text{for any } k : \langle \mathcal{R}_k, \mathcal{R}_k \rangle > 0, \qquad (14)$$

where each entry of $\tilde{\mathcal{Y}}_t$ is L_1-normalized according to the rule:

$$\tilde{y}_{i_1,\ldots,i_{N-1},t} = \frac{y_{i_1,\ldots,i_{N-1},t}}{\sum_{i_1=1}^{I_1} \cdots \sum_{i_{N-1}=1}^{I_{N-1}} y_{i_1,\ldots,i_{N-1},t}}. \qquad (15)$$

Similar rules as for the standard XRAY algorithm can be used for selecting the sample k that determines the face of $\mathcal{C}(\mathcal{Y}_A)$ to which the residual tensor $\mathcal{R}_k$ is normal. Thus, the rules (4) and (5) take the respective forms: $k = \arg \max_t \langle \mathcal{R}_t, \mathcal{R}_t \rangle$ and

$$k = \arg \max_v \left\| \left[\sum_{i_1=1}^{I_1} \cdots \sum_{i_{N-1}=1}^{I_{N-1}} r_{i_1,\ldots,i_{N-1},v} y_{i_1,\ldots,i_N} \right]_+ \right\|_2 .$$

The projection stage is performed by solving the following nonnegative least squares (NNLS) problem:

$$C^* = \arg \min_{C \geq 0} \|\mathcal{Y} - \mathcal{Y}_A \times_N C\|_2^2, \qquad (16)$$

which can be readily done with many numerical algorithms, e.g. the HALS-CPD [2]. If the size of $\mathcal{Y}$ is not very large, the problem (16) can be rewritten by unfolding the tensors with respect to their N-th mode. Thus:

$$C^* = \arg \min_{C \geq 0} \|Y_{(N)} - CY_{A(N)}\|_2^2, \qquad (17)$$

where $Y_{(N)} \in \mathbb{R}_+^{I_N \times \prod_{p=1}^{N-1} I_p}$ and $Y_{A(N)} \in \mathbb{R}_+^{|\mathcal{K}| \times \prod_{p=1}^{N-1} I_p}$ are unfolded versions of $\mathcal{Y}$ and $\mathcal{Y}_A$, respectively. The problem (17) can be solved with any NNLS solver. The residual tensor is given by $\mathcal{R} = \mathcal{Y} - \mathcal{Y}_A \times_N C^*$.

Remark 1. The selection stage (14) with any rule for choosing the face of $\mathcal{C}(\mathcal{Y}_A)$ has the computational complexity $O\left(\prod_{n=1}^N I_n\right)$. For the projection, we have at most $O\left(J \prod_{n=1}^N I_n\right)$, assuming that $J \ll I_N$.

---

[1] The inner product $< \mathcal{A}, \mathcal{B} >$ between the tensors $\mathcal{A} = [a_{i_1,\ldots,i_N}] \in \mathbb{R}^{I_1 \times \ldots \times I_N}$ and $\mathcal{B} = [b_{i_1,\ldots,i_N}] \in \mathbb{R}^{I_1 \times \ldots \times I_N}$ is defined as follows: $< \mathcal{A}, \mathcal{B} >= \sum_{i_1=1}^{I_1} \cdots \sum_{i_N=1}^{I_N} a_{i_1,\ldots,i_N} b_{i_1,\ldots,i_N}$.

The proposed algorithm assumes that the N-th order tensor $\mathcal{Y}$ is observed and each subtensor $\mathcal{Y}_t = \mathcal{Y}(:,\ldots,:,t)$ along the N-th mode is represented by a linear mixture of other (source) subtensors. The mixing model is formulated in (10), where $\mathcal{A}$ represents the source tensors, and C contains the mixing coefficients. Note that the model (10) assumes that the sources are mixed along the N-th mode but it can be applied to tensors mixed along any mode using the permutation of modes, i.e. by rearranging the dimensions of $\mathcal{Y}$.

Due to the assumption that the model (10) represents linear mixtures, its intrinsic application seems to be a Blind Source Separation (BSS) problem. However, the condition of the separability in (11) may be too strong in many real BSS problems. Many research papers [10,13,14] reported that the separable factorization models can be useful in other machine learning problems, such as clustering or classification. Since the computational complexity of (10) along the N-th mode is similar to other modes (see Remark 1), the proposed model can also be useful for clustering of video sequences or other multi-way arrays, e.g. spectrograms.

4   Experiments

The proposed algorithms were tested for solving a synthetic BSS problem. We assumed that the mixtures are modeled by (10), where $\mathcal{A}$ contains the sources, and C expresses the mixing operator. Thus, the sources are mixed along one mode. The nonnegativity constraint is imposed onto both the sources and the mixing matrix. We selected six natural source images of the Tatra mountains which are illustrated in Fig. 1(a). Each image is represented in the space of RGB colors by the tensor $\mathcal{A}_j \in \mathbb{R}_+^{256 \times 256 \times 3}$. The mixing matrix $C \in \mathbb{R}_+^{T \times 6}$ was generated in various ways: (a) *case 1*: $c_{jt} = \max\{0, \check{c}_{jt}\}$ with $\forall t, j : \check{c}_{jt} \sim \mathcal{N}(0,1)$ (normal distribution); (b) *case 2*: the first 6 signals taken from the benchmark Sparse10sinc4.mat that is freely accessible in the Matlab toolbox *NMFLAB for Signal Processing* [23]. The benchmark contains sparse upper half-sine waveforms, which are all nonnegative and locally smooth. The number of samples is equal to 1000. For each case, the mixed images are stored in the 4D tensor $\mathcal{Y} \in \mathbb{R}_+^{256 \times 256 \times 3 \times T}$. For the case 2, the mixed gray-scale images are illustrated as a 3D plot in Fig. 1(b).

The mixing matrix C should satisfy the condition (11) in order to the observed tensor $\mathcal{Y}$ be separable. For the case 1, the number of extreme rays in $\mathcal{C}(\mathcal{Y})$ depends on T, and it can be modeled by the binomial distribution. In the experiment, we set $T = 5000$, because for this number and $J = 6$, $\mathcal{Y}$ is separable (with probability equal to nearly one) according to Definition 1. The matrix C in the case 2 is sparse, however, it cannot be presented in the form (11). From its structure, the number of the extreme rays in $\mathcal{C}(\mathcal{Y})$ is equal to only two. Hence, this is a non-separable case.

We tested the following algorithms: M1 – XRAY for tensor decomposition with the random selection rule, M2 – XRAY for tensor decomposition with the maximum residual rule, M3 – XRAY for tensor decomposition with the maximum distance rule, and M4 – the HALS-NTF (the nonnegatively constrained

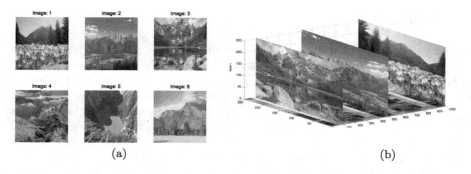

Fig. 1. Test images: (a) sources; (b) 3D grey-scale mixtures.

CP decomposition with the HALS algorithm for factor updating) [2]. The algorithms M1–M3 do not need any parameters to set in advance but the parameters in M4 were set accordingly [2]. All the algorithms were run several times, where in each run a new sample of C (for the case 1) or a new initializer was generated (for the M1 and M4). Note that if C in the model (10) does not satisfy the linear separability property, the intrinsic scale and permutation ambiguities are not the only ones that occur in this model. The problem was discussed by Donoho and Stodden [4], regarding the NMF model. Thus, the usage of Monte Carlo (MC) approach for validating the algorithms is important, especially for the random selection rule and the case 2, where the separability property is weakly satisfied.

The results are evaluated with the Signal-to-Interference Ratio (SIR) measure [2], both between the original and estimated tensors $\{\mathcal{A}_i\}$ as well as the matrices C. Each algorithm was run 100 times, and the worst case was selected from multiple SIR-values.

Table 1 presents the minimal SIR values for estimating the sources $\mathcal{A}$ as well as the mixing matrix C. It also contains the averaged elapsed time of running each algorithm. Additionally, the quality of the results can be inferred from the estimated matrix C by finding the rows that give the anchors or the extreme rays. The averaged numbers of estimated extreme rays for the M1–M3 are also listed in Table 1. We do not provide the information on the extreme rays for the M4 because this is not geometry-based algorithm.

Table 1. Results obtained for testing the algorithms M1–M4 for both cases.

Algorithms:	Case 1				Case 2			
	M1	M2	M3	M4	M1	M2	M3	M4
Min. SIR for $\{\mathcal{A}_i\}$ [dB]	273.3	273.1	273.1	−0.15	0	113.9	113.9	0.97
Min. SIR for C [dB]	−2.23	207.3	207.3	0	−1.52	112.2	112.2	0.18
Elapsed time [sec.]	97.44	90.15	122.6	2968	19.61	19.28	25.18	527.5
Avg. No. Extr. Rays	5.6	6	6	-	2	2	2	-

5   Conclusions

In the paper, we extended the XRAY algorithm to the separable NTF model. In the numerical experiments, the discussed algorithms are validated in terms of the SIR-measure and the elapsed time, assuming the data satisfies the intrinsic properties of the model, i.e. nonnegativity of the latent factors and linear separability along a given mode. The experiments show that the best results for the BSS problem with 3D images can be obtained with the algorithms: M2 and M3. All the proposed algorithms work very well for the fully separable BSS problem as in the case 1. Surprisingly, the M2 and M3 give very good estimates (with $SIR > 100\,$dB, both for the sources and the mixing matrix) even for the non-separable case 2 for which the number of estimated extreme rays is equal to two. The XRAY with the random selection rule (M1) works worse than with M2 and M3. For comparison, we also used the standard CP algorithm with the nonnegative constraints. Unfortunately, the CP algorithm is not able to provide satisfactory results, mostly because $I_3 < J$, which leads to underdetermined estimation. Furthermore, all the proposed algorithms provided the satisfactory result within the time much shorter than the standard CP algorithm.

Summing up, the proposed algorithms are very efficient for separable NTF. Despite, we only demonstrated their robustness for the synthetic and noise-free BSS problems, their application potential is large. Recently, many research papers [24–29] have reported the usefulness of the geometry-based NMF and multi-way array decomposition algorithms in many applications, including BSS and clustering problems. In the further research, we plan to extend the area of their applications to online clustering of video sequences and video tracking in the sense of finding the video frames that present the end-positions of a moving object. Such the frames can be regarded as extreme rays in the geometric representation of video sequences.

Furthermore, the proposed algorithms can also be used for estimating any factor matrix in the model (9), assuming that the selected factor is represented by (11). If so, the other factor matrices can be estimated using the SVD-based algorithms that were proposed by Zhou and Cichocki [30].

Acknowledgment. This work was partially supported by the grant 2015/17/B/ST6/01865 funded by National Science Center (NCN) in Poland.

References

1. Lee, D.D., Seung, H.S.: Learning the parts of objects by non-negative matrix factorization. Nature **401**, 788–791 (1999)
2. Cichocki, A., Zdunek, R., Phan, A.H., Amari, S.I.: Nonnegative Matrix and Tensor Factorizations: Applications to Exploratory Multi-way Data Analysis and Blind Source Separation. Wiley, Hoboken (2009)
3. Vavasis, S.A.: On the complexity of nonnegative matrix factorization. SIAM J. Optim. **20**(3), 1364–1377 (2009)

4. Donoho, D., Stodden, V.: When does non-negative matrix factorization give a correct decomposition into parts? In: Proceedings of NIPS, vol. 16. MIT Press (2004)
5. Wang, Y.X., Zhang, Y.J.: Nonnegative matrix factorization: a comprehensive review. IEEE Trans. Knowl. Data Eng. **25**(6), 1336–1353 (2013)
6. Ding, C., Li, T., Jordan, M.I.: Convex and semi-nonnegative matrix factorizations. IEEE Trans. Pattern Anal. Mach. Intell. **32**(1), 45–55 (2010)
7. Thurau, C., Kersting, K., Bauckhage, C.: Convex non-negative matrix factorization in the wild. In: Proceedings of 9th IEEE International Conference on Data Mining, ICDM 2009, Washington, D.C., USA, pp. 523–532 (2009)
8. Esser, E., Möller, M., Osher, S., Sapiro, G., Xin, J.: A convex model for nonnegative matrix factorization and dimensionality reduction on physical space. IEEE Trans. Image Process. **21**(7), 3239–3252 (2012)
9. Arora, S., Ge, R., Kannan, R., Moitra, A.: Computing a nonnegative matrix factorization - provably. SIAM J. Comput. **45**(4), 1582–1611 (2016)
10. Kumar, A., Sindhwani, V., Kambadur, P.: Fast conical hull algorithms for near-separable non-negative matrix factorization. In: Proceedings of the 30th International Conference on Machine Learning (ICML), Atlanta, Georgia, USA, vol. 28, pp. 231–239 (2013)
11. Bioucas-Dias, J.M., Plaza, A., Dobigeon, N., Parente, M., Du, Q., Gader, P., Chanussot, J.: Hyperspectral unmixing overview: geometrical, statistical, and sparse regression-based approaches. IEEE J. Sel. Topics Appl. Earth Obs. Remote Sens. **5**(2), 354–379 (2012)
12. Chan, T.H., Ma, W.K., Ambikapathi, A.M., Chi, C.Y.: A simplex volume maximization framework for hyperspectral endmember extraction. IEEE Trans. Geosci. Remote Sens. **49**(11), 4177–4193 (2011)
13. Gillis, N.: Successive nonnegative projection algorithm for robust nonnegative blind source separation. SIAM J. Imaging Sci. **7**(2), 1420–1450 (2014)
14. Arora, S., Ge, R., Halpern, Y., Mimno, D.M., Moitra, A., Sontag, D., Wu, Y., Zhu, M.: A practical algorithm for topic modeling with provable guarantees. In: Proceedings of ICML. JMLR Workshop and Conference Proceedings, vol. 28, pp. 280–288 (2013). www.JMLR.org
15. Ding, W., Rohban, M.H., Ishwar, P., Saligrama, V.: Topic discovery through data dependent and random projections. In: Proceedings of ICML, vol. 28, pp. 471–479 (2013)
16. Bittorf, V., Recht, B., Re, C., Tropp, J.: Factoring nonnegative matrices with linear programs. In: Pereira, F., Burges, C.J.C., Bottou, L., Weinberger, K.Q. (eds.) Advances in Neural Information Processing Systems, vol. 25, pp. 1214–1222 (2012)
17. Gillis, N., Luce, R.: Robust near-separable nonnegative matrix factorization using linear optimization. J. Mach. Learn. Res. **15**, 1249–1280 (2014)
18. Huang, K., Sidiropoulos, N., Swami, A.: Non-negative matrix factorization revisited: uniqueness and algorithm for symmetric decomposition. IEEE Trans. Sig. Process. **62**(1), 211–224 (2014)
19. Ouedraogo, W.S.B., Souloumiac, A., Jaïdane, M., Jutten, C.: Simplicial cone shrinking algorithm for unmixing nonnegative sources. In: ICASSP, pp. 2405–2408. IEEE (2012)
20. Zdunek, R.: Initialization of nonnegative matrix factorization with vertices of convex polytope. In: Rutkowski, L., Korytkowski, M., Scherer, R., Tadeusiewicz, R., Zadeh, L.A., Zurada, J.M. (eds.) ICAISC 2012. LNCS (LNAI), vol. 7267, pp. 448–455. Springer, Heidelberg (2012). doi:10.1007/978-3-642-29347-4_52

21. Araujo, M.C.U., Saldanha, T.C.B., Galvao, R.K.H., Yoneyama, T., Chame, H.C., Visani, V.: The successive projections algorithm for variable selection in spectroscopic multicomponent. Chemometr. Intell. Lab. Syst. **57**(2), 65–73 (2001)

22. Gillis, N.: Robustness analysis of hottopixx, a linear programming model for factoring nonnegative matrices. SIAM J. Mat. Anal. Appl. **34**(3), 1189–1212 (2013)

23. Cichocki, A., Zdunek, R.: NMFLAB for signal and image processing. Technical report, Laboratory for Advanced Brain Signal Processing, BSI, RIKEN, Saitama, Japan (2006)

24. Fu, X., Ma, W.K., Huang, K., Sidiropoulos, N.D.: Blind separation of quasi-stationary sources: exploiting convex geometry in covariance domain. IEEE Trans. Sig. Process. **63**(9), 2306–2320 (2015)

25. Fu, X., Sidiropoulos, N.D., Ma, W.K.: Power spectra separation via structured matrix factorization. IEEE Trans. Sig. Process. **64**(17), 4592–4605 (2016)

26. Fu, X., Huang, K., Yang, B., Ma, W.K., Sidiropoulos, N.D.: Robust volume minimization-based matrix factorization for remote sensing and document clustering. IEEE Trans. Sig. Process. **64**(23), 6254–6268 (2016)

27. Yang, Z., Xiang, Y., Rong, Y., Xie, K.: A convex geometry-based blind source separation method for separating nonnegative sources. IEEE Trans. Neural Netw. Learn. Syst. **26**(8), 1635–1644 (2015)

28. Yin, P., Sun, Y., Xin, J.: A geometric blind source separation method based on facet component analysis. Sig. Image Video Process. **10**(1), 19–28 (2016)

29. Zhu, Y., Wang, N., Miller, D.J., Wang, Y.: Convex analysis of mixtures for separating non-negative well-grounded sources. Sci. Rep. **6**(38350), 1–13 (2016). doi:10.1038/srep38350

30. Zhou, G., Cichocki, A.: Canonical polyadic decomposition based on a single mode blind source separation. IEEE Sig. Process. Lett. **19**(8), 523–526 (2012)

Automatic Learning of Gait Signatures
for People Identification

Francisco Manuel Castro[1]([✉]), Manuel J. Marín-Jiménez[2], Nicolás Guil[1],
and Nicolás Pérez de la Blanca[3]

[1] Computer Architecture Department, University of Málaga, Málaga, Spain
fcastro@uma.es
[2] Computing and Numerical Analysis Department,
University of Córdoba, Córdoba, Spain
[3] Computer Science and Artificial Intelligence Department,
University of Granada, Granada, Spain

Abstract. This work targets people identification in video based on
the way they walk (i.e. gait). While classical methods typically derive
gait signatures from sequences of binary silhouettes, in this work we
explore the use of convolutional neural networks (CNN) for learning
high-level descriptors from low-level motion features (i.e. optical flow
components). We carry out a thorough experimental evaluation of the
proposed CNN architecture on the challenging TUM-GAID dataset. The
experimental results indicate that using spatio-temporal cuboids of optical flow as input data for CNN allows to obtain state-of-the-art results
on the gait task, both for identification and gender recognition, with an
image resolution eight times lower than the previously reported results
(i.e. 80×60 pixels).

1 Introduction

The goal of *gait recognition* is to identify people by the way they walk. This
type of biometric approach is considered non-invasive, since it is performed at
a distance, and does not require the cooperation of the subject that has to be
identified, in contrast to other methods as iris- or fingerprint-based approaches.
Gait recognition has application in the context of video surveillance, ranging
from control access in restricted areas to early detection of persons of interest
as, for example, v.i.p. customers in a bank office.

From a computer vision point of view, gait recognition could be seen as a particular case of human action recognition. However, gait recognition requires more
fine-grained features than action recognition, as differences between different gait
styles are usually much more subtle than between common action categories (i.e.
'high jump' vs. 'javelin throw') included in state-of-the-art datasets [18].

In last years, great effort has been put into the problem of people identification based on gait recognition [13]. However, previous approaches have mostly
used hand-crafted features for representing the human gait, which are not easily scalable to diverse datasets. Therefore, we propose an end-to-end approach
based on convolutional neural networks that given low-level optical flow maps,

© Springer International Publishing AG 2017
I. Rojas et al. (Eds.): IWANN 2017, Part II, LNCS 10306, pp. 257–270, 2017.
DOI: 10.1007/978-3-319-59147-6_23

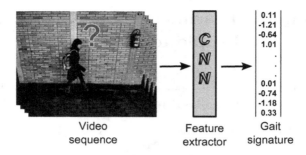

Fig. 1. **Goal of this paper.** We aim at automatically learning gait signatures from video sequences of people walking to identify them.

directly extracted from video frames (see Fig. 1), is able to learn and extract higher-level features suitable for representing human gait: *gait signature*.

To the best of our knowledge, this is the first work where convolutional neural networks are applied to the problem of gait identification using as input optical flow features. Therefore, our main contributions are: (i) a preprocessing stage to extract, organize and normalize low-level motion features for defining the input data; (ii) a convolutional neural network architecture to extract discriminative gait signatures from low-level motion features; and, (iii) a thorough experimental study to validate the proposed framework on the standard TUM-GAID dataset for gait identification and gender identification, obtaining state-of-the-art results with video frames whose size is eight times smaller than the ones used in previously reported results.

The rest of the paper is organized as follows. We start by reviewing related work in Sect. 2. Section 3 explains our approach for learning gait signatures and identifying people. Section 4 contains the experiments and results. Finally, we present the conclusions and future work in Sect. 5.

2 Related Work

Traditionally, deep learning approaches based on Convolutional Neural Networks (CNN) have been used in image-based tasks with great success [16]. In the last years, deep architectures for video have appeared, specially focused on action recognition, where the inputs of the CNN are subsequences of stacked frames. In [17], Simonyan and Zisserman proposed to use as input to a CNN a volume obtained as the concatenation of frames with two channels that contain the optical flow in the x-axis and y-axis respectively. To normalize the size of the inputs, they split the original sequence in subsequences of 10 frames, considering each subsample independently. A natural modification is presented by Ji *et al.* [14], where a 3D convolutional network is developed to capture temporal information from multiple frames. Then, Tran *et al.* [19] propose a new 3D network which uses raw videos as input, instead of preprocessed inputs. Recently, a new approach has been developed by He *et al.* [10]. They propose a new kind of CNN which has a large number of layers and residual connections to avoid the vanishing gradient problem. Although several papers can be found for the task of human action recognition using deep learning techniques, it is hard to find such type of approaches applied to the problem of gait recognition. In [12], Hossain and Chetty propose the use of Restricted Boltzmann Machines to extract gait features from binary silhouettes, but a very small probe set (*i.e.* only ten different subjects) was used for validating their approach. A more recent work [23], uses a random set of binary silhouettes from a sequence to train a CNN that accumulates the calculated features to achieve a global representation of the dataset. In [1], raw 2D GEI [9] (Gait Energy Image: the main idea is to compute a temporal averaging of the binary silhouettes of the target subject) are employed to train a simple CNN for gait recognition. A more complex work is presented in [7] where GEI are used to train an ensemble of CNN and a Multilayer Perceptron is employed as classifier. In [22], given two GEI descriptors, they learn a metric to decide whether both descriptors belong to the same subject or not. All those previous CNN-based approaches propose precomputed GEI descriptors as input features. In contrast, our approach builds a spatio-temporal volume of optical flow [17] as input to a CNN specially designed for gait recognition, what will allow the CNN to learn characteristic gait patterns directly from the source, *i.e.* the motion.

3   Proposed Approach

In this section we describe our proposed framework to address the problem of gait recognition using CNN. The pipeline proposed for gait recognition based on CNN is represented in Fig. 2: *(i)* compute optical flow (OF) along the whole

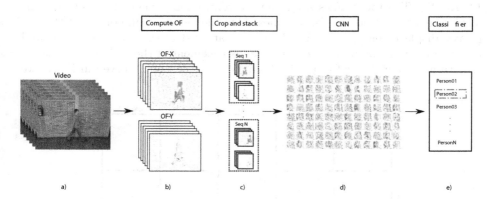

Fig. 2. Pipeline for gait recognition. (a) The input is a sequence of RGB video frames. (b) Optical flow is computed along the sequence. (c) Optical flow maps are cropped and stacked in subsequences of L maps. (d) Optical flow subsequences are passed through the CNN to obtain gait signatures. (e) Classification of the extracted gait signatures. Note: positive flows are displayed in pink and negative flows in blue (best viewed in color). (Color figure online)

sequence; *(ii)* build up a data cuboid from consecutive OF maps; *(iii)* feed the CNN with OF cuboid to extract the gait signature; and, *(iv)* apply a classifier to decide the subject identity.

3.1   Input Data

The use of optical flow (OF) as input data for action representation in video with CNN has already shown excellent results [17]. Nevertheless human action is represented by a wide, and usually well defined, set of local motions. In our case, the set of motions differentiating one gait style from another is much more subtle and local. An important question here is whether the gait information can be decoded from simple and low resolution (*e.g.* 80×60) optical flow. With this resolution we choose a sufficiently small scale to make our results useful for most of the standard video cameras and recording scenarios. Note that higher resolutions could achieve better results but at the price of higher computing time and lesser applicability.

Let F_t be an OF map computed at time t and, therefore, $F_t(x, y, c)$ be the value of the OF vector component c located at coordinates (x, y), where c can be either the horizontal or vertical component of the corresponding OF vector. The input data I_L for the CNN are cuboids built by stacking L consecutive OF maps F_t, where $I_L(x, y, 2k - 1)$ and $I_L(x, y, 2k)$ corresponds to the value of the horizontal and vertical OF components located at spatial position (x, y) and time k, respectively, ranging k in the interval $[1, L]$.

Since each original video sequence will probably have a different temporal length, and CNN requires a fixed size input, we extract subsequences of L frames from the full-length sequences. In Fig. 3 we show five frames distributed every six frames along a subsequence of twenty-five frames in total (i.e. frames 1, 7, 13, 19, 25). Top row frames show the horizontal component of the OF (*x*-axis displacement) and bottom row frames show the vertical component of the OF (*y*-axis displacement). It can be observed that most of the flow is concentrated in the horizontal component, due to the displacement of the person. In order to remove noisy OF located in the background, as it can be observed in Fig. 3, we

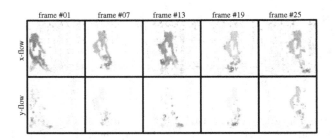

Fig. 3. Input data. Frames extracted from a subsequence of 25 frames. (top) Optical flow in *x*-axis. (bottom) Optical flow in *y*-axis. Note: positive flows are displayed in pink and negative flows in blue (best viewed in color). (Color figure online)

might think in applying a preprocessing step for filtering out those vectors whose magnitude is out of a given interval. However, since our goal in this work is to minimize the manual intervention in the process of gait signature extraction, we will use those OF maps as returned by the OF algorithm.

Implementation details. First of all, we resize the RGB video frames to a common size of 80 × 60 pixels, keeping the original aspect ratio of the video frames. Then, we compute dense OF on pairs of frames by using the method of Farneback [6] implemented in OpenCV library. In parallel, people are located in a rough manner along the video sequences by background substraction [15]. Then, we crop the video frames to remove part of the background, obtaining video frames of 60 × 60 pixels (full height is kept) and to align the subsequences (people are x-located in the middle of the central frame, #13) as in Fig. 3.

Finally, from the cropped OF maps, we build subsequences of 25 frames by stacking OF maps with an overlap of $\mathcal{O}\%$ frames. In our case, we chose $\mathcal{O} = 80\%$, that is, to build a new subsequence, we use 20 frames of the previous subsequence and 5 new frames. For most state-of-the-start datasets, 25 frames cover almost one complete gait cycle, as stated by other authors [2].

In order to increase the number of samples available for training, we compute 8 spatial displacements of ±5 pixels in all directions, $i.e.$ (−5, −5), (−5, 0), (0, −5), ..., (0, 5). Then, the corresponding mirror sequences are computed. These procedure allows us to obtain about 270 k training samples. Finally, before feeding each sample into the CNN, the mean value of the whole training dataset is subtracted.

3.2 CNN Architecture for Gait Signature Extraction

The CNN architecture we propose for gait recognition is based on the one described in [17] for general action recognition in video. However, in our case, the input has a size of 60 × 60 × 50, obtained from the sequence of 25 OF frames with their corresponding two channels, as explained in the previous section.

The proposed CNN is composed by the following sequence of layers (Fig. 4): '$conv1$', 96 filters of size 7 × 7 applied with stride 1 followed by a normalization and max pooling 2 × 2; '$conv2$', 192 filters of size 5 × 5 applied with stride 2 followed by max pooling 2 × 2; '$conv3$', 512 filters of size 3 × 3 applied with stride 1 followed by max pooling 2 × 2; '$conv4$', 4096 filters of size 2 × 2 applied

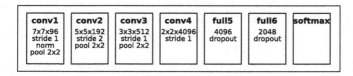

Fig. 4. Proposed CNN architecture for gait signature extraction. Four convolutional layers are followed by two fully connected layers. The top layer is a softmax classifier that can be used to directly derive an identity.

with stride 1; '*full5*', fully-connected layer with 4096 units and dropout; '*full6*', fully-connected layer with 2048 units and dropout; and, '*softmax*', softmax layer with as many units as subject identities. All convolutional layers use the rectification (ReLU) activation function.

Implementation details. We use the implementation of CNN provided in Mat-ConvNet library [20]. This library allows to develop CNN architectures in an easy and fast manner using the Matlab environment. In addition, it takes advantage of CUDA and cuDNN [5] to improve the performance of the algorithms.

We perform CNN training following an incremental process to speed up and to facilitate the convergence. In this incremental process, initially, we train a simplified version of our CNN (i.e. '*conv1*' without normalization, '*conv4*' 512 filters, '*full5*' 512 units, '*full6*' 256 units and no dropout) and, then, we use its weights for initializing the layers of a more complex version of that simpler CNN (i.e. adding normalization, 0.1 dropout and more filters and units). By this way, we train four incremental CNN versions using the previous weights until we obtain the final CNN architecture represented in Fig. 4. During the training of the CNN, the weights are learnt using mini-batch stochastic descent algorithm with momentum equal to 0.9 in the first three CNN version iterations, and 0.95 during the last one. We set weight decay to $5 \cdot 10^{-4}$ and dropout to 0.4. The learning rate is initially set to 10^{-2} and divided by 10 when the validation error become stagnant. At each epoch, a mini-batch of 150 samples is constructed by random selection over a balanced training set (i.e. almost same proportion of samples per class).

3.3   Classification Strategies

Once we have obtained the gait signatures, the final stage consists in classifying those signatures to derive a subject identity. Although the softmax layer of the CNN is already a classifier (i.e. each unit represents the probability of belonging to a class), the fully-connected layers can play the role of gait signatures that can be used as input of a Support Vector Machine (SVM) classifier. Since we are dealing with a multiclass problem, we define an ensemble of C binary SVM classifiers with linear kernel in an 'one-vs-all' fashion, where C is the number of possible subject identities. Previous works (*e.g.* [4]) indicate that this configuration of binary classifiers is suitable to obtain top-tier results in this problem. Note that we $L2$-normalize the top fully-connected layer before using it as feature vector, as early experiments supported this normalization.

A classical alternative to discriminative classifiers is the nearest neighbour (NN) classifier, which does not require any training step. Actually, we can easily extend our gait recognition system by just adding samples of the new subjects to our *gallery set* (i.e. the models).

Note that in Sect. 3.1, we split the whole video sequence into overlapping subsequences of a fixed length, and those subsequences are classified independently. Therefore, in order to derive a final identity for the subject walking along the whole sequence, we apply a *majority voting* strategy on the labels assigned to each subsequence.

4   Experiments and Results

We present here the experiments designed to validate our approach and the results obtained on the selected dataset for gait recognition.

4.1   Dataset

We run our experiments on the recent 'TUM Gait from Audio, Image and Depth' (TUM-GAID) dataset [11] for gait recognition. In TUM-GAID 305 subjects perform two walking trajectories (left to right and vice versa) in an indoor environment. Two recording sessions were performed, one in January and the second in April, where subjects wore different clothes. The action is captured by a Microsoft Kinect sensor which provides a video stream with a resolution of 640×480 pixels with a frame rate of approximately 30 fps. Some examples can be seen in Fig. 5 depicting the different conditions included in the dataset.

Hereinafter the following nomenclature is used to refer each of the four walking conditions considered: *normal* walk (N), carrying a *backpack* of approximately 5 kg (B), wearing coating *shoes* (S), as used in clean rooms for hygiene conditions, and *elapsed time* $(TN\text{-}TB\text{-}TS)$. Each subject of the dataset is composed of: six sequences of normal walking, two sequences carrying a bag and two sequences wearing coating shoes. In addition, 32 subjects were recorded in both sessions (i.e. January and April) so they have 10 additional sequences. Therefore, the overall amount of videos is 3400.

To standardize the experiments performed on the dataset, the authors have defined three subsets of subjects: training, validation and testing. The training set is used for obtaining a robust model against the different covariates of the dataset. This partition is composed of 100 subjects and the sequences N, B and S. The validation set is used for validation purposes and contains 50 different subjects. Finally, the test set contains other 155 different subjects used in the test

Fig. 5. TUM-GAID dataset. People walking indoors under four walking conditions: normal walking, wearing coats, carrying a bag and wearing coating shoes. Top and bottom rows show the same set of subjects but in different months of the same year.

phase. As the set of subjects is different between the test set and the training set, a new training of the identification model must be performed using four normal (N) sequences per subject. The rest of sequences are used for testing and to obtain the accuracy of the model.

For the viability of our experiments with CNN, we resized all the videos to a resolution of 80×60 pixels (i.e. 8 times lower resolution). Nevertheless, we will show in the experimental results (Sect. 4.3), that we obtain state-of-the-art results with such low resolution, what, in our opinion, highlights the potential of CNN for gait recognition.

4.2 Experimental Setup

We describe here the three types of experiments we carried out on the dataset with the proposed approach: *experiment A* to validate our approach; *experiment B* to validate that our method is able to identify people wearing different clothes between training and test sets; and, *experiment C* to assess that our gait signatures are useful for other gait biometrics such as gender recognition.

Experiment A: gait recognition with clothing and carrying conditions. This is the core experiment of this paper, where we aim at evaluating the capacity of the proposed CNN model to extract gait signatures robust enough to deal with covariate factors as clothing changes (*e.g.* long coats or coating shoes) or carrying conditions (*e.g.* backpacks). In fact, the CNN model trained here will be used for the subsequent experiments.

Training of the CNN convolutional filters is carried out by using only sequences of the standard training and validation subject partitions (i.e. $100+50$ subjects) of TUM-GAID, including the three scenarios. Once the CNN model is trained with those samples, the learnt weights from layers '*conv1*' to '*full6*' are frozen (i.e. not modified any more). In order to evaluate the performance of the CNN-based gait signatures on the test subject partition (i.e. 155 subjects), only the softmax layer will be fine-tuned by using the training sequences of scenario 'N' from the test subject partition, as the subject identities have changed. However, when we use SVM or NN classifiers, no CNN fine-tuning is needed, as we will use the output of layer '*full6*' directly as our gait signature (i.e. the automatic gait descriptor extracted from the input sequence).

The results of this experiment are summarized in Table 1, where each row corresponds to a different combination of features and classifiers: softmax 'SM', support vector machine 'SVM' and nearest neighbour 'NN'. Each column contains the recognition results of the diverse scenarios included in the dataset (N, B, S) plus the average on the three scenarios ('Avg'). For completeness, we report rank-1 ('R1') and rank-5 ('R5') results. Metric *rank-1* measures the percentage of test samples where the top one assigned identity corresponds to the right one. Metric *rank-5* measures the percentage of test samples where the ground truth identity is included in the first five ranked identities for the corresponding test sample.

Moreover, for comparison purposes, we have implemented the 'Pyramidal Fisher Motion' (PFM) descriptor, as described in [4], since it does not need

Table 1. Experiment A. Percentage of correct recognition on scenarios *N-B-S* of TUM-GAID dataset by using *rank-1* (R1) and *rank-5* (R5) metrics. Each row corresponds to a different combination of features and classifiers. Best average results are marked in bold.

	Method/Rank	*N*		*B*		*S*		Avg	
		R1	R5	R1	R5	R1	R5	R1	R5
80 × 60	CNN-SM	99.4	100	94.5	99.4	94.2	98.7	96.0	99.4
	CNN-SVM	99.7	100	97.1	99.4	97.1	99.4	**98.0**	**99.6**
	CNN-NN+PCA256	99.4	99.7	97.7	98.7	96.1	97.7	97.7	98.7
	CNN-NN+PCA128	99.7	100	98.1	98.4	95.8	97.1	97.9	98.5
	CNN-NN+PCA064	99.7	100	98.1	98.4	94.8	96.4	97.5	98.3
	PFM	75.8	93.2	70.3	91.3	32.3	62.9	59.5	82.5
	PF@640 × 480	99.7	99.7	99.0	99.4	99.0	99.4	**99.2**	**99.5**

binary silhouettes as input for its computation and has previously reported state-of-the-art results for the problem of gait recognition [3]. Note that we have used the PFM descriptor both in the original resolution video sequences (row 'PFM@640 × 480') and in the low resolution version of the sequences (row 'PFM@80 × 60'), to allow a fair comparison with our CNN-based gait signatures that use the low resolution version. For 'PFM@640 × 480', we have used the whole video sequence to compute a single descriptor, as in the original paper [4]. Whereas in 'PFM@80 × 60', we have computed several PFM using the same set of subsequences extracted for CNN, making an even much fairer comparison. After the classification of each PFM of the sequence, majority voting was applied to obtain a final identity.

Experiment B: elapsed time. The goal of this experiment is to evaluate the robustness of the CNN-based gait signatures against changes of people appearance at different periods of time. In this experiment, we apply the CNN model trained in 'Experiment A' on the 'elapsed time' subset of TUM-GAID (Sect. 4.1), which is composed of 16 subjects for training and validation, and 16 for testing. From the training sequences of the 'normal' scenario *TN* of the 16 test subjects, we obtained 10620 samples that were used to fine-tune the softmax layer of the CNN trained in the previous experiment, as the subject identities changed. Then, we used the test sequences of the three *elapsed time* scenarios to evaluate the performance.

The results of this experiment are summarized in Table 2, where each row corresponds to a different combination of features and classifiers, including PFM. Each column presents the recognition results of the diverse scenarios included in the *elapsed time* subset (*TN*, *TB*, *TS*) plus the average on the three scenarios ('Avg'). For completeness, we report rank-1 ('R1') and rank-5 ('R5') results.

Experiment C: gait-based gender recognition. Gender recognition based on gait signatures is considered a kind of soft biometric, which allows to prune a

Table 2. Experiment B. Percentage of correct recognition on scenarios *TN-TB-TS* of TUM-GAID dataset by using *rank-1* (R1) and *rank-5* (R5) metrics. Each row corresponds to a different combination of features and classifiers. Best average results are marked in bold.

	Method/Rank	*TN*		*TB*		*TS*		Avg	
		R1	R5	R1	R5	R1	R5	R1	R5
80 × 60	CNN-SM	53.1	87.5	40.6	90.6	50.0	90.6	47.9	**89.6**
	CNN-SVM	59.4	87.5	50.0	78.1	62.5	93.8	57.3	86.5
	CNN-NN+PCA256	59.4	71.9	56.3	65.6	56.3	65.6	57.3	67.7
	CNN-NN+PCA128	62.5	71.9	56.3	65.6	59.4	68.8	**59.4**	68.8
	CNN-NN+PCA064	62.5	68.8	53.1	62.5	59.4	68.8	58.3	66.7
	PFM	50.0	84.4	40.6	81.3	25.0	75.0	38.5	80.2
	PFM@640 × 480	78.1	87.5	56.3	87.5	46.9	87.5	**60.4**	87.5

subset of subjects for a subsequent finer identification. The goal of this experiment is to validate the quality of the gait signatures learnt in the first experiment to train a binary linear SVM for gender classification. For evaluation purposes, we train the gender classifier only on the gait sequences included in the training and validation subject partitions. In TUM-GAID, which provides labels at video level for this task, the proportion of male and female subjects in the test set is 62.6% and 37.4%, respectively.

The results of this experiment are summarized in Table 4, where we show both the confusion matrices for each scenario, plus the overall accuracy of the classifier. For comparison purposes, bottom row contains the accuracy reported for this task in paper [11].

4.3 Results and Discussion

We ran our experiments on a computer with 32 cores at 2 GHz, 256 GB of RAM and a GPU Nvidia Tesla K40c, with MatConvNet library running on Matlab 2014b for Ubuntu 14.04. After splitting the training sequences (of the training subjects) into subsequences, we got a training set composed of 269352 samples used for learning the filters from '*conv1*' to '*full6*' layers (see Fig. 4); and a second training set composed of 108522 samples for training the softmax layer from the subset of test subjects[1]. With all these samples, the whole training process (from the first CNN model until the fine-tuning of the softmax layer of the final model) took about 60 h.

Due to the specificity of the dataset, in the training step we had to balance the number of samples of the different kinds of walking (i.e. normal, carrying a bag and wearing coating shoes). To do this, we defined different training subsets

---

[1] Note that TUM-GAID distinguishes between training/test *subjects* and training/test *sequences*. Test sequences are never used for training or validation of the model.

Table 3. State-of-the-art on TUM GAID. Percentage of correct recognition on TUM-GAID for diverse methods published in the literature. Bottom row corresponds to our proposal, where instead of using video frames at 640 × 480, a resolution of 80 × 60 is used. Each column corresponds to a different scenario. Best results are marked in bold. (See main text for further details).

	Method	N	B	S	Avg	TN	TB	TS	Avg
640 × 480	SDL [24]	-	-	-	-	**96.9**	-	-	-
	GEI [11]	99.4	27.1	52.6	59.7	44.0	6.0	9.0	19.7
	SEIM [21]	99.0	18.4	96.1	71.2	15.6	3.1	28.1	15.6
	GVI [21]	99.0	47.7	94.5	80.4	62.5	15.6	**62.5**	46.9
	SVIM [21]	98.4	64.2	91.6	84.7	65.6	31.3	50.0	49.0
	RSM [8]	**100**	79.0	97.0	92.0	58.0	38.0	57.0	51.3
	PFM [4]	99.7	**99.0**	**99.0**	**99.2**	78.1	**56.3**	46.9	**60.4**
80 × 60 (ours)	CNN-SVM	99.7	97.1	97.1	98.0	59.4	50.0	**62.5**	57.3
	CNN-NN128	99.7	98.1	95.8	97.9	62.5	**56.3**	59.4	59.4

with the same number of samples of each walking scenario, and when the CNN converged, we continued the training with a different subset. At the end of the training phase, all samples of the original training set had been passed through the CNN at least twice to guarantee a good performance of the model. If this step is not performed, the CNN would learn mainly specialized filters for 'normal' walk, as we have four times more samples of this kind than the others.

Focusing on 'Experiment A', the results in Table 3 (columns 'N', 'B' and 'S') indicate that from low resolution frames (i.e. 80 × 60) the trained CNN model is able to extract gait signatures that used in combination with standard SVM classifiers, it is attained an average of 98% rank-1 correct recognition (see row 'CNN-SVM'), and 99.6% of rank-5 accuracy. Comparing SVM with SM, we can see that the obtained results are quite similar, although SVM accuracy is slightly better, indicating a good linear separability of the test subjects given the extracted gait signatures. For speeding-up the NN classifier, the 2048-dimensional gait descriptors were compressed with the standard principal components analysis (PCA) algorithm – vectors are $L2$-normalized and mean is subtracted before PCA – obtaining compact signatures of 64, 128 and 256 dimensions. The average results reported in rows 'CNN-NN+PCAx' are comparable to the ones yielded by the parametric classifiers (i.e. SVM and SM), making attractive the use of NN in combination with these CNN-based signatures as no training stage is needed if adding new identities to our recognition system is required. Furthermore, our proposal outperforms PFM descriptor when used on the same low resolution video sequences (i.e. 59.5% vs. 98%), although average rank-1 accuracy for PFM at full resolution is around 1% better that CNN. Nevertheless, our CNN-based signature extractor has been trained in a fully automatic manner, in contrast to the hand-crafted steps need for computing PFM. Focusing on the results on scenarios 'B' and 'S', we can conclude that our CNN signatures are able to successfully represent the discriminative motion patterns that characterize the different subjects regardless the clothing or shoes worn or the bags carried. Remember that we used a set of individuals totally different for training the CNN filters than the one used for testing the signatures obtained with them.

Moving to 'Experiment B', the 'elapsed time' experiment proposed in TUM-GAID is more challenging than the previous one, as there is a temporal gap of months between recordings of the same subjects. This higher level of difficulty is reflected in the results of Table 2, where we directly use the CNN previously trained in 'Experiment A'. In terms of rank-1 accuracy, SVM behaves on average better than SM, as previously reflected in 'Experiment A'. However, NN classifiers improves on SVM, what suggests that this set of subjects is not linearly separable given the gait signatures. Comparing to PFM, only the full resolution version obtains average results slightly better than CNN, due to the good results achieved in the normal (TN) scenario. Note that results reported for TB and TS are equal or lower than the CNN ones. Comparing our best results with previously published ones, we observe in Table 3 that our accuracy (rank-1) is on a par with those methods, even though we are using video frames with a resolution eight times lower than the others. Note that our average accuracy (columns 'Avg') in both sets of experiments is greater than the ones reported in all the compared papers, but PFM, what emphasizes the quality of the gait signatures returned by the proposed CNN.

Finally, the results in Table 4 suggest that the problem of gender recognition can be successfully addressed based on just motion features (i.e. optical flow). Regardless the scenario, accuracy for female recognition is lower than male recognition (i.e. $\approx$77% vs. $\approx$96%). This can be due to the ratio among female and male samples in the dataset. Comparing to the results reported by Hofmann *et al.* in [11], the average on the three scenarios (N, B, S) for our method is 88.9%, whereas the average for their method is 87.8%, despite the lower resolution of our video inputs.

Table 4. Experiment C. Confusion matrices for gender recognition based on convolutional gait signatures. For each scenario, each cell in 'CM' contains the percentage of probe samples assigned to each gender. Rows in 'Acc' contain the overall accuracy per scenario.

		N		B		S		AVG
		Female	Male	Female	Male	Female	Male	
CM	Female	**78.4**	21.6	**77.6**	22.4	**76.7**	23.3	
	Male	4.6	**95.4**	4.6	**95.4**	3.6	**96.4**	
Acc	CNN+SVM (ours, 80 × 60)	89.0%		88.7%		89.0%		**88.9**
	Hofmann *et al.* [11] (640 × 480)	95.8%		74.8%		92.9%		87.8

5   Conclusions and Future Work

This paper has presented a thorough study of convolutional neural networks applied to the demanding problem of people identification based on gait. The experimental validation has been carried out on the challenging dataset TUM-GAID, by using a low resolution version of the original video sequences (i.e. eight times lower). The results indicate that starting from just sequences of optical flow, the proposed CNN is able to extract meaningful gait signatures (i.e. $L2$-normalized top fully-connected layer) that allow to obtain high recognition rates on the available scenarios (i.e. different clothing and wearing bags), achieving state-of-the-art results, in contrast to classical approaches for gait recognition that use hand-crafted features, mainly based on binary silhouettes or dense tracklets. In terms of classification strategies, an ensemble of 'one-vs-all' linear SVM is a good choice, although a NN approach on PCA compressed descriptors offers similar accuracy, not requiring any training step. Finally, we have shown that our automatically learnt gait signatures are suitable for gender recognition, what might allow to filter out a set of individuals before running a finer identification procedure. As future work, we plan to extend our study to other datasets for gait recognition where multiple viewpoints are available and other CNN architectures combining OF with RGB data. Also, as different kind of labels are available in TUM-GAID (gender, age, shoe type) we plan to explore multi-task architectures. Finally, we also plan to experiment with higher resolution samples to find out the dependence between resolution and accuracy.

References

1. Alotaibi, M., Mahmood, A.: Improved gait recognition based on specialized deep convolutional neural networks. In: 2015 IEEE Applied Imagery Pattern Recognition Workshop (AIPR), pp. 1–7, October 2015
2. Barnich, O., Droogenbroeck, M.V.: Frontal-view gait recognition by intra- and inter-frame rectangle size distribution. Pattern Recogn. Lett. **30**(10), 893–901 (2009)
3. Castro, F.M., Marín-Jiménez, M.J., Guil, N.: Empirical study of audio-visual features fusion for gait recognition. In: Azzopardi, G., Petkov, N. (eds.) CAIP 2015. LNCS, vol. 9256, pp. 727–739. Springer, Cham (2015). doi:10.1007/978-3-319-23192-1_61
4. Castro, F.M., Marín-Jiménez, M., Guil Mata, N., Muñoz Salinas, R.: Fisher motion descriptor for multiview gait recognition. Int. J. Patt. Recogn. Artif. Intell. **31**(1) (2017)
5. Chetlur, S., Woolley, C., Vandermersch, P., Cohen, J., Tran, J., Catanzaro, B., Shelhamer, E.: cudnn: Efficient primitives for deep learning. CoRR abs/1410.0759 (2014)
6. Farnebäck, G.: Two-frame motion estimation based on polynomial expansion. In: Bigun, J., Gustavsson, T. (eds.) SCIA 2003. LNCS, vol. 2749, pp. 363–370. Springer, Heidelberg (2003). doi:10.1007/3-540-45103-X_50
7. Gálai, B., Benedek, C.: Feature selection for lidar-based gait recognition. In: 2015 International Workshop on Computational Intelligence for Multimedia Understanding (IWCIM), pp. 1–5 (2015)

8. Guan, Y., Li, C.T.: A robust speed-invariant gait recognition system for walker and runner identification. In: International Conference on Biometrics (ICB), pp. 1–8 (2013)

9. Han, J., Bhanu, B.: Individual recognition using gait energy image. IEEE PAMI **28**(2), 316–322 (2006)

10. He, K., Zhang, X., Ren, S., Sun, J.: Deep residual learning for image recognition. In: CVPR, pp. 770–778, June 2016

11. Hofmann, M., Geiger, J., Bachmann, S., Schuller, B., Rigoll, G.: The TUM gait from audio, image and depth (GAID) database: multimodal recognition of subjects and traits. J. Vis. Commun. Image Represent. **25**(1), 195–206 (2014)

12. Hossain, E., Chetty, G.: Multimodal feature learning for gait biometric based human identity recognition. In: Neural Information Processing, pp. 721–728 (2013)

13. Hu, W., Tan, T., Wang, L., Maybank, S.: A survey on visual surveillance of object motion and behaviors. IEEE Trans. Syst. Man Cybern. Part C Appl. Rev. **34**(3), 334–352 (2004)

14. Ji, S., Xu, W., Yang, M., Yu, K.: 3D Convolutional Neural Networks for human action recognition. IEEE PAMI **35**(1), 221–231 (2013)

15. KaewTraKulPong, P., Bowden, R.: An improved adaptive background mixture model for real-time tracking with shadow detection. In: Remagnino, P., Jones, G.A., Paragios, N., Regazzoni, C.S. (eds.) Video-Based Surveillance Systems, pp. 135–144. Springer, New York (2002)

16. Krizhevsky, A., Sutskever, I., Hinton, G.E.: Imagenet classification with deep convolutional neural networks. In: NIPS, pp. 1097–1105 (2012)

17. Simonyan, K., Zisserman, A.: Two-stream convolutional networks for action recognition in videos. In: NIPS, pp. 568–576 (2014)

18. Soomro, K., Zamir, A.R., Shah, M.: UCF101: a dataset of 101 human action classes from videos in the wild. In: CRCV-TR-12-01, November 2012

19. Tran, D., Bourdev, L.D., Fergus, R., Torresani, L., Paluri, M.: Learning spatiotemporal features with 3D convolutional networks. In: ICCV. IEEE (2015)

20. Vedaldi, A., Lenc, K.: MatConvNet - convolutional neural networks for MATLAB. In: Proceedings of the ACM International Conference on Multimedia (2015)

21. Whytock, T., Belyaev, A., Robertson, N.: Dynamic distance-based shape features for gait recognition. J. Math. Imaging Vis. **50**(3), 314–326 (2014)

22. Wu, Z., Huang, Y., Wang, L., Wang, X., Tan, T.: A comprehensive study on cross-view gait based human identification with deep CNNs. IEEE PAMI **PP**(99) (2016)

23. Wu, Z., Huang, Y., Wang, L.: Learning representative deep features for image set analysis. IEEE Trans. Multimedia **17**(11), 1960–1968 (2015)

24. Zeng, W., Wang, C., Yang, F.: Silhouette-based gait recognition via deterministic learning. Pattern Recogn. **47**(11), 3568–3584 (2014)

Comprehensive Evaluation of OpenCL-Based CNN Implementations for FPGAs

Ricardo Tapiador-Morales[1], Antonio Rios-Navarro[1],
Alejandro Linares-Barranco[1(✉)], Minkyu Kim[2], Deepak Kadetotad[2],
and Jae-sun Seo[2]

[1] Robotic and Technology of Computers Lab, University of Seville, Seville, Spain
`alinares@atc.us.es`
[2] School of Electrical, Computer and Energy Engineering,
Arizona State University, Tempe, AZ, USA
`jaesun.seo@asu.edu`

Abstract. Deep learning has significantly advanced the state of the art in artificial intelligence, gaining wide popularity from both industry and academia. Special interest is around Convolutional Neural Networks (CNN), which take inspiration from the hierarchical structure of the visual cortex, to form deep layers of convolutional operations, along with fully connected classifiers. Hardware implementations of these deep CNN architectures are challenged with memory bottlenecks that require many convolution and fully-connected layers demanding large amount of communication for parallel computation. Multi-core CPU based solutions have demonstrated their inadequacy for this problem due to the memory wall and low parallelism. Many-core GPU architectures show superior performance but they consume high power and also have memory constraints due to inconsistencies between cache and main memory. OpenCL is commonly used to describe these architectures for their execution on GPGPUs or FPGAs. FPGA design solutions are also actively being explored, which allow implementing the memory hierarchy using embedded parallel BlockRAMs. This boosts the parallel use of shared memory elements between multiple processing units, avoiding data replicability and inconsistencies. This makes FPGAs potentially powerful solutions for real-time classification of CNNs. In this paper both Altera and Xilinx adopted OpenCL co-design frameworks for pseudo-automatic development solutions are evaluated. A comprehensive evaluation and comparison for a 5-layer deep CNN is presented. Hardware resources, temporal performance and the OpenCL architecture for CNNs are discussed. Xilinx demonstrates faster synthesis, better FPGA resource utilization and more compact boards. Altera provides multi-platforms tools, mature design community and better execution times.

Keywords: Deep learning · Convolutional Neural Network · Hardware acceleration · OpenCL · FPGA · Caffe · Xilinx · Altera

© Springer International Publishing AG 2017
I. Rojas et al. (Eds.): IWANN 2017, Part II, LNCS 10306, pp. 271–282, 2017.
DOI: 10.1007/978-3-319-59147-6_24

1   Introduction

In recent years, throughout a series of breakthrough algorithms [1–5], convolutional neural networks significantly improved the state-of-the-art in large-scale image recognition tasks. Driven by such success, CNNs have become widespread across a broad range of applications including vision, object detection, speech recognition, autonomous driving, image captioning, etc.

Typically CNNs consists of a large number of deep layers, and could involve hundreds of millions of parameters (i.e. convolution kernels, bias). Using high-end GPGPUs (General Purpose Graphic Processing Units), the networks are trained iteratively using back-propagation algorithm for days or weeks, and then the networks with trained weights can be deployed onto hardware for classification tasks. There has been a number of prior works [6–12] that built hardware on different platforms for efficient CNN implementation (as accelerators or complete architecture on hardware), such as FPGA [6–9] and ASIC (application-specific integrated circuits) [10–12]. ASIC or custom chip designs show better energy-efficiency, but may not flexibly map various CNN algorithms easily with the rigid circuits. On the other hand, FPGA platforms are much more flexible and could easily map any given CNN algorithm with hardware optimizations. For FPGAs, the designers could perform manual RTL designs [7], but using high-level synthesis tools could prove effective [8,9] in terms of design time and wide design space exploration. The authors in [8] employed HLS tools in Xilinx framework to optimize CNN implementation, while the authors in [9] explored Open Computing Language (OpenCL) based implementation in Altera framework for throughput optimization of CNNs.

Since the high-level synthesis tools are developed differently within different frameworks of Xilinx and Altera, it is difficult to determine which option or FPGA chip would be the best candidate for certain objectives (area, speed, etc.) from the designers point of view. In this paper, we provide a comprehensive evaluation and comparison of the same CNN using both Xilinx and Alteras OpenCL-based high-level synthesis tool flows.

The remainder of the paper is organized as follows. In Sect. 2, the OpenCL programming and models are described. In Sect. 3, Alteras OpenCL design flow and hardware system is discussed, while Xilinxs SDAaccel design flow and hardware platform is presented in Sect. 4. LeNet-5 ConvNet [13] for MNIST database digits classification scenario is presented in Sect. 5. In Sect. 6, the hardware results and implementation are compared between the two designs from different vendors in a comprehensive manner. The paper is concluded in Sect. 7.

2   OpenCL for FPGA

Parallel computing has considerably improved in the last years thanks to the technology scaling favors. In the past decade, parallelism improvement started to be oriented to multi-core architectures for general purpose computers, or many-core to more specific solutions, as GPGPUs. For example, the Tesla K80 accelerator has 4,992 cores with a dual-GPU design that allows up to 2.9 double precision TFLOPS or 8.73 single-precision TFLOPS [14]. Special interest has existed for the FPGAs and SoC framework that includes programmable logic cells oriented to embedded co-design solutions.

To implement a given CNN model onto FPGA hardware, we start from the publicly available codes in the Caffe framework [15]. The input image for the CNN model is converted to a text file from Python interface in Caffe and the text file is read from OpenCL host code. Using the Python interface in Caffe, both the input data and weights are extracted and fed to the OpenCL host code, on a batch of input images. The hardware implementation computes till the last inner product layer output and compares it to the expected output from Caffe, to ensure correct functionality. Typically, the CNN models in Caffe are realized using double-precision values for the nodes and the weights. Considering efficient hardware implementation, we first find out how much precision reduction could be achieved while having minimal degradation in the final classification accuracy, and this reduced precision will be used when the OpenCL codes are written.

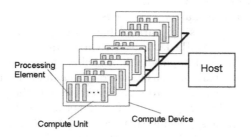

Fig. 1. OpenCL platform model.

The OpenCL *platform model* [16] consists of a host computer connected to several devices. Each OpenCL device is divided into compute units (CUs). Each CU is divided into processing elements (PE), where computations occur (Fig. 1). The OpenCL application is implemented as both host code and device kernel code. Each part will run in their specific hardware. The host code submits each kernel code as commands from the host to PEs through the memory hierarchy.

Regarding to the *execution model*, OpenCL has two units of execution: kernels that execute in one or more platforms, and a host program that executes on a host computer. Kernels execute the computation through work-items (with an associated ID), which are executed in groups (work-groups). The context of what the kernel executes is defined by the host. The host program uses the OpenCL API to create and manage the context. This API has a set of functions that enable the host interaction with devices through a command-queue. There are three main commands: kernel (to order the kernel execution), memory (data transfer between host and devices) and synchronization (synchronize points for order definition between commands). When a kernel-enqueue command submits a kernel for execution, an index space is defined. This index space is called NDRange in OpenCL, which corresponds to an N-dimensional index space. N is 1, 2 or 3. The NDRange is decomposed into work-groups forming blocks. It is defined by three integer arrays: global size (the extent of the index space in each dimension), an offset index (initial value of indices in each dimension), and the local-size (size of the work-group in each dimension).

The *memory model* consists of four regions in OpenCL for work-item execution. *Global memory* is where all work-items of all work-groups have to read and write data. These accesses must be cached. *Constant memory* is a region that remains without changes during the kernel execution. The host initializes this memory. *Local memory* is the one used by work-groups locally. It is shared by all work-items. It can be mapped into regions of the global memory. And *private memory* is a memory region that is only visible for a work-item, such that any other work-item cannot access this memory of a particular work-item. Data flow between memory regions is controlled through commands that the host enqueues. The memory consistency is guaranteed in a work-item and between a work-group and all its work-items, but there is no guarantee of memory consistency between different work-groups executing a kernel.

In OpenCL, there are mainly two *ways to parallelize a kernel*: (1) using multiple compute-units (CUs) in parallel (see Fig. 2, left), and (2) vectorising data processing through SIMD kernels with a unique CU (see Fig. 2, right). When multiple CUs are used in parallel, a kernel is replicated and the replicated kernels work simultaneously, increasing throughput and, therefore, consuming global memory bandwidth and hardware resources. On the other hand, SIMD vectorization increases throughput by vectoring kernels, which allows processing multiple work items in a single instruction (SIMD). This alternative is more efficient than using several CUs because it only duplicates the data paths. In this paper, SIMD experiments are presented because the use of replicated CUs generates global memory bottlenecks due to many parallel accesses during the execution.

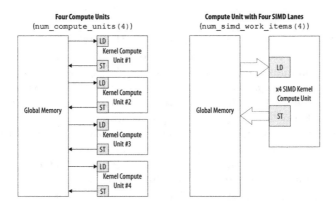

Fig. 2. Multiple parallel CU (left) versus single CU with SIMD (right).

Finally, there are two supported *programming models*: data parallel and task parallel. The data parallel model defines a computation as a sequence of instructions applied to multiple elements of a memory. On the other hand, the task parallel model requires the kernel to be executed in a single work-item of a work-group. In this case, several kernels can be executed in parallel. Synchronization is possible between work-items of a work-group or through two types of enqueued commands: barrier (it ensures all previous commands have been executed) or wait-on-an-event (the command to be executed waits for a particular event in memory before executing itself).

3   Altera OpenCL

Altera OpenCL (AOCL) is a framework for developing host applications that send kernels to be executed in parallel in FPGA resources. Work-groups, their work-items and memory models are implemented automatically in FPGA resources from an OpenCL description of the kernels and a C++ host application calling different functions from API libraries, such as: set buffers, call kernels, synchronize through events and read results. AOCL allows users to abstract the traditional hardware FPGA development flow and instead work with a much faster and higher-level software development flow. Using this design flow, it is possible to emulate OpenCL code in a FPGA, generating synthesis report files as timing or resources summaries.

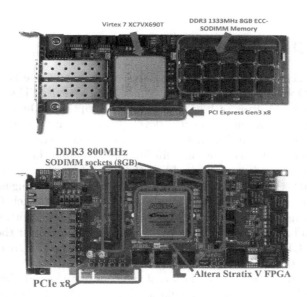

Fig. 3. Bottom Terasic DE5-Net Altera platform. Top Alpha-Data ADM-PCIE-7V3 Xilinx platform.

The design flow consists of two parts: host software application and kernel accelerator hardware on FPGA. The concept is that host sends data to the kernels, where complex calculations are accelerated. The design flow for an Altera board using OpenCL consists of several steps. The first step is to describe the functionality of the kernels using C/C++ and then to optimize each kernel applying OpenCL directives to generate a *.cl file. In addition, a host application must be written in C/C++ using the recommended environment. The OpenCL implementation of the LeNet-5 CNN for MNIST handwritten digit recognition [13] has been deployed on a Terasic DE5-Net (see Fig. 3). This board supports Altera OpenCL and its main characteristics include up to 8 GB DDR3 RAM memory running at 800 MHz, 72 Mb SRAM running at 550 MHz, PCIe-x8 and Altera Stratix V GX-5SGXEA7N-2F45C2 FPGA, which features are shown in Table 1.

Table 1. Right resources of Alpha-Data ADM-PCIE-7V3 Xilinx platform. Left resources of Terasic DE5-Net Altera platform

Resource	Amount	Resource	Amount
Logic elements (K)	622	Logic cells (K)	693
M20K (Blocks/Mbits)	2560/50	Slices (K)	108
18-bit 18-bit Multipliers	512	Distributed RAM(Kb)	10
27-bit 27-bit Multipliers	256	DSP slices	3600
		Block RAM (#/Mb)	2940/53

4   Xilinx SDAccel

The SDAccel [17] development environment is a command line based tool suite for compiling OpenCL programs into a Xilinx FPGA device. The design flow is similar to AOCL in terms of host and kernel descriptions. Directives and API must be replaced when same project is developed for both vendor environments. SDAccel runs on RedHat Linux OS in a batch mode using a command file. These commands allow to define the solution name, adding the target device and host files, creating the kernels and adding the files where they are implemented, creating the Xilinx OpenCL compute unit binary file, and building and packaging the systems. Several CUs per kernel can be implemented. Each CU can have several PEs, which emulates the SIMD architecture. One important advantage over the Altera tool is that SDAccel lets the programmer to test the application (emulation on CPU) before compiling and generating the FPGA binary file. A disadvantage is that SDAccel is less mature than AOCL.

SDAccel allows hardware emulation of the codesign program before building the system for FPGA. Hardware emulation is slower than CPU emulation since it uses a hardware simulator, but this emulation reproduces the final design on FPGA. The main advantage of using hardware emulation is to avoid the long implementation times (8 h on average for this work).

Table 2. Test results: No parallelism/Unroll/SIMD

Kernel Name	Exec. time (ms)	Logic Cells / Elem. (K)	DSP slices	BRAM (Kb)
Xilinx Virtex 7 690T				
conv_pool1	3.63 / 1.96 / 1.96	4.9 / 6.2 / 5.1	11 / 11 / 11	180 / 216 / 216
conv2	7.62 / 4.92 / 4.92	4.8 / 4.8 / 4.9	11 / 11 / 11	108 / 144 / 144
pool2	0.03 / 0.06 / 0.06	3.0 / 4.0 / 3.0	4 / 4 / 4	72 / 144 / 144
ip1_relu	0.55 / 0.55 / 0.55	4.2 / 4.2 / 4.2	11 / 11 / 11	72 / 72 / 72
ip2	0.35 / 0.35 / 0.35	4.0 / 3.0 / 4.0	9 / 9 / 9	72 / 72 / 72
Altera Stratix V GXA-7				
conv_pool1	1.01 / 1.01 / 0.98	145.7 / 42.3 / 73.7	8 / 31 / 57	5225 / 6205 / 11200
conv2	3.95 / 3.96 / 4.27	300.5 / 34.0 / 34.0	8 / 31 / 31	3207 / 4882 / 4900
pool2	0.08 / 0.07 / 0.13	6.9 / 6.9 / 6.8	2 / 2 / 2	279 / 273 / 279
ip1_relu	1.01 / 1.81 / 2.02	5.8 / 5.8 / 5.8	4 / 4 / 4	1471 / 1470 / 1500
ip2	0.15 / 0.14 / 0.13	5.7 / 5.7 / 5.7	4 / 4 / 4	1471 / 1470 / 1500

After emulation, the final step is to build the system in the real hardware of the target device. When the compilation/implementation is completed, a number of files have been created to run the application, such as the executable, the Xilinx OpenCL binary container (*.xclbin) and the FPGA programming file. For these experiments, the Alpha Data ADM-PCIE-7V3 [18] board (see Fig. 3) has been used under the CentOS 6.6 operating system. The main features include two 8 GB ECC-SODIMM memory up to 1333 MT/s (faster than Altera DE5 platform), one PCI Express Gen3 x8 and Xilinx Virtex 7 XC7VX690T-2FFG1157C. The features of the FPGA are listed in Table 2. Besides the DSP slices, the specification of the FPGA is similar to that of Alteras Stratix V-GXA7.

5   LeNet-5 and MNIST Scenario

LeNet-5 CNN [13] architecture (shown in Fig. 5) serves as the baseline for many recent CNN-based classification algorithms. It combines three architectural ideas to ensure a certain degree of shift, scale and distortion invariance: local receptive fields, shared weights and spatial sub-sampling. The input layer represents a size-normalized and centered image. In this case, the size corresponds to the size of MNIST database digits (28×28). The first layer (C1) is the result of a set of convolutions over the input image. Each pixel in C1 receives inputs from a set of units located in a small neighborhood of the previous layer. This represents the kernel of the convolution (5×5 in our case). These operations are able to learn and extract elementary visual features, such as edges, end-points, and corners. The combination of these features by subsequent layers are able to detect higher-order features. C1 in this example extracts 20 features from the input image. S2 performs a sub-sampling operation of local averaging, reducing the resolution of the feature maps where distinctive features are encoded. Typically, these convolution and sub-sampling layers are sequentially instantiated for feature map combinations. They are implemented in a bi-pyramid way: at each layer, the number of feature maps is increased as the spatial resolution is decreased. C3 is a convolution layer for 50 smaller feature maps and S4 is the corresponding sub-sampling layer that performs the same operation as that in S2. C3 combines all of the S2 features. The last layer of this CNN is a fully-connected classifier with 500 input neurons and 10 output neurons, which also includes a Rectification Linear Unit (ReLU).

6 Comparison Study

The implementation of this Le-Net5 using the OpenCL framework impose some restrictions. Figure 4 (bottom) shows the block diagram of the OpenCL solution. It can be seen that the host application, running on a computer, sends input images and kernel weights to the logic through PCIe interface. Data is then stored in the DDR memory in the platform, called Global Memory. This memory is continuously and iteratively accessed by the logic (FPGA) through all the

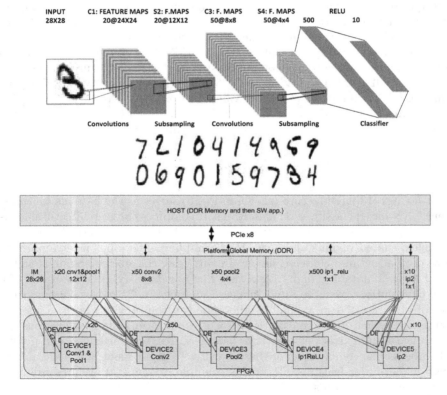

Fig. 4. LeNet-5 ConvNet architecture (top) for MNIST digit recognition (middle) and its OpenCL based hardware block diagram (bottom).

parallel devices physically implemented in hardware. The CNN is structured in 5 kernels (stages), where first kernel implements first layer convolutions and their subsampling operations (conv_pool1); second kernel performs the second layer convolutions (conv2), which is more complex since it has to take results from 20 instances of previous layer, and perform convolutions for 50 instances of this second layer. Then, the third kernel implements the second subsampling operations (pool2). The forth kernel has 500 instances for the classifier unit, whose inputs are the outputs of the previous 50 instances (ip1_relu). Each of these 500 devices send their output to a final layer with 10 instances (one per digit) to categorize the winner digit in the classification (ip2). Each of these devices read the global memory, process the corresponding operation, and then write back the results to the global memory. Consecutive kernels (stages or layers of the CNN) execute in-order, which are controlled by special events included during OpenCL compilations. This architecture needs a high bandwidth DDR memory interface to support all required parallel instances. OpenCL can implement each kernel in a replicated manner as many times in parallel as possible, or it can execute one after the other sequentially if no parallelism can be implemented. As more parallelism is employed, the global memory behavior worsens. The main difference between Altera and Xilinx platforms is the DDR3 on-board memory speed (800 MHz for Altera and 1333 MHz for Xilinx) as mentioned previously. OpenCL allows other memory implementations to avoid this shared memory bottleneck, like local pipes that connect two devices directly in the logic. Each of these pipes is implemented through small FIFOs as a point-to-point communication channel between two devices. For CNNs, these pipes do not represent a feasible solution because internal convolution layers, such as C3 in this case, have to read all the S2 outputs and combine them into each of the 50 C3 outputs. This represents 50 pipes at C3 per for each of the 20 S2 units, which is not viable, in terms of resource consumption, for the selected platforms. Therefore, we selected the global memory interface as the possible solution to work for both platforms, and we provide a comprehensive comparison. Three different tests have been developed for these platforms with the Le-Net5. The first test consists of comparing each FPGA executing each layer of the CNN without any kind of parallelism. The second test aims to do same measurements when loops are unrolled. For the last test, SIMD directives have been included to vectorise each layer. Table 3

Table 3. Acceleration comparison

Kernel Name	Xilinx vs Altera Acceleration	% Acceleration
conv_pool1	3.59 / 1.94 / 2	259 / 94 / 100
conv2	1.92 / 1.24 / 1.15	92 / 24 / 15
pool2	-2.66 / 1.16 / 2.16	-166 / -16 / -116
ip1_relu	-1.83 / 3.29 / 3.67	-83 / -229 / -267
ip2	2.33 / 2.5 / 2.69	133 / 150 / 169

shows the results of these three experiments. Execution times, logic resources, the number of DSP units and needed blockRAM are shown per kernel. In general, execution time is improved upon employing more parallelism up to a limit. The limit occurs due to the bottleneck that the global memory accesses impose. As expected, the usage of logic gates and DSP units increases when parallelism is increased. Altera tools are able to extract much more parallelism than Xilinx, as it can be seen on logic elements/cells and DSP utilization. There are very small differences between unrolling and SIMD for both platforms for this experiment. Altera tool is able to extract more aggregation for SIMD than Xilinx. In fact, for Xilinx, both unrolling and SIMD have almost same results.

7   Real Time Experiment

In order to demonstrate the different DDR memory bandwidth limits of these two platforms, the same real-time experiment has been performed in both platforms. The experiment consists of connecting a webcam to the host application, which continuously reads in an image frame, normalizing it and resizing to 28×28 pixels using OpenCV libraries. The host sends kernels parameters in the beginning and then it iterates the process of acquiring an image frame, pre-processing it, sending it to the platform and checking the final classification results. The onboard DDR in the Altera platform could not support the memory bandwidth required by this demonstration and the time per frame is continuously increasing (starting at 10 ms per frame). In contrast, Xilinx platform supported this real-time experiment owing to the higher DDR bandwidth. Results show that time increases when parallelism is applied. This is due to memory bandwidth when multiple access to global memory are done. Bottlenecks slow down kernel increasing execution time. Figure 5 shows a screen-shot of the real time running

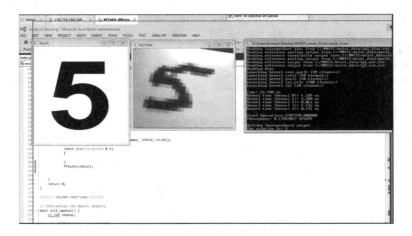

Fig. 5. LeNet-5 ConvNet architecture (top) for MNIST digit recognition (middle) and its OpenCL based hardware block diagram (bottom).

demonstration. In general, logic elements, DSP and BlockRAM have increased when parallelized directives are applied. However, the time does not get better due to bottleneck generated by DDR memory bandwidth. Table 3 represents the acceleration between vendors. Execution times for Xilinx are much better than Altera except for pool2 and ip1_relu stages.

8   Conclusions

This work presents a comparison between two OpenCL FPGA-based platforms (Altera and Xilinx) executing a convolutional neural network. Results show that the Altera platform has better execution time for each kernel than the Xilinx platform for all test scenarios. However, the Xilinx platform requires less FPGA resources than the Altera counterpart to execute the same CNN model. The real-time experiment developed for both platforms has demonstrated that the DDR memory bandwidth is crucial for the global memory communication architecture. Other memory architectures, such as pipes, were implemented internally to the FPGA without requiring any off-chip memory bandwidth, but it was insufficient for CNNs because of their point-to-point connections. A new memory model that allows having double-buffered memory spread on the FPGA blockRAM will avoid the bottlenecks identified in this work. This will allow having more CUs in parallel to further improve the performance. Therefore, beyond the differences between the platforms, this presented exploratory work for implementing full CNNs architectures on FPGA with OpenCL, shows that proposed codesing architecture lacks on memory bandwidth because of the dense connections between layers.

Acknowledgments. This work has been partially supported by Xilinx and Altera University Programs (through platform donations); by the NPP project funded by SAIT (2015–2018); and by the Spanish government grant (with support from the European Regional Development Fund) COFNET (TEC2016-77785-P).

References

1. Krizhevsky, A., et al.: ImageNet classification with deep convolutional neural networks. In: Advances in Neural Information Processing Systems (2012)
2. Simonyan, K., Zisserman, K.: Very deep convolutional networks for large-scale image recognition. In: International Conference on Learning Representations (2015)
3. Szegedy, C., et al.: Going deeper with convolutions. In: IEEE Conference on Computer Vision and Pattern Recognition (CVPR) (2015)
4. Ioffe, S., Szegedy, C.: Batch normalization: accelerating deep network training by reducing internal covariate shift. In: International Conference on Machine Learning (ICML) (2015)
5. He, K., Zhang, X., Ren, S., Sun, J.: Deep Residual Learning for Image Recognition (2015). arXiv:1512.03385

6. Farabet, C., et al.: Hardware accelerated convolutional neural networks for synthetic vision systems. In: IEEE International Symposium on Circuits and Systems (2010)
7. Gokhale, V., et al.: A 240 G-ops/s mobile coprocessor for deep neural networks. In: IEEE Conference on Computer Vision and Pattern Recognition Workshops (2014)
8. Zhang, C., et al.: Optimizing FPGA-based accelerator design for deep convolutional neural networks. In: ACM International Symposium on Field-Programmable Gate Arrays (FPGA) (2015)
9. Suda, N., Chandra, V., Dasika, G., Mohanty, A., Ma, Y., Vrudhula, S., Seo, J., Cao, Y.: Throughput-optimized opencl-based fpga accelerator for large-scale convolutional neural networks. In: ACM International Symposium on Field-Programmable Gate Arrays (FPGA) (2016)
10. Chen, Y., Luo, T., Liu, S., Zhang, S., He, L., Wang, J., Li, L., Chen, T., Xu, Z., Sun, N., Temam, O.: DaDianNao: a machine-learning supercomputer. In: IEEE/ACM International Symposium on Microarchitecture (MICRO) (2014)
11. Chen, Y.-H., Krishna, T., Emer, J., Sze, V.: Eyeriss: an energy-efficient reconfigurable accelerator for deep convolutional neural networks. In: IEEE International Solid-State Circuits Conference (ISSCC) (2016)
12. Sim, J., Park, J.-S., Kim, M., Bae, D., Choi, Y., Kim, L.-S.: A 1.42TOPS/W deep convolutional neural network recognition processor for intelligent IoE systems. In: IEEE International Solid-State Circuits Conference (ISSCC) (2016)
13. LeCun, Y., Bottou, L., Bengio, Y., Haffner, P.: Gradient-based learning applied to document recognition. Proc. IEEE **86**(11), 2278–2324 (1998)
14. Tesla K80 GPU Accelerator, Board Specification BD_07317_001_V05, January 2015. http://images.nvidia.com/content/pdf/kepler/Tesla-K80-BoardSpec-07317-001-v05.pdf
15. Jia, Y., et al.: Caffe: convolutional architecture for fast feature embedding. In: ACM International Conference on Multimedia (2014)
16. Howes, L., Munshi, A. (eds.) The OpenCL Specification Version: 2.0. Khronos OpenCL Working Group. https://www.khronos.org/
17. Xilinx SDAccel development environment user guide. http://www.xilinx.com/products/design-tools/software-zone/sdaccel.html
18. Alpha Data ADM-PCIE-7V3 user manual. http://www.alpha-data.com/pdfs/adm-pcie-7v3%20user%20manual.pdf

Machine Learning Improves Human-Robot Interaction in Productive Environments: A Review

Mauricio Zamora[1], Eldon Caldwell[1], Jose Garcia-Rodriguez[2(✉)],
Jorge Azorin-Lopez[2], and Miguel Cazorla[2]

[1] University of Costa Rica, San José, Costa Rica
{mauricio.zamora,eldon.caldwel}@ucr.ac.cr
[2] University of Alicante, Alicante, Spain
{jgr,jazorin,mcazorla}@ua.es

Abstract. In the new generation of industries, including all the advances introduced by Industry 4.0, human robot interaction (HRI), by means of automatic learning and computer vision, become an important element to accomplish. HRI allows to create collaborative environments between people and robots, avoiding the latter generating a risk of occupational safety. In addition to the automatic systems, the interaction by mean of automated learning processes provides necessary information to increase productivity and minimize delivery response times by helping to optimize complex production planning processes. In this paper, it is presented a review of the technologies necessary to be considered as basic elements in all processes of industry 4.0 as a crucial linking element between humans, robots, intelligent and traditional machines.

Keywords: Machine learning · Manufacturing · Computer Vision · Robotics · Augmented reality

1 Introduction

From the first industrial revolution there was a process of economic, social and technological transformation that began in England in the second half of the eighteenth century, with the steam engine, mass production. In the nineteenth century, the commercialization of electricity begasn, the development of communication technologies and the beginnings of automation, marked the guidelines of what became known as the second industrial revolution [1].

At the end of the twentieth century, the technological advances allowed the beginning of an era in which computer science plays a crucial role: the third industrial revolution. There was given incipient use of robotic technologies of repetition of task with little interaction with the production systems. Nowadays, the fourth industrial revolution or also known as Industry 4.0 arrived, where production processes are linked to the customization of production, which increases

© Springer International Publishing AG 2017
I. Rojas et al. (Eds.): IWANN 2017, Part II, LNCS 10306, pp. 283–293, 2017.
DOI: 10.1007/978-3-319-59147-6_25

the complexity of manufacturing, where communication and control are one of the pillars [1, 2].

In consequence, in today's world, companies are experiencing a strong transformation in their production processes, due to the upgrade in Industry 4.0. The concept of automation and data exchange as core in manufacturing technologies are strongly involved. Moreover, productive environments can be seen as an ecosystem of technologies that include robotics, cyberphysical systems, Big Data and the Internet of things embedded in a collaborative environment with people [3–5].

Cybernetics is the study of complex systems principles which involves the control and exchange of information between machines and humans [6]. One of the criteria to be considered in the automated industries is the interaction of humans with robotic systems in a more natural and safe way, minimizing stress to operators by the use of robots [7].

To achieve this natural interaction between robots and humans in manufacturing environments, robots need to behave as a partner in the performance of a work, rather than a tool to get the job done. To get the robot to be a co-worker, automatic learning must be involved so that they can learn the behavior patterns of the user and have the ability to adapt to the tasks of collaborative work. This learning takes place through initial training by operators [8].

In this paper, we present an approach to the technologies necessary to be considered as basic elements in the implementation of manufacturing environments in Industry 4.0 focusing on machine learning, computer vision and augmented reality.

The remaining of the paper is organized as follows: Sect. 2 presents the basics of machine learning applied to cybernetics. In Sect. 3, we introduce the topic of Cybermanufacturing, describing the relation among machine learning and computer vision working together to create cognitive systems. Section 4 reviews the literature related with human-robot interaction and machine learning to provide the reader with a summary of the current research in the area. Finally, in Sect. 5 we draw some conclusion and further works.

2   Machine Learning in Cybernetic Systems

In order to make robots better adapted to interaction with people, the application of automatic learning is required, both in tasks where robots can perform with full autonomy, and in those where they need to learn from humans. The most important learning algorithms used in social robotics for HRI are reviewed, where supervised and unsupervised learning techniques will be used to analyze information from cyber environments in order to generate high level information [9].

In the area of automatic learning applied to cybernetics, we are interested in establishing the possible applications that can be made for each type of algorithm. The main types of algorithms are: supervised and unsupervised.

– **Supervised learning**, is enployed when training with datasets with known inputs and outputs. It is extremely useful in industrial processes where quality control is sought at workstations, as Ferreiro and Sierra indicate in [10]. These quality inspection processes can be carried out by simple sensors (e.g. weight, color or size [11]) as well as by computer vision which allows the identification of quality with respect to the shape of the products processed in other workstations [5]. This is important if the plant (industry) is seen as a system interconnected between all its workstations. This conception allows time and material savings by avoiding process pieces that will not meet the proposed quality standards. In the case of a particular case of computer vision, it is easier to use because it fits a variety of products or processes without having to invest in new components (i.e. it would be only necessary to re-train with known data). In this process, the intervention of operators interacting with each other and with the machines is needed. Hence, mechanisms of control and exchange of information should be established [12]. Computer vision is also used in manufacturing for quality control, collision detection [13], navigation [14] and augmented reality [15].

– **Unsupervised learning**. In productive environments, many contexts are found in which relations between inputs and outputs are unknown, unsupervised learning algorithms allow to organize information or find patterns in them [16]. One of the main points to consider is the protection of the integrity of the operators. For example in the human-robot interaction, learning from cameras can be used to track the movements of operators so that robots can predict intentions and recognize the behavior of people with whom it collaborates [17]. This allows to create more flexible working environments for the performance of human tasks, but this requires that the algorithms of automatic learning make decisions from sets of data of partially known environments having to establish control policies to protect people. In this context, Santoro and Tamburrini [9] proposed to use a mixed scheme, using supervised and unsupervised learning, where the sets of data used for learning of behavior patterns were performed by a "trainer".

2.1 Current Challenges in Manufacturing

With the high level of worldwide competition, manufacturing has to be well planned to respond quickly with high quality products [18]. In order to manage this situation, pressure must be exerted on each of the processes of production engineering: from the design stages [16], process planning, complex calculations to production cell modifications [18].

These processes generate and consume high volumes of data from the definition of customer needs, planning and floor programming. All data is synthesized in information to support decision making [19]. The process is very complex, Monostori [20] recommends using pattern recognition techniques, expert systems, artificial neural networks and fuzzy systems generating mixed systems between manufacturing techniques and Artificial Intelligence.

Many of these processes are so complex because of the large number of parameters needed to calculate production plans and capacity calculations in production plants. These calculations require Artificial Intelligence techniques to support real-time operator decision making [20]. In Industry 4.0 some challenges are found in many aspects of the manufacturing process: definition of demand, design and interaction.

Definition of Demand. The initial problem for a production process is the definition of demand and the particular needs of each product. A novel technique known as social manufacturing is applied today to use cross-enterprise information by leveraging the benefits of information and communications technologies [21], as well as obtaining information from the Internet of Things, Big Data, Cloud technologies and Advanced manufacturing processes [22].

The social manufacturing and cloud manufacturing environments generate large volumes of unstructured data, which are not suitable for data mining type processing, so progress has been made in extracting information using neural networks based on cases and rules [21].

Automated learning is being applied in dynamic programming, specifically with Neural Networks of Retro-propagation and Case-Based Reasoning (CBR) techniques, in order to set the rules for starting production to decrease the time needed for the production [23].

Design. In traditional manufacturing environments, product engineers designed a product that was created massively. Currently, there are new developments in production, where either the designers of the company or the end users themselves can request the manufacture of products through models created by CAD applications [24]. It generates a level of stress in manufacturing companies since they must restart production processes for a new product, which may not be for mass production, with and increase in their operating costs.

A key element in manufacturing after the design phase of a new product is the generation of a pilot production program that provides information to create the model used in mass manufacturing. This initial manufacturing execution generates small datasets, Tsai and Li [25] proposed to use neural networks with diffusion-neural network (DNN) learning function to optimize initial production models.

Robotic Processes. Industrial processes are intricately linked with technology. At present, we are talking about cyber physical systems, which are an evolution of mechatronics systems, where the main difference is that mechatronics systems are the integration of mechanics, electronics and computing in the creation of devices that are used with integrated control systems [26]. These systems are usually part of processes that do not compare information to dynamically react to changes. Cybernetic systems are mechatronic systems that strongly integrate computer technologies, networks and physical processes, through the use of intelligent objects that are interconnected locally or globally through the Internet of things, using data and services for making autonomous decisions taking advantage of automatic learning [26].

Interaction in Productive Environments. In current industries, the goal is to create environments where human and robots can interact is increasing. However, it generates a series of complex problems due to the interaction to establish communication mechanisms between humans and robots [8]. The complexity is increased in industrial environments due to the noisy produced by the machinery of the factories. In industrial HRI, an important point to consider is the occupational safety. It could be more important where there are young and inexperienced operators who usually do not read safety standards carefully and do not follow instructions to avoid injuries [27].

Mobile robots, fixed robots and recently social robots are different types of robots that can be found in an industrial environment. The mobile robots serve to move through the plant elements such as: raw material, supplies, finished product among others. Currently, they follow predefined routes with colored lines. The task is to recognize lines through color sensors or cameras. In situations with a more dynamic production scheme, greater interaction of humans with robots and greater sharing of physical space is produced. Wang et al. [13] proposed a model of a plant using 3D sensors for the detection, linking real time motion sensors to imitate the models with the elements of the reality in order to calculate the minimum distance between the human and the robot to setting a collision detection system.

In an industrial environment, HRI can be improved using computer vision and augmented reality. The computer vision allows to help in the self-localization and mapping (SLAM), detection and tracking of people, identification of human activities and facial expressions, as example. Digital content can be overlaid on images from the environment on mobile devices to create augmented reality systems that help operators interact in real time. In addition, wearable cameras can be used to facilitate human activities using first-person vision (FPV) [28]. On the other hand, as negative effect, Mehlmann et al. [29] determined that using eye tracking approach could cause problem over time.

3   Human-Robot Interaction and Machine Learning Classification

As shown in Table 1, most of the interaction was usually made through mobile mechanisms, especially tablets, in order to facilitate the movement of people. The interaction is carried out through buttons and showing graphical information about the processes and statistics of use of the devices (both intelligent and regular machines). In another significant amount of occasions a traditional computer is used to show similar information to workers. It should be noted that in some cases the use of tablets was to implement augmented reality systems, which also showed instructions to the operators to carry out their work.

Table 1. Human Robot Interaction

Ref.	Display	Inf. exchange	Autonomy (LOA)	Applications
[27]	Mobile	Visual	Reduced computational decision	Detecting events
[4]	Computer	Visual	Complete human decision	Controlling processes
[13]	Mobile	Touch	Complete human desicion	Navigation, Detecting events, Interaction, Automatic inspection
[28]	Mobile	Touch	Reduced computational decision	Navigation, Detecting events, Interaction, Automatic inspection
[29]	Computer	Voice	Complete human desicion	Controlling processes
[9]	Mobile	Visual	Reduced computational decision	Interaction
[7]	Mobile	Voice	Reduced human decision	Navigation

The level of autonomy of robots, as shown in Table 1, is low. Most of the analyzed teams are controlled mainly by humans, very few provide capabilities of total autonomy control by artificial intelligence or where artificial intelligence controls most of decisions. Much of the automatic decision-making capability is being used to determine events (related to stimuli on each machine, events generated by interaction processes with other intelligent machines in complex processes are not being controlled). In addition, automatic learning capability is used in pattern recognition for navigation in controlled environments.

Regarding the robot interaction, voice and visual form are the most used information exchange. They are more related to automatic learning according to the data identified in Table 1. Systems interact naturally with people. In the case of the visual form, most cases referred to computer vision systems, where it is being used to create augmented reality systems in order to facilitate interaction with the operators. This interaction is achieved by using ocular projection systems or through mobile devices, providing additional information about the work environment. Experiments were done to interact with people through eye tracking but it was determined that their continued use in long lapses of time could cause problems for users [29].

In the case of machine learning methods, the majority of cases identified in the state of the art, as shown in Table 2, are artificial neural networks, with different types of learning algorithms that vary according to the problem (being Backpropagation the most used). These neural networks fed on data from sensors to provide dynamic decision-making environments.

Table 2. Machine Learning

Ref.	Applications	Approaches	Algorithms type
[20]	Manufacturing	Neural networks	Supervised, Unsupervised
[21]	Manufacturing	Neural networks	Supervised, Unsupervised
[10]	Manufacturing	Neural networks	Supervised, Unsupervised
[27]	Bioinformatics	Clustering	Supervised
[23]	Manufacturing	Neural networks	Supervised
[9]	Social	Neural networks	Supervised, Unsupervised
[7]	Navigation	Neural networks	Supervised
[25]	Manufacturing	Bootstrapping	Supervised
[17]	Interaction	Non parametric	Supervised
[30]	Bioinformatics	Support vector machines	Unsupervised
[31]	Cooperative robotics	Neural networks	Supervised, Unsupervised
[32]	Social robotics	Neural networks	Supervised
[33]	Cognitive robotics	Genetic algorithms	Supervised
[34]	Navigation	Genetic algorithms	Supervised
[14]	Navigation	Neural networks	Supervised, Unsupervised
[35]	Optimization and metaheuristic	Neural networks	Unsupervised
[36]	Optimization and metaheuristic	Neural networks	Supervised
[37]	Machine perception	Neural networks	Supervised
[38]	Optimization and metaheuristic	Neural networks	Supervised

Most of the scenarios described were trained to artificial neural networks with known data sets and results, as shown in Table 2 (supervised algorithm). However, in many industrial processes it is necessary to generate algorithms that can provide decision making under conditions of small or limited data sets (several authors recommend using statistical techniques in these cases).

An adequate training of neural networks is required in new industries, because the interaction of robots and operators is sought in the most natural way. In addition to collaborating with automatic decisions in production processes. Intelligent systems will be the basis on which the new industry will be based.

Although the new industry will focus on intelligent systems, interaction with people will always be present at some point in the overall process. It can be seen that most of the interaction is done by obtaining information through sensors, as can be seen in the Table 1. It seeks to establish an adequate interconnection beyond the use of sensors with SME, traditional machinery and operators with real-time monitoring and monitoring by artificial intelligence, to maximize production, improve human robot interaction and minimize occupational health risks to operators and the physical risks to the productive elements.

4   Discussion

We have revised research works that provide human robot interaction through Computer Vision (CV) and Machine Learning (ML) methods in order to reflect how to use such technologies in an industrial environment. These tecniques provides machinery with autonomous working capabilities and minimize the interaction gap between the operators and the Smart Manufacturing Equipment (SME). The workers, in some cases, have not developed the same level of technical development, either because they are novel workers or old employees. so, the use of mechanisms that use the CV and ML plays an important role in this situation.

It is proposed that SME considers real-time collision detection scheme through RGB-D cameras detection of the people around them, so that in a first instance protects humans from injury. This could be considered as the basic level of interaction in a collaborative environment of industrial robotics. A second level is to provide natural interface mechanisms with operators, where robots through automatic learning would be able to recognize and learn human behaviors and activities to serve them as collaborations without having to use programming of activities of the traditional robotics. Robots, in turn, can predict needs for tools or materials to increase human productivity.

Due to the fact that in many industries there are limitations of voice communication by manufacturing noise, it is proposed to use "wearable" type sensors that allow communication with SME. Giving the definition of a third level, which seeks the implementation of lens-type devices, which allow the operator to receive information through real time augmented reality (AR) of SME, including instructions for assembly or safety for the operator. In addition, it is sought that SME uses CV through ANN training to learn how to recognize products and raw material according to standards. Moreover, in order to serve as mechanisms to control the quality of production before being sent to other machines or the operators, these techniques would avoid "scrap" or reprocessing, with the reduction of production time.

Another recurring problem in the industry, by including a source of customization factor in Industry 4.0, is determining the Capacity Resource Planning (CRP). It makes very difficult to define Product Quantity Routes (PQR) and Bottle Neck (BN) due to the dynamics of the production. In consequence, it has been demonstrated that the SME provides greater autonomy, when using

ML to determine the sequence of processes and, in turn, to provide the human operators with the necessary instructions for product assembly customizing it through AR.

5   Conclusions

This paper is mainly focused on the use of artificial intelligence in industry, exploring the topics through the central idea of cybermanufacturing including robotics, computer vision, manufacturing, artificial intelligence and human robot interaction. The paper reviews them from the perspective of an entire industrial production process, starting at the definition of the idea and the needs of a product to its production; emphasizing on how Smart Manufacturing Equipment (SME) interacts with each other and the operators, seeking the creation of an intelligent environment for the development of an industrial production.

The main purpose is to discuss the necessity to use artificial intelligence as the core element for designing future cyberfactories based on the concepts of Industry 4.0. It will allow the interaction of intelligent and traditional equipment and the collaborative work between operators, considering elements such as computer vision to intrinsically promote the quality of the products and the occupational safety of people. Moreover, it will allow to consider the evolutionary proposal of production systems to make them more dynamic in response to times and flexibility of production of new products with variants caused by specific needs of consumers.

Future research lines propose the integration of intelligent logistic systems. The production planning in real time will allow intelligent distribution channels to integrate distribution schemes with robots directly from the factory. While HRI has made significant advances there are several challenges, including the use of two-way communication natural interfaces in real time between humans and robots. In this way, humans would feel that devices are really workmates rather than collaborative tools.

Acknowledgements. This work has been funded by the Spanish Government TIN2016-76515-R grant for the COMBAHO project, supported with Feder funds.

References

1. Kang, H.S., Lee, J.Y., Choi, S., Kim, H., Park, J.H., Son, J.Y., Kim, B.H., Noh, S.D.: Smart manufacturing: Past research, present findings, and future directions. Int. J. Precis. Eng. Manuf. - Green Technol. **3**(1), 111–128 (2016)
2. Drath, R., Horch, A.: Industrie 4.0: hit or hype? [Industry Forum]. IEEE Ind. Electron. Mag. **8**(2), 56–58 (2014)
3. Hermann, M., Pentek, T., Otto, B.: Design principles for industrie 4.0 scenarios. In: Proceedings of the Annual Hawaii International Conference on System Sciences, pp. 3928–3937, March 2016

4. Lee, J., Bagheri, B., Kao, H.A.: Recent advances and trends of cyber-physical systems and big data analytics in industrial informatics. In: International Conference on Industrial Informatics (INDIN) 2014, November 2015 (2014)
5. Hedelind, M., Jackson, M.: How to improve the use of industrial robots in lean manufacturing systems. J. Manuf. Technol. Manag. **22**(7), 891–905 (2011)
6. Siddique, N.H., Mitchell, R., O'Grady, M., Jahankhani, H.: Cybernetic approaches to robotics. Paladyn **2**(3), 109–110 (2011)
7. Meisner, E., Isler, V., Trinkle, J.: Controller design for human-robot interaction. Auton. Rob. **24**(2), 123–134 (2008)
8. Goodrich, M.A., Schultz, A.C.: Human-robot interaction: a survey. Found. Trends® Hum. Comput. Interact. **1**(3), 203–275 (2007)
9. Santoro, M., Marino, D., Tamburrini, G.: Learning robots interacting with humans: from epistemic risk to responsibility. AI Soc. **22**(3), 301–314 (2008)
10. Ferreiro, S., Sierra, B.: Comparison of machine learning algorithms for optimization and improvement of process quality in conventional metallic materials. Int. J. Adv. Manuf. Technol. **60**(1–4), 237–249 (2012)
11. Fast-Berglund, Å., Fässberg, T., Hellman, F., Davidsson, A., Stahre, J.: Relations between complexity, quality and cognitive automation in mixed-model assembly. J. Manuf. Syst. **32**(3), 449–455 (2013)
12. Lee, J., Bagheri, B., Jin, C.: Introduction to cyber manufacturing. Manuf. Lett. **8**, 11–15 (2016)
13. Wang, L., Schmidt, B., Nee, A.Y.C.: Vision-guided active collision avoidance for human-robot collaborations. Manuf. Lett. **1**(1), 5–8 (2013)
14. Hornung, A., Bennewitz, M., Strasdat, H.: Efficient vision-based navigation. Auton. Rob. **29**(2), 137–149 (2010)
15. Makris, S., Karagiannis, P., Koukas, S., Matthaiakis, A.S.: Augmented reality system for operator support in human–robot collaborative assembly. CIRP Ann. Manuf. Technol. **65**(1), 61–64 (2016)
16. Ericson, G., Franks, L., Rohrer, B.: How to choose algorithms for Microsoft Azure Machine Learning (2016)
17. Xiao, S., Wang, Z., Folkesson, J.: Unsupervised robot learning to predict person motion. In: 2015 IEEE International Conference on Robotics and Automation (ICRA), pp. 691–696 (2015)
18. Erdin, M.E., Atmaca, A.: Implementation of an overall design of a flexible manufacturing system. Procedia Technol. **19**, 185–192 (2015)
19. Puik, E., Telgen, D., Moergestel, L., Ceglarek, D.: Assessment of reconfiguration schemes for Reconfigurable Manufacturing Systems based on resources and lead time. Robot. Comput.-Integr. Manuf. **43**, 30–38 (2017)
20. Monostori, L.: AI and machine learning techniques for managing complexity, changes and uncertainties in manufacturing. IFAC Proc. Volumes (IFAC-Papers Online) **15**(1), 119–130 (2002)
21. Leng, J., Jiang, P.: A deep learning approach for relationship extraction from interaction context in social manufacturing paradigm. Knowl.-Based Syst. **100**, 188–199 (2015)
22. Cheng, Y., Tao, F., Zhao, D., Zhang, L.: Modeling of manufacturing service supply-demand matching hypernetwork in service-oriented manufacturing systems. Robot. Comput.-Integr. Manuf. **45**, 59–72 (2015)
23. Priore, P., Fuente, D., Puente, J., Parreño, J.: A comparison of machine-learning algorithms for dynamic scheduling of flexible manufacturing systems. Eng. Appl. Artif. Intell. **19**(3), 247–255 (2006)

24. Gao, W., Zhang, Y., Ramanujan, D., Ramani, K., Chen, Y., Williams, C.B., Wang, C.C., Shin, Y.C., Zhang, S., Zavattieri, P.D.: The status, challenges, and future of additive manufacturing in engineering. Comput.-Aided Des. **69**, 65–89 (2015)
25. Tsai, T.I., Li, D.C.: Utilize bootstrap in small data set learning for pilot run modeling of manufacturing systems. Expert Syst. Appl. **35**(3), 1293–1300 (2008)
26. Penas, O., Plateaux, R., Patalano, S., Hammadi, M.: Multi-scale approach from mechatronic to Cyber-Physical Systems for the design of manufacturing systems. Comput. Ind. (2016)
27. Tatic, D., Tesic, B.: The application of augmented reality technologies for the improvement of occupational safety in an industrial environment. Comput. Ind. **85**, 1–10 (2017)
28. Leo, M., Medioni, G., Trivedi, M., Kanade, T., Farinella, G.M.: Computer vision for assistive technologies. Comput. Vis. Image Underst. **154**, 1–15 (2015)
29. Mehlmann, G., Häring, M., Janowski, K., Baur, T., Gebhard, P., André, E.: Exploring a model of gaze for grounding in multimodal HRI. In: Proceedings of the 16th International Conference on Multimodal Interaction - ICMI 2014, pp. 247–254 (2014)
30. Rani, P., Liu, C., Sarkar, N., Vanman, E.: An empirical study of machine learning techniques for affect recognition in human-robot interaction. Pattern Anal. Appl. **9**(1), 58–69 (2006)
31. Panait, L., Panait, L., Luke, S.: Cooperative multi-agent learning: the state of the art. Auton. Agents Multi-Agent Syst. **3**(11), 387–434 (2005)
32. Mohammad, Y., Nishida, T.: Toward combining autonomy and interactivity for social robots. AI Soc. **24**(1), 35–49 (2009)
33. Ramík, D.M., Madani, K., Sabourin, C.: A Soft-Computing basis for robots' cognitive autonomous learning. Soft Comput. **19**(9), 2407–2421 (2014)
34. Vlassis, N., Toussaint, M., Kontes, G., Piperidis, S.: Learning model-free robot control by a Monte Carlo em algorithm. Auton. Rob. **27**(2), 123–130 (2009)
35. Guo, L., Hao, J.H., Liu, M.: An incremental extreme learning machine for online sequential learning problems. Neurocomputing **128**, 50–58 (2014)
36. Li, D.C., Yeh, C.W.: A non-parametric learning algorithm for small manufacturing data sets. Expert Syst. Appl. **34**(1), 391–398 (2008)
37. Lee, J.H., Ha, S.H.: Recognizing yield patterns through hybrid applications of machine learning techniques. Inf. Sci. **179**(6), 844–850 (2009)
38. Sudha, L., Dillibabu, R., Srivatsa Srinivas, S., Annamalai, A.: Optimization of process parameters in feed manufacturing using artificial neural network. Comput. Electr. Agric. **120**, 1–6 (2016)

Machine Learning Methods from Group to Crowd Behaviour Analysis

Luis Felipe Borja-Borja[1], Marcelo Saval-Calvo[2], and Jorge Azorin-Lopez[2($\boxtimes$)]

[1] Universidad Central del Ecuador,
Ciudadela Universitaria Av. América, Quito, Ecuador
[2] Computer Technology Department, University of Alicante,
Carretera San Vicente s/n, 03690 San Vicente del Raspeig, Spain
`jazorin@dtic.ua.es`

Abstract. The human behaviour analysis has been a subject of study in various fields of science (e.g. sociology, psychology, computer science). Specifically, the automated understanding of the behaviour of both individuals and groups remains a very challenging problem from the sensor systems to artificial intelligence techniques. Being aware of the extent of the topic, the objective of this paper is to review the state of the art focusing on machine learning techniques and computer vision as sensor system to the artificial intelligence techniques. Moreover, a lack of review comparing the level of abstraction in terms of activities duration is found in the literature. In this paper, a review of the methods and techniques based on machine learning to classify group behaviour in sequence of images is presented. The review take into account the different levels of understanding and the number of people in the group.

Keywords: Human behavior analysis · Motion analysis · Trajectory analysis · Machine learning · Crowd automated analysis · Computer vision

1 Introduction

Nowadays, video surveillance of people is a widely used tool because there are many cameras that facilitate the capture and storage of video. Most of these products are dependant on an operator to analyze the content of stored information. Knowing this limitation it is necessary to provide systems of video surveillance that make possible the automatic identification of behavior. These types of system can be carried out using computer vision techniques, since they allow the identification of patterns of people behavior in an unsupervise manner as gestures, movements and activities among others. In general terms, machine learning, it is possible to model the behavior of people in open or closed spaces such as universities, shopping malls, parks or streets, and then analyze them using automatic learning methods.

There are currently many researches on Human Behavior Analysis such as, [1] that have resulted in the identification of various types of people's behavior

© Springer International Publishing AG 2017
I. Rojas et al. (Eds.): IWANN 2017, Part II, LNCS 10306, pp. 294–305, 2017.
DOI: 10.1007/978-3-319-59147-6_26

in video sequences. These behaviors have been classified from the simplest to the most complex taking into account their duration, from seconds to hours. For these behaviors a classification has been proposed in [2].

The objective of this paper is to provide a classification of human behavior analysis proposals taking into account the size of the group or crowd, identifying the number of people that comprises it, the type of behavior detected, the level of abstraction (from simple actions to complex behaviors) and the techniques used for its treatment and analysis. The most important public datasets are also reviewed which are used to test algorithms there exist several studies on the identification of human behaviors such as [2–5]. In [6] a taxonomy of groups with fewer and more members is established, in addition the methods to analyze them are specified. There are works such as [7], where it is proposed to analyze the behavior of crowds by classifying them into two levels, macro and micro. Despite research efforts to analyze behavior in groups and crowds, we still have many fronts on this subject for researchers.

According to the above the objectives of this paper are: to propose a classification of group and crowd behaviour analysis proposals according to the number of members and the level of abstraction regarding the duration of behavior detected.

2   Aspects of Human Behavior Analysis

In this section the main aspects of the human behavior analysis are explained. First we will present the different levels of understanding and later the main datasets available for experimentation.

2.1   Description of Human Behavior Types and Semantics (Gesture, Motions, Activities, Behavior)

In order to identify human behavior according to the level of abstraction and understanding the data has to be classified depending on the meaning, duration and complexity of tasks performed by humans.

Classifications of activities has taking as its main reference the level of complexity of them, from the easiest to the most complex. The complexity factor is directly related to the time duration of the activity, generally, an activity is considered complex if it has a longer duration. In [8] four levels related to their semantics:

- **Level 1 (Gestures):** Basic movements of parts of the body that last a time. Examples of gestures can be movements of the hand, arm, foot or head among others.
- **Level 2 (Actions):** Also called atomic, consists of actions performed by a single person, their duration is larger than a gesture. An example of actions could be walking, running, jumping.

- **Level 3 (Interaction):** In this category human-human or human-object interaction activities are performed. Examples of these interactions can be two people dancing, kissing, running one behind another, children playing, people cycling.
- **Level 4 (Group Activity):** At this level of description it conforms to two or more groups of people, one or more objects can intervene in the scene. An athletic race, basketball team forwarding, pedestrians crossing a street, a football game, a fight in a stadium can be examples of group activities.

Another taxonomy of human behavior that classifies it according to the complexity and duration time is proposed in [2]. In this approach, the analysis is classified on the degree of semantics in four levels:

- **Level 1 (Motion):** Detection in seconds or frames.
- **Level 2 (Action):** Detection of simple tasks in terms of seconds. The human can interact with objects, or be sitting, standing, walking.
- **Level 3 (Activity):** These are tasks from of minutes to hours. They constitute the sequence of actions, such as cleaning a room, washing a vehicle.
- **Level 4 (Behavior):** This is the higher level of understanding since its duration time can be hours and days. Example behavior can be daily routines of a person, personal habits, mix of two activities in logical sequence.

Both taxonomies described above are based on the daily activities of people, taking into account important factors such as the level of semantics, the duration and the activities composed of other simpler parts such as movements and actions. They described the levels/orders of behavior from the simple movements lasting seconds, to complex activities performed by people for several minutes, hours and even days. The aim of the researchers has been to propose a general classification human behaviour. There are other classification, however, in this work we are going to base our proposal on these focused on group and crowd behavior classification.

2.2 Specialized Datasets

In [9] Blusden and Fisher presented a set of datasets which include sequences for individual and group behavior which are part of the BEHAVE project and include some form of ground truth. Since this paper is focused on group and crowd analysis, the individual datasets are not studied, but authors refer to the original paper for further details.

In group analysis, there are three datasets belonging to BEHAVE project: CAVIAR, CVBASE, ETISEO. Examples of behavior detected in these datasets are: InGroup (The people are in a group and not moving very much), Approach (Two people or groups with one (or both) approaching the other), WalkTogether (People walking together), Meet (Two or more people meeting one another), Split (Two or more people splitting from one another), Ignore (Ignoring of one another), Chase (One group chasing another), Fight (Two or more groups fighting), RunTogether (The group is running together), Following (Being followed).

In this paper we analyze the behavior of groups and crowds such as pedestrians, crowds in public places such as stadiums or squares, interactions of large and small groups, sport actions such as soccer and basketball, and others. The datasets used by the researchers are numerous, being the main ones the following: BEHAVE, BIWI, VSPETS, ETH, DGPI, UHD, HMDB, SportsVU, PETS, UNM, ViF, Bus STATIONS, Subway STATIONS, others. Also in some cases the researchers use their own datasets or videos obtained on YouTube. In [10] it is proposed a study and dataset classification taking into account the behaviors, number of people involved, techniques used to recognize behaviors, types of scene, year of publication, among other characteristics. From this study, an absense of RGB-D (Color and depth) datasets is shown.

With the objective of studying human behavior, in the last years several public datasets have been created. In these dataset, video sequences with contents of several activities in different scenarios and situations are stored. There are also sites dedicated to study particular activities such as a movement or action of a sport, identification of abandoned objects, or daily activities (ADL) such as having a cup of coffee, detection of falls of human, gait study, gesture analysis.

These studies are directly related to public datasets, where tests of the algorithms and techniques used in each case are performed, in certain studies more than one dataset is used to check the accuracy of the recognition systems developed, in other cases it is used custom datasets or the researcher's own, video sequences obtained in public places like bus stations or trains, also of people who carry out activities in squares, streets and commercial centers of a city, are also used. There are very few studies that use YouTube as a source for video footage for research.

Video analysis to perform such a study requires effort and time for researchers, thousands of man-hours are used for the labeling of the different situations that need to be identified in a video. Currently, in cities, it is common to find camcorders capturing video that are later stored. However, all this large amount of information is not available for public access and experimentation.

3   Classification of the Level of Understanding of Groups

To analyze human behavior by using video surveillance cameras, a system based on computer vision requires following a series of ordered steps as suggested in [11]. This paper aims to organize a classification of human behavior according to the number of people that make up a group or crowd, and the techniques, algorithms or frameworks used for analysis.

Human behaviour analysis (HBA) investigations have different applications: improving the quality of life of human beings, in aspects such as support in the health area to detect unusual behaviors, for example falls of elderly people in assisted living environments (AAL) [3,11,12]; surveillance of pedestrians, fights, people running, assaults, ingesting liquor in public places, for example.

The classification of tasks performed by humans described in the previous section are analyzed in [3] according to the level of semantics (in ascending order according to the duration time of this is): Movement (seconds), actions (seconds, minutes), activity (minutes, hours), behavior (hours, days). Each of these tasks must be recognized and modeled, using different techniques, algorithms and other tools suitable for this task.

Turaga et al. [4] proposed a scale of recognition of human activities from simple (actions) to complex (activities), for actions called simple uses (Non-Parametric, Volumetric, and Parametric), for activities called complex uses (Graphical Models, Syntactic, Knowledge Based). Another organization proposal for recognition of activities is set out in [11], where it proposes the Chain of Activity Recognition. This approach divides the recognition process into different procedures, which are: Data Acquisition, Preprocessing, Segmentation, Characteristic extraction, Classification, Decision. Most current research focuses on the last two procedures of this proposal and is often referred to as the learning and decision phases.

In the studies about human behavior of groups and crowds analyzed, it was found that there are few works dealing with RGBD cameras and analysis of human behavior using 3D information. It is important to highlight the work of Wu et al. [13]. They proposed the MoSIF method is combined with HMM [13] to analyze video sequences obtained from a Microsoft Kinect RGBD device. The accuracy obtained is 60% for 3600 video sequences. However, according to the authors, a better result could be obtained if they used more videos to improve learning.

The methods of classification can be supervised and not supervised, and can be used individually or combined using boosting techniques.

On the subject of behavior and trajectories of groups of people there are also some approaches that are based on (HBA) study individually, for example: to recognize activities of groups of people we use the Group Activity Descriptor Vector (GADV) Proposed in [14]. This method has as its predecessor the Activity Description Vector (AVD) revised in [14,15], and aims to recognize human behavior in advance.

3.1   Features of a Groups and Crowds

For example, Andrade et al. [16] detected behaviour of a crowd in different scenarios considered unusual or an emergency, usually provoked by a minority of people in the crowd. These behaviors are coded in Hidden Markov Models (HMM) with mixture of Gaussians output (MOGHMMs), detecting within the different scenes according to their density of people that conform it. It should be considered that the system must be previously trained to detect a type of behavior considered normal that usually have the majority of members of a crowd analyzed. Analyzing specifically the modeling of dense crowds is still an open problem of researchers.

In a public space, where there are a lot of people, the behaviour could be analysed by two variables: actions and duration. Its behavior and its duration. A general trend could be noticed and described as the actions considered normal ones have an extended duration,... a general trend that would be described as that the behaviors considered normal ones have an extended duration, in which most people make up the crowd, while the behaviors considered abnormal are caused by few people in the crowd and in short times of duration. For the study of these types of behaviors, Hu et al. [17] proposed to use a statistical exploration method analyzing the video in a separate way as sliding windows in which the behaviors considered anomalous are detected, taking into account that the algorithm used in this technique requires monitoring.

As we have previously described in order to understand the behavior of crowd, we must take into account the social behavior of the masses, since in this one can observe patterns of behavior that can be modeled by computer studying their structure and special characteristics as proposed in [18]. This study analyzes the human activity of medium level in the granularity, that is to say in the number of people that conform it based on algorithms for the detection of pedestrians and tracking of several moving objects. A particular fact is that the study considers small groups of people traveling together considering the hierarchy of smaller to larger size of the group. It takes into account the proximity of pairs of people and their speed when walking in a particular scene. According to [18], a group is formed from two people, in addition it must feet other parameters such as: if they are within 2,13 meters of each other and not separated by another in the middle, have the same speed up to within 0,15 m/s, and is traveling in the same direction within 3o. When a member of the group stops fulfilling these characteristics or complies with them, it can be said that he or she is inside or outside the group.

The datasets can be chosen by the researchers according to their criteria, taking into account the suitability of their objective. The data are grouped into two categories the heterogeneous, referring to the general activities and the specific when these actions have a special treatment. A third category is included in [10], which specify techniques for motion capture such as the use of infrared, thermal and motion capture (MOCAP).

3.2 Behavior of Groups and Crowds

This paper shows in Tables 1 and 2 a classification of the group size according to the number of members and the activities that each type of group performs, besides specifying the methods, algorithms and forms of recognition that can be used for their study. We can see the following analyzed fields: Ref = Reference to the article, CL = Classification (number of people if exist), TE = Technique, D = Dataset, LA = Level Abstraction. In the column LA = Level Abstraction we show three levels of abstraction: Mot = Motion, Act = Action, Actv = Activity, also two automatic tasks, CP = Count-People and Tra = Tracking.

The classification according to the number of people is in two main sets GROUP and CROWD. Group is defined as the compound of two or more people in a given site and performing an action or activity. Crowd is a composition of people larger that a group that performs simultaneous activities.

The types of behaviors analyzed using video surveillance are limited and specific. The most frequently studied behaviors are the following: Tracking,

Table 1. Classification of proposals reviewed

AR	CL	Technique	D	LA
[15]	G	Self-Organizing Map (SOM) Supervised Self-Organizing Map (SSOM) Neural GAS (NGAS) Linear Discriminant Analysis (LDA) k-Nearest Neighbour (kNN) Multiclassifier (MC)	CAVIAR	Actv
[19]	G	Convolutional Neural Networks (CNN) Long Short-Term Memory (LSTM)	UAV	Actv
[20]	G	Multiple Object Tracking Accuracy (MOTA) K-Shortest Pats Optimization(KSP) Markov Decision Process(MDP) Recurrent Neural Networks(RNN)	TOWN ETH HOTEL STATION	
[21]	C	Collective Transition priors (CT) Mixture of dynamic texture (DTM) Hierarchical clustering (HC) Coherent filtering (CF)	CUHK	Mot
[22]	C	Pedestrian Simulation (PS) Person re-identification (PT) Pedestrian tracking (MPF)	NY station Shanghai-Expo	Mot
[23]	C	Motion Pattern Features (MDA)	N	Mot
[24]	G	Stability Features (HDP)	BEHAVE	Actv
[25]	G	Hidden Markov Models (HMM) Dynamic Probabilistic Networks (DPN)	HMDB BEHAVE	Mot
[26]	G(50)	Inter-Relation Pattern Matrix(IRPM) Game-Theoric Conversational Groups (GTCG) Spectral Clustering (R-GTCG SC)	DGPI	Actv
[27]	C	Model Dynamic Textures Temporal (MDT-temp) Local Motion Histogram (LMH) Spatail (MDT-spat)	UNM UCSD	Mot
[28]	G(25)	Markov Chain Monte Carlo (MCMC) Gaussian Mixture Model (GMM)	FIFA WC 2006	Tra
[29]	G	Category Feature Vectors (CFVs) Gaussian Mixture Models (GMM) Recognizing algorithm (CFR)	N	Actv
[30]	G	Convolutional Neural Network (CNN) Feed Forward Network (FFN)	SportsVU	Actv
[31]	G	Multiple Human Tracking (MHT) Correct Detected Tracks (CDT) False Alarm Tracks (FAT) Track Detection Failure (TDF)	ETH UHD	Tra
[32]	G	Neural Network (NN)	N	Act
[33]	C	Histogram of Oriented Gradients (HOG) Histogram of Optical Flow (HOF) Motion Boundary Histogram (MBH)	UMN UCSD CUHK PETS2009 ViF Rodriguez's UCF Own dataset	Act
[34]	C	Hidden Markov Models (HMM) Support Vector Machine (SVM) Robust Local Optical Flow (RLOF)	PETS UMN	Act

Table 2. Classification of proposals reviewed

AR	CL	Technique	D	LA
[35]	G	Cumulative Match Characteristic (CMC) Synthetic Disambiguation Rate (SDR) Center Rectangular Ring Ratio-Occurrence (CRRO) Block based Ratio-Occurrence (BRO)	2008 i-LIDS MCTS	Act
[36]	C	Accumulated Mosaic Image Difference (AMID) OpticalFlow+BackgroundModel (OFBM) Markov Random Fields (MRF) Support Vector Machine (SVM)	Subway Station Bus Station Plaza	Mot
[37]	C	Support Vector Machine (SVM) Library for Support Vector Machines (LIBSVM) Basis Radial Function (BRF) Block Matching Algorithm (BMA)	UMN	Act
[38]	C	Fast Corner Detect (FAST) Support Vector Machine (SVM)	BEHAVE	Act
[39]	G	Evolving Networks (EN) Monte Carlo (MC)	N	Mot
[40]	G	Linear Trajectory Avoidance (LTA)	N	Mot
[7]	G(20)	Bag of words modelling (BoW)	Novel dataset	Mot
[41]	G	Gaussian Mixture Model (GMM) EM algorithm	N	Actv
[42]	G	Minimum Description Length (MDL)	COLLECTIVE ACTIVITY BEHAVE	Actv
[43]	G	Hidden Markov Models (HMM) Dynamic Bayes Networks (DBN)	BIWI	Tra
[44]	G(20)	Multi-model MHT	Own	Tra
[45]	G	Voronoi Diagrams Model (VDM)	N	Mot
[5]	G	Dynamic Probabilistic Networks (DPNs) Dynamically Multi-Linked (DML) Hidden Markov Model (HMM)	PETS 2004 YouTube	Mot
[46]	G(25)	Support Vector Machines (SVM)		Act
[47]	C	Hidden Markov Model (HMM)	N	Mot
[43]	G	Sampling Importance Resampling (SIR) Discrete Choice Model (DCM) Multi Hypothesis Tracking (MHT) Statistical Shape Modeling (SSM)	BIWI	Tra
[48]	G(90)	Heuristic learned (HL)	N	CP
[49]	C	Bag of Words (BoW) Locality-constrained Linear Coding (LLC) Vector Quantization (VQ)	BEHAVE	Mot
[50]	C	Unsupervised Bayesian Clustering Framework (UBCF)	N	Mot
[51]	C	Bayesian Marked Point Process (MPP)	CAVIAR VSPETS SOCCER	CP
[52]	C	Social Force Model (SFM) Pure Optical Flow(POF)	UNM	
[53]	C	Detection of moving regions	METRO	Tra
[54]	C	Linear Fitting (LF) Unpervised Neural Network (UNN)	N	CP

trajectories, bicyclist, pedestrian, skateboarders, count people in a group or crowd, street fights, interaction objects-people, motions or actions in sports, human actions (walking, jogging, running, boxing, hand waving and hand clapping).

The dataset frequently used for experimentation to analyze the behavior of groups and crowds are the following: BEHAVE, BIWI, CAVIAR, VSPETS, ETH, DGPI, UHD, HMDB, SportsVU, PETS, UNM, ViF, Bus STATIONS, Subway STATIONS, others. Also in some cases the researchers use their own dataset or videos obtained on YouTube.

Based on the information analyzed in the papers, it is possible to propose a classification according to the level of abstraction of the analyzed human behavior of groups and crowds according to the case, in order of shortest to longest duration of behavior we propose three levels of abstraction: Motion, Action, Activity, also two automatic tasks, Count-People and Tracking.

The techniques or methods frequently used to analyze human behavior of groups and crowds using video surveillance are the following: Bag of Words, Deep Neural Networks, Hidden Markov Models, Monte Carlo, Gaussian Mixture Model, Multiple Human Tracking, Support Vector Machines. Many authors use these methods including specific tunings and mask the name with a slight modification.

4   Conclusions and Future Directions

In this work the human behavior of groups and crowds has been approached taking into account the degree of semantics and especially the size of people that integrate the group or crowd, in addition has been considered behaviors like; Sports teams of soccer and basketball, pedestrians, groups of people in metro and bus stations, people grouped in parks and squares. We propose a classification of behavior of groups and crowds according to degree of semantics has been carried out in three types; Motion, Action, Activity, also two automatic tasks, Count-People and Tracking. It has included techniques and algorithms that researchers use for analysis, and has included the datasets used, which in most of the investigations are traditional and in a few cases custom datasets or YouTube videos are used.

As future work, it is important to address the issue of video sequences using RGBD cameras, as this type of technology is currently in increasing use.

References

1. Azorin-Lopez, J., Saval-Calvo, M., Fuster-Guillo, A., Garcia-Rodriguez, J., Orts-Escolano, S.: Self-organizing activity description map to represent and classify human behaviour. In: IJCNN 2015 (2015)
2. Chaaraoui, A.A., Climent-Pérez, P., Flórez-Revuelta, F.: A review on vision techniques applied to human behaviour analysis for ambient-assisted living. Expert Syst. Appl. **39**(12), 10873–10888 (2012)

3. Cardinaux, F., Deepayan, B., Charith, A., Hawley, M.S., Mark, S., Bhowmik, D., Abhayaratne, C.: Video based technology for ambient assisted living: a review of the literature. J. Ambient Intell. Smart Environ. (JAISE) **1364**(3), 253–269 (2011)
4. Turaga, P., Chellappa, R., Subrahmanian, V., Udrea, O.: Machine recognition of human activities: a survey. IEEE Trans. Circ. Syst. Video Technol. **18**, 1473–1488 (2008)
5. Ryoo, M.S., Aggarwal, J.K.: Recognition of high-level group activities based on activities of individual members. In: 2008 IEEE Workshop on Motion and Video Computing, WMVC 2008, January 2008
6. Mihaylova, L., Carmi, A.Y., Septier, F., Gning, A., Pang, S.K., Godsill, S.: Overview of Bayesian sequential Monte Carlo methods for group and extended object tracking. Digit. Sig. Process.: Rev. J. **25**(1), 1–16 (2014)
7. Climent-Pérez, P., Mauduit, A., Monekosso, D.N., Remagnino, P.: Detecting events in crowded scenes using tracklet plots. In: Proceedings of the International Conference on Computer Vision Theory and Applications (2014)
8. Vishwakarma, S., Agrawal, A.: A survey on activity recognition and behavior understanding in video surveillance. Vis. Comput. **29**(10), 983–1009 (2013)
9. Blunsden, S., Fisher, R.B.: The BEHAVE video dataset: ground truthed video for multi-person behavior classification. Ann. BMVA **2010**(4), 1–11 (2010)
10. Chaquet, J.M., Carmona, E.J., Fernández-Caballero, A.: A survey of video datasets for human action and activity recognition. Comput. Vis. Image Underst. **117**(6), 633–659 (2013)
11. Banos, O., Damas, M., Pomares, H., Rojas, F., Delgado-Marquez, B., Valenzuela, O.: Human activity recognition based on a sensor weighting hierarchical classifier. Soft Comput. **17**(2), 333–343 (2013)
12. Bruckner, D., Yin, G.Q., Faltinger, A.: Relieved commissioning and human behavior detection in ambient assisted living systems. Elektrotechnik und Informationstechnik **129**(4), 293–298 (2012)
13. Wu, Y., Jia, Z., Ming, Y., Sun, J., Cao, L.: Human behavior recognition based on 3D features and hidden markov models. Sign. Image Video Process. **10**(3), 495–502 (2015)
14. Azorin-Lopez, J., Saval-Calvo, M., Fuster-Guillo, A., Garcia-Rodriguez, J., Cazorla, M., Signes-Pont, M.T.: Group activity description and recognition based on trajectory analysis and neural networks, pp. 1585–1592 (2016)
15. Azorin-Lopez, J., Saval-Calvo, M., Fuster-Guillo, A., Oliver-Albert, A.: A predictive model for recognizing human behaviour based on trajectory representation. In: International Joint Conference on Neural Networks (IJCNN), pp. 1494–1501 (2014)
16. Andrade, E., Blunsden, S., Fisher, R.: Hidden markov models for optical flow analysis in crowds. In: 18th International Conference on Pattern Recognition, pp. 460–463, January 2006
17. Hu, Y., Zhang, Y., Davis, L.S.: Unsupervised abnormal crowd activity detection using semiparametric scan statistic. In: IEEE Computer Society Conference on Computer Vision and Pattern Recognition Workshops, vol. 1, pp. 767–774 (2013)
18. Ge, W., Collins, R.T., Ruback, B.: Automatically detecting the small group structure of a crowd. In: 2009 Workshop on Applications of Computer Vision, WACV 2009 (2009)
19. Goel, K., Robicquet, A.: Learning causalities behind human trajectories (2015)
20. Maksai, A., Wang, X., Fua, P.: Globally consistent multi-people tracking using motion patterns, vol. 1 (2016). arXiv preprint arXiv:1612.00604

21. Shao, J., Loy, C.C., Wang, X.: Scene-independent group profiling in crowd, pp. 2219–2226 (2014)
22. Yi, S., Li, H., Wang, X.: Pedestrian travel time estimation in crowded scenes. Shenzhen Institutes of Advanced Technology, Chinese Academy of Sciences, pp. 3137–3145 (2015)
23. Yi, S., Li, H., Wang, X.: Understanding pedestrian behaviors from stationary crowd groups. In: Proceedings of the IEEE Computer Society Conference on Computer Vision and Pattern Recognition, 07–12 June 2015, pp. 3488–3496 (2015)
24. Al-Raziqi, A., Denzler, J.: Unsupervised framework for interactions modeling between multiple objects. In: Proceedings of the 11th Joint Conference on Computer Vision, Imaging and Computer Graphics Theory and Applications, VISI-GRAPP, vol. 4, pp. 509–516 (2016)
25. Shen, C., Xie, R., Zhang, L., Song, L.: Small group people behavior analysis based on temporal recursive trajectory identification. Institute of Image Communication and Network Engineering, Shanghai Jiao Tong University, Cooperative Medianet Innovation Center (2015)
26. Vascon, S., Mequanint, E.Z., Cristani, M., Hung, H., Pelillo, M., Murino, V.: Detecting conversational groups in images, sequences: a robust game-theoretic approach. Comput. Vis. Image Underst. **143**, 11–24 (2016)
27. Li, W., Mahadevan, V., Vasconcelos, N.: Anomaly detection and localization in crowded scenes. IEEE Trans. Pattern Anal. Mach. Intell. **36**(1), 18–32 (2014)
28. Liu, J., Tong, X., Li, W., Wang, T., Zhang, Y., Wang, H., Yang, B., Sun, L., Yang, S.: Automatic player detection, labeling and tracking in broadcast soccer video. In: Proceedings of the British Machine Vision Conference 2007, pp. 3.1–3.10 (2007)
29. Lin, W., Sun, M.-T., Poovandran, R., Zhang, Z.: Human activity recognition for video surveillance. In: IEEE International Symposium on Circuits and Systems, pp. 2737–2740, June 2008
30. Harmon, M., Lucey, P., Klabjan, D.: Predicting shot making in basketball using convolutional neural networks learnt from adversarial multiagent trajectories (2016). arXiv preprint arXiv:1609.04849
31. Camplani, M., Paiement, A., Mirmehdi, M., Damen, D., Hannuna, S., Burghardt, T., Tao, L.: Multiple human tracking in RGB-D data: a survey, June 2016
32. Wickramaratna, K., Chen, M., Chen, S.C., Shyu, M.L.: Neural network based framework for goal event detection in soccer videos. In: Proceedings - Seventh IEEE International Symposium on Multimedia, ISM 2005, vol. 2005, pp. 21–28 (2005)
33. Hamidreza Rabiee, H.M., Haddadnia, J.: Emotion-based crowd representation for abnormality detection hamidreza. Int. J. Artif. Intell. Tools (2016)
34. Fradi, H., Dugelay, J.L.: Spatial and temporal variations of feature tracks for crowd behavior analysis. J. Multimodal User Interfaces **10**(4), 307–317 (2016)
35. Gong, S., Cristani, M., Yan, S., Loy, C.C.: Person re-identification. In: Gong, S., Cristani, M., Yan, S., Loy, C.C. (eds.) ACVPR. Springer, London (2014)
36. Cao, L., Huang, K.: Video-based crowd density estimation and prediction system for wide-area surveillance. China Commun. **10**(5), 79–88 (2013)
37. Liao, H., Xiang, J., Sun, W., Feng, Q., Dai, J.: An abnormal event recognition in crowd scene. In: Proceedings - 6th International Conference on Image and Graphics, ICIG 2011, pp. 731–736, September 2011
38. Chang, M.C., Krahnstoever, N., Lim, S., Yu, T.: Group level activity recognition in crowded environments across multiple cameras. In: Proceedings - IEEE International Conference on Advanced Video and Signal Based Surveillance, AVSS 2010, pp. 56–63, February 2010

39. Gning, A., Mihaylova, L., Maskell, S., Pang, S.K., Godsill, S.: Group object structure and state estimation with evolving networks and Monte Carlo methods. IEEE Trans. Sig. Process. **59**(4), 1383–1396 (2011)
40. Pellegrini, S., Ess, A., Tanaskovic, M.: Wrong turn-no dead end: a stochastic pedestrian motion model. In: 2010 IEEE Computer Society Conference on CVPRW (2010)
41. Perše, M., Kristan, M., Kovačič, S., Vučkovič, G., Perš, J.: A trajectory-based analysis of coordinated team activity in a basketball game. Comput. Vis. Image Underst. **113**(5), 612–621 (2009)
42. Yin, Y., Yang, G., Man, H.: Small human group detection and event representation based on cognitive semantics. In: Proceedings - 2013 IEEE 7th International Conference on Semantic Computing, ICSC 2013, pp. 64–69, September 2013
43. Ge, W., Collins, R.T., Ruback, R.B.: Vision-based analysis of small groups in pedestrian crowds. IEEE Trans. Pattern Anal. Mach. Intell. **34**(5), 1003–1016 (2012)
44. Lau, B., Arras, K.O., Burgard, W.: Multi-model hypothesis group tracking and group size estimation. Int. J. Soc. Robot. **2**(1), 19–30 (2010)
45. Jacques, J.C.S., Braun, A., Soldera, J., Musse, S.R., Jung, C.R.: Understanding people motion in video sequences using Voronoi diagrams. Pattern Anal. Appl. **10**(4), 321–332 (2007)
46. Schuldt, C., Barbara, L., Stockholm, S.: Rcognizing human actions: a local SVM approach. Department of Numerical Analysis and Computer Science. In: Pattern Recognition, Proceedings of the 17th International Conference on ICPPR 2004, vol. 3, pp. 32–36 (2004)
47. Andrade, E.L., Blunsden, S., Fisher, R.B.: Modelling crowd scenes for event detection (2006)
48. Kilambi, P., Ribnick, E., Joshi, A.J., Masoud, O., Papanikolopoulos, N.: Estimating pedestrian counts in groups. Comput. Vis. Image Underst. **110**(1), 43–59 (2008)
49. Zhang, C., Yang, X., Lin, W., Zhu, J.: Recognizing human group behaviors with multi-group causalities. In: Proceedings of the IEEE/WIC/ACM International Conference on Web Intelligence and Intelligent Agent Technology Workshops, WIIAT 2012, pp. 44–48 (2012)
50. Brostow, G.J., Cipolla, R.: Brostow: unsupervised bayesian detection of independent motion in crowds (2006)
51. Ge, W., Collins, R.T.: Marked point processes for crowd counting. In: IEEE Computer Vision and Pattern Recognition 2009, pp. 2913–2920 (2009)
52. Mehran, R., Oyama, A., Shah, M.: Abnormal crowd behaviour detection using social force model. In: IEEE Conference on Computer Vision and Pattern Recognition, no. 2, pp. 935–942 (2009)
53. Cupillard, F., Brémond, F., Thonnat, M.: Tracking groups of people for video surveillance. In: Remagnino, P., Jones, G.A., Paragios, N., Regazzoni, C.S. (eds.) Video-Based Surveillance Systems, pp. 89–100. Springer, USA (2002)
54. Kong, D., Gray, D., Tao, H.: A viewpoint invariant approach for crowd counting. In: Proceedings - International Conference on Pattern Recognition, vol. 3, pp. 1187–1190 (2006)

Unsupervised Color Quantization
with the Growing Neural Forest

Esteban José Palomo[1,2]([✉]), Jesús Benito-Picazo[1], Ezequiel López-Rubio[1],
and Enrique Domínguez[1]

[1] Department of Computer Languages and Computer Science, University of Málaga,
Bulevar Louis Pasteur, 35, 29071 Málaga, Spain
{ejpalomo,jpicazo,ezeqlr,enriqued}@lcc.uma.es
[2] School of Mathematical Sciences and Information Technology,
University of Yachay Tech, Hacienda San José s/n, Urcuquí, Ecuador
epalomo@yachaytech.edu.ec

Abstract. Image processing has become a very common application for
artificial intelligence-based algorithms. More precisely, color quantization
has become an important issue when it comes to supply efficient transmission and storage for digital images, which consists of color indexing for
minimal perceptual distortion image compression. Artificial Neural Networks have been consolidated as a powerful tool for unsupervised tasks,
and therefore, for color quantization purposes. In this work we present
a novel color quantization approach based on the Growing Neural Forest (GNF), which is a Growing Neural Gas (GNG) variation where a
set of trees is learnt instead of a general graph. Its suitability for color
quantization processes is supported by experimental results obtained,
where the GNF outperforms other self-organizing models such as the
GNG, GHSOM and SOM. As future work, more datasets and competitive models will be taken into account.

Keywords: Color quantization · Self-organization · Tree-structured
model · Unsupervised learning

1 Introduction

Color quantization is a common image processing task, which consists of selecting
a small but representative set of indexed colors (codebook) for coding the original digital image with minimum perceptual distortion. This process is used to
reduce the storage requirements and the transmission bandwidth of color images
while maintaining an acceptable image fidelity. Moreover, it is considered as a
useful lossy compression method to find an acceptable set of colors (codebook)
that can be used to represent the original colors of a digital image. The quality
of the codebook depends on the error between the original image and the resultant image. An optimal codebook aims to minimize this error, which is usually
measured by a mean square error criterion.

© Springer International Publishing AG 2017
I. Rojas et al. (Eds.): IWANN 2017, Part II, LNCS 10306, pp. 306–316, 2017.
DOI: 10.1007/978-3-319-59147-6_27

There are several well-known codebook design algorithms such as k-means algorithm [10], fuzzy c-means [1], competitive learning [5], self-organizing map [7], and their variants. The Self-Organizing Map (SOM) [6] was the starting point for the development of many self-organizing models [11,12,14]. Some of them try to face some drawbacks of the original SOM regarding its pre-established network architecture, i.e. topology and number of neurons [8,13]. The Growing Hierarchical Self-Organizing Map (GHSOM) [20] represents a hierarchical extension of the SOM to reflect hierarchical data, where the entire architecture of the neural network is automatically determined during the unsupervised learning process. The Growing Neural Gas (GNG) [4] constitutes a successful self-organizing neural network model that solves the fixed-network architecture problem of the SOM. The GNG is based on the Neural Gas (NG) model [15], but the GNG provides a neuron growth and removal mechanism to automatically determine the number of neurons during the unsupervised learning process according to the input data.

The Growing Neural Forest (GNF) was recently proposed in [17] as an improved version of the GNG, where a spanning tree for each connected component (subgraph) of the overall graph is computed. This way only those units which are connected to the winning unit in a spanning tree are updated. Therefore, the GNF learns a set of trees (forest) so that each tree represents a connected cluster of data, presenting a better adaptation to input data that the GNG. Since many self-organizing models were successfully applied in the past to color quantization [3,18,19,21], in this work the GNF has also been used for this task in order to demonstrate its goodness in this application.

The reminder of the paper is organized as follows[1]. The GNF model is explained in Sect. 2. Experimental results on color quantization are detailed in Sect. 3. Section 4 concludes this paper.

2   The Model

In this section a brief summary of the Growing Neural Forest self-organizing model is provided. Please refer to [17] for more details. A Growing Neural Forest (GNF) comprises a graph with an adaptive number of nodes (neurons) and edges (links). The nodes and the edges are inserted and removed from the graph as necessary along the adaptation procedure. The current number of neurons is noted H. Let the training set be $\mathcal{S}$, where $\mathcal{S} \subset \mathbb{R}^D$, and D is the input space dimension. Every neuron $i \in \{1, ..., H\}$ contains a prototype $\mathbf{w}_i \in \mathbb{R}^D$ and an error variable $e_i \in \mathbb{R}$, $e_i \geq 0$. Every connection is associated to an age, which is a natural number. Let the set of connections be $A \subseteq \{1, ..., H\} \times \{1, ..., H\}$, so that no links are allowed between a neuron and itself, $(i, i) \notin A$. Here A is defined as an undirected graph, i.e. $(i, j) \in A$ if and only if $(j, i) \in A$.

Given the current set of links, the Q connected components of the graph are called subgraphs, so that the number of subgraphs is less or equal then

---

[1] The present paper is an extension of the extended abstract entitled "A Color Quantization Approach based on the Growing Neural Forest", which was presented at the 2016 Latin American Conference on Computational Intelligence [16].

the number of neurons, $Q \leq H$. Each subgraph has an associated spanning tree. Hence a forest is formed from the set of the Q spanning trees. Let $\hat{A} \subseteq \{1, ..., H\} \times \{1, ..., H\}$ be the set of links of the spanning trees, where $\hat{A} \subseteq A$. Since A is an undirected graph, $\hat{A}$ is undirected too. Figure 1 depicts a possible GNF structure.

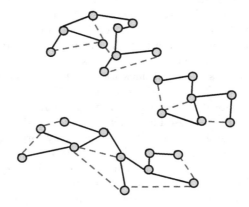

Fig. 1. Example GNF model with three subgraphs. The links which do not belong to the spanning trees are depicted with dashed lines.

The learning rule for the GNF is based on the Growing Neural Gas [4]. It finds a spanning tree for each connected component (subgraph) of the overall graph, so that only those neurons which are linked to the winning neuron in a spanning tree are modified. The learning algorithm is given by these steps:

1. Start with two neurons ($H = 2$) joined by a link, so that the graph contains only one subgraph which comprises both neurons. The link set A only contains the connection between both neurons, and the set of the links of the spanning trees $\hat{A}$ is initialized to A. The prototype of each neuron is initialized to an input sample drawn at random from the training set $\mathcal{S}$. The error variables are initialized to zero. The age of the connections are initialized to zero, too.
2. Draw a training sample $\mathbf{x}_n \in \mathbb{R}^D$ at random from the training set $\mathcal{S}$. Please note that there can be a training sample which is drawn more than once. This is harmless and in fact it will happen repeatedly for small datasets.
3. Find the nearest neuron q and the second nearest neuron s in terms of Euclidean distance:

$$q = \arg \min_{i \in \{1, ..., H\}} \|\mathbf{w}_i(n) - \mathbf{x}_n\| \tag{1}$$

$$s = \arg \min_{i \in \{1, ..., H\} - \{q\}} \|\mathbf{w}_i(n) - \mathbf{x}_n\| \tag{2}$$

If there is a tie in the computation of the first and second nearest neurons, then it is broken by selecting one of the tied neurons uniformly at random.

4. Increment by one the age of all edges of A which depart from q.
5. Add the squared Euclidean distance between $\mathbf{x}_n$ and the nearest neuron q to the error variable e_q:

$$e_q(n+1) = e_q(n) + \|\mathbf{w}_q(n) - \mathbf{x}_n\|^2 \tag{3}$$

6. Modify q and all its direct topological neighbors in its spanning tree with step size ϵ_b for unit q and ϵ_n for the neighbors, where $\epsilon_b > \epsilon_n > 0$:

$$\epsilon(n,i) = \begin{cases} \epsilon_b & \text{iff } n = q \\ \epsilon_n & \text{iff } (n \neq q) \wedge (n,q) \in \hat{A} \\ 0 & \text{iff } (n \neq q) \wedge (n,q) \notin \hat{A} \end{cases} \tag{4}$$

$$\mathbf{w}_i(n+1) = (1 - \epsilon(n,i))\,\mathbf{w}_i(n) + \epsilon(n,i)\,\mathbf{x}_n \tag{5}$$

It must be noted that $(n,q) \in \hat{A} \Leftrightarrow (q,n) \in \hat{A}$ since $\hat{A}$ is an undirected graph.

7. If q and s are connected by a link in A, then set the age of this link to zero. Otherwise, create a link between q and s.
8. Remove the links in A with an age larger than a_{max}. Then remove all neurons which have no outgoing links in A.
9. If the current time step n is an integer multiple of a parameter λ and the current number of units is lower than the maximum number of neurons H_{max}, then insert a new neuron as follows. First find the neuron r with the maximum error among all neurons. After that, determine the neuron z with the maximum error among all direct neighbors of r in A. Then create a new neuron k, insert links connecting k with r and z, and remove the original link between r and z. After that, decrease the error variables e_r and e_z by multiplying them with a constant α, and initialize the error variable e_k to the new value of e_r. Setup the prototype of neuron k to be midway between those of r and z, as follows:

$$\mathbf{w}_k(n) = \frac{1}{2}\left(\mathbf{w}_r(n) + \mathbf{w}_z(n)\right) \tag{6}$$

Finally, recompute the spanning trees by Kruskal algorithm [9] and update $\hat{A}$ to the set of connections which belong to spanning trees.
10. Decrease all error variables e_i by multiplying them by a predefined constant d with $0 < d < 1$.
11. If the maximum number of time steps has been reached, then halt. Otherwise, go to step 2.

It is worth noting that the algorithm can easily be adapted to online learning by obtaining each new incoming sample in Step 2. The learning rates ϵ_b and ϵ_n can also be set to decay as learning progresses. Linear or exponential decay rates could be employed for this purpose.

The parameters of the above learning algorithm are the following:

– The step size for the winning neuron ϵ_b is a positive real number which controls how much the prototype of winning neuron moves to the input sample.
– The step size for the neighbor neurons ϵ_n is a positive real number which controls how much the prototypes of the neighbors of the winning neuron move to the input sample. We require $\epsilon_b > \epsilon_n$ so that the neighbors move less than the winning neuron.
– The a_{max} parameter is a positive integer which specifies how long a link must have existed without having any input for which its ends are the first and second best matching neurons. The smaller a_{max}, the faster the old links are pruned. The case $Q = H$ is possible, i.e. there is one connected component per neuron. However, it is undesirable because the model would not have learned any connectivity among the neurons. As the value of the a_{max} parameter increases this case becomes unlikely because very few connections are removed. So the $Q = H$ case can be avoided by choosing a large enough value for a_{max}.
– The λ parameter is a positive integer which establishes how often a new neuron can be inserted into the model. The lower λ, the faster the model can grow.
– The maximum number of neurons H_{max} establishes a limit on the overall size of the model.
– The α parameter is a positive real number with $0 < \alpha < 1$. It controls the reduction in the error variables of the parent neurons r and z which give rise to a new neuron. This is done so that r and z are not chosen again to create a new neuron. The lower α, the larger the reduction of the error variables.
– The error variable decay parameter d is a positive real number with $0 < d < 1$. It controls how fast the quantization errors of the neurons are forgotten. The lower d, the faster the past errors are forgotten so that the error variables adapt faster to the new state of the model.

The GNF is suitable for a wide range of unsupervised learning tasks, since it can be trained online or offline depending on the application. In what follows it is employed for color quantization of natural images.

3   Experimental Results

In order to demonstrate the capabilities of the GNF to achieve color quantization tasks, it has been tested against other three competitive unsupervised learning neural network models, challenging them through a battery of tests involving eight images commonly used in image processing literature which can be checked in Fig. 2. The chosen research method is quasi-experimental, which is adequate for our exploratory purposes in this paper.

First model is the popular and widely used in clustering processes Kohonen's Self-Organizing Map (SOM). This neural network has been trained for 5 different map sizes, namely 3×4, 4×5, 5×6, 6×7, and 7×8.

Fig. 2. Color images used for color quantization experiments: (a) baboon, (b) lake, (c) Lena, (d) house, (e) milk, (f) peppers, (g) plane and (f) Tiffany.

Second one is the Growing Hierarchical Self-Organizing Map (GHSOM) that consists of an artificial neural network which represents an evolution of the traditional SOM by incorporating a hierarchical architecture featuring a set of independent growing self-organizing maps. This model has two parameters for controlling the growth of the maps (τ_1) and the hierarchical expansion (τ_2). For our experiments we fixed $\tau_1 = 0.5$ and we set five different values for the τ_2 parameter (0.5, 0.05, 0.025, 0.01, 0.005).

The last neural network model GNF has been compared to the Growing Neural Gas (GNG), whose features were briefly explained in the introductory section of this document. Both neural networks have been trained with the same parameters obtained through an automatized empiric trial and error-based optimization process. Those parameters appear in Table 1.

In order to perform the training process, each image was converted into a dataset to be fed to the neural network model, where each pixel of the image represents a data sample with three features ($D = 3$) corresponding to the RGB components. The number of samples of each dataset is $M = 262,144$ as these images have a resolution of 512×512 pixels. Training of all four models was done for 10 runs and during 2 epochs assigning five different values (10, 20, 30, 40 and 50) to the H_{max} parameter for the GNF and GNG for comparison purposes. Note that in the SOM model their map sizes have been chosen to be as similar as possible to the previous number of neurons, whereas for the GHSOM model this number will be approximate because of the particular way the network structure evolves through the training process.

Table 1. Selected parameter values for the GNF and GNG models.

Parameter description	Values
Step size for the winning unit	$\epsilon_b = 0.1$
Step size for the neighbor unit	$\epsilon_n = 0.007$
Maximum edge for an edge	$a_{max} = 48$
New units insertion	$\lambda = 19$
Maximum number of units	$H_{max} = 10-50$
Reduction of the error variables	$\alpha = 0.1$
Error variable decay	$d = 0.987$

Aiming to obtain a clear representation of the color quantization results yielded by the GNF model, differences images between the original and the quantized images for each network model and 50 neurons approximately, are presented in Figs. 3 and 4. Note that the SOM and GHSOM models reflect more errors in the image quantization process.

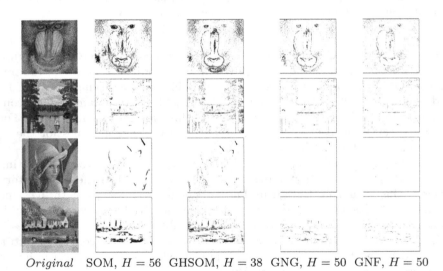

Original   SOM, $H = 56$ GHSOM, $H = 38$ GNG, $H = 50$ GNF, $H = 50$

Fig. 3. Differences images for the 4 network models results and indicated number of neurons. Rows from top to bottom: Baboon, Lake, Lena and House images. The Darker the color, the most difference.

Quantitative results have been assessed with three popular performance measures such as the Peak Signal-to-Noise Ratio (PSNR), which is expressed as follows (in dB, higher is better):

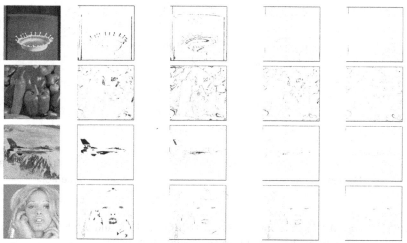

Original   SOM, $H = 56$ GHSOM, $H = 38$ GNG, $H = 50$ GNF, $H = 50$

Fig. 4. Differences images for the 4 network models results and indicated number of neurons. Rows from top to bottom: Milk, Peppers, Plane and Tiffany images. The Darker the color, the most difference.

$$PSNR = 10 \log_{10} \left(\frac{3 \times 255^2}{MSE} \right) \tag{7}$$

The Mean Squared Error (MSE), which is defined as follows (lower is better):

$$MSE = \frac{1}{M} \sum_{i=1}^{M} \|\mathbf{w}_i - \mathbf{x}_i\|^2 \tag{8}$$

where M is the number of samples in the dataset, $\mathbf{x}_i$ is the i-th input sample and $\mathbf{w}_i$ is the prototype of the winning neuron corresponding to x_i.

Last one is Normalized Cross-Correlation (NCC) [2] (Eq. 9) (the closer to 1, the better):

$$NCC = \frac{\sum_{i=1}^{M} (\mathbf{w}_i * \mathbf{x}_i)}{\sum_{i=1}^{M} \mathbf{x}_i^2} \tag{9}$$

where M is the number of samples in the dataset, $\mathbf{x}_i$ is the i-th input sample and $\mathbf{w}_i$ is the prototype of the winning neuron corresponding to x_i.

PSNR, MSE and NCC results obtained after training the 4 models with the eight images mentioned at the beginning of this section are provided in Figs. 5 and 6 for values between 10 and 60 neurons. Each plot has the number of neurons in its horizontal axis, the value of the performance measure in its vertical axis, and the data series represent the mean value of that particular

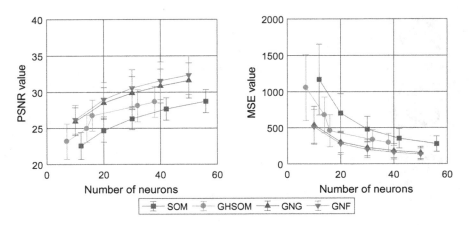

Fig. 5. PSNR and MSE values for the SOM, GHSOM, GNG and GNF models considering a number of neurons for each between 10 and 56 Neurons.

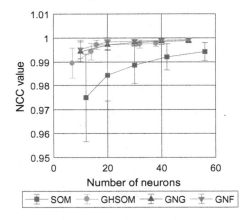

Fig. 6. NCC values for the SOM, GHSOM, GNG and GNF models considering a number of neurons for each between 10 and 56 Neurons.

performance measure for 10 runs and all the images considered for a particular number of neurons. The number of neurons of the GHSOM is the mean number of neurons obtained after 10 runs. These figures reveal that GNF MSE and PSNR values are always better than GNG network values with the same training parameters. They also illustrate that NCC values for GNF model are always higher or equal than GNG network values with the same training configuration parameters. When it comes to SOM and GHSOM models, figures illustrate that results obtained are even further away from the results obtained by GNF model than GNG results are.

4   Conclusions

In this document, a new approach to digital image color quantization processes by means of the Growing Neural Forest (GNF) has been proposed. The GNF supposes an evolution of the Growing Neural Gas insofar as it incorporates connected component detection in the connection graph and associated spanning tree computing. This allows tree structure tracing and, therefore, the learning of the connection graph's set of trees through the connection graph.

Color quantization process is been achieved by training a GNF using several color images. Experimental results obtained by performing different test batteries to a fixed set of well-known color images, as opposed to results obtained from tests performed to the GNG, GHSOM and SOM models with the same set of images, point out that GNF model described in this paper overtakes the aforementioned models in color quantization tasks performance, standing as a suitable model for this purpose. As future work, more datasets and competitive models will be taken into account.

Acknowledgments. This work is partially supported by the Ministry of Economy and Competitiveness of Spain under grant TIN2014-53465-R, project name Video surveillance by active search of anomalous events. It is also partially supported by the Autonomous Government of Andalusia (Spain) under projects TIC-6213, project name Development of Self-Organizing Neural Networks for Information Technologies; and TIC-657, project name Self-organizing systems and robust estimators for video surveillance. Finally, it is partially supported by the Autonomous Government of Extremadura (Spain) under the project IB13113. All of them include funds from the European Regional Development Fund (ERDF). The authors thankfully acknowledge the computer resources, technical expertise and assistance provided by the SCBI (Supercomputing and Bioinformatics) center of the University of Málaga. They also gratefully acknowledge the support of NVIDIA Corporation with the donation of the Titan X GPU used for this research.

References

1. Bezdek, J.C.: Pattern Recognition with Fuzzy Objective Function Algoritms. Plenum Press, New York (1981)
2. Chokkalingam, S.P., Komathy, K., Vinoth, M., Saravanan, R.: Performance analysis of bone images using various edge detection algorithms and denoising filters. Int. J. Pharma Bio Sci. **5**(2), B943–B954 (2014)
3. Dekker, A.: Kohonen neural networks for optimal color quantization. Netw.: Comput. Neural Syst. **5**, 351–367 (1994)
4. Fritzke, B.: A growing neural gas network learns topologies. Adv. Neural Inf. Process. Syst. **7**, 625–632 (1995)
5. Hertz, J., Krogh, A., Palmer, R.: Introduction to the Theory of Neural Computation. Addison-Wesley, Redwood City (1991)
6. Kohonen, T.: Self-organized formation of topologically correct feature maps. Biol. Cybern. **43**(1), 59–69 (1982)
7. Kohonen, T.: The self-organizing map. Proc. IEEE **78**(9), 1464–1480 (1990)
8. Kohonen, T.: Essentials of the self-organizing map. Neural Netw. **37**, 52–65 (2013)

9. Kruskal, J.B.: On the shortest spanning subtree of a graph and the traveling sales-man problem. Proc. Am. Math. Soc. **7**, 48–50 (1956)
10. Linde, Y., Buzo, A., Gray, R.: An algorithm for vector quantizer design. IEEE Trans. Commun. **28**(1), 84–95 (1980)
11. López-Rubio, E.: Probabilistic self-organizing maps for continuous data. IEEE Trans. Neural Netw. **21**(10), 1543–1554 (2010)
12. López-Rubio, E.: Probabilistic self-organizing maps for qualitative data. Neural Netw. **23**(10), 1208–1225 (2010)
13. López-Rubio, E., Palomo-Ferrer, E.: Growing hierarchical probabilistic self-organizing graphs. IEEE Trans. Neural Netw. **22**(7), 997–1008 (2011)
14. López-Rubio, E., Palomo-Ferrer, E.J., de Lazcano-Lobato, J.M.O., Vargas-González, M.C.: Dynamic topology learning with the probabilistic self-organizing graph. Neurocomputing **74**(16), 2633–2648 (2011). http://www.sciencedirect.com/science/article/pii/S0925231211002360
15. Martinetz, T., Schulten, K.: A "Neural-Gas" network learns topologies. Artif. Neural Netw. **I**, 397–402 (1991)
16. Palomo, E.J., Lpez-Rubio, E.: Extended abstract: a color quantization approach based on the growing neural forest. In: 2016 IEEE Latin American Conference on Computational Intelligence (LA-CCI), pp. 1–2, November 2016
17. Palomo, E., López-Rubio, E.: Learning topologies with the growing neural forest. Int. J. Neural Syst. **26**(4), 1650019 (2016)
18. Palomo, E.J., Domínguez, E.: Hierarchical color quantization based on self-organization. J. Math. Imag. Vis. **49**(1), 1–19 (2013)
19. Papamarkos, N.: Color reduction using local features and a sofm neural network. J. Imag. Syst. Technol. **10**(5), 404–409 (1999)
20. Rauber, A., Merkl, D., Dittenbach, M.: The growing hierarchical self-organizing map: Exploratory analysis of high-dimensional data. IEEE Trans. Neural Netw. **13**(6), 1331–1341 (2002)
21. Xiao, Y., Leung, C.S., Lam, P.M., Ho, T.Y.: Self-organizing map-based color palette for high-dynamic range texture compression. Neural Comput. Appl. **21**(4), 639–647 (2012)

3D Body Registration from RGB-D Data with Unconstrained Movements and Single Sensor

Victor Villena-Martinez, Andres Fuster-Guillo, Marcelo Saval-Calvo$^{(\boxtimes)}$,
and Jorge Azorin-Lopez

Department of Computer Technology, University of Alicante, Alicante, Spain
{vvillena,fuster,msaval,jazorin}@dtic.ua.es

Abstract. In this paper, the problem of 3D body registration using a single RGB-D sensor is approached. It has been guided by three main requirements: low-cost, unconstrained movement and accuracy. In order to fit them, an iterative registration method for accurately aligning data from single RGB-D sensor is proposed. The data is acquired while a person rotates in front of the camera, without the need of any external marker or constraint about its pose. The articulated alignment is carried out in a model-free approach in order to be more consistent with the real data. The iterative method is divided in stages, contributing to each other by the refinement of a specific part of the acquired data. The exploratory results validate the proposed method that is able to feed on itself in each iteration improving the final result by a progressive iteration, with the required precision under the conditions of affordability and unconstrained movement acquisition.

1 Introduction

Nowadays, there are several techniques to obtain a 3D model of the human body. This model is useful in many applications of different scopes: medical, textile, shoes, etc. The tandem of 3D technology and medicine has a long trajectory, 3D models have been used to help experts study patients and make decisions [23], e.g.: dietetic treatment, child growth monitoring, podiatry, orthopedics, among others There are several solutions to obtain a high definition 3D model of the human body, but they are expensive and bulky, making its portability more difficult, with a prohibitive cost. The usefulness of these models has made more attractive the development of inexpensive systems with few restrictions for the subject pose during the acquisition. Consumer RGB-D, low-cost RGB-D, or just RGB-D sensors have become popular in fitting those requirements [9] due to their combination of affordability and portability. However, the accuracy of them is not high enough for some applications.

Therefore, in this paper we face the problem of providing a 3D body representation using RGB-D sensors. The proposal has to fit some requirements: (i) affordability and flexibility for broad transferring solutions, (ii) unconstrained

© Springer International Publishing AG 2017
I. Rojas et al. (Eds.): IWANN 2017, Part II, LNCS 10306, pp. 317–329, 2017.
DOI: 10.1007/978-3-319-59147-6_28

method in terms of free movement of the subject in front of the sensor, (iii) accurate alignment of the data to create an accurate representation.

Low-cost RGB-D sensors are considered a new type sensor due to the large usage they are receiving in the scientific and industrial community. This kind of sensors combine color and depth information, the later one estimated using Time of Flight (ToF) or structured-light techniques. In this work, we propose a method for RGB-D devices, due to they fit the low-cost requirement, and their accuracy is appropriate for body modeling since they were originally developed for body movement tracking [17]. However, it could be possible to use our approach with any other sensor which provides color and depth information.

To develope a full model of bodies using RGB-D data, our proposal faces the registration or alignment of 3D points. The registration is the process to align one data set onto one or more data sets. It could be performed rigidly and non-rigidly, which refer to transform the data all with the same transformation, or by applying different transformations to each data point [18]. Registration of 3D data is a widely studied problem. Henry et al. [8] made a 3D reconstruction of indoor environments with an RGB-D sensor. The work of [11] carried out an accurate 3D registration of the foot using a PrimeSense rotating around it. These sensors could be used in combination with augmented reality markers in the scene in order to estimate more accurately the transformation to obtain the 3D model [12].

The 3D registration of the human body is complex due to its articulated nature, and the impossibility to keep the same pose in different instants of time. These inconveniences could be approached by different acquisition approximations to obtain a model:

- Single view, performing a partial reconstruction of the subject of interest.
- Adding sensors to acquire the subject from different angles in the same instant of time, avoiding any movement of the subject.
- Controlled environment, using external elements to the sensor, like augmented reality markers.
- Using articulated/isometric techniques for the registration. These techniques take into account the movements performed by the subject during the acquisition. This work is focused in this approach.

The rest of the paper is organised as follows: a study of the background is presented in Sect. 1.1; an explanation of the proposed methiod is done in Sect. 2; the experiments are presented in Sect. 3; finally, the conclusion are presented in Sect. 4.

1.1 Related Works

Various proposals for body scanning could be found, either related to the sensor system or to the registration. There exist commercial body scanners, like *Vitus BodyScan* or the *Cyberware Whole Body Scanner*. However, these systems are expensive. Some works carried out the reconstruction of body introducing multiple RGB-D devices [10,20,22]. Nevertheless, the affordability and flexibility are required in this work. One restriction is the use of a single RGB-D sensor, avoiding complex instrumentation, but forcing the rotation of the body in front of the camera. Then, the use of articulated or isometric techniques are mandatory [14] to overcome misalignment problems.

The articulated alignment is currently under intense study. In human bodies it consists in non-rigid alignment of the different parts in the body, taking into account the joints, allowing pose variations during the acquisition. It divides the body into different parts defined by pairs of joints, and rigidly register each part. To determine each part the work [4] differentiates each one from a single view obtained with an RGB-D sensor. Shotton et al. [21] proposed a method to estimate the parts of the body an its pose in real time from a single frame using random forest. Schwarz er al. [19] estimated the pose from depth data, using geodesic distances and optical flow for occlusions. Other works use Convolution Neural Networks (CNNs) for pose and skeleton estimation [5,13,25].

Focused on the articulated 3D reconstruction of human body, two approaches have been used, model-free and model-based techniques [3]. The former needs multiple depth frames from different viewpoints to get information of the whole body and perform the 3D reconstruction. The latter techniques fit the data points on a generic model as [26], which uses the SCAPE [1].

Model-free approaches use different techniques to get the 3D model. Barmpotis et al. [2] carried out a real-time reconstruction of a body of a subject in front of the camera through parametrization of cylindrical objects using Cartesian tensor and *b-splines*. Cui et al. [6] proposed a method to obtain a full body model using a single depth camera with a Kinect sensor and employing an articulated model to describe de non-rigid motion of the body forcing the subject placed in "T" pose. Wang et al. [24] proposed a cylindrical representation of the body model, performing the registration in a top-bottom-top manner and using the cylinders to simplify the registration to a bi-dimensional problem.

Our main goal is to develop a system for 3D body registration of subjects. The system aims to be flexible, accurate, and affordable. In order to accomplish this objective, we propose to use a single RGB-D sensor, due to its affordability and flexibility, being the subject who rotates in front of the camera. Furthermore, the registration is performed in a model-free approach using all the acquired information from a large set of frames. To the best of our knowledge, there exist the need to propose a system which accurately registers full body fitting the requirements above mentioned. To achieve the accuracy of the problem, we propose an iterative method for 3D reconstruction of the human body, based on an initial estimation of the skeleton. This method is divided in three main stages, each one contributing to the others by the refinement of a specific part of the data, performing an iterative strategy fitting the accuracy of the problem.

2   Method Overview

In this paper, an iterative method for 3D registration of the human body is proposed. This method is based on the initial estimation of a 3D medial axis (i.e. a skeleton of the subject), which is iteratively improved by the registration process itself. Initially, the proposed method expects as input a set of views, being each one a 3D point cloud aligned with the color information of the scene. In these set of views, the subject is rotating by itself in front of the camera in a "tired T" pose (see Fig. 1) that is appropriate for perceiving the whole body along the different views.

Figure 1 shows the outline of the method composed by three stages that are fed back in each iteration in order to improve the final result.

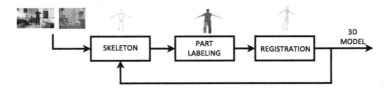

Fig. 1. Iterative scheme of the method

The first stage estimates the skeleton of the subject (position of body joints related to the 3D point set). Moreover, in each iteration, the information of previous skeletons and registered points are used to overcome possible problems (see Sect. 2.1). At the labeling stage, the different parts of the body are detected and each point in the view is accordingly labeled (forearm, chest...). The points labels are used in the isometric alignment of the data. Finally, the registration stage performs the articulated alignment of the 3D data having as output a set of aligned views that are, moreover, used as feedback for the next iteration of the process.

2.1 Skeleton Estimation

The skeleton model to be estimated is composed by a set of *joints* and *bones*. The *joints* are a set of 3D points in the location of the body joints related to the input point set. The *bones* of the skeleton are segments connecting the joints keeping the shape model of a human body. As a case of study in this paper, the skeleton model used is composed by 23 joints (see Table 1) and 25 bones but any other approach can be used.

Table 1. Set of joints model used in this paper

N	Joint	N	Joint	N	Joint	N	Joint
1	Head	7	Left wrist	13	Right hand	19	Left knee
2	Neck	8	Left elbow	14	Right thumb	20	Left ankle
3	Spine shoulder	9	Left shoulder	15	Right hand tip	21	Right hip
4	Left hand tip	10	Right shoulder	16	Middle spine	22	Right knee
5	Left thumb	11	Right elbow	17	Spine base	23	Right ankle
6	Left hand	12	Right wrist	18	Left hip		

The skeleton estimation is a challenging problem, mainly when the subject is not facing the camera and becoming very difficult when the body is positioned from the side with respect to the camera. In order to correct the estimation errors, different refinements have been proposed:

- re-estimating left and right joints and bones in case the subject back towards the camera (see Sect. 2.1).
- detecting and discarding wrongly estimated joints by overlapping (see Sect. 2.1). For example, when the subject is seen from the side, both arms are detected in the same position (see Fig. 2).
- re-estimating joints placed in wrong locations, discarded, or those that have not been estimated (see Sect. 2.1).

Backward Correction. One of the usual problems in body skeleton estimation algorithms become when the subject is facing backwards to the camera due to the calculated joints are not correct being exchanging the left and right ones.

The proposed method checks distances between the same joints in different views. The joints are clustered into four groups: right arm (10–14), left arm (4–9), right leg (21–23) and left leg (18–20). Then, the view where the subject is facing the camera is taken as reference (i.e. view with best skeleton). For each view, the clusters are anyalized so as each joint is compared to its corresponding and its opposite (other side). If the distance is larger in the correspoding, an exchange is performed to correct it. It is important to remark the need of a prior alignment of the torso (the most static part of the body).

Overlapped Joints. Another common problem in skeleton estimation algorithms is that, in some cases, the joints are overlapped. Figure 2 shows the estimation of the skeleton when the subject is seen from the side with respect to the camera. Figure 2-left shows the problem, the joints of both arms (4–8 and 11–15 from Table 1) are overlapped. This problem will result in a wrong labeling of the body parts. Again, this correction also needs that the torso is previously registered.

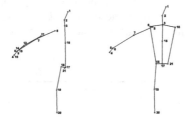

Fig. 2. Overlapped joints problem. Left: Wrong joints in an arm. Both are detected in the same side. Right: Skeleton correction after correcting wrong joints.

First, the method improves the estimation of joints by positioning those from the reference frame into the remaining views once pre-aligned. Then, the overlapped joints are detected assuming that the euclidean distance between hands should be greater than the distance between shoulders (joints 9 and 10).

Next, it is necessary to decide which of the two groups of joints (left or right) is wrong. Hence, we calculate the distance between the hand and both its corresponding shoulder and the opposite one. If the former is larger, those joints are incorrect and a re-estimation is required as explained in Sect. 2.1.

Re-estimation. After the correction of the overlapped joints, the skeleton model may have some non-detected joints, or wrongly positioned ones. For example, in Fig. 2-right image, the wrist joint is placed at the elbow, and the elbow joint at the shoulder, producing errors in the labeling stage. To solve this problem, the first step is to detect the frames where the joints are wrong estimated by calculating the standard deviation of the distances between the joints showed in Table 2. Then, the distances for each frame are calculated and compared with the reference frame, checking the following condition: $|distance - reference\_distance| > std$, where std is the standard deviation of all the distances. In those frames were the condition is true, a re-estimation of the wrist and elbow joints is needed.

Table 2. Joint distances calculated for correct position of joint in the arms

Joint	
6–7	Left hand-wrist distance
9–8	Reft shoulder-elbow distance
10–11	Right shoulder-elbow distance
12–13	Right hand-wrist distance

Assuming a "tired T" pose of the subject, the problem could be solved as basic geometry problem. Given two points A and B, being the hand and shoulder as shown in Fig. 3, and the segment which connects them, we find a third point C (elbow or wrist) that belongs to the segment and is placed at a certain distance from A or B. These distances, from the hand to the wrist and from the shoulder to the elbow, are recalculated taking into account those from the rest of views.

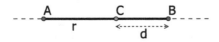

Fig. 3. Segment from A, hand, to B, shoulder, to find C elbow or wrist.

2.2   Part Labeling

The articulated registration of the human body needs to know to which part of the body each point of the point cloud belongs to. These parts are defined taking into account the joints of the skeleton and the bones, which are the segments defined by pairs of joints following human body shape. With the information of the previous stage, each 3D point of the point cloud per view is classified in a part of the body. In the experiments, 12 parts have been considered (Table 3) but this could be different depending on the requirements of the problem.

Table 3. Labels and the associated part of the body

Part		Part		Part	
1	Head	5	Left hand	9	Left thigh
2	Torso	6	Right biceps	10	Left calf
3	Left biceps	7	Right forearm	11	Right thigh
4	Left forearm	8	Right hand	12	Right calf

To classify the points, the distance from each 3D point to each bone of the skeleton is computed. The closest bone to an specific point determines the part in which will be classified. The distance from a point to a line is given by the orthogonal projection of the point in the line, so it could be calculated with basic three dimensional geometry.

Next, the topology defined by a part is validated, checking the connectivity of the points under the same label. Each part or label only contains connected points. Sometimes, if two parts are close, points from one could be labeled as the other. In this case, the clusters are analyzed selecting one of them as correct, and classifying again the remaining groups.

2.3   Registration

The registration stage performs the alignment of different views to obtain the whole 3D model. This alignment is usually performed with rigid registration techniques. Rigid registration algorithms could be classified in terms of [16]: precision (coarse and fine) and registration strategy (pairwise and multiview). However, rigid registration techniques are not enough, such as in the problem of body modeling due to the acquisition of the different viewpoints are not simultaneously taken and the subject is in motion. For that reason, it is necessary to take advantage of the topological information provided in previous stages, i.e. joints, bones and body parts labeling.

In this paper, the articulated registration is approached by dividing the model in parts and aligning each one independently with a rigid registration method. Then, the transformation has to be propagated to the dependent parts, e.g.: in case of the arm (Fig. 4), a transformation to the biceps must be applied to the

Fig. 4. Arm articulated registration process. From left to right, first the biceps is registered propagating to the rest. Second, the forearm is registered propagating to the hand. Thirdly, the hand is registered.

forearm and hand, and so on. The dependencies between the different part of the body are defined by the graph in Fig. 5.

Hence, the registration is divided into different steps (see Fig. 6). First, the registration of the torso is performed to coarsely align the different views of the subject. Second, the views are filtered to determine which one could be used. Finally, the registration of the limbs is carried out.

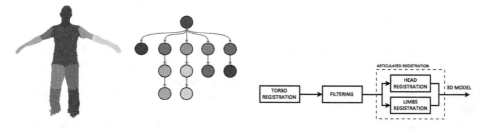

Fig. 5. Dependencies between body parts   **Fig. 6.** Schematic of articulated registration

Torso Registration. The torso is fine registered using Iterative Closest Point (ICP) [15] with a pairwise strategy. The main objective of this step is to perform the alignment of all views based on the torso, due to the remaining parts depend on it and it is the most static part of the body. This implies that a transformation of the torso must be applied to other parts.

In this case, it is possible to use a fine approximation without a previous alignment due to the closeness between consecutive frames having a minimum variation between consecutive views. If the views are not close each other, a pre-alignment could be used.

Filtering Views. The views are filtered to select and remove the parts that are not used to be registered. The method discard the parts that lie in one of the next situations: the part is occluded (side views); the number of points of a part is widely different that the average in the rest of views; and, the distance between joints in the skeleton is widely different that the same in the rest of frames.

Limbs Registration. In this step, each part is aligned individually, taking into account the dependencies with other parts in order to apply the transformations (see Fig. 5). The head usually has few points and a uniform geometry mainly in the back, which difficult its alignment. To register this part, first, a coarse alignment is performed using Random Sample Consensus (RANSAC) [7] with visual features as SIFT and SURF. Then, a fine aligned it is done with ICP.

The arms and legs are composed by different parts, so the transformations are applied according to the dependencies of each part (see Fig. 4). In this step, a pairwise strategy is used between consecutive frames with a coarse registration based on Principal Component Analysis (PCA) and ICP as fine refinement. After that, a multiview strategy is applied to each part for improving the final result. This strategy consists in align each view with the whole model in order to polish the imperfections produced by the registration of consecutive views with a pairwise strategy.

3   Experiments

In this section a set of experiments have been carried out to verify the results provided by our method. This exploration results are evaluated using visual inspection. The setup includes an RGB-D sensor Microsoft Kinect V2 and the Microsoft Kinect SDK. The SDK provides the skeleton of the subject, estimated with a variation of the method proposed by [21] that provides joints for thumbs tracking, which are not useful for this method.

The experimentation is divided into three parts: overlapped joints detection and correction; joints re-estimation when they are wrongly estimated; 3D registration of body subjects with the proposed iterative method.

3.1   Overlapped Joints

The proposed method can solve problems where both arms have been detected in the same side, shown in Fig. 7a. This problem has direct consequences in the labeling and registration stages, as can be seen in Fig. 7b which shows the result of the labeling stage with the previous wrong skeleton where the arm is divided into regions that do not correspond to any correct human model. With the proposed method in Sect. 2.1, the skeleton is corrected by removing the wrong joints, and re-estimating some of them as can be seen in Fig. 7c. Then, the 3D data can be properly labeled (Fig. 7d) to provide useful information for the articulated registration step.

3.2   Joints Re-estimation

The second experiment studies the joints re-estimation, when the skeleton estimation missed some joint. For example, Fig. 8a shows a skeleton where the joint 9, corresponding with the left shoulder, has not been detected. This problem produces erroneous detection of body parts, in this case the left biceps and shoulder

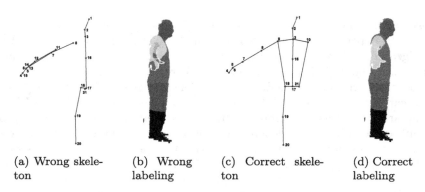

(a) Wrong skele-   (b) Wrong   (c) Correct skele-   (d) Correct
ton   labeling   ton   labeling

Fig. 7. (a) shows the wrong joints estimation due to the arms overlapping, and (b) shows the problems in labeling of the arm. (c) shows the joint correction and (d) shows the adequate labeling after applying the proposed method.

are equally labeled (Fig. 8b). The proposed method can solve this situation by providing the skeleton showed in Fig. 8c making the left biceps distinguishable in the labeling stage.

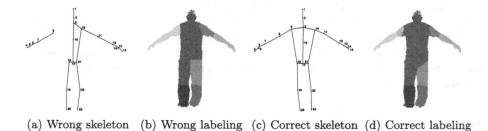

(a) Wrong skeleton (b) Wrong labeling (c) Correct skeleton (d) Correct labeling

Fig. 8. (a) shows the skeleton model in which the joint 9 has not been estimated; (b) shows the wrong labeling of the data where shoulder and biceps are the same part; (c) presents the corrected skeleton with the proposed method, which leads to an adequate labeling shown in (d)

3.3   Articulated Registration

In order to evaluate the articulated registration, a sequence of frames of a man facing the camera making free movements with his arms is used. A subset of these frames is showed in Fig. 9. Figure 9-bottom-right depicts the 3D points of all frames superimposed. After applying the proposed method, the articulated alignment achieves an accurate registration of the views represented in Fig. 10.

Fig. 9. First row and three left figures in bottom row are a subset of frames of free movements with the arms. Bottom right shows 3D data overlapped of the movement

Fig. 10. Result of the articulated alignment

Fig. 11. 3D model obtained with the proposed iterative method

3.4   3D Body Model

The purpose of the proposed method is to obtain a 3D registered model of a human body. To evaluate the overall performance of the proposal, various experiments have been carried out. Because of the similarity between the results, only one example is shown in this paper, but similar accuracy has been achieved for the rest of models. The model is obtained from a sequence of 281 images of the subject rotating by itself in front the camera. Then, the whole method has been applied to those frames, obtaining an accurate result, as shown in Fig. 11 in frontal and profile view.

4   Conclusion

In this paper, an iterative method for 3D body registration is presented aiming to be low-cost, dealing with unconstrained movements and accuracy. In line of these objectives, RGB-D sensors have been selected for the acquisition of 3D data to fit the low-cost requirement. The acquisition of this data is performed while the subject rotates in front of the camera, without the need of any external device or forcing the subject to be in a specific pose. The alignment of the different views is performed in a model-free approach in order to be more consistent with the real data.

The iterative method for 3D registration of the human body is based on an initial estimation of the skeleton. The method is divided in three stages, contributing to each other by the refinement of a specific part of the data. Moreover, the method feeds back in each iteration to improve the final result.

The experimentation with real data allows to validate the proposed method. It has been shown how it can progressively improve the acquired data to achieve the more accurate registration. The articulated alignment have been tested performing free arm movements, proving that the method can take into account those movements aligning all views into a specific pose. Finally, the main purpose of the method have been carried out performing the 3D registration of a body rotating in front of the camera satisfying the requirements of low-cost, unconstrained movements and accuracy.

As future lines, it is considered to introduce non-rigid registration methods to study the shape variation of subjects by obtaining 3D models in different instants of time in order to quantify that variation.

References

1. Anguelov, D., Srinivasan, P., Koller, D.: Scape: shape completion and animation of people. ACM Trans. Graph. (TOG) **24**(3), 408–416 (2005)
2. Barmpoutis, A.: Tensor body: real-time reconstruction of the human body and avatar synthesis from RGB-D. IEEE Trans. Cybern. **43**(5), 1347–1356 (2013)
3. Bogo, F., Black, M.J., Loper, M., Romero, J.: Detailed full-body reconstructions of moving people from monocular RGB-D sequences. In: ICCV, pp. 2300–2308 (2015)
4. Charles, J., Everingham, M.: Learning shape models for monocular human pose estimation from the Microsoft Xbox Kinect. In: 2011 IEEE International Conference on Computer Vision Workshops, pp. 1202–1208. IEEE (2011)
5. Chen, W., Wang, H., Li, Y., Su, H., Tu, C., Lischinsk, D., Cohen-Or, D., Chen, B.: Synthesizing training images for boosting human 3D pose estimation. CoRR (2016)
6. Cui, Y., Chang, W., Nöll, T., Stricker, D.: KinectAvatar: fully automatic body capture using a single kinect. In: Park, J.-I., Kim, J. (eds.) ACCV 2012. LNCS, vol. 7729, pp. 133–147. Springer, Heidelberg (2013). doi:10.1007/978-3-642-37484-5_12
7. Fischler, M.A., Bolles, R.C.: Random sample consensus: a paradigm for model fitting with applications to image analysis and automated cartography. Commun. ACM **24**(6), 381–395 (1981)
8. Henry, P., Krainin, M., Herbst, E., Ren, X., Fox, D.: Experimental Robotics, vol. 79. Springer, Heidelberg (2014)
9. Lai, K., Bo, L., Ren, X., Fox, D.: Consumer depth cameras for computer vision. In: Fossati, A., Gall, J., Grabner, H., Ren, X., Konolige, K. (eds.) Consumer Depth Cameras for Computer Vision, p. 167. Springer, London (2013)
10. Lin, S., Chen, Y., Lai, Y.-K., Martin, R.R., Cheng, Z.-Q.: Fast capture of textured full-body avatar with RGB-D cameras (2016)

11. Lovato, C., Bissolo, E., Lanza, N., Stella, A., Giachetti, A.: A low cost and easy to use setup for foot scanning (2014)
12. Mihalyi, R.G., Pathak, K., Vaskevicius, N., Fromm, T., Birk, A.: Robust 3D object modeling with a low-cost RGBD-sensor and AR-markers for applications with untrained end-users. Robot. Auton. Syst. **66**, 1–17 (2015)
13. Oliveira, G.L., Valada, A., Bollen, C., Burgard, W., Brox, T.: Deep learning for human part discovery in images. In: IEEE International Conference on Robotics and Automation (2016)
14. Pellegrini, S., Schindler, K., Nardi, D.: A generalisation of the ICP algorithm for articulated bodies. Proc. BMVC **3**, 4 (2008)
15. Rusinkiewicz, S., Levoy, M.: Efficient variants of the ICP algorithm. In: Proceedings Third International Conference on 3-D Digital Imaging and Modeling, pp. 145–152. IEEE Computer Society (2001)
16. Salvi, J., Matabosch, C., Fofi, D., Forest, J.: A review of recent range image registration methods with accuracy evaluation. Image Vis. Comput. **25**(5), 578 (2007)
17. Saval-Calvo, M., Azorin-Lopez, J., Fuster-Guillo, A., Garcia-Rodriguez, J., OrtsEscolano, S., Garcia-Garcia, A.: Evaluation of sampling method effects in 3D nonrigid registration. Neural Comput. Appl. 1–15 (2016)
18. Saval-Calvo, M., Azorin-Lopez, J., Fuster-Guillo, A., Mora-Mora, H.: μ-mar: Multiplane 3D marker based registration for depth-sensing cameras. Expert Syst. Appl. **42**(23), 9353–9365 (2015)
19. Schwarz, L.A., Mkhitaryan, A., Mateus, D., Navab, N.: Human skeleton tracking from depth data using geodesic distances and optical flow. Image Vis. Comput. **30**(3), 217–226 (2012)
20. Shapiro, A., Feng, A., Wang, R., Li, H., Bolas, M., Medioni, G., Suma, E.: Rapid avatar capture and simulation using commodity depth sensors. Comput. Animation Virtual Worlds **25**, 201–211 (2014)
21. Shotton, J., Sharp, T., Kipman, A., Fitzgibbon, A., Finocchio, M., Blake, A., Cook, M., Moore, R.: Real-time human pose recognition in parts from single depth images. Commun. ACM **56**(1), 116 (2013)
22. Tong, J., Zhou, J., Liu, L., Pan, Z., Yan, H.: Scanning 3D full human bodies using kinects. IEEE Trans. Vis. Comput. Graph. **18**, 643–650 (2012)
23. Treleaven, P., Wells, J.: 3D body scanning and healthcare applications. Computer **40**(7), 28–34 (2007)
24. Wang, R., Choi, J., Medioni, G.: Accurate full body scanning from a single fixed 3D camera. In: Proceedings - 2nd Joint 3DIM/3DPVT Conference: 3D Imaging, Modeling, Processing, Visualization and Transmission, 3DIMPVT 2012, pp. 432–439 (2012)
25. Wei, S.E., Ramakrishna, V., Kanade, T., Sheikh, Y.: Convolutional pose machines. In: Proceedings of CVPR, pp. 4724–4732 (2016)
26. Weiss, A., Hirshberg, D., Black, M.: Home 3D body scans from noisy image and range data. In: Computer Vision (ICCV), pp. 1951–1958 (2011)

Human Activity Recognition for Health and Well-being Applications

Posture Transitions Identification
Based on a Triaxial Accelerometer
and a Barometer Sensor

Daniel Rodríguez-Martín$^{(\boxtimes)}$, Albert Samà, Carlos Pérez-López,
and Andreu Català

Technical Research Centre for Dependency Care and Autonomous Living
(CETpD), Universitat Politècnica de Catalunya – BarcelonaTech (UPC),
Barcelona, Spain
{daniel.rodriguez-martin,albert.sama,
carlos.perez-lopez,andreu.catala}@upc.edu

Abstract. Posture transitions (PT) are important movements among the activities performed in daily life of older adults. Their analysis provides information related to the amount of activity performed by a patient over a day and, furthermore, they are useful for assessing symptoms in some movement disorders such as Parkinson's disease. Many research works have attempted to automatically identify PT relying on the use of machine learning algorithms and light and small accelerometers, since they might be embedded into wearable systems, being unobtrusive for the users. However, distinguishing PTs through a single sensor results in complex classifiers requiring high computational resources, since some PT (such as Stand-to-Sit and Sit-to-Stand PT) may provide very similar acceleration signals. In this paper, we propose a barometer sensor with the aim of complementing the information provided by accelerometers. In addition, a hierarchical algorithm is presented, which is based on Support Vector Machines to detect PT including falls and Lying-to-Stand PT through a single sensor device. Results in 14 users show that the use of a barometer sensor enables the hierarchical algorithm to distinguish Sit-to-Stand from Stand-to-Sit transitions, and Falls from Lying-to-Stand with accuracies over 99%.

Keywords: Posture transitions · Barometer · Accelerometer · Support Vector Machines

1 Introduction

Human activity recognition with monitoring purposes can be a tool of great interest for clinics in order to improve therapeutic strategies [1]. Within the field of Parkinson's disease (PD), or other diseases with motor disorders, it can help to evaluate symptomatology and enhance medication adjustment [2–4]. Among the different movements and activities a person can perform, posture transitions (PT) are of special interest; mainly Sit-to-Stand (*SiSt*) and Stand-to-Sit (*StSi*) transitions. These PT are one of the most mechanically demanding activities, being difficult to be executed by people with

© Springer International Publishing AG 2017
I. Rojas et al. (Eds.): IWANN 2017, Part II, LNCS 10306, pp. 333–343, 2017.
DOI: 10.1007/978-3-319-59147-6_29

mobility impairments, leading to falls [5]. In PD patients, posture transitions are considered as a pre-requisite to walk and a propitious moment for the appearance of Freezing of Gait (FoG). These episodes, before initiating gait, can cause falls that entail serious consequences [6, 7]. The detection of FoG, in ambulatory conditions, is a highly complex task due to the great similarity between the episodes of FoG and some natural movements of patients. However, the detection of SiSt and StSi transitions has shown to enhance the detection of FoG by means of eliminating false positives [8].

In order to monitor these PT, researchers have employed many systems, such as video, photography, and walkways with pressure. However, wearables systems based in Micro-Electro-Mechanical-Systems (MEMS) technology have been widely used due to their small size and low invasiveness [9]. The main MEMS sensors used for posture analysis are accelerometers, which provide movement linear acceleration and the orientation of the sensor with respect to gravity. Nonetheless, other systems might help to complement this information. In a previous work, barometers have shown to be a good tool in order to detect movements with a vertical component such as *SiSt* and *StSi* transitions. For example, Massé et al. showed the performance of different barometers for distinguishing *SiSt* from *StSi* [10]. Only one barometer showed acceptable measurement changes enough to differentiate a *SiSt* from a *StSi*. Besides, barometers have also been used to detect falls with promising results such as those from Tolkiehn et al. [11], Bianchi et al. [12], and Moncada et al. [13].

This paper presents a system to detect *SiSt* and *StSi* transitions, based on the information provided by an accelerometer and an up-to-date barometric pressure sensor. Besides, a fall detector has been also implemented since they are a common consequence in some PD symptoms such as FoG [14]. As stated before, falls can be also detected through barometers due to the loss of altitude. Furthermore, other activities such as using an elevator or going up and down stairs, where the barometer signal is sensitive, have also been studied. The algorithm relies on a hierarchical structure of classifiers based on thresholds and support vector machine (SVM) algorithms.

This work has been performed under the "Freezing in Parkinson's Disease: Improving Quality of Life with an Automatic Control System" (MASPARK) project [15], which aims to detect and contextualize Freezing of Gait (FoG) symptom.

2   Postures and Falls Detection Algorithm

In this section, we present the methodology for detecting posture transitions and falls combining the information from a triaxial accelerometer and a barometer.

The algorithm to detect PT and falls is presented in Fig. 1 as a block diagram. Through this hierarchical scheme, the algorithm is capable of determining whether a person, who is wearing an inertial system, has fallen or has done a posture transition among *SiSt*, *StSi* and a *Lying-to-Stand* (*LySt*). There is a state called *'Other activity'* in which other detected actions are labelled, such as walking, sit, stand, lying on a bed, and walking upstairs and downstairs.

To distinguish among these PT, the algorithm is composed of two supervised-learning classifiers called C1 and C2 (shown in orange) and two threshold-based classifiers C3 and

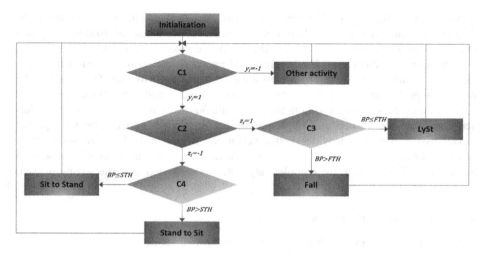

Fig. 1. Posture transition algorithm block diagram. Initialization is indicated as the green block. Yellow blocks correspond to the final states of the algorithm (posture transitions or *fall*). Orange blocks correspond to supervised-learning classifiers based on accelerometer features and, finally, blue boxes are the threshold-based classifiers, which relies on the barometer sensor response. (Color figure online)

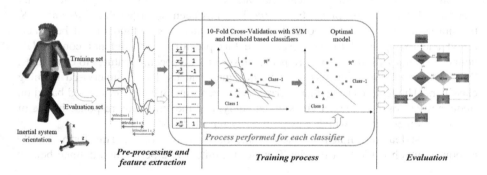

Fig. 2. Main scheme to train and evaluate the classifiers involved in the posture transition and *fall* detection algorithm.

C4 (shown in blue). C1 and C2 classifiers determine the type of posture transition through accelerometer features while C3 and C4 classifiers detect the direction of movement by means of the barometer, identifying the event.

The method shown in Fig. 1 is applied over the signals acquired by the triaxial accelerometer and the barometric pressure sensor. The approach followed to obtain and treat the signals is represented in Fig. 2. First, signals are filtered and divided into windows 50% overlapped. Each window is 3.2 s long and the sampling frequency is 40 Hz given posture transition temporal requirements described in [9]; in consequence, windows are composed of 128 samples. Then, each signal window is characterized through specific features, according to the requirements of the supervised-learning and threshold-based classifiers, which are detailed in the next subsections.

The four classifiers used by the presented approach are trained through the same process. Training and validation datasets are established to be the data from 70% of the users who participated in the database collection, who are randomly selected among all users. Testing dataset correspond to the remaining 30% of users. From the training data set, classifiers are trained and hyper-parameters are found, considering the optimal model as the one that provides better performance with a 10-fold cross-validation process. Once the four classifiers are trained, they are hierarchically structured according to Fig. 1, then, the complete algorithm is evaluated with the testing dataset.

2.1   C1 Classifier Description

This classifier distinguishes between the following two states:

1. The user has performed either a PT (*SiSt* or *StSi*), a *LySt* or a *Fall* event.
2. The user has performed a different activity such as walking, lying, sitting, standing, going up/down stairs or lift.

This classifier depends on accelerometer features and takes a previous work as the starting point [9], where it was determined that the sum of harmonics below 0.68 Hz and above 0 Hz from the vertical and lateral axes of a waist-worn accelerometer identifies that a PT took place. In this work, it has been observed that *falls* and *LySt* transitions have a similar response in this band. Thus, this method cannot be generalised and extrapolated to our case, since this previous work did not contemplate neither *falls* nor *LySt* transitions. Hence, a classifier with specific features to cover falls, SiSt, StSi and LySt transitions must be designed and used with supervised learning classifiers. The features used characterize these events and have been selected based on frequency response of the activities under study, and the amplitude response in one of the accelerometer axes, which was previously reported in [9].

The first feature employed relies on the Short Time Fourier Transform (STFT) response, which is the Fast Fourier Transform limited by a period and defined below:

$$X_h^W = \sum\nolimits_{n=1}^{N} x_n e^{-i2\pi h \frac{n}{N}} \qquad (1)$$

where $h = 1,...,N$; X_h^W is the output of the STFT at the W window. X_h^W is composed of a set of N complex numbers that represents the amplitude and phase of the harmonic (N is set by the length of W).

With the aim of computing a specific frequency band, let $\left|X_{f1,f2}^W\right|$ be the amplitudes of the harmonics between the $f1$ and $f2$ harmonics of the x-axis. The first feature is then composed of the sum of the harmonics at the three axes as shown in Eq. (2).

$$Acc.Energy_{f_1,f_2} = \sum\nolimits_{h=f_1}^{f_2} |X_h^W| + \sum\nolimits_{h=f_1}^{f_2} |Y_h^W| + \sum\nolimits_{h=f_1}^{f_2} |Z_h^W| \qquad (2)$$

where X_h^W, Y_h^W and Z_h^W corresponds to the frequency response of the x axis, y axis and z axis, respectively.

The second feature is the mean of the vertical axis (x-axis) given by $\bar{x}^W = \frac{1}{N}\sum_{i=1}^{N} x_i^w$. Finally, the third feature is the walking index described in a previous work [9]. This index quantifies the amount of energy corresponding to walking activity by dividing the walking band energy ($Acc.Energy_{0.68,3}$) with the amount of energy obtained within the PT band ($Acc.Energy_{0.01,0.68}$).

The features used to characterise acceleration windows result into vectors that are also employed to train a SVM with radial basis function kernel, in which the accuracy is maximised to select the hyper-parameters C and γ. The obtained model is, then, selected and tested within the hierarchical structure of classifiers.

2.2   C2 Classifier Description

Once the previous classifier (C1) confirms that the current window is a candidate to be *StSi, SiSt, LySt* transitions or *Fall,* this classifier determines whether the person has performed a *StSi* or *SiSt* transition or, on the contrary, a fall event (*Fall* or *LySt*). We present five features that could characterize and distinguish these actions. These features are inspired by the work in [9] and are based on the accelerometer response.

The first feature is the sum of the harmonics amplitudes between 0 and 0.68 Hz from the vertical axis of the accelerometer, excluding the 0 Hz value. The second is the sum of amplitudes of the three axes between 0 and 0.68 Hz excluding the 0 Hz value. These two features aim to detect significant movements within a specific band that is sensitive to PT's. The third is the mean of the vertical axis of the accelerometer, used to know the orientation of the trunk regarding the ground. The fourth and fifth features are the difference between the accelerometer axes means in the current and the previous windows. These two features provides information about the change of orientation in a PT of *SiSt* and *StSi* [16]. Table 1 summarizes the five inputs for the C2 classifier.

Table 1. Features description for the C2 classifier

Feature	Description
STFT of the vertical axis between 0 and 0.68 Hz	$\sum_{h=f_1}^{f_2} |X_h^W|,\quad f_1 = 0.01, f_2 = 0.68$
Sum of the STFT response of the 3 axes between 0 and 0.68 Hz	$\sum_{h=f_1}^{f_2} |X_h^W| + \sum_{h=f_1}^{f_2} |Y_h^W| + \sum_{h=f_1}^{f_2} |Z_h^W|,\quad f_1 = 0.01, f_2 = 0.68$
Means of the vertical axis	$\bar{x}^W = \frac{1}{N}\sum_{i=1}^{N} x_i^w$
y axis and x axis difference	$\overline{yx}^w = \left(y'^W - x'^W\right) - \left(y'^{W-1} - x'^{W-1}\right)$, where $\bar{x}'^w = \bar{x}^w - \bar{x}^{w-1}$ and $\bar{y}'^w = \bar{y}^w - \bar{y}^{w-1}$
y axis and z axis difference	$\overline{yz}^w = \left(z'^W - y'^W\right) - \left(z'^{W-1} - y'^{W-1}\right)$, where $\bar{z}'^w = z^W - z^{W-1}$

As well as in the previous classifier, these features are trained with a SVM with a RBF kernel. The model is the one that maximizes the accuracy obtained for different hyper-parameters C and γ.

2.3 Barometer-Based Classifiers (C3 and C4)

One of the inconveniences of the PT classifiers has been the complexity for distinguishing *SiSt* and *StSi* transitions [16], which implies the use of several features and many computational resources. In [16], 13 features and 23 support vectors are employed to distinguish a SiSt from a StSi, while in this problem only one feature is needed and a threshold-based classifier is proposed. *SiSt* and *StSi* inertial signals yield very similar features, leading to a very complex classifier with several parameters to compute [9]. We propose to solve this issue with the use of a MEMS barometer with the aim of simplifying the algorithm.

MEMS barometers provide reliable values of the atmospheric pressure. Dramatic changes in pressure can be found by comparing the density of air at sea level compared to the top of a mountain [17]. However, in short distances, it is also possible to detect small change in altitude in the barometric signal [10]. As Massé et al. explained, in order to detect a *StSi* or *SiSt* transition, the barometer requires to have a resolution of 0.061 mbar.

In this work, we are using the BMP280 sensor, which has a resolution of 0.0016 mbar with a pressure noise of 0.0013 mbar, which is enough for detecting such small changes. However, barometers are sensitive to many external agents, such as temperature, although BMP280 includes a temperature compensation algorithm and it does not affect to the measures obtained in the algorithm [18]. Barometers are also sensitives to significant atmospheric pressure changes such as opening or closing doors since pressures could be different between adjacent rooms. Thus, once we know that a given window contains either a *StSi* or a *SiSt* transition, barometers are essential to easily identify the direction of the transition. The same occurs with *falls* and *LySt* transitions.

C3 and C4 classifiers are set to depend on a single feature, given that barometers provide clear signals for distinguishing PT's. In consequence, instead of using machine learning classifiers, a threshold is used to identify PT's. We propose to find these thresholds by maximizing the margin between the two classes by means of a linear kernel SVM.

The selected feature for both classifiers (C3 and C4) is the slope of a line fitted through linear regression on the values of the barometric pressure sensor at the current window W. Let be, then, r the slope obtained from the regression line obtained from BMP_1^N values. Then, within the C4 classifier, if $r > STH$, we consider the window to be a *StSi* since the barometric pressure increases, denoting that there is more air above the sensor at the end of the window than at the beginning. In contrast, when $r < STH$, the barometric pressure decreases during the window leading to an increase of the altitude, denoting a SiSt transition.

In the case of the C3 classifier we use the same reasoning to determine whether a *Fall* or a *LySt* took place, since the main feature is a change of altitude. Then, if $r < FTH$, a *LySt* is determined; oppositely, if $r > FTH$, a *Fall* is established.

3 Data Collection

Data base has been collected within the Technical Research Centre for Dependency Care and Autonomous Living. The database is composed of triaxial acceleration and atmospheric pressure signals obtained from 14 healthy users (6 female/8 male) from 21 to 60 years old (Average: 34.36 ± 9.5 years old). The protocol test that each user performed is composed of 9 different simulated falls, 6 *SiSt* and 6 *StSi* transitions on 3 different seats: mattress (28 cm height), office chair with wheels (43 cm chair) and standard chair (45 cm height). Users also walked for several seconds during two bouts. Going upstairs and downstairs by walking were also included, as well as using the elevator. The protocol took around 10 min to each user Table 2.

Table 2. Set of activities from the database protocol test

Database protocol test	
Front fall	$2 \times StSi$ and *SiSt* (on a sofa)
Back fall	$2 \times StSi$ and *SiSt* (on a static chair)
Left side with turn fall	$2 \times StSi$ and *SiSt* (on an office chair)
Right side with turn fall	Walk 20 m
Left side fall	Go upstairs
Right side fall	Go downstairs
Fall after wall shock	Use elevator (up one floor)
Front fall with knees	Use elevator (down one floor)
Fall from bed	Walk 20 m

The protocol was captured with a new version of the 9×2 sensor [19], which incorporates an accelerometer (LSM9DS0) and the BMP280 barometer, which has an accuracy of 0.12 mbar, resolution of 0.0016 mbar with a pressure noise of 0.0013 mbar. The system stores the inertial data into a microSD card in order to be treated afterwards. All the tests were video recorded in order to have a gold standard, which was obtained by labelling the videos.

4 Results and Discussions

As shown in Fig. 2, each of the classifiers was trained individually with data from 70% of the users and, then, the structure of classifiers was executed with the data from the remaining 30% of users. In this section, first, the results obtained in the 10-cross fold-validation from the four classifiers are presented. Then, the results of the hierarchical structure of classifiers are shown.

A total of 126 simulated *falls*, 84 *StSi*, 126 *LySt*, 84 *SiSt* were obtained in the data collection. From the 14 users, 11 (randomly selected) have been employed to train and the remaining 3 users to evaluate. This process has been executed 10 times to prevent specific training/test divisions from biasing the results. Results from the 10-fold Cross Validation are shown in Table 3. The four classifiers show accuracies above 95%.

Table 3. Average 10-fold cross-validation results among 10 random selections of training and testing datasets

Classifier	Average accuracy
C1 (*PT or Fall*/Other activity)	97.5%
C2 (*Fall* or LySt/*StSi-SiSt*)	95.1%
C3 (*Fall/LySt*)	98.6%
C4 (*StSi/SiSt*)	100%

The evaluation of the algorithm with the testing data was performed following the rules presented in [8]. That is, instead of evaluating windows, each activity described in the test protocol was considered as an evaluation unit, regardless its length. For example, when evaluating the detection of *StSi* transitions, a true positive is considered to be a *StSi* transition correctly detected; a true negative is defined as any other activity (e.g. *Stand*) that was rejected correctly. A false positive is defined as an activity that is detected as the activity under analysis (e.g. *StSi* transition) but corresponds to another type. Finally, false negative are activities of the type under analysis that were missed by the classifier. This way, true negatives were balanced in contrast to a classical window analysis, where number of true negatives are overestimated, which leads to also overestimating specificity. During the evaluation process, it has been observed that most users stood up from the lying position in the same way that they stood up from sitting. This was obtained because *LySt* were preceded by *falls*, and users fell over a mattress. In consequence, given that *LySt* transitions and *SiSt* were performed in an extremely similar way, their labels have been merged. Results from the evaluation process are shown in Table 4.

As shown in Table 4 on the left, results are very high in the collected datasets since most measurements are above 95%. However, it can be observed that sensitivity of *falls* as well as specificity of *StSi* are under 90%. This is due to the confusion gotten in the evaluation of two of the falls. Left and right side falls described in the test protocol imply that the user hits a wall with her/his left shoulder and then the user slides through the wall to the ground. These falls always end with the patient sitting on the ground and, in few cases, the patient ends lying on the ground. Furthermore, this fall is executed slowly since the slide through the wall provokes a friction that impedes falling abruptly, as occurs with the other falls. This movement is extremely similar to a *StSi* transition and, thus, results of these states are not so high. Given this rationale, we can confirm the good performance of the posture detection algorithm.

On the right of Table 4, we can observe the evaluation with a matrix confusion involving the barometer-dependent classifiers and excluding the rest of activities. In

Table 4. Evaluation results from the testing dataset. On the left, results' average along the 10 repetitions performed on the process depicted in Fig. 2. Results of the hierarchical structure of classifiers for each final state are presented. On the right, the confusion matrix (window evaluation) of the classifiers where the barometer determines the state is shown. Results are the sum of states after having executed the evaluation algorithm 10 times.

Evaluation results		Predicted	
Sensitivity *Falls*	0.866	*Fall*	*LySt*
Specificity *Falls*	0.989	*Fall*	303	1
Sensitivity *LySt*	0.974	*LySt*	0	391
Specificity *LySt*	0.970	Predicted	
Sensitivity *Stand to Sit*	0.971	*Stand to Sit*	*Sit to Stand*
Specificity *Stand to Sit*	0.834	*Stand to Sit*	255	1
Sensitivity *Sit to Stand*	0.974	*Sit to Stand*	2	207
Specificity *Sit to Stand*	0.970		

this case, we have used the window evaluation since all the activities have the same duration and can be evaluated in this way. Results show that barometer can almost perfectly determine an event by analysing the behaviour of its signal in a PT detection.

5 Conclusions

The improvement of new technologies enables the non-invasive monitoring by means of wearable inertial systems. One of the most employed are accelerometers but, in the detection of PT such as *SiSt* and *StSi*, the performance is not very high and classifiers with a high computational burden are necessary to identify these transitions with acceptable results [9]. In this work, we introduce the barometer in order to contextualize the information of the accelerometer and provide, with a simple algorithm, an excellent performance. Results shown prove that barometers are a promising tool to identify PT's including falls in which a changes of altitude are observed. Although the use of barometers might benefit in many applications, this algorithm is expected to be useful to contextualize FoG within Parkinson's Disease, since, as shown in [9], similar pattern are found between PD patients and healthy users. This way, future work should focus on providing an accurate monitoring of the daily living activities with the aim of contextualizing the Parkinson's disease symptomatology.

Acknowledgements. Part of this project has been performed within the framework of the MASPARK project which is funded by La Fundació La Marató de TV3 20140431. The authors, thus, would like to acknowledge the contributions of their colleagues from MASPARK consortium.

References

1. Salarian, A., Russmann, H., Vingerhoets, F.J.G., Burkhard, P.R., Aminian, K.: Ambulatory monitoring of physical activities in patients with Parkinson's disease. IEEE Trans. Biomed. Eng. **54**, 2296–2299 (2007)
2. Pérez-López, C., Samà, A., Rodríguez-Martín, D., Moreno-Aróstegui, J.M., Cabestany, J., Bayes, A., Mestre, B., Alcaine, S., Quispe, P., Laighin, G.Ó., Sweeney, D., Quinlan, L.R., Counihan, T.J., Browne, P., Annicchiarico, R., Costa, A., Lewy, H., Rodríguez-Molinero, A.: Dopaminergic-induced dyskinesia assessment based on a single belt-worn accelerometer. Artif. Intell. Med. **67**, 47–56 (2016)
3. Salarian, A., Russmann, H., Wider, C., Burkhard, P.R., Vingerhoets, F.J.G., Aminian, K.: Quantification of tremor and bradykinesia in Parkinson's disease using a novel ambulatory monitoring system. IEEE Trans. Biomed. Eng. **54**, 313–322 (2007)
4. Pérez-López, C., Samà, A., Rodríguez-Martín, D., Català, A., Cabestany, J., Moreno-Arostegui, J., de Mingo, E., Rodríguez-Molinero, A.: Assessing motor fluctuations in Parkinson's disease patients based on a single inertial sensor. Sensors. **16**, 2132 (2016)
5. Cheng, P.-T., Liaw, M.-Y., Wong, M.-K., Tang, F.-T., Lee, M.-Y., Lin, P.-S.: The sit-to-stand movement in stroke patients and its correlation with falling. Arch. Phys. Med. Rehabil. **79**, 1043–1046 (1998)
6. Schaafsma, J.D., Balash, Y., Gurevich, T., Bartels, A.L., Hausdorff, J.M., Giladi, N.: Characterization of freezing of gait subtypes and the response of each to levodopa in Parkinson's disease. Eur. J. Neurol. **10**, 391–398 (2003)
7. Bartels, A., Balash, Y., Gurevich, T., Schaafsma, J., Hausdorff, J., Giladi, N.: Relationship between freezing of gait (FOG) and other features of Parkinson's: FOG is not correlated with bradykinesia. J. Clin. Neurosci. **10**, 584–588 (2003)
8. Rodríguez-Martín, D., Samà, A., Pérez-López, C., Català, A., Cabestany, J., Rodríguez-Molinero, A.: Posture detection with waist-worn accelerometer: an application to improve freezing of Gait detection in Parkinson' s disease patients. In: Chen, W., Augusto, J.C., Seoane, F., Lehocki, F., Wolf, K.-H., Arends, J., Ungureanu, C., Wichert, R. (eds.) Recent Advances in Ambient Assisted Living – Bridging Assistive Technologies, e-Health and Personalized Health Care, pp. 3–17 (2015)
9. Rodriguez-Martin, D., Samà, A., Perez-Lopez, C., Català, A., Cabestany, J., Rodriguez-Molinero, A.: SVM-based posture identification with a single waist-located triaxial accelerometer. Expert Syst. Appl. **40**, 7203–7211 (2013)
10. Massé, F., Bourke, A.K., Chardonnens, J., Paraschiv-Ionescu, A., Aminian, K.: Suitability of commercial barometric pressure sensors to distinguish sitting and standing activities for wearable monitoring. Med. Eng. Phys. **36**, 739–744 (2014)
11. Tolkiehn, M., Atallah, L., Lo, B., Yang, G.-Z.: Direction sensitive fall detection using a triaxial accelerometer and a barometric pressure sensor. In: 33rd Annual International Conference of the IEEE Engineering in Medicine and Biology Society, pp. 369–372 (2011)
12. Bianchi, F., Redmond, S.J., Narayanan, M.R., Cerutti, S., Lovell, N.H.: Barometric pressure and triaxial accelerometry-based falls event detection. IEEE Trans. Neural Syst. Rehabil. Eng. **18**, 619–627 (2010)
13. Moncada-Torres, A., Leuenberger, K., Gonzenbach, R., Luft, A., Gassert, R.: Activity classification based on inertial and barometric pressure sensors at different anatomical locations. Physiol. Meas. **35**, 1245–1263 (2014)
14. Bloem, B.R., Hausdorff, J.M., Visser, J.E., Giladi, N.: Falls and freezing of gait in Parkinson's disease: a review of two interconnected, episodic phenomena. Mov. Disord. **19**, 871–884 (2004)

15. MASPARK: Freezing in Parkinson's disease: improving quality of life with an automatic control system. La Marató de TV3-20140431 (2015–2017)
16. Rodríguez-Martín, D., Samà, A., Pérez-López, C., Cabestany, J., Català, A., Rodríguez-Molinero, A.: Posture transition identification on PD patients through a SVM-based technique and a single waist-worn accelerometer. Neurocomputing **164**, 144–153 (2015)
17. West, J.B., Lahiri, S., Maret, K.H., Peters, R.M., Pizzo, C.J.: Barometric pressures at extreme altitudes on Mt. Everest: physiological significance. J. Appl. Physiol. **54**, 1188–1194 (1983)
18. Robert Bosch GmbH: Datasheet BMP280 digital pressure sensor. https://www.bosch-sensortec.com/bst/products/all_products/bmp280
19. Rodríguez-Martín, D., Pérez-López, C., Samà, A., Cabestany, J., Català, A.: A wearable inertial measurement unit for long-term monitoring in the dependency care area. Sensors **13**, 14079–14104 (2013)

Deep Learning for Detecting Freezing of Gait Episodes in Parkinson's Disease Based on Accelerometers

Julià Camps[1]([✉]), Albert Samà[1], Mario Martín[2], Daniel Rodríguez-Martín[1],
Carlos Pérez-López[1], Sheila Alcaine[3], Berta Mestre[3], Anna Prats[3],
M. Cruz Crespo[3], Joan Cabestany[1], Àngels Bayés[3], and Andreu Català[1]

[1] Technical Research Centre for Dependency Care and Autonomous Living,
CETPD, Universitat Politècnica de Catalunya, Barcelona Tech.,
Rambla de l'Exposició 59-69, 08800 Vilanova i la Geltrú, Spain
`julia.camps.sereix@est.fib.upc.edu`
[2] Knowledge Engineering and Machine Learning Group,
Universitat Politècnica de Catalunya, Barcelona Tech.,
C/ Jordi Girona 1-3, 08034 Barcelona, Spain
[3] Unidad de Parkinson y trastornos del movimiento (UParkinson), Barcelona, Spain

Abstract. Freezing of gait (FOG) is one of the most incapacitating symptoms among the motor alterations of Parkinson's disease (PD). Manifesting FOG episodes reduce patients' quality of life and their autonomy to perform daily living activities, while it may provoke falls. Accurate ambulatory FOG assessment would enable non-pharmacologic support based on cues and would provide relevant information to neurologists on the disease evolution.

This paper presents a method for FOG detection based on deep learning and signal processing techniques. This is, to the best of our knowledge, the first time that FOG detection is addressed with deep learning. The evaluation of the model has been done based on the data from 15 PD patients who manifested FOG. An inertial measurement unit placed at the left side of the waist recorded tri-axial accelerometer, gyroscope and magnetometer signals. Our approach achieved comparable results to the state-of-the-art, reaching validation performances of 88.6% and 78% for sensitivity and specificity respectively.

Keywords: Freezing of gait · Parkinson's disease · Deep learning ·
Signal processing · Inertial measurement unit

1 Introduction

Parkinson's disease (PD), with a prevalence of approximately 1% among people of age above 65, is the second most common neurodegenerative disorder [15,16,23,26]. PD patients manifest several motor and non-motor symptoms. Accurate automatic symptoms detection in PD patients' provides relevant indicators about their condition [4]. Clinicians by disposing of these indicators can

© Springer International Publishing AG 2017
I. Rojas et al. (Eds.): IWANN 2017, Part II, LNCS 10306, pp. 344–355, 2017.
DOI: 10.1007/978-3-319-59147-6_30

maintain an updated assessment over the patient's regimen, which permits to improve the patient's quality of life.

Freezing of gait (FOG) is a symptom associated with PD condition. FOG is usually manifested in episodes shorter than 10 s [20]. Suffering FOG episodes may provoke falling accidents when patients are willing to perform walking related actions [3]. According to Nieuwboer and Giladi [14], FOG might be defined as an inability to deal with concurrent cognitive, limbic, and motor inputs, causing an interruption of locomotion. Many medical research studies have been carried out to discover strategies to combat this symptom. These studies have proved that techniques such as to induce an auditory or visual external a rhythm stimulus to PD patients to improve their walking capacity while minimising FOG's incisive frequency [2]. Automatic FOG monitoring would permit to provide online support to patients through rhythmic auditory cues, which may significantly enhance the patients' autonomy during their activities of daily living (ADL) [2,28].

Recently, with the increase of computing power of small devices and the adoption of wearable sensors for biomedical research, wearable sensors are increasingly becoming a common practice for detecting motor symptoms in PD patients within the research community. The state-of-the-art on algorithms for automatic FOG detection is shallow machine learning (ML) algorithms applied to signals acquired from inertial wearable sensors [12,13,18]. The state-of-the-art performance for FOG detection is defined by performances within the range [85%, 95%] for the geometric mean (GM) between sensitivity and specificity. However, the complexity in designing handcrafted features and the scarcity of data from PD patients collected under real-life-like conditions for developing reliable solutions for monitoring FOG in naturalistic environments, are the major impediments preventing the research community from mastering the problem.

Feature learning is a set of techniques that learns a transformation of raw data input to a representation that can be exploited by ML methods. Deep learning (DL) methods are feature learning methods with multiple levels of representation. DL models can learn feature extractions that can easily handle multimodal data, missing information and high dimensional feature spaces. Thus, when working with DL methods, the manual feature engineering can be obviated, which is otherwise necessary for traditional ML methods. Furthermore, DL models can outperform shallow ML algorithms when enough data to represent the complexity of a target problem are provided adequately.

To the best of our knowledge, this is the first paper to present a DL model for addressing FOG detection in PD patients from inertial sensors data. Our approach implements an 8-layer convolutional neural network (ConvNet), which is composed of: 5 convolutional layers, 2 dense (i.e. fully-connected) layers and an output layer. In the experiments performed, this model achieved comparable performances to the state-of-the-art. Concretely, the presented models were able to achieve performances of 88.6% and 78% for sensitivity and specificity respectively. The main contribution of this study is to serve as a basis from which designing DL models capable of mastering FOG detection from inertial data collected in naturalistic environments.

2   Related Work

Automatic FOG detection is an open research issue which has been widely addressed by several combinations of devices and algorithms. This section reviews some of these approaches.

In 2008, Moore et al. [13] presented the Moore-Bächlin FOG Algorithm (MBFA), a novel method for automatic FOG detection in PD patients that manifest FOG. The MBFA mainly consists of a freeze index (FI) threshold, where FI is defined as the ratio between the power spectral density in the gait freezing band (i.e. 3–8 Hz) and in the locomotion band (i.e. 0.5–3 Hz). This approach was able to achieve highly accurate results, considering the simplicity of the method. Concretely, they were able to detect 78% of FOG events correctly. These results were retrieved by a general threshold across all patients data. However, they also proved that calibrating the threshold to each patient increased the method's performance about an 11%.

In 2012, Mazilu et al. [12] presented a novel approach for monitoring FOG in PD patients, which combines the usage of smartphones and wearable accelerometers as devices, while using, for the first time, ML algorithms for the online FOG detection task. Some of the ML algorithms they tested were: random forests (RF), decision trees (C4.5), naive Bayes (NB) and k-nearest neighbours (k-NN). They reported top results of 66.25% and 95.38% for sensitivity and specificity, respectively, using user-independent settings.

In 2016, Rodríguez et al. [18] presented a study aiming at FOG detection in PD patients during their ADL, and adopting the support vector machines (SVMs) for the FOG bi-classification task. They proposed an innovative feature extraction which is designed to be implementable in low-power consumption wearables for online FOG detection. The data, which was composed of inertial signal recordings at 40 Hz from a single inertial measurement unit (IMU) placed at the left side of the waist, was acquired following the same conventions reviewed in [13]. However, laboratory data acquisition biases the data with information related to the experiment characteristics as in [12,13]. Thus, they collected the data at the patients' homes, configuring each test to adapt to the real activities in which the patient would experience FOG, rather than employing homogeneous lab settings to force patients to trigger FOG events. Although they have not reported test error results in this work, they performed a comparative study of the state-of-the-art feature extraction techniques for FOG detection, while reporting cross-validation error when training a model for each combination of ML algorithm (e.g. k-NN, RF, NB and SVM) and feature extraction strategy. Furthermore, they considered different window sizes (i.e. ranging from 0.8 to 6.4 s) to maximise the representation power of each configuration. Their results suggested that SVMs with their proposed feature generation are powerful strategies for FOG detection since the cross-validation performance for this configurations were the most accurate among all regardless of the window size. Concretely, highest results were achieved by using a window duration of 1.6 s, for which they reported 89.77% as the GM between sensitivity and specificity.

Recently, in 2017, Rodríguez et al. [19] presented an extension of [18] and [17]. In this study, they showed performance results for the GM between sensitivity and specificity of 76%, which was determined by using an episode-based evaluation strategy, instead of the window-based strategy.

3 Deep Learning for FOG Detection

3.1 Feature Extraction with ConvNets

A ConvNet is a type of feed-forward deep neural network, which typically combines convolutional layers with traditional dense layers to reduce the number of weights composing the model. Convolutional layers enforce local connectivity between neurones of adjacent layers to exploit spatially local correlation. Concretely, convolutional layers are formed by kernels that share weights and, thus, permit to learn position invariant features from the input data.

Therefore, convolutional layers can extract features from data that have underlying spatial or temporal patterns, such as images or signal data. Furthermore, stacking these layers permits to extract progressively more abstract patterns.

While traditional DL models are composed of stacked dense layers, which lead to an overwhelming number of weights, ConvNets implement a powerful and efficient alternative if the target data present underlying spatial patterns.

3.2 Architecture

The presented approach is a one-dimensional ConvNet, which is described as

$$C(16|3) - C(16|3) - C(16|3) - C(16|3) - C(16|3) - D(32) - D(32) - L$$

where $C(x|y)$ corresponds to a convolutional layer of x kernels of length y, $D(z)$ corresponds to a dense layer of z neurons, and L is the last layer of the network.

FOG events detection was treated as a bi-classification task, such that FOG instances were labelled as positive values (i.e. 1) whereas non-FOG instances were labelled as negative values (i.e. -1). The models, thus, should be able to retrieve negative and positive values. A linear function with L2 weight regularisation penalty coefficient set to 0.01 was, thus, implemented as the activation function of the last layer, which was set to a dense layer of 1 neurone.

3.3 Data Representation and Augmentation

A common practice to enhance ML models' training quality is to normalise the data. Data were, thus, normalised by the precomputed sample standard deviation from the overall training dataset.

The most common technique to deal with classification tasks in time-series data is to use a windowing strategy. Windowing consists of splitting the data into equally-sized consecutive parts to address the classification task window-wisely instead of instance-wisely. The classification task was, therefore, addressed window-wisely.

Data augmentation techniques permit to increase the knowledge extracted from data by performing replications on the data that are consistent with the task's domain. Data augmentation strategies implemented were:

– To randomly shift the window starting points.
– To rotate each windowed signal by simulating a rotation on the waist-sensor through a rotation matrix generated by sampling angles (see Fig. 1 for the axis reference) over a distribution defined by the following ranges: x-axis: $[-30°, +30°]$; y-axis: $[-45°, +45°]$; and z-axis: $[-10°, 10°]$; which were designed to resemble naturally introduced rotations due to the patient's waist form and movements.

DL models are powerful feature extractors, however, if being provided with insufficient information, these models would fail to solve the classification task. Rodríguez et al. [19] performs a feature extraction which is computed from the current and previous windows. This window transitional information usage suggested that data from one single window could be insufficient to succeed in the FOG detection task. To provide the DL models with a sample representation from which they could learn which part of sample data belonged to the current window and which to the transition between windows, a novel strategy was implemented which is hereinafter referred as stacking. The stacking strategy can be seen as a function which outputs the window to fed to the model from the current and the previous window data plus a value for the stacking parameter p; thus, $S_n = stacking(W_n, W_{n-1}, p)$ where: S_n refers to the nth stacked window used for feeding the model; W_n and W_{n-1} refer to the current and previous windowed data, respectively; and p is a trade-off parameter of the $stacking(W_n, W_{n-1}, p)$ function, which is hereafter defined:

1. The previous and the current windows are split into p equally-sized parts.
2. Each ith part of the current window is paired with the $(i-1)^{th}$ part, even if this one belongs to previous or to the same window, in which case that part is replicated. The parts from the previous window which were not paired are removed.
3. Each remaining part is transformed by applying the fast Fourier transform (FFT).
4. Each part of the current window is complemented by its predecessor part by just concatenating the columns of the current part to the difference between both parts, which produces a new extended part composed of 18 columns (i.e. 9 of the current window + 9 of the predecessor part of this concrete part).
5. Finally all parts are re-stacked together maintaining the initial temporal order.

4   Experiments

4.1   Data Collection

The dataset [18] employed is composed of inertial signals from 15 PD patients recorded by a single IMU placed at the left side of the patient's waist, as shown in Fig. 1. This IMU generated 9 signals sampled at 200 Hz as output. The 9 signals represented the measurements of 3 tri-axial sensors: gyroscope, accelerometer and magnetometer.

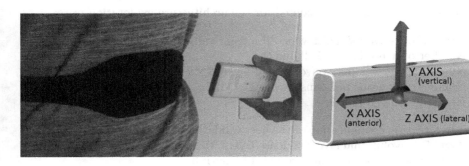

Fig. 1. The data collector device and its location on body [18].

The data collection was performed within the scope of the Freezing in Parkinson's Disease: Improving Quality of Life with an Automatic Control System (MASPARK)[1] project. Inclusion criteria were: (I) being diagnosed with PD according to the UK Brain Bank; (II) having Hoehn & Yahr stage above 2 in OFF state; (III) not having dementia according to DSM-IV criteria; (IV) and giving their written informed consent for using the collected data in the research carried out. All data was, furthermore, gathered at the patients' homes to increase the resemblance of targeting the same problems in real-life environments. The patients performed a set of activities such as showing their place and carrying an object from one room to another. These activities were afterwards labelled by clinicians relying only on the video recording. The data collection protocol included some activities which were specially introduced to increase the difficulty of the FOG detection task. Concretely, adding data from activities such as brushing their teeth, painting and erasing in a sheet of paper, was endeavoured to force models to learn robust representations of FOG.

4.2   Training and Tested Configurations

A complete exhaustive search of the hyperparameter space was infeasible due to computational and time constraints. The hyperparameters exploration was

---

[1] http://futur.upc.edu/15557508.

undertaken following an iterative semi-heuristic strategy formed by several stages, in each of which some hyperparameters were set to the optimal configurations tested; thus reducing significantly the combinations of configurations to consider. Note that the adoption of this strategy implies that the presented approaches constitute only local minima over the overall hyperparameters space considered.

The model's hyperparameters are hereafter described while discussing the ranges of values contemplated and the suboptimal configurations selected for each of them.

Architecture. The values considered for the architecture exploration were:

- Number of convolutional layers: all integer values within the range $[2, 10]$.
- Number of dense layers: 1, 2 and 3.
- Number of kernels per convolutional layer: 8, 16, 32, 64, 128, 256 and 512.
- Number of neurons per dense layer: 16, 32, 64, 128, 256, 512 and 1024.
- Kernel lengths: 3 and 5.

The architecture configuration of the best validation models trained were defined as: 4 and 5 convolutional layers, 2 dense layers, 16 kernels per convolutional layer, 32 neurons per dense layer and kernels of length 3.

Data Augmentation. Data augmentation parameters that determine the number of shifts and rotations to be performed were set to:

- Number of shifts = 4 (data is always shifted).
- Number of added rotations = 1 (original data was included).

From the augmentation process, the amount of training data was, thus, 8 times higher.

Stacking Parameter p. Values considered for the p stacking parameter were: 1, 2, 3, 4 and 8. Finally, p was set to values 1, 2 and 3.

Sampling Frequency and Window Duration. Rodríguez et al. [18] performed successfully FOG detection on a subset of the same dataset here considered by adopting a subsampling frequency of 40 Hz and window duration of 3.2 s This previous study suggested, thus, that that windows of at least 3.2 s and a sampling frequency at least of 40 Hz were sufficient values to address the detection task. Therefore, the values tested were:

- Sampling frequency (Hz): 40, 50, 100 and 200.
- Window duration (s): 2.56, 3, 3.2, 5.12 and 10.24.

Finally, these hyperparameters were set to 100 Hz and 2.56 s for sampling frequency and window duration respectively.

Batch Size. DL models are trained batch-wisely instead of sample-wisely, being each sample composed by a windowed group of instances already processed (i.e. normalised and stacked). A batch is a fraction of the overall processed training data to be fed into the DL model, which is generated from splitting these data into equally sized parts. Then, DL models compute the gradients and perform the weight corrections batch-wisely; thus, applying fewer corrections during the training process.

According to Goodfellow et al. [7], generalisation error is often best for a batch size of 1. However, this strategy is time-consuming. Batch sizes tested, thus, were: 1, 8, 16, 32, 64 and 128. From these values, 16 was selected.

Number of Epochs. DL models are trained iteratively with the aim of approaching relevant local optima on the neurones' weights space to successfully represent the target problem. These iterations are denoted as epochs. Correctly establishing the number of epochs is important to avoid useless computation when the model has converged while preventing overfitting. The number of epochs by which a model may reach convergence will usually be correlated to several other characteristics, such as the model's architecture, the data, the optimisation method and its internal parameters (e.g. learning rate), and the regularisation strategies being employed. The hyperparameters process performed kept the models' architectures and the dropout indexes as variable throughout the overall study. An early-stopping strategy was, thus, implemented instead of fixing the number of epochs.

Activations. The most widely exploited activation function for DL methods is the rectified linear unit (ReLU) [11,22]. Besides, training DL models with ReLU activations increases the training time efficiency significantly, which is a major bottleneck when working with DL techniques. Therefore, all activations, except for the activation function of the *last layer*, are set to ReLU functions.

Error Loss. FOG detection is addressed as a binary classification tasks. Hinge loss algorithm is a loss error method specialised for bi-classification problems, which is defined as:

$$L_{hinge} = mean(max(0, 1 - \boldsymbol{y}_{true} * \boldsymbol{y}_{pred})) \tag{1}$$

where $\boldsymbol{y}_{true}$ are the real labels of the data, which can either be -1 or 1, while $\boldsymbol{y}_{pred}$ are the model label predictions which can adopt real values in the range $[-\infty, +\infty]$.

However, it was noted that the class imbalance in the training data was preventing the model from learning strong FOG representations. Therefore, the weighted version of the hinge loss function, which is hereafter defined, was implemented in the final models.

$$\boldsymbol{L}_{w-hinge}^{+} = max(0, 1 - \boldsymbol{y}_{true}^{+} * \boldsymbol{y}_{pred}) * P_{no-FOG}$$
$$\boldsymbol{L}_{w-hinge}^{-} = max(0, 1 - \boldsymbol{y}_{true}^{-} * \boldsymbol{y}_{pred}) * P_{FOG}$$
$$L_{w-hinge} = mean(\boldsymbol{L}_{w-hinge}^{+} + \boldsymbol{L}_{w-hinge}^{-})$$

where $\boldsymbol{y}_{true}^{+}$ defines the real positive samples in the data while $\boldsymbol{y}_{true}^{-}$ defines the negative ones, P_{FOG} is the prior of FOG samples in the training data (i.e. the percentage of fog samples in the training dataset) and P_{no-FOG} is the prior of non-FOG samples in the training data.

Optimizer. In DL models the loss optimisation responsibility is delegated to a stochastic gradient-based algorithm. The algorithms considered were: root mean square propagation (RMSProp) [24]; AdaDelta [29], which is an extension of adaptive gradient algorithm (AdaGrad) [5]; and adaptive momentum (Adam) [9]. However, no significant difference was found from the initial results between the tested algorithms. Finally, it was decided to select Adam, due to taking into account the momentum which adds robustness to the gradients and lowers the effect of outlier batches on the weight updates [9]. Moreover, Adam is the most widely implemented of all algorithms considered [1,8,10,25,27]. Therefore, the models presented were trained via backpropagation and Adam algorithm as the optimisation method.

Learning Rate. In DL models, the learning rate is a parameter of the optimizer algorithm which indicates the size of the step to be applied when correcting the model's weights by the newly computed gradient. Finally, the tested values for the learning rate were within the range $[5 \cdot 10^{-3}, 5 \cdot 10^{-5}]$.

Weight Initialization. Although being a crucial hyperparameter for successfully training DL models, it has been stated that several initializations will usually allow a DL model to train in a proper way [7]. The weight initialization strategy was, thus, set to the method presented by Glorot et al. [6], which, indeed, permitted to train our models successfully.

Regularization. Models trained initially were prone to overfit on the training data; thus, dropout strategies were implemented after every layer of the models' [21]. Dropout indexes considered were: 0, 0.1, 0.15, 0.2, 0.25 and 0.5.

5   Results

Table 1 presents the models' designs that achieved the highest validation GM values between sensitivity and specificity. From the top configuration results, it was observed that only models trained with windows of 2.56 s achieved validation GM above 80%.

From Table 1 it can be observed that the best models' configurations presented are defined by 4 and 5 convolutional layers, each of which was composed of 16 kernels of length 3; 2 dense layers of 32 neurons; a dropout measure within the range $[0.1, 0.25]$; stacking p parameter equals 1 and 2; and the remaining common characteristics for all trained models which are commented in Sect. 4.2.

The performance of ML algorithms is usually compared using benchmarking datasets. Rodríguez et al. [18] replicate other authors' feature extraction strategies

Table 1. Top validation models. Columns naming description: '# c-layers' → number of convolutional layers; '# kernels' → number of kernels per layer; '# d-neuron' → number of neurons of each dense layer; 'stacking' → number of parts that each window was split before re-stacking (p stacking parameter); 'dropout' → dropout index; 'val_sen' and 'val_spe' → validation values of sensitivity and specificity.

# c-layer	# kernel	# d-neuron	stacking	dropout	val_sen	val_spe
5	16	32	1	0.25	86.5%	79.2%
4	16	32	1	0.10	80.2%	83.8%
4	16	32	1	0.20	82.8%	81.4%
4	32	64	2	0.20	85.2%	78.1%
5	16	32	2	0.20	85.2%	78.4%
4	16	32	2	0.20	78.3%	85.0%

to compare a novel SVM-based method on 6 patients' data to the state-of-the-art for FOG detection. These data matches to a subset of the data considered here. Thus, our approach was compared to results reported by Rodríguez et al. [18]. Hence the results table in Rodríguez et al. [18] was adapted to include our approach, producing Table 2. However, Rodríguez et al. [18] perform a 10-fold cross-validation over 6 patients' data, while our results from Table 1 were produced from 4 validation patients which were never used for training the models. Note that this disadvantage advocates our approach.

Table 2. State-of-the-art for FOG detection evaluated on a subset of 6 patients' according to [18]. Values illustrated correspond to the best validation GM result reported in [18] for that ML algorithm, while in column 'DL' shows the top validation GM achieved.

	k-NN	RF	NB	SVM	DL
Rodriguez et al. [18]	83.71	84.36	79.99	89.77	-
Mazilu et al. [18]	81.85	82.9	82.08	84.24	-
Tripoliti et al. [18]	75.04	62.25	79.15	79.57	-
Moore et al. [18]	71.07	70.46	71.39	78.43	-
FFT + stacking	-	-	-	-	82.1

Table 2 indicates that our approach exhibited comparable validation performance to the state-of-the-art results replicated by Rodríguez et al. [18] on data from the same distribution. However, some of the methods to which our approach was compared to, were specially designed for being implementable in real-time. These other methods are intended to be implementable in low-power consumption devices, while our approach requires significant memory and computational resources.

6   Conclusion

This paper is, to the best of our knowledge, the first study to present a method for FOG detection based on DL models. However, our approach was just a first attempt to tackle the FOG detection problem by employing DL techniques which achieved comparable results to the state-of-the-art methods referred from [18].

Interesting extensions of our approach are:

- To replace the last dense layers of the model by recurrent neural networks (RNNs) or long short term memory (LSTM) layers to enhance the current representation of temporal data.
- To implement personalization strategies, such as to retrain the model with partial information on each patient before evaluating on it. These techniques discern from the common practices adopted in the DL literature. However, they are usually considered by clinicians, who prioritise robustness of models over scalability.

Finally, our results suggest that exploring time-series endeavoured DL techniques (e.g. RNNs and LSTMs) could lead to outperforming the state-of-the-art for automatic FOG detection.

Acknowledgements. Part of this project was performed within the framework of the MASPARK project which is funded by La Fundació La Marató de TV3 20140431. The authors, thus, would like to acknowledge the contributions of their colleagues from MASPARK.

References

1. Ankit, K., et al.: Ask me anything: dynamic memory networks for natural language processing. CoRR abs/1506.07285 (2015)
2. Arias, P., et al.: Effect of rhythmic auditory stimulation on gait in parkinsonian patients with and without freezing of gait. PloS One **5**(3), e9675 (2010)
3. Bloem, B.R., et al.: Falls and freezing of gait in Parkinson's disease: a review of two interconnected, episodic phenomena. Mov. Disord. **19**(8), 871–884 (2004)
4. Del Din, S., et al.: Free-living monitoring of Parkinson's disease: lessons from the field. Mov. Disord. **31**(9), 1293–1313 (2016)
5. Duchi, J., Hazan, E., Singer, Y.: Adaptive subgradient methods for online learning and stochastic optimization. J. Mach. Learn. Res. **12**(Jul), 2121–2159 (2011)
6. Glorot, X., Bengio, Y.: Understanding the difficulty of training deep feedforward neural networks. In: Aistats, vol. 9, pp. 249–256 (2010)
7. Goodfellow, I., Bengio, Y., Courville, A.: Deep Learning. MIT Press, Cambridge (2016)
8. Gregor, K., et al.: Draw: a recurrent neural network for image generation. arXiv preprint arXiv:1502.04623 (2015)
9. Kingma, D.P., Ba, J.: Adam: a method for stochastic optimization. CoRR abs/1412.6980 (2014)
10. Kiros, R., et al.: Skip-thought vectors. In: 28th Advances in Neural Information Processing Systems, pp. 3294–3302 (2015)

11. Krizhevsky, A., et al.: Imagenet classification with deep convolutional neural networks. In: Advances in Neural Information Processing Systems, pp. 1097–1105 (2012)
12. Mazilu, S., et al.: Online detection of freezing of gait with smartphones and machine learning techniques. In: 2012 6th International PervasiveHealth and Workshops, pp. 123–130 (2012)
13. Moore, S.T., MacDougall, H.G., Ondo, W.G.: Ambulatory monitoring of freezing of gait in Parkinson's disease. J. Neurosci. Meth. **167**(2), 340–348 (2008)
14. Nieuwboer, A., Giladi, N.: Characterizing freezing of gait in Parkinson's disease: models of an episodic phenomenon. Mov. Disord. **28**(11), 1509–1519 (2013)
15. Nussbaum, R.L., Ellis, C.E.: Alzheimer's disease and Parkinson's disease. New Engl. J. Med. **348**(14), 1356–1364 (2003). pMID: 12672864
16. Pringsheim, T., et al.: The prevalence of Parkinson's disease: a systematic review and meta-analysis. Mov. Disord. **29**, 1583–1590 (2014)
17. Rodríguez-Martín, D., Samà, A., et al.: Posture detection based on a waist-worn accelerometer: an application to improve freezing of gait detection in Parkinson's disease patients. Recent Adv. Ambient Assist. Living-Bridging Assistive Technol. E-Health Personalized Health Care **20**, 3 (2015)
18. Rodríguez-Martín, D., Samà, A., et al.: Comparison of features, window sizes and classifiers in detecting freezing of gait in patients with parkinson's disease through a waist-worn accelerometer. In: Frontiers in Artificial Intelligence and Applications, vol. 288 (2016)
19. Rodríguez-Martín, D., Samà, A., Pérez-López, C., Català, A., Arostegui, J.M.M., Cabestany, J., Bayés, À., Alcaine, S., Mestre, B., Prats, A., et al.: Home detection of freezing of gait using support vector machines through a single waist-worn triaxial accelerometer. PLoS One **12**(2), e0171764 (2017)
20. Schaafsma, J.D., et al.: Characterization of freezing of gait subtypes and the response of each to levodopa in Parkinson's disease. Eur. J. Neurol. **10**(4), 391–398 (2003)
21. Srivastava, N., et al.: Dropout: a simple way to prevent neural networks from overfitting. J. Mach. Learn. Res. **15**(1), 1929–1958 (2014)
22. Szegedy, C., et al.: Going deeper with convolutions. In: CVPR (2015)
23. Tanner, C.M., Goldman, S.M.: Epidemiology of Parkinson's disease. Neurol. Clin. **14**(2), 317–335 (1996)
24. Tieleman, T., et al.: Lecture 6.5-rmsprop: divide the gradient by a running average of its recent magnitude. COURSERA: Neural Netw. Mach. Learn. 4 (2012)
25. Venugopalan, S., et al.: Sequence to sequence - video to text. In: The ICCV (2015)
26. WHO: Neurological disorders: public health challenges (2006)
27. Xu, K., et al.: Show, attend and tell: neural image caption generation with visual attention **2**(3), 5 (2015) arXiv preprint arXiv:1502.03044
28. Young, W.R., Shreve, L., Quinn, E.J., Craig, C., Bronte-Stewart, H.: Auditory cueing in Parkinson's patients with freezing of gait. What matters most: action-relevance or cue-continuity? Neuropsychologia **87**, 54–62 (2016)
29. Zeiler: ADADELTA: an adaptive learning rate method. CoRR abs/1212.5701(2012)

Presenting a Real-Time Activity-Based Bidirectional Framework for Improving Social Connectedness

Kadian Davis[1,3]($\boxtimes$), Evans Owusu[2], Geert van den Boomen[1],
Henk Apeldoorn[1], Lucio Marcenaro[3], Carlo Regazzoni[3], Loe Feijs[1],
and Jun Hu[1]

[1] Department of Industrial Design, Eindhoven University of Technology,
5612 AZ Eindhoven, The Netherlands
{k.a.davis,g.j.a.v.d.Boomen,h.apeldoorn,l.m.g.feijs,j.hu}@tue.nl
[2] Independent Researcher, Eindhoven, The Netherlands
owboateng@gmail.com
[3] DITEN - Università degli Studi di Genova, 16145 Genoa, Italy
{lucio.marcenaro,carlo.regazzoni}@unige.it

Abstract. New research on ambient displays within ambient assisted living (AAL) environments, demonstrates solid potential for the application of bidirectional activity-based context aware systems for promoting social connectedness between the elderly and their caregivers. Using visual, auditory or tactile modalities, such systems can reveal subtle information concerning health and well-being and stimulate co-presence between the pair. In this paper, we present the design and development of an activity-based framework aimed at enabling the real-time viewing of bidirectional activity states between the elderly and their caregivers. This framework seeks to overcome the limitations of existing ambient displays deployed in AAL settings, which are in most cases unidirectional and confined to the homes of its users. Our bidirectional activity-based framework, is based on an extensive literature review, expert advice and user feedback, which informed the design decisions about the product features and functionality. The system exploits a highly accurate activity recognition model to facilitate real-time activity awareness and an "always connected" service through portable interactive devices for stimulating social connectedness within the AAL domain.

Keywords: Ambient assisted living · Social connectedness · Human activity recognition · Ambient displays · Internet of Things

1 Introduction

On a global scale, there has been an unprecedented rise in human life expectancy [9]. This trend towards increased longevity could have profound social, political and economic implications for ageing societies. As a result of these challenges, policy makers, not-for-profit organizations and major industry partners have

© Springer International Publishing AG 2017
I. Rojas et al. (Eds.): IWANN 2017, Part II, LNCS 10306, pp. 356–367, 2017.
DOI: 10.1007/978-3-319-59147-6_31

sparked heightened interest in the development of cutting-edge interventions directed toward usability, accessibility, and safety, to provide social support, and enable older adults to maintain their independence and age in place.

To a great extent, previous ambient assisted living (AAL) technologies have explored remote monitoring, indoor positioning, and fall detection to inform caregivers and emergency services of abnormal events in the activities of daily living (ADLs) of older adults [1]. Recent AAL research corroborates the significance of the exchange of human activity information for improving social connectedness between the elderly and their caregivers [6,16]. However, the design, development, deployment and evaluation of such activity-based bidirectional systems is mostly left unexplored.

In the current era of connected devices, the combination of ubiquitous sensing technologies, the Internet and everyday objects could support older adults and their caregivers to maintain social connectivity despite separation by geographical distance. In particular, advanced human activity recognition (HAR) algorithms allow accurate detection of physical states while light's dynamic characteristics has shown potential for rendering activity information on the periphery of the users' attention to improve social connectedness [5,16].

In this paper, we present the design and development of a real-time activity-based bidirectional framework for improving social connectedness between the elderly and their caregivers. In subsequent sections, we give an overview of the related work, discuss our design rationale and provide a detailed description of the system's architecture. Finally, we discuss the limitations of the system and our plans for future work.

2   Related Work

Social connectedness refers to the "sense of belonging" or the "feeling of being in touch" and having enough meaningful social relationships [14]. Social support is a critical component of healthy and active ageing. While loneliness is common among all age groups, the elderly are more prone to experience this social phenomenon [8]. Notably, caregivers whether family, friends, neighbours or professional helpers are key pillars of social support for older adults.

The pervasive nature of mobile devices and the Internet of Things (IOT), provide game-changing opportunities for developing smart solutions for enhancing social connectivity between older adults and their caregivers in computer-mediated environments. Moreover, the emergence of social awareness technologies is a catalyst for a paradigm shift from direct communication channels such as social media, Skype and email to maintain awareness of people around us indirectly [11]. Inspired by the principle of calm technology [17], ambient displays beautifully illustrate awareness information in the periphery of the user's attention so one can access this information at their own convenience [10].

Formerly, most classical ambient displays such as the Digital Family Portrait [13] and the CareNet Display [3] were entirely focused on furnishing caregivers with context information to support the remote monitoring of the older adults and provide peace of mind to their family members. Conversely, a few contemporary ambient displays such as those presented in [6,16] are geared toward interpersonal awareness using physical activity information in mediated environments. However, in both cases, the ambient displays are positioned as decorative objects in a fixed location inside the users' home or work environments. Thus, prohibiting access to the counterpart's activity information outdoors. Moreover, the aforementioned ambient display studies [3,6,13,16] demonstrated awareness information using one decorative object namely ambient photo frames or a snowglobe lighting device.

This work presents a novel user-centered approach to the design and development of an amalgamation of interactive tools to enhance social connectedness between the elderly and their caregivers. To address the limitations of the previous research, we have developed a bidirectional framework, which receives input from embedded smartphone accelerometer and gyroscope sensors and renders information through the interactive materiality of everyday artifacts such as Philips Hue light orbs, a portable LED walking cane and wallet. Consequently, enabling an "always connected" communication channel through pervasive interactive devices.

3   Design Rationale

Our bidirectional activity-based framework is an ambient lighting system that detects human activities and provides visual feedback through a LED cane, LED wallet and Philips hue light orbs to create a sense of awareness and social connectedness between older adults and their caregivers. In general, physical activity data has been studied about health issues, e.g., for motivating people to exercise. However, Consolvo et al. indicates that the exchange of physical activity information can create awareness and influence social cohesion in technology-mediated environments [2]. These findings were used to drive our decision to use physical activity data for our social connectedness framework.

We were guided by the following design heuristics obtained through a thorough review of the literature [10,12,15] and our own findings from previous research [4,6,7] using ambient displays.

- The system should be practical, not distracting, portable, perceptible, comfortable, meaningful, reliable, subtle, discrete, aesthetically pleasing, accessible and safe.
- The system should accommodate the vision and motor impairments of the elderly population and should appeal to the intrinsic motivation to share knowledge.
- The system should support ease of use, affordance and learnability bearing in mind that the elderly are susceptible to cognitive impairments, which affects their attention and memory.
- The system should support the elderly's autonomy and should seamlessly fit into their existing lifestyle patterns.

Motivated by the central goal of designing usable, acceptable and accessible products for the elderly and their caregiver counterparts we sought to determine appropriate everyday objects for conveying activity information that would meet our design criteria. This was done over the course of several brainstorming sessions. Notably, to provide an "always connected" service, we were interested in complementing our already existing Hue lighting system with portable ambient lighting devices. As such, we consulted the following prospective users: one biomedical engineer, two industrial designers and two gerontechnology researchers, two electrical engineers, one embedded software engineer, a retired professor, an elderly professional in the medical industry, one retired engineer and two professors to capture their likes and preferences for interactive products to convey activity information. We encountered various suggestions including the following LED objects: a walking cane, wallet, bracelet, wrist-watch, ball, portable speakers and other wearables such as clothing and shoes. After much deliberation, we decided that the led cane and wallet were most suited for conveying activity information while simultaneously adhering to the design heuristics.

4   System Overview

The entire system is composed of 5 major subsystems as illustrated in Fig. 1. A remote server subsystem resides in the central part of the system and is responsible for classifying human activities and relaying detected activities to other subsystems. A LED and Hue subsystem are located on each side of the remote server subsystem respectively. Each LED subsystem consists of a waist-mounted smartphone, an ESP microcontroller with Wi-Fi capability and an LED ring or strip. The waist-mounted phone is equipped with an accelerometer and a gyroscope for measuring the proper acceleration and orientation of the body respectively [5]. A custom built Android application (LED controller app.), collects the accelerometer and gyroscope readings (sensor data) at a frequency of 50 Hz (see [5]) and sends it to the remote server subsystem for classification. The Android application maintains two socket connections to the central remote server, one for sending sensor data to the server for classification and the other for receiving the classified activities of the counterpart. Subsequently, the classified activities obtained are transformed into lighting property encodings and then broadcasted to the cane's led strip or the wallet's led ring via the ESP microcontroller Wi-Fi module. To achieve this, the waist mounted phone requires a 3G/4G internet connection by which data is streamed to the remote server and a portable Wi-Fi hotspot to provide internet connection to the ESP Wi-Fi module.

Besides, the Hue subsystem consists of a mobile phone with Wi-Fi internet connection and a Philips Hue bridge and bulb. Another custom-built Android application (i.e. the Hue controller), maintains a single socket connection to the central server subsystem for receiving the classified activities of the partner. The Hue controller then relays this information to the hue bulbs as light property encodings via the hue bridge. The Hue subsystems are deployed indoors to convey bidirectional activity information while users are situated in the comfort of their

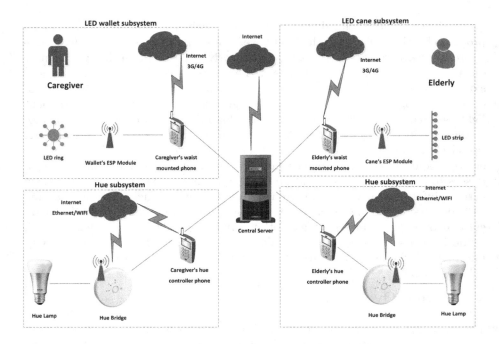

Fig. 1. Overview of the bidirectional activity-based system

homes while the LED devices are carried when users are outdoors. This enables an "always connected" system to users.

5   Central Server Subsystem

The central server subsystem is composed of a multi-threaded socket server implemented in the Python programming language, a hybrid of Support Vector Machine and a Hidden Markov Model (SVM-HMM) classifier implemented in Matlab and a RabbitMQ message queuing server. Figure 2, illustrates the server subsystem.

The multi-threaded server allows concurrent requests to be handled in parallel to reduce response delays. It receives continuous streams of raw sensor data from the waist-mounted mobile phones from the LED subsystems, which is passed to the SVM-HMM classifier for cleaning, feature extraction and selection, and classification. It uses a trained hybrid SVM-HMM activity classification model with an accuracy of 99.7% [5], for detecting one of six basic activities (laying, sitting, standing, walking, walking upstairs, walking downstairs) every 2.5 s. Please refer to [5] for more details on the classification model. Since the SVM-HMM model was trained with feature sets computed on fixed length windows of 2.5 s, a maximum of one activity can be determined in a window. In the server's implementation, however, sensor data is passed to the classifier, every

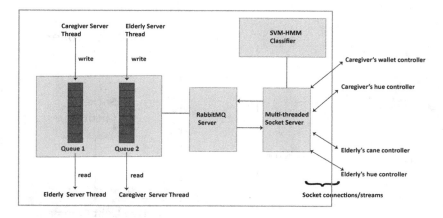

Fig. 2. Central server subsystem

two windows instead of one. This decision was based on the fact that the classifier requires a minimum amount of sensor data within a window to correctly extract and select features. In rare cases, the data received in a window was not sufficient for classification. Therefore, passing two windows of data to the classifier allowed a minimum of one activity to be detected without errors.

Thus, a minimum delay of 5 s was introduced. Moreover, an additional delay of 1 s was spawned by the actual activity classification process and the relay of the activity information resulting in an average system delay of 6 s between the time a participant changes an activity state and the time their partner sees this change in the peripheral displays (LED and Hue client subsystems).

To achieve bidirectional exchange of activity information, client subsystems must uniquely identify themselves and their partners. With this information, the RabbitMQ server, a robust and high-performance implementation of the Advanced Message Queuing Protocol (AMQP), creates two queues for each elderly-caregiver pair. The caregiver server thread then writes to one of the queues, which is read by the elderly server thread and is later read from the queue written by the elderly server thread and vice-versa, thereby facilitating the bidirectional exchange of data.

6   The Hue Subsystem

The Hue subsystem consists of a mobile phone with Wi-Fi internet connection, a Philips Hue bridge and two Hue bulbs infixed in orbs. The Philips Hue is a connected lighting system that enables lighting properties such intensity, colour and brightness to be manipulated over a network. It is furnished with a network bridge for establishing an internet connection and provides an API for building custom applications to control the light over a network. The bulbs connect to the bridge via the ZigBee Light Link open standards protocol[1]. On the left side

---

[1] https://www.developers.meethue.com/documentation/how-hue-works.

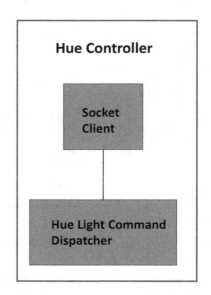

Fig. 3. Hue subsystem

of Fig. 3, the components of the Philips lighting system are illustrated, which includes a Hue bridge, a Hue bulb, and three example light orbs in different colours. The major software constituents of the Hue controller Android application is displayed on the right of the diagram. The Hue bridge connects to the internet via an Ethernet cable and each Hue bulb connected to the hue bridge is identifiable by a unique ID and name on the network. However, the accompanying API is local to the bridge, i.e. it is only accessible on a local Wi-Fi network and is therefore unreachable outside this network. Accordingly, all applications or interfaces that send commands to the bridge should be on the same network as the bridge. This is a security requirement implemented by Philips to prevent Hue lights from being controlled from outside a user's home network. In consequence, we built a Hue controller Android application on a dedicated phone, which connects to the same Wi-Fi network as the Hue bridge. The Hue controller application makes a socket connection to the central remote server and waits for streams of detected activities from the counterpart. Each received activity value is mapped to a set lighting colour properties and relayed to the bridge, which then forwards these commands to the connected light bulb.

Cognizance must be taken of the fact that the mapping between the six basic activities and activity levels (active, passive, resting) and those of activity levels and lighting encodings (colour and brightness) are not arbitrary. As discussed in [7], a person walking or walking upstairs or downstairs is said to be in an "active state", one sitting or standing is in a "passive state" while another who is laying is in a "resting state". Figure 4, depicts these mappings. The mappings were generated to conceal participants actual states for privacy purposes and to

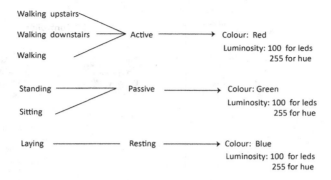

Fig. 4. Mapping between basic activities, activity levels and light configurations

mentally stimulate participants to decipher the actual activities and activity patterns as articulated in [6]. Furthermore, the choice of red, green and blue lighting colours with moderate brightness to represent active, passive and resting states respectively, is by virtue of an exploratory study discussed in [7].

7   The LED Cane Subsystem

To enable elderly-caregiver pairs to have continuous receipt of activity informa- tion outdoors, in a similar mode (i.e. display via lights), we envisioned using portable everyday objects as forms of display. After careful consultation with design experts and potential users, we agreed upon a cane adorned with the Adafruit NeoPixel Digital RGB LED strip for the elderly. The Adafruit NeoPixel Digital RGB LED Strip is a set of individually addressable RGB LEDs encased in a strip as shown in the left of Fig. 5. The LED subsystem controls the NeoPixel LEDs based on the sequence of detected activity information received from the remote server subsystem, which is later transformed into lighting colour prop- erties. To achieve this, the LED strip's microcontroller requires a communica- tion module to receive input from the LED controller Android application on the waist-mounted smartphone. Since the LED Cane subsystem is designed for outdoor use, a wireless connection between the mobile phone and the cane's microcontroller module was the most feasible approach. In retrospect, we basi- cally had two options available: a Wi-Fi or Bluetooth connection. For connection reliability and speed, the Wi-Fi option was chosen.

We selected the ESP8266 based Wi-Fi board, a low-cost Wi-Fi chip with TCP/IP stack and a microcontroller from a Chinese manufacturer, Espressif Systems. The board not only enables Wi-Fi but also comes with an on board Li- ion battery charger and a voltage regulator circuit. To power the microcontroller and the LED strip, a 3.7 V Samsung 18650 Lithium-ion battery of 2600 mAh capacity was used. Batteries are charged via the inbuilt micro USB connection. Embedded software is uploaded to the microcontroller via the serial port, using

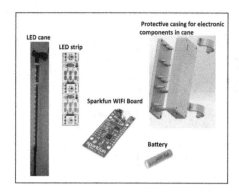 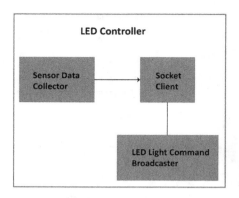

Fig. 5. LED cane subsystem

a so-called FTDI board (a USB-to-serial adapter). Also, development of the embedded software was done within the popular Arduino environment.

Recall that the waist-mounted smartphone has two main responsibilities with respect to the LED subsystems; i.e. (1) sending collected sensor (accelerometer and gyroscope) data to the remote server for activity classification, and (2) relaying the received partner's activity data to the cane's microcontroller so as to change the LEDs' colour and brightness. To enable this functionality, the LED controller application running on the waist-mounted phone requires a cellular internet connection (3G/4G) to communicate with the remote server and a portable Wi-Fi network to communicate with the Sparkfun microcontroller. The main software components of the LED controller application is illustrated on the right of Fig. 5.

The power consumption of the Wi-Fi board, with the LEDs connected, is a limitation to the maximum amount of time the system can be deployed outdoors. We determined the current consumption of the Wi-Fi board, with the LEDs connected using a stable 3.3 V power supply. When all LEDs are at full brightness, and only one colour is "active" the total current consumption is almost 200 mA with 16 LEDs and when all LEDs were off the current consumption was no more than 88 mA.

According to the datasheet[2], the battery has a capacity of 2600 mAh when discharging with a constant current of 520 mA down to a voltage of 2.75 V (cut-off voltage) after the cell is charged up to 4.2 V. Also, the Wi-Fi module is expected to have at least a supply voltage of 3.0 V. With the voltage regulator in between the battery and the Wi-Fi module, having a dropout voltage of 125 mV, the minimal battery voltage needed is 3.125 V. That means that we cannot use the full capacity of the cell since we cannot drain it down to 2.75 V. With a discharge current of 0.2 A the available operating time is 12.5 h. However, the rate of discharge of the battery does not only depend on the consumption of

---

[2] http://www.batteryspace.com/samsung-lithium-18650-rechargeable-cell-3-7v-2600 mah-9-62wh---icr18650-26f---un38-3-passed.aspx.

the LEDs but also on the speed of communication between the microcontroller's Wi-Fi module, the LED controller app. and the LED strip. Overall, we recorded an average operating battery life of 4 h in a full outdoor deployment.

8   The LED Wallet Subsystem

The components of the wallet subsystem are similar to the LED cane subsystem excluding the shape of the led strip (ring-shaped LEDs) and the rechargeable lithium battery utilized. The wallet subsystem components are portrayed in Fig. 6. Basically, the communication protocol between the wallet's Wi-Fi micro-controller, LED controller app and LED ring is identical to that of the cane subsystem. However, we chose a different form factor for the wallet subsystem, as the battery needed to be as flat as possible to not show any protrusions. Thus, a MikroElektronika 3.7 V rechargeable lithium battery of 2000 mAh capacity was installed. With a lower battery capacity, we recorded an average operating battery life of 3 h in a full outdoor deployment.

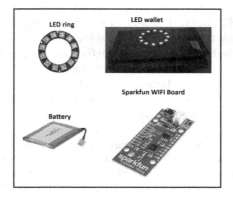

Fig. 6. LED wallet subsystem

9   Discussions and Conclusions

In this paper, we present the design and development of a real-time activity-based bidirectional framework aimed at improving social connectedness between the elderly and their caregivers. Leveraging the internet of things (IoT), this system can be easily deployed and promises to transform the way the elderly and their caregivers communicate and maintain awareness of each other's activities in daily life. Unlike its predecessors e.g., the Digital Family Portrait [13], CareNet Display [3] and the Snowglobe [16], we strive to present an accessible form of technology designed for use in and outside the home environment. By adopting a user-centric approach, we assessed the users' needs and expectations, their context of use and feedback in the design loop. The process culminated with the realization of intelligent interactive product suited for use in the ambient assisted living domain. In particular, the Hue lighting system, LED wallet and LED cane may have considerable added value or complement traditional approaches to maintain awareness and stimulate social connectedness. Through an in-depth description of the technical processes and system components, we aspire to encourage researchers to design innovative and usable systems that enhance social relationships in AAL environments.

Despite the considerable potential of our system, we acknowledge some limitations and challenges we have encountered during the design, implementation and test deployment. Firstly, efficient battery power is the prime limitation of our system's deployment outdoors. In future work, we will consider optimizing computation and communication within the system and also use LEDs of higher resistance to help reduce power consumption. Moreover, introducing a sleep mode based on a pre-defined stimulus will also aid in reducing battery consumption. Secondly, the average delay of 6 s through the exchange of activity cues might be unacceptable in situations where feedback is safety-critical. Even though, the bidirectional activity-based system is not intended as a safety monitoring system, we will investigate methods of reducing the classification window size of our SVM-HMM model in order to reduce the overall response time. Finally, although the system has been implemented, deployed and tested, an experimental study has not been conducted to evaluate general user acceptance and its implications on social connectedness in AAL. Hereafter, we plan to implement the system on a larger scale and perform an experimental study in order to analyse in depth the objective, behavioural and social connectedness implications of the system and its acceptance.

Acknowledgments. This work was supported in part by the Erasmus Mundus Joint Doctorate (EMJD) in Interactive and Cognitive Environments (ICE), which is funded by Erasmus Mundus under the FPA no. 2010–2012. Many thanks to the participants who gave useful insights and perspectives toward the design and implementation of our system. Also, we would like to express our appreciation to Mr. Danny Jansen and Mr. Jasper Sterk for their assistance with this project.

References

1. Bennett, T.R., Wu, J., Kehtarnavaz, N., Jafari, R.: Inertial measurement unit-based wearable computers for assisted living applications: a signal processing perspective. IEEE Sig. Process. Mag. **33**(2), 28–35 (2016)
2. Consolvo, S., Everitt, K., Smith, I., Landay, J.A.: Design requirements for technologies that encourage physical activity. In: Proceedings of the SIGCHI Conference on Human Factors in Computing Systems, pp. 457–466. ACM (2006)
3. Consolvo, S., Roessler, P., Shelton, B.E.: The carenet display: lessons learned from an in home evaluation of an ambient display. In: Davies, N., Mynatt, E.D., Siio, I. (eds.) UbiComp 2004. LNCS, vol. 3205, pp. 1–17. Springer, Heidelberg (2004). doi:10.1007/978-3-540-30119-6_1

4. Davis, K., Feijs, L., Hu, J., Marcenaro, L., Regazzoni, C.: Improving awareness and social connectedness through the social hue: insights and perspectives. In: Proceedings of the International Symposium on Interactive Technology and Ageing Populations, pp. 12–23. ACM (2016)
5. Davis, K., Owusu, E., Bastani, V., Marcenaro, L., Hu, J., Regazzoni, C., Feijs, L.: Activity recognition based on inertial sensors for ambient assisted living. In: 2016 19th International Conference on Information Fusion (FUSION), pp. 371–378. IEEE (2016)
6. Davis, K., Owusu, E., Hu, J., Marcenaro, L., Regazzoni, C., Feijs, L.: Promoting social connectedness through human activity-based ambient displays. In: Proceedings of the International Symposium on Interactive Technology and Ageing Populations, pp. 64–76. ACM (2016)
7. Davis, K., Owusu, E., Marcenaro, L., Feijs, L., Regazzoni, C., Hu, J.: Evaluating human activity-based ambient lighting displays for effective peripheral communication. In: Proceedings of the 11th EAI International Conference on Body Area Networks, BodyNets 2016, pp. 148–154. ICST (Institute for Computer Sciences, Social-Informatics and Telecommunications Engineering), Brussels (2016)
8. Hazer, O., Boylu, A.A.: The examination of the factors affecting the feeling of loneliness of the elderly. Procedia-Soc. Behav. Sci. **9**, 2083–2089 (2010)
9. He, W., Goodkind, D., Kowal, P.: An aging world: 2015, pp. 1–165. US Census Bureau (2016)
10. Mankoff, J., Dey, A.K., Hsieh, G., Kientz, J., Lederer, S., Ames, M.: Heuristic evaluation of ambient displays. In: Proceedings of the SIGCHI Conference on Human Factors in Computing Systems, pp. 169–176. ACM (2003)
11. Markopoulos, P., Ruyter, B.D., Mackay, W.: Awareness Systems: Advances in Theory Methodology and Design, 1st edn. Springer, London (2009)
12. Matthews, T., Rattenbury, T., Carter, S.: Defining, designing, and evaluating peripheral displays: an analysis using activity theory. Hum.-Comput. Interact. **22**(1–2), 221–261 (2007)
13. Rowan, J., Mynatt, E.D.: Digital family portrait field trial: support for aging in place. In: Proceedings of the SIGCHI Conference on Human Factors in Computing Systems, pp. 521–530. ACM (2005)
14. Van Bel, D.T., IJsselsteijn, W.A., de Kort, Y.A.: Interpersonal connectedness: conceptualization and directions for a measurement instrument. In: CHI 2008 Extended Abstracts on Human Factors in Computing Systems, pp. 3129–3134. ACM (2008)
15. Vastenburg, M.H., Visser, T., Vermaas, M., Keyson, D.V.: Designing acceptable assisted living services for elderly users. In: Aarts, E., Crowley, J.L., Ruyter, B., Gerhäuser, H., Pflaum, A., Schmidt, J., Wichert, R. (eds.) AmI 2008. LNCS, vol. 5355, pp. 1–12. Springer, Heidelberg (2008). doi:10.1007/978-3-540-89617-3_1
16. Visser, T., Vastenburg, M.H., Keyson, D.V.: Designing to support social connectedness: The case of snowglobe. Int. J. Des. **5**(3), 129–142 (2011)
17. Weiser, M., Brown, J.S.: The coming age of calm technology. In: Weiser, M., Brown, J.S. (eds.) Beyond Calculation, pp. 75–85. Springer, New York (1997)

Software Testing and Intelligent Systems

Using Ants to Fight Wildfire

Pablo C. Cañizares$^{(\boxtimes)}$, Mercedes G. Merayo, and Alberto Núñez

Departamento de Sistemas Informáticos y Computación,
Universidad Complutense de Madrid, Madrid, Spain
{pablocc,mlmgarci,albenune}@ucm.es

Abstract. The control of fire spreading is a (research) challenge. The impact of the fire in the environment makes essential the study and analysis of fire spread with the goal of designing new tools that help to mitigate the wildfire expansion and, as a consequence, their effects. In this work we introduce a platform to deploy an algorithm, based on Ant Colony Optimization, to determine the best plan to attack fire focus. The framework is based on a theoretical model that allows us to represent the main elements of the environment in which fire evolves. The tool provides a visualisation component to model realistic landscapes.

1 Introduction

Forests play an essential role on the biological diversity. With the 9% total surface of the earth, it is known that the 50% existing species lives in forest ecosystem [11]. Therefore, forests make the planet more habitable, emitting oxygen and cleaning the air by the absorption of contamination. In addition, the wooded areas have an ecological function, preserving and regulating the climate, contributing to limit the effects of storms and acting as a heat buffer [8].

Unfortunately, an average of one percent of world's forests is reported to be affected each year by fires. The global annual area burned in the last years has reached an average of 377 Mha [9]. Wildfires can have deep impact on forest ecosystem, destroying the native species and leading to longer-term indirect effects such as loss of habitat, affecting to the nutrient retention and water infiltration. The loss of organisms, such as invertebrates, pollinators and decomposers, which are a fundamental element in forest ecosystems, can slow down the growth of the forest [13]. Another consequence of fires is the impact generated in the global CO_2 emissions and global warming. It is shown that fires play a key role in the CO_2 levels. Fires produce CO_2 emissions equal to 50% of those originated from humans [6]. Additionally, wildfires can cause several damages to houses and buildings, as well as, in extreme situations, human deaths. The forest fires in Australia (2009) and Greece (2007) caused 173 and 80 victims, respectively.

Research partially supported by the Spanish MEC project DArDOS (TIN2015-65845-C3-1-R) and the Comunidad de Madrid project SICOMORo-CM (S2013/ICE-3006).

© Springer International Publishing AG 2017
I. Rojas et al. (Eds.): IWANN 2017, Part II, LNCS 10306, pp. 371–380, 2017.
DOI: 10.1007/978-3-319-59147-6_32

Besides to the damages caused by the fire, it is worth noting that millions of dollars are spent to extinguish and reconstruct the affected areas. A full accounting considers long-term and complex costs, including impact on watersheds, ecosystems, infrastructure, businesses, individuals, and the local and national economy. As an example, the estimated cost of a massive wildfire that devastated 125,000 acres in California (2003) came to 1.2 billion$ and required the evacuation of 100,000 residents [7].

The deep impact caused by wildfires on the ecosystem have raised the interest of the scientific community. During the last years, several approaches have been introduced to minimize the risks caused by the fire disasters. Among them, it can be found learning methodologies to build decision making systems [17], techniques based on artificial intelligence to predict the beginning of a fire focus in an specific zone using a knowledge-based system [3,15] and schedule approaches to planning the extinction [4].

In this paper we propose a forest fire modelling framework, biologically inspired, to detect and mitigate the wildfires expansion. The framework includes a flexible and scalable platform to deploy algorithms based on swarm intelligence, specifically, it includes one Ant Colony Algorithm used to optimize the detection of fire focus. We have developed a theoretical model that allows us to represent the main elements of the environment in which fire evolves. In order to simulate the expansion of the fire we have use a fire spread model based on cellular automata [10], in which all the elements of our model, affecting the forest fire spreading, can easily be incorporated. The platform has been developed using an advanced simulation engine, known as OMNeT++, that provides a visualisation platform to model realistic landscapes.

The rest of the paper is structured as follows. Section 2 presents the forest wildfire model. Next, in Sect. 3 we describe the swarm intelligence algorithm used to solve the problem. Section 4 presents some experiments. Finally, in Sect. 5, we present the conclusions and some lines of future work.

2   Forest Fire Model

One of the main goals of this paper is to provide some mechanisms which assists in detecting, preventing and mitigating the effects of wildfires in the natural ecosystems. With this aim, we have developed a framework to model, with a realistic detail level, the most important elements which are involved in a forest fire.

One of the main actors of this phenomenon are the *forests*. It is well known the wide spectrum, diversity and extension of the existing wooded areas. Therefore, it is usual that these areas are composed by several species of trees, brushwood and other geographical elements such as mountains, rivers and grasslands. In order to deal with all these aspects, we decided to model the global *surface* of a forest by a grid, in which each square corresponds to a *region*. The user can model the forest in regions which must have the same size but the dimension can range from centimeters to hectares. These regions include information related to the most relevant characteristics of the land, such as the vegetation volume, humidity, temperature, elevation, inclination and probability of fire propagation. Another factor that must be considered, due to its relevant influence in the propagation of fires, is the evolution of *wind flows* that affect the different areas of the forest. Therefore, our model also include data corresponding to the resultant of all wind speeds and directions that act over each of the regions.

Definition 1. Let $\mathcal{G} = \{0, 1, ..., n\} \times \{0, 1, ..., m\}$ where $n, m \in \mathbb{N}$. We define a *region* of $\mathcal{G}$ as a tuple $r = (c, v, p, st, l, i, t, hum, wf)$ where $c \in \mathcal{G}$ represents the coordinates of the region, v is the total volume of vegetation measured in $\frac{m^3}{ha}$, $p \in [0, 1]$ is the probability of fire propagation in this area, st indicates the state of the region, $l \in \mathbb{R}$ is the elevation of the land measured in meters, $i \in [0, 90]$ represents the inclination angle, $t \in \mathbb{R}$ is the average temperature measured in Celsius degrees, $hum \in [0, 100]$ represents the humidity level of the area and $wf \in [0, 360) \times [0, 12]$ indicates the direction measured in degrees and the wind speed based on Beaufort scale [5] that affect the region.

The state of the region st takes values in $\{Healthy, OnFire, Burned\}$. *Healthy* means that the region is clear of fire, *OnFire* indicates that the region is partially on fire and *Burned* indicates that the fire burnt down the region to the ground. Given a region $r = (c, v, p, st, l, i, t, hum, wf)$ we let $pos(r)$ be equal to c.

A *surface* modeled by $\mathcal{G}$ is a tuple $\mathcal{S}_\mathcal{G} = (h, w, R, ld)$ where $h, w \in \mathbb{N}_+$ correspond to the height and weight of a forest measured in meters, respectively, R is a set of regions of $\mathcal{G}$ and $ld : R \times R \longrightarrow \mathbb{R}_> \cup \{\infty\}$ is the difficulty level of access function. In order to ensure that the set of regions completely covers the surface we need two conditions hold:

1. For all distinct $r, r' \in R$ we have $pos(r) \neq pos(r')$.
2. For all $c \in \mathcal{G}$ there exists $r \in R$ such that $pos(r) = c$.

Given $r_1, r_2 \in R$ such that $pos(r_1) = (x_1, y_1)$ and $pos(r_2) = (x_2, y_2)$ we say that r_1 and r_2 are *neighbors*, denoted by $neigh(r_1, r_2)$, if and only if $|x_1 - x_2| \leq 1 \wedge |y_1 - y_2| \leq 1 \wedge (x_1 = x_2 \oplus y_1 = y_2)$.

The function $ld(r_1, r_2)$ returns a value that represents the difficulty level to access between r_1 and r_2. The function returns ∞ for all the pairs (r_1, r_2) such that $neigh(r_1, r_2)$ does not hold. We say that $\sigma = <r_1, ..., r_w>$ is a *path* of $\mathcal{S}_\mathcal{G}$ if and only if for all $1 \leq i < w, neigh(r_i, r_{i+1})$ and for all $1 \leq i < j \leq w$ we have $r_i \neq r_j$. Abusing the notation we will write $r \in \sigma$ if there exists $1 \leq i \leq w$ such that $r = r_i$. □

Simulating the behavior of wildfires over a surface requires the application of wildfire models [2,12,14,16], a collection of equations that allow to calculate rate of spread, fireline intensity, fuel consumption and fire effects among others. The application of these models is indispensable for forest fire management and they are essential in the operating tools used in forestry agencies. In this work, due to the features of our approach, we have applied a model based on cellular automata (CA) [10] in which all the factors affecting the forest fire spreading can easily be incorporated. The model leads to algorithms which can exploit the inherent parallelism of the CA structure. Cellular automata consist of a grid of cells These cells are in a specific state that changes over time on the basis of its neighbors states and a function. Despite the simplicity of this structure, it allows to model complex systems. The simulation of the wildfire corresponds to the evolution of the states of the regions define in the considered surface along the time.

3   Swarm Intelligence Algorithm

In this work we propose a parallel algorithm inspired on swarm intelligence to mitigate the effects of the wildfires in the natural ecosystems. Given a surface in which different fire focuses have been detected the algorithm searches for the shortest paths to reach each of them from a specific point where the firefighting brigade is located. In this way, we try to help to determine the best strategy for fire control. Specifically, our approach follows the Ant Colony Optimization (ACO) metaheuristic, which imitates the behaviour of real ants to solve complex problems in a distributed way. In order to apply the classical ACO paradigm to the problem faced in this work, the different features considered in our model have been integrated in the path selection and pheromone update functions.

Definition 2. Let $S_{\mathcal{G}} = (h, w, R, ld)$ be a surface, $F \subseteq R$ be the set of the regions where the fire focuses are located and $r_0 \in R$ be the region considered the starting point. Let $\sigma_k = < r_0, r_1^k, ..., r_{l_k}^k >$ the path traversed by an ant k from the starting region. We define the probability of an ant k moving from region r_i to region r_j having as target a region $r_t \in F$ as:

$$p_{ij}^{kt} = \begin{cases} \dfrac{V_{ij}^t}{\sum_{neigh(r_i,r_j) \wedge r_j \notin \sigma_k} V_{ij}^t} & \text{if } r_i = r_{l_k}^k \wedge neigh(r_i, r_j) \wedge r_j \notin \sigma_k \\ 0 & \text{otherwise} \end{cases} \tag{1}$$

where $V_{ij}^t = \left[\tau_{ij}^t\right]^{\alpha} \cdot \left[\dfrac{1}{ld(r_i,r_j)}\right]^{\beta}$

The component τ_{ij}^t represents the pheromone value associated with the connection between the involved regions. Initially, the level of pheromone between any two regions is 0. As we previously said, the difficulty level of access function $ld(r_i, r_j)$ included in the model of the surface takes into account the different parameters associated with the regions (elevation, inclination, temperature...) that can affect the accessibility. Finally α, β are constants used to assign a weigh to each of the considered parameters.

The update of τ_{ij}^t, that indicates the pheromone level deposited by ants in the transition between regions r_i and r_j on the way to the target r_t, is performed as follows:

$$\tau_{ij}^t \leftarrow (1 - \rho) \cdot \tau_{ij}^t + \sum_{k=1}^{m} \Delta\tau_{ij}^{tk} \tag{2}$$

Algorithm Ants_Mitigate_Fire(*focuses*, *nAnts*, *nIter*, *startReg*)
while *(termination condition does not hold)* **do**
    Solution ← ∅;
    Targets ← *focuses* ∪ *starReg*;
    foreach *origReg* in *Targets* **do**
        foreach *destReg* in *Targets* **do**
            lIter ← *nIter*;
            while *lIter* > 0 **do**
                SolutionTarget ← ∅;
                Release ants;
                foreach *ant* **do**
                    Path ← *origReg*;
                    while *(destReg not reached* ∧ *antIsAlive)* **do**
                        step ← calculateNextStep();
                        if *step* ≠ ∅ **then**
                          Path ← *Path* ∪ *step*;
                        else
                          killAnt();
                        end
                    end
                    SolutionTarget ← *SolutionTarget* ∪ *Path*;
                end
                updatePheromones();
                lIter ← *lIter* − 1;
            end
            Solution ← *Solution* ∪ shortestPath(origReg, destReg);
        end
    end
    buildHamiltonGraph(Solution);
    spreadFire();
end

Algorithm 1. Algorithm schema for mitigate wildfire inspired by ACO.

where ρ denotes the pheromone evaporation rate, m is the number of ants and $\Delta\tau_{ij}^{tk}$ is the amount of pheromone deposited by ant k in the transition from r_i to r_j, which is given by the following equation:

$$\Delta\tau_{ij}^{kt} = \begin{cases} Q/l_k & \text{if } \exists\ 1 \leq h < l_k : r_h = r_i \wedge r_{h+1} = r_j \\ 0 & \text{otherwise} \end{cases} \qquad (3)$$

where Q is a parameter of the model.   ☐

The proposed algorithm is shown in Algorithm 1. Intuitively, the algorithm calculates the shortest paths between the starting region and each fire focus. The process consists in the application of the ACO metaheuristic, using the path selection and pheromone update function previously defined. It is worth noting that if any ant is not able to continue on its way to the target focus, this ant is killed and its local solution does not affect to the pheromone update. The path construction process is performed a predefined number of iterations until a final solution is obtained. At this point, each region containing a fire focus and the starting region are connected by a path. In order to select the shortest path that covers all the fire focuses, a hamiltonian graph is built using the provided solution. Afterwards, the fire evolves by applying the fire spreading model proposed in the previous section. This process does not stop until the predefined number of iterations have been performed.

4   Experiments

In this section we provide several experiments used to evaluate our proposal for mitigating wildfire in natural ecosystems. In order to accomplish this evaluation, we have implemented Algorithm 1 using the simulation framework OMNeT++ 5.0 [1]. The implementation of Algorithm 1 and the GUI to execute experiments are available at http://antares.sip.ucm.es/tools/ants/index.html.

In order to analyse both the usability and accuracy of Algorithm 1, 3 different surfaces have been modelled. These surfaces have been configured with a

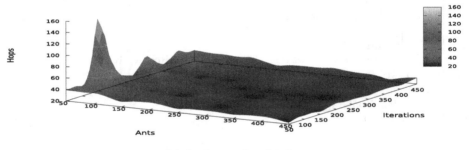

(a) Surface with 1 fire focus.

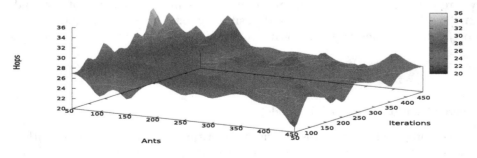

(b) Surface with 3 fire focuses.

Fig. 1. Number of hops to mitigate wildfire in an surface of 25×25 regions

pre-defined number of fire focuses, which represents an active wildfire. In these experiments, we use 1 and 3 fire focuses for each surface. Also, we have used a range of values between 50 and 450 for both the number of ants and the number of iterations. Each region used to model these surfaces covers an area of 10m². The proposed algorithm (see Algorithm 1) has been executed on these surfaces with the aim of finding the best path to mitigate the propagation of the existing fire focuses.

The first surface consists of a grid of 25×25 regions. The size of this surface is equivalent to almost 9 soccer fields. Figure 1(a) shows the results obtained by executing Algorithm 1 in this surface with 1 fire focus, while Fig. 1(b) shows the results for the same surface with 3 fire focuses. These charts depict the length of the shortest path to reach all the fire focuses from an initial point in the grid. In general, the length of the paths obtained when 1 fire focus is used ranges between 23 and 39. There is one exception when 50 ants and 150 iterations are used. In this case, the ants are not able to find the best path and they build a path containing a significant number of regions compared with the rest of the solutions. This fact is clearly reflected in the chart when 3 fire focuses are used. In this case, using a low number of ants has a direct impact on the quality of

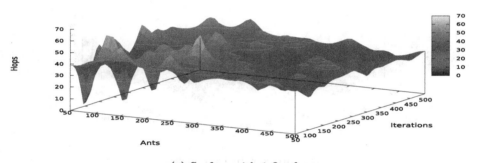

(a) Surface with 1 fire focus.

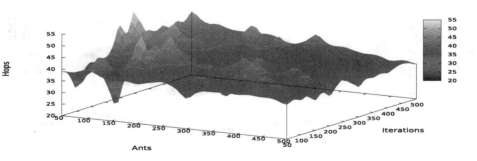

(b) Surface with 3 fire focuses.

Fig. 2. Number of hops to mitigate wildfire in an area of 50×50 regions

the obtained solution. Hence, increasing the number of ants slightly reduce the length of the obtained solution. However, the reduced dimensions of the terrain does not provide a significant reduction in the total length of the path when increasing the number of ants and iterations.

The second surface consists of a grid of 50 × 50 regions, which approximately represents an area 4 times larger than the gardens of Versalles. Figure 2(a) shows the results when 1 fire focus is used. Similarly to the previous surface, it can be seen a slight reduction in the length of the obtained path when the number of ants and the number of iterations increase. Moreover, due to the specific characteristic of the terrain, the algorithm executed with a low number of ants and iterations sometimes does not find a valid solution. This is reflected in the three peaks reaching zero. However, increasing the number of ants or the number of iterations seems to solve this problem, providing a suitable solution. Figure 2(b) shows the results when 3 fire focuses are used. In this case, the results obtained are similar to the ones obtained when 1 fire focus is used.

The last modelled surface consists of a grid of 100 × 100 regions, having a total extension nearly double of Vatican City. Figure 3(a) shows the results when

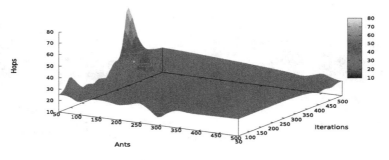

(a) Surface with 1 fire focus.

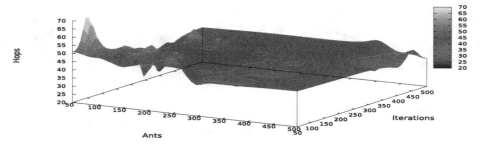

(b) Surface with 3 fire focuses.

Fig. 3. Number of hops to mitigate wildfire in an area of 100 × 100 regions

1 fire focus is used. In this case we can observe two interesting facts. First, there is a peak that represent a path much larger than the rest of the solutions when a low number of ants are used. Similarly than the first surface, in this case the ants were unable to find a suitable solution. Second, there are some variations in the length of the solution when the number of ants and iterations are below 350. This fact is mainly caused by the main characteristics of the terrain. However, when a high number of ants or iterations are used, the length of the solutions seems to be stabilized. Figure 3(a) shows the results when 3 fire focuses are used. Similarly, there are some variations in the length of the solutions. However, in this case, these variations occur when the number of ants used are less than 250. In the rest of the experiments, the solution seems to be stabilised. Using more than 450 ants and more than 450 iterations reflects a slight reduction in the length of the obtained path.

As a conclusion, increasing the number of ants executed in these modelled surfaces provides better solutions than using a low number of ants. This is mainly caused because increasing the number of executed ants also increases the probability of finding the best solution. When using a low number of ants, the solution can be improved by increasing the number of iterations. In some cases, using the maximum values for the number of ants and iterations may provide a slight improvement in the solution, which is more noticeable in large surfaces.

5   Conclusions and Future Work

In this paper we have developed a framework to model wildfire scenarios and simulate their evolution in natural ecosystems. First, we have modelled the main characteristics of the terrain for representing a wide variety of natural surfaces. Second, we provide a theoretical model that allows us to represent the fire in this environment. Finally, we have developed an algorithm, based on Ant Colony Optimization, to mitigate the wildfire spread in natural ecosystems.

In order to check both the usability and accuracy of this algorithm, we have carried out some experiments by modelling three different surfaces. Thus, the proposed algorithm were executed in each one the these surfaces. We can observe that increasing the number of ants provides better solutions than using a low number of ants. This means that using a high number of ants increase the probability of finding the best solution to reach all the fire focuses. Additionally, when using a low number of ants, the quality of the obtained solution can be improved by increasing the number of iterations.

Future work include modelling larger surfaces to analyse the scalability of the proposed algorithm. Also, we plan to extend the provided language to model a wider spectrum of natural ecosystems.

References

1. OMNeT++ 5.0 (2016). https://omnetpp.org/
2. Albini, F.A.: A model for fire spread in wildland fuels by-radiation. Combust. Sci. Technol. **42**(5–6), 229–258 (1985)

3. Alonso-Betanzos, A., Fontenla-Romero, O., Guijarro-Berdiñas, B., Hernández-Pereira, E., Andrade, M.I.P., Jiménez, E., Soto, J.L., Carballas, T.: An intelligent system for forest fire risk prediction and fire fighting management in Galicia. Expert Syst. Appl. **25**(4), 545–554 (2003)

4. Avesani, P., Perini, A., Ricci, F.: Combining CBR and constraints reasoning in planning forest fire fighting. In: 1st European Workshop on Case-based reasoning, pp. 235–239 (1993)

5. Beaufort, F.: Beaufort wind scale. British Rea-Admiral (1805)

6. Bowman, D., Balch, J., Artaxo, P., Bond, W., Carlsonand, J.M., Cochrane, M.A., D'Antonio, C.M., DeFries, R.S., Doyle, J., Harrison, S., Johnston, F., Keeley, J., Krawchuk, M., Kull, C., Marstond, J.B., Moritz, M., Prentice, I., Roos, C., Scott, A., Swetnam, T.W., van der Werf, G., Pyne, S.: Fire in the earth system. Science **324**(5926), 481–484 (2009)

7. Dale, L.: The true cost of wildfire in the Western US. Western Forestry Leadership Coalition (2009)

8. FAO (Food and Agriculture Organization of the United Nations). Global forest resources assessment 2010 (2010)

9. Giglio, L., Randerson, J.T., Werf, G.R.: Analysis of daily, monthly, and annual burned area using the fourth-generation global fire emissions database (gfed4). J. Geophys. Res.: Biogeosci. **118**(1), 317–328 (2013)

10. Karafyllidis, I., Thanailakis, A.: A model for predicting forest fire spreading using cellular automata. Ecol. Model. **99**(1), 87–97 (1997)

11. Mora, C., Tittensor, D.P., Adl, S., Simpson, A., Worm, B.: How many species are there on earth and in the ocean? PLoS Biol. **9**(8), e1001127 (2011)

12. Morvan, D., Dupuy, J.L.: Modeling of fire spread through a forest fuel bed using a multiphase formulation. Combust. Flame **127**(1–2), 1981–1994 (2001)

13. Nasi, R., Meijaard, E., Applegate, G., Moore, P.: Forest fire and biological diversity. Unasylva **53**(209) (2002)

14. R.C. Rothermel. A mathematical model for predicting fire spread in wildland fuels. Forest Services, (1972)

15. Sakr, G., Elhajjand, I.H., Mitriand, G., Wejinya, U.C.: Artificial intelligence for forest fire prediction. In: International Conference on Advanced Intelligent Mechatronics (AIM), pp. 1311–1316. IEEE (2010)

16. Weber, R.O.: Modelling fire spread through fuel beds. Prog. Energy Combust. Sci. **17**(1), 67–82 (1991)

17. Wiering, M.A., Dorigo, M.: Learning to control forest fires. In: 12th International Symposium on Computer Science for Environmental Protection, pp. 378–388. Metropolis Verlag (1998)

Using Evolutionary Computation to Improve Mutation Testing

Pedro Delgado-Pérez[1]([⊠]), Inmaculada Medina-Bulo[1],
and Mercedes G. Merayo[2]

[1] Departamento de Ingeniería Informática, Escuela Superior de Ingeniería,
Universidad de Cádiz, Cádiz, Spain
{pedro.delgado,inmaculada.medina}@uca.es
[2] Departamento de Sistemas Informáticos y Computación,
Universidad Complutense de Madrid, Madrid, Spain
mgmerayo@fdi.ucm.es

Abstract. The work on mutation testing has attracted a lot of attention during the last decades. Mutation testing is a powerful mechanism to improve the quality of test suites based on the injection of syntactic changes into the code of the original program. Several studies have focused on reducing the high computational cost of applying this technique and increasing its efficiency. Only some of them have tried to do it through the application of genetic algorithms. Genetic algorithms can guide through the generation of a reduced subset of mutants without significant loss of information. In this paper, we analyse recent advances in mutation testing that contribute to reduce the cost associated to this technique and propose to apply them for addressing current drawbacks in Evolutionary Mutation Testing (EMT), a genetic algorithm based technique with promising experimental results so far.

Keywords: Software testing · Mutation testing · Genetic algorithms

1 Introduction

Mutation testing [20] is an effective technique to help improve the fault detection capability of a test suite. In order to estimate this capability, we measure the extent to which the developed test suite is able to detect different faults injected into the program. These faults are introduced by means of *mutation operators* that generate *mutants*. Detecting a mutation means that the test suite can distinguish the behaviour of the mutant and the original program. In this case, the mutant is said to be *killed*. Otherwise, when the outputs of the original and the mutant are equal, we say that the mutant is *alive*. However, some alive mutants are *equivalent* to the original program and cannot be killed by any test case. Mutation testing is a powerful testing technique but it has a high computational cost due to the potentially large number of mutants that can be generated, in particular, when the program requires a high compilation and test execution

© Springer International Publishing AG 2017
I. Rojas et al. (Eds.): IWANN 2017, Part II, LNCS 10306, pp. 381–391, 2017.
DOI: 10.1007/978-3-319-59147-6_33

time. Therefore, it is required to propose approaches that, on the one hand, reduce the size of the set of mutants and, on the other hand, provide a high level of fault detection. As a result, several methods have been proposed to alleviate as much as possible this problem [19], including search-based techniques.

In this paper we focus on the application of the approaches proposed in mutation testing to Evolutionary Mutation Testing, a technique for the selection of a reduced set of mutants based on a genetic algorithm [8]. It will allow us to improve the promising results reported by this methodology to date for WS-BPEL compositions [6] and C++ object-oriented systems [3]. We propose to integrate the following approaches:

- **Selective mutation based on the quality metric** [7], which aims at identifying mutants that could be excluded without losing effectiveness of the test suite. We can choose either to discard a subset of mutation operators (operator-based selective mutation) or give preference to the selection of mutants from the most valuable operators (rank-based mutant selection).
- **A multi-objective approach** that drives the search towards finding, not only undetected mutants, but also mutants with a great coverage impact [14] and well spread through the code [18].
- **Trivial Compiler Equivalence** [15], a mechanism to detect equivalent mutants when they have identical machine code as the original program.

The rest of the paper is organized as follows: Sect. 2 surveys the use of genetic algorithms in mutation testing, and more specifically in EMT. Then, we explain the proposals based on each of these findings and comment the benefits of their application in Sects. 3, 4 and 5. Finally, we present the conclusions derived from this paper.

2   Genetic Algorithms in Mutation Testing

Genetic algorithms have been applied to software testing since many years ago [21]. With regard to mutation testing, genetic algorithms have been successfully investigated and incorporated into different systems for the generation of test cases [16]. However, they have also shown to be useful in mutant generation to increase the efficiency of mutation testing. We can remark the following advances:

- The evolution in parallel of the population of mutants and test cases using different genetic operators [1,12].
- Several studies have shown that genetic algorithms can provide better results than other search-based techniques [10,13] in finding interesting higher order mutants.
- Genetic algorithms have been used to select a subset of mutants [18] and also a subset of mutation operators [2] with the goal of minimising the loss of information.

Evolutionary Mutation Testing [6] was proposed to select a subset of mutants with the help of an evolutionary algorithm. Specifically, a genetic algorithm was implemented in the *GAmera* tool [5] to search for *strong mutants* that can help derive new test cases. Strong mutants fall in one of these categories:

- *Potentially equivalent mutants*: mutants that are not detected by the test suite.
- *Difficult to kill mutants*: mutants that are only killed by a specific test case that does not kill any other mutants.

The fitness function gives the highest value to those mutants killed by few test cases and those test cases killing few other mutants at the same time. In this way, strong mutants are the best valued by the fitness function. The mutants in the new generations are then:

- Derived from those mutants with a high fitness thanks to reproductive operators (mutation and crossover). This should lead to the generation of strong mutants in new generations.
- Produced by a random algorithm.

The studies originally conducted on EMT [6] reported that this technique is able to find all strong mutants generating 15% less mutants than random selection for WS-BPEL compositions. Recently, the experiments performed with C++ systems and class-level mutation operators [3] supported that EMT is, in fact, more effective than random mutant selection, though the difference was not as significant as in the experiments with WS-BPEL. This fact might be motivated by the way that strong mutants are distributed over the search space, and reveals that further research is required to enhance the effectiveness of EMT.

3 Quality Metric for Selective Mutation

3.1 Background

Some mutants have greater potential than others to improve the quality of the test suite. They can guide the tester through the definition of *high-quality test cases* [4], that is, test cases that are able to find non-trivial faults. Regarding mutation operators, those ones that generate mutants that can be killed, mostly, by a default "happy path" test case, are not useful. These considerations are embodied in the quality metric devised in [7]. The formula of the quality metric of a mutant Q_m is defined as follows:

$$Q_m = \begin{cases} 0, & m \in E \\ 1 - \dfrac{1}{(|M| - |E|) \cdot |T|} \displaystyle\sum_{t \in K_m} |C_t|, & m \in D \end{cases} \tag{1}$$

where:

- M is the set of valid mutants.
- E is the set of equivalent mutants.
- D is the set of killed mutants.
- T is an adequate and minimal test suite, i.e., a test suite of the minimal size that kills all non-equivalent mutants.
- K_m is the set of test cases that kill the mutant m.
- C_t is the set of mutants killed by the test case t.

This quality metric punishes the existence of equivalent mutants as well as takes into account a twofold criteria regarding killed mutants:

- The number of test cases that kill a mutant: the fewer test cases kill the mutant, the better.
- The number of mutants that a particular test case kill: the fewer the mutants killed by a test case, the better. This property is valuable because we only have a reduced subset of mutants that can induce the generation of that test case.

Next, we find the definition of the quality of a mutation operator Q_o as the mean of the quality of the set of mutants generated by the operator o (M_o):

$$Q_o = \frac{1}{|M_o|} \sum_{m \in M_o} Q_m \qquad (2)$$

Recently, this metric has been used to determine the capacity of mutation operators to help the tester enhance the fault detection power of the test suite with high-quality test cases. The mutation operators were sorted in a ranking according to their quality. Then, two different selective strategies were applied taking into account the ranking:

- **Operator-based selective mutation**: this strategy selected mutants generated by a subset of mutation operators.
- **Rank-based mutant selection**: this strategy favoured the selection of the mutants generated by the top ranked operators. The results of the rank-based strategy showed that this method for the selection of mutants offers a better outcome than the random selection of mutants.

Both strategies were evaluated by measuring the percentage of test cases that would not be generated due to the dismissed mutants. This approach was successful in reducing the number of mutants without a meaningful loss of effectiveness.

3.2 Improvements Applying Selective Mutation Based on Quality Metrics

Selective mutation following operator rankings improves the test suite through the selection of a subset of all the mutants. Initially, we use it to analyse and determine which are the best and the worst-valued operators in order to incorporate the gained knowledge into the mutation tool. Then, EMT can take advantage of this information to reduce the set of operators to be applied and, therefore, the cost. As a result, the combination of both techniques can be used to further reduce the cost. These results can be used in two different ways:

- *First proposal*: Instead of generating mutants from all mutation operators, EMT only generates mutants from the best-valued operators following the ranking based on the quality metric. This proposal requires to determine if a mutation operator should be discarded (based on previous studies with the set of mutation operators applied).
- *Second proposal:* Instead of generating mutants with the same probability, the mutants are selected in a rank-based manner: the probability of selecting mutants from the best-valued operators is higher. This criterium is applied to the subset of mutants randomly produced in each generation of the genetic algorithm.

Mutant-based selective mutation and operator-based selective mutation have shown to be useful to reduce the cost of mutation testing while retaining effectiveness. However, in the experiments presented in [4], mutant-based selection yielded better results than operator-based selection when applied to mutation operators at the class-level in C++. However, the analysis of traditional operators has usually revealed great redundancy among operators and, consequently, a subset of operators could subsume the rest. As a consequence, it is unclear which strategy shows a better performance, so the proposal should be selected depending on the nature of the set of mutation operators based on previous empirical results.

3.3 Benefits

Both proposals modify the performance of EMT in the following ways:

- *First proposal*: By removing the operators at the bottom of the classification, the efficiency of EMT is improved because it will avoid that the genetic algorithm produces low-quality mutants. Since the quality metric punishes equivalent mutants, which decreases the value of a mutation operator, we will be also preventing the selection of equivalent mutants. In general, the genetic algorithm finds quicker those mutants that can lead to the enhancement of the test suite. These mutants can be labelled as *resistant mutants.*
  The reason that non-equivalent mutants remain alive is either (a) the test suite does not cover the mutant or (b) the test suite covers the mutant but it is not able to reveal its mutation. Since the quality metric is devoted to the generation of high-quality test cases, the refinement is achieved with a large proportion of mutants that fall in the case (b), which are the most interesting because they are not easy to design.

- *Second proposal:* The first proposal involves the risk of eliminating the possibility that some mutants of a high quality derived from low-quality operators are generated. The application of the rank-based strategy removes that risk, especially in those cases when it is not easy to find a sufficient set of mutation operators. The same benefits mentioned for the first proposal hold, but there is a higher probability that some low-quality mutants are generated.

4   Multi-objective Evolutionary Mutation Testing

4.1   Background

One of the major drawbacks of mutation testing is the presence of equivalent mutants. They cannot be completely discarded in general because this is an undecidable problem. Therefore, they have to be detected when the alive mutants are analysed, which is a time-consuming task. Several researchers have proposed different techniques to identify and remove equivalent mutants [9,11]. Recently, several works have studied the impact that mutations have on the code coverage in order to mitigate the effects of the equivalence [14,17]. Intuitively, those mutations which cause a great impact on the coverage of the test suite execution are less likely to produce an equivalent mutant. In addition, there should not be parts of the code without mutations, because it would avoid that those fragments are analysed with mutation testing [18].

The results obtained when applying an operator-based and a mutant-based selective strategy have been compared [23]. Namely, two different mutant-based strategies are applied: *One-round* random selection (mutants selected with the same probability) and *Two-round* random selection (the operator that generates the mutant is selected with the same probability). Given that Two-round random selection yielded better results than One-round random selection, it is plausible to think that each mutation operator is useful to address a different feature. Similar results were reported in other experiments using class-level mutation operators [4]. The experiments reported in [22] also suggest that sampling mutants from each method of the program and from each mutation operator performs better than One-round random selection.

4.2   New Objectives

Based on the aforementioned studies and their results, EMT should follow a multi-objective approach. We have detected three aspects that should be considered in the fitness function of the genetic algorithm:

1. **Maximise the coverage impact.** EMT currently assigns the highest fitness to potentially equivalent mutants, but it cannot distinguish between equivalent mutants and resistant mutants. Analysing the coverage impact of the mutants will help during the selection of non-equivalent mutants with the highest probability.

2. **Maximise the scattering in the code.** The genetic algorithm currently selects mutations without taking care of the location in which it is injected. As a consequence, the reduced subset of mutants might not be dealing with some parts of the code appropriately. Therefore, it is preferable spreading mutations all over the code so that all code items are covered [18,22].
3. **Maximise the scattering in the set of operators.** We foster the generation of mutants from all mutation operators, favouring the selection of mutants from operators barely applied so far. This aspect is not currently taken into account by the genetic algorithm. This objective aligns with recent findings in the mutation testing literature [4,22,23].

4.3 Benefits

Next, we present the benefits of incorporating each of these aspects into EMT:

1. *Coverage impact*: Penalisation of those mutants with the highest probability of resulting in equivalent mutants prevents from selecting equivalent mutants for reproduction and new equivalent mutants being generated.
2. *Scattering in the code*: A program can be divided into different classes, which count with different methods comprised of multiple statements. Furthermore, in an object-oriented program, some methods or even blocks are directly associated to specific object-oriented features, such as constructors or exceptions. Therefore, a fitness function aware of the coverage will avoid that the subset of selected mutants concentrates in the same area of the code.
3. *Scattering in the set of operators*: Each of the operators affects to different features, especially in the case of class-level operators. Consequently, strong mutants might be generated by a large subset of mutation operators. As a result of this improvement, the gap between the percentage of mutants generated by each of the operators is less significant.

It is worth noting that the weight of the application of each of aspects can be parametrised in order to prioritise the most convenient for the specific context.

5   Trivial Compiler Equivalence

5.1   Background

As aforementioned, the existence of equivalent mutants is one of the main problems in mutation testing. Recently, a new technique, called Trivial Compiler Equivalence (TCE), has been proposed to detect some equivalent mutants automatically [15]. This technique has shown the ability to reduce an average of 30% the set of equivalent mutants in programs coded in C.

TCE combines the use of the compiler *gcc* and the utility *diff*: *gcc* is used to compile the code and generate an optimised executable, and *diff* allows to compare these executables to search for equivalences between different versions of the program. This mechanism is able to detect two types of mutants:

- **A subset of equivalent mutants**: a mutant is identified as equivalent when there is no difference between the binary file of the original program and a mutant.
- **A subset of duplicated mutants**: a mutant is identified as duplicated when the binary file derived from the mutant is equal to the the binary file derived from another mutant.

The experimental results showed that the application of this technique is affordable in terms of execution time. However, the most expensive task in TCE is the compilation time, which increases with the level of optimisation selected. The detection of equivalent and duplicated mutants is not significant when compared to the compilation time.

5.2 Improvements Applying TCE

The genetic algorithm in EMT searches for potentially equivalent mutants, which have not been detected by the current test suite. Hopefully, this type of mutants will help derive new test cases. However, potentially equivalent mutants can also turn out to be equivalent, being unable to distinguish them from those that lead us to improve the test suite. Therefore, some of those equivalent mutants can be automatically detected when EMT is combined with TCE. Given that TCE is based on the compiler *gcc*, the technique is available for those mutation tools that apply to programming languages within the collection of compilers in *gcc*, like *GiGAn* for C++ [3].

We propose two different options for applying EMT in conjunction with TCE:

- *First proposal*: Applying TCE before the execution of EMT. In this case, all mutants are compiled to create an executable and TCE compares them with the executable of the original program. Those mutants detected by TCE as equivalent are then marked to avoid that they can be used during the execution of EMT.
- *Second proposal*: Applying TCE during the execution of EMT. In this case, the mutants selected in each generation are analysed using TCE. Those mutants identified as equivalent are penalised by means of the fitness value assigned to them.

The detection of duplicated mutants could also be beneficial to avoid the selection of mutants that can lead to the generation of the same test case. However, for the manual generation of test cases, it might be useful sometimes to count with redundant mutants: a tester may find difficult to kill a mutant but may find more feasible to produce a test case for another mutant which kills both mutants.

5.3   Benefits

The benefits of implementing these strategies are:

- *First proposal*: Applying TCE to all mutants from the beginning, we avoid that EMT selects equivalent mutants that can be detected by TCE. However, we are wasting time trying to detect equivalent mutants which might never be selected by EMT, so this proposal is not recommended for programs with a high compilation time. On the contrary, it is very convenient when there is evidence that TCE is able to detect a high percentage of equivalent mutants for the set of mutation operators applied.
- *Second proposal*: In this case, we only apply TCE when finding a potentially equivalent mutant during the execution of EMT. Although this proposal decreases the computational cost of applying TCE, it does not avoid selecting equivalent mutants occasionally.

Nevertheless, we have to take into account the restrictions imposed by TCE:

- TCE has only been applied to C programs. Thus, it is not clear that, if it is applied to other languages or other type of mutation operators, we will obtain the same detection power.
- Under the premise that an equivalent mutant and a resistant mutant are slightly different, it is possible that some equivalent mutants can guide the genetic algorithm on the selection of resistant mutants in new generations. Therefore, removing or penalising equivalent mutants can impact the search for potentially equivalent mutants.

6   Conclusions

In this paper, we have analysed the current operation mode of the technique called Evolutionary Mutation Testing. Despite evidence of its usefulness, based on the empirical results obtained from its application to programs in different languages, from a review of the mutation literature it becomes clear that further improvements can be made. In particular, EMT does not solve the hardest task of mutation testing: the detection of equivalent mutants. With the goal of overcoming this drawback, we propose to consider the use of additional information in the calculation of the fitness function and the application of TCE, a technique for the automated detection of some equivalent mutants. The new fitness function also promotes the generation of mutants from all mutation operators and covering all the code.

We also propose the generation of mutants following an operator-based and/or a rank-based selective approach, which also penalises the generation of mutants from operators usually producing many equivalent mutants. However, the application of these two selective strategies should be explored further with different sets of mutation operators to know the extent to which mutant-based selection is superior to operator-based selection, as previous studies suggest.

Rank-based selective mutation is based on the rank selection method used in genetic algorithms; it would also be interesting to compare its results with different mutant-based selection techniques, such as roulette wheel selection, tournament selection or stochastic universal sampling.

In spite of proposing several improvements for EMT, there is still room for other ones. For instance, we should try to find a way to avoid that all test cases are executed to calculate the fitness function, especially in those cases when the test execution time is high.

Acknowledgements. Paper partially funded by the research scholarship PU-EPIF-FPI-PPI-BC 2012-037 (University of Cádiz) and by Spanish government projects DAr-DOS (TIN2015-65845-C3-3-R (MINECO/FEDER)), SICOMORo-CM (S2013/ICE-3006) and the Excellence Network SEBASENet (TIN2015-71841-REDT (MINECO)).

References

1. Adamopoulos, K., Harman, M., Hierons, R.M.: How to overcome the equivalent mutant problem and achieve tailored selective mutation using co-evolution. In: Deb, K. (ed.) GECCO 2004. LNCS, vol. 3103, pp. 1338–1349. Springer, Heidelberg (2004). doi:10.1007/978-3-540-24855-2_155
2. Banzi, A.S., Nobre, T., Pinheiro, G.B., Árias, J.C.G., Pozo, A., Vergilio, S.R.: Selecting mutation operators with a multiobjective approach. Expert Syst. Appl. **39**(15), 12131–12142 (2012). http://dx.doi.org/10.1016/j.eswa.2012.04.041
3. Delgado-Pérez, P., Medina-Bulo, I., Segura, S., Domínguez-Jiménez, J.J., García-Domínguez, A.: GiGAn: evolutionary mutation testing for C++ object-oriented systems. In: The 32nd ACM Symposium on Applied Computing (SAC 2017) (2017)
4. Delgado-Pérez, P., Segura, S., Medina-Bulo, I.: Assessment of C++ object-oriented mutation operators: a selective mutation approach. Softw. Test. Verif. Reliab. (2017). http://dx.doi.org/10.1002/stvr.1630
5. Domínguez-Jiménez, J.J., Estero-Botaro, A., García-Domínguez, A., Medina-Bulo, I.: GAmera: an automatic mutant generation system for WS-BPEL compositions. In: Proceedings of the 7th IEEE European Conference on Web Services, pp. 97–106. IEEE Computer Society Press, Eindhoven, November 2009. http://dx.doi.org/10.1109/ECOWS.2009.18
6. Domínguez-Jiménez, J.J., Estero-Botaro, A., García-Domínguez, A., Medina-Bulo, I.: Evolutionary mutation testing. Inf. Softw. Technol. **53**(10), 1108–1123 (2011). http://dx.doi.org/10.1016/j.infsof.2011.03.008
7. Estero-Botaro, A., Palomo-Lozano, F., Medina-Bulo, I., Domínguez-Jiménez, J.J., García-Domínguez, A.: Quality metrics for mutation testing with applications to WS-BPEL compositions. Softw. Test. Verif. Reliab. **25**(5–7), 536–571 (2015). http://dx.doi.org/10.1002/stvr.1528
8. Goldberg, D.E.: Genetic Algorithms in Search, Optimization and Machine Learning, 1st edn. Addison-Wesley Longman Publishing Co. Inc., Boston (1989)
9. Hierons, R., Harman, M., Danicic, S.: Using program slicing to assist in the detection of equivalent mutants. Softw. Test. Verif. Reliab. **9**(4), 233–262 (1999). http://dx.doi.org/10.1002/(SICI)1099-1689(199912)9:4⟨233::AID-STVR191⟩3.0.CO;2-3

10. Jia, Y., Harman, M.: Constructing subtle faults using higher order mutation testing. In: Proceedings of the Eighth IEEE International Working Conference on Source Code Analysis and Manipulation, 2008, pp. 249–258, September 2008. http://dx.doi.org/10.1109/SCAM.2008.36

11. Offutt, A.J., Pan, J.: Detecting equivalent mutants and the feasible path problem. In: Proceedings of the Eleventh Annual Conference on Computer Assurance, Systems Integrity. Software Safety. Process Security (COMPASS 1996), pp. 224–236, June 1996. http://dx.doi.org/10.1109/CMPASS.1996.507890

12. de Oliveira, A.A.L., Camilo-Junior, C.G., Vincenzi, A.M.R.: A coevolutionary algorithm to automatic test case selection and mutant in mutation testing. In: Proceedings of the IEEE Congress on Evolutionary Computation (CEC 2013), pp. 829–836, June 2013. http://dx.doi.org/10.1109/CEC.2013.6557654

13. Omar, E., Ghosh, S., Whitley, D.: HOMAJ: a tool for higher order mutation testing in AspectJ and Java. In: Proceedings of the IEEE Seventh International Conference on Software Testing, Verification and Validation Workshops (ICSTW 2014), pp. 165–170, March 2014. http://dx.doi.org/10.1109/ICSTW.2014.19

14. Papadakis, M., Delamaro, M., Traon, Y.L.: Mitigating the effects of equivalent mutants with mutant classification strategies. Sci. Comput. Program. **95**(Part 3), 298–319 (2014). http://dx.doi.org/10.1016/j.scico.2014.05.012, special Section: ACM SAC-SVT 2013 + Bytecode 2013

15. Papadakis, M., Jia, Y., Harman, M., Le Traon, Y.: Trivial compiler equivalence: a large scale empirical study of a simple, fast and effective equivalent mutant detection technique. In: Proceedings of the 37th International Conference on Software Engineering - Volume 1 (ICSE 2015), pp. 936–946. IEEE Press, Piscataway (2015). http://dx.doi.org/10.1109/ICSE.2015.103

16. Pargas, R.P., Harrold, M.J., Peck, R.R.: Test-data generation using genetic algorithms. Softw. Test. Verif. Reliab. **9**(4), 263–282 (1999). http://dx.doi.org/10.1002/(SICI)1099-1689(199912)9:4⟨263::AIDSTVR190⟩3.0.CO;2-Y

17. Schuler, D., Zeller, A.: Covering and uncovering equivalent mutants. Softw. Test. Verif. Reliab. **23**(5), 353–374 (2013). http://dx.doi.org/10.1002/stvr.1473

18. Schwarz, B., Schuler, D., Zeller, A.: Breeding high-impact mutations. In: Proceedings of the 4th IEEE International Conference on Software Testing, Verification, and Validation Workshops (ICSTW 2011), pp. 382–387 (2011). http://dx.doi.org/10.1109/ICSTW.2011.56

19. Usaola, M., Mateo, P.: Mutation testing cost reduction techniques: a survey. IEEE Softw. **27**(3), 80–86 (2010). http://dx.doi.org/10.1109/MS.2010.79

20. Woodward, M.R.: Mutation testing - its origin and evolution. Inf. Softw. Technol. **35**(3), 163–169 (1993). http://dx.doi.org/10.1016/0950-5849(93)90053-6

21. Xanthakis, S., Ellis, C., Skourlas, C., Le Gall, A., Katsikas, S., Karapoulios, K.: Application of genetic algorithms to software testing. In: Proceedings of the 5th International Conference on Software Engineering and Applications, pp. 625–636 (1992)

22. Zhang, L., Gligoric, M., Marinov, D., Khurshid, S.: Operator-based and random mutant selection: better together. In: Proceedings of the IEEE/ACM 28th International Conference on Automated Software Engineering (ASE 2013), pp. 92–102, November 2013. http://dx.doi.org/10.1109/ASE.2013.6693070

23. Zhang, L., Hou, S.S., Hu, J.J., Xie, T., Mei, H.: Is operator-based mutant selection superior to random mutant selection? In: Proceedings of the 32nd ACM/IEEE International Conference on Software Engineering - Volume 1 (ICSE 2010), pp. 435–444, ACM, New York (2010). http://dx.doi.org/10.1145/1806799.1806863

Towards Deterministic and Stochastic Computations with the Izhikevich Spiking-Neuron Model

Ramin M. Hasani$^{(\boxtimes)}$, Guodong Wang, and Radu Grosu

Cyber Physical Systems Group, Institute of Computer Engineering,
Vienna University of Technology, Vienna, Austria
{ramin.hasani,guodong.wang,radu.grosu}@tuwien.ac.at

Abstract. In this paper we analyze simple computations with spiking neural networks (SNN), laying the foundation for more sophisticated calculations. We consider both a deterministic and a stochastic computation framework with SNNs, by utilizing the Izhikevich neuron model in various simulated experiments. Within the deterministic-computation framework, we design and implement fundamental mathematical operators such as addition, subtraction, multiplexing and multiplication. We show that cross-inhibition of groups of neurons in a winner-takes-all (WTA) network-configuration produces considerable computation power and results in the generation of selective behavior that can be exploited in various robotic control tasks. In the stochastic-computation framework, we discuss an alternative computation paradigm to the classic von Neumann architecture, which supports information storage and decision making. This paradigm uses the experimentally-verified property of networks of randomly connected spiking neurons, of storing information as a stationary probability distribution in each of the sub-network of the SNNs. We reproduce this property by simulating the behavior of a toy-network of randomly-connected stochastic Izhikevich neurons.

1 Introduction

Contemporary models of neural-computation are based on mathematical abstractions capturing the main principles underlying information processing within the nervous system of animals [1]. These models have also to consider other criteria, such as the speed of information processing, robustness, plausibility of parameters, and complexity of parameter optimization. For the development of neural-computation blocks, two key modeling components are required: (1) A suitable model for neurons, and (2) A proper model for their connectivity.

Suitable neuron model. Various single-compartment models for the neurons building up a spiking neural network (SNN) have been proposed, imitating the way the brain processes information [1–3]. Examples include the biologically-inspired Hodgkin-Huxley model (HH) [4], the leaky integrate-and-fire model (I&F) [5], and the Izhikevich neuron model (INM) [6]. HH has biologically meaningful and physically measurable parameters. It can reproduce all the neuro-computational features of real neurons, while being computationally expensive.

© Springer International Publishing AG 2017
I. Rojas et al. (Eds.): IWANN 2017, Part II, LNCS 10306, pp. 392–402, 2017.
DOI: 10.1007/978-3-319-59147-6_34

I&F on the other hand, is the simplest way of modeling spiking activities, where an action potential is generated by accumulating a graded potential that is thresholded with a firing function. Although I&F is computationally cost-efficient, it fails to reproduce most of the spiking patterns generated by real neurons. INM benefits from both efficiency and generality. It consists of two ordinary differential equations (ODE), including four tuneable parameters, with which various firing patterns of the biological neurons are reproducible.

Connectivity model. Synapses are information-transmission relays, from pre-synaptic, to post-synaptic neurons, respectively. The dynamics of the information transmission depends on several mechanisms, such as, the neurotransmitter concentration at the pre-synaptic neuron, the receptors concentration at the post-synaptic neuron, the state of activation of the pre-synaptic neuron, as well as the neurotransmitters-secretion dynamics [7]. In the next section, we design a synapse model (SM) which includes such dependencies.

Using the (energy-efficient) neural-processing blocks, we define small-scale neural networks, underlying more complicated computations. Arithmetic using spiking neurons, can be performed in either a deterministic- [8–10], or in a stochastic-computation paradigm [11–13]. Within the deterministic paradigm, we show how several simple, rate-based mathematical operations, such as, addition, subtraction, multiplexing, and multiplication, can be performed efficiently within the INM and the SM. Addition and subtraction are designed inspired by the temporal summation of the graded potential, at the junction of a group of pre-synaptic neurons, with a post-synaptic neuron [14]. We design a selective filter, by employing a lateral cross-inhibition of a region of neurons, in a winner takes all (WTA) architecture [15]. We then employ this structure, to design a novel spike-rate neural-multiplier circuit. Finally, we suggest small-scale neural nets that can be feasibly employed in the robotic control processes.

Within the stochastic-computation paradigm, we reproduce a major property of the stochastic, randomly-structured SNNs, initially suggested, and analytically proved, by Habenschuss et al. in [16]: Within an arbitrarily structured stochastic SNN, subjected to external inputs, there exists a stationary, spiking-activity probability distribution of the SNN, which resembles a mechanism of knowledge storing, within the firing states of the network. Here, we repeat this experiment by using the INM, and recapitulate the achieved results.

The rest of the paper is organized as follows: In Sect. 2, we describe the deterministic computational model, based on the INM and the SM. Within the same section, we design the mathematical operators, such as addition, subtraction, selective element, and multiplication. In Sect. 3, we discuss the stochastic computational power of the spiking neurons on storing knowledge through a small-scale SNN. Section 4 concludes with a review of our results.

2 Deterministic Computations with INM and SM

In this section we design fundamental mathematical operators, including addition, subtraction, multiplication and multiplexing, by using the INM and the SM.

First, we briefly introduce the neuron model. Then, we describe a synapse model implementation and correspondingly, we develop the arithmetic networks.

2.1   Izhikevich Neuron Model (INM)

INM [6], is a computationally cost-efficient model, which can reproduce many biologically realistic spiking patterns generated by the nerve cells. The model comprises a quadratic ODE modeling the *membrane potential* v, and a linear ODE representing a *membrane recovery* variable u, as follows:

$$\frac{dv}{dt} = 0.04v^2 + 5v + 140 - u + I \tag{1}$$

$$\frac{du}{dt} = a(bv - u), \tag{2}$$

where v and u get *reset* after each spike with the following rule:

$$\text{if } v \geq 30 \text{ mV, then} \begin{cases} v \longleftarrow c \\ u \longleftarrow u + d. \end{cases} \tag{3}$$

and a, b, c and d are dimensionless parameters standing for the *time constant* of the recovery variable u, *sub-threshold oscillations* of the membrane potential, *after-spike reset* value for the membrane potential and *after-spike reset* value of the recovery variable u, respectively. By tuning these parameters one can recreate various neuro-computational properties of a real nerve cell, including tonic spiking, fast spiking, chattering (see [3] for a complete list of spiking patterns).

2.2   Synapse Model (SM)

Within a neural network, information is transferred through *synapses*. They allow the pre-synaptic neurons to control the membrane potential of the post-synaptic neurons. If a synapse causes an increase of the post-synaptic membrane potential, it is called *excitatory*, otherwise it is called *inhibitory*.

An action potential from the pre-synaptic neuron results in the secretion of neurotransmitters that induces a change in conductance of the post-synaptic neuron [5]. Based on Ohm's law, such current can be modeled as:

$$I_{syn} = G(v_{pre})(E_{syn} - v_{post}), \tag{4}$$

where $G(v_{pre})$ is the *conductance* of the synapse which depends on the *pre-synaptic potential* v_{pre} and E_{syn} is the membrane *reversal potential*. If $E_{syn} \geq 0$, the synapse is excitatory, otherwise the synapse is inhibitory.

A proper approximation of $G(v_{pre})$ is the total *concentration of neurotransmitter* release as a result of a pre-synaptic action potential:

$$G(v_{pre}) = \frac{G_{max}}{1 + e^{-(v_{pre} - v_{th})/v_{st}}}, \tag{5}$$

where G_{max} is the *maximum conductance* of a synapse representing the maximum concentration of the available neurotransmitters. And v_{th} and v_{st} determine the *threshold* and *stiffness* of the release, respectively [5].

We have implemented a Simulink model of the described synapse. In this synapse model the parameters are chosen as follows: G_{max} = weight of the synapse, $v_{th} = -45$ mV, $v_{st} = -4$ mV, $E_{syn} = 0$ mV for excitatory synapses and $E_{syn} = -90$ mV for inhibitory synapses.

2.3 Mathematical-Operators Implementation

In this section we use the INM and the SM in order to design task-specific networks of spiking neurons. For this purpose, we implement a Simulink model of the INM for creating basic math operators such as addition, subtraction, multiplexing and multiplication. First, we use three particular sets of parameters of the INM for designing operator networks which results in producing tonic spiking, fast spiking and chattering spiking patterns, presented in Table 1.

Table 1. The parameters a, b, c, and d for three spiking patterns.

Spiking pattern	a	b	c	d
Tonic Spiking (TS)	0.02	0.2	−65	8
Fast Spiking (FS)	0.1	0.2	−65	2
Chattering (CH)	0.02	0.02	−50	2

Addition and Subtraction Implementation. Given the spiking patterns defined above, we implement addition (subtraction) by summing up (subtracting) the rates of the spikes generated from two pre-synaptic neurons, within a time interval τ. To this end, we introduce a third neuron with TS activity, which is a post-synaptic neuron to those input neurons.

If both input neurons excite the post-synaptic neuron then the spiking activity of this neuron, given proper synaptic weights, will be the sum of that of pre-synaptic neurons. Figure 1A illustrates the addition network architecture together with an example of its input/output spiking activities.

With the same network architecture, one can model subtraction, if one of the two pre-synaptic input neurons inhibits the post-synaptic neuron. By choosing the correct weight for the SM, the network outputs the difference of the rates of the spiking activity of the two input neurons, (See Fig. 1B for a subtraction network realization together with an example of input/output spiking activities).

The proposed architecture is generalizable to addition or subtraction of the spiking activity of n input neurons. In the general case, for each input number we need an input neuron. Therefore, the total number of neurons for addition of n spike-rates, is $n+1$. The CPU time for the simulation of adding two spike-rates within a 100 s time interval is calculated as 0.81 s. Within a 100 s calculation time

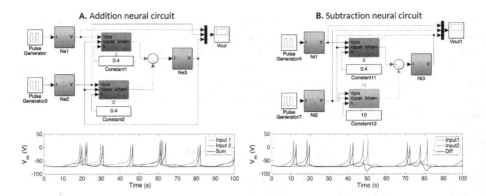

Fig. 1. Addition and subtraction computational networks within a 100 s time-interval τ. (A) Two input neurons' spike-rates are summed up by merging their outputs to a single post-synaptic neuron. Simulation results are shown at the bottom. (B) Subtraction of two input firing-rates is performed by excitation through the first neuron and inhibition through the second neuron. Simulation results of such network are shown on the bottom.

interval each input neuron can have a maximum of 50 spikes in order to deliver the correct output results. This is due to the refractory period of individual neurons. However, the addition of larger numbers than 50 is possible in two different ways: (1) Using neurons with a FS configuration where more spikes can be generated within the same time interval. (2) Increasing the calculation time interval τ. Figure 2 depicts the relationship between the calculation time interval τ and the cpu time of the Simulink simulations.

Selective-Element (SE) Implementation. A winner takes all (WTA) configuration of neurons, where neurons laterally cross-inhibit each other creates a selective behavior for the network. Figure 3 shows the SE structure: Spiking neurons are cross-inhibiting each other. As long as the inputs to all neurons are

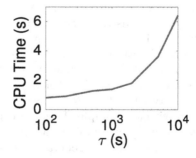

Fig. 2. MATLAB/Simulink CPU time as a function of the calculation time interval τ. Note that τ-axis is plotted in log-scale.

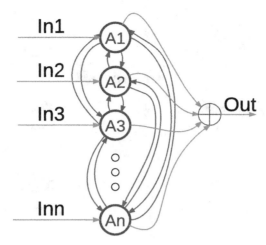

Fig. 3. Winner takes all (WTA) structure of the SE. Blue circles are spiking neurons. Green (red) arrows represent excitatory (inhibitory) synapses. (Color figure online)

the same, the output stays at its equilibrium state. Once one of the inputs varies from the others, the SE outputs the response of that specific neuron. Depending on the number of inputs n, we need $n+1$ neurons, $2n+1$ excitatory synapses and $n^2 - n$ inhibitory synapses, in order to create the SE. We employ our selective structure for designing a spike-rate-multiplier network.

Multiplication Implementation. Multiplication of spike rates is a costly operation within the INM and SM models. This should be no surprise, as multiplication is a costly operation in current micro-processors, too. In most cases, it is reduced to a repeated addition, where one of the inputs dictates the number of additions. Moreover, while amplitude is encoded in hardware in space (number of bits), this is encoded in neural circuits in time (number of spikes in a given interval). As a consequence, in order to know the magnitude of the operands, one needs first to wait for the given interval, before performing multiplication.

For full generality, if the inputs are stochastic, we have to initially set a time interval τ_1, where we read the spike-rate of the second neuron m_2. The first input neuron m_1 is pre-synaptic to all the inputs of an $n \times 1$ SE designed as above. The neurons of the SE are configured such that A_1, in case of activation, produces one spike for every pre-synaptic action potential, neuron A_2 produces two, neuron A_3 three, and neuron A_n produces n spikes upon arrival of a pre-synaptic action potential. This is achieved by setting up the neurons in TS, FS and CH spiking configurations. Within a calculation time interval τ_2, through a readout circuit consisting of a spike-counter and a demultiplexer, we vary the synaptic connection of the A_i neuron of the SE corresponding to the spike-rate of the second neuron m_2. The SE therefore chooses to multiply the spike-rate of neuron m_1 with m_2 which is the desired operation.

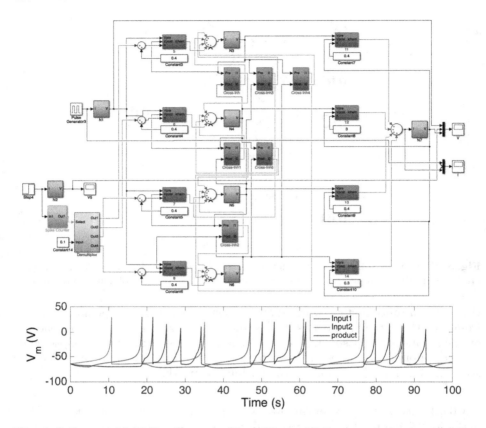

Fig. 4. Spike-rate Multiplier Network. (Top) The multiplier neural circuit multiplies the rate of the firing of two input neurons. The first input neuron is pre-synaptic to the neurons of the SE and the rate of firing of the first neuron is calculated by using a readout circuit that selects what the desired multiplication path is. (Bottom) An example of the performance of the multiplier network. Firing-rate of 3 of the neuron N_1 (blue), is multiplied with that of neuron N_2 (red), which is 4. As a consequence, the network produces 12 spikes (black) which is the desired result. The time interval is set to 100 s. (Color figure online)

Figure 4 shows the structure of the multiplier in Simulink. The circuit is able to multiply any spike-rate of neuron N_1, with 1, 2, 3 or 4 which are the allowed spike-rates neuron N_2. This limitation is due to the SE, which includes only 4 neurons. Neuron N_1 simultaneously branches to the neurons of the SE. The spike rate of neuron N_2, defines the path selected by a readout demultiplexer. It makes the synaptic weight of the target multiplication slightly higher than the other inputs to the SE. Finally, at the output we see the result of the multiplications as the spike-rate of neuron N_7. The CPU time for the simulation of the multiplier block within a 100 s time interval is calculated as 1.3 s.

In general, for multiplying two spike-rates, m and n, one needs 2 input neurons, $n + 1$ neurons within the SE, $2n$ excitatory, and $n^2 - n$ inhibitory synapses, respectively. If the input to the network is deterministic, the circuit estimates the number of firing activities of the second neuron (which is configured in a TS mode). Therefore there is no need for the initial time interval τ_1.

It is worth noting that more and more experimental evidence supports the existence of neurons capable of exponentiation and logarithm [17]. Using such neurons multiplication becomes very cheap: $m * n = exp(ln(m) + ln(n))$.

3 Stochastic Computations: Storing Knowledge in the Form of a Probability Distribution Using the INM

In this section we use the INM and the SM for reproducing the following hypothesis: *There exists a stationary probability distribution, of firing activity of neurons, in a randomly-connected recurrent SNN* [16].

The hypothesis suggests that a significant amount of information is stored in regions of the SNN forming a stationary probability distribution [16]. Such structure in the brain presumably results in fast decoding of information, when necessary, and facilitates the process of decision making (see e.g. [18,19]).

This stationary-probability model defines an alternative computation model to conventional computers, which sequentially store the data on a hard disk [11]. Below, we create a small SNN with the INM and the SM, and capture such a stationary-probability property in a networks of spiking neurons.

3.1 Convergence of Network-States to a Stationary Distribution

It has been observed that in a network of randomly connected spiking neurons, in presence of noise, the state of the spiking activity of the neurons, within a predefined time interval, follows a stationary probability distribution, for the entire network, and for any subgroup of the network [16]. More precisely:

Theorem: Considering an arbitrary SNN, C, in presence of noise, and a set of input neurons $I = 1 \ldots N$, which stimulate the network with different rates $x_i(t)$. For an input with constant firing-rate or any generated input which obeys Markov properties, the firing activity of the neurons of the network follows a stationary probability distribution $p_c(y|x)$, of network states y. This is true for any subgroup of C, where a marginal stationary distribution $p_c(\hat{y}|x)$, of network states $\hat{y}$ exists, to which the sub-network exponentially converges [16].

In the following, we design a small network and investigate its spiking activities. Figure 5A represents a network of 10 spiking neurons, stimulated by two input neurons, I_1 and I_2, which fire at a constant rate. The network comprises seven excitatory neurons and three inhibitory neurons hardwired through synapses. Each Izhikevich neuron (tuned for tonic spiking (TS) configuration) is fed by a white noise source which makes the neuron spike stochastically. The firing activity of the neurons in a 200 ms time-window taken from a 2000 ms simulation of the recurrent SNN is shown in Fig. 5B.

Let *a simple-state* of the firing activity of a subgroup C of neurons be defined as their firing activity (regardless of the rate of spikes) within a time interval θ.

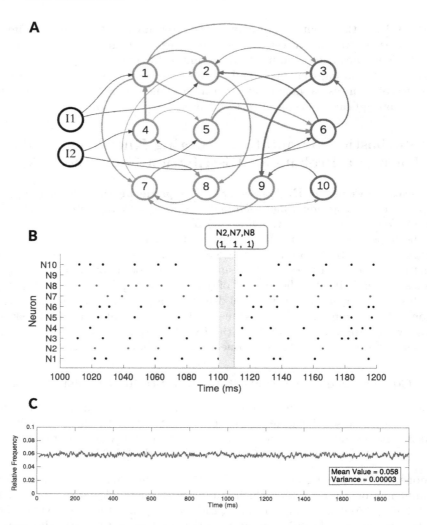

Fig. 5. Inference in a randomly structured recurrent SNN. (A) Network of 12 spiking neurons wired in a random topology. *Input* neurons are shown in black, *excitatory* neurons and synapses in blue and *inhibitory* ones are colored in red. (B) Firing activity of 10 neurons of a random recurrent SNN in a 200 ms time-window from a 2000 ms-simulation of the network. The *green interval* (10 ms) indicates a simple state in which we observe as the firing activity of the subgroup of neurons N_2, N_7, and N_8. Their representative state at this interval is $(1, 1, 1)$, where they all fire action potentials. (C) Stationary distribution of spiking activity of the sub-network N_2, N_7, N_8, in the state in which all of them have spiking activities, $(1, 1, 1)$, within a time interval 10 μs. (Color figure online)

For instance, if we take the firing activity of neurons N_2, N_2 and N_8 in the green time interval shown in Fig. 5B, their corresponding state is determined as $(1, 1, 1)$ where they are all generating action potentials. Therefore, for this subset of three neurons, seven possible states can be determined, as we do not count the silence state $(0, 0, 0)$.

We now quantify the frequency of such a state during a simulation, and perform sampling to compute its probability distribution. For instance, we sample the simulation described above for 10000 times and calculate the probability of the occurrence of the state $(1, 1, 1)$ for the marginal network consisting of neurons N_2, N_7 and N_8 over the simulation time. Figure 5C depicts the normalized frequency of the state. The frequency remains constant over the simulation time.

This result indicates that certain input data to a hardwired recurrent SNN can be stored in the form of a probability distribution of spiking activity of the neurons. Potential applications of such a powerful property of spiking neural networks include, solving constraint satisfaction problems, which are out of the scope of this paper. The interested reader is referred to [11,20,21].

4 Conclusions

This paper described deterministic and stochastic computations with small-scale spiking neural networks. In the deterministic-computation framework, we created addition and subtraction circuits with the Izhikevich spiking neuron model. We continued by exploring the power of mutual inhibition of groups of spiking neurons, by defining a winner-takes-all (WTA) configuration. We showed that WTA exhibits a multiplexing behavior, that can be useful for many control tasks. In particular, we used WTA for designing a multiplication circuit.

In the stochastic-computation framework, we reproduced an important property of networks of randomly connected spiking neurons: The storage of knowledge within the states of the network, as the firing shape of a stationary probability distribution. We employed stochastic Izhikevich neurons for designing an SNN in a random topology, and computed the frequency-rate of a marginal state of the network, in a simulated experiment. We observed that the probability of the occurrence of such a state, remains constant over the simulation time, which illustrates the correctness of the theorem.

For future work, we aim to utilize our computational elements, in the control of common robotic tasks, such as moving objects with a robotic arm, motion planning with rover robots, and stabilization of quad-copters.

Acknowledgments. This work was partially supported by the NSF-Frontiers Cyber-Cardia Award, FWF-NFN RiSE Award, FWF-DC LMCS Award, FFG Harmonia Award, FFG Em2Apps Award, and the TUW CPPS-DK Award.

References

1. Herz, A.V.M., Gollisch, T., Machens, C.K., Jaeger, D.: Modeling single-neuron dynamics and computations: a balance of detail and abstraction. Science **314**(5796), 80–85 (2006)
2. Abbott, L.F., Kepler, T.B.: Model neurons: from Hodgkin-huxley to hopfield. In: Garrido, L. (ed.) Statistical Mechanics of Neural Networks. LNP, vol. 368, pp. 5–18. Springer, Heidelberg (1990)

3. Izhikevich, E.M.: Which model to use for cortical spiking neurons. IEEE Trans. Neural Netw. **15**, 1063–1070 (2004)
4. Hodgkin, A.L., Huxley, A.F.: A quantitative description of membrane current and its application to conduction and excitation in nerve. J. Physiol. **117**(4), 500 (1952)
5. Koch, C., Segev, I.: Methods in Neuronal Modeling: From Ions to Networks. MIT Press, Cambridge (1998)
6. Izhikevich, E.M.: Simple model of spiking neurons. IEEE Trans. Neural Netw. **14**, 1569–1572 (2003)
7. Schutter, E.D.: Computational Modeling Methods for Neuroscientists. The MIT Press, Cambridge (2009)
8. Pfeil, T., Grubl, A., Jeltsch, S., Muller, E., Muller, P., Petrovici, M.A., Schmuker, M., Bruderle, D., Schemmel, J., Meier, K.: Six networks on a universal neuromorphic computing substrate. arXiv preprint arXiv:1210.7083 (2012)
9. McDonnell, M.D., Boahen, K., Ijspeert, A., Sejnowski, T.J.: Engineering intelligent electronic systems based on computational neuroscience [scanning the issue]. Proc. IEEE **102**(5), 646–651 (2014)
10. Benjamin, B.V., Gao, P., McQuinn, E., Choudhary, S., Chandrasekaran, A.R., Bussat, J.-M., Alvarez-Icaza, R., Arthur, J.V., Merolla, P.A., Boahen, K.: Neurogrid: a mixed-analog-digital multichip system for large-scale neural simulations. Proc. IEEE **102**(5), 699–716 (2014)
11. Maass, W.: Noise as a resource for computation and learning in networks of spiking neurons. Proc. IEEE **102**(5), 860–880 (2014)
12. Hasani, R.M.: Design of CMOS silicon neurons for noise assisted computations in spiking neural networks. Politesi Digital Library of PhD and Post Graduate Theses, Politecnico di Milano (2015)
13. Hasani, R.M., Ferrari, G., Yamamoto, H., Kono, S., Ishihara, K., Fujimori, S., Tanii, T., Prati, E.: Control of the correlation of spontaneous neuron activity in biological and noise-activated CMOS artificial neural icrocircuits. arXiv preprint arXiv:1702.07426 (2017)
14. Magee, J.C.: Dendritic ih normalizes temporal summation in hippocampal CA1 neurons. Nat. Neurosci. **2**(6), 508–514 (1999)
15. Maass, W.: On the computational power of winner-take-all. Neural Comput. **12**(11), 2519–2535 (2000)
16. Habenschuss, S., Jonke, Z., Maass, W.: Stochastic computations in cortical microcircuit models. PLoS Comput. Biol. **9**(11), e1003311 (2013)
17. Jones, P.W., Gabbiani, F.: Logarithmic compression of sensory signals within the dendritic tree of a collision-sensitive neuron. J. Neurosci. **32**(14), 4923–4934 (2012)
18. Vul, E., Pashler, H.: Measuring the crowd within probabilistic representations within individuals. Psychol. Sci. **19**(7), 645–647 (2008)
19. Tenenbaum, J.B., Kemp, C., Griffiths, T.L., Goodman, N.D.: How to grow a mind: statistics, structure, and abstraction. Science **331**(6022), 1279–1285 (2011)
20. Jonke, Z., Habenschuss, S., Maass, W.: Solving constraint satisfaction problems with networks of spiking neurons. Front. Neurosci. **10**, 118 (2016)
21. Binas, J., Indiveri, G., Pfeiffer, M.: Spiking analog VLSI neuron assemblies as constraint satisfaction problem solvers. In: IEEE International Symposium on Circuits and Systems (ISCAS), pp. 2094–2097. IEEE (2016)

A Formal Framework to Specify and Test Systems with Fuzzy-Time Information

Juan Boubeta-Puig[1], Azahara Camacho[2], Luis Llana[2], and Manuel Núñez[2]($\boxtimes$)

[1] Department of Computer Science and Engineering, School of Engineering,
University of Cadiz, Cadiz, Spain
juan.boubeta@uca.es
[2] Departamento de Sistemas Informáticos y Computación,
Universidad Complutense de Madrid, Madrid, Spain
{mariaazc,llana,manuelnu}@ucm.es

Abstract. We specify the behavior of a sensor network with different sensor stations distributed all along the region of Andalusia (South of Spain). The main goal of this network is the measure of air quality taking into account the maximum levels of certain pollutants. The problem that we try to solve with this formalization is the management of time inaccuracies between the components of a network with the aim of avoiding the malfunctions that are derived from them. We present the formal syntax and semantics of our variant of fuzzy-timed automata and define all the automata corresponding to the different parts of the netowrk.

1 Introduction

The use of formal methods for the specification and analysis of network systems and of the protocols running over them is not new [3,4]. However, formal methods are not widely used to guide the development of computer systems. This is an anomalous situation in engineering, where the simplest project includes a detailed design of the artifact to be built. Therefore, formal methods should use blueprints to guide the *construction* of computer systems [8], in particular, to formally implement the validation process via testing [5,7].

In this paper we present a formal design of a non-trivial system: the measure of air quality through a sensors network, with sensor stations distributed all along Andalusia (a region of Spain). First, we had to find a formalism to represent the network. The problem with well-established formal methods is that they are oriented to systems where uncertainty usually does not play a relevant

The research presented in this paper has been partially supported by the Spanish MINECO/FEDER project DArDOS (TIN2015-65845-C3-1-R and TIN2015-65845-C3-3-R), the Comunidad de Madrid project SICOMORo-CM (S2013/ICE-3006) and the University of Cádiz project (PR2016-032). The second author is supported by a Universidad Complutense de Madrid - Santander Universidades grant. The first author would like to thank the ho spitality of the Design and testing of reliable systems research group, at Universidad Complutense de Madrid, during his stay when this research was carried out.

I. Rojas et al. (Eds.): IWANN 2017, Part II, LNCS 10306, pp. 403–414, 2017.
DOI: 10.1007/978-3-319-59147-6_35

role. In fact, they are good to cope with time information, parallelism and non-determinism, features required in our study, but approximate reasoning is out of the scope of these languages. Specifically, they are not good to deal with *fuzzy time*. Suppose that a certain signal should be received every 10 min and there is a lapse of 10 min and 0.001 s since the reception of the last message then we should not raise an alarm because this *error* might be due to several reasons, none of them showing a failure of the device sending the signal. For example, the clock of the receiver might be wrong or the message might have been delayed while being retransmitted in a point between the sender and the receiver. We decided to *adapt* an existing formalism, timed automata [1], to consider fuzzy time. Although there are many proposals to define fuzzy automata [2,6,11,12], we considered timed automata because it is much easier to reuse its existing tools to deal with fuzzy time, most notably UPPAAL [9], than build tools from scratch. Technically, our proposal is a hybrid between timed automata [1] and our proposal of fuzzy automata [2].

We introduce a formalism to formally design complex systems where inaccuracies in the measure of time are managed via fuzzy logic. In addition, we introduce an implementation relation so that we can match the expected behavior (the one given by the specification) and the one of the studied system. The idea is to analyze mismatches in order to detect potential malfunctioning. In order to show the usefulness of our formalism, we will formally specify a sensors network formed by a central server connected to 61 sensor stations, each of them conformed by 6 sensors, used to capture the amount of different pollutants in the air of Andalusia.

The rest of the paper is structured as follows. In Sect. 2 we present some basic concepts of fuzzy logic and introduce our fuzzy adaption of timed automata. In Sect. 3 we present our case study and formally specify the sensors network. Finally, in Sect. 4 we present our conclusions.

2   Introducing Fuzzy-Timed Automata

In this section we introduce some basic concepts of fuzzy logic that will be used afterwards to define our *fuzzy* version of timed automata.

In ordinary logic, a set or a relation is determined by its characteristic function: a function that returns true if the element is in the set (or if some elements are related) and false otherwise. In the fuzzy framework we have a complete range of values in the interval $[0, 1]$; the larger is the value, the more confidence we have in the assessment. We consider relations over the set of real numbers $\mathbb{R}$. Therefore, a fuzzy relation is a mapping from the Cartesian product $\mathbb{R}^n$ into the interval $[0, 1]$.

In this paper, we consider the fuzzy relations given in Fig. 1. These relations depend on a non negative real number $\lambda \geq 0$. We use these functions to define fuzzy-timed automata. The only property we use, which simplifies the writing, is the commutativity of equality $\overline{x = y}^{\lambda} = \overline{y = x}^{\lambda}$, viewed as a binary operation over $\mathbb{R}$.

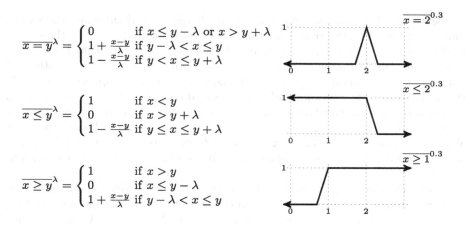

$$\overline{x=y}^{\lambda} = \begin{cases} 0 & \text{if } x \le y - \lambda \text{ or } x > y + \lambda \\ 1 + \frac{x-y}{\lambda} & \text{if } y - \lambda < x \le y \\ 1 - \frac{x-y}{\lambda} & \text{if } y < x \le y + \lambda \end{cases}$$

$$\overline{x \le y}^{\lambda} = \begin{cases} 1 & \text{if } x < y \\ 0 & \text{if } x > y + \lambda \\ 1 - \frac{x-y}{\lambda} & \text{if } y \le x \le y + \lambda \end{cases}$$

$$\overline{x \ge y}^{\lambda} = \begin{cases} 1 & \text{if } x > y \\ 0 & \text{if } x \le y - \lambda \\ 1 + \frac{x-y}{\lambda} & \text{if } y - \lambda < x \le y \end{cases}$$

Fig. 1. Fuzzy order relations

A triangular norm (abbreviated t-norm) is a binary operation used in fuzzy logic to generalize the *conjunction* in propositional logic. Therefore, we require a t-norm to satisfy similar properties. We also require an extra property: monotonicity. Intuitively, the resulting truth value does not decrease if the truth values of the arguments increase.

Definition 1. *A t-norm is a function $T : [0,1] \times [0,1] \mapsto [0,1]$ which satisfies the following properties:*

- *Commutativity: $T(x,y) = T(y,x)$.*
- *Monotonicity: $T(x,y) \le T(z,u)$ if $x \le z$ and $y \le u$.*
- *Associativity: $T(x,T(y,z)) = T(T(x,y),z)$.*
- *Number 1 is the identity element: $T(x,1) = x$.*
- *Number 0 is nilpotent: $T(x,0) = 0$.*

Since t-norms are associative, we can generalize them to take as parameter a list of values:

$$T(x_1, x_2, \ldots, x_{n-1}, x_n) = T(x_1, T(x_2, \ldots, T(x_{n-1}, x_n) \ldots))$$

The following t-norms are often used:

Łukasiewicz t-norm: $T(x,y) = \max(0, x + y - 1)$. We represent this t-norm with the symbol $\curlywedge$.

Gödel t-norm: $T(x,y) = \min(x,y)$. We represent this t-norm with the symbol $\overline{\wedge}$.

Product t-norm: $T(x,y) = x \cdot y$ (real number multiplication). We represent this t-norm with the symbol $\bigstar$.

Hamacher product t-norm: $T(\delta_1, \delta_2) = \frac{\delta_1 \cdot \delta_2}{\delta_1 + \delta_2 - \delta_1 \cdot \delta_2}$. We represent this t-norm with the symbol $*$.

Next, we define *fuzzy-timed* automata. In classical timed automata theory [1], time is expressed in the time constraints. Hence, we need to modify these constraints in order to be able to introduce fuzziness. In ordinary timed automata theory, the time constraints consist of conjunctions of inequalities. Instead of ordinary crisp inequalities we use their fuzzy counterparts appearing in Fig. 1. We could have more freedom in allowing general convex fuzzy sets, but we have preferred to keep our constraints close to the original ones so we can use the theory developed for timed automata. The role of a conjunction in Fuzzy Theory is played by *t*-norms. There is not a *canonical* *t*-norm: we have presented 4 of the more used *t*-norms and the designer can specify which one is more appropriate in each situation.

In order to define fuzzy-timed automata we need some additional ingredients. First, we need a set of *variables* denoted by Vars. We denote the elements of Vars by $x, y, z, \ldots$. These variables will take values in real numbers. We assume that there is a special variable ts that represents a clock. This variable can only take non negative values. An *environment* is a function $e : \text{Vars} \mapsto \mathbb{R}$ that maps variables and clocks into real numbers. We denote the set of environments by Env. We extend the environments to sequences of variables in the natural way: $e(x_1, \ldots, x_n) = (e(x_1), \ldots, e(x_n))$. The modification of a variable is denoted by $e[\overline{v}/\overline{x}]$, meaning that the variables in $\overline{x}$ take the values indicated in $\overline{v}$. We consider a set of *channel labels*, denoted by A. From this set we define the set of *output channels*: an output channel is defined by $a!(id, ID, \overline{v})$, where a is a channel label, id is the identifier of the sender of the message, ID is the set of identifiers of the receivers, and $\overline{v}$ is the tuple of sent data. Formally, $A!$ is the set

$$\{a!(id, ID, \overline{v}) \mid a \in A, id \in \mathbb{N}, ID \subseteq \mathbb{N}, \ \overline{v} \in \bigcup_{i=1}^{|\text{Vars}|} \mathbb{R}^i\}$$

We will also consider a set of *input channels* $a?(id, \overline{x})$ where $a \in A$ and id is the identifier of the component receiving the value. Formally, $A?$ is the set

$$\{a?(id, \overline{x}) \mid a \in A, id \in \mathbb{N}, \ \overline{x} \in \bigcup_{i=1}^{|\text{Vars}|} \text{Vars}^i\}$$

Finally, we consider the set of actions $\text{Acts} = A \cup A! \cup A?$. Next we define the formal syntax of our fuzzy constraints using Backus-Naur Form (BNF).

Definition 2. *A* fuzzy constraint *is a formula built from the following BNF:*

$$C ::= \text{True} \mid C_1 \triangle C_2 \mid \overline{x \bowtie n}^\lambda \mid \overline{x_1 - x_2 \bowtie n}^\lambda$$

where $\triangle$ *is a t-norm,* $\bowtie \in \{\leq, =, \geq\}$, $x, x_1, x_2 \in \text{Vars}$, $\lambda \in \mathbb{R}^+$, *and* $n \in \mathbb{N}$. *We denote the set of fuzzy constraints by* $\mathcal{FC}$.

In timed automata theory, constraints are used to decide if the automata can stay in a location and to decide if a transition can be executed. All this is done by checking if a valuation satisfies the corresponding constraint. In fuzzy theory the notion of satisfaction is not crisp: we do not have a boolean answer but a value in the interval $[0, 1]$. Therefore, we do not have that a constraint is true or false but a *satisfaction grade* of a constraint.

Definition 3. *Let* $e \in$ Env *be an environment and* $C \in \mathcal{FC}$ *be a fuzzy constraint. We inductively define the* satisfaction grade *of* C *in* e, *written* $\mu_C(e)$, *as*

$$
\begin{cases}
1 & \text{if } C = \text{True} \\
\overline{e(x) \bowtie n}^{\lambda} & \text{if } C = \overline{x \bowtie n}^{\lambda}, \; \bowtie \in \{\leq, =, \geq\} \\
\overline{e(x_1) - e(x_2) \bowtie n}^{\lambda} & \text{if } C = \overline{x_1 - x_2 \bowtie n}^{\lambda}, \; \bowtie \in \{\leq, =, \geq\} \\
\triangle\left(\mu_{C_1}(e), \mu_{C_2}(e)\right) & \text{if } C = C_1 \triangle C_2
\end{cases}
$$

Let us remark that $\mu_C(e) \in [0, 1]$.

Definition 4. *A* fuzzy-timed automaton *is a tuple* (L, l_0, n, E, I) *where:*

- L *is a finite set of locations.*
- $l_0 \in L$ *is the initial location.*
- $n \in \mathbb{N}$ *is the identifier of the automata.*
- $E \subseteq L \times \text{Acts} \times \mathcal{FC} \times L$ *is the set of edges; we write* $l \xrightarrow{a,C} l'$ *whenever* $(l, a, C, l') \in E$.
- $I : L \mapsto \mathcal{FC}$ *is a function that assigns invariants to locations.*

Next we are going to define the operational semantics of fuzzy-timed automata. We need it to obtain the fuzzy traces that are used for the conformance relations. This operational semantics is given in terms of transitions, which are enabled when time constraints hold. Since we do not have crisp time constraints, transitions must be decorated with a real number $\alpha \in [0, 1]$. This number indicates its certainty. In order to define the operational semantics we need a t-norm. Let us explain the reason. In the definition of the operational semantics of an ordinary timed automaton, action transitions require the condition "$e \models C$ and $e[r] \models I(l')$". This conjunction must be transformed into its fuzzy version: a t-norm. Our operational semantics is a labeled transition system. The states are triples (l, e), where l is a location of the automaton and e is the environment. This operational semantics has two kind of transitions: timed transitions and action transitions. The former represents the passing of time whereas the later represents the execution of an action. The actions are basically those of the set Acts with one exception: when an input action is performed we need to know the data transmitted in order to update the environment. So we consider the set:

$$
\text{Acts}_v = A \cup A! \cup \{a?(id, \overline{v}) \mid a \in A, \; \overline{v} \in \bigcup_{i=1}^{|\text{Vars}|} \mathbb{R}^i\}
$$

We also consider the initial environment e_{None} where all variables are set to 0.

Definition 5. *Let $fA = (L, l_0, id, E, I)$ be a fuzzy-timed automaton and $\triangle$ be a t-norm. The $\triangle$-operational semantics of fA is the probabilistic labeled transition system whose set of states is $L \times$ Env, the initial state is (l_0, e_{None}), and the set of transitions is given by the following two rules:*

1. $(l, e) \xrightarrow{d} \alpha(l, e[ts + d/ts])$, $\mu_{I(l)}(e[ts + d/ts]) = \alpha$.
2. $(l, e) \xrightarrow{a} \alpha(l', e')$,, *being* $\alpha =\triangle (\mu_{I(l')}(e'), \mu_C(e))$, *and such that one of the following conditions hold:*
   (a) $l \xrightarrow{a,C} l'$ *with* $e' = e$.
   (b) $l \xrightarrow{a!(id, ID, \overline{v}), C} l'$ *with* $a = a!(id, ID, \overline{v})$ *and* $e' = e$.
   (c) $l \xrightarrow{a?(id, \overline{x}), C} l'$ *with* $a = a?(id, \overline{v})$ *and* $e' = e[\overline{x}/\overline{v}]$.

We are going to represent systems in which internal clocks are not synchronized. In fact, it is very likely, and realistic, that there will be either delays or advances in the internal clocks. Thereby, in order to define transitions that allow the communication we need to remove timed transitions. So, we define a new kind of *untimed* transition.

Definition 6. *Let $(l, e), (l', e') \in L \times$ Env, $\alpha \in [0, 1]$ and $a \in$ Acts$_v$. We write $(l, e) \overset{a}{\Longrightarrow} \alpha(l', e')$ if there exist two consecutive transitions $(l, e) \xrightarrow{d} \alpha_1(l_1, e_1)$ and $(l_1, e_1) \xrightarrow{a} \alpha_2(l', e')$ such that $\alpha = \alpha_1 \cdot \alpha_2$.*

The transitions $(l, e) \overset{a}{\Longrightarrow} \alpha(l', e')$ are defined for the labeled transition system that is derived from an automaton A.

Next we are going to consider a network of automata.

Definition 7. *An automata network is given by the following BNF.*

$$N ::= (A, l, e) \mid \|_S(N_1, \ldots, N_k)$$

where $k \geq 2$ is a natural number, $S \subseteq A$ is the synchronization alphabet, e is an environment, and l is a location of the fuzzy automaton A.

Given $\triangle$ a t-norm, the operational semantics of an automata network is given by the rules in Fig. 2.

Once we have defined the operational semantics of an automata network, we can define a conformance relation. In a conformance relation we have a specification, that in our case is an automata network, and an implementation. The implementation is a real system. However, we make a (realistic) assumption to compare specifications and implementations: the latter can produce sequences of actions belonging to the same set of actions as the automata network representing the specification.

Definition 8. *Let N be an automata network, and $\triangle$ be a $t - norm$. We say that $(w, p) \in$ Acts$_v{}^* \times [0, 1]$ is a fuzzy trace of N according to $\triangle$, denoted by $(w, p) \in$ ftr$_\triangle(N)$, if there exist a sequence of transitions*

$$N = N_0 \overset{a_1,}{\Longrightarrow} p_1 N_1 \ldots \overset{a_k}{\Longrightarrow} p_n N_k$$

such that $w = a_1 \ldots a_k$ and $p =\triangle (p_1, \ldots p_k)$.

$$\frac{(l, e) \xLongrightarrow{\beta} _\alpha (l', e')}{(A, l, e) \xLongrightarrow{\beta} _\alpha (A, l', e')} \beta \in \mathsf{Acts_v}$$

$$\frac{a \in S,\ N_{i_0} \xLongrightarrow{a!(id_{i_0}, ID, v)} _{\alpha_{i_0}} N'_{i_0}, \forall j \in ID : N_j \xLongrightarrow{a?(j, v)} _{\alpha_j} N'_j, \forall j \notin ID \cup \{i_0\} : N'_j = N_j}{\|s(N_1, \ldots, N_k) \xLongrightarrow{a} _{\triangle\{\alpha_i \mid i \in ID \cup \{i_0\}\}} \|s N'_1, \ldots, N'_k}$$

$$\frac{N_i \xLongrightarrow{\beta} _\alpha N'_i, \beta \in \{b, b!(id, ID, \overline{v}), b?(id, \overline{v})\}, b \notin S}{\|s(N_1, \ldots, N_i, \ldots N_k) \xLongrightarrow{\beta} _\alpha \|s(N_1, \ldots, N'_i, \ldots N_k)}$$

Fig. 2. Automata network semantics

Let s (the specification) and i (the implementation) be two automata networks and $\triangle$ be a t-norm. We say the i conforms to s under the t-norm $\triangle$, denoted by i $\mathsf{conf}_\triangle$ s, if and only if for all trace $(w, p) \in \mathsf{ftr}_\triangle(i)$ there exists a trace $(w, p') \in \mathsf{ftr}_\triangle(s)$ such that $p \geq p'$.

3   Case Study: A Sensors Network Controlling Air Quality

In this section, we present our case study about a sensors network for controlling air quality. First, we present the main indicatives of air quality used during our study. Next, we give the formal specification, using the automata introduced in the previous section, of all the components of the network. The final goal of our research is to use the specification to derive expected properties of the system and check whether the real system fulfill them.

3.1   Measuring Air Quality

Our case study is the sensor network implemented by the Andalusian regional government to control air quality across the region. The total area of the region is 87,268 Km^2 (33,694 square miles) and it currently has a population of around 8.4 million people. The network has 61 sensor stations located all around the region. Each station measures the key air pollutants $PM_{2.5}$, PM_{10}, CO, O_3, NO_2 and SO_2 every 10 min. In addition, some of them measure other pollutants not relevant for this case study, such as NO and SH_2. In order to validate our proposed system, we use all the real sensing data collected during December 2016.

In order to establish the bounds to determine *good* air quality, we use the index proposed by the U.S. Environmental Protection Agency (EPA) [10] because this agency provides complete information about how every air quality level affects different risk groups. EPA provides the value interval for detecting a specific air quality level for every isolated pollutant (see Table 1). As it can been seen in this table, an average of the last 1, 8 or 24 hour-values must be calculated depending on the considered air pollutant. For example, the average value in 1-hour period is required for NO_2, while 8-hour period is needed for CO. Once the average has been calculated, the air quality level can be reported by choosing the range to which the value belongs to. EPA defines the so-called *Air Quality Index* (AQI), a general level for air quality based on the maximum level of each pollutant. There are six possible air quality levels: *Good, Moderate, Unhealthy for Sensitive Groups, Unhealthy, Very Unhealthy, Hazardous.* The corresponding health risks are given for each air pollutant and level. For example, having a NO_2 1 hour-avg value of 55 ppb (parts per billion) and a *CO* 8 hour-avg value of 13.7 ppm (parts per million) implies the detection of the following air quality levels: *Moderate* for NO_2 and *Unhealthy* for *CO*. As the *Unhealthy* level is more dangerous than the *Moderate* one, the air quality level (AQI) for this location will be *Unhealthy*.

Table 1. AQI categories.

Air quality category		Pollutants					
Name	Level	NO_2 (*ppb*) 1 h	SO_2 (*ppb*) 1 h	CO (*ppm*) 8 h	O_3 (*ppm*) 8 h	$PM_{2.5}$ ($\mu g/m^3$) 24 h	PM_{10} ($\mu g/m^3$) 24 h
Good	1	0–53	0–35	0.0–4.4	0.000–0.054	0.0–12.0	0–54
Moderate	2	54–100	36–75	4.5–9.4	0.055–0.070	12.1–35.4	55–154
Unhealthy for sensitive groups	3	101–360	76–185	9.5–12.4	0.071–0.085	35.5–55.4	155–254
Unhealthy	4	361–649	186–304	12.5–15.4	0.086–0.105	55.5–150.4	255–354
Very unhealthy	5	650–1249	305–604	15.5–30.4	0.106–0.200	150.5–250.4	355–424
Hazardous	6	1250–2049	605–1004	30.5–50.4	>0.200	250.5–500.4	425–604

3.2 Formal Specification

In this section, we present the specification, using our fuzzy-timed automata, of all the components of the network. We have three main entities. First, *sensors* are the small devices collecting information about a particular pollutant. These sensors are associated with *sensor stations*, containing one sensor per pollutant (in our case, 6 different sensors). Our network has 61 sensor stations that communicate with a *server* where the information is processed and alarms, if needed, are produced. Servers and stations also communicate to deal with potential errors, both malfunctioning of a station and wrong adjustments of local clocks. First, we present the variables used in these automata to store and manage the collected data:

- u: variable for saving the daily report of a sensor station.
- v: variable for saving the value obtained by the sensor of one air pollutant.
- x: 6-tuple for saving the value of each air pollutant measured in a sensor station.

- y: 3-tuple for saving the information obtained from checking the clocks of the stations.
- z: variable to store the mean of delays of the clock of a sensor station during an hour.
- w: variable to store the *state* of a sensor station; 1 denotes *erroneous* state while 0 denotes *normal* state.
- ts: variable for storing the time value of the clock.
- ts_{aux}: auxiliary variable for storing time that will be helpful to the trigger of some actions.
- ts_{rep}: auxiliary variable for storing the last time when a report was sent to the stations of the network.

Next, we describe each of the three components that we mentioned before. The most basic elements of the system are *sensors*. We have a different type of sensor for each air pollutant that they measure: $PM_{2.5}$, PM_{10}, CO, O_3, NO_2 and SO_2. Since our network has 61 sensor stations, we have a total of 366 sensors. The functionality of sensors includes two actions:

- Data Capture: Register the current level of the pollutant that the sensor measures. This is performed by the action *getInfo*.
- Data Sending: Send to the sensor station the info collected about the amount of pollutants. Depending on the measured pollutant, the action performed is *sensor1* for $PM_{2.5}$, *sensor2* for PM_{10}, *sensor3* for CO, *sensor4* for O_3, *sensor5* for NO_2 and *sensor6* for SO_2.

In Fig. 3 we give a graphical representation of the fuzzy-timed automata corresponding to the first sensor (S_{11}) of the first sensor station (M_1). The definition of all the sensors follow the same pattern.

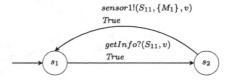

Fig. 3. Sensor S_{11} automaton.

Sensors are collected in group of 6 (one per every pollutant) and they conform *sensor stations*. A total of 61 stations are distributed along the Andalusian territory for measuring air quality. The main functionality of stations is implemented with the following actions:

- Data Capture: Receive the current level of the pollutant from the sensor and save this information for sending it to the central server every 10 min. This is performed by the actions *sensor1* to *sensor6*, receiving an input with the information of each sensor.

- Data Sending: Send to the central server the information collected, from all its associated sensors, about the amount of pollutants. This is performed by the action *air*, but firstly, data is accumulated in the station with the action *getData*.
- Report an error: Send to the central server the abnormal state of the station because the clock is not set to the right time. This is performed by the action *error*.
- Recover from an error: Apply the solution sent by the server in order to recover the behavior of the station. This is performed by the action *reset*, giving the value of ts from the server to the ts of the station.
- Save delay: Receive the mean of delays that suffer the clock of the station every hour. This is obtained from the central server and it is performed by the actions *warning*, capturing the information from the server, and *corrections*, saving the delay that should be corrected.
- Save daily report: Receive the daily report generated by the central server with the information obtained from the station. This is performed by the action *report*.

In Fig. 4 we present a graphical view of the automaton corresponding to the sensor station M_1. Again, the automaton corresponding to other sensor stations are similar.

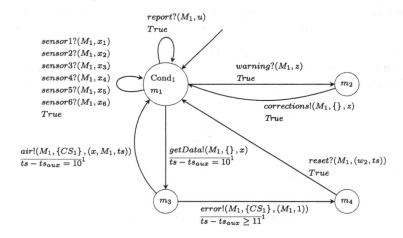

Fig. 4. Sensor station M_1 automaton.

The *central server* collects the information captured by all the sensors of the sensor stations. In particular, it controls the delays of the clocks of the sensor stations. The functionality of this element is implemented with the following actions:

- Data Reception: Receive the information of each sensor station and process it. This is performed by the actions *air*, receiving an input with the information of the station, and the action *checkPat*, applying the patterns each time that information is received.

- Sending of information: Send to the corresponding sensor station the information obtained from the evaluation of their clocks. Firstly, we save the errors of the clocks with the action *getErrors* when the delay is greater or equal to 5 and they are saved with *saveErrors*. Then, with the action *warning* we send the errors detected every hour and with *report*, we send the information collected of the station during a day.
- Reset from an error: Receive an abnormal state of a sensor station and the server sends the measure to apply with the aim of recovering that station from an error state. This is performed by the actions *error* and *reset*.

In Fig. 5 we present a graphical representation of the automaton defining the central server CS_1.

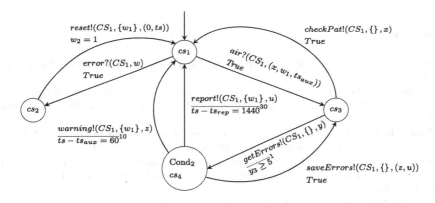

Fig. 5. Central server CS_1 automaton.

The network requires the synchronization of all the elements in two levels. First, we define the synchronization of the sensors with their corresponding station. The system corresponding to the sensor station i, where $1 \leq i \leq 61$, is:

$$SS_i = \|_A(S_{i1}, S_{i2}, S_{i3}, S_{i4}, S_{i5}, S_{i6}, M_i)$$

where $A = \{sensor1, sensor2, sensor3, sensor4, sensor5, sensor6\}$. Second, we define the synchronization of the sensor stations with the central server. This is formally defined as:

$$N = \|_B(SS_1, SS_2, \ldots, SS_{61}, CS_1)$$

where $B = \{air, error, reset, warning, report\}$.

4   Conclusions

We have introduced a formalism to specify systems with uncertain information. We have defined the syntax and operational semantics of our *fuzzy* version of the classical timed automata formalism. In order to show the usefulness of our formalism, we have used it to formally represent a real system. Specifically, we present the formal specification of a sensor network conformed by a central server, 61 sensor stations and 366 sensors devoted to the study of air quality. The measurement of these levels are performed periodically and some reports and warnings are generated according to some timing requirements. The main problem, concerning time, is that lacks of synchronization between local clocks and the clock of the server may provoke the malfunction of the network. Although the measure of the quality of the air is the first goal of the network, it is essential to check that the time in this scenario is correct in order to assure that the timing triggers of the actions are accurate.

References

1. Alur, R., Dill, D.: A theory of timed automata. Theoret. Comput. Sci. **126**, 183–235 (1994)
2. Andrés, C., Llana, L., Núñez M.: Self-adaptive fuzzy-timed systems. In: 13th IEEE Congress on Evolutionary Computation, CEC 2011, pp. 115–122. IEEE Computer Society (2011)
3. Bosik, B.S., Uyar, M.Ü.: Finite state machine based formal methods in protocol conformance testing. Comput. Netw. ISDN Syst. **22**, 7–33 (1991)
4. Cavalli, A., Favreau, J.P., Phalippou, M.: Standardization of formal methods in conformance testing of communication protocols. Comput. Netw. ISDN Syst. **29**, 3–14 (1996)
5. Cavalli, A.R., Higashino, T., Núñez, M.: A survey on formal active and passive testing with applications to the cloud. Ann. Telecommun. **70**(3–4), 85–93 (2015)
6. Doostfatemeh, M., Kremer, S.C.: New directions in fuzzy automata. Int. J. Approximate Reasoning **38**(2), 175–214 (2005)
7. Hierons, R.M., Bogdanov, K., Bowen, J.P., Cleaveland, R., Derrick, J., Dick, J., Gheorghe, M., Harman, M., Kapoor, K., Krause, P., Luettgen, G., Simons, A.J.H., Vilkomir, S., Woodward, M.R., Zedan, H.: Using formal specifications to support testing. ACM Comput. Surv. **41**(2), 1–9 (2009)
8. Lamport, L.: Who builds a house without drawing blueprints? Commun. ACM **58**(4), 38–41 (2015)
9. Larsen, K.G., Pettersson, P., Yi, W.: Uppaal in a nutshell. Int. J. Softw. Tools Technol. Transf. **1**(1–2), 134–152 (1997)
10. Mintz D.: Technical assistance document for the reporting of daily air quality - the Air Quality Index (AQI). Technical report EPA-454/B-16-002, U.S. Environmental Protection Agency (2016)
11. Mordeson, J.N., Malik, D.S.: Fuzzy Automata and Languages: Theory and Applications. Chapman & Hall/CRC, Boca Raton (2002)
12. Wee, W.G., Fu, K.S.: A formulation of fuzzy automata and its application as a model of learning systems. IEEE Trans. Syst. Sci. Cybern. **5**(3), 215–223 (1969)

Intelligent Transportation System to Control Air Pollution in Cities Using Complex Event Processing and Colored Petri Nets

Gregorio Díaz, Hermenegilda Macià$^{(\boxtimes)}$, Valentín Valero,
and Fernando Cuartero

School of Computer Science, University of Castilla-La Mancha,
02071 Albacete, Spain
{Gregorio.Diaz,Hermenegilda.Macia,Valentin.Valero,
Fernando.Cuartero}@uclm.es

Abstract. Pollution due to road traffic in big cities is an important problem in our society, with consequences for human health. In this paper we deal with this problem, proposing an Intelligent Transportation System (ITS) model based on Complex Event Processes (CEP) technologies and Petri nets that takes into account the levels of environmental pollution according to the air quality levels accepted by the international recommendations. Thus, we are tackling a rather common problem in big cities nowadays, where traffic restrictions must be applied due the pollution. Petri nets are then used in this paper as a tool to make decisions about traffic regulations, so as to reduce pollution levels.

Keywords: Complex event processing · Event processing languages · Formal methods · Petri nets

1 Introduction

The increase of vehicles in the road traffic is a characteristic phenomenon of our society, which means that related problems such as traffic accidents, pollution (air and noise), long travel times, etc., are increasing in the same way. Numerous reports have been published in order to know the extension of this problem, and in particular this has motivated the development of a new area of study, such as Intelligent Transportation Systems (ITS) [11], in which numerous papers have been published in recent years. ITS has become an important element, both for improving human life and modern economy [3], with the main objective of optimizing road traffic, managing the capacity of the roads, improving drivers safety, reducing energy consumption and improving the quality of the environment, among many others.

For the study and development of ITS, a key component is traffic modeling, which provides a framework to better investigate and test the state of the road

---

This work was supported in part by the Spanish Ministry of Science and Innovation and the European Union FEDER Funds under Grant TIN2015-65845-C3-2-R.

I. Rojas et al. (Eds.): IWANN 2017, Part II, LNCS 10306, pp. 415–426, 2017.
DOI: 10.1007/978-3-319-59147-6_36

in real time and accurately predict future traffic. In general, a desirable model must meet the following requirements:

- It must be consistent with traffic flow.
- It must be flexible, by using parameters that characterize the traffic flow, and be able to represent different situations and random changes in the traffic flow.
- It should be simple, but capturing the information required in order to take decisions about traffic regulations.

In this context, we focus on traffic control in cities, taking into account the levels of environmental pollution according to the air quality levels accepted by the international recommendations [13]. Thus, we are tackling a rather common problem in big cities, where traffic restrictions must be applied due to pollution.

The methodology we use to design ITS is Complex Event Processing (CEP) [7] in combination with formal methods to model and test the proposed solutions. CEP provides users with facilities for analyzing and correlating large volumes of data in the form of events with the aim of detecting relevant or critical situations for a particular domain in real time. To fulfill this objective, the conditions describing the situations of interest to be detected must be specified as event patterns. Patterns are implemented by using the languages provided by CEP engines, the so-called Event Processing Languages (EPLs), and once the patterns are defined they can be deployed in the CEP engine in question.

Additionally, Petri Nets (PNs) [10] is a formalism which provides mathematical rigor and a graphical representation of the model, which allows us to have a better comprehension from a visual model, but also to profit from the mathematical underlying model in order obtain important results about all its possible behaviors. Furthermore, Petri nets are supported by tools, which allow us the simulation and analysis of the behaviour of a given system in a suitable manner. There are some related works in the literature. Grunder et al. [8] have defined a Petri-Net based modelling and analysing of public transportation systems by using Vod Petri Nets, a formalism derived from Colored Petri Nets. Čapkovič [4] uses three different models of Petri nets to model segments of a transport network, and Riouali et al. [12] use hybrid Petri Nets to model discrete and continuous aspects of traffic flow dynamics.

To sum up, our proposal is to combine the use of EPL and Petri Nets to model and test ITS and, specifically, the city traffic flow taking into account air pollution conditions. Thus, the contributions stated by this paper are:

- Definition of EPL patterns to detect risky situations produced by air pollutants.
- A model of traffic flow using Petri Nets.
- A combination of EPL and Petri Nets to provide an ITS.
- A methodology to test an ITS by using a sound theoretical model (PNs).

The structure of the paper is as follows. Section 2 presents a brief background about CEP technologies and the specific model of Petri nets we use, Colored Petri

Nets (CPNs). The city road model by using CPNs is presented in Sect. 3 and the AQI modeling by EPL patterns in Sect. 4. Section 5 presents the ITS system by the integration of both the CPN model and the EPL patterns. Finally, Sect. 6 presents our conclusions and lines of future work.

2 Background

We explain the background for both the CEP technology that we use for the event patterns that capture air quality level conditions and the CPN formalism.

2.1 Complex Event Processing

In CEP, a *situation* is an event occurrence or an event sequence that requires an immediate reaction. Events can be classified into two main categories: *simple events*, which are indivisible and happen at a point in time and *complex events*, which usually contain more semantic meaning and are obtained by the processing of a set of other events. Complex events can be derived from other events by applying or matching *event patterns*, i.e. templates where the conditions describing situations to be detected are specified. A *CEP engine* is the software used to match these patterns over continuous and heterogeneous event streams, and to raise real-time alerts after detecting them. These event patterns are implemented by using EPLs. Further information about existing EPLs can be found in the survey by Cugola and Margara [5]. CEP is performed in 3 stages: (1) event capture –it receives events to be analyzed by CEP technology, (2) analysis –based on the event patterns previously defined in the CEP engine, the latest will process and correlate the information in the form of events in order to detect critical or relevant situations in real time, and (3) response –after detecting a particular situation, it will be notified to the system, software or device in question. The main advantage of using this technology is that such relevant or critical situations can be identified and reported in real time, thus reducing latency in decision making. It is noteworthy that we have chosen Esper EPL [2] as EPL in this work, since this rich high level processing language is more complete than others are, providing more temporal and pattern operators for defining the situations of interest. For the sake of brevity, we refer to the particular language Esper EPL simply as EPL through the rest of the paper.

There are three important structural elements to consider from EPL in this work: the schema, which defines the event type structure, an *every pattern*, which provides us with all the events fulfilling a certain condition from the input flow, *sliding time windows pattern operators* for processing event information in time slides, as well as the arithmetic operators such as average, counter, etc. Figure 1 shows three EPL extracts, where *(a)* specifies an event schema with two properties *propname1* and *propname2*, both of type *double*, *(b)* specifies an event pattern *Pattern1*, which detects the input events whose *propname1* is greater than 10.0 and *(c)* calculates the average of *propname2* values over the last 8 h.

418   G. Díaz et al.

```
(a)          create schema event_name ( propname1 double, propname2 double );

(b)          @Name('Pattern1')
             insert into Pattern1
             select a1.propname1
             from pattern [every a1 = event_name(a1.propname1>10.0)]

(c)          @Name('Pattern2')
             insert into Pattern2
             select avg(a1.propname2) as average
             from pattern [(every a1 = event_name)].win:time(8 hours)
```

Fig. 1. EPL basic schema and two patterns.

2.2 Colored Petri Nets

A Petri Net (PN) is a bipartite directed graph, with two types of node, places (circles) and transitions (rectangles). Places and Transitions can be connected by arcs, either place-transition (PT) or transition-place (TP) arcs (Fig. 2).

Places usually represent states or system conditions while transitions are the actions or events that produce changes in the system state. For instance, places *Start* and *End* in Fig. 2 represent an initial and final state, respectively, whereas places *L1, L2* represent system conditions. Places are then annotated by tokens to indicate system states. These tokens are usually depicted by dots or the number of them on the corresponding place. For example, a token on a place can indicate that the condition represented by this place is currently satisfied. The current state is thus defined by the set of tokens on every place, called the Petri Net marking, and a firing rule determines the conditions under which transitions are fired (executed) in order to change the current marking (see [10] for a complete formal definition of the model).

There is a well known extension of Petri nets, Colored Petri Nets (CPNs), which extend the basic model with both data and time. CPNs are supported by a very well-known tool, CPN Tools [1], which allows us to create, edit, simulate and analyze CPNs, so the notation described below is that used in this specific tool. In this paper, we only present an informal description of the CPN dynamical behavior, so we omit the formal definitions, which can be found in [9]. In CPNs places have an associated *color set* (a data type), which specifies the set of allowed token colors at this place, so that tokens bring some data information, according to the data type of its associated place. Furthermore, tokens can have

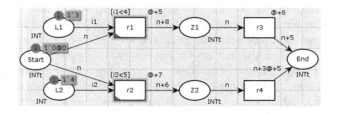

Fig. 2. A Colored Petri Net.

a timestamp associated, indicating in that case the time at which they will be available in order to fire some transitions. We will consider a discrete time model, but it would not be a problem to consider a continuous time model. Places are then graphically annotated with their corresponding color sets, which are indicated in the bottom right-hand side of them. For instance, places *L1, L2* have the color set *INT*, which represents the integer set, whereas places *Start* and *End* have the color set *INTt*, which represents a timed color set of integer numbers. The notation to specify the tokens on an untimed place is $n`x + +m`y$, which indicates that we have n tokens with color value x and m tokens with color y. For timed places the corresponding notation is $n`x@t + + + m`y@u$, where we now have n timed tokens with color x and timestamp t and m tokens with color y and timestamp u. Notice that the markings on the places are depicted in green beside their places, together with the number of them in a green circle.

There is also a discrete global clock which represents the total time elapsed in the system model, which is initialized to 0. The clock only moves forward when no transition can be fired at the current time, and it will move forward to the next time at which some transition can be fired, as indicated below.

Arcs can have inscriptions (*arc expressions*), constructed using variables, constants, operators and functions. To evaluate an arc expression we need to bind the variables, which consists of assigning a value to the variables that appear in the arc inscription. These values are then used to select the token colors that must be removed or added when firing the corresponding transition. For example, the variable n attached to the arc from *Start* to *r2* will be binded to the value of the token in the place *Start*. Arc expressions can have associated time information. We will only use time information in the output arcs from transitions, which is used to indicate a time that must be added to the current global clock value in order to get the timestamp of the tokens produced by these arcs. Actually, when all the output arcs of a transition have the same time inscription, there is a shorthand notation in CPN Tools by which this time information is associated with the transition instead of the output arcs (labels "$@+x$" in the upper right-hand side of transitions).

Transitions can also have associated guards and priorities. Guards are Boolean expressions that can prevent their firing. These guards are indicated in the upper left-hand side of transitions. When a transition has a guard, it must evaluate to true for the binding to be enabled, otherwise the binding is disabled and the transition cannot be fired. Furthermore, we can use transition priorities, so that in the event of a conflict between two transitions that can be fired (executed) at a given time, the transition with the highest level of priority is fired first, where smaller values of priorities correspond to higher levels of priority. Thus, a transition can only be fired at the current time if we have at least one available token[1] on each precondition place for which a binding allows the firing of the transition and no other transition with a higher level of priority can be fired at the current time. A token is considered available when its timestamp is less than or equal to the current system time.

---

[1] We only consider CPNs with weights 1 in the arcs.

For instance, for the CPN of Fig. 2 both transitions $r1$ and $r2$ can be fired at time 0 with the marking indicated in the figure. The firing of $r1$ removes the token on the place $L1$ and produces a new timed token on $z1$, so the marking on $z1$ is $1`8@5$. From this new marking transition $r3$ can be fired, which removes the token on $z1$ and produces the following marking on the timed place $End$: $1`13@11$. Should we have fired $r2$ initially, instead of $r1$, then we would have the marking $1`6@7$ on $z2$, and then firing $r4$ we could reach the marking $1`9@12$ on End.

Following this same procedure for every CPN N with a given initial marking M_0 we can obtain all the reachable markings from it. We call $Reach(N, M_0)$ the set of all reachable markings from M_0 (*state space* of (N, M_0)). This set is of particular interest because it provides us with information about all the events that can occur in a system modeled by the considered net.

3   Modeling the City Map with Petri Nets

A city will be divided into zones, which are represented by places, labelled with the zone names: $Zonei$, $i = 1, 2, \ldots, n$, where n is the number of zones. Our goal is to obtain different routes to travel from $Zone1$ to $Zonen$. Transitions will represent movements from one zone to another by traversing some street in which a sensor station has been placed in order to get air quality level information. Thus, transition $stationij$ captures the movement from $Zonei$ to $Zonej$. These transitions have a boolean guard, which will be modified depending on the pollutant levels obtained from their corresponding sensor station. The arc from a transition $stationij$ to the place $Zonej$ will have an associated delay, which corresponds to the average time required to travel from $Zonei$ to $Zonej$ using this route. As illustration, we show a simple city map in Fig. 3, in which each transition only reads air quality level data from its corresponding sensor station, but notice that two transitions could get their information from a same sensor station.

Places $Zonei$ will have INTt as color set, so the initial marking will only have one token on the place $Zone1$ ($1`1@0$), and we expect to reach a final marking in which there is a single token on the place $Zonen$: $1`1@t$, where t will be the time elapsed to go from $Zone1$ to $Zonen$, according to the followed route. Figure 4 shows the state space corresponding to the CPN of Fig. 3, with the following values: $c12 = 3$, $c13 = 9$, $c14 = 3$, $c15 = 8$, $c23 = 4$, $c36 = 2$, $c45 = 6$, $c56 = 2$.

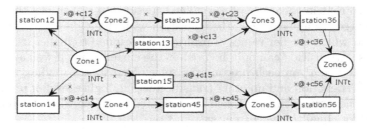

Fig. 3. PN representing a city map.

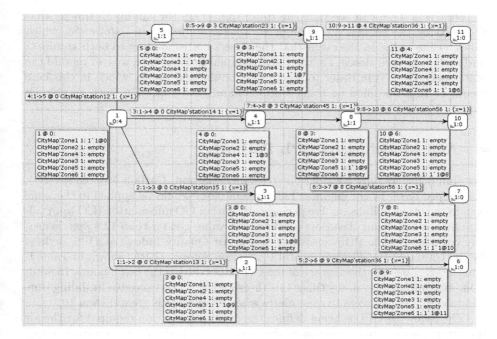

Fig. 4. Stace spate for the PN modeling the city map.

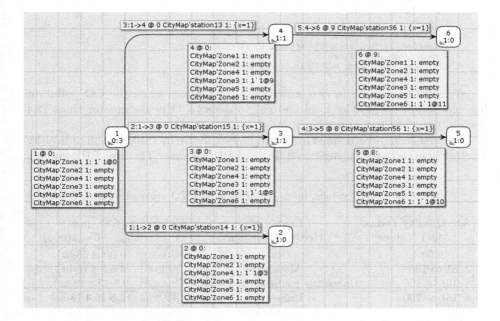

Fig. 5. Stace spate for the PN modeling the city map with closed streets.

From Fig. 4 we can see that the route that takes the minimum time is that obtained by the sequence of firings: *station12*, *station23* and *station36*, which takes 6 time units. In the event that the streets corresponding to both *station12* and *station45* were closed, i.e. their guards equal to false, then we can rebuild the state space (Fig. 5) in order to get the minimum time route, which in this case would be that obtained by firing *station15* and then *station56*, with an elapsed time of 10 time units.

4   AQI Modeling by EPL Patterns

For simplicity, we only consider one relevant pollutant, Carbon Monoxide (CO), mainly produced by car engines. This methodology would be applied in the same way to the other pollutants. We periodically gather CO information (CO ppm values) from a collection of sensors (stations). A value greater than 12.4 *ppm* is usually considered Unhealthy according to the Air Quality Index (AQI) [6] defined by the United States Environmental Protection Agency (EPA) shown in Table 1. This index considers the average of the CO values during the last 8 hours. In addition, we compute how many times this threshold of 12.4 is exceeded in sliding time windows of 3 hours, so as to raise alerts when its number is greater than 10.

The following EPL schema defines the event information required for the air quality measurements. It contains the time at which the measurement is done, the station and location identifiers and the pollutants included in the AQI index ($PM_{2.5}$, PM_{10}, O_3, NO_2, SO_2 and CO). Remind that we only consider in this paper the CO pollutant.

```
create schema AirMeasurement
    (timestamp string, stationId string, location string,
    pm2_5 float, pm10 float,
    o3 float, no2 float, so2 float, co float);
```

We use this schema in two scenarios: (a) to detect when the CO pollutant has reached an unhealthy level and (b) to detect CO peaks greater than 12.4 ppm in the last three hours.

The first scenario is defined in the AQI index shown in Table 1 as an unhealthy level. It is detected by using two different EPL patterns. The first EPL pattern

Table 1. AQI values, levels, color code and CO average ranges.

Air Quality Index (AQI) Values	Levels of Health Concern	Colors	CO (ppm) $C_{low} - C_{high}(avg)$
1: 0 to 50	Good	Green	0.0-4.4 (8-hr)
2: 51 to 100	Moderate	Yellow	4.5-9.4 (8-hr)
3: 101 to 150	Unhealthy for Sensitive Groups	Orange	9.5-12.4 (8-hr)
4: 151 to 200	Unhealthy	Red	12.5-15.4 (8-hr)
5: 201 to 300	Very Unhealthy	Purple	15.5-30.4 (8-hr)
6: 301 to 500	Hazardous	Maroon	30.5-50.4 (8-hr)

calculates the *CO* pollutant average for the last eight hours, grouping it by each station.

```
@Name('CO_Avg')
insert into CO_Avg
select a1.stationId as stationId, avg(a1.co) as value
from pattern [(every a1 = AirMeasurement)].win:time(8 hours)
group by a1.stationId
```

The second pattern checks whether the average, obtained by the previous pattern, is higher than the threshold specified by the AQI index as unhealthy, that is, higher than 12.4 *ppm* (level 4 in Table 1). Similar patterns can be created to detect the other levels by modifying the condition.

```
@Name('CO_PollutantLevel')
insert into CO_PollutantLevel
select a1.stationId as stationId, 4 as levelNumber,
       'CO_Unhealthy' as levelName
from pattern [(every a1 = CO_Avg((a1.value > 12.4 and a1.value <= 15.4)))]
```

In the second scenario we compute the number of times that CO values are greater that 12.4 in time sliding windows of three hours. With both patterns we will be able to make decisions about traffic regulations.

```
@Name('Threshold4CO')
insert into Threshold4CO
select a1.stationId as stationId, count(a1.co) as num
from pattern [every a1 = AirMeasurement(a1.co>12.4)].win:time(3 hours)
group by a1.stationId;
```

5   ITS for Traffic Control by Using CEP and CPNs

We start from the CPN model presented in Fig. 3, which is now enriched in order to manage the two situations of interest indicated in the previous section, with the intention of making decisions about traffic control by closing and opening the connections between two zones. Thus, we consider that connections are closed for four hours when one of the following situations occurs:

1. The pollution level is higher than 3, i.e. Unhealthy, Very Unhealthy or Hazardous.
2. The level is Unhealthy for Sensitive Groups (3) and 10 peaks reaching values higher than 12.4 *ppm* have at least occurred.

These situations are encoded in our CPN model by the boolean function **close** shown in the third part of Fig. 6, where parameter y represents the output

```
▼fun geny(y)=            ▼fun genz(z)=              ▼fun genPoll(y,z)=(geny(y),genz(z));
 (case y of              (case z of                ▼fun close(y,z)=
 1 => y+discrete(0,1)    0 => z+discrete(0,5)       if y>3 orelse (y=3 andalso z>=10)
 |5=> y-discrete(0,1)    |20=> Int.max (0,z-discrete(0,5))   then true else false;
 |_ => y+discrete (0,1)-discrete(0,1));  |_ => Int.max(0,z+discrete (0,20-z)-discrete(0,5)));
```

Fig. 6. Function declarations in CPNTools.

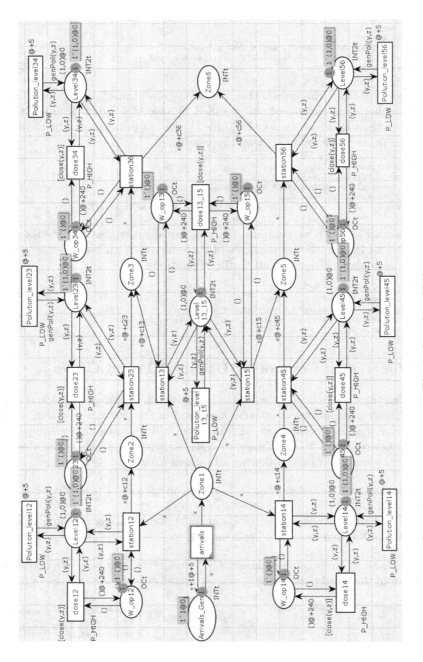

Fig. 7. Complete CPN.

levelNumber obtained when applying pattern CO_PollutantLevel and parameter z represents the output num obtained when applying pattern Threshold4CO.

Figure 7 shows the CPN for our ITS. Markings in Places named *Levelij* are obtained from the sensor stations associated to *stationij*.[2] In a real scenario they will be automatically provided, but in this simple approach we generate these values by using the random function genPoll shown in the third part of Fig. 6, which uses two auxiliary functions geny, genz, which use discrete distributions to produce the values. We have considered uniform distributions, but other distributions could also be applied in order to simulate the pollution evolution. Transition *Pollution_levelij* will be fired every 5 min, thus producing a new CO measure from the previous function genPoll, which updates the information stored in the level places.

Thus, notice that transitions *closeij* will fire and therefore close a zone connection when the conditions described above by function close are fulfilled. In this case, when *closeij* is fired, this zone connection will be closed for a period of 4 hours, which is captured by the place *W_opij*, whose token timestamp is then increased by 240 min, so as to prevent a new firing of *stationij* for this period of time.

Furthermore, notice the use of priorities in the transitions in order to enforce a correct order in their execution, so transitions *Pollution_levelij* have the minimum priority, transitions *stationij* have an intermediate priority, whereas *closeij* must have the maximum priority.

Now, in order to feed the model with cars and see a dynamic evolution of the system we use a transition called *arrivals* (left part of Fig. 7), which in this case generates car inputs at periods of five minutes. Each car enters into *Zone1* and it wants to reach *Zone6* as final destination.

6   Conclusions and Future Work

In this paper we have proposed an ITS model for traffic control considering air pollutant levels. The proposed model conforms the three aspects that an ITS should satisfy, that is, to be consistent with traffic flow, flexible to characterize a dynamic flow and simple but at the same time capable of representing changing situations.

Our proposal captures traffic flows in terms of Colored Petri Nets and air quality conditions are gathered by a collection of sensor stations. Then, we have applied CEP technologies in order to detect critical situations on the basis of the AQI standards. Those techniques together, Colored Petri Nets and CEP, thus allow us to define an extended CPN model in which we can make decisions about traffic regulations.

In this paper, then, we have presented the model structure. As future work, we intend to gather real information about traffic flows and air pollution. The first objective will be to obtain the best distribution to characterize this real

---

[2] Notice that *station13* and *station15* have the same sensor station associated.

426   G. Díaz et al.

information. This objective will allow us to check if the functions defined to control the traffic flows using the closed-open mechanism are conformant to the real situations, or should we consider new ones, otherwise.

A final step would be to deploy this solution by using a CEP engine and an Enterprise Business System (EBS) to test this approach in a real scenario. These tests will be performed by comparing the data obtained from simulations with those obtained in the real scenario.

References

1. CPNTools HomePage, February 2017. http://www.cpntools.org/
2. Esper - Complex Event Processing, February 2017. http://www.espertech.com/esper/
3. Bekiaris, E., Nakanishi, J.: Economic Impacts of Intelligent Transportation Systems, vol. 8. Elsevier, Amsterdam (2004)
4. Čapkovič, F.: Petri net-based modelling and simulation of transport network segments. In: Król, D., Fay, D., Gabryś, B. (eds.) Propagation Phenomena in Real World Networks. ISRL, vol. 85, pp. 135–154. Springer, Cham (2015). doi:10.1007/978-3-319-15916-4_6
5. Cugola, G., Margara, A.: Processing flows of information: from data stream to complex event processing. ACM Comput. Surv. **44**(3), 15:1–15:62 (2012)
6. EPA. Technical assistance document for the reporting of daily air quality - the Air Quality Index (AQI). Technical report EPA-454/B-16-002, U.S. Environmental Protection Agency, North Carolina, US, May 2016
7. Etzion, O., Niblett, P.: Event Processing in Action, 1st edn. Manning Publications Co., Greenwich (2010)
8. Grunder, O., Elmoudni, A., Abbas-Turki, A.: Public transportation systems: modeling and analysis, based on a new Petri net approach. In: Proceedings of the Second IEEE International Conference on Systems, Man and Cybernetics (SMC 2002) (2002)
9. Jensen, K., Kristensen, L.M.: Coloured Petri Nets: Modelling and Validation of Concurrent Systems, 1st edn. Springer Publishing Company, Heidelberg (2009)
10. Peterson, J.L.: Petri Net Theory and the Modeling of Systems. Prentice Hall PTR, Upper Saddle River (1981)
11. Picone, M., Busanelli, S., Amoretti, M., Zanichelli, F., Ferrari, G.: Advanced Technologies for Intelligent Transportation Systems. Springer Nature, Heidelberg (2015)
12. Riouali, Y., Benhlima, L., Bah, S.: Petri net extension for traffic road modelling. Int. J. Sci. Eng. Res. **7**(11), 282–299 (2016)
13. World Health Organization (WHO). Health risk assessment of air pollution. general principles (2016)

Heuristics for ROSA's LTS Searching

Fernando López Pelayo[1]($\boxtimes$), Fernando Cuartero Gomez[1], Diego Cazorla[1],
Pedro Valero-Lara[2], and Mercedes Garcia Merayo[3]

[1] Departamento de Sistemas Informáticos, E. S. I. I. - U. C. L. M.,
02071 Albacete, Spain
{FernandoL.Pelayo,Fernando.Cuartero,Diego.Cazorla}@uclm.es
[2] Centro Nacional de Supercomputación, BSC, 08034 Barcelona, Spain
pedro.valero@bsc.es
[3] Departamento de Sistemas Informáticos y Computación,
Universidad Complutense, 28040 Madrid, Spain
mgmerayo@fdi.ucm.es

Abstract. The authors have been aimed by the goal of reducing the computational cost of searching the Labeled Transition Systems, LTSs, generated by Process Algebras. In particular, we have been following the idea of moving the order of the computational cost required to find/reach a desired node/state from the exponential of the classical Breadth-First Search fashion to the polynomial produced either by the Depth-First Search or by the A^* algorithm [6]. As usual, they both take as size of the problem the branching factor of the LTS.

This paper, first, presents the Normal Formed ROSA processes required to, second, define a sound topological structure over this Process Algebra. The underlying notion of distance from this topology can be taken as the heuristics to guide the search for whatever node which reachability want to be studied, by means of an A^* algorithm.

Keywords: Process Algebra · AI searching · Computational complexity

1 Introduction

It is very well known the usefulness of Process Algebras, PAs, for easily specifying and afterwards capturing the behaviour of a computing system in spite of their too high computational cost to develop an analysis with. Nevertheless, one (maybe the most important) of the reasons of not being so much used as these features indicate is the computational cost of traversing the LTS when studying the reachability of any node.

The authors presented the paper entitled *Looking for a Cheaper ROSA* [9] in this same forum 6 years ago so contributing in this direction. In that work we presented a brand new metric structure over the set of ROSA processes, its main contribution was the novelty of the structure that was supported by

---

F.L. Pelayo—Research supported by project TIN2015-65845-C3-2-R.

I. Rojas et al. (Eds.): IWANN 2017, Part II, LNCS 10306, pp. 427–437, 2017.
DOI: 10.1007/978-3-319-59147-6_37

a means-ends policy regardless of any other consideration. Now, we make an improvement on these results mainly because of two reasons; first we present the detailed definition of the Normal Formed ROSA processes, and second we define a new topological structure far better than the one sketched then, since the latter fits better with the true future behaviour of the processes so providing a sound and, perhaps, optimum heuristics to make finding whatever node a shorter task.

A very clear example of this huge computational cost we previously refer can be found in a previous work of some of the authors of this paper [10], where the LTS generated by the specification of the MPEG2 (IBBP) encoding algorithm (the Fig. 11 of this reference [10] clearly shows it) exhibits a lot of paths from the root node that encodes the starting point of the running of the system. The shortest path from the initial state to the final one is the result we are interested in (Fig. 12 of this same work [10] captures it).

When we deal with such a type of scenarios an 'intelligent' agent that by perceiving the features of this domain would be able to take actions improving the chances of success, i.e., to discard some fruitless paths, or even better to only choose the best of them, would be very welcome.

Edelkamp et al. provide several Heuristics for Analysing Graph Transition Systems [4], almost all of them based on structural issues or features of the morphology of the graphs to be analyzed, and the rest based on classical distance notions as Hamming or Levenshtein ones.

Gradara et al. use an heuristic search in an A^* algorithm for finding deadlocks over CCS [5]. In essence they only visit the most promising state to deduce tentative deadlock conditions based on considering both safe and critical points for possible deadlocks. These considerations has to be helped by temporal logic formulae.

Artificial Intelligence A.I. agents can be classified into several kinds according to the type of problems to solve or according to the strategies to follow. One of the typical problems to work in, is searching for a particular state among a lot of them. The most common cases in the field literature of P.As. assume this searching process as blind/non-guided because there is no enough knowledge of the environment to drive it.

We claim to have the required knowledge of the environment as being able to develop a guided searching by an heuristic function that estimates very accurately how good a state is to pass through for finding a goal node.

ROSA is a Markovian process algebra that took PNAL as starting point. Markovian time is added by means of the inclusion of both actions which duration are modelled by Exponentially distributed random variables of parameters $\lambda \in \mathbb{R}^+ - \{0\}$, and immediate actions which duration can be modelled by $Exp[\infty]$. There is another big difference between **ROSA** and PNAL associated with the non-determinism handling. While in PNAL probabilistic choices have precedence against non-deterministic ones, in **ROSA** happens the contrary. This fact implies immediate consequences in the Normal Forms required for both languages that make them different. **ROSA** goes beyond that in the sense that pure non-determinism is assumed when cooperating, up to certain point, some type of actions.

ROSA has its own place among the probabilistic, non-deterministic and timed models as it does not impose any syntactical restrictions on the components of a parallel operator, and thus, the specification task becomes easier and more straightforward than in the most of other models.

The usefulness of **ROSA**, as well as of so many PAs is out of any doubt, but as exposed, the computational cost of the fully unfolding of the LTS is unreachable from a practical perspective, so that we propose a way to only unfold the most promising state among the reachable (through a single transition) set of states from a given one. This, of course, would mean a very big saving on the computational cost of producing the part of the LTS required to study the reachability of a given state by the semantics of **ROSA**.

This paper is structured as follows: next Section provides a brief description of the Markovian process Algebra **ROSA** and justifies why closer syntactical and semantic definitions are required. Section 3 describes the Lexicographical Normal Formed **ROSA**'s syntax and semantics. Then a new topological structure over the set of Lexicographical Normal Formed **ROSA** processes is defined. The provided metrics is the basis of the heuristic evaluation function of a namely A^* algorithm to search the LTS generated by the operational semantics of **ROSA** processes. The conclusions and future work Section finishes the paper.

2   The Markovian Process Algebra ROSA

Let $\Delta = \{a, b, c, \ldots\}$ be an ordered finite set of action types.

Let $Id = \{X, Y, Z, \ldots\}$ be a finite set of variables of process.

We will denote by the last letters of the latin alphabet $r, s, t, \ldots$ probabilities.

We will denote by greek letters $\alpha, \beta, \gamma, \ldots$ time parameters for actions.

Terms of **ROSA** are defined by the following BNF expression:

$$P ::= \mathbf{0} \mid X \mid a.P \mid \langle a, \lambda \rangle.P \mid P \oplus P \mid P + P \mid P \oplus_r P \mid P\|_A P \mid recX : P$$

where $\langle a, \lambda \rangle$ represents an action labelled by a which execution time is modeled by a random Negative Exponential distribution of parameter λ ($Exp[\lambda]$), $\lambda \in \mathbb{R}^+ - \{0\}$ so that, immediate action labeled by a should be denoted by $\langle a, \infty \rangle$ but as an abbreviation we also denote them it as just a, $A \subseteq \Delta$, $a \in \Delta$, $X \in Id$, . is concatenation, $\oplus$, $+$ and $\oplus_r$ are internal, external and probabilistic choices, $r \in [0, 1]$, $\|$ is parallel, rec stands for recursion and P is a process of **ROSA**.

The Algebra induced by this expression makes up the set of **ROSA** processes. A detailed description of the operational semantics and the performance evaluation algorithm of **ROSA** can be found in [11], where we began to search the way of making **ROSA** a very usable formalism therefore some steps have been done in the line of fully automatize its analyzing skills.

As we are aimed to make **ROSA** usage as cheap (in the computational cost field) as possible, we have decided to provide the set of **ROSA** processes with a topological structure. The basic idea is taking this metric as the heuristic evaluation function to be minimized so playing the same role that h function does within the A^* searching algorithm.

The fist property to be held by every metric is

$$\forall P, Q \in \{\textbf{ROSA } processes\}.D(P,Q) = 0 \Leftrightarrow P = Q$$

In this case both P and Q are just **ROSA**-syntactical expressions denoting processes, so it is clear that if P and Q are syntactically (according to **ROSA**) identical expressions the distance between them must be 0. Nevertheless it is also easy to see that in some cases two different processes according to ROSA syntax, not always represent two actual different processes in terms of their behaviours, let see some examples:

Example 1. Let P and Q be a pair of **ROSA** processes, we need that processes $P \oplus Q$ and $Q \oplus P$ have distance 0, because in whatever interpretation of the semantics of processes, they should be equivalent. The same could be said about the processes $P + Q$ and $Q + P$, so this commutative property must be preserved. Moreover, the weighted commutative property of $\oplus_r$ should be also fulfilled, thus $P \oplus_r Q$, has to be equivalent to $Q \oplus_{1-r} P$, or more precisely the distance between them must be 0 in a correct definition of distance. □

Example 2. Furthermore, the definition of distance should be consistent with the associativity of processes so that, given P, Q and R three **ROSA** *processes* we want that $d((P \oplus Q) \oplus R, P \oplus (Q \oplus R)) = 0$. In this line the associativity of $+$ and the weighted associativity of $\oplus_r$ has to be preserved. □

Example 3. Also, there are some cases in which distributive property must be satisfied. For instance, let us take P, Q and R as **ROSA** *processes*, then we want that $d((P \oplus Q) + R, (P + R) \oplus (Q + R)) = 0$.

Distributive is a difficult property to be studied and guaranteed, thus, we will follow the results presented in [2] on this issue, and the corresponding distributive laws. □

Example 4. Finally, we want that derivative operators could be removed, and then, the equivalent expression without them should have distance 0 with the former. For instance we want that $d(a.0||_{\emptyset}b.0, a.b.0 + b.a.0) = 0$. □

In fact, we think that in an appropriate semantics, two equivalent processes must have distance 0 between them. The main objective of this paper is not the study of a theoretical semantics, such as denotational or axiomatic semantics. Of course, with the basis of our operational semantics, we could define a notion of bisimulation [1,3], and take this equivalence as the basis. But this is a considerable amount of effort, and this work have been already done. In fact, in [2] a **Proof System** is defined, and it is demonstrated the equivalence of a denotational semantics and a set of axioms and inference rules, in the sense that this system is sound and complete. That is, if two processes have the same denotational semantics, then, it can be proved by using the proof system that they are equivalent, and on the contrary, if the equivalence may be proved in the proof system, then, the processes have the same denotational semantics.

In order to solve all the cases shown in the above examples we have decided to introduce a sort of normal forms for **ROSA** processes.

3 Lexicographical Normal Formed ROSA, LNF-ROSA

3.1 Syntax of LNF-ROSA

Let $\Delta = \{a, b, c, \ldots\}$ be an ordered finite set of action types.

We will denote by the last letters of the latin alphabet $r, s, t, \ldots$ probabilities.

We will denote by greek letters $\alpha, \beta, \gamma, \ldots, \lambda, \ldots, \mu, \ldots$ time parameters for actions. They could be positive reals or even ∞, as they are representing the parameter of the negative exponential random distribution which models the duration of the corresponding action (type).

$\langle a, \lambda \rangle$ denotes the action a whose execution takes a time modelled by an exponential of parameter λ (∞ is the parameter for immediate actions).

Terms of *LNF-ROSA* are defined by the following grammar:

$\forall P \in \{LNF\text{-}ROSA\ processes\}:$

$$P ::= \mathbf{0} \mid \bigoplus_{A_i \in \mathcal{A}} Q_i$$

where:

- O is the Stop process.
- Q_i is a Probabilistic LNF-ROSA process (PLNFR).
- $\mathcal{A}$ is a convex family of sets of actions.
- At internal choice level, sets must appear in their action-induced lexicographic order.

$\forall Q \in \{PLNFR\ processes\}:$

$$Q ::= \bigotimes_{j} [q_j] T_j$$

where:

- T_j is an Action LNF-ROSA process (ALNFR).
- $\forall j. q_j \geq q_{j+1}$. At probabilistic level, probabilities must appear in decreasing order. If two had the same probability then they would be ordered by their ALNFRs.
- $\sum_j q_j = 1$.

$\forall T \in \{ALNFR\ processes\}:$

$$T ::= \sum_{a \in \mathbf{Type}(A)} \langle a, \lambda_A \rangle . V_a$$

where:

- $V_a \in \{LNF\text{-}ROSA\ processes\}$.
- Appearances of action types are ordered as the set they belong to.
- **Type** is a function defined on the set of multisets of actions which associates to every multiset of actions, the set consisting of their action types:

$$\textbf{Type} : \mathcal{M}[\Delta \times \mathbb{R}^+ - \{0\} \cup \{\infty\}] \longrightarrow \mathcal{P}[\Delta]$$

The Algebra induced by this expression makes up the set $\{LNF\text{-}ROSA\ processes\}$.

3.2 Operational Semantics of LNF-ROSA

Operational Semantics gives to a process its meaning represented by a labeled transition system which shows how the process is transformed into other(s) process(es), thus an operational semantics is made up by every possible transformation of all the language terms.

On the other hand, the operational semantics gives a precise interpretation of the operators presented when defining the syntax of the language.

The operational semantics is defined in a sort of ordered labeled transition system (Plotkin [12] and Milner's [7] style) in which non-deterministic behaviours are first solved by means of **non-deterministic transitions**.

These are tuples $< P, Q >$ where:

- $P \in \{LNF\text{-}ROSA\ processes\}$.
- $Q \in \{PLNFR\ processes\}$.

These transitions will be represented by

$$P \longrightarrow Q$$

It means that P can evolve immediately (they do not take any time at all) and behaves as Q. We will also say that P can evolve internally to Q.

These transitions capture the internal decisions that the system makes for resolving the non-deterministic evolutions which can come either from pure internal/non-deterministic behaviours or from competitions between actions with similar rates (an action demanded by the environment can be attended in more than one manner but either all of them would be immediate, their temporal parameters are all ∞, or, none of them have temporal parameters equal to ∞); $\forall A_j \in \mathcal{A}$:

$$\bigoplus_{A_j \in \mathcal{A}} T_j \longrightarrow T_j$$

Afterwards the probabilistic evolutions of the system are captured by means of **probabilistic transitions**.

These are tuples $< P, Q, r >$ where:

- $P \in \{PLNFR \; processes\}$.
- $Q \in \{ALNFR \; processes\}$.
- $r \in [0,1]$.

These transitions will be represented by

$$P \longrightarrow_r Q$$

This means that P can evolve immediately and behaves as Q with probability r. We will also say "P can evolve probabilistically to Q with probability r" or "P can evolve with probability r to Q"; $\forall i$:

$$\bigotimes_i [q_i] Q_i \longrightarrow_{q_i} Q_i$$

These transitions capture the pure probabilistic behaviour of the system (generative interpretation).

To finish with, **action transitions** will take place.

These are tuples $< P, Q, a, \lambda >$ where:

- $P \in \{ALNFR \; processes\}$.
- $Q \in \{LNF\text{-}ROSA \; processes\}$.
- $r \in [0,1]$.
- $a \in \Delta$.
- $\lambda \in \mathbb{R}^+ - \{0\} \cup \{\infty\}$.

These transitions will be represented by

$$P \xrightarrow{a,\lambda} Q$$

It means that P can evolve by executing the action labeled by a, taking a time described by an Exponential random distribution with parameter λ, and then behaves as Q. We will also say that P can evolve to Q by executing the action $\langle a, \lambda \rangle$; $\forall a \in \mathbf{Type}(A)$:

$$\sum_{a \in \mathbf{Type}(A)} \langle a, \lambda_A \rangle.V_a \xrightarrow{a,\lambda_A} V_a$$

The **Operational Semantics** of $LNF\text{-}ROSA$ is defined as the set of transitions (non-deterministic t., probabilistic t. and action t.) we can derive by using the previously stated rules.

Once Syntax and Semantics of LNF-ROSA processes have been defined we want to guarantee that some syntactically different ROSA processes having the same behaviour also share their syntactical expressions. This is why we use their LNF-ROSA expression instead. The following example illustrates what we mean.

Let P and Q be ROSA processes:

$$P = (((\langle d, 1 \rangle.0 + \langle a, 2 \rangle.0) \oplus (\langle b, 1 \rangle.0 + \langle a, 3 \rangle.0)) \oplus_{0.3} (\langle f, \infty \rangle.0 \oplus \langle e, 1 \rangle.0)$$

$$Q = (\langle e, 1 \rangle.0 \oplus \langle f, \infty \rangle.0) \oplus_{0.7} ((\langle a, 3 \rangle.0 + \langle b, 1 \rangle.0) \oplus (\langle d, 1 \rangle.0 + \langle a, 2 \rangle.0))$$

Notation: $\forall P \in \{\textbf{ROSA } processes\}$ its lexicographical normal form is $\|P\|$
Therefore, P and Q share their *lexicographical* normal forms $\|P\| = \|Q\|$

$$= [0.7]\langle e, 1\rangle.0 \otimes [0.3]\langle a, 3\rangle.0 + \langle b, 1\rangle.0 \oplus [0.7]\langle e, 1\rangle.0 \otimes [0.3]\langle a, 2\rangle.0 + \langle d, 1\rangle.0$$
$$\oplus [0.7]\langle f, \infty\rangle.0 \otimes [0.3]\langle a, 3\rangle.0 + \langle b, 1\rangle.0 \oplus [0.7]\langle f, \infty\rangle.0 \otimes [0.3]\langle a, 2\rangle.0 + \langle d, 1\rangle.0$$

because the notion of distance we keep in mind is mainly based on the syntactical expressions of the processes.

We assume as equal **ROSA** processes, every pair of them having the same Lexicographical Normal Forms:
$\forall P, Q \in \{\textbf{ROSA } processes\}.P = Q \Leftrightarrow \|P\| = \|Q\|$
This is a sound assumption since in [2] an equivalent proof of the pure functional behaviour of **ROSA** can be found. Although there is a difference of precedence between non-deterministic and probabilistic behaviours of PNAL and **ROSA** this fact has not been used to prove the soundness. Concerning the timed behaviour [8] presents a complete Proof System for Timed Observations.

In this version we will omit the treatment of recursion, because it implies an important mathematical apparatus so requiring a considerable amount of space, and the result does not justify this effort. This is due to the fact that for defining correctly a (lexicographical) normal form for infinite processes, we need a power domain, as well as an order relation, so that, an infinite process would be the limit of a chain of ascending finite processes, each of them, an approximation of this limit. In order to guarantee the existence of this limit, we need both to introduce a fixed point theory, and to prove that every operator is continuous.

Since we think that this considerable work is not interesting in our study, we leave for a future work the completion of this operator, and we address the interested reader to the paper [2], where the semantics for infinite processes in a syntax similar to that of **ROSA** is described. Thus, from now on, operator $recX : P$ is not considered.

Once the notion of *lexicographical normal form* is defined we are able to provide the metric in charge of supporting the heuristic to search the LTS.

4   Topological Structure for LNF-ROSA Processes

Definition 1. *Given a pair of* **ROSA** *processes P and Q the distance between them is $D(P, Q)$*

$$D(P, Q) = \frac{1}{2^{l(\|P\| \sqcup \|Q\|)}} - \frac{1}{2^N}$$

where:

- $l(string)$ *is the length of the string being part of a LNF-ROSA process $\|P\|$ and is defined inductively over the syntactic structure of $\{LNF\text{-}ROSA processes\}$, as follows*

$$l : \{LNF - \textbf{ROSA } processes\} \longrightarrow \mathbb{N}$$

$$
\begin{aligned}
\textbf{0} &\mapsto 0 \\
a.P &\mapsto 2 + l(P) \\
\langle a, \lambda \rangle.P &\mapsto 2 + l(P) \\
\textstyle+_{a \in A_j} P_a &\mapsto m - 1 + \textstyle\sum_{a \in A_j} l(P_a)(m = |A_j|) \\
\textstyle\bigoplus_{A_j \in \mathcal{A}_i} P_j &\mapsto k - 1 + \textstyle\sum_{A_j \in \mathcal{A}_i} l(P_j)(k = |\mathcal{A}_i|) \\
\textstyle\bigotimes_{i \in \{1...n\}} [q_i] P_i &\mapsto 2n - 1 + \textstyle\sum_{i \in \{1...n\}} l(P_i) \\
(P) &\mapsto 2 + l(P)
\end{aligned}
$$

- $N = max\{l(\|P\|), l(\|Q\|)\}$
- $P \sqcup Q$ is the longest common 0-ending string of both processes P and Q

Once defined the target state S_F (the one to achieve), the distance D between a given state and S_F is taken as the heuristic value of that state. The lower the better. It plays the role of the usual heuristic function $\mathbf{h}$ of the A^* search algorithm.

$$
\begin{aligned}
\mathbf{h_{S_F}} : \{\textbf{ROSA } procs.\} &\longrightarrow (0, 1] \\
P &\mapsto D(P, S_F)
\end{aligned}
$$

Finally, our proposal is, given an initial state S_0 and a final one S_F, to apply all the rules of the operational semantics of LNF-**ROSA** to S_0 so generating a set of processes, and only keep unfolding the state belonging to this set that minimizes h_{S_F}. Therefore, the computational cost of searching the LTS is moved from exponential to polynomial.

Theorem 1. *The function D so defined is a metric.*

Proof. "D is a metric over $\{LNF - \textbf{ROSA } processes\} \Leftrightarrow D$ holds $(1) \wedge (2) \wedge (3)$" where:

1. $\forall P, Q \in \{LNF - \textbf{ROSA } processes\}.D(P, Q) = 0 \Leftrightarrow P = Q$.
2. $\forall P, Q \in \{LNF - \textbf{ROSA } processes\}.D(P, Q) = D(Q, P)$.
3. $\forall P, Q, T \in \{LNF - \textbf{ROSA } processes\}.D(P, Q) \leq D(P, T) + D(T, Q)$.

All along the proof of this theorem processes P, Q and T are assumed to fulfill LNF-ROSA syntax.

1. $D(P, Q) = 0 \Leftrightarrow \frac{1}{2^{l(P \sqcup Q)}} - \frac{1}{2^N} = 0 \Leftrightarrow l(P \sqcup Q) = N \Leftrightarrow P = Q$.

2. $D(P, Q) = D(Q, P) \Leftrightarrow \dfrac{1}{2^{l(P \sqcup Q)}} - \dfrac{1}{2^N} = \dfrac{1}{2^{l(Q \sqcup P)}} - \dfrac{1}{2^N}$

   $\Leftrightarrow P \sqcup Q = Q \sqcup P \Leftrightarrow \sqcup$ symmetry.

3. $D(P, Q) \leq D(P, T) + D(T, Q) \Leftrightarrow D(P, T) + D(T, Q) - D(P, Q) \geq 0$

   $\Leftrightarrow \dfrac{1}{2^{l(P \sqcup T)}} - \dfrac{1}{2^m} + \dfrac{1}{2^{l(T \sqcup Q)}} - \dfrac{1}{2^o} - \dfrac{1}{2^{l(P \sqcup Q)}} + \dfrac{1}{2^n} \geq 0$

   $\Leftrightarrow (\dfrac{1}{2^n} - \dfrac{1}{2^m} - \dfrac{1}{2^o}) + (\dfrac{1}{2^{l(P \sqcup T)}} + \dfrac{1}{2^{l(T \sqcup Q)}} - \dfrac{1}{2^{l(P \sqcup Q)}}) \geq 0.$

where:

- $m = max\{l(P), l(T)\}$.
- $o = max\{l(T), l(Q)\}$.
- $n = max\{l(P), l(Q)\}$.

where either one can be less than the other two (**A**), or all the same (**B**)

A : **Let's assume** $n < (m = o)$

$$n < m \Leftrightarrow n \leq m - 1 \Leftrightarrow 2^n \leq 2^{m-1} \Leftrightarrow \frac{1}{2^n} \geq \frac{1}{2^{m-1}}$$

$$\Leftrightarrow \frac{1}{2^n} - \frac{1}{2^{m-1}} \geq 0 \Leftrightarrow \frac{1}{2^n} - \frac{1}{2^m} - \frac{1}{2^m} \geq 0 \Leftrightarrow \frac{1}{2^n} - \frac{1}{2^m} - \frac{1}{2^o} \geq 0$$

$$\sqcup \; transitivity \Leftrightarrow l(P \sqcup Q) \geq min\{l(P \sqcup T), l(T \sqcup Q)\}$$

$$\Leftrightarrow (l(P \sqcup Q) \geq l(P \sqcup T)) \vee (l(P \sqcup Q) \geq l(T \sqcup Q))$$

$$\Leftrightarrow (2^{l(P \sqcup Q)} \geq 2^{l(P \sqcup T)}) \vee (2^{l(P \sqcup Q)} \geq 2^{l(T \sqcup Q)})$$

$$\Leftrightarrow (\frac{1}{2^{l(P \sqcup Q)}} \leq \frac{1}{2^{l(P \sqcup T)}}) \vee (\frac{1}{2^{l(P \sqcup Q)}} \leq \frac{1}{2^{l(T \sqcup Q)}})$$

$$\Leftrightarrow (\frac{1}{2^{l(P \sqcup T)}} - \frac{1}{2^{l(P \sqcup Q)}} \geq 0) \vee (\frac{1}{2^{l(T \sqcup Q)}} - \frac{1}{2^{l(P \sqcup Q)}} \geq 0)$$

$$\Rightarrow \frac{1}{2^{l(P \sqcup T)}} + \frac{1}{2^{l(T \sqcup Q)}} - \frac{1}{2^{l(P \sqcup Q)}} \geq 0$$

The cases where $m < (n = o)$ or equivalently $o < (m = n)$, have very similar proof.

B : The proof is also valid here.

5   Conclusions and Future Work

In this paper we have provided the set of **ROSA** processes with a metric structure which allows to define a *heuristic function* for the sake of (computationally) improving the task of searching for 'a goal node' by means of taking this heuristic function to guide an A^* algorithm.

This heuristics provides a way to follow in order to make formalisms as Process Algebras a little easier and cheaper to be used. In fact, it would be only required to visit one node per stage/layer of the LTS when studying reachability.

Our future work in this line is mainly concerned with reducing the cost to get this sort of standardized shape required for ROSA processes to became suitable to identify identical behaviours.

References

1. Bloom, B.: Ready simulation, bisimulation, and the semantics of CCS-like languages. Ph.D. thesis, Department of Electrical Engineering and Computer Science, MIT (1989)

2. Cazorla, D., Cuartero, F., Valero, V., Pelayo, F.L., Pardo, J.J.: Algebriac theory of probabilistic and nondeterministic processes. J. Logic Algebraic Program. **55**(1–2), 57–103 (2003)

3. Cleaveland, R., Hennessy, M.: Testing equivalence as a bisimulation equivalence. Formal Aspects Comput. **5**, 1–20 (1993)

4. Edelkamp, S., Jabbar, S., Lafuente, A.L.: Heuristic search for the analysis of graph transition systems. In: Corradini, A., Ehrig, H., Montanari, U., Ribeiro, L., Rozenberg, G. (eds.) ICGT 2006. LNCS, vol. 4178, pp. 414–429. Springer, Heidelberg (2006). doi:10.1007/11841883_29

5. Gradara, S., Santone, A., Villani, M.L.: Using heuristic search for finding deadlocks in concurrent systems. Inf. Comput. **202**(2), 191–226 (2005)

6. Hart, P.E., Nilsson, N.J., Raphael, B.: A formal basis for the heuristic determination of minimum cost paths. IEEE Trans. Syst. Sci. Cybern. **4**(2), 100–107 (1968)

7. Milner, R.: Communication and Concurrency. Prentice Hall, Upper Saddle River (1989)

8. Ortega-Mallén, Y., Frutos-Escrig, D.: A complete proof system for timed observations. In: Abramsky, S., Maibaum, T.S.E. (eds.) CAAP 1991. LNCS, vol. 493, pp. 412–440. Springer, Heidelberg (1991). doi:10.1007/3-540-53982-4_23

9. Pelayo, F.L., Cuartero, F., Cazorla, D.: Looking for a cheaper ROSA. In: Cabestany, J., Rojas, I., Joya, G. (eds.) IWANN 2011. LNCS, vol. 6692, pp. 380–387. Springer, Heidelberg (2011). doi:10.1007/978-3-642-21498-1_48

10. Pelayo, F.L., Cuartero, F., Valero, V., Cazorla, D.: Analysis of the MPEG-2 encoding algorithm with ROSA. Electron. Notes Theoret. Comput. Sci. **80**(1), 185–202 (2003). https://doi.org/10.1016/S1571-0661(04)80818-1

11. Pelayo, F.L., Pelayo, M.L., Guirao, J.G.: Generating the syntactic and semantics graphs for a Markovian process algebra. J. Comput. Appl. Math. **204**, 38–47 (2007)

12. Plotkin, G.D.: A structural approach to operational semantics. Technical report DAIMI FN-19, Computer Science Department, Aarhus University (1981)

Real World applications of BCI Systems

Suitable Number of Visual Stimuli for SSVEP-Based BCI Spelling Applications

Felix Gembler, Piotr Stawicki, and Ivan Volosyak$^{(\boxtimes)}$

Faculty of Technology and Bionics, Rhine-Waal University of Applied Sciences,
47533 Kleve, Germany
ivan.volosyak@hochschule-rhein-waal.de
http://www.hochschule-rhein-waal.de

Abstract. Steady state visual evoked potentials (SSVEPs)-based Brain-Computer interfaces (BCIs) provide a pathway for re-establishing communication to people with severe disabilities. In the presented study, we compared accuracy and speed of three SSVEP-based BCI spelling applications in order to investigate the influence of the number of visual stimuli on the BCI performance. Three systems with four, six and 28 stimulating frequencies were tested. Ten subjects (one female) participated in this study. The highest ITR achieved in the experiment was 51.77 bpm. It is interesting, that it was achieved with the system based on six flickering targets. Our results confirm that the number of stimuli has high impact on classification accuracy and BCI literacy of SSVEP-based BCIs.

Keywords: Brain-computer interface (BCI) · Electroencephalogram (EEG) · Steady state visual evoked potential (SSVEP) · Visual stimuli

1 Introduction

A brain-computer interface (BCI) allows direct acquisition of human brain activity patterns and translates them into control commands [4,10,23]. This article focuses on steady state visual evoked potential (SSVEP)-based BCIs, based on a neural response which is evoked by repetitive visual stimuli [1].

The implementation of SSVEP-based BCIs as spelling interfaces has been a major research field in BCI community. An important issue preventing a broader use of BCIs is so-called BCI illiteracy (also synonymously called BCI deficiency), basically describing the fact that for some users the BCI cannot detect their intentions accurately. That also takes into account the situations if classification accuracy cannot surpass a certain threshold of e.g. 70%. The BCI literacy rate is defined reciprocally as the percentage of users who are able to achieve effective control over the BCI.

Generally speaking, a relatively high amount of targets can be implemented with SSVEP-based BCIs. Hwang et al. developed a SSVEP-based BCI spelling

© Springer International Publishing AG 2017
I. Rojas et al. (Eds.): IWANN 2017, Part II, LNCS 10306, pp. 441–452, 2017.
DOI: 10.1007/978-3-319-59147-6_38

system adopting a QWERTY-style LED keyboard [8]. Such multitarget application can also be implemented on computer screens using the frequency approximation methods [21]. Though multitarget BCIs usually allow higher speed, some articles have previously reported slightly worse BCI performance with a higher number of stimuli for some users [2,5,22]. The literacy rate is generally higher with BCIs implementing a low number of visual stimuli; some larger BCI studies with only four targets reported that even all users were able to gain control over the application [6,7,14,19].

A major disadvantage of spelling applications with smaller number of simultaneously presented targets is that typically several steps are necessary in order to select a desired character, due to the fact that the English alphabet consists of 26 letters (which is far more than the number of simultaneously presented targets). Additional commands for error correction and special characters also need to be considered. On the other hand, the time required for a reliable single selection is shorter, as a smaller number of frequencies needs to be distinguished. Such systems also tend to be more user friendly due to the reduced cognitive load, and cause less frustration of potential false classifications.

It should also be noted, that the careful design of the interface can directly improve the spelling speed as well. Some research groups developed spellers that feature dictionary support, allowing users to spell multiple-word texts faster (see e.g. [16,20]).

The aims of the presented work were to explore how many targets can be reliably distinguished from each other using the SSVEP paradigm in the series of online experiments and to find out the optimal number of visual stimuli in terms of user friendliness and performance. For this purpose, we tested three different custom-made applications utilizing different amounts of simultaneously flickering targets. It is important to investigate the number of BCI targets under practical real life conditions, as in our previous studies a big difference between online and offline results was observed. In the formerly developed *Three-step speller* four distinct stimulation frequencies were used [6]; the here presented *Two-step speller* was implemented with six stimuli, similar to [20]. Performance of these spellers was compared to the multitarget *QWERTZ-speller* which resembled a German keyboard [12]. For this speller 28 frequencies were used, which allowed character selection in one single step.

2 Methods and Materials

2.1 Subjects

Ten subjects (one female) with a mean (SD) age 25.5 (4.03) years participated in the study, all students or employees of the Rhine-Waal University of Applied Sciences in Kleve. All subjects (healthy adult volunteers) gave written informed consent in accordance with the Declaration of Helsinki. Information needed for the analysis of the experiments was stored anonymously, and cannot be traced back to the participant. The EEG recording took place in a normal laboratory room. Spectacles were worn when appropriate. Subjects did not receive any financial reward for participation in this study.

2.2 Signal Acquisition

Subjects were seated in front of a LCD screen (BenQ XL2420T, resolution: 1920 × 1080 pixels, vertical refresh rate: 120 Hz) at a distance of about 60 cm. The used computer system operated on Microsoft Windows 7 Enterprise running on an Intel processor (Intel Core i7, 3.40 GHz). Standard Ag/AgCl electrodes were used to acquire the signals from the surface of the scalp. The ground electrode was placed over AF_Z, the reference electrode over C_Z, and the eight signal electrodes were placed at predefined locations on the EEG-cap marked with $P_Z, PO_3, PO_4, O_1, O_2, O_Z, O_9$ and O_{10} in accordance with the international system of EEG electrode placement. Standard abrasive electrolytic electrode gel was applied between the electrodes and the scalp to bring impedances below 5 kΩ. An EEG amplifier, g.USBamp (Guger Technologies, Graz, Austria), was utilized. The sampling frequency was set to 128 Hz. During the EEG signal acquisition, an analogue band pass filter (between 2 and 30 Hz) and a notch filter (around 50 Hz) were applied directly in the amplifier.

2.3 Signal Processing

Minimum energy combination method (MEC) [3,17] was used for SSVEP signal classification. To detect a specific frequency in the spatially filtered signals the SSVEP power estimations for all N_f frequencies were normalized into probabilities,

$$p_i = \frac{\hat{P}_i}{\sum_{j=1}^{N_f} \hat{P}_j} \text{ with } \sum_{i=1}^{N_f} p_i = 1 \tag{1}$$

where $\hat{P}_i$ is the ith power estimation, $1 \leq i \leq N_f$.

Further, in order to increase the difference between probabilities, a Softmax function was applied:

$$p'_i = \frac{e^{\alpha p_i}}{\sum_{j=1}^{j=N_f} e^{\alpha p_j}} \text{ with } \sum_{i=1}^{i=N_f} p'_i = 1 \tag{2}$$

with $\alpha = 0.25$.

All classifications were performed on the basis of the hardware synchronization of the EEG amplifier (g.USBamp); the new EEG data were transferred to the PC in blocks of 13 samples (101.5625 ms with the sampling rate of 128 Hz). The classification was performed with a stepwise increasing sliding window (see [17] for more details). If the ith stimulation frequency had the highest probability p'_i and exceeded certain predefined thresholds β_i the corresponding command was classified. After each classification the classifier output was rejected for the duration of about 914 ms (9 blocks). During this gaze shifting period, the targets did not flicker and the user changed his focus to another target unhindered (please also refer to [17] for more details). In this study the minimal classification time window was set to 2 s for all three spellers.

Stimulation frequencies for the presented *Two-step speller* and the *Three-step speller* were determined in a short calibration session with the previously developed SSVEP-BCI wizard software [6].

To implement the 28 targets needed for the *QWERTZ-speller* frame-based stimulus approximation method as proposed by Wang et al. [21] was used. For the *Two-step speller* and *Three-step speller* the frequencies were selected as divisors of the vertical refresh rate of the monitor, ensuring the constant number of frames in each cycle. Four or six frequencies were used for stimulation in these cases. In the frame-based stimulus approximation method a varying number of frames is used in each cycle. The stimulus signal at frequency f is generated by

$$\text{stim}(f, i) = \text{square}[2\pi f(i/\text{RefreshRate})],$$

where $\text{square}(2\pi f t)$ generates a square wave with frequency f and i is the frame index. E.g., the black/white reversing interval for the approximated frequency 17 Hz includes 17 cycles of varying length (three or four frames). By using the formula above, the one-second stimulus sequence of 17 Hz can be generated: (*4* 4 3 4 3 4 3 4 3 4 3 4 3 4 3 *4* *4* 3 4 3 4 3 4 3 4 3 4 3 4 3 4 3). For the online spelling task with the *QWERTZ speller* approximated frequencies between 6.1 and 11.7 Hz (resolution 0.2 Hz) were used to avoid overlapping in the harmonics.

2.4   Software

Three different spelling applications were tested in the online experiments. The *Three-step Speller* and the *Two-step Speller* resemble previously developed graphical user interface (GUI) layouts [5,20] and allow selection of single letters and complete words, which can be selected in two or three steps, respectively.

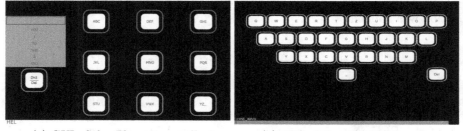

(a) GUI of the *Three-step speller*   (b) GUI of the *QWERTZ speller*

Fig. 1. (a) The graphical user interface of the *Three-step speller*. In the *Three-step speller* only four stimulation frequencies were used. (b) The *QWERTZ speller*, which allows direct selection of letters (in one step only). (Color figure online)

In each speller, stimulation frequencies were presented as flickering boxes on the computer screen. The size of the boxes varied in relation to the SSVEP amplitude during the experiment as described in [17]. Each box was outlined by a frame which determined the maximum size a box could reach. In order to increase the user friendliness further, every command classification was followed by an audio feedback with the name of the selected command or the letter spelled.

Two-Step Speller. Initially a matrix of 30 white stimulation boxes and 6 additional gray boxes containing dictionary entries and a "back" button were displayed (see Fig. 2a). The size of the white boxes varied between 140×90 and 170×130 pixels; the size of the grey boxes varied between 370×90 and 400×130 pixels. Twenty-seven of the white stimulation boxes contained single characters (26 letters plus the white space). Through selection of the box "Dictionary" the user was able to select one of up to five suggested words. The boxes "Delete" and "Clear word" allowed the user to delete either the previously selected character or the whole lastly spelled word. As only six distinct frequencies were used, each row of white boxes presented one stimulation frequency. E.g. the boxes "F","G","H","I", and "J" all flickered in unison and by gazing at a particular box the user selected the row containing the desired letter. After each selection, the boxes of the selected row were displayed flickering with individual frequencies while the other rows were grayed out. Now again only six boxes were flickering: the five boxes from the selected row and also the "back" button (see Fig. 2b). The output of the *Three-step Speller* (spelled text) was displayed at the center of the screen. Figure 3 shows the necessary steps to select word suggestions from the included dictionary.

(a) Selection of letter "H" - 1st step   (b) Selection of letter "H" - 2nd step

Fig. 2. The *Two-step speller* during the online copy spelling task. The subject had to spell the letter "H". Every row of the stimulation matrix flickered with a different frequency (same frequencies for columns). In order to select the desired letter, the subject needed to select the corresponding row first (a). In the second step all characters of the selected row flickered with different frequencies and the desired letter "H" was selected. In total, two correct commands were necessary to select any desired character.

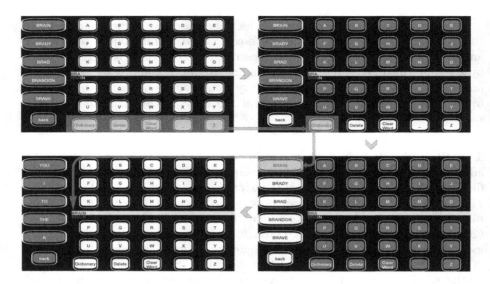

Fig. 3. Dictionary mode of the *Two-step speller*. At first the subject selected the row containing the box "Dictionary" (top left) and in the second step the box itself (top right). Then one of five suggested words were displayed (bottom right). The user chose the word "BRAIN" (bottom left).

Three-Step Speller. The *Three-step speller* was realized similarly. Three successive correct commands were necessary in order to select a desired character. Initially a matrix of nine boxes, each containing three letters of the alphabet (26 letters plus the white space), was presented. The frames of the boxes were colored differently, with each color corresponding to one of the stimulation frequencies (see Fig. 1a); an additional 10th box, containing the command "Dict/Del" (delete the last spelled character or switch to the *dictionary mode*) was located on the left side of the screen. Box sizes varied between the default 140×130 and 230×210 pixels. After the first performed selection, the boxes of the selected row were highlighted with individual colors (green, red, and blue), while the other rows were grayed out. As only the frequencies (and frame colors) changed, but not the position of the target letter, no gaze shifting was necessary. The user was able to select one out of three boxes, containing three letters each. Afterwards, the boxes were rearranged and again three individual boxes, each containing one of the characters from the previously selected box, were presented. Finally, the user was able to select the desired character. The role of the yellow-framed box changed depending on the current step of the selection procedure. Initially the user could either delete the last selected letter or select one of the word suggestions which were positioned above the "Dict/Del" button. After the 1st and 2nd step the yellow framed box contained the command "back" which gave the user the opportunity to go to the previous screen, e.g. in case of misclassification.

QWERTZ Speller. The *QWERTZ speller* (as presented in [12]) displayed 28 buttons which were arranged into four rows resembling a German QWERTZ-keyboard layout (see Fig. 1b). The box sizes varied between 130×90 and 170×120 pixels. The rows contained 10, 9, 7, and 2 buttons from the top down. A "space" and a "delete" button occupied the bottom row. Each box flickered with a specific frequency, using previously described frame-based stimulus approximation method. Therefore, any desired character was selected in a single step.

2.5   Experimental Setup

After signing the consent form, each subject was prepared for the EEG recording. At first subjects went through the steps of a Wizard software which determined the stimulation frequencies for the *Three-step speller* and the *Two-step speller*. Afterwards, subjects tested the spelling applications as follows: Initially, subjects participated in a familiarization run, spelling the word "BCI" and a word of their own choice (e.g. their first name). Next, each subject used each GUI in random order to spell the phrase "RHINE WAAL" (names of two rivers determining location of our University). The spelling phase ended automatically when the phrase was spelled correctly. In case a subject was not able to execute a desired classification within a certain time frame, or if repeated false classifications occurred, the experiment was stopped manually. Spelling errors were corrected via the "delete" button. Information needed for the further analysis was stored anonymously during the experiment. After the test phase the subjects completed a post-questionnaire, answering questions regarding the preferred spelling application. The entire session took on average about 40 min for each subject. Subjects had the opportunity to opt-out of the study at any time.

3   Results

The overall BCI performance for the three tested spelling applications is given in Table 1. Provided are the time needed to complete the task, the command accuracy and the commonly used information transfer rate (ITR) in bits/min (see e.g. [23]). The overall number of possible choices depended on the spelling application; 4, 6 and 28 for the *Three-step speller*, *Two-step speller* and the *QWERTZ speller*, respectively.

A comparison of the tested spelling applications based on the results is provided in Table 2. The accuracy was calculated based on the number of correct command classifications divided by the total number of classified commands C_n. To obtain the average command classification time, the total time needed for the spelling task, was divided by C_n.

Table 3 summarizes results from the post questionnaire for all subjects.

Table 1. Results for ten subjects (copy spelling of the phrase "RHINE WAAL"). Subjects that were not able to successfully control one of the tested spellers were excluded from the calculation of mean values and represented with the "−" sign. For each subject the highest achieved ITR is shown in bold.

Subject [#]	Three-step speller			Two-step speller			QWERTZ speller		
	Time [sec]	Acc. [%]	ITR [bpm]	Time [sec]	Acc. [%]	ITR [bpm]	Time [sec]	Acc. [%]	ITR [bpm]
1	164.125	91.89	19.82	59.922	100	**51.77**	79.625	85.71	37.31
2	128.883	100	27.93	175.094	89.66	18.53	60.633	91.67	**47.47**
3	179.867	94.29	18.60	174.484	85.71	20.00	83.484	76.47	**35.45**
4	182.203	100	19.76	113.547	95.45	**25.72**	-	-	-
5	299.203	100	12.03	141.781	100	**21.88**	118.523	83.33	20.44
6	227.602	89.4	13.50	-	-	-	-	-	-
7	167.781	100	21.46	207.797	88.46	13.52	153.461	75.00	**21.95**
8	268.430	91.67	11.70	-	-	-	-	-	-
9	171.031	96.88	19.64	229.531	92.31	13.70	114.867	100	**25.11**
10	149.195	97.14	**24.88**	238.977	95.83	13.49	160.063	75.00	21.05
Mean	193.83	96.13	18.93	167.64	93.43	22.33	110.09	83.88	29.82
SD	54.19	4.04	5.31	60.75	5.29	12.70	37.76	9.46	10.39

Table 2. Overview of the tested spelling applications. Subjects that were not able to successfully control a spelling interface were excluded from the calculation of mean values for that particular system.

	Number of displayed stimuli	Literacy rate [%]	Max ITR [bpm]	Mean ITR [bpm]	Mean Acc [%]	Classification time (average) [sec]
Three-step speller	28	100	27.93	18.93	96.13	5.856
Two-step speller	6	80	51.77	22.33	93.43	6.631
QWERTZ speller	4	70	47.47	29.23	83.88	7.284

Table 3. Questionnaire results about the tested spellers.

Subject	Gender	Age	PC usage [h/week]	BCI-experience	Most exhausting speller	Most intuitive speller	Relative change in tiredness
1	M	28	25	Yes	QWERTZ	Three-step	Neutral
2	M	21	15	No	Two-step	Three-step	Neutral
3	M	31	40	Yes	QWERTZ	QWERTZ	More tired
4	M	21	35	No	Three-step	Two-step	More tired
5	F	26	70	Yes	QWERTZ	Three-step	Neutral
6	M	26	70	Yes	Two-step	Two-step	More tired
7	M	22	35	No	QWERTZ	QWERTZ	Neutral
8*	M	21	10	No	-	-	Neutral
9	M	28	50	Yes	QWERTZ	QWERTZ	Neutral
10	M	32	65	Yes	Two-step	QWERTZ	Neutral

*Subject 8 only gained control over the Three-step speller.

4 Discussion

It is important to mention that this study was performed only with three graphical user interfaces on a very small user group, so no general conclusions can be drawn. However, some certain trend can be observed. As it can be seen in Table 1, BCI performance varied considerably between subjects. While some subjects performed best with the *QWERTZ speller*, others were faster with the *Two-step speller*. Interestingly, one subject (subject 10) reached peak ITR with the *Three-step speller*. All subjects were able to control the *Three-step speller*, reaching an average accuracy of 96.13%. For the BCI spellers with higher number of targets, BCI literacy rate and average classification accuracy dropped (see Table 2). The BCI illiterate subjects were excluded from further calculation of mean values. While more than 80% of the subjects were able to control the *Two-step speller*, reaching a mean accuracy of 93.43%, only 70% of all subjects completed the spelling task with the *QWERTZ speller*. For the *QWERTZ speller* the frequency difference between two flickering targets was as small as 0.2 Hz (in general, even denser resolutions can be realized as e.g. 0.05 Hz, for further details refer to [13]). In terms of ITR, four subjects achieved peak performance with the *QWERTZ speller*. Interestingly, the highest ITR (51.77 bpm) was achieved with the *Two-step speller*. This can be explained by the fact that this subject had a relatively low classification accuracy with the *QWERTZ speller*.

Indeed, as displayed in Table 2, BCI-accuracy generally drops with an increasing number of targets for most subjects. This is in line with other observations in BCI research (see e.g. [24]).

The high classification accuracies of low target SSVEP-based BCIs make them a valuable tool for hybrid BCIs, which combine input signals of different brain patterns, or biosignals such as eye gaze (see e.g. [9]). Apart form the higher literacy rate, a further advantage of BCIs with fewer stimuli is that they seem to be less stressful for the user (see e.g. [5]). The majority of the subjects stated that the *QWERTZ speller* was the most exhausting one to look at. The comparably low accuracy also caused frustration for some subjects. In addition to that, the time the subject had to look at a flickering target was generally larger for the *QWERTZ speller*. The average command classification time (including the gaze shifting period) was 7.3 s for the *QWERTZ speller*, which is considerably longer than the mean classification times for the systems with smaller number of targets, 5.9 s and 6.6 s for the *Two-step speller* and *Three-step speller*, respectively. This demonstrates that the time segment length for obtaining accurate classification of SSVEP responses is generally larger, if more stimuli are used. The importance of the of appropriate time window length has already been discussed in [18].

Nevertheless, the performance with the *QWERTZ speller* is quite promising; a mean ITR of 29.82 bpm was achieved. These results indicate the potential use of noninvasive SSVEP-based BCIs as a standalone high-speed communication tool. It is important to mention, however, that all subjects were familiar with the QWERTZ-layout, as all of them stated that they use computers in daily life quite often (see Table 3).

All in all, subjects gave generally positive feedback regarding the user friendliness of all tested systems. However, three subjects reported an increased level of tiredness after the experiment (see Table 3). Among these subjects were two subjects, who were unable to gain control over the *QWERTZ speller*. In order to keep the experiment duration short, we used only several words as test phrases. Therefore, the integrated dictionaries were used very sparsely. In earlier practical experiments we found that the integrated dictionary generally doubles the amount of spelled characters per minutes. Further improvements could be made regarding the calibration software used in the experiment.

Most SSVEP based BCIs depend to some extend on oculomotor control. However, some SSVEP applications designed to measure attentional states do not depend on the user's ability to shift his or her gaze [11]. Some effort has also been made toward the development of gaze independent SSVEP-based BCI communication tools [15]. The systems in the presented study still relied on the user's eye movements. For the *QWERTZ speller* the dependence on eye gaze control is strongest because of the close proximity of targets.

An additional calibration phase could be integrated in order to determine the optimal number of stimuli for each user. The optimal number of targets in terms of accuracy as well as ITR varied strongly between subjects. The comparison of the mean accuracy and ITR values for the different application shows that ITR rises with increasing number of targets, while accuracy drops. Hence optimal number of stimuli needs to be determined according to individual performance, user friendliness, and application type. While speed attracts much attention in development of BCI application, high accuracies are the priority for control applications and also tend to provide the highest literacy rate. Tests with a larger number of subjects are necessary to confirm the results. Also, further tests with brain-injured patients are necessary, as these may differ from findings of this study. Our future work will focus on the further development of low target SSVEP-based BCIs and data fusion with eye tracking devices to enhance the speed of such hybrid systems.

5   Conclusions

Three SSVEP-based spelling applications offering four, six, and 28 selection targets, respectively, were tested with 10 healthy subjects in order to explore optimal number of stimuli for SSVEP-based BCI. All subjects achieved reliable control over the 4-target system, whilst 80% effectively controlled the 6-target BCI system and only 70% were successful with the speller based on 28 targets. Subjects that were not able to successfully control one of the spellers were excluded from the calculation of the corresponding mean values. The comparison of mean values for BCI-literate subjects shows that ITR increases with the number of targets in the used speller, while classification accuracy decreases. The comparably low accuracy of the system with the highest number of targets caused frustration for some subjects. The results show that the optimal number of stimuli in a SSVEP-based BCI depends on the application it is designed for, and must be investigated further.

Acknowledgment. This research was supported by the European Fund for Regional Development under Grant GE-1-1-047. We also thank the participants of this study.

References

1. Bin, G., Gao, X., Yan, Z., Hong, B., Gao, S.: An online multi-channel SSVEP-based brain-computer interface using a canonical correlation analysis method. J. Neural Eng. **6**(4), 046002 (2009)
2. Carvalho, S.N., Costa, T.B., Uribe, L.F., Soriano, D.C., Almeida, S.R., Min, L.L., Castellano, G., Attux, R.: Effect of the combination of different numbers of flickering frequencies in an SSVEP-BCI for healthy volunteers and stroke patients. In: 2015 7th International IEEE/EMBS Conference on Neural Engineering (NER), pp. 78–81. IEEE (2015)
3. Friman, O., Volosyak, I., Gräser, A.: Multiple channel detection of steady-state visual evoked potentials for brain-computer interfaces. IEEE Trans. Biomed. Eng. **54**(4), 742–750 (2007)
4. Gao, S., Wang, Y., Gao, X., Hong, B.: Visual and auditory brain-computer interfaces. IEEE Trans. Biomed. Eng. **61**(5), 1436–1447 (2014)
5. Gembler, F., Stawicki, P., Volosyak, I.: Towards a user-friendly BCI for elderly people. In: Proceedings of the 6th International Brain-Computer Interface Conference Graz (2014)
6. Gembler, F., Stawicki, P., Volosyak, I.: Autonomous parameter adjustment for SSVEP-based BCIs with a novel BCI wizard. Front. Neurosci. **9**, 1–12 (2015)
7. Guger, C., Allison, B.Z., Großwindhager, B., Prückl, R., Hintermüller, C., Kapeller, C., Bruckner, M., Krausz, G., Edlinger, G.: How many people could use an SSVEP BCI? Front. Neurosci. **6**, 1–6 (2012)
8. Hwang, H.J., Lim, J.H., Jung, Y.J., Choi, H., Lee, S.W., Im, C.H.: Development of an SSVEP-based BCI spelling system adopting a QWERTY-style LED keyboard. J. Neurosci. Methods **208**(1), 59–65 (2012)
9. Liu, Y.H., Wang, S.H., Hu, M.R.: A self-paced p300 healthcare brain-computer interface system with SSVEP-based switching control and kernel FDA+ SVM-based detector. Appl. Sci. **6**(5), 142 (2016)
10. Nicolas-Alonso, L.F., Gomez-Gil, J.: Brain computer interfaces, a review. Sensors **12**(2), 1211–1279 (2012)
11. Norcia, A.M., Appelbaum, L.G., Ales, J.M., Cottereau, B.R., Rossion, B.: The steady-state visual evoked potential in vision research: a review. J. Vis. **15**(6), 4 (2015)
12. Nunez, F., Gembler, F., Stawicki, P.: Information transfer rate differences in SSVEP BCI: alphabetical and keyboard layouts in a GUI. In: Volosyak, I. (ed.) EEG-Based Brain-Computer Interfaces for Healthcare Applications, pp. 101–110. Shaker Verlag (2016)
13. Stawicki, P., Gembler, F., Volosyak, I.: Evaluation of suitable frequency differences in SSVEP-based BCIs. In: Blankertz, B., Jacucci, G., Gamberini, L., Spagnolli, A., Freeman, J. (eds.) Symbiotic 2015. LNCS, vol. 9359, pp. 159–165. Springer, Cham (2015). doi:10.1007/978-3-319-24917-9_17
14. Stawicki, P., Gembler, F., Volosyak, I.: Driving a semi-autonomous mobile robotic car controlled by a SSVEP-based BCI. Comput. Intell. Neurosci. **2016**, 5 (2016). Hindawi

15. Treder, M.S., Schmidt, N.M., Blankertz, B.: Gaze-independent brain-computer interfaces based on covert attention and feature attention. J. Neural Eng. **8**(6), 066003 (2011)
16. Vilic, A., Kjaer, T.W., Thomsen, C.E., Puthusserypady, S., Sorensen, H.B.D.: DTU BCI speller: an SSVEP-based spelling system with dictionary support. In: 2013 35th Annual International Conference of the IEEE Engineering in Medicine and Biology Society (EMBC), pp. 2212–2215. IEEE (2013)
17. Volosyak, I.: SSVEP-based Bremen-BCI interface - boosting information transfer rates. J. Neural Eng. **8**(3), 036020 (2011)
18. Volosyak, I., Cecotti, H., Gräser, A.: Steady-state visual evoked potential response - impact of the time segment length. In: Proceedings of the 7th International Conference on Biomedical Engineering BioMed2010, Innsbruck, Austria, 17–19 February, pp. 288–292 (2010)
19. Volosyak, I., Gembler, F., Stawicki, P.: Age-related differences in SSVEP-based BCI performance. Neurocomputing (2017, in press)
20. Volosyak, I., Moor, A., Gräser, A.: A dictionary-driven SSVEP speller with a modified graphical user interface. In: Cabestany, J., Rojas, I., Joya, G. (eds.) IWANN 2011. LNCS, vol. 6691, pp. 353–361. Springer, Heidelberg (2011). doi:10.1007/978-3-642-21501-8_44
21. Wang, Y., Wang, Y.T., Jung, T.P.: Visual stimulus design for high-rate SSVEP BCI. Electron. Lett. **46**(15), 1057–1058 (2010)
22. Wei, Q., Feng, S., Lu, Z.: Stimulus specificity of brain-computer interfaces based on code modulation visual evoked potentials. PloS One **11**(5), e0156416 (2016)
23. Wolpaw, J., Birbaumer, N., McFarland, D., Pfurtscheller, G., Vaughan, T.: Brain-computer interfaces for communication and control. Clin. Neurophysiol. **113**, 767–791 (2002)
24. Xing, S., McCardle, R., Xie, S.: Reading the mind: the potential of electroencephalography in brain computer interfaces. In: 2012 19th International Conference Mechatronics and Machine Vision in Practice (M2VIP), pp. 275–280. IEEE (2012)

A Binary Bees Algorithm for P300-Based Brain-Computer Interfaces Channel Selection

Víctor Martínez-Cagigal[1]([✉]) and Roberto Hornero[1,2,3]

[1] Biomedical Engineering Group, E.T.S.I. de Telecomunicación,
Universidad de Valladolid, Valladolid, Spain
victor.martinez@gib.tel.uva.es, robhor@tel.uva.es
[2] IMUVA, Instituto de Investigación en Matemáticas, Universidad de Valladolid,
Valladolid, Spain
[3] INCYL, Instituto de Neurociencias de Castilla y León, Salamanca, Spain

Abstract. Brain-computer interface (BCI) systems need to work in real-time with large amounts of data, which makes the channel selection procedures essential to reduce over-fitting and to increase users' comfort. In that sense, metaheuristics based on swarm intelligence (SI) have demonstrated excellent performances solving complex optimization problems and, to the best of our knowledge, they have not been fully exploited in P300-BCI systems. In this study, we propose a modified SI method, called binary bees algorithm (b-BA), that allows users to select the most relevant channels in an evolutionary way. This method has been compared to particle swarm optimization (PSO) and tested with the 'III BCI Competition 2005' dataset II. Results show that b-BA is suitable for use in this kind of systems, reaching higher accuracies (mean of $96.0 \pm 0.0\%$) than PSO (mean of $93.5 \pm 2.1\%$) and the original ones (mean of $94.0 \pm 2.8\%$) using less than the half of the initial channels.

Keywords: Brain-computer interfaces · P300 evoked potentials · Channel selection · Swarm intelligence · Bees algorithm · Particle swarm optimization

1 Introduction

A brain-computer interface (BCI) is a communication system between the brain and the environment that translates the users' intentions into device control commands in real-time [15]. Due to its portability, low cost and non-invasive nature, electroencephalography (EEG) is commonly used for monitoring the brain activity of the user. Thus, electric potentials, which are used as reliable control signals, are recorded placing several electrodes on the scalp [15].

In particular, P300 evoked potentials are positive peaks mainly produced in the parietal cortex in response to infrequent and particularly significant stimuli at about 300 ms after their onset [15]. One of the most popular applications based on these potentials is the P300 Speller, which was originally developed by Farwell and Donchin [5] and allows users to spell words and select certain

© Springer International Publishing AG 2017
I. Rojas et al. (Eds.): IWANN 2017, Part II, LNCS 10306, pp. 453–463, 2017.
DOI: 10.1007/978-3-319-59147-6_39

commands. The user focuses attention on one of the character cells of a displayed matrix while its rows and columns are randomly intensified. When the target's row or column are illuminated, a P300 potential is generated on the user's scalp [10]. Therefore, the attended character can be determined as the intersection of the row and the column where these P300 potentials were found.

However, due to the high inter-session variability and the low signal-to-noise ratio, there is a need to compute an average of several repetitions (i.e., sequences) in order to detect a reliable P300 potential. Thus, the system should analyze high dimensional data in real-time, which can easily produce over-fitting [3]. Channel selection procedures can reduce the curse of dimensionality, as well as reduce the power consumption in wireless EEG caps, increase the users' comfort and, in some cases, increase the system performance [3]. Nevertheless, this selection is not trivial: for an N-channel set, there are 2^N possible subsets; making the exhaustive search intractable [3]. There are previous studies that have addressed this problem in P300-BCI systems. Most of them have used backward elimination strategies with different selection criteria [3,13,14] or similar approaches [4,16,19]. In addition, several variants of particle swarm optimization (PSO) have been applied to the same problem [6,8,11].

Nevertheless, swarm intelligence (SI) is a growing research field that integrates a large amount of different methods from PSO that, to the best of our knowledge, have not been applied to the channel selection problem in P300-BCI systems yet. Metaheuristics based on SI have demonstrated excellent performances solving complex optimization problems by means of integrating the collective behavior of decentralized and self-organized systems, whose population is formed by agents that interact among themselves and the environment [18].

In this study, a modified binary version of the bees algorithm (b-BA) is proposed to its application in the P300-BCI channel selection procedure. The algorithm is based on a wrapper method, which is dependent on a linear discriminant analysis (LDA) classifier, with a multi-objective purpose: (i) to maximize the system's accuracy; and (ii) to minimize the required number of channels.

2   Subjects

The proposed method have been tested with the 'III BCI Competition 2005' dataset II, which has been recorded from two different subjects (named A and B) in five spelling sessions [1]. In each session, subjects were asked to spell several words. EEG signals were recorded using a 64-channel cap with a sampling frequency of 240 Hz. Then, signals were bandpass-filtered from 0.1–60 Hz. For each subject, training and testing sets were made of 85 and 100 characters, respectively. Further information can be found in [1].

3   Methods

Figure 1 shows the signal processing pipeline that was followed in this study. Firstly, a feature extraction stage was applied, in which: (i) a 0–700 ms window

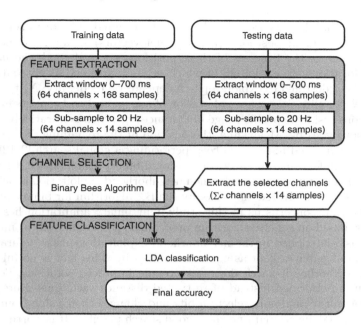

Fig. 1. Detailed signal processing pipeline employed in this paper, composed of three stages: feature extraction, channel selection and feature classification.

from the stimuli onset was extracted; and (ii) a sub-sampling to $f = 20\,\text{Hz}$ was performed (i.e., 14 samples for each stimulus in each channel) [10]. Later, the SI-based method was applied, returning the final selected channel vector $c = [c_1, c_2, \ldots, c_{N_d}]$, where $c_n = \{0, 1\}$ for $n = 1, 2, \ldots, N_d$, and $N_d = 64$ is the original number of channels. Note that $c_n = 0$ means that channel n is not selected, whereas $c_n = 1$ means that it is selected. Finally, the solution c was evaluated in the testing subset and final accuracy (defined as the number of correct predicted characters to the total number of characters) was calculated. It is worthy to note that this pipeline was repeated 10 times in order to avoid local minima.

3.1 Binary Bees Algorithm

The original bees algorithm (BA), developed by Pham et al. in 2006 [12], is based on the foraging behavior of honey bees. Initially, the hive sends scout bees that move randomly looking for nectar from flowers. When a bee discovers a promising patch, it returns to the hive and reports the direction, distance and quality of the patch to other bees. Its quality (i.e., fitness) is proportional to the number of bees destined to exploit it. Meanwhile, a reduced number of bees continue searching for new locations [12].

Based on the aforementioned behavior, the original BA starts sending N_b scout bees to random positions (i.e., channel vectors) in an N_d-dimensional search space. Once the fitnesses of these positions are evaluated, the m_s most promising locations are exploited, of which m_{es} are assigned as elite sites. The number of bees destined to exploit a patch depends on its quality: N_o for normal sites, and N_e for elite sites. The exploitation of a neighborhood consists in testing near locations inside a specific range of distances η, which is iteratively reduced as generations goes by. Meanwhile, the position of the remaining $N_b - m_s$ bees is randomly generated in order to keep performing a global search [2,12].

Since we are focusing on an N_d-dimensional binary search space, there is no point in using continuous distances when neighborhood exploitation is being performed. This fact motivated us to modify the BA algorithm for binary problems like the one that concerns us in this study. Our binary adaptation lies on two approaches, based on genetic algorithm procedures: (i) elitism and (ii) mutation. Firstly, elitism is applied in the neighborhood exploitation phase, guaranteeing that the initial patch is kept as generations goes by. This fact is useful in case that the neighborhood search does not provide improved solutions. Secondly, a mutation is performed instead of restricted distance variations. Here, the η-mutation operator randomly selects η bits and changes their value from 0 to 1, or vice versa, making a binary neighborhood search possible. It is worthy to note that the amount of mutation η is iteratively decreased by a $\lambda \in [0,1]$ factor. The b-BA algorithm is detailed in the Fig. 2.

3.2   Multi-objective Cost Function

The fitness function employed in this study to test the quality of a possible solution is based on a multi-objective approach that (i) maximizes the system accuracy, and (ii) minimizes the required number of channels.

In order to estimate the system accuracy, we have employed a cross-validation approach with $K = 5$ iterations over the training dataset. In other words, training data is divided into 5 equally-sized subsets. An LDA classifier is trained with each of these subsets and tested with the rest, returning 5 different accuracies. Lastly, f_α final accuracy is calculated as the average of all of them.

The function that evaluates the amount of channels must be null when it is required only one channel and maximum when N_d channels are required. We consider that a polynomial function is suitable for this problem, which favors the reduction of channels when the amount is large. Thus, the final multi-objective cost function is the following:

$$F(\boldsymbol{c}) = \omega_\alpha(1 - f_\alpha) + \omega_c\left(\frac{\sum \boldsymbol{c} - 1}{N_d - 1}\right)^3, \text{ with } \omega_\alpha + \omega_c = 1, \tag{1}$$

where ω_α and ω_c are weighting coefficients.

Pseudocode: Binary Bees Algorithm.

Input : $N_d, N_b, N_o, N_e, m_s, m_{es}, \eta, \lambda$
Output: c

1 $X \leftarrow$ RandomPositions$(N_b, N_d) \in \{0,1\}$;
2 **while** $\neg$EndOfCondition **do**
3 $\quad F \in \mathbb{R}^{N_b \times 1} \leftarrow$ EvaluateFitness(X);
4 $\quad g \leftarrow \arg\min_X(F)$;
5 $\quad$ NewGen $\leftarrow$ Zeros(N_b, N_d);
6 $\quad \eta \leftarrow$ round$(\lambda \cdot \eta)$;
7 $\quad X_s \leftarrow$ SortPositions(X, F);
8 $\quad$ Sites $\in \mathbb{Z}^{m_s \times N_d} \leftarrow$ BestPositions(X_s, m_s);
9 $\quad$ **foreach** Sites$_i$ **do**
10 $\quad\quad$ **if** $i < m_{es}$ **then**
11 $\quad\quad\quad$ $nr = N_e$;   % Elite site;
12 $\quad\quad$ **else**
13 $\quad\quad\quad$ $nr = N_o$;   % Normal site;
14 $\quad\quad$ **end**
15 $\quad\quad$ $N \leftarrow$ Zeros$(nr + 1, N_d)$;
16 $\quad\quad$ $N_1 \leftarrow$ Sites$_i$ % Elitism;
17 $\quad\quad$ **for** $j = 2$ **to** $nr + 1$ **do**
18 $\quad\quad\quad$ $N_j \leftarrow$ Sites$_i$;
19 $\quad\quad\quad$ $N_j \leftarrow$ Mutation(N_j, η);
20 $\quad\quad$ **end**
21 $\quad\quad$ $F_n \in \mathbb{R}^{nr+1 \times 1} \leftarrow$ EvaluateFitness(N);
22 $\quad\quad$ NewGen$_i \leftarrow \arg\min_N(F_n)$;
23 $\quad$ **end**
24 $\quad$ **for** $r = 1$ **to** $(N_b - m_s)$ **do**
25 $\quad\quad$ NewGen$_{m_s + r} \leftarrow$ RandomPositions$(1, N_d)$;
26 $\quad$ **end**
27 $\quad X \leftarrow$ NewGen;
28 **end**
29 **return** $c \leftarrow g$;

Fig. 2. Detailed b-BA pseudocode that is employed in this study.

4   Results

In order to avoid local minima, the pipeline shown in Fig. 1 have been repeated 10 times for each method. In order to make a comparison between the proposed algorithm b-BA and a binary PSO algorithm, both of them have computed 100 generations. The binary PSO approach have been employed by modifying particle positions using a sigmoid function [6]. Multi-objective cost function coefficients were set to $\omega_\alpha = 0.7$ and $\omega_c = 0.3$, since we have considered that reaching suitable accuracies is more important than drastically reducing the number of required channels. Method-specific parameters, which are detailed in the Table 1, were set following the recommendations of the literature [2,9,12].

Table 1. Specific hyperparameters of each method.

Method	Hyperparameters		Value
PSO	N_p	Number of particles	20
	α_1	Individual confidence coefficient	2
	α_2	Swarm confidence coefficient	2
b-BA	N_b	Number of scout bees	10
	N_o	Number of normal bees	2
	N_e	Number of elite bees	5
	m_s	Number of sites to exploit	5
	m_{es}	Number of elite sites	1
	η	Initial amount of mutation	32
	λ	Decreasing mutation factor	0.95

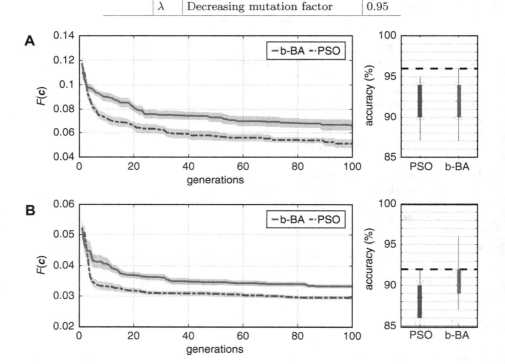

Fig. 3. (Left) Average convergence curves among the repetitions of each method for subject A (top) and B (bottom). Note that the lines correspond to the mean and the shaded areas portray the standard deviation. (Right) Testing accuracies distributions among the repetitions of each method for 15 sequences. Original accuracies using the 64-channel set are indicated with a black dashed line.

Figure 3 shows the average convergence curves among the 10 repetitions of both methods for each subject, as well as the boxplots that illustrates the testing accuracies distributions in comparison with the 64-channel set ones. As can be

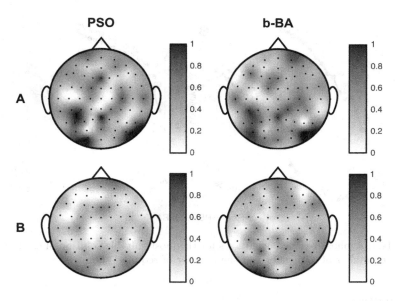

Fig. 4. Rank of each channel for PSO (left) and b-BA (right) methods in subject A (top) and B (bottom).

Table 2. Highest reached accuracies for 5 and 15 sequences in testing phase.

	Method sequences	Raw		PSO		b-BA	
		5	15	5	15	5	15
Subject A	Accuracy	49%	96%	42%	95%	57%	96%
	No. channels	64	64	25	25	29	29
Subject B	Accuracy	77%	92%	72%	92%	78%	96%
	No. channels	64	64	17	17	19	19
Mean	Accuracy	**63.0%**	**94.0%**	**57.0%**	**93.5%**	**67.5%**	**96.0%**
	No. channels	**64.0**	**64.0**	**21.0**	**21.0**	**24.0**	**24.0**

seen, PSO converges faster than b-BA, even though the reached accuracies are smaller. The rank of each channel for both methods and subjects is shown in Fig. 4. Rank is defined as the normalized number of times that a specific channel is selected among the repetitions of a desired method. As can be noticed, selected channel distributions over the scalp are similar for both methods. The reached accuracies for the most suitable solution of each method are also shown in the Table 2. Finally, Fig. 5 shows the testing phase reached accuracies for the most suitable solution of each method in function of the number of sequences and channels.

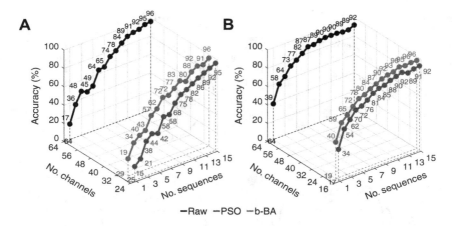

Fig. 5. Testing phase reached accuracies for the most suitable solution of each method versus the number of sequences and the number of required channels.

5   Discussion

As shown in Fig. 3, the designed multi-objective fitness function decreases as generations goes by. Although the convergence curve of PSO reaches faster the minimal values of $F(c)$ than b-BA, the obtained accuracies does not improve the original 64-channel set solutions. It is worthy to note that the boxplots shows that b-BA is able (1) to reach the original accuracy for subject A, and (2) to improve it for subject B. In fact, Table 2 exposes that the proposed algorithm, b-BA, improves the solutions of the 64-channel set using less than the half part of the original channels for both 5 and 15 sequences. However, PSO is not able to improve the original solutions, sacrificing slightly the performance in exchange for a drastically reduced channel set. In particular, original mean accuracies for the 64-channel set are 63% and 94% for 5 and 15 sequences, respectively. PSO reaches almost similar accuracies with a mean of 21 channels: 57% and 93.5%; and b-BA improves them with a mean of 24 channels: 67.5% and 96%. In addition, Fig. 5 shows that the aforementioned behavior is also present for a variable number of sequences. As can be noticed from the curves, b-BA is able to reach similar or even better accuracies than the 64-channel set, whereas PSO is not. It is worthy to note that the results with few number of sequences are relevant due to the fact that reaching higher accuracies using less number of sequences is one of the current challenges in BCI systems. The less sequences are required, the higher the speed of the system.

As can be seen in Fig. 4, both SI methods recurrently select a specific set of channels. An accumulation on the parietal and occipital cortex, mainly for subject A, is clearly observed. This fact reinforce the study of Krusienski et al. [10], which stated that the P300 evoked potentials are mainly produced in the parietal cortex and recommended a channel set dispersed over these locations, commonly used in P300-BCI studies.

Despite the fact that PSO reached smaller accuracies than b-BA, it also selected less number of channels. Therefore, these results are likely produced due to the channel-accuracy tradeoff that is reflected in the expression (1). This evinces that both methods are suitable for use in P300-BCI systems, even though the results can slightly vary between repetitions, owing to the fact that SI methods does not guarantee a global optimum, but assure a sub-optimal solution in a proper amount of time [17]. In addition, SI methods usually require fixing a large amount of hyperparameters that cannot be optimized by validation procedures, which makes the experience of the user a valuable competence. Despite of these disadvantages, SI methods present an excellent performance solving complex optimization problems due to their: (1) stochasticity, useful when the cost function is multimodal and discontinuous; (2) self-organization, by removing the need of supervision; (3) flexibility, due to their easy adaptation to variable problems; (4) robustness, owing to the fact that the fail on an individual does not affect the global performance at all; and (5) scalability, because their architecture do not depend on the number of agents [2, 17, 18].

It is also worthy to note that we have employed a wrapper selection method [7]. In other words, the fitness of each possible solution is evaluated according to the performance of the LDA classifier. Because of the selection depends on the classifier, high accuracies are guaranteed in testing phase. However, cross-validation, training and testing procedures are needed in order to compute a single fitness evaluation, which makes the proposed algorithm a computationally expensive approach [7].

There are few studies that have applied SI methods to P300-BCI channel selection. In fact, all of them have used PSO, which reinforces the fact that these metaheuristics have not been fully exploited yet. Jin et al. [8] used a wrapper approach that integrated PSO and bayesian LDA (BLDA) over a custom dataset. Due to the fact that the objective of the paper was to create a P300 Speller with Chinese characters, channel selection procedure was not deeply explained. However, it was reported that the selected subsets by PSO were able to improve the common 8 and 16-channel sets performances. In addition, Perseh and Sharafat [11] achieved a mean accuracy of 97.5% in the 'III BCI Competition 2005' dataset using a combination of discrete wavelet transform (DWT), BLDA, and a modified version of PSO with Bhattacharyya distances. Lastly, Gonzalez et al. [6] were able to reach perfect accuracies with 33 channels in the same dataset using their proposed method, which combined: lifting wavelet transform (LWT) features, support vector machine (SVM) classification and PSO. All of these were wrapper approaches that combined PSO with sophisticated feature extraction and classification methods. In this study, we have tried to simplify both procedures using the state-of-the-art solutions: sub-sampling as feature extraction, and LDA as feature classification; in order to focus on the individual performance of both SI methods. Based on the obtained results, we can speculate than the combination of b-BA with more sophisticated feature extraction and classification methods could reach even higher accuracies.

Even though the results show that both methods, specially b-BA, are suitable for use in P300-based BCI systems channel selection procedures, we can point out some limitations. As previously stated, b-BA requires fixing a high amount of hyperparameters, which makes their optimization computationally intractable using validation procedures. Although there are several guidelines, its optimization lies on the user experience to a large extent [2]. Also, SI-based methods are self-organized, meaning that a selected solution may not be necessarily the same than that obtained in a different iteration. For this reason, it is recommended to perform several repetitions of the algorithm in order to avoid local minima. In addition, we have used a wrapper approach that depends on the LDA classifier, which increases the computational cost substantially [7]. Lastly, it is worthy to note that the dataset only includes EEG signals from two subjects and thus, the generalization of the results is obviously limited. In order to overcome these limitations, we contemplate the following future research lines: (i) reducing the number of hyperparameters, (ii) getting rid of the classifier dependence using alternative metrics that could be able to evaluate properly the predictive power of a solution, and (iii) testing the methods with larger P300-based BCI databases.

6   Conclusion

Channel selection procedures in P300-BCI systems are essential to reduce the curse of dimensionality and increase the users' comfort. Furthermore, nature-inspired metaheuristics, which have demonstrated excellent performances solving complex optimization problems, have not been fully exploited in this kind of systems yet. Although there are several previous approaches that have used PSO, SI is a research growing field that integrates a large amount of methods. The objective of this study was to propose an alternative method from PSO, named b-BA, that is able to successfully select the most relevant channels without compromising the system's performance. The method has been tested with the 'III BCI Competition 2005' dataset II and compared with a common state-of-the-art PSO. Results not only show that b-BA has reached higher accuracies (mean of $96.0 \pm 0.0\%$) than PSO (mean of $93.5 \pm 2.1\%$) and that obtained with the original 64-channel set (mean of $94.0 \pm 2.8\%$), but also has selected a mean of 24 channels, less than the half of the original ones. Despite the dominance of b-BA, both PSO and b-BA are suitable for use in P300-based BCI systems channel selection phase.

Acknowledgments. This work was partially supported by the project TEC2014-53196-R of 'Ministerio de Economía y Competitividad' (MINECO) and FEDER. In addition, V. Martínez-Cagigal was in receipt of a 'Promoción de Empleo Joven e Implantación de la Garanta Juvenil en I+D+i' grant from MINECO and the University of Valladolid.

References

1. Blankertz, B., Müller, K.R., Krusienski, D.J., Schalk, G., Wolpaw, J.R., Schlögl, A., Pfurtscheller, G., Millán, J.D.R., Schröder, M., Birbaumer, N.: The BCI competition III: validating alternative approaches to actual BCI problems. IEEE Trans. Neural Syst. Rehabil. Eng. **14**(2), 153–159 (2006)

2. Brownlee, J.: Clever Algorithms: Nature-Inspired Programming Recipes, 2 edn. (2011). http://www.cleveralgorithms.com
3. Cecotti, H., Rivet, B., Congedo, M., Jutten, C., Bertrand, O., Maby, E., Mattout, J.: A robust sensor-selection method for P300 brain-computer interfaces. J. Neural Eng. **8**(1), 016001 (2011)
4. Colwell, K.A., Ryan, D.B., Throckmorton, C.S., Sellers, E.W., Collins, L.M.: Channel selection methods for the P300 Speller. J. Neurosci. Methods **232**, 6–15 (2014)
5. Farwell, L.A., Donchin, E.: Talking off the top of your head: toward a mental prosthesis utilizing event-related brain potentials. Electroencephalogr. Clin. Neurophysiol. **70**(6), 510–523 (1988)
6. Gonzalez, A., Nambu, I., Hokari, H., Iwahashi, M., Wada, Y.: Towards the classification of single-trial event-related potentials using adapted wavelets and particle swarm optimization. In: Proceedings of the 2013 IEEE International Conference Systems, Man, Cybernetics, SMC 2013, pp. 3089–3094 (2013)
7. Guyon, I., Elisseeff, A.: An introduction to variable and feature selection. J. Mach. Learn. Res. **3**, 1157–1182 (2003)
8. Jin, J., Allison, B.Z., Brunner, C., Wang, B., Wang, X., Zhang, J., Neuper, C., Pfurtscheller, G.: P300 Chinese input system based on Bayesian LDA. Biomed. Tech. **55**(1), 5–18 (2010)
9. Kennedy, J., Eberhart, R.: A discrete binary version of the particle swarm algorithm. In: 1997 IEEE International Conference on Systems, Man, and Cybernetics, Computational Cybernetics and Simulation, vol. 5, pp. 4–8 (1997)
10. Krusienski, D., Sellers, E., McFarland, D., Vaughan, T., Wolpaw, J.: Toward enhanced P300 speller performance. J. Neurosci. Methods **167**(1), 15–21 (2008)
11. Perseh, B., Sharafat, A.R.: An efficient P300-based BCI using wavelet features and IBPSO-based channel selection. J. Med. Signals Sens. **2**(3), 128–143 (2012)
12. Pham, D.T., Ghanbarzadeh, A., Koç, E., Otri, S., Rahim, S., Zaidi, M.: The bees algorithm - a novel tool for complex optimisation problems. In: Intelligent Production Machines and Systems - 2nd I*PROMS Virtual International Conference, pp. 454–459 (2006)
13. Rakotomamonjy, A., Guigue, V.: BCI competition III: dataset II - ensemble of SVMs for BCI P300 speller. IEEE Trans. Biomed. Eng. **55**(3), 1147–1154 (2008)
14. Rivet, B., Cecotti, H., Maby, E., Mattout, J.: Impact of spatial filters during sensor selection in a visual P300 brain-computer interface. Brain Topogr. **25**(1), 55–63 (2012)
15. Wolpaw, J.R., Birbaumer, N., McFarland, D.J., Pfurtscheller, G., Vaughan, T.M.: Brain-computer interfaces for communication and control. Clin. Neurophysiol. **113**(6), 767–791 (2002)
16. Xu, M., Qi, H., Ma, L., Sun, C., Zhang, L., Wan, B., Yin, T., Ming, D.: Channel selection based on phase measurement in P300-based brain-computer interface. PLoS ONE **8**(4), 1–9 (2013)
17. Yang, X.S.: Nature-Inspired Optimization Algorithms, 1st edn. Elsevier Inc., Amsterdam (2014)
18. Yang, X.S., Cui, Z., Xiao, R., Gandomi, A.H., Karamanoglu, M.: Swarm Intelligence and Bio-Inspired Computation: Theory and Applications, 1st edn. Elsevier Inc., Amsterdam (2013)
19. Yu, T., Yu, Z., Gu, Z., Li, Y.: Grouped automatic relevance determination and its application in channel selection for P300 BCIs. IEEE Trans. Neural Syst. Rehabil. Eng. **23**(6), 1068–1077 (2015)

A Comparison of a Brain-Computer Interface and an Eye Tracker: Is There a More Appropriate Technology for Controlling a Virtual Keyboard in an ALS Patient?

Liliana García[1]([✉]), Ricardo Ron-Angevin[2], Bertrand Loubière[3],
Loïc Renault[4], Gwendal Le Masson[3], Véronique Lespinet-Najib[1],
and Jean Marc André[1]

[1] IMS UMR 5218, CIH, ENSC-BordeauxINP, Bordeaux, France
{liliana.garcia,veronique.lespinet,
jean-marc.andre}@ensc.fr
[2] Dpto. Tecnología Electrónica, Universidad de Málaga, Málaga, Spain
rra@dte.uma.es
[3] Centre SLA, CHU, Bordeaux, France
bertrand.loubiere@permobil.com,
gwendal.le-masson@u-bordeaux.fr
[4] ENSC - Bordeaux INP, Bordeaux, France
loic.renault@ensc.fr

Abstract. The ability of people affected by amyotrophic lateral sclerosis (ALS), muscular dystrophy or spinal cord injuries to physically interact with the environment, is usually reduced. In some cases, these patients suffer from a syndrome known as locked-in syndrome (LIS), defined by the patient's inability to make any movement but blinks and eye movements. Tech communication systems available for people in LIS are very limited, being those based on eye-tracking and brain-computer interface (BCI) the most useful for these patients. A comparative study between both technologies in an ALS patient is carried out: an eye tracker and a visual P300-based BCI. The purpose of the study presented in this paper is to show that the choice of the technology could depend on user's preference. The evaluation of performance, workload and other subjective measures will allow us to determine the usability of the systems. The obtained results suggest that, even if for this patient the BCI technology is more appropriate, the technology should be always tested and adapted for each user.

Keywords: Brain-Computer Interface (BCI) · Eye tracker · ALS · Usability · Speller

1 Introduction

Several of the neurological diseases that human can suffer result in severe disabilities. For instance, the ability of people affected by amyotrophic lateral sclerosis (ALS), muscular dystrophy or spinal cord injuries to physically interact with the environment, is usually reduced, and they may even lose it completely.

© Springer International Publishing AG 2017
I. Rojas et al. (Eds.): IWANN 2017, Part II, LNCS 10306, pp. 464–473, 2017.
DOI: 10.1007/978-3-319-59147-6_40

ALS patients suffer from a syndrome known as locked-in syndrome (LIS). In its classical modality, this syndrome is defined by the patient's inability to make any movement but blinks and eye movements, despite being still conscious. This renders them completely dependent not only on their close family, but also on ventilatory machines to remain alive. If the disease draws on, the patient is bound to be unable to make even those residual movements, thus remaining completely isolated.

If there are several tech communication systems available for people who has residual muscular control, for people in LIS, these technologies are very limited. Probably, the technologies based on eye-tracking, electrooculography (EOG) and brain-computer interface (BCI) are the only useful for these patients [1].

As much the eye tracker as the electrooculography are based on the measurement of eye activity, and both rely on the users' abilities to control their eye-muscles.

The eye trackers, generally, include two components: a light source and a camera. The camera tracks the reflection of the light source along with visible ocular features such as the pupil. Electrooculography is a technique for measuring, though electrodes placed around the eyes of the user, the corneo-retinal standing potential that exists between the front and the back of the human eye. The resulting signal is called the electrooculogram.

A brain-computer interface (BCI) is based on the analysis of the brain activity recorded during certain mental activities, in order to control an external device. Currently, the most commonly used BCI systems are those based on electroencephalographic (EEG) signals, mainly because they can be recorded in a non-invasive manner and show adequate temporal resolution. Among them, those based on the P300 event-related potential (ERP) are very common due to easiness with which this ERP can be elicited. Specifically, the P300 is a positive deflection in voltage occurring about 300 ms after an infrequent or significant stimulus is perceived [2]. P300 wave amplitude is typically between 2 μV and 5 μV and is symmetrically distributed around central scalp areas, showing greater amplitude in occipital rather than frontal regions [3].

From those technologies, the eye-tracking is, at present, one the most advanced devices for communication in patients in LIS and, specially, in patients with ALS. A recent study tried to explore the effectiveness of communication and the variable affecting the eye-tracking computer system utilization in patients ALS [4]. The study was carried out on 30 patients with advanced ALS and 19 showed a high acceptance and average daily eye-tracking system utilization of 300 min. However, the remaining 11 subjects reported limited and irregular daily use of the device, being the reported causes, gaze fatigue (8 subjects), oculomotor impairment, i.e., inability to properly move the eyes (2 subjects) and difficulty to keep the head still (1 subject). Finally, authors concluded that limitation of the eye-tracking is given by the fact that it actually relies on eye movements. For patients with oculomotor dysfunction, the use of an eye-tracking is uncomfortable or even impossible, being necessary to provide other technologies, such are those based on brain-computer interface (BCI).

Another study [1] compares three technologies to provide binary communication: eye-tracking, electrooculography (EOG) and auditory brain-computer interface. The participant of the study was a patient with ALS who had been in the LIS for 6 years. He was able to communicate with slow residual eye movements, restricting the number of choices. With the EOG based system, the user reached an accuracy mean of 71% with

5 choices. With the eye tracking based system, the user had difficulties looking at a particular direction and only two choices were provided however, the reached accuracy was 100%, being all selections classified correctly. Finally, with the auditory BCI, only two choices were provided and participant reached accuracies above 75%. In this study, the reduced number of choices provided was due to the slow residual eye movements of the participant.

Other BCIs used for communication purpose are those based on visual P300 signal. They are based on the P300 speller first developed by Farwell and Donchin [5], which is still referenced and intensely studied [3, 6–8]. In this BCI, a 6 × 6 matrix of letters, arranged in rows and columns, is shown to the subject. The user focuses his/her attention on the matrix element he/she wishes to select as each row and column is flashed (i.e., intensified) randomly, one after the other. After a number of flashes, the symbol that the user has supposedly chosen is presented on screen.

Recently, a study compares a visual P300-based brain-computer interface and an eye-tracking for controlling an Internet browser [9]. A total of 12 patients with severe motor impairment (11 affected by ALS, and 1 affected by Duchenne muscular dystrophy, participated in this study). According to the obtained results in this study, the performance measures showed the advantages of using the eye tracker as a communication device. Besides, participants rated the eye tracker as a more satisfying device and considered the BCI as a technology requiring more effort and that was more time-consuming than the eye tracker. The conclusion of this study was that if users can rely on eye movements, they tend to consider the eye tracker as a superior technology.

The purpose of the study presented in this paper is to show that the choice of a technology could depend on user's preference, and not all the users have the same preference. To this end, the usability of the two technologies [10, 11], the eye tracker and a visual P300-based BCI, will be evaluated.

2   Methods

2.1   Participants

One French man, 57-years-old, diagnosed with amyotrophic lateral sclerosis in 2010 and without any impairment of cognitive functions, participated in this study. The patient, with severe motor impairment, was naïve to both technologies: the eye tracker and the BCI. He was able to move his eyes and had difficulty to communicate through the voice. He gave informed consent through a protocol reviewed by the ENSC-IMS Cognitive team. The experiment was carried out at the CHU (centre Hospitalier Universitaire) at Bordeaux.

2.2   Procedure

On the same day, the participant tested the visual P300-based BCI (session 1) and the eye tracker (session 2). During the experiment, the participant sat in his wheelchair in a reclining position at a distance about 60 cm from the screen. Before the beginning of each session, instructions regarding the procedure and the device (BCI and eye tracker) management were given in verbal form. The experiment was conducted in accordance with standard ethical guidelines as defined by the Declaration of Helsinki and the study was approved by the Ethics Committee of the University of Málaga.

Because the objective of the study was to compare the usability of the eye tracker and the visual BCI in a communication task, the speller size was the same for both technologies. The speller used was based on the classical Farwell and Donchin [5] speller, which consists on a 6 × 6 matrix of symbols (36 alphanumeric letters and numbers) arranged within rows and columns (see Fig. 1). The matrix size was 14.69 cm, being the symbols size of 1.17 cm and the distance between symbols of 1.53 cm.

Fig. 1. Schematic representation of a classical P300 speller BCI

Each session consisted of a calibration phase and an evaluation phase. The purpose of the calibration phase was to adapt the technology to the user. Once the calibration was done, the subject participated in the evaluation phase to copy-spell the sentence "il fait beau" (i.e., "the weather is nice"). The participant was allowed to correct each error only once. After the copy-spelling tasks, he was asked to complete a visual analogue scale (VAS) of: fatigue, difficulty, stress and difficulty to perceive the characters, and the NASA-TLX test [12] to evaluate the subjective cognitive workload.

At the end of the last session, the participant was asked to express his preference between the two technologies. A comparative questionnaire adapted from the SUS (System Usability Scale) allowed to evaluate six dimensions: favourite, complex, comfortable, stressful, controllable, tiring.

2.3   Equipment and Tasks

Brain-Computer Interface. EEG was recorded using gold electrodes placed at positions Fz, Cz, Pz, Oz, P3, P4, PO7 and PO8, according to the 10/20 international system. All channels were referenced to the left earlobe, using FPz as ground. The EEG was amplified through a 16 channel biosignal amplifier (g.BSamp, Guger Technologies). The amplifier settings were 0.5 and 100 Hz for the band-pass filter, the notch (50 Hz) was on, and the sensitivity was 500 μV. The EEG was then digitized at a rate of 256 Hz by a 12-bit resolution NI-USB-6210 data acquisition card (National Instruments). All aspects of EEG data collection and processing were controlled by the BCI2000 system.

During the calibration phase, each row and column was randomly flashed 10 times. Therefore, each character was randomly intensified 20 times. The duration of each flash was 125 ms and the inter-stimulus interval (ISI) between flashes was also 125 ms. There was a pause of 6 s after each sequence of flashes (i.e., after a character had been selected). The calibration consisted in spelling the words "lune", "feux" and "kilo" and the number "2015". It is important to mention that the time required for the calibration phase for the BCI system depends on the number of words to spell (4 in this experiment).

After these runs, we performed a stepwise linear discriminant analysis (SWLDA) of the data from the last three runs to obtain the weights for the on-line P300 classifier.

After calibration and training of the classifier, the evaluation phase started (see Fig. 2). We set the number of intensification sequences to the minimum number need to reach 100% accuracy off-line.

Fig. 2. Participant during the evaluation phase of the BCI system

Eye Tracker. The experiment was carried out using the Tobii C15 (Tobii Technology, Sweden) [13]. The eye tracker interface speller was configured to be identical to the one used for the BCI. The calibration phase consisted in fixating 9 targets located on different positions of the screen. The time required for this calibration phase should be, usually, very short. The evaluation phase started only once the operator considered an acceptable calibration. For the evaluation phase (see Fig. 3), participant could select a symbol by gazing at the intended target for 1.6 s.

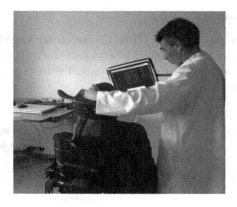

Fig. 3. Participant during the evaluation phase of the Tobii communicator system

2.4 Objective and Subjective Measures

To compare the performance of the BCI and the eye tracker, different objective measured were considered: the time required for the calibration phase, the time required for the evaluation phase, the final written sentence and the number of errors.

Regarding the subjective measures, we analyze the NASA-TLX for each experiment in order to evaluate the subjective workload. The different Visual Analog Scales (VAS), added to the comparative questionnaire adapted from the SUS (System Usability Scale), allow us to obtain a global subjective assessment of usability.

3 Results

3.1 Objective Measures

The different objective measures obtained for each technology are shown in Table 1.

Table 1. Objective measures (times, final sentence, errors) obtained for BCI and Tobii system.

	BCI	Tobii
Time for calibration phase	8 min	5 min
Time for evaluation phase	7 min 34 s	7 min 47 s
Final written sentence	IL FAIT BEA7L	IL LFAI7_BEAP65
Number of error	3	7

The user had high difficulty to fix the targets and the calibration failed several times. Finally, the calibration was done considering only the right eye.

Regarding the BCI, the participant required 7 flashes (one flash is the intensification of one row and one column) to obtain 100% of accuracy. Finally, we configured 8 flashes to select a letter during the evaluation phase.

3.2   Subjective Measures

The total subjective workload (NASA-TLX- Global score; ranged from 0 to 100) and dimensions contributions to the subjective workload (mental, physical and temporal demand, performance, effort and frustration; ranged from 0 to 33.3) for each technology is sowed in Fig. 4.

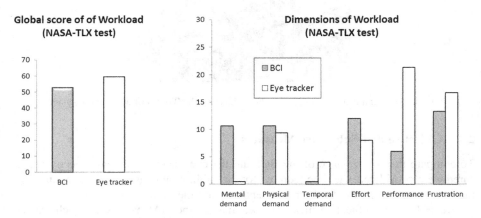

Fig. 4. NASA-TLX scores obtained for BCI and Tobii

The obtained values of the different Visual Analog Scales (VAS) for each technology is shown in Fig. 5.

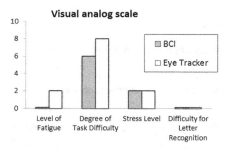

Fig. 5. VAS obtained for BCI and Tobii

Finally, the obtained results of the comparative questionnaire are shown in Table 2.

Table 2. Comparatives results between BCI and Tobii.

	BCI	Tobii
Positive view	The most controllable	
	The most comfortable	
	The favourite	
Negative view	The most stressful	
		The most complex
		The most tired

4   Discussion and Conclusion

In this study, two different technologies for communication purpose designed for people with severe motor impairment have been compared through objective and subjective measures. The experiment has been carried out in a participant with ALS. The obtained results show that, for this participant, the visual BCI has been considered more advantageous that the eye tracker as a communication device.

During the calibration phase, the participant had, not only a high difficulty to gaze the different targets presented by the eye tracker, but also high difficulty to keep this eyes open, being this an obstacle to calibrate the system. Finally, after several runs, the calibration was carried out with the right eye and, even in this condition, the calibration was not really satisfactory for the operator. This longer time required for the calibration of the Tobi could affect the perceived fatigue and affected the usability of the system. However, for the BCI, the participant did not have any problem during the calibration phase, getting 100% of performance with only 7 flashes. Besides, even if the time required for the evaluation phase was similar for both technologies, the number of error was higher with the Tobii system. These results show a better performance with the BCI system.

Regarding the subjective measures, the global score of workload of the visual BCI was lower than of the Tobii system. This result seems to be contrary to that other study [1, 9] which suggest that the workload of the eye tracker was lower that the workload of the BCI. The mental demand and the effort subscale were considered higher in the BCI system compare to the Tobii System. The higher time needed to select a letter during the evaluation phase with the BCI system (24 s) was, probably, contributed to this increase in both dimensions (mental demand and effort). However, the BCI system requires less temporal demand, makes easier the development of the task and produces less frustration. Probably, the fact that the demand temporal contribution was so low, could have a positive effect on performance and frustration.

The obtained results in the different VAS show no level of fatigue for the BCI. Although the degree of task difficulty was lower for the BCI, the obtained values were important for both technologies (6 for the BCI and 8 for the Tobii). None of the technology were considered stressful and the subject did not present any difficulty to recognize the characters.

The obtained results in the comparative questionnaire allow to summarize which technology has been considered as most advantageous for the participant. Undoubtedly, the visual BCI has been select as the favourite, being the stressful the only dimension with a negative point of view for the BCI however, as it was mentioned before, the level of stress was very low for both technologies.

The preference to BCI system can be due to ability to achieve control of BCI over Eye-tracing. By the way, the participant verbally reported feeling satisfaction particularly related with the BCI experience, possibly because it could represent a greater challenge that needed to be raised.

Although a recent study concluded that an eye tracker system is more advantageous than a P300-based BCI for communication purpose, from these results we conclude that a BCI system may be a not negligible alternative solution for some patients with special difficulties, not only for controlling their eye movements, but who present difficulties to manage an eye tracker. Even if the results have been obtained with only one patient, these show that the technology should be always tested and adapted for each user, not being able to stablish a specific technology as the most appropriate without, previously, testing it.

Acknowledgements. We would like to dedicate this work to the memory of the participant involved in this study. This work was partially supported by the University of Málaga, by the Spanish Ministry of Economy and Competitiveness through the project LICOM (DPI2015-67064-R) and by the European Regional Development Fund (ERDF).

References

1. Käthner, I., Kübler, A., Halder, S.: Comparison of eye tracking, electrooculography and an auditory brain-computer interface for binary communication: a case study with a participant in the locked-in state. J. Neuroeng. Rehabil. **12**(1), 76, 1–11 (2015)
2. Wolpaw, J.R., Birbaumer, N., McFarland, D.J., Pfurtscheller, G., Vaughan, T.M.: Brain-computer interfaces for communication and control. Clin. Neurophysiol. **113**(6), 767–791 (2002)
3. Krusienski, D.J., Sellers, E.W., McFarland, D.J., Vaughan, T.M., Wolpaw, J.R.: Toward enhanced P300 speller performance. J. Neurosci. Methods **167**(1), 15–21 (2008)
4. Spataro, R., Ciriacono, M., Manno, C., La Bella, V.: The eye-tracking computer device for communication in amyotrophic lateral sclerosis. Acta Neurol. Scand. **130**(1), 40–45 (2014)
5. Farwell, L., Donchin, E.: Talking off the top of your head: toward a mental prosthesis utilizing event related brain potentials. Electroencephalogr. Clin. Neurophysiol. **70**(6), 510–523 (1988)
6. Bianchi, L., Sami, S., Hillebrand, A., Fawcett, I., Quitadamo, L., Seri, S.: Which physiological components are more suitable for visual ERP based brain-computer interface? A preliminary MEG/EEG study. Brain Topogr. **23**(2), 180–185 (2010)
7. Kleih, S., Nijboer, F., Halder, S., Kübler, A.: Motivation modulates the P300 amplitude during brain- computer interface use. Clin. Neurophysiol. **121**(7), 1023–1031 (2010)
8. Sellers, E.W., Krusienski, D.J., McFarland, D.J., Vaughan, T.M., Wolpaw, J.R.: A P300 event-related potential brain-computer interface (BCI): the effects of matrix size and inter stimulus interval on performance. Biol. Psychol. **73**(3), 242–252 (2006)
9. Pasqualotto, E., Matuz, T., Federici, S., Ruf, C.A., Bartl, M., Belardinelli, M.O., Birbaumer, N., Halder, S.: Usability and workload of access technology for people with severe motor impairment a comparison of brain-computer interfacing and eye tracking. Neurorehabil. Neural Repair **29**(10), 950–957 (2015)

10. ISO 9241-11: Ergonomic requirements for office work with visual display terminals (VDTs) – Part 11: guidance on usability (1998)
11. Nielsen, J.: What is usability? In: Usability Engineering. Academic Press, Cambridge, pp. 23–48 (1993)
12. Hart, S.G., Staveland, L.E.: Development of NASA-TLX (Task Load Index): results of empirical and theoretical research. Adv. Psychol. **52**, 139–183 (1988)
13. Tobii ATI – Tobii C15 Augmentative and Alternative Communication Device, May 2012. http://www.tobii.com/assistive-technology/northamerica/products/hardware/tobii-C15/

SSVEP-Based BCI in a Smart Home Scenario

Abdul Saboor[1], Aya Rezeika[2], Piotr Stawicki[2], Felix Gembler[2],
Mihaly Benda[2], Thomas Grunenberg[2], and Ivan Volosyak[2(✉)]

[1] Faculty of Communication and Environment, Rhine-Waal University of Applied
Sciences, 47533 Kleve, Germany
[2] Faculty of Technology and Bionics, Rhine-Waal University of Applied Sciences,
47533 Kleve, Germany
ivan.volosyak@hochschule-rhein-waal.de
http://www.hochschule-rhein-waal.de

Abstract. Steady state visual evoked potentials (SSVEPs)-based Brain-Computer Interfaces (BCIs) can provide hand-free human interaction with the environment. In the presented study, visual stimuli were displayed on Epson Moverio BT-200 augmented reality glasses, which can be easily used in smart homes. QR codes were used to identify the devices to be controlled with the BCI. In order to simulate a real life scenario, participants were instructed to go out of the lab to get a coffee. During this task light switches, elevator and a coffee machine were controlled by focusing on SSVEP stimuli displayed on the smart glasses. An average accuracy of 85.70% was achieved, which suggests that augmented reality may be used together with SSVEP to control external devices.

Keywords: Brain-Computer Interface (BCI) · Electroencephalogram (EEG) · Steady state visual evoked potential (SSVEP) · Visual stimuli · Smart glasses · Augmented reality · QR-code

1 Introduction

Brain-Computer Interface (BCI) is a field of Human Computer Interaction (HCI) in which brain activity is recorded using signal processing devices and analyzed using signal processing algorithms in real time. The captured signals are used as input signals to the physical world, thus, providing human interaction to the surroundings without any muscular movement [20].

For BCI systems, several non-invasive BCI methods are in practice, including P300, motor imagery (MI) and steady sate visually evoked potential (SSVEP) [5]. However, according to Amiri et al. 2013, SSVEP-based BCI system are more accurate and have higher information transfer rate (ITR) [1], which makes SSVEP a great opportunity for research and development.

The SSVEP-based BCI systems have various areas of applications, including the control of wheelchairs, robots, gaming environments and many more [10,13,15].

© Springer International Publishing AG 2017
I. Rojas et al. (Eds.): IWANN 2017, Part II, LNCS 10306, pp. 474–485, 2017.
DOI: 10.1007/978-3-319-59147-6_41

Lately, the smart glasses technology also proved to be very useful. Some studies were conducted using applications with different brands of smart glasses. These applications belong to various fields, for example, medical and surgical [12], facial recognition [16], objects identification [14]. Some studies combined smart glasses with EEG recording, e.g. [8,9,17].

E.g. a study conducted by Tsuru et al. used SSVEP with smart glasses (Epson Moverio BT-200, Seiko Epson Corporation, Suwa, Nagano, Japan) [17]. The brain activity was recorded with an Emotiv EEG neuroheadset. The study was performed with two selectable classes during real-life situations i.e. resting, walking, and talking. In contrast, we used a state-of-art battery powered actiCHamp (Brain Products GmbH, Gilching, Germany) amplifier and actiCAP (Brain Products GmbH, Gilching, Germany) active electrodes. While the low cost device is more suitable for mobile solutions, the actiChamp allowed a higher number of selectable classes.

Most of the VEP based BCI systems used today are limited in terms of portability. Overall, the BCI systems are bulky due to use of monitors or LCD panels [21]. Therefore, it is needed to develop systems which can be carried easily, making truly portable BCI systems.

The study explores the usage of BCI along with internet of things (IOT). In this respect, the concept of portable BCI system is tested with the half-transparent smart glasses Epson Moverio BT-200, presenting the visual stimuli in front of the real world environment.

To simulate a smart home scenario, a quick response technology QR-code is used to identify controllable items in the environment such as lights, coffee machine, and an elevator.

2   Methods and Materials

articipants did not receive any financial reward.

2.1   Participants

To perform the test run with the system, seven volunteers participated in the study (one female). Participants had a mean age of 25.7 (range 22–39). All of the participants had previous experience with BCI, except one. The participants were students or employees of the Rhine-Waal University of Applied Sciences. All subjects (healthy adult volunteers) gave written informed consent in accordance with the Declaration of Helsinki. Information needed for the analysis of the experiments was stored anonymously, and cannot be traced back to the participant. This research was approved by the Ethical Review Board of the Medical Faculty of the University Duisburg-Essen. Subjects did not receive any financial reward for participation in this study.

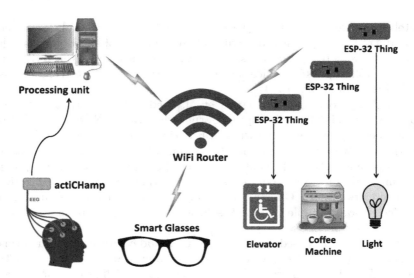

Fig. 1. An overview of the used system. Only key components of the smart home scenario are presented.

2.2   Hardware

The experimental setup consisted of:

- EPSON Moverio BT-200 Smart glasses - presentation of the SSVEP stimuli,
- BrainProducts actiCHamp EEG amplifier and actiCAP active electrodes - signal acquisition,
- Dell Latitude E6410 (Dell Inc., Round Rock, TX, USA) with i5 2.4 GHz CPU Laptop-PC - SSVEP signal classification and UDP server,
- Hama (2in1-WLAN-Adapter) (Hama GmbH & Co KG, Manheim, Germany) portable WiFi router - managing the wireless connection,
- Four ESP32 Thing (SparkFun Electronics, Niwot, CO, USA) microcontroller boards - execution of the classified tasks.

The smart glasses, the processing unit and microcontrollers were connected to the WiFi router using a secure (WPA2 encrypted) protocol to control the external home appliances like light, coffee machine or elevator (see Fig. 1).

Moverio BT-200: The graphical user interface was displayed on the Moverio BT-200 smart glasses which had a fully transparent display. The glasses provide polysilicon TFT active matrix display with LCD size of 0.42 in. with aspect ratio of 16:9. The 24 bit color display had a refresh rate of 60 Hz. These smart glasses also had an integrated VGA camera and provide WiFi and bluetooth connectivity. The operating system of the smart glasses was Android version 4.0.4.

ESP32 Thing: The ESP32 Thing is a WiFi compatible microcontroller development board. It provided nearly 30 I/O pins and bluetooth connectivity, thus making it an ideal choice for IOT applications. The operating voltage range is between 2.2 V and 3.6 V. Thanks to the integrated LiPo battery it can be used wireless.

actiCHamp and actiCAP Electrodes: For EEG signal acquisition, the 24 bit active channel amplifier was used. The 6 volt DC rechargeable battery (actiPOWER, manufactured by Brain Products, Gilching, Germany) was used to power up the amplifier allowing a portable system. The actiCHamp used 8 EEG channels, including reference (8 actiCAP electrodes with a built in noise subtraction unit). Electrode gel was applied between the electrodes and the scalp in order to bring impedances below 15 $k\Omega$. The following electrode locations were used: F_{PZ} as ground, C_Z as reference, and CP_1, P_Z, P_3, P_4, O_1, O_Z, O_2 as EEG signal channels. The EEG data was recorded with a sampling frequency of 500 Hz, downsampled to 200 Hz. During the EEG signal acquisition, an analog notch filter around 50 Hz was used to negate the noise originating from nearby power cables, and a digital band pass filter was applied between 3 and 60 Hz. The DELL Laptop used for EEG signal classification operated on Microsoft Windows 7 Enterprise running on an Intel processor (Intel Core i5, 3.40 GHz).

2.3 Signal Processing

For SSVEP signal classification, the minimum energy combination method (MEC) was used [4] which creates a set of N_s channels that minimize the nuisance signals. Considering N_t samples of EEG data, recorded for each of N_y signal electrodes, the SSVEP response for a flickering stimuli of f Hz, measured with the i-th electrode, can be described as function of the frequency f and its harmonics k, with corresponding amplitudes $a_{i,k}$ and $b_{i,k}$:

$$y_i(t) \;=\; \sum_{k=1}^{N_h} a_{i,k} \sin(2\pi k f t) + b_{i,k} \cos(2\pi k f t) + E_{i,t}, \tag{1}$$

where $E_{i,t}$ represents the noise component of the electrode i. For a time segment length of T_s, acquired with a sampling frequency of F_E Hz, the model can be described in vector form as $y_i = X\tau_i + E_i$ where $y_i = [y_i(1), \dots, y_i(N_t)]^T$. The vector τ_i contains the corresponding amplitudes $a_{i,k}$ and $b_{i,k}$ and X describes the $N_t \times 2N_h$ SSVEP model matrix containing the sine and cosine components, associated with the N_h harmonics of the stimulus frequency f. In more detail, the model matrix is composed as followed: For $k = 1, \dots, N_h$ the $N_t \times 2$ matrices X_k are defined by

$$\begin{aligned}
X_k(t,1) &= \sin(2\pi k f t), \\
X_k(t,2) &= \cos(2\pi k f t), \quad \text{for } t = 1, \dots, N_h;
\end{aligned} \tag{2}$$

the model matrix X is then composed by concatenating X_k, $k = 1, \ldots N_h$

$$X = [X_1, \ldots, X_{N_h}]. \tag{3}$$

A set of channels vectors, weighted combinations of the electrode signals, is created to minimize the nuisance signals. In this respect N_s channel vectors $s_i, i = 1, \ldots, N_s$ of length N_t are defined as linear combination of the electrode signals; the $N_t \times N_s$ matrix

$$S = [s_1, \ldots, s_{N_s}] \tag{4}$$

can be written as $S = XW$, where the $N_t \times N_s$ matrix W contains the corresponding weights. The noise and nuisance signal can be estimated by removing the SSVEP components from the signal. In this respect, the $N_t \times N_y$ matrix Y containing the electrode signals is projected on the orthogonal complement of the SSVEP signal model matrix X,

$$\tilde{Y} = Y - X(X^T X)^{-1} X^T Y. \tag{5}$$

A weights vector $\hat{w}$ that minimizes remaining signal $\tilde{Y}$ needs to be found: An optimal weight combination for the electrode signals can then be found by calculating the eigenvectors of the symmetric matrix $\tilde{Y}^T \tilde{Y}$ (please refer to [18] for more details):

The solution of the optimization problem

$$\min_{\hat{w}} \| \tilde{Y}\hat{w} \|^2 = \min_{\hat{w}} \hat{w}^T \tilde{Y}^T \tilde{Y}\hat{w} \tag{6}$$

is given by the eigenvector v_1 corresponding to the minimal eigenvalue λ_1. Additional uncorrelated channels can be added by choosing the next smallest eigenvalue and its corresponding eigenvalue. The weight matrix is then set to

$$W = \left[\frac{v_1}{\sqrt{\lambda_1}}, \ldots, \frac{v_{N_s}}{\sqrt{\lambda_{N_s}}} \right], \tag{7}$$

where $\lambda_1 \leq \lambda_2 \leq \ldots \lambda_{N_s}$ are the calculated eigenvalues.

In order to cancel 90 % of the nuisance signal the total number of channels N_s is selected by finding the smallest value for N_s satisfying

$$\frac{\sum_{i=1}^{N_s} \lambda_i}{\sum_{j=1}^{N_y} \lambda_j} > 0.1. \tag{8}$$

To detect the SSVEP response for a specific frequency, the power of that frequency and its N_h harmonics is estimated by

$$\hat{P}_i = \frac{1}{N_s N_h} \sum_{l=1}^{N_s} \sum_{k=1}^{N_h} \| X_k^T s_l \|^2, \tag{9}$$

where X_k are defined in (2). For all frequencies $\hat{P}_i, i = 1, \ldots, N_f$ is calculated according to (9).

The SSVEP power estimations of all N_f considered frequencies are then normalized,

$$p_i = \frac{\hat{P}_i}{\sum_{j=1}^{N_f} \hat{P}_j},$$

and, to increase the difference between the values, a Softmax function

$$p'_i = \frac{e^{\alpha p_i}}{\sum_{j=1}^{j=N_f} e^{\alpha p_j}}, \qquad (10)$$

with $\alpha = 0.25$ was applied (see [18]).

For the system implementation in this paper, N_h was set to 2, in order to avoid overlapping between the frequencies. N_f was determined by the number of stimulation frequencies. The sampling rate of the amplifier, F_E was set to 128 Hz. EEG data were evaluated in blocks of 13 samples (approximately 0.1 s).

Only if a particular stimulation frequency had the highest probability, exceeded a certain predefined threshold and the classification time window exceeded a certain minimum time period, the corresponding command was classified. The minimum SSVEP classification time window was set to 30 blocks ($N_t = 390$ samples, approximately 3 seconds) and increased block-wise if no command was classified on the basis of the collected data. After 60 blocks ($N_t = 780$), a sliding classification window was used.

2.4 Software

The Epson smart glasses BT-200 operated on android 4.0.4, the GUI displayed on the smart glasses was implemented using Android Studio 2.2. To realize constantly flickering objects, a Handler class is used which schedules the execution of tasks at certain time intervals. The GUI offered two phases, the QR-code scanning phase and the Visual Stimulation Phase.

QR-code Scanning Phase: At the beginning of the experiment, a welcome screen was shown and after a certain time interval, which was required for application launch, the QR-code scanner (see Fig. 2a) was loaded. The live camera image was displayed on the glasses and the application waited until a relevant QR-code was scanned and recognized (see Fig. 2b).

Visual Stimulation Phase: The visual stimuli screen was activated once a QR-code was scanned. The options were displayed according to the scanned QR-code. If e.g., the user scanned the QR-code in the elevator, the visual stimulation phase displayed the following four options: Ground, 1st floor, 2nd floor and 3rd floor (see Fig. 2c). The scanned QR-code data was also sent to the processing unit. The flickering on the smart glasses started after receiving the corresponding

Table 1. Overview of the QR codes and options associated with corresponding frequencies available.

	Location	Frequencies [Hz]	Options
QR-Light1	Lab1 (1st floor)	6	On
		7	Off
QR-Elevator	Floor	6	Ground floor
		7	1st floor
		8	2nd floor
		9	3rd floor
QR-Light2	Lab2 (Ground floor)	6	On
		7	Off
QR-Coffe machine	Lab2 (Ground floor)	6	Espresso
		7	Capuccino
		8	Caffee crema

feedback from the processing unit. The progress bars reflected the current power estimations of the corresponding frequencies (see Fig. 2d). During the experiment three different QR codes were used: One was placed in the lab (first floor) where the experiment started (QR-Light1), one inside of an elevator (QR-Elevator), two codes were placed in a second Lab (QR-Light2) and (QR-Coffee machine). The selection options associated with each of the codes are displayed in Table 1. Note that the number of available options also depended on the QR-code selection; between two and four frequencies were used.

The information about the selection was sent to the smart-glass via the UDP protocol (see Fig. 2e) and to the ESP 32 units which handled the hardware via I/O pins. This included performing the selection, e.g. turning the light on/off or pressing the desired button in the lift (see Fig. 2f).

To control the lights, a set of two wireless socket switches (Funkschalter FS20 ST-2, Conrad, Germany) were remotely controlled through ATmega328 development board (myAVR, Laser & Co. Solutions, Germany) with an TX868-75 transmitter (ELV, Germany) operating on FS20 protocol. The FS20 are ready to use adapter devices with operating frequency 868.35 MHz (AM band). These devices were used for switching the lights on and off. Thanks to the FS20 wireless switch connection, the own build circuits were working with secure voltages of 5 V or less. The elevator buttons were pressed using a servomechanism controlled with the ESP32 internal PWN generator. The coffee-machine selection was simulated with an Adafruit BiColor LED matrix connected directly to the ESP32 using an I2C connection, displaying the number of the classified selection.

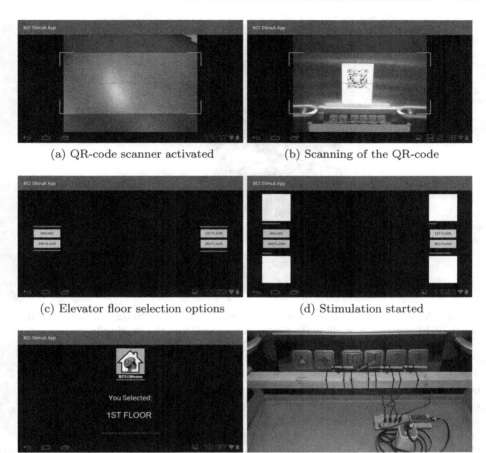

(a) QR-code scanner activated   (b) Scanning of the QR-code

(c) Elevator floor selection options   (d) Stimulation started

(e) Selection feedback   (f) Execution of the selected option

Fig. 2. Screenshots of smart glasses GUI application during QR-code scanning (a, b), the feedback (d, e), and a photo of a custom-made lift hardware interface (f). The user's task was to scan the QR-code inside the elevator (b) and choose the option *1ST FLOOR* by gazing at the proper stimuli (c, d). After the system has made the selection, the feedback was sent to the stimulation interface on the smart glasses (e) and the command was sent to the ESP32 board in the elevator to execute the proper floor selection (f).

3   Experimental Setup

The experiment began in the BCI laboratory (Lab 1, first floor). After signing the consent form and reading an information sheet, participants were prepared for the EEG recording. Afterwards, participants put on the smart-glasses. The actiChamp amplifier with actiPOWER battery and the laptop were placed in a basket which was carried by the subject (see Fig. 3). Initially a test QR code was scanned, so that participants got familiar with the functionality of the system.

Fig. 3. A participant during the experiment. The actiCHamp amplifier with actiPOWER battery as well as the laptop were carried by the participant.

The task for the participants was to go down to a second lab, get a coffee and come back to the BCI-lab. During the experiment, several selections had to be made with the BCI system. When arriving to or leaving a lab, a light switch was controlled to turn the lights on/off; the elevator buttons as well as the desired coffee needed to be selected with the system. If a wrong classification was made the user continued the experiment without correcting the error. The detailed sequence for the task is given in Table 2. The entire session took on average

Table 2. Tasks of the Experiment, each subject had to choose the following options: turning the light on or off, select the proper floor in the elevator (ground floor or first floor), and choosing one of the three coffee options (simulation).

Location	Task	QR code	Correct selection	Associated frequency [Hz]
Lab1	Turn off light	QR-Light1	Off	7
1st floor	Select Ground	QR-Elevator	Ground	6
Lab2	Turn on light	QR-Light2	On	7
Lab2	Select coffee	QR-Coffee machine	Cafe crema	8
Lab2	Turn off light	QR-Light2	Off	7
Ground	Select 1st floor	QR-Elevator	1st floor	7
Lab1	Turn on light	QR-Light1	On	6

about 20 min for each subject. Subjects had the opportunity to opt-out of the study at any time.

4 Results

BCI performance was accessed by calculating the accuracy, the number of correct command classifications divided by the total number of classified commands. In total seven SSVEP commands needed to be selected. The overall results are provided in Table 3. An average command classification of 85.70% was reached.

Table 3. Results of the experiment. The selection accuracy (Acc.) for each subject is calculated based on their correct selections.

User	Light (Room1)	Elevator	Light (Room 2)	Coffee machine (Room 2)	Light (Room 2)	Elevator	Light (Room 1)	Acc.
1	Correct	Correct	Correct	Wrong	Correct	Correct	Correct	6/7
2	Correct	Wrong	Correct	Correct	Correct	Correct	Wrong	5/7
3	Correct	Correct	Correct	Correct	Correct	Wrong	Correct	6/7
4	Correct	Correct	Correct	Wrong	Correct	Correct	Correct	6/7
5	Correct	Correct	Correct	Wrong	Correct	Correct	Correct	6/7
6	Correct	Correct	Correct	Correct	Correct	Correct	Correct	7/7
7	Correct	Correct	Correct	Wrong	Correct	Correct	Correct	6/7
Mean	7/7	6/7	7/7	3/7	7/7	6/7	6/7	85.70%

5 Discussion and Conclusion

This study was designed to simulate SSVEP-based BCI control combined with smart glasses in a simulated smart home environment. The majority of participants controlled the system with accuracies above 80%. Interestingly, the lowest selection accuracy occurred when operating the simulated coffee machine.

The different illumination conditions in the second room might be one of the affecting parameters, which opens an opportunity for further research.

Smart glasses seamlessly blend digital content into the real world. QR codes containing information about the item to which they were attached, were used to identify the devices that could be controlled with the BCI.

Build in error correction allow the implementation of logos. Each of the QR codes we used contained a logo (*BCI@Home*) making it more attractive to the user on one hand and on the other hand, providing the information that the code belongs to the interactive smarthome/BCI system.

The ESP32 boards were used to communicate with controllable devices (such as light switches, elevator and coffee machine). Wireless security is an important aspect in smart homes. Therefore, WiFi Protected Access II (WPA2) security protocols were utilized.

Up to four stimulation frequencies were used. Systems with four targets using computer monitors or LED panels allow average classification accuracies above 90% [6,7]. In the presented paper, the achieved accuracies were not as high. There were several factors that complicated signal classification. One aspect that needs to be investigated further is the contrast. Brighter colors yield a stronger SSVEP response [3,17]. The Epson glasses featured two UV shades for added contrast. In this experiment, the darkest shades were used. Some participants intentionally focused on a darker surfaces in their environment to increase the contrast further. Other participants could select the desired options easily, even without any shades in the testing phase.

Another obstacle might be movement artifacts due to the recording conditions. As observed by Lin et al., SSVEP detectability for standing participants, are comparable with seated participants and deteriorates when changing their position from standing to walking [11]. In this study, participants were usually standing still during SSVEP stimulation phase.

Due to their strong SSVEP response, we used low stimulation frequencies (6–9 Hz). Higher stimulation frequencies on the other hand are more subtle and less tiring (see e.g. [2,19]). Recently, promising results with embedded high-frequency polychrmomatic SSVEP stimuli were achieved [2], which could be used in interactive display systems as presented here.

The results are promising and indicate that the combination of augmented reality and SSVEP could provide intuitive and reliable smart home control.

Acknowledgment. This research was supported by the European Fund for Regional Development under Grant GE-1-1-047.

References

1. Amiri, S., Fazel-Rezai, R., Asadpour, V.: A review of hybrid brain-computer interface systems. Adv. Hum.-Comput. Interact. **2013**, 1 (2013)
2. Chien, Y.Y., Lin, F.C., Zao, J., Chou, C.C., Huang, Y.P., Kuo, H.Y., Wang, Y., Jung, T.P., Shieh, H.P.D.: Polychromatic SSVEP stimuli with subtle flickering adapted to brain-display interactions. J. Neural Eng. **14**(1), 016018 (2016)
3. Duszyk, A., Bierzyńska, M., Radzikowska, Z., Milanowski, P., Kuś, R., Suffczyński, P., Michalska, M., Łabécki, M., Zwoliński, P., Durka, P.: Towards an optimization of stimulus parameters for brain-computer interfaces based on steady state visual evoked potentials. PLoS ONE **9**(11), 1–11 (2014)
4. Friman, O., Volosyak, I., Gräser, A.: Multiple channel detection of steady-state visual evoked potentials for brain-computer interfaces. IEEE Trans. Biomed. Eng. **54**(4), 742–750 (2007)
5. Gao, S., Wang, Y., Gao, X., Hong, B.: Visual and auditory brain-computer interfaces. IEEE Trans. Biomed. Eng. **61**(5), 1436–1447 (2014)
6. Gembler, F., Stawicki, P., Volosyak, I.: Autonomous parameter adjustment for SSVEP-based BCIs with a novel BCI wizard. Front. Neurosci. **9**, 7–12 (2015)

7. Guger, C., Allison, B.Z., Growindhager, B., Prückl, R., Hintermüller, C., Kapeller, C., Bruckner, M., Krausz, G., Edlinger, G.: How many people could use an SSVEP BCI? Front. Neurosci. **6**, 1–6 (2012)
8. Jiang, S., Zhou, P., Li, Z., Li, M.: Poster abstract: emotion-driven lifelogging with wearables. In: 2016 IEEE Conference on Computer Communications Workshops (INFOCOM WKSHPS), pp. 1091–1092, April 2016
9. Kim, Y., Kaongoen, N., Jo, S.: Hybrid-BCI smart glasses for controlling electrical devices. In: 2015 54th Annual Conference of the Society of Instrument and Control Engineers of Japan (SICE), pp. 1162–1166, July 2015
10. Lalor, E.C., Kelly, S.P., Finucane, C., Burke, R., Smith, R., Reilly, R.B., Mcdarby, G.: Steady-state vep-based brain-computer interface control in an immersive 3D gaming environment. EURASIP J. Appl. Sig. Process. **2005**, 3156–3164 (2005)
11. Lin, Y.P., Wang, Y., Wei, C.S., Jung, T.P.: Assessing the quality of steady-state visual-evoked potentials for moving humans using a mobile electroencephalogram headset. Front. Hum. Neurosci. **8**, 1–10 (2014)
12. Mitrasinovic, S., Camacho, E., Trivedi, N., Logan, J., Campbell, C., Zilinyi, R., Lieber, B., Bruce, E., Taylor, B., Martineau, D., et al.: Clinical and surgical applications of smart glasses. Technol. Health Care **23**(4), 381–401 (2015)
13. Muller, S., Celeste, W.C., Bastos, T., Sarcinelli, M.: Brain-computer interface based on visual evoked potentials to command autonomous robotic wheelchair. J. Med. Biol. Eng. **30**(6), 407–415 (2010)
14. Rumiński, J., Bujnowski, A., Kocejko, T., Wtorek, J., Andrushevich, A., Biallas, M., Kistler, R.: Performance analysis of interaction between smart glasses and smart objects using image-based object identification. Int. J. Distrib. Sens. Netw. **12**(3), 6254827 (2016)
15. Stawicki, P., Gembler, F., Volosyak, I.: Driving a semiautonomous mobile robotic car controlled by an SSVEP-based BCI. Comput. Intell. Neurosci. **2016**, 5 (2016)
16. Sumi, N., Moshnyaga, V.: A novel face recognition for smart glasses. In: 2016 IEEE Region 10 Symposium (TENSYMP), pp. 115–118. IEEE (2016)
17. Tsuru, K., Miura, H., Matsui, D.: A new stimulation for steady-state visually evoked potentials based brain-computer interface using semi-transmissive patterns with smartglasses. In: 2015 International Conference on Cyberworlds (CW), pp. 165–168, October 2015
18. Volosyak, I.: SSVEP-based Bremen-BCI interface - boosting information transfer rates. J. Neural Eng. **8**(3), 036020 (2011)
19. Volosyak, I., Valbuena, D., Lüth, T., Malechka, T., Gräser, A.: BCI demographics II: how many (and what kinds of) people can use an SSVEP BCI? IEEE Trans. Neural Syst. Rehabil. Eng. **19**(3), 232–239 (2011)
20. Wolpaw, J.R., Birbaumer, N., McFarland, D.J., Pfurtscheller, G., Vaughan, T.M.: Brain-computer interfaces for communication and control. Clin. Neurophysiol. **113**(6), 767–791 (2002)
21. Wu, Z., Lai, Y., Xia, Y., Wu, D., Yao, D.: Stimulator selection in SSVEP-based BCI. Med. Eng. Phys. **30**(8), 1079–1088 (2008)

How to Reduce Classification Error in ERP-Based BCI: Maximum Relative Areas as a Feature for P300 Detection

Vinicio Changoluisa[1,2]([✉]), Pablo Varona[1], and Francisco B. Rodriguez[1]([✉])

[1] Grupo de Neurocomputación Biológica, Dpto. de Ingeniería Informática,
Escuela Politécnica Superior, Universidad Autónoma de Madrid, 28049 Madrid, Spain
{pablo.varona,f.rodriguez}@uam.es
[2] Universidad Politécnica Salesiana, Quito, Ecuador
fchangoluisa@ups.edu.ec

Abstract. Currently, one of the challenges in a Brain Computer Interface (BCI) technologies is the improvement real-time event-related potential (ERP) detection. Variability and low signal-to-noise ratio (SNR) impair detection methods. We hypothesized that if in a P300-based BCI we find the electrodes with the maximum relative voltage area (the "maximum relative" term refers to the area within each trial, but not between trials) where a P300 can be located, we will improve the performance of a classifier and reduce the number of trials necessary to achieve 100% success. We propose a method that calculates successively the maximum relative voltage areas in the P300 region of the EEG signal for each stimulus. In this way, differences between a target and a non-target stimulus are maximized. This method was tested with a linear classifier (LDA), known for its good performance and low computational cost. We observed that a single electrode with maximum relative voltage area in a P300 region can give more information than the traditional 4 electrode measurement. The preliminary results show that by detecting appropriate characteristics in the EEG signal, we can reduce the error by trial as well as the number of electrodes. The detection of the maximum relative voltage area in the EEG electrodes is a characteristic that can contribute to increase the SNR and decrease the prediction error with the smallest number of trials in the P300-based BCI systems. This type of methods that seek specific characteristics in the signals can also contribute to the management of the variability present in the BCI systems. This method can be used both for an online and offline analysis.

Keywords: Brain Computer Interface · Event-related potentials · P300 · Online · LDA

1 Introduction

A BCI is a technology that allows interaction with external devices without making use of peripheral muscles or nerves. There are different ways to implement a BCI, one of them is through P300 ERP. P300 component is the name

© Springer International Publishing AG 2017
I. Rojas et al. (Eds.): IWANN 2017, Part II, LNCS 10306, pp. 486–497, 2017.
DOI: 10.1007/978-3-319-59147-6_42

given to a positive signal deflection appearing at approximately 300 ms after a stimulus [16]. To evoke a P300 in a subject it is necessary to present a specific stimulus, usually both frequent and infrequent stimuli are presented. The subject is instructed to respond to infrequent stimuli, this procedure is known as the oddball paradigm [15]. There are several ways of presenting this stimulus, although the majority is based on the P300 speller paradigm of Farwell and Donchin [5]. One of the techniques used for its detection is electroencephalography (EEG), among its advantages are its high temporal resolution and relative low cost.

One of the biggest challenges for the P300-based BCI is to improve ERP detection in real-time [10]. Variability of the signal is an obstacle that does not allow achieving this goal. The variability of amplitude and latency impair the detection of the P300 component, these are affected by the attention of the subject or the difficulty of the task [6,12]. The high inter-trial variability impairs the early detection of this ERP. In order to overcome those problems, the stimulus is repeated several times, although repetitions can cause fatigue [14] and a decrease in task performance [3,8]. This is why most investigations seek directly or indirectly to improve the accuracy of the system with the least number of trials and thus improve the Information Transfer Rate ITR [10]. Therefore, achieving high accuracy with a single trial is a challenge. Another intrinsic disadvantage of this BCI modality is the low SNR, and thus the difficulty to differentiate between the P300 signal and neural background activity [2].

It is well known that P300 components are more prominent in the central, parietal and occipital lobes [12], although it is not clear yet what characteristics of the signal of each lobe can be measured to help the detection of this ERP [13]. For example, ERP signal has been parameterized with the amplitude peak or with the area under the curve (AUC) in a P300 window, as described by Farwell and Donchin [5] without conclusive results when these parameters were used as unique characteristic for the detection of P300. However, we suggest that by measuring the maximum AUC in the P300 region of each stimulus can be used to locate the electrodes that improve the SNR in a P300-based BCI.

Our method is based on the fact that if we know a priori the instant that the target stimulus is presented in a P300-based BCI we will be able to look for where there is an amplitude increase and therefore an increase of AUC in the EEG signal. In other words, we believe that, while the target stimuli the AUC increases in the P300 region, the non-target stimulus does not increase as much. Thus, electrodes having the highest AUC in the P300 region are potentially likely to have better characteristics for detecting P300 effectively. Here we propose a new method to detect the maximum relative AUC in a P300 region. We analyze how the detection of this AUC at each stimulation contributes to improve SNR in P300-based BCI. We want to know how maximum relative AUC parameter helps the detection of P300 component with a minimum number of trials. We believe that by finding electrodes with the maximum relative AUC at each stimulation, we will improve the prediction of a classifier.

2   Materials and Methods

In this section, we describe the dataset used for the analysis, the data preprocessing, we detail the proposed method and the procedure for classification.

2.1   DataSet

Second Competition BCI dataset IIb [1] was used for this offline analysis. P300 speller paradigm based on a paradigm described by Donchin et al. [4] was used to collect data from a single subject. These data were divided into three sessions, two sessions with data to train a classifier (42 characters) and one remaining session (with 31 characters) to predict the target characters.

P300 Speller Paradigm. In this paradigm, the subject was asked to concentrate on a letter that was inside a matrix with 36 alphanumerical characters, 6 rows and 6 columns. Each row and column flashed randomly one at a time, so each character will have 12 stimuli. The user must mentally count the times the row or column with the target character flashed. This presentation of 12 stimuli per character was based on an oddball paradigm [15], in which when two stimuli were presented: one frequent and one infrequent, the latter elicit a P300 component [15]. In this case, the row and column with the target character were the infrequent stimuli so they will contain a P300 component.

Data Collection. To spell one character, each row and column (of the 12) were flashed for 100 ms, randomly, one at a time. After each flash all rows and columns were blank for 75 ms, as explained in the competition description. In the end, there were 180 stimuli or flashing per character (12 row/column × 15 times). The set of 12 stimuli was called trial and are repeated 15 times per character.

The data of the subject were collected at a sampling frequency of 240 Hz in three hierarchies: sessions, runs and characters. Each session contained runs, each runs contained a set of characters that form a word. In each run the subject focused on one character at a time.

2.2   Preprocessing

For our offline analysis we took 600 ms (144 samples) from the onset of the stimulus. Each of these segment was called epoch. Each trial was bandpass filtered 0.5–30 Hz and normalized to an interval of $[-1\ 1]$. Although this method was used for offline processing, in the future, our goal is to use it for online processing.

2.3   Area Calculation Method

The main objective of our method was to find the electrodes with maximum relative AUC in the zone where a P300 component can appear. The relative term was used since the maximum P300 AUC of the stimulus of one trial may be less compared to the maximum AUC of another stimulus of a different trial.

In general terms, the method calculated the AUC of the voltage signal generated by each stimulus (target and non-target) in the place where P300 was presumed exist, near the 300 ms (we will call them P300 window). The stimulus that contained the largest AUC was assumed to be the one that contained a P300 component. Subsequently, we evaluated the number of hits per electrode, which contained the maximum number of hits will be the maximum score and the rest of hits per electrode was normalized in function of this one. A hit means having detected all P300 components to accomplish the objective task. We used the training data to calibrate the P300 window and obtained the greatest number of hits. The electrodes with the highest score (or hits) will be considered for the classifier. We present this method of the maximum relative AUC with the P300 speller paradigm by Donchin [4], the method is general enough to be used in any other ERP paradigm.

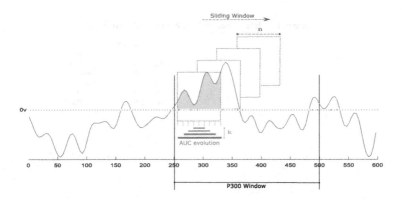

Fig. 1. Diagram of method with P300 window. First, we determine the limits of the P300 window, the difference between the upper and lower limit of P300 window will determine the number of sliding windows (SW). Then, the size of the SW (n) is established and within it, we calculate the evolution of the AUC k times. k depends on SW

In what follows, we illustrate how the method was applied in the P300 speller paradigm previously described (character matrix 6-by-6). In this paradigm, it is considered that only two out of twelve stimuli (6 rows and 6 columns) should have increased AUC approximately to 300 ms from the onset of the stimulus, while the remaining ten (stimuli) would maintain their AUC or at least they would not change as much as the previous ones. However, it is known that the variability between stimulus and stimulus makes it difficult to differentiate them [10], which makes it a problem to overcome.

First, we established a P300 window where it is known that this component is generated, normally between 250 and 500 ms from the onset of the stimulus [12],

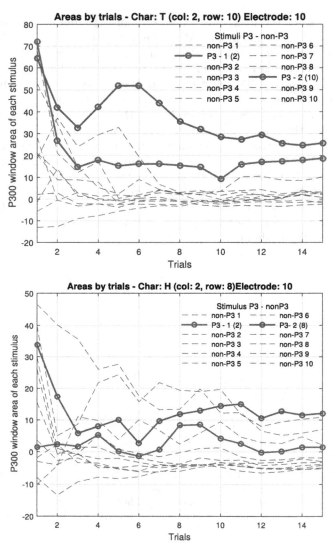

Fig. 2. Evolution of the AUC by trial. Top panel: In all trials to detect the letter 'T', two target stimuli have the largest AUC compared to the ten non-target stimuli. Bottom panel: Non-target stimuli has the largest area in almost all trials.

although it was adjusted according to the number of hits. Once established the P300 window our method was in charge of calculating the evolution of the AUC. This evolution implies that within the P300 window a sliding window is established and within it a small initial interval (of n samples) is configured to calculate the AUC, see Fig. 1. As the interval grows we continue to calculate the area until reaching the limits of the sliding window that will be the largest interval calculated. This sliding window advances in steps of 1 sample until it reaches the limit of the P300 window.

In an ideal case, the calculated AUC of the target stimulus would evolve much faster than the non-target stimulus. That is, in one trial of 12 stimuli, 2 stimuli (1 row and 1 column) will have larger AUC values than the remaining ten, as shown in the top panel of Fig. 2. With this in each trial, the maximum area of the rows and columns was chosen to result in a single character of the stimulation matrix.

The detection of the target character was considered a success, otherwise, a failure. In each character, we calculated the relative error by dividing the number of hits by the number of trials. At the end, each electrode had a relative error depending on the number of hits. The electrodes that had the highest number of hits was the one that for us more information contributes.

In the Fig. 2 we can observe the evolution of the areas by trial of two characters, with good and poor results. In the top panel, there is a character of session 10 that was successful in all the trials, the maximum areas corresponds to the target stimuli. On the other hand, in the bottom panel, one target character of session 11 was not selected correctly in any trial since the maximum relative AUC does not correspond to the target stimuli.

2.4 Classification

For the prediction of the target character we have chosen Krusienski method [9]. This is a fast method to classify with LDA. In this method, 800 ms (192 samples) were taken after each intensification which we call epoch. Each epoch was filtered with a moving average and then decimated with a factor of 12, leaving a total of 16 samples (192/12) per electrode. Eight electrodes: Fz, Cz, Pz, $P3$, $P4$, $PO7$, $PO8$, Oz; were selected and concatenated. In the end each epoch has 128 samples (16*8). Two training sessions with 42 characters were used to train the LDA model. The prediction of the class was made trial to trial. 0 was assigned for non-P300 class and 1 for P300 class. To infer the target character from test set was used:

$$predicted_{row} = argmax_{rows} \left[\sum_{i_{row}} w \cdot f(x_{i_{row}}) \right] \tag{1}$$

$$predicted_{col} = argmax_{cols} \left[\sum_{i_{col}} w \cdot f(x_{i_{col}}) \right] \tag{2}$$

where x is the feature vector, $f(\cdot)$ is a transformation function and w is a vector of classification weights [9].

3   Results

Our method allowed to identify the electrodes with maximum relative AUC in the P300 window that helped us to improve the classifier's success using

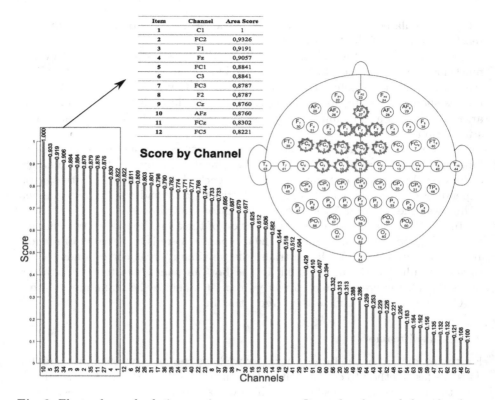

Fig. 3. Electrodes and relative maximum area score. It can be observed that the electrodes with greater relative AUC in the P300 window are in the central and frontal lobes. Histogram of area-by-electrodes scores: the scores were obtained according to the number of hits that were achieved in each trial.

the fewest trials. The central and frontal lobes were the electrodes with higher relative AUC (see Fig. 3) and they help to decrease the percentage of error, as shown in Fig. 4. Our results were compared with those obtained by applying the Krusienski method and with the results of the BCI competition 2003 winners. Although the winners of the competition used a Gaussian SVM kernel to achieve 100% success from 5 trial, the results are comparable since they use the same dataset. To facilitate the description of the results we present 2 approaches for our analysis. In the first, we show the maximum percentage of hits with the lowest number of trials. In the second approach, we demonstrate that with the selection of the appropriate electrodes we can improve the number of hits and decrease the computational cost for the detection of P300.

3.1   Reduction of Error and Number of Trials

A performance of 97.41% of the total hits (15 trials) was achieved with 10 electrodes, 6 electrodes of the occipital and parietal lobe of the Krusienski method:

Pz, Oz, $P3$, $P4$, $PO7$, $PO8$; plus 4 electrodes with higher AUC: $C1$, $FC2$, Fz, $F1$. With this combination of electrodes 100% of hits from the fourth trial was achieved, overcoming the Krusienski method that reaches 100% from the ninth trial and total hits of 92.25%. Our results also surpassed the winners of the BCI competition 2003 [7] that achieved 100% success from the fifth trial, who use 10 electrodes: Fz, Cz, Pz, Oz, $C3$, $C4$, $P3$, $P4$, $PO7$, $PO8$.

Figure 4 presents the percentage of hits for each trial. From the first trial the classifier results improved with the electrodes of maximum relative AUC. In this trial the hits are increased from 17 characters (from 31) with the Krusienski method to 23 hits with the method of the areas. In the third trial our method of the AUC only fails in 1 characters, then from the fourth trial all the characters are correct.

In the figures below (Figs. 4 and 5), a 10–20 system brain mapping was used to illustrate the results and the electrodes used in each configuration. Each electrodes configuration has a different color and marker that determines which electrodes were used. For example, the green color and circle marker represents

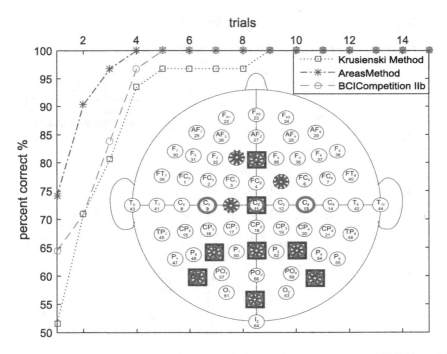

Fig. 4. Comparison of hits by trials. Method of the relative maximum AUC (Asterisk) with 10 electrodes: 6 occipital/parietal lobes and 4 central/frontal lobes reach 100% accuracy from the fourth trial. Krusienski method (squares) with 8 electrodes reach 100% error from the ninth trials and BCI Competition 2003 Dataset IIb (circle) from the fifth trial with 10 electrodes (Color figure online)

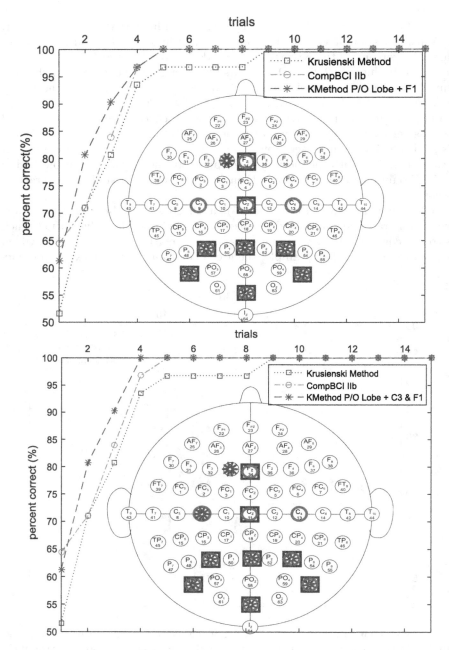

Fig. 5. Influence of the central or frontal lobe electrodes in the classifier. Top Panel: The trials required for a 100% success decrease from 9 to 5 when one electrode of the central (C3) or frontal (F1) lobe with maximum relative AUC is added to the electrodes of the occipital and parietal lobes of Krusienski method. Bottom panel: C3 and F1 electrodes plus occipital and parietal lobes electrodes of Krusienski method reach 100% success from the fourth trial. (Color figure online)

the results BCI competition, in the brain mapping, each electrode is marked with green and circle.

3.2   The Importance of Selecting the Appropriate Electrode

A single electrode of the central or frontal lobe with maximum relative AUC plus 6 of the occipital and parietal lobes was sufficient to reduce the number of trials necessary to achieve 100% success from the fifth trial. Our method reduces to 7 the number of electrodes necessary to achieve 100% of success from the fifth trial surpassing the BCI competition 2003 winners who used 4 electrodes of the central and frontal lobes and the results of the Krusienski method using 2 (see Fig. 5).

We wanted to quantify how much the electrodes of the central and frontal lobes contribute in hits rate. For this, we evaluated the results of the classifier taking into account only the electrodes of the occipital and parietal lobes used in the Krusienski method, without taking into account Fz and Cz electrodes. We observed that the classifier started with a low percentage of success and only in the ninth trial reached a total success. While adding a single electrode of the central or frontal lobe ($C3$, $F1$), the classifier to hit 100% from the fifth trial Fig. 5. The results even improve with the addition of F1 and C3 to the electrodes of the parietal and occipital lobes, see bottom panel of Fig. 5.

4   Discussion and Conclusions

This paper proposes a new methodology to extract more information from EEG electrodes in a P300-based BCI. We present the relative maximum AUC as a characteristic that contributes to the detection of the P300 component. This will help the detection of the electrodes with higher SNR that contribute to the reduction of the number of trials necessary for the detection of P300 and to improve the prediction error of the classifier.

We observed that with the use of a single electrode of the central or frontal lobe, with maximum relative AUC, plus 6 of the parietal and occipital lobes the results of the classifier can be improved. Figure 5 shows that the electrode $F1$ contribute more to the classifier than CZ and FZ used in the Krusienski method [9]. Our results also surpass the winners of the BCI competition II [7], who use 4 electrodes from these brain regions: FZ, CZ, $C3$, $C4$. Electrode selection methods allow using fewer electrodes and reducing setup time and cost for an EEG-based P300 spellers [11].

Similarly, the results of the Fig. 4 show that using 4 electrodes with higher relative AUC and 6 electrodes of the parietal and occipital lobes we can maximize the percentage of total hits to 97.41% and hit with a smaller number of trials. We understand that there are other electrodes of these lobes (central/frontal) that provide information but to a lesser degree that even some of them may even impair the detection of P300.

It is important to emphasize that what we are detecting is not necessarily the maximum AUC in a electrode, rather we are detecting how different AUC's are in this brain region. We are showing that the AUC differences between the target and non-target stimuli of the central and frontal area greater than in other brain regions in this subject. With this work we are looking for AUC to maximize the differences between a target and a non-target stimulus. It is still unclear which features are specifically involved in the detection of P300. However it would seem that the AUC in the P300 window is a feature that could help detect this ERP with few trials. It is necessary to consider that there are other characteristics that contribute to the detection of P300 like N200, N100, among others.

One of the weaknesses that we observe in this method derives from external noise component that only damages an epoch, fluctuations that alter the calculation of the areas and therefore the maximum relative AUC. We consider that the search for other characteristics, such as those mentioned in the previous paragraph, can help to overcome it.

This method aims to help the adaptability of BCI in each subject detecting a specific feature in each stimulus of each electrode. We believe that these types of methods, which look for specific characteristics, can help overcome the inter- and intra-subject variability that affect BCI. This is why in the future we will study how these characteristics vary between subject and subject.

Acknowledgments. This work was funded by Spanish projects of Ministerio de Economía y Competitividad/FEDER TIN2014-54580-R, DPI2015-65833-P (http://www.mineco.gob.es/) and Predoctoral Research Grants 2015-AR2Q9086 of the Government of Ecuador through the Secretaría de Educación Superior, Ciencia, Tecnología e Innovación (SENESCYT).

References

1. Blankertz, B., Muller, K.R., Curio, G., Vaughan, T.M., Schalk, G., Wolpaw, J.R., Schlogl, A., Neuper, C., Pfurtscheller, G., Hinterberger, T., Schroder, M., Birbaumer, N.: The BCI competition 2003: progress and perspectives in detection and discrimination of EEG single trials. IEEE Trans. Biomed. Eng. **51**(6), 1044–1051 (2004)
2. Blankertz, B., Lemm, S., Treder, M., Haufe, S., Müller, K.-R.: Single-trial analysis and classification of ERP components–a tutorial. NeuroImage **56**(2), 814–825 (2011)
3. Boksem, M.A.S., Meijman, T.F., Lorist, M.M.: Effects of mental fatigue on attention: an ERP study. Cogn. Brain. Res. **25**(1), 107–116 (2005)
4. Donchin, E., Spencer, K.M., Wijesinghe, R.: The mental prosthesis: assessing the speed of a P300-based brain-computer interface. IEEE Trans. Rehabil. Eng. **8**(2), 174–179 (2000)
5. Farwell, L.A., Donchin, E.: Talking off the top of your head: toward a mental prosthesis utilizing event-related brain potentials. Electroencephalogr. Clin. Neurophysiol. **70**(6), 510–523 (1988)
6. Kahneman, D.: Attention and effort. Citeseer (1973)

7. Kaper, M., Meinicke, P., Grossekathoefer, U., Lingner, T., Ritter, H.: BCI competition 2003-data set IIB: support vector machines for the P300 speller paradigm. IEEE Trans. Biomed. Eng. **51**(6), 1073–1076 (2004)
8. Käthner, I., Wriessnegger, S.C., Müller-Putz, G.R., Kübler, A., Halder, S.: Effects of mental workload and fatigue on the P300, alpha and theta band power during operation of an ERP (P300) brain-computer interface. Biol. Psychol. **102**, 118–129 (2014)
9. Krusienski, D.J., Sellers, E.W., Cabestaing, F., Bayoudh, S., McFarland, D.J., Vaughan, T.M., Wolpaw, J.R.: A comparison of classification techniques for the P300 speller. J. Neural Eng. **3**(4), 299–305 (2006)
10. Mak, J.N., Arbel, Y., Minett, J.W., McCane, L.M., Yuksel, B., Ryan, D., Thompson, D., Bianchi, L., Erdogmus, D.: Optimizing the P300-based brain-computer interface: current status, limitations and future directions. J. Neural Eng. **8**(2), 025003 (2011)
11. McCann, M.T., Thompson, D.E., Syed, Z.H., Huggins, J.E.: Electrode subset selection methods for an EEG-based P300 brain-computer interface. Disabil. Rehabil. Assistive Technol. **10**(3), 216–220 (2015)
12. Polich, J.: Updating P300: An integrative theory of P3a and P3b. Clin. Neurophysiol. **118**(10), 2128–2148 (2007)
13. Rakotomamonjy, A., Guigue, V.: BCI competition III: dataset II- ensemble of SVMs for BCI P300 speller. IEEE Trans. Biomed. Eng. **55**(3), 1147–1154 (2008)
14. Salvaris, M., Sepulveda, F.: Visual modifications on the P300 speller BCI paradigm. J. Neural Eng. **6**(4), 046011 (2009)
15. Squires, N.K., Squires, K.C., Hillyard, S.A.: Two varieties of long-latency positive waves evoked by unpredictable auditory stimuli in man. Electroencephalogr. Clin. Neurophysiol. **38**(4), 387–401 (1975)
16. Sutton, S., Braren, M., Joseph, Z., John, E.R.: Evoked-potential correlates of stimulus uncertainty. Science **150**(3700), 1187–1188 (1965)

Machine Learning in Imbalanced Domains

Deep Fisher Discriminant Analysis

David Díaz-Vico[✉], Adil Omari, Alberto Torres-Barrán,
and José Ramón Dorronsoro

Dpto. Ing. Informática and Instituto de Ingeniería del Conocimiento,
Universidad Autónoma de Madrid, Madrid, Spain
`david.diazv@estudiante.uam.es`

Abstract. Fisher Discriminant Analysis' linear nature and the usual
eigen-analysis approach to its solution have limited the application of
its underlying elegant idea. In this work we will take advantage of some
recent partially equivalent formulations based on standard least squares
regression to develop a simple Deep Neural Network (DNN) extension
of Fisher's analysis that greatly improves on its ability to cluster sample
projections around their class means while keeping these apart. This is
shown by the much better accuracies and g scores of class mean classifiers
when applied to the features provided by simple DNN architectures than
what can be achieved using Fisher's linear ones.

Keywords: Linear Discriminant Analysis · Deep Neural Networks ·
Non-linear classifiers

1 Introduction

Fisher' Linear Discriminant Analysis (FLDA from now on) is a very well known
linear dimensionality reduction/feature extraction technique that, while able to
provide useful data representations, does not intend, in principle, to solve a
given classification problem and, thus, it has known only a limited use as a tool
to build classifiers. There may be two main reasons for this. The first one is
its linear nature. In fact, while quite attractive, its main goal of concentrat-
ing the projected features around their class means while keeping those means
apart can, for most problems, only be partially achieved by FLDA's linear projections. Moreover, in order to build a powerful classifier, we would most likely
need to apply a non-linear classifier to the FLDA features, but this combination of a linear projection followed by a non-linear classifier may at best be
only competitive with the direct application of the non-linear classifier over the
initial features.

In any case, FLDA has been successfully applied in a number of problems,
most notably on face recognition where the original Fisher Face method [1] has
been progressively improved to become the state of the art in this area. A natural
idea is thus to somehow extend FLDA to a nonlinear procedure by applying it
after some non-linear pre-processing of the original features. This is the goal of
Kernel Discriminant Analysis (KDA, [12]), which addresses binary problems and

© Springer International Publishing AG 2017
I. Rojas et al. (Eds.): IWANN 2017, Part II, LNCS 10306, pp. 501–512, 2017.
DOI: 10.1007/978-3-319-59147-6_43

where the well known reduction for such problems of FLDA to a linear regression problem [4, Chap. 5], is extended into a kernel setting.

As just said, KDA only is available in principle for binary problems. A different non-linear extension that works for multiclass problems was proposed in [8,14], where FLDA is applied on the nonlinear features obtained after processing the original features by the hidden layers of a standard feed-forward multilayer perceptron (MLP). For this, the non-linear z features are first obtained on the MLP's last hidden layer as $z = f(x, W)$, where W denotes the MLP weights and biases up to the last hidden layer and $f(x, W)$ the effect of the MLP forward pass on the original features x. Then, FLDA's standard criterion function is used on these z to get FLDA's projecting matrix A by minimizing one of the several criterion functions J proposed in FLDA. Thus, we can view the overall cost function $J(A, W)$ as depending separately on the FLDA's projection matrix A and on the MLP's weight and bias set W, which suggests to optimize $J(A, W)$ alternating the minimization on W and A. More precisely, for a given W_k and $z^k = f(x, W_k)$, we first derive the A_k matrix by minimizing $J(A, W_k)$ by FLDA's standard eigenprocedure. Then, the new W_{k+1} are derived minimizing $\mathcal{J}(W) = J(A_k, W)$; as shown in [8], the gradient $\nabla_W \mathcal{J}$ can be explicitly computed by backpropagation. Notice that this ensures $J(A_k, W_{k+1}) < J(A_k, W_k) < J(A_{k-1}, W_k)$ and this alternating two-step process can be iterated towards a minimum (A^*, W^*) of J. As in [8,14], we shall refer to this procedure as Non-linear Discriminant Analysis (NLDA).

While in principle any number of hidden layers could be considered, only a single hidden layer was used in [8,14], as was customarily done before the advent of deep neural networks (DNN). These have provided two main insights. The first one is a better understanding of network initialization plus efficient minimization and regularization procedures, which have made largely routine the previously near impossible training of many layered networks. The second one is the availability of symbolic gradient computation in platforms such as Theano [2] or TensorFlow [9] that make it possible the consideration of cost functions much more general than the square error or cross entropy usually applied in neural network-based regression or classification.

Both could be applied to improve on NLDA, either by keeping the alternating minimization of $J(A, W)$ but working with deeper networks or, simply by applying symbolic differentiation on A, W directly to the joint $J(A, W)$ cost function. Here, however, we will follow a much more direct approach by taking advantage of the results in [13,15,16], where a link is established between a concrete formulation of FLDA and a related Least Squares Regression (LSR) problem with a particular, class-based target choice. We shall make extensive use of this approach which we will call Least Squares Discriminant Analysis (LSDA). More precisely, it is shown in [15,16] that there is an isometry between projections derived from a specific FLDA formulation and those derived from the solution of the LSR problem. In turn, this implies that if distance based classifiers such as k-Nearest Neighbors or (as done here) minimum class-mean distances are used, either a renormalized FLDA or LSR approaches result in equivalent classifiers.

Once the previous set up is available, it is straightforward to carry the LSR problem into a DNN setting, working with the same targets as in the linear case but which now are to be approximated by the outputs of a suitable DNN. This is the approach we shall follow here and, besides a short, self-contained presentation of the LSR and FLDA equivalence in [15,16] our contributions are the following:

- The proposal of Deep Fisher Discriminant Analysis, DFDA, along the lines just summarized.
- A comparison of DFDA with classical FLDA over several, large size, binary and multiclass datasets, that shows a much better performance of DFDA.

The paper is organized as follows. In Sect. 2 we shall review classical FLDA as well as the distance-based classifier equivalence established in [15,16] between classical FLDA and a concrete LSR problem. Deep Fisher Discriminant Analysis is introduced in Sect. 3 and in Sect. 4 we will compare its performance with that of classical LFDA over several relatively large multiclass and, in some cases, imbalanced problems. As we shall see, the accuracies and g scores of the DFDA classifiers are substantially better. Finally the paper closes with a brief discussion and pointers to further work.

2 Fisher's Linear Discriminant Analysis and Least Squares Counterparts

2.1 Fisher's Linear Discriminant Analysis

We first briefly review classical FLDA. As mentioned, its goal is to linearly project the original patterns in such a way that these projections are close to their class means while these class means are kept apart. Several criterion functions can be used for this goal and many of them are in fact equivalent; see [6, Sect. 10.2]. Here we will seek to maximize the trace criterion

$$g(W) = \mathrm{trace}(s_T^{-1} s_B) = \mathrm{trace}\left((A^t S_T A)^{-1}(A^t S_B A)\right), \qquad (1)$$

where A is the projection matrix, S_B and S_T denote the between-class and total covariance matrices respectively of the sample patterns and s_B and s_T are their counterparts for the projections $z = Ax$; see [16], Subsect. 2.2 for more details. Solving $\nabla_A g = 0$ leads to

$$0 = -2 S_T A s_T^{-1} s_B s_T^{-1} + 2 S_B A s_T^{-1}$$

i.e., $S_T^{-1} S_B A = A s_T^{-1} s_B$ or, up to an invertible transformation of A (which won't change the cost function $g(W)$), to

$$S_T^{-1} S_B A = A\Lambda, \qquad (2)$$

with Λ the non-zero eigenvalues of $S_T^{-1} S_B$ (and of $s_T^{-1} s_B$). Thus, for such an A we have

$$g(A) = \mathrm{trace}(s_T^{-1} s_W) = \mathrm{trace}\,\Lambda = \lambda_1 + \ldots + \lambda_q,$$

which we maximize by sorting the eigenvalues in Λ in descending order and selecting the $q = \min\{d, C - 1\}$ largest ones and some conveniently normalized associated eigenvectors; here d is pattern dimension and C the number of classes; q is then the rank of S_B. Notice that the minimizer of (1) is not uniquely defined, and some normalization has to be introduced; an usual choice is to impose $A^t S_T A = I_q$.

In some problems S_T may be ill conditioned and not have full rank. One possibility in this case is to use the Moore-Penrose inverse of S_T; another, and the one we follow here, is Regularized Discriminant Analysis [5], where we work with $S_t + \lambda I$ for an appropriate $\lambda > 0$.

2.2   Least Squares Regression and Fisher's Linear Discriminant Analysis

It is a very well known result [4] that for 2-class problems, a solution to FLDA can be obtained solving a Least Squares Regression (LSR) problem

$$\min \frac{1}{2}\|Y - 1_n w_0 - Xw\|^2 \tag{3}$$

where 1_n is the all ones vector, X is the $n \times d$ data matrix and Y is an appropriate target matrix defined by settingbreak $Y_p = n/n_1$ if x_p belongs to class 1 and $Y_p = -n/n_2$ if x_p belongs to class 2. If m_1 and m_2 denote the class means on the original features, it can be then checked that $w = S_T^{-1}(m_1 - m_2)$. Since now $S_B = (m_1 - m_2)(m_1 - m_2)^t$, it follows that $S_T^{-1} S_B w = w\gamma$, with $\gamma = (m_1 - m_2)^t S_T^{-1}(m_1 - m_2)$, i.e., w is an eigen-solution of (2) and, hence, a dilation of a FLDA's projection vector computed as in the previous subsection.

There have been several attempts to carry this result to a multiclass setting. Among the most successful ones are the proposals by Park and Park [13] and the somewhat simpler one in Zhang et $al.$ [16], which we follow here and briefly explain next in a much more concise way.

Consider again the LSR problem (3), where the target matrix Y to be adequately chosen. For simplicity we assume that S_T is regular; if not, we can simply replace it with $S_T + \lambda I$ for some $\lambda > 0$. The optimal LSR solution is then

$$w = S_T^{-1} X^t HY,$$

where H is the **centering matrix**

$$H = I_n - \frac{1}{n} 1_n 1_n^t;$$

in particular, $X^t H$ is the $n \times d$ matrix whose p-th row equals $x_p^t - m^t$. The key assumption is now that we can choose a target matrix Y such that we can write S_B as

$$S_B = X^t HY Y^t HX.$$

Assuming this, let us write $Q = X^t HY$ (and, thus, $S_B = QQ^t$); we show next how can we transform the solution $w = S_T^{-1} X^t HY = S_T^{-1} Q$ of (3) into an eigensolution A of (2). Consider the semidefinite positive matrix $R = Q^t S_T^{-1} Q$ whose

SVD decomposition is $R = \tilde{V}\tilde{\Gamma}\tilde{V}^t$. Assuming for simplicity that rank$(Q) =$ rank$(S_B) = C-1$, at least one of the diagonal elements in $\tilde{\Gamma}$ will be zero; reordering $\tilde{\Gamma}$ if necessary, we assume it to be the element in the (C, C) matrix entry. We can thus drop the last row and column of $\tilde{\Gamma}$ (that are 0) to get a $(C-1) \times (C-1)$ diagonal matrix Γ, and the last column of $\tilde{V}$ to get a $C \times (C-1)$ matrix V that verifies $V^tV = I_{C-1}$ and for which we can also write R as $R = V\Gamma V^t$.

Now it is easy to check that the pair $(A = wV\Gamma^{-1/2}, \Gamma)$ is an eigensolution of (2) with normalization $A^t S_T A = I_q$, for we have

$$S_T^{-1}S_B A = S_T^{-1}QQ^t wV\Gamma^{-1/2} = wQ^t S_T^{-1}QV\Gamma^{-1/2} = wRV\Gamma^{-1/2}$$
$$= wV\Gamma V^t V\Gamma^{-1/2} = A\Gamma.$$

In other words, if we choose Y adequately, from $Q = X^t HY$ we can derive the LSR solution w, the SVD decomposition (V, Γ) of $Q^t S_T^{-1}Q$ and the FLDA eigensolution $A = wV\Gamma^{-1/2}$. This combines Algorithm 4 in [16] and the discussion in its Sect. 6 to derive a FLDA solution A from the LSR solution w.

As mentioned, one thing to consider is the possibility of S_T being singular. This can be easily handled now by working with a Ridge Regression problem, i.e., solving for an appropriate $\lambda > 0$

$$\min \frac{1}{2}\|Y - 1_{nC}\, w_0^t - Xw\|^2 + \frac{\lambda}{2}\text{trace}(w^t w),$$

where 1_{nC} denotes the $n \times C$ all ones matrix and w_0 is a C-dimensional vector.

2.3   Equivalence of Distance Classifiers

As in [16], let's consider the projection matrix $B = A\Gamma^{1/2} = wV$ instead of FLDA's standard A. It is easy to see that B is also an eigen-solution of (2) associated to the normalization $B^t S_T B = \Gamma$. Let $y = w_0 + w^t x$ and $\omega = B^t x$ be the LSR and B-eigen-projections of a pattern x, respectively. We then have

$$\|\omega - \omega'\|^2 = (x - x')^t BB^t(x - x') = (x - x')^t wVV^t w^t(x - x') = \|y - y'\|^2.$$

Thus, any Euclidean distance-based classifier will give the same results when applied to the B-eigen-projections ω than when applied to the LSR ones y. This will be the case for a k-Nearest Neighbor classifier and also for the nearest class mean classifier

$$\delta_{NCM}(x) = \arg\min_c \|w^t x - w^t \overline{x}_c\| = \arg\min_c \|y - \overline{y}_c\|,$$

which we will use here. In other words, we can obtain a FLDA-like nearest class mean classifier directly from the LSR solution, without having to perform the eigen-analysis that FLDA requires. We will call this procedure, proposed in [15,16], Least Squares Discriminant Analysis, LSDA, and take advantage of this in Sect. 3 to define our deep Fisher classifiers but, before that, we close this section with two examples of suitable target matrices Y.

2.4 Two Examples

The well known relationship between the LSR and FLDA solutions for 2 class problems mentioned above also follows easily from the previous discussion. In fact, using the target vector $Y_p = n/n_1$ if x_p is in class 1 and $Y_p = -n/n_2$ if x_p is in class 2, then we have $Q = X^t HY = m_1 - m_2$ and, hence, $S_B = (m_1 - m_2)(m_1 - m_2)^t = QQ^t$. Besides, $R = QS_T^{-1}Q = (m_1 - m_2)^t S_T^{-1}(m_1 - m_2) = \gamma$, with a trivial SVD decomposition $R = 1\gamma 1$ and, thus, we have here $A = w \cdot 1 \cdot \gamma^{-1/2} = S_T^{-1}(m_1 - m_2)\gamma^{-1/2}$ and $B = A\gamma^{1/2} = S_T^{-1}(m_1 - m_2) = w$, i.e., the w and y projections now coincide.

For the general multi-class case, it is shown in [16, Eq. (4)], that we can write

$$S_B = X^t HE\Pi^{-1}E^t HX = X^t HE\Pi^{-1/2}\Pi^{-1/2}E^t HX,$$

where Π is the diagonal matrix with $\Pi_{cc} = n_c$ and E is the $n \times C$ indicator matrix with rows e_p such that if x_p is in class c, $e_{pc} = 1$ and $e_{pc'} = 0$ for $c' \neq c$. Thus here we can take $Q = X^t HE\Pi^{-1/2}$ and it is also shown in [16, Eq. (25)], that we can write the LSR solution $w = S_T^{-1}X^t HY$ as

$$w = S_T^{-1}X^t HE\Pi^{-1/2} = S_T^{-1}Q$$

if we use $Y = HE\Pi^{-1/2}$ as the target matrix. It is now easy to see that for such Y and x_p in class c, we have $Y_{pc} = \frac{n-n_c}{n\sqrt{n_c}}$, and $Y_{pc'} = -\frac{\sqrt{n_c}}{n}$ otherwise. These are the targets we shall use in the next section.

3 Deep Fisher Discriminant Analysis

We have just argued how we can obtain for a general C class problem a nearest class-mean classifier equivalent to one acting on the B-based Fisher projections by performing the following steps:

1. For a given training set D_{tr}, solve for a data matrix X_{tr}, class indicator matrix E_{tr} and targets $Y_{tr} = HE_{tr}\Pi^{-1/2}$ the LSR problem

$$\min \frac{1}{2}\|Y_{tr} - \mathbf{1}_{nC}\, w_0^t - X_{tr}w\|^2,$$

   obtaining the optimal $d \times C$ matrix w^* and C-dimensional vector w_0^*.
2. Compute the projections $y = w_0^* + (w^*)^t x$ for $x \in D_{tr}$ and their class means $\overline{y}_c = w_0^* + (w^*)^t \overline{x}_c$.
3. Assign an x in a test sample D_{ts} to the class whose mean the projection $y = w_0^* + (w^*)^t x$ it is closest to; that is, to the class c^* for which

$$c^* = \arg \min_c\{\|y - \overline{y}_c\|\}.$$

Now, a natural idea to extend this to a non linear setting is to perform the LSR computations on features z obtained by a non-linear processing $z = \Phi(x)$ of the original features x. An example of this is Kernel Discriminant Analysis,

KDA [12,16], where a certain generalized eigenvalue problem involving the kernel matrix $K = ZZ^t$ is solved, with the matrix Z being $Z = \Phi(X)$; in particular the projections z are not needed explicitly as they enter the computations through a kernel k such that $z \cdot z' = \Phi(x) \cdot \Phi(x') = k(x, x')$. As it is often the case in kernel methods, handling the $n \times n$ matrix K can be too costly in large sample problems and some suitable low rank approximation would have to be used.

A simpler alternative, better suited in principle for large sample problems, is to derive the z features using a straightforward DNN extension of the previous linear setup; more precisely, in a DNN setting we would

1. Solve over a training set D_{tr} the LSR problem

$$\min \frac{1}{2}\|Y_{tr} - f(X_{tr}, \mathcal{W})\|^2$$

   with Y_{tr} the previous training target matrix and the p-th row of the matrix $f(X_{tr}, \mathcal{W})_p$ is given by $f(X_{tr}, \mathcal{W})_p = f(x_p, \mathcal{W})$, where $f(x, \mathcal{W})$ is the transfer function of a deep network with linear outputs and overall weight set $\mathcal{W}$; we thus obtain an optimal DNN weight set $\mathcal{W}^*$.
2. Compute the projections $y_p = f(x_p, \mathcal{W}^*)$ over D_{tr}.
3. Assign a new $x \in D_{ts}$ to the class whose mean the projection $y = f(x, \mathcal{W}^*)$ is closest to.

Writing $\mathcal{W}^* = (w_0^*, w^*, W^*)$ with w_0^*, w^* the linear output weights, these optimal w_0^*, w^* solve the LSR problem (3) over the last hidden layer features $z = \Phi(x, W^*)$, with Φ the DNN transfer function up to the last hidden layer. Thus, the class mean classifier of the full DNN is equivalent to a class mean classifier over some FLDA projections of the z patterns in the last hidden layer which, in turn, are also learned by tuning the W component of the overall weight $\mathcal{W}$. In other words, the DNN also performs a particular kind of representation learning, as in this case it learns in its last hidden layer new features that have been optimized to perform FLDA on them. We will call this Deep Fisher Discriminant Analysis, or DFDA.

As in the linear case, we may avoid singularity issues here by adding a regularization term, i.e., solving for instance

$$\min_{w,W} \frac{1}{2}\|Y - f(X, w_0, w, W)\|^2 + \frac{\lambda}{2}\text{trace}(w^t w + \widetilde{W}^t \widetilde{W}),$$

with $\widetilde{W}$ the components of W excluding the biases at each hidden layer. This is the cost function we will use. Of course, in the deep case one may use other regularization terms for the W components of the overall weight structure $\mathcal{W}$ (such as, for instance, dropout), but the term $\frac{\lambda}{2}\text{trace}(w^t w)$ should be kept in any case for the linear weights w.

4 Numerical Experiments

4.1 Datasets and Quality Measures

We will consider the datasets `SensIT Vehicle (combined)`, `dna`, `ijcnn1`, `letter`, `mnist`, `pendigits`, `satimage`, `shuttle`, `w7a`, `w8a` and `usps`. All

Table 1. Train sample size, dimension, number of classed and ratio between the maximum and minimum class sizes for the considered datasets.

Problem	N. patterns	Dimension	N. classes	Class ratios
combined	78,823	100	3	2.156
dna	2,000	180	3	2.265
ijcnn1	49,990	22	2	9.301
letter	15,000	16	26	1.128
mnist	60,000	784	10	1.244
pendigits	7,494	16	10	1.085
satimage	4,435	36	6	2.583
shuttle	43,500	9	7	5,684.667
w7a	24,692	300	2	32.368
w8a	49,749	300	2	32.637
usps	7,291	256	10	2.203

of them also have separate, well defined train-test splits and, except mnist, are available on the Datasets section of the LIBSVM web site; for mnist we have used Scikit-Learn to fetch it from mldata. We have put an emphasis in relatively large, multiclass datasets; in Table 1 we give their dimension, total number of train patterns, number of classes and their maximum class size ratios, i.e., the ratio of the maximum class size to the minimum one. Data sizes go from 2,000 (dna) to 78,823 (combined) and the number of classes ranges from 2 to 26 (letter); while some of them are quite balanced, others (w7a, w8a and particularly shuttle) present large class imbalances, having class size ratios $\gg 1$. Because of this, the main quality measure we will use for model evaluation will be the g-score, i.e., the geometric mean of the different class sensitivities:

$$g = \left(\prod_{c=1}^{C} S_c \right)^{1/C} = \left(\prod_{1}^{C} \frac{m_{cc}}{\sum_j m_{cj}} \right)^{1/C} ,$$

where m_{cj} is the (c, j) entry of the confusion matrix, that is, the number of class c patterns that are assigned to class j. The g-score measure is often used in imbalanced classification as it is more robust to markedly different class sizes than accuracy, easily achieved by assigning small class patterns to the largest class. Because of this we will also use g as the merit function for hyper-parameter selection of both linear and deep models. Nevertheless, we shall also report the accuracies $a = \frac{\sum_c m_{cc}}{\sum_{c,j} m_{cj}}$.

4.2 Deep Model Universal Approximation Capabilities

The goal when using deep versions of FLDA is to obtain better representations in the last hidden layer of which the final, Fisher-like, linear transform can take advantage. In an extreme perspective, overfitting should concentrate each class around its mean while keeping these classes far apart. Once vanishing gradients are avoided and proper training is possible, the simplest way to overfit a dataset is to work with a deep enough network with rather large hidden layers, which ensures a large number of weights. We will do so here, building for each dataset four DFDA models having between 2 and 5 fully connected hidden layers with 100 units each but controlling overfit with a proper penalty. For most problems the number of weights approximately varies thus between 20,000 and 50,000; on the other hand, given its pattern dimension, for mnist the first hidden layer already has 78,400 weights. We stress that our main goal here is not to obtain top quality models; for instance, for mnist, convolutional networks would be needed for this and the fully connected layers be much larger than the ones considered here. Instead, our main goal here is to measure DFDA's performance and compare it against that of FLDA.

As a benchmark reference we will build an LSDA model for each dataset using the class Ridge in scikit-learn; for DFDA models we will use the MLPRegressor class also in scikit-learn. MLPRegressor only allows for deep MLPs with fully connected layers and L_2 penalization; we will use relu activations and the adam solver. This solver is a faster, more stable version of gradient descent but convergence may still be slow. Moreover, the targets Y_{tr} are rather small to begin with, so the convergence tolerance should also be small. Because of this we will work with a maximum number of 20,000 iterations and use a tolerance of 10^{-9}; other solver parameters are left at their default values. A more powerful alternative could have been to work with a general DNN framework such as Keras [3] that has Theano or TensorFlow as backends and offers a much wider range of network architectures (including for instance convolutional layers) or penalties (L_1 or dropout). However, most of the problems considered do not lend themselves to, say, using convolutional layers and, on the other hand, the structural simplicity of MLPRegressor models results in a much faster training.

We will work with mini-batch sizes min(200, num_patterns), i.e., the default for the MLPRegressor class. Therefore, the only hyper-parameter we have to set is the L_2 penalty alpha of the LSDA and DFDA models. For both cases we will select it using the RandomizedSearchCV model selection framework in scikit-learn. To select the optimal alpha values, we will perform 100 uniform random searches of alpha values in a range $(0, \alpha_{\max})$, averaging for each one its g scores over 10 cross validation folds built on the training set and retaining the value giving the largest validation score. The test g-scores are reported in Table 2 and the accuracies in Table 3. In each case the reported test g and accuracy values are obtained training 10 DLDA models with different random initializations, averaging their outputs as well as those of the corresponding test patterns and computing the class predictions and the test confusion matrix over these averages. The tables also give the rankings of each model over the different problems.

As it can be seen, the deep DFDA models clearly improve on the LSDA ones in terms of g scores and accuracies, with the best results usually obtained with the largest 5-hidden layer network. This is particularly remarkable on the

Table 2. Test g scores of the LSDA and of the DFDA models with 2, 3, 4 and 5 100 unit, hidden layers. We write in parenthesis the g score ranking of each model over the different problems and the corresponding ranking means in the last line.

Problem	LSDA	DFDA			
		2 HL	3 HL	4 HL	5 HL
combined	0.780 (5)	0.800 (3)	0.791 (4)	0.817 (2)	**0.831** (1)
dna	0.940 (5)	0.952 (3)	0.952 (3)	**0.955** (1)	0.953 (2)
ijcnn1	0.772 (5)	0.826 (3)	0.809 (4)	0.946 (2)	**0.955** (1)
letter	0.683 (5)	0.924 (4)	0.936 (3)	0.952 (2)	**0.957** (1)
mnist	0.870 (5)	0.929 (4)	0.947 (3)	0.952 (2)	**0.962** (1)
pendigits	0.805 (5)	0.975 (4)	0.976 (3)	0.978 (2)	**0.981** (1)
satimage	0.808 (5)	0.883 (2)	**0.887** (1)	0.876 (3)	0.873 (4)
shuttle	0.565 (3)	0.000 (5)	0.414 (4)	0.838 (2)	**0.979** (1)
w7a	0.794 (3)	0.694 (5)	0.776 (4)	**0.852** (1)	0.849 (2)
w8a	0.781 (5)	**0.837** (1)	0.827 (3)	0.827 (3)	0.831 (2)
usps	0.870 (4)	0.943 (2)	**0.949** (1)	0.939 (3)	0.939 (3)
rank mean	4.55	3.27	3.00	2.09	1.73

Table 3. Test accuracies of the LSDA and of the DFDA models with 2, 3, 4 and 5 100 unit, hidden layers. Again, we write in parenthesis the accuracy ranking of each model over the different problems and the corresponding ranking means in the last line.

Problem	LSDA	DFDA			
		2 HL	3 HL	4 HL	5 HL
combined	0.770 (5)	0.791 (3)	0.782 (4)	0.808 (2)	**0.819** (1)
dna	0.927 (5)	0.957 (3)	0.954 (4)	**0.960** (1)	0.958 (2)
ijcnn1	0.855 (5)	0.891 (3)	0.883 (4)	0.977 (2)	**0.984** (1)
letter	0.694 (5)	0.925 (4)	0.937 (3)	0.952 (2)	**0.957** (1)
mnist	0.873 (5)	0.931 (4)	0.947 (3)	0.952 (2)	**0.962** (1)
pendigits	0.825 (5)	0.975 (4)	0.976 (3)	0.978 (2)	**0.981** (1)
satimage	0.835 (5)	0.891 (2)	**0.897** (1)	0.885 (3)	0.882 (4)
shuttle	0.913 (5)	0.943 (3)	0.934 (4)	**0.989** (1)	0.986 (2)
w7a	0.984 (3)	0.843 (5)	0.978 (4)	0.988 (2)	**0.989** (1)
w8a	0.984 (5)	0.988 (2)	0.988 (2)	0.987 (4)	**0.989** (1)
usps	0.883 (5)	0.949 (2)	**0.954** (1)	0.946 (3)	0.944 (4)
rank mean	4.82	3.18	3.00	2.18	1.73

`shuttle` problem. Notice in Table 2 the g score of the 2-hidden layer network is 0, due to no pattern in the smallest class being correctly classified (this class has about 50,000 times less patterns than the biggest one).

5   Discussion and Further Work

While elegant and enticing, classical Fisher Linear Discriminant Analysis (FLDA) has fallen into some disuse, partly because of its linear nature but also because of the eigenanalysis it requires, which doesn't lend itself to be considered over very large datasets or to be learned in an iterative basis. Most of these difficulties are greatly alleviated when instead of a "pure" FLDA approach, one follows the equivalent LSR set-up proposed in [15,16] and discussed above. Moreover, this lends itself into a natural extension to a Deep Neural Network setting, pairing the LSR target matrix with a highly complex deep pattern processing.

This is our approach here, where we have shown how simple 2-to-5 layer networks can noticeably improve the performance of FLDA. We have applied some of the latest tools in deep networks, such as Glorot initialization [7], RELU activations [11] or ADAM optimizers [10] but, in any case, the networks considered are relatively small and rather simple. There are thus several venues we can follow to improve on the results reported here. For instance, we can use other, more specialized, DNN architectures, easily available through the keras wrapper for Theano or TensorFlow, which may include convolutional layers for highly structured inputs such as mnist. These layers could also be helpful in problems such as person identification, where the current state of the art are the Fisher Face procedures, that apply a suitable version of Fisher analysis over face images. We could also consider DFDA networks as representation learners, using the network outputs of the last hidden layer as features upon which classifiers stronger than nearest neighbors or class mean distances could be applied. This may be particularly suitable for imbalanced problems. In this line it is also interesting to compare the performance of Deep Fisher networks with that of other non linear Fisher extensions, such as KDA. We are currently studying these issues.

Acknowledgments. With partial support from Spain's grants TIN2013-42351-P, TIN2016-76406-P, TIN2015-70308-REDT and S2013/ICE-2845 CASI-CAM-CM. Work supported also by project FACIL-Ayudas Fundación BBVA a Equipos de Investigación Científica 2016, the UAM-ADIC Chair for Data Science and Machine Learning and Instituto de Ingeniería del Conocimiento. The third author is also supported by the FPU-MEC grant AP-2012-5163. We gratefully acknowledge the use of the facilities of Centro de Computación Científica (CCC) at UAM.

References

1. Belhumeur, P.N., Hespanha, J.P., Kriegman, D.J.: Eigenfaces vs. fisherfaces: recognition using class specific linear projection. IEEE Trans. Pattern Anal. Mach. Intell. **19**(7), 711–720 (1997). http://dx.doi.org/10.1109/34.598228
2. Bergstra, J., Breuleux, O., Bastien, F., Lamblin, P., Pascanu, R., Desjardins, G., Turian, J., Warde-Farley, D., Bengio, Y.: Theano: a CPU and GPU math expression compiler. In: Proceedings of Python for Scientific Computing Conference (SciPy), Oral Presentation, Jun 2010
3. Chollet, F.: Keras: deep learning library for Theano and TensorFlow (2015). https://github.com/fchollet/keras

4. Duda, R.O., Hart, P.E., Stork, D.G.: Pattern Classification, 2nd edn. Wiley, Hoboken (2000)
5. Friedman, J.H.: Regularized discriminant analysis. J. Am. Stat. Assoc. **84**(405), 165–175 (1989)
6. Fukunaga, K.: Introduction to Statistical Pattern Recognition. Computer Science and Scientific Computing. Academic Press, Boston (1990)
7. Glorot, X., Bengio, Y.: Understanding the difficulty of training deep feedforward neural networks. In: JMLR W&CP: Proceedings of 13th International Conference on Artificial Intelligence and Statistics (AISTATS 2010), vol. 9, pp. 249–256, May 2010
8. González, A.M., Dorronsoro, J.R.: Natural learning in NLDA networks. Neural Netw. **20**(5), 610–620 (2007). http://dx.doi.org/10.1016/j.neunet.2006.09.014
9. Google: Tensorflow, an open source software library for machine intelligence. https://www.tensorflow.org/
10. Kingma, D.P., Ba, J.: Adam: a method for stochastic optimization. CoRR abs/1412.6980 (2014). http://arxiv.org/abs/1412.6980
11. Krizhevsky, A., Sutskever, I., Hinton, G.E.: ImageNet classification with deep convolutional neural networks. In: Pereira, F., Burges, C., Bottou, L., Weinberger, K. (eds.) Advances in Neural Information Processing Systems, vol. 25, pp. 1097–1105. Curran Associates, Inc. (2012). http://papers.nips.cc/paper/4824-imagenet-classification-with-deep-convolutional-neural-networks.pdf
12. Mika, S., Rätsch, G., Weston, J., Schölkopf, B., Smola, A.J., Müller, K.: Invariant feature extraction and classification in kernel spaces. In: Advances in Neural Information Processing Systems, NIPS Conference, Denver, Colorado, USA, 29 November – 4 December 1999, vol. 12, pp. 526–532 (1999)
13. Park, C.H., Park, H.: A relationship between linear discriminant analysis and the generalized minimum squared error solution. SIAM J. Matrix Anal. Appl. **27**(2), 474–492 (2005). http://dx.doi.org/10.1137/040607599
14. Santa Cruz, C., Dorronsoro, J.R.: A nonlinear discriminant algorithm for feature extraction and data classification. IEEE Trans. Neural Netw. **9**(6), 1370–1376 (1998). http://dx.doi.org/10.1109/72.728388
15. Ye, J.: Least squares linear discriminant analysis. In: Proceedings of 24th International Conference on Machine Learning (ICML 2007), Corvallis, Oregon, USA, 20–24 June 2007, pp. 1087–1093 (2007)
16. Zhang, Z., Dai, G., Xu, C., Jordan, M.I.: Regularized discriminant analysis, ridge regression and beyond. J. Mach. Learn. Res. **11**, 2199–2228 (2010)

An Iterated Greedy Algorithm for Improving the Generation of Synthetic Patterns in Imbalanced Learning

Francisco Javier Maestre-García[1], Carlos García-Martínez[1],
María Pérez-Ortiz[2], and Pedro Antonio Gutiérrez[1(✉)]

[1] Department of Computer Science and Numerical Analysis, University of Córdoba,
Campus de Rabanales, C2 building, 14071 Córdoba, Spain
{i12magaf,cgarcia,pagutierrez}@uco.es
[2] Department of Quantitative Methods, Universidad Loyola Andalucía,
Córdoba, Spain
mariaperez@uloyola.es

Abstract. Real-world classification datasets often present a skewed distribution of patterns, where one or more classes are under-represented with respect to the rest. One of the most successful approaches for alleviating this problem is the generation of synthetic minority samples by convex combination of available ones. Within this framework, adaptive synthetic (ADASYN) sampling is a relatively new method which imposes weights on minority examples according to their learning complexity, in such a way that difficult examples are more prone to be over-sampled. This paper proposes an improvement of the ADASYN method, where the learning complexity of these patterns is also used to decide which sample of the neighbourhood is selected. Moreover, to avoid suboptimal results when performing the random convex combination, this paper explores the application of an iterative greedy algorithm which refines the synthetic patterns by repeatedly replacing a part of them. For the experiments, six binary datasets and four over-sampling methods are considered. The results show that the new version of ADASYN leads to more robust results and that the application of the iterative greedy metaheuristic significantly improves the quality of the generated patterns, presenting a positive effect on the final classification model.

Keywords: Over-sampling · Imbalanced classification · ADASYN · Iterative greedy algorithm · Metaheuristics

1 Introduction

Learning from imbalanced data represents one of the current challenges in machine learning. In classification domains, imbalanced distributions occur when

This work has been partially subsidised by the TIN2014-54583-C2-1-R, TIN2015-70308-REDT, and TIN2014-55252-P projects of the Spanish Ministerial Commission of Science and Technology (MINECO, Spain) and FEDER funds (EU).

I. Rojas et al. (Eds.): IWANN 2017, Part II, LNCS 10306, pp. 513–524, 2017.
DOI: 10.1007/978-3-319-59147-6_44

one or more classes have a significantly higher a-priori-probability [16,17]. The difficulty of learning from imbalanced data is that classifiers will often assume that the classes are equally represented in the dataset [26]. Consequently, standard classifiers will be biased towards the majority class, significantly harming the performance of the minority one [8]. Nonetheless, in most cases, rare objects (or minority class samples) will be of great interest and should be the focus of machine learning algorithms [15], e.g. in financial engineering, where it is crucial to detect fraudulent credit card activities from a pool of large transactions [3]. Note however, that an uneven data distribution is not the only factor that hinders the learning in these cases [16,17], the complexity of the data or the size of the training set being also determining factors.

Imbalanced learning is an active field, and there is a wide range of techniques proposed for improving the minority class sensitivity. For a detailed survey of these, we refer the reader to [16]. Two main groups of methods can be emphasized: (1) data preprocessing techniques, where the class priors are changed by under-sampling the majority class, over-sampling the minority one [2,4] or modifying class priors by changing class labels [6]; or (2) specific learners, where the classifier is forced to pay more attention to the minority class [7,29]. Hybrid methods can also be found in the literature and have shown an outstanding performance for this matter. These methods usually combine data and algorithm-level techniques [8,21], e.g. traditional ensemble methods (such as bagging and boosting) [8] which combine resampling with multiple learning models. Although these techniques were not originally proposed to address imbalanced problems, their relatively superior performance [8,22,28,30] enabled ensemble methods to gain attention within this topic. Finally, there have been some recent and successful attempts to tackle class imbalance using ranking algorithms [5].

The analysis made in this paper is mainly contextualised on data approaches, as these present several advantages over algorithmic solutions [11], all basically stemming from the fact that this approach does not rely on the reformulation of a certain classification algorithm. The most straightforward resampling idea would be repetitive over-sampling, which simply replicates existing data points (in a similar way to cost-sensitive learning). However, generative sampling is usually preferred, where the sparse data space is populated with new data points, producing ideally a more dense, smooth and uniformly distributed dataset [21]. Although both over-sampling and under-sampling are widely used, some studies [17] emphasize over-sampling methods for complex and highly imbalanced datasets.

Concerning generative sampling, the most popular method is the Synthetic Minority Over-sampling TEchnique (SMOTE) [4], which is based on a random interpolation between minority class data points (between a randomly chosen pattern and one of its k nearest neighbours). This algorithm presents several important handicaps, all of them related to the omission of the majority class in the over-sampling process, which, in some cases, depending on the data distribution, can result in a set of synthetic patterns that lie in the majority class region and that hinders the learning process. Because of this, there have been different proposals over the years to improve SMOTE: borderline-SMOTE [13], which focuses on sampling only those minority data points close to the classification decision boundary; cluster-oversampling [18], which considers the so-called 'rare' regions, which are resampled individually; or safe-level SMOTE [2] and LN-SMOTE [24], which generate new synthetic examples in the direction of the regions populated by the minority class, to avoid introducing artificial examples within majority class regions.

ADASYN [15] is also a relatively new approach which refines SMOTE by focusing on those patterns with a higher learning complexity (i.e. those closer to the class boundary). As a result, ADASYN adaptively shifts the focus of the classification model towards more difficult examples, showing promising results [15]. Although more complex examples are resampled with a higher probability, the interpolation process of ADASYN still relies on a random selection of one of the k nearest neighbours of the pattern, which, depending on the dataset and the choice of k, can lead to the generation of minority samples in the majority-class region (see Fig. 1), i.e. one the main problems with SMOTE still remains partially unsolved when using ADASYN. As can be checked in this Figure, new synthetic patterns could be generated in regions associated to the majority class, depending on the choice of the k parameter for the nearest neighbour analysis. Different k parameters could be chosen for different patterns, which emphasizes the necessity of using an optimisation algorithm.

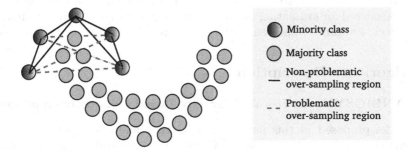

Fig. 1. Imbalanced toy dataset with problems associated to standard oversampling.

In this sense, over-sampling data-level approaches can be formulated as an optimisation problem to further refine the data distribution and avoid the above-mentioned class inconsistencies. In particular, the generation of synthetic samples can be defined as follows: given an imbalanced dataset $D = \{D^m \cup D^M\}$, with D^m being the set of examples of the minority class and D^M the set of examples of the majority class (or examples of other classes; $|D^m| \ll |D^M|$), the goal is to generate a set of synthetic examples S such that the performance of a classifier trained on $D \cup S$ is maximised. This problem can be effectively addressed by metaheuristic algorithms. However, to the best of our knowledge, there are very few works in this direction. For example, the work of

Ghazikhani et al. [12] presents an evolutionary algorithm that evolves the optimal regions to apply over-sampling. On the other hand, under-sampling can be also approached by using a binary genetic algorithm for selecting which majority patterns are used for learning, and there are different works exploiting this idea [23]. Finally, Wong et al. [32] used an evolutionary algorithm for simultaneously applying under-sampling and over-sampling.

Iterated Greedy (IG) [10,27] is a simple metaheuristic that, in contrast to evolutionary methods, specifically incorporates heuristic information to address the problem at hand. This metaheuristic is based on the application of two operations on the best found solution: *destruction* and *construction*. During the former one, some solution components are removed, generating a partial solution. In the latter, a heuristic greedy procedure completes the partial solution. The process is repeated until a stopping condition is met. This paper proposes an iterated greedy model to improve the solution of over-sampling algorithms in imbalanced datasets. Our method encodes candidate solutions as sets of synthetic samples, where destruction and construction operators apply on. Its heuristic construction operator is an enhanced version of ADASYN that considers the complexity of the patterns also when selecting the neighbour for the convex combination, but could also be applied to other over-sampling strategies. More specifically, from the k nearest neighbours, we select a complex pattern that is, at the same time, close to the pattern to be resampled.

The rest of this paper is structured as follows: Sect. 2 presents the description of the over-sampling algorithm proposed. Section 3 describes the different experiments considered for evaluating the proposal and the corresponding results. And finally, Sect. 4 outlines some concluding remarks.

2   Algorithm Description

2.1   ANEIGSYN: An Iterated Greedy Based Generative Sampler

The model proposed in this paper is an IG that addresses the task of over-sampling as an optimisation problem. Given an imbalanced dataset $D = D^M \cup D^m$, the method searches over the candidate sets of synthetic examples for those from which better classifiers can be induced. The process is divided in four stages:

- *Initialisation:* D is divided into three sets that are subsequently used in their corresponding stages: Tr from which synthetic examples are generated, with 50% of the examples in D, V for guiding the search process by validating the synthetic sets produced, with 30% of the examples in D, and Te for testing the final classifier, with rest of D (20%). Te is never considered for inducing the classifier, but just for evaluation once the search process has finished.
- *Construction:* A candidate set of synthetic examples S is built according to the heuristic described in Sect. 2.2 and using the examples in $Tr \cup V$. This stage generates new synthetic examples, as many as to balance the number of examples of majority and minority classes in $Tr \cup S$, that are included into S. This stage can also be applied from scratch, i.e. using an empty candidate

set S, or over a partial candidate set that is the output of the Destruction
stage.

- *Validation:* Any new candidate set S is evaluated according to the performance of the classifier induced from $Tr \cup S$ when predicting the label of the examples in V. The considered performance metric is the geometric mean (see Sect. 3 for more details).
- *Destruction:* The best so far candidate set is partially destroyed by removing 15% randomly chosen examples. Then, this partially destroyed solution becomes the input for the construction stage of the next iteration.

Finally, the process finishes after a maximum number of iterations, 100 in our case, and the best obtained classifier is evaluated on Te.

2.2 Improved Version of ADASYN: Adaptive NEighbours Synthetic Sampling (ANESYN)

The heuristic procedure that generates the initial solution and reconstructs every partial one is at the core of every IG algorithm, given that it exploits problem knowledge to produce better solutions than random ones [9].

As stated before, one of the most useful sources of information to generate synthetic examples is the proximity of the solutions to the decision boundary from which synthetic patterns are generated, which can help the learner to approximate it accurately, and how the selected generative solutions are combined, because inappropriate pairings may favour the generation of synthetic examples in non-appropriate regions (see Fig. 1). Therefore, our proposed improved version of ADASYN emphasizes these two aspects, producing synthetic solutions from original solutions that are expected to be close to the decision boundary and close to each other. Its main scheme is:

1. The proximity to the decision boundary of each minority example $s_i \in G^m$ is estimated according to the equation ($G = Tr \cup V$):

$$\widehat{r_i} = \frac{r_i}{\sum_j^{|G^m|} r_j}, \text{ where } r_i = |\{s : s \in G^M \text{ and } s \in N_{G \cup S_p}(s_i, k)\}|/k,$$

   where $N_{G \cup S_p}(s_i, k)$ is the set of the k samples from $G \cup S_p$ (i.e. the set of samples used for generation of patterns plus the synthetic patterns already generated but not removed from the solution) closest to s_i.
2. The number of times each minority sample will participate as the main generative sample is computed according to its estimated proximity to the decision boundary, as $g_i = \lfloor \widehat{r_i} \cdot N \rfloor$. Given that we use the floor operator, $\lfloor \cdot \rfloor$, the required number of synthetic patterns is probabilistically completed increasing the number of times minority examples participate according to the remainders ($r_i \cdot N \mod \sum_j^{|G^m|} r_j$).
3. For each time a minority sample s_i is used for generating a synthetic sample, a secondary original pattern $s_j \in N_G(s_i, k)$ (note that in this case we

discard already generated synthetic patterns) is selected in accordance to the probabilities p_j:

$$p_j = \varepsilon_j / \sum_k^{|G^m|} \varepsilon_k, \text{ and } \varepsilon_j = \alpha \widehat{r'_j} + (1 - \alpha)/d(s_i, s_j),$$

where $\widehat{r'_j}$ is $\widehat{r_j}$ normalised according to the sum of $\widehat{r_j}$ for those patterns in the vicinity $N_G(s_i, k)$, and $d(s_i, s_j)$ is the euclidean distance between the examples, also normalised by the sum of the same quantity for the elements in $N_G(s_i, k)$.

4. Finally, a new synthetic example is generated per pair of generative samples $\{s_i, s_j\}$ as the linear interpolation $s_i + \lambda \cdot (s_j - s_i)$, where λ is a random value in $(0, 1)$.

Note that the proximity of every minority example (Step 2) is computed considering both the original examples and the synthetic ones that have not been removed from the current solution ($G \cup S_p$), so minority examples whose vicinity has been populated with many synthetic ones will be nominated as main generative examples less times than previously.

Additionally, instead of fixing the number of nearest neighbours analysed (see Step 1, parameter k) for all datasets, we have considered an adaptive method. The value of k considered for the algorithm is the minimum value for which there is at least one majority sample among the k closest ones for each minority example.

3 Experiments

This section presents the different experiments considered for evaluating the performance of the method proposed. We have tested our algorithm on six real-world machine learning datasets, which can be downloaded from the UCI machine learning repository[1] [20] and Keel repository[2] [1]. The battery of datasets is described in Table 1 and is similar to that considered in the original ADASYN paper [15], but we have also included the **page** dataset. Table 1 includes the imbalance ratio (IR), which is the ratio between the number of patterns of the majority class and that of the minority one.

For the experimental design a holdout procedure is performed 100 times, i.e. 100 training/test partitions are randomly performed, where, as previously discussed, 50% of data is used as training while the remaining 50% is considered for test purposes. All the partitions are stratified, in such a way that the original class distribution of the dataset is approximately maintained in the partitions. Since the ANEIGSYN algorithm is based on a validation set, the 50% of training data is also partitioned into two stratified sets, as detailed in Sect. 2.1.

---

[1] https://archive.ics.uci.edu/ml/datasets.html.
[2] http://sci2s.ugr.es/keel/imbalanced.php.

Table 1. Characteristics of the datasets used in this paper

Dataset	# Patterns	# Minority patterns	# Majority patterns	IR	# Attributes
abalone	731	42	689	16.40	7
ionos	351	126	225	1.79	34
page	5472	559	4913	8.79	10
pima	768	268	500	1.87	8
vowel	990	90	900	10.00	10
vehicle	846	199	647	3.25	18

All the experiments have been performed using a decision tree as base classifier, given that it is one of the most popular classification methods and has been also considered for evaluating other over-sampling algorithms [13,15]. We have used the implementation included in the Python scikit-learn machine learning framework [25], where an optimised version of the CART algorithm is considered for the tree induction. This implementation includes heuristic algorithms, where locally optimal decisions are made at each node. This makes the induction process non deterministic, and different trees can be obtained according to the seed used for random number generation.

Different preprocessing methods have been compared in this paper:

- A decision tree directly obtained from the original dataset, without preprocessing (Original).
- The well known SMOTE algorithm [4].
- The standard ADASYN method, as described in [15], where the difficulty of classifying each minority class pattern is considered for selecting the patterns to be resampled.
- The ANESYN method consists of the application of the improved version of ADASYN, without using the proposed IG algorithm (see Sect. 2.2).
- The application of the improved ADASYN together with the iterative refinement of the generated interpolation (ANEIGSYN, see Sect. 2.1).

The number of nearest neighbours considered is $k = 5$ for all methods, except ANESYN and ANEIGSYN, which automatically select the value of k as explained in Sect. 2.2.

The results have been reported in terms of two metrics, both specially designed to deal with imbalanced data (in both cases, the positive class is the minority one):

- The Geometric Mean of the sensitivities ($GM = \sqrt{S_p \cdot S_n}$) [19], where S_p is the sensitivity for the positive class (ratio of correctly classified patterns considering only this class) and S_n is the sensitivity for the negative one.
- The Area Under the ROC curve (AUC) [14]. Although receiver operating characteristic (ROC) graphs are useful tools for analysing the performance of classifiers, these graphs need to be reduced to single scalar values to ease their comparison. The most common method is to compute the area under the curve (AUC), which has been shown to be equivalent to the probability

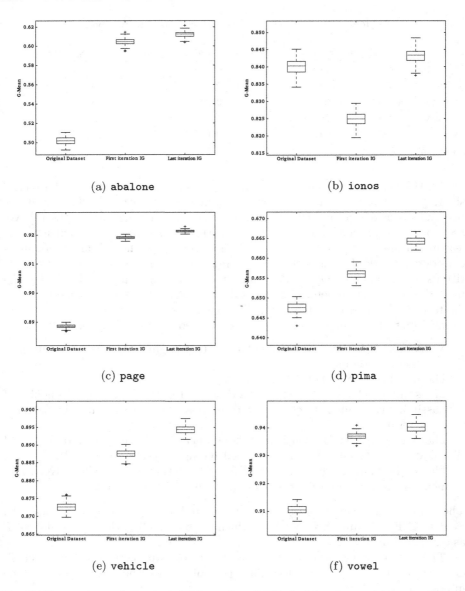

Fig. 2. Comparison of the test GM results of 150 decision trees obtained without preprocessing the data (Original) or by including the synthetic data generated by ANEIGSYN for the first and the last iterations. These results are averaged over 100 executions of the algorithm.

that a classifier assigns a higher score to a randomly chosen positive pattern than to a negative one.

A first experiment was performed to check whether the iterative process of ANEIGSYN resulted in better suited synthetic patterns. Figure 2 includes different boxplots comparing the test GM results obtained by generating 150 decision trees using the original dataset (Original) and complementing this dataset with the synthetic data generated by ANEIGSYN, before applying the destruction/construction iterative process ('First iteration IG' in the figure), and after the iterative process ('Last iteration IG' in the figure). As can be checked, the generation of synthetic data improves the results obtained by using the original data for all datasets, except for `ionos`, probably due to the relatively low degree of imbalance of this dataset ($IR = 1.79$). However, the IG algorithm improves the results of the trees consistently for all datasets if we compare the first and last iterations. These results confirm that the good performance obtained by the IG algorithm is not due to the lucky chance of the seed considered for the generation of the decision tree but to the quality of the synthetic patterns generated by the method.

Additionally, the results of the different algorithms can be checked in Table 2, where the average test GM and AUC performances are included. We also obtain the ranking of each method in each dataset according to test GM and AUC ($R = 1$ for the best performing method and $R = 5$ for the worst one) and the average rankings are included in Table 2. In order to check if the differences

Table 2. Average test performance values obtained from the 100 executions and Wilcoxon statistical test results

	Original	SMOTE	ADASYN	ANESYN	ANEIGSYN
GM					
abalone	0.54222	*0.58094*	0.57606	0.56608	**0.60565**
ionos	0.85164	**0.85273**	0.84114	0.84343	*0.85212*
page	0.89739	0.91108	0.90436	*0.91825*	**0.92186**
pima	0.65718	0.65778	0.65964	*0.66284*	**0.66618**
vehicle	0.89159	0.89003	*0.89741*	0.89516	**0.89772**
vowel	0.93069	0.93156	0.92885	*0.93799*	**0.94233**
$\overline{R}_{GM}$	4.33	3.00	3.67	*2.83*	**1.17**
Wilcoxon p	0.028	0.028	0.046	0.028	–
AUC					
abalone	0.63602	0.64108	*0.64777*	0.64159	**0.65864**
ionos	*0.85471*	**0.85588**	0.83731	0.84536	0.85396
page	0.90107	0.91145	0.90686	*0.91913*	**0.92230**
pima	0.66566	0.66599	0.66507	*0.66660*	**0.66880**
vehicle	0.89385	0.89228	0.89132	*0.89679*	**0.89896**
vowel	0.93328	0.93341	0.93102	*0.93939*	**0.94342**
$\overline{R}_{AUC}$	3.83	3.00	4.33	*2.50*	**1.33**
Wilcoxon p	0.046	0.046	0.028	0.028	–

The best result is in bold face and the second one in italics

522   F.J. Maestre-García et al.

found are significant, a Wilcoxon statistical test [31] has been considered. The
corresponding p-values of the tests comparing ANEIGSYN against each of the
other methods are also shown in Table 2.

From the results of this Table, we can first conclude that better results are
obtained by applying SMOTE than by applying ADASYN. Moreover, the two
proposals of this paper (ANESYN and ANEIGSYN) generally improved the
performance of ADASYN. Firstly, the improvement of the ADASYN method
implemented (ANESYN) results in much better ranking values than standard
ADASYN, both for GM and AUC. Indeed, the results of ANESYN are the
second best ones from all the methods compared. Secondly, we can clearly conclude that the ANEIGSYN method is the best performing one in GM and AUC,
although slightly better results are obtained for GM. Moreover, the statistical
tests confirm these results (all differences favouring ANEIGSYN are significant,
p-values < 0.05), showing that the differences are not due to random nature of
the algorithms.

4   Conclusions

This paper presents two main contributions: (1) a new improved version of the
ADASYN method [15], based on a better selection of the secondary pattern from
which the synthetic example is generated, and (2) an iterated greedy algorithm
for repeatedly destructing and reconstructing the set of synthetic patterns to
improve the performance of the over-sampling process. From the results obtained
by evaluating six benchmark imbalanced datasets and two baseline over-sampling
methods (SMOTE [4] and ADASYN [15]), the improved version of ADASYN is
seen to yield superior performance than the original one, and the iterated greedy
algorithm is able to improve all the results. As future research lines, we would like
to test the use of probabilistic classifiers together with continuous performance
metrics for evaluating the quality of the synthetic patterns, which we think could
enhance the convergence of the algorithm.

References

1. Alcalá, J., Fernández, A., Luengo, J., Derrac, J., García, S., Sánchez, L., Herrera,
   F.: Keel data-mining software tool: data set repository, integration of algorithms
   and experimental analysis framework. J. Multiple-Valued Log. Soft Comput.
   17(2–3), 255–287 (2010)
2. Bunkhumpornpat, C., Sinapiromsaran, K., Lursinsap, C.: Safe-level-SMOTE: safe-
   level-synthetic minority over-sampling technique for handling the class imbalanced
   problem. In: Theeramunkong, T., Kijsirikul, B., Cercone, N., Ho, T.-B. (eds.)
   PAKDD 2009. LNCS (LNAI), vol. 5476, pp. 475–482. Springer, Heidelberg (2009).
   doi:10.1007/978-3-642-01307-2_43
3. Chan, P.K., Fan, W., Prodromidis, A.L., Stolfo, S.J.: Distributed data mining in
   credit card fraud detection. IEEE Intell. Syst. Appl. **14**(6), 67–74 (1999)
4. Chawla, N.V., Bowyer, K.W., Hall, L.O., Kegelmeyer, W.P.: SMOTE: synthetic
   minority over-sampling technique. J. Artif. Intell. Res. **16**, 321–357 (2002)

5. Cruz, R., Fernandes, K., Cardoso, J.S., Costa, J.F.P.: Tackling class imbalance with ranking. In: 2016 International Joint Conference on Neural Networks (IJCNN), pp. 2182–2187. IEEE (2016)

6. Domingos, P.: Metacost: a general method for making classifiers cost-sensitive. In: Proceedings of 5th ACM SIGKDD International Conference on Knowledge Discovery and Data Mining, pp. 155–164. ACM (1999)

7. Fernández-Caballero, J.C., Martínez-Estudillo, F.J., Hervás-Martínez, C., Gutiérrez, P.A.: Sensitivity versus accuracy in multiclass problems using memetic pareto evolutionary neural networks. IEEE Trans. Neural Netw. 21(5), 750–770 (2010)

8. Galar, M., Fernández, A., Barrenechea, E., Bustince, H., Herrera, F.: A review on ensembles for the class imbalance problem: bagging-, boosting-, and hybrid-based approaches. IEEE Trans. Syst. Man Cybern. Part C Appl. Rev. 42(4), 463–484 (2012)

9. García-Martínez, C., Lozano, M., Rodriguez, F.J.: Arbitrary function optimization. No free lunch and real-world problems. Soft. Comput. 16(12), 2115–2133 (2012)

10. García-Martínez, C., Rodriguez, F.J., Lozano, M.: Tabu-enhanced iterated greedy algorithm: a case study in the quadratic multiple knapsack problem. Eur. J. Oper. Res. 232, 454–463 (2014)

11. Garcia-Pedrajas, N., Pérez-Rodríguez, J., de Haro-García, A.: OligoIS: scalable instance selection for class-imbalanced data sets. IEEE Trans. Cybern. 43(1), 332–346 (2013)

12. Ghazikhani, A., Yazdi, H.S., Monsefi, R.: Class imbalance handling using wrapper-based random oversampling. In: 20th Iranian Conference on Electrical Engineering (ICEE 2012), pp. 611–616. IEEE (2012)

13. Han, H., Wang, W.-Y., Mao, B.-H.: Borderline-SMOTE: a new over-sampling method in imbalanced data sets learning. In: Huang, D.-S., Zhang, X.-P., Huang, G.-B. (eds.) ICIC 2005. LNCS, vol. 3644, pp. 878–887. Springer, Heidelberg (2005). doi:10.1007/11538059_91

14. Hanley, J.A., McNeil, B.J.: The meaning and use of the area under a receiver operating characteristic (ROC) curve. Radiology 143(1), 29–36 (1982)

15. He, H., Bai, Y., Garcia, E.A., Li, S.: ADASYN: Adaptive synthetic sampling approach for imbalanced learning. In: International Joint Conference on Neural Networks (IJCNN), pp. 1322–1328 (2008)

16. He, H., Garcia, E.A.: Learning from imbalanced data. IEEE Trans. Knowl. Data Eng. 21(9), 1263–1284 (2009)

17. Japkowicz, N., Stephen, S.: The class imbalance problem: a systematic study. Intell. Data Anal. 6(5), 429–449 (2002)

18. Jo, T., Japkowicz, N.: Class imbalances versus small disjuncts. ACM SIGKDD Explor. Newsl. 6(1), 40–49 (2004)

19. Kubat, M., Matwin, S.: Addressing the curse of imbalanced training sets: one-sided selection. In: Proceedings of 14th International Conference on Machine Learning, pp. 179–186. Morgan Kaufmann (1997)

20. Lichman, M.: UCI machine learning repository (2013). http://archive.ics.uci.edu/ml

21. Lim, P., Goh, C.K., Tan, K.C.: Evolutionary cluster-based synthetic oversampling ensemble (eco-ensemble) for imbalance learning. IEEE Trans. Cybern. 99, 1–12 (2016)

22. Liu, X.Y., Wu, J., Zhou, Z.H.: Exploratory undersampling for class-imbalance learning. IEEE Trans. Syst. Man Cybern. Part B 39(2), 539–550 (2009)

23. Luengo, J., Fernández, A., García, S., Herrera, F.: Addressing data complexity for imbalanced data sets: analysis of smote-based oversampling and evolutionary undersampling. Soft. Comput. **15**(10), 1909–1936 (2011)
24. Maciejewski, T., Stefanowski, J.: Local neighbourhood extension of smote for mining imbalanced data. In: 2011 IEEE Symposium on Computational Intelligence and Data Mining (CIDM), pp. 104–111. IEEE (2011)
25. Pedregosa, F., Varoquaux, G., Gramfort, A., Michel, V., Thirion, B., Grisel, O., Blondel, M., Prettenhofer, P., Weiss, R., Dubourg, V., et al.: Scikit-learn: machine learning in Python. J. Mach. Learn. Res. **12**(October), 2825–2830 (2011)
26. Pérez-Ortiz, M., Gutiérrez, P.A., Tino, P., Hervás-Martínez, C.: Oversampling the minority class in the feature space. IEEE Trans. Neural Netw. Learn. Syst. **27**(9), 1947–1961 (2016)
27. Ruiz, R., Stützle, T.: A simple and effective iterated greedy algorithm for the permutation flowshop scheduling problem. Eur. J. Oper. Res. **177**, 2033–2049 (2007)
28. Seiffert, C., Khoshgoftaar, T.M., Van Hulse, J., Napolitano, A.: Rusboost: a hybrid approach to alleviating class imbalance. IEEE Trans. Syst. Man Cybern.-Part A: Syst. Hum. **40**(1), 185–197 (2010)
29. Thai-Nghe, N., Gantner, Z., Schmidt-Thieme, L.: Cost-sensitive learning methods for imbalanced data. In: International Joint Conference on Neural Networks (IJCNN), pp. 1–8. IEEE (2010)
30. Wang, S., Minku, L.L., Yao, X.: Resampling-based ensemble methods for online class imbalance learning. IEEE Trans. Knowl. Data Eng. **27**(5), 1356–1368 (2015)
31. Wilcoxon, F.: Individual comparisons by ranking methods. Biom. Bull. **1**(6), 80–83 (1945)
32. Wong, G.Y., Leung, F.H., Ling, S.H.: A novel evolutionary preprocessing method based on over-sampling and under-sampling for imbalanced datasets. In: Industrial Electronics Society, IECON 2013–39th Annual Conference of the IEEE, pp. 2354–2359. IEEE (2013)

Fine-to-Coarse Ranking in Ordinal and Imbalanced Domains: An Application to Liver Transplantation

María Pérez-Ortiz[1]([✉]), Kelwin Fernandes[2,3], Ricardo Cruz[3],
Jaime S. Cardoso[2,3], Javier Briceño[4], and César Hervás-Martínez[5]

[1] Department of Quantitative Methods, Universidad Loyola Andalucía, Córdoba,
Spain
mariaperez@uloyola.es
[2] Faculty of Engineering, University of Porto, Porto, Portugal
kafc@inesctec.pt, jaime.cardoso@inesctec.pt

[3] INESC TEC, Porto, Portugal
rpcruz@inesctec.pt
[4] Liver Transplantation Unit, Reina Sofia Hospital, Córdoba, Spain
[5] Department of Computer Science and Numerical Analysis, University of Córdoba,
Córdoba, Spain
chervas@uco.es

Abstract. Nowadays imbalanced learning represents one of the most vividly discussed challenges in machine learning. In these scenarios, one or some of the classes in the problem have a significantly lower a priori probability, usually leading to trivial or non-desirable classifiers. Because of this, imbalanced learning has been researched to a great extent by means of different approaches. Recently, the focus has switched from binary classification to other paradigms where imbalanced data also arise, such as ordinal classification. This paper tests the application of learning pairwise ranking with multiple granularity levels in an ordinal and imbalanced classification problem where the aim is to construct an accurate model for donor-recipient allocation in liver transplantation. Our experiments show that approaching the problem as ranking solves the imbalance issue and leads to a competitive performance.

Keywords: Imbalanced data · Ranking · Ordinal classification · Oversampling

This work has been subsidized by the TIN2014-54583-C2-1-R and the TIN2015-70308-REDT projects of the Spanish Ministerial Commission of Science and Technology (MINECO, Spain), FEDER funds (EU), the PI-0312-2014 project of the "Fundación pública andaluza progreso y salud" (Spain), the PI15/01570 project ("Proyectos de Investigación en Salud"), and also by NanoSTIMA: Macro-to-Nano Human Sensing: Towards Integrated Multimodal Health Monitoring and Analytics/NORTE-01-0145-FEDER-000016.

© Springer International Publishing AG 2017
I. Rojas et al. (Eds.): IWANN 2017, Part II, LNCS 10306, pp. 525–537, 2017.
DOI: 10.1007/978-3-319-59147-6_45

1   Introduction

Liver transplantation, although representing nowadays a widely-accepted and successful treatment for patients who present terminal liver disease, is hampered by the low availability of suitable donors. Several strategies have been considered over the years to construct a system to prioritize recipients on the waiting list and optimise the utility of the organ, but most of them only consider characteristics of either donors or recipients, therefore obviating the potential compatibility between these. This paper proposes a novel donor-recipient liver allocation system based on machine learning by means of a dataset comprised of donor-recipients pairs from different centres (seven Spanish transplantation units and the London King's College hospital).

The problem tackled in this paper is imbalanced, as it is common in biomedicine and real-world applications, meaning this that there exist one or several classes that are under-represented in the dataset, which usually leads the classification strategy to a non-appropriate trivial final model [4,22]. In this case, the dependent variable is the time leading up to graft loss, where patients are monitored for a year after transplantation. The classes of the problem are: less than 15 days, between 15 and 90 days, between 90 and 365 days and more than a year (showing then an ordinal nature). The number of transplants that result in graft loss are in this case, a significant minority with respect to successful transplants, however, these cases are the most interesting ones for constructing a proper model for organ allocation.

To deal with class imbalance, different approaches have been proposed over the years, mainly data-based approaches [4,12] (such as over-sampling the minority class or under-sampling the majority one) or algorithmic approaches [3] (e.g. cost-sensitive learning or post-processing strategies). Since imbalanced data does not only arise in binary standard classification domains, there are different extensions of these initially proposed algorithms to other learning paradigms being researched [19,23].

In the case of ordinal classification, the different categories of the class to predict follow a given ranking order, but not a cardinal order. This must be considered in the different stages of the learning process; such as training the algorithm, measuring the performance and applying preprocessing techniques (e.g. over-sampling or under-sampling).

As said, from all the techniques that have been proposed to deal with class imbalance, over-sampling and under-sampling approaches can be highlighted, because they have been shown to improve classifier performance over imbalanced datasets and do not depend on a specific classification approach. In this sense, previous studies suggest that over-sampling could be more powerful than under-sampling [16], specially for highly imbalanced and small datasets. This could be due to the potential loss of useful information when performing under-sampling. The main problem, however, with over-sampling approaches is that most of these methods assume a distribution of minority samples which may not hold in reality, resulting in synthetic patterns lying in inadequate areas of the dataset [20]. Concerning algorithmic approaches, cost-sensitive learning has been shown to result in over-fitting [10, 20] and although hybrid and ensemble algorithms are common and usually successful, they rely on a specific method and are difficult to optimise. Recently, a novel proposal has joined the set of strategies available to deal with class imbalance, the use of rank learners, which has shown very promising results for this matter [6, 7]. This paper explores the use of this type of algorithms in the context of ordinal and imbalanced data and compares it to other proposals in the literature.

In our case, the use of rankers is natural, given that: (1) the ordinal classes to predict (i.e. that represent the time leading up to graft failure) come from a continuous latent variable which represents the exact number of days[1] and (2) from the definition of the problem tackled our aim is to construct an accurate organ allocation system, that ranks patients according to their suitability for the transplant. This information can be used to construct the model, introducing valuable information about the rank of the patterns, which is lost when using coarse labels. The experiments performed in this paper compare different rank-based learners with several ordinal classifiers and techniques to deal with class imbalance, showing that: (1) the problems that present such highly imbalanced nature require the use of specially designed techniques to avoid trivial classifiers; (2) rank-based learners represent a suitable option to deal with imbalanced data; (3) these strategies show comparable results to the use of other approaches to deal with class imbalance, and (4) the use of fine labels (as opposed to coarse labels) can satisfactorily complement the classification leading to more robust results.

The paper is organised as follows: Sect. 2 shows a description of the dataset and methodology used; Sect. 3 describes the experimental study and analyses the results obtained; and finally, Sect. 4 outlines some conclusions.

2   Materials and Methodology

2.1   Dataset Description

A multi-centred retrospective analysis was made of 7 Spanish Liver transplant units. Recipient, donor, retrieval and transplant characteristics were reported at the time of transplant. Patients undergoing partial, split or living-donor liver transplantation and patients undergoing combined or multi-visceral transplants were excluded from the study. Liver transplantation units were homogeneously distributed throughout Spain. The Spanish dataset constructed has 634 patterns (donor-recipient pairs) corresponding to the years 2007 and 2008. The proportion of combined transplant was 2.3% in both cohorts. The proportion of partial grafts was 0.9% and 9.1% in the Spanish and British cohort, respectively. The few cases of combined transplantation were those of liver and kidney, which, in several series, have been reported to not decrease the outcome of the liver graft.

---

[1] Note, however, that a simple regression analysis is not feasible because of the high number of organs which survived the 365 day threshold (for which, we do not have more information).

In addition, the dataset was completed with information about donor-recipient pairs from the King's College Hospital (London), to perform a supranational study of donor-recipient allocation in liver transplantation. To obtain a similar number of patterns, only reported pairs of recipients over eighteen years of age between January 2002 and December 2010 were included. A dataset containing 858 English donor-recipient pairs was collected. In order to merge the datasets, several variables were selected, 16 recipient variables, 17 donor variables and 5 surgically related variables, as can be seen in [18]. All patients were followed from the date of transplant until either death, graft loss or completion of the first year after the liver transplant. The final dataset was comprised of 1406 patterns.

To solve the donor-recipient matching problem, the dependent variable is the class label which is equal to 1 when representing graft loss up to the first 15 days after the transplant, equal to 2 if the loss occurs between 15 days and 3 months, equal to 3 when the loss is after 3 months and before a year, and, finally, the last class corresponds to the patterns which do not present graft loss after the first year and is represented by label 4. The variables selected for the dataset can be seen in [18].

The choice of class limits for the dataset were not arbitrary (15 days, 3 months and a year); in addition to being considered as the most pertinent, [18] shows that the cumulative frequency slope of the graft loss curve changes strongly somewhere around those class limits. An important point is the limit located at 15 days since it is defined by experts as a critical point for survival or loss. In this case, the application of a regression-based technique is not suitable for the problem, due to the high number of points belonging to the more than 1 year category, which do not incorporate any knowledge about the real value of the number of days until either graft loss or death. The class distribution of the dataset is {76,76,62,1223}, which shows a highly imbalanced nature.

2.2   Methodology

The ordinal regression problem consists of predicting the label y for an input vector $\mathbf{x}$, where $\mathbf{x} \in \mathcal{X} \subseteq \mathbb{R}^d$ and $y \in \mathcal{Y} = \{\mathcal{C}_1, \mathcal{C}_2, \ldots, \mathcal{C}_Q\}$, i.e. $\mathbf{x}$ is in a K-dimensional input space and y is in a label space of Q different labels corresponding to the categories. The objective is to find a classification rule or function $f \colon \mathcal{X} \to \mathcal{Y}$ to predict the labels of new patterns, given a training set of N points, $D = \{(\mathbf{x}_i, y_i), i = 1, \ldots, N\}$. A natural label ordering is included for ordinal regression, $\mathcal{C}_1 \prec \mathcal{C}_2 \prec \ldots \prec \mathcal{C}_Q$, where $\prec$ is an order relation given by the nature of the classification problem. Many ordinal regression measures and algorithms consider the rank of the label, i.e. the position of the label in the ordinal scale, which can be expressed by the function $\mathcal{O}(\cdot)$, in such a way that $\mathcal{O}(\mathcal{C}_q) = q$, $q = 1, \ldots, Q$. The assumption of an order between class labels makes that two different elements of $\mathcal{Y}$ could always be compared by using the relation $\prec$, which is not possible under the nominal classification setting.

Rank Learners for Ordinal Classification. The *multipartite ranking* problem is a generalisation of the well-known bipartite ranking one. ROC analysis, which evaluates the ability of classifiers to sort positive and negative instances in terms of the area under the ROC curve, is a clear example of training a binary classifier to perform well in a bipartite ranking problem. Multipartite ranking can be seen as an intermediate point between ranking and sorting. It is similar to ranking because training patterns are labelled with one of Q ordered ratings ($Q = 2$ for bipartite ranking), but here the goal is to learn from them a ranking function able to induce a total order in accordance with the given training ratings [9,21], which is similar to sorting. The objective of multipartite ranking is to obtain a classifier which ranks "high" classes ahead of "low" classes (in the ordinal scale), being this a refinement of the order information provided by an ordinal classifier, as the latter does not distinguish between objects within the same category. The relationship between multipartite ranking and ordinal classification is discussed in [9]. An ordinal regression classifier can be used as a ranking function by interpreting the class labels as scores [13]. However, this type of scoring will produce a large number of ties (which is not desirable for multipartite ranking). On the other hand, a multipartite ranking function $f(\cdot)$ can be turned into an ordinal classifier by deriving thresholds to define an interval for each class, but how to find the optimal thresholds is an open issue.

We propose to consider pairwise scoring rankers for the class imbalance problem. Ranking algorithms have been found to be very competitive classifiers for classification [6]. And the crucial property is that there is no class imbalance when performing pairwise ranking, since each observation of a class is compared to every observation of the other class, being the associated learning process balanced.

Scoring pairwise rankers are a type of rankers, where each observation $\mathbf{x}_i$ is compared against all others $\mathbf{x}_j$, and if $\mathbf{x}_i \succ \mathbf{x}_j$, then the model learns a scoring function s so that if $\mathbf{x}_i \succ \mathbf{x}_j$ then $s(\mathbf{x}_i) > s(\mathbf{x}_j)$, with $s \colon X \to \mathbb{R}$.

This scoring function then needs to be converted back to ordinal classes. A threshold strategy is proposed. Let s_i be the ordered score of observation i and q_i be the true class, we search the threshold left-to-right by invoking the function min_error with initial parameters $(s_0, q_0, 0)$, where:

$$\text{min\_error}(s_i, q_i, \hat{q}) = \begin{cases} \varepsilon_{q_i \hat{q}}, & \text{when } i = N, \\ \min\left\{\varepsilon_{q_i \hat{q}} + \text{min\_error}(s_{i+1}, q_{i+1}, \hat{q}), \ \text{min\_error}(s_i, q_i, \hat{q}_{i+1})\right\} \end{cases}$$

Several cost matrices ε can be used. For instance:

- **Homogeneous:** $\varepsilon_{q\hat{q}} = \{1 \text{ if } q \neq \hat{q} \text{ and } 0 \text{ otherwise} \mid \forall q, \hat{q}\}$
- **Absolute costs:** $\varepsilon_{q\hat{q}} = \{|q - \hat{q}| \mid \forall q, \hat{q}\}$
- **Inverse class frequency:** $\varepsilon_{q\hat{q}} = \left\{\frac{N}{QN_q + 1} \text{ if } q \neq \hat{q} \text{ and } 0 \text{ otherwise} \mid \forall q, \hat{q}\right\}$,
  where N_q is the number of patterns in class $\mathcal{C}_q$.

So far the underlying model has not been discussed; several adaptations have been presented in the literature: from neural networks [2] to gradient boosting [24]. In this work, the underlying model used is an SVM trained in the space of differences, an adaptation of [14]. The space of differences is constructed by

$\mathbf{x}_{mn}^{(ab)} = \mathbf{x}_m^{(a)} - \mathbf{x}_n^{(b)}$ with $\mathcal{C}_a < \mathcal{C}_b$, for every two observations indexed by m and n. The pairs are imbalance however, so each pair is re-weighted by a factor of $\prod_{q=1}^{Q} N_q/(N_a N_b)$, which corresponds to the inverse class frequency.

Fine-to-Coarse Ranking. Our learning problem is framed as an ordinal regression problem with two levels of granularity on the target variable $\mathcal{Y}_{\text{fine}} = \{\mathcal{C}_1^{\text{fine}}, \mathcal{C}_2^{\text{fine}}, \ldots, \mathcal{C}_{Qf}^{\text{fine}}\}$ and $\mathcal{Y}_{\text{coarse}} = \{\mathcal{C}_1^{\text{coarse}}, \mathcal{C}_2^{\text{coarse}}, \ldots, \mathcal{C}_{Qc}^{\text{coarse}}\}$, such that $\mathcal{Y}_{\text{fine}}$ can be monotonously partitioned in Qc disjoint groups.

This is a very common setting and can be seen in a wide range of scenarios. For instance, stundents receive a continuous fine-grained mark (e.g. 0–100%) which is usually under-segmented into a smaller number of intervals (e.g. fail, pass, pass with honors). Another example can be observed in ranking companies being broadly categorized as small-medium-large enterprises based on a fine-grained scale given by the number of employees, annual turnover, etc.

In this sense, coarse labels are a semantic abstraction of the original phenomenon being quantified. Despite this abstraction may simplify the analysis of each group, it may impose a significant loss of information when building rankers. For instance, for the specific problem tackled in this work, the number of comparisons used for training in a coarse scheme is restricted to patients on different intervals, while on a fine scheme we can distinguish patients that survived n days from patients that survived $n + 1$ days (see Fig. 1).

Thus, we propose to train the ranking model with the entire original information (fine labels) in order to obtain a sound and stable ranker. On the other hand, the transformation from ranking to ordinal classification is done using the ordinal classes (coarse labels).

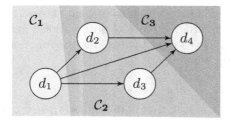 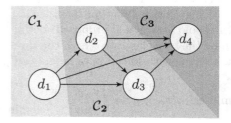

Fig. 1. Comparison of Coarse Ranking and Fine-to-Coarse Ranking. Transitive edges are not shown to improve readability.

Over-Sampling in Ordinal Domains. As stated before, the ordinal nature of the classes in ordinal classification should be taken into account for all the stages of the learning machine (e.g. for preprocessing the data). To compare the impact of rank-based learners, this paper also considers two over-sampling techniques in the context of ordinal classification and regression [19,23]. The main intuition behind these methods is that the ordering structure of the classes can be exploited when generating new synthetic patterns. A synthetic pattern in this case corresponds to a virtual donor-recipient pair, which we create to balance the class distribution and make the classifier pay more attention to minority classes. These virtual pairs are created using the information of other pairs, so new synthetic patterns are not totally virtual, but rather based on the combination of two donors and two recipients.

The strategies tested in this paper are:

- Over-sampling for regression problems (OR): The idea of using over-sampling in the presence of real-valued outputs presented in [23] is used in this paper, as an alternative when fine labels are given (instead of coarse categories). In this case, new synthetic patterns are created using a convex combination of two neighbours, where the new label is created also by convex combination of the label of the two patterns.
- Ordinal graph-based over-sampling (OGO): The ordinal nature of the data is exploited considering a neighborhood graph between the classes, which aims to capture the underlying manifold of the ordinal labelling space. New patterns are generated on the paths that preserve the ordinal structure of this manifold and create a spatial continuity on the input space.

Further details about these methods can be consulted in [19,23].

3   Experiments

3.1   Methodologies Tested

The experiments in this paper have been designed to compare several methodologies with our proposal of using rank-based learners as an alternative to imbalanced classification:

- Support vector machine using the one-versus-one approach (SVM) [15], a state-of-the-art nominal classification methodology. Both linear (LSVM) and non-linear (SVM) versions are employed.
- Cost-sensitive SVM [3], which poses a higher cost to minority classes (using the imbalance ratio of the class).
- Support Vector Machine for Ordinal Regression with Implicit Constraints (SVORIM) [5], a state-of-the-art SVM-based technique for ordinal classification.
- The proportional odds model (POM) [17], i.e. a standard ordinal logistic regression method which is widely used in the literature.
- Rank, the strategy presented in Sect. 2.2, where the ordinal classification problem is transformed to a learning to rank problem.
- F2C-Rank (also presented in Sect. 2.2) where the days leading up to graft failure are used to construct the ranking between the patterns.

As said, different threshold optimisation techniques are also tested for both the Rank and F2C-Rank strategies, in order to transform the ranking to discrete labels, as well as two over-sampling approaches (OGO and OR, depending on the nature of the label used).

3.2   Evaluation Metrics

Several measures can be considered for evaluating ordinal classifiers. The most common ones in machine learning are the Mean absolute error (MAE) and the accuracy (Acc) [11]. However, these measures may not be the best option, for example, when measuring performance in the presence of class imbalance [1], and/or when the costs of different errors vary markedly. The accuracy (Acc) is defined by:

$$Acc = \frac{100}{N} \sum_{i=1}^{N} (I(y_i^* = y_i)),$$

where $I(\cdot)$ is the zero-one loss function, y_i is the desired output for pattern $\mathbf{x}_i$, y_i^* is the prediction of the model and N is the total number of patterns in the dataset. However, this metric does not take the order of the categories into account, and it is not recommended for imbalanced datasets.

The average mean absolute error ($AMAE$) [1] is the mean of MAE classification errors throughout the classes, where MAE is the average absolute deviation of the predicted class from the true class (in number of categories on the ordinal scale). It is able to mitigate the effect of imbalanced class distributions. Let MAE_j be the MAE for a given j-th class:

$$MAE_j = \frac{1}{N_j} \sum_{i=1}^{N_j} |\mathcal{O}(y_i) - \mathcal{O}(y_i^*)|, \ 1 \leq j \leq Q,$$

where $\mathcal{O}(\mathcal{C}_j) = j$, $1 \leq j \leq Q$, i.e. $\mathcal{O}(y_j)$ is the order of class label y_j. Then, the $AMAE$ measure can be defined in the following way:

$$AMAE = \frac{1}{Q} \sum_{j=1}^{Q} MAE_j.$$

The Maximum Mean Absolute Error ($MMAE$), which corresponds to the MAE value considering only the patterns from the class with the greatest difference between true values as compared to the predicted ones:

$$MMAE = \max \{MAE_j; j = 1, \ldots, Q\},$$

where MAE_j is the MAE value considering only the patterns from the j-th class. This measure was recently proposed [8] and it is very interesting, since a low $MMAE$ represents a low error for all classes. MAE values are between 0 and $Q - 1$, and so are $AMAE$ and $MMAE$.

These four performance metrics briefly summarise the most important aspects of confusion matrices when dealing with ordinal and imbalanced data: *Acc* giving an idea of the global performance, *MAE* representing overall ordinal errors and *AMAE* and *MMAE* reflecting the magnitude of errors in the ordinal scale (the former the mean deviation and the latter the deviation associated to the worst classified class).

3.3   Experimental Setting

For evaluating the results, a stratified 10-fold technique has been used to divide the data, and the results have been taken as the mean and standard deviation.

The parameters for all methods have been chosen using a nested 5-fold cross-validation over the training set (independently of the 10-fold technique). The final parameter combination was the one which obtained, in mean, the best average performance for the 5 validation sets of this nested 5-fold cross-validation, where the metric used was the *AMAE*. The test sets were never used during model selection. The kernel selected for all the non-linear kernel methods was the Gaussian one, $K(\mathbf{x}, \mathbf{y}) = \exp\left(-\frac{\|\mathbf{x}-\mathbf{y}\|^2}{\sigma^2}\right)$, where σ is the kernel width. For every tested kernel method, the kernel width was tuned within the range $\sigma \in \{10^{-3}, 10^{-2}, \ldots, 10^3\}$, as well as the cost parameter associated to SVM-based methods, $C \in \{10^{-3}, 10^{-2}, \ldots, 10^3\}$.

3.4   Results

Table 1 shows the results obtained for the dataset considered with the methods previously mentioned. Several conclusions can be extracted from this table: Firstly, the complexity of the problem considered can be appreciated, as most methods obtain trivial models (accurate but that almost always predict the survival class, as indicated by the values of AMAE and MMAE of 1.5 and 3 respectively). Secondly, the choice of a linear or nonlinear model does not seem to have an important impact on the performance (as can be seen comparing the results of LSVM and SVM or CS-LSVM and CS-SVM). Thirdly, cost-sensitive approaches do not present an acceptable performance, confirming this that over-sampling should be preferred. In this regard, over-sampling helps the method to focus on the classification of the minority class and avoid trivial classifiers. The sole use of an ordinal method is not enough to obtain an appropriate classifier, as can be seen when analysing the results of LSVORIM (very competitive results for imbalance-nature metrics but a extremely poor overall performance, even worse than a random classifier). Concerning rank-based learners, these are seen to obtain a competitive performance and reach the results of other state-of-the-art classifiers (such as POM when combined with an ordinal over-sampling strategy). The use of fine labels help the method to optimise the results even further, specially for *Acc*, *MAE* and *AMAE*, obtaining worse results for *MMAE*, but a promising trade-off in all metrics. The combination of a rank-based learner with over-sampling is not satisfactory, which could mean that the sole use of a rank

Table 1. Average and standard deviation results for the test sets

Method	Acc	MAE	$AMAE$	$MMAE$
Nominal and ordinal classifiers				
LSVM	**85.11 ± 0.34**	*0.304 ± 0.004*	1.500 ± 0.000	3.000 ± 0.000
CS-LSVM	85.04 ± 0.47	0.306 ± 0.008	1.501 ± 0.002	3.000 ± 0.000
SVM	**85.11 ± 0.34**	*0.304 ± 0.004*	1.500 ± 0.000	3.000 ± 0.000
CS-SVM	**85.11 ± 0.34**	**0.303 ± 0.005**	1.500 ± 0.000	3.000 ± 0.000
LSVM+OGO	**85.11 ± 0.79**	**0.303 ± 0.017**	1.500 ± 0.000	3.000 ± 0.000
LSVORIM	20.20 ± 31.69	1.435 ± 0.196	**1.102 ± 0.196**	**2.185 ± 0.393**
LSVORIM+OGO	54.82 ± 31.97	0.994 ± 0.822	1.420 ± 0.179	2.566 ± 0.402
POM	*84.97 ± 0.34*	0.304 ± 0.006	1.500 ± 0.001	3.000 ± 0.000
POM+OGO	63.33 ± 3.12	0.540 ± 0.049	1.410 ± 0.086	2.562 ± 0.179
Rank-based learners				
Rank	64.37 ± 19.29	0.804 ± 0.439	1.422 ± 0.081	*2.416 ± 0.541*
F2C-Rank	67.75 ± 12.85	0.669 ± 0.281	*1.407 ± 0.096*	2.452 ± 0.511
Rank+OGO	69.74 ± 20.14	0.600 ± 0.399	1.443 ± 0.109	2.624 ± 0.497
F2C-Rank+OR	84.69 ± 0.45	0.310 ± 0.012	1.493 ± 0.021	2.975 ± 0.079
Comparison between threshold optimisation strategies for ranking				
Rank-Inv	64.37 ± 19.29	0.804 ± 0.439	1.422 ± 0.081	2.416 ± 0.541
Rank-Unif	84.55 ± 1.05	0.314 ± 0.018	1.495 ± 0.022	3.000 ± 0.000
Rank-Abs	85.04 ± 0.36	0.306 ± 0.010	1.489 ± 0.026	3.000 ± 0.000
F2C-Rank-Inv	67.75 ± 12.85	0.669 ± 0.281	1.407 ± 0.096	2.452 ± 0.511
F2C-Rank-Unif	*84.97 ± 0.32*	0.306 ± 0.009	1.495 ± 0.020	3.000 ± 0.000
F2C-Rank-Abs	84.83 ± 0.26	0.310 ± 0.010	1.499 ± 0.011	3.000 ± 0.000

The best result is in **bold** face and the second best result is in *italics*.

learner is enough to deal with the imbalanced nature of the data. Specially, the use of OR (as opposed to F2C-Rank) deteriorates the results and leads almost to a trivial classifier, which could be due to the over-sampling strategy itself. Finally, considering the use of different threshold optimisation strategies, the use of the inverse function (which takes into account the imbalanced nature of the data) presents the best results (which are the ones used for comparison in the rest of the experiments).

Note that the developed computational models are used as a decision support system. With MELD (one of the current assignation methodologies used world-wide) donors are generally assigned to the candidates at greatest-risk, a policy that does not allow the transplant team to do the matching according to the principles of fairness and survival benefit [18]. The method proposed here for organ allocation seeks to minimise futile liver transplantation, giving primary attention to patients with the best predicted lifetime gained due to transplantation. Note that although rankers have been tested in this case using classification metrics, they have shown to produce a better ordering between the classes, meaning that the output of the ranking algorithm could be used for the construction of the allocation system.

4   Conclusions

A practical case of imbalance ordinal classification is analysed in this paper: a dataset consisting of liver transplant survival information from eight transplantation units. The survival information is given in the number of days up to either graft loss or death or one year (365). A simple regression would not be appropriate due to the non-linear relation between the independent and dependent variables, and the boundedness nature of the problem (0–365). Typically, survival information would be discretised in classes prior to the application of classification classifier.

The traditional approach is compared against ranking, in particular pairwise scoring ranking. This model family has been found to produce good results in an imbalance context [6]. The expressiveness of ranking avoids the prior discretisation of the survival variable, which has been named as fine labels. A scoring ranking model predicts a score for each observation, which indicates the ranking order against the others. This ranking score must then be converted back to the original discretised class-space, which we have called coarse labels.

The suggested approach is contrasted with traditional ordinal classification methods: vanilla SVM, SVORIM and POM, with both linear and RBF kernels. Since the dataset is imbalanced, and the classifier is evaluated using imbalance metrics, these classifiers are also evaluated by prior over-sampling and by introducing cost-sensitivity. The proposed method is also tested against ranking trained only using coarse labels.

In the experimental setting, the proposed ranking method is found to be highly competitive, especially when considering the MMAE and AMAE metrics, which are imbalance-sensitive metrics. It was found that over-sampling is always preferred to introducing inverse frequency costs in order to balance the classes. Interestingly, this was not verified for ranking. Ranking performs generally competitively against traditional approaches, even without over-sampling. The proposed ranking approach, which makes use of fine labels, presents a promising trade-off in all metrics resulting in a final model that could be useful as a decision support system for the medical community.

References

1. Baccianella, S., Esuli, A., Sebastiani, F.: Evaluation measures for ordinal regression. In: Proceedings of the Ninth International Conference on Intelligent Systems Design and Applications, ISDA 2009, pp. 283–287 (2009)
2. Burges, C., Shaked, T., Renshaw, E., Lazier, A., Deeds, M., Hamilton, N., Hullender, G.: Learning to rank using gradient descent. In: Proceedings of the 22nd international Conference on Machine Learning - ICML 2005, pp. 89–96. ACM, New York (2005)

3. Chang, C.C., Lin, C.J.: LIBSVM: a library for support vector machines. ACM Trans. Intell. Syst. Technol. **2**(3), 27 (2011)
4. Chawla, N.V., Bowyer, K.W., Hall, L.O., Kegelmeyer, W.P.: SMOTE: synthetic minority over-sampling technique. J. Artif. Intell. Res. **16**, 321–357 (2002)
5. Chu, W., Keerthi, S.S.: Support vector ordinal regression. Neural Comput. **19**, 792–815 (2007)
6. Cruz, R., Fernandes, K., Cardoso, J.S., Costa, J.F.P.: Tackling class imbalance with ranking. In: 2016 International Joint Conference on Neural Networks (IJCNN), pp. 2182–2187, July 2016
7. Cruz, R., Fernandes, K., Pinto Costa, J.F., Perez Ortiz, M., Cardoso, J.S.: Ordinal class imbalance with ranking. In: Rojas, I., et al. (eds.) IWANN 2017, Part II. LNCS, vol. 10306, pp. 538–548. Springer, Cham (2017)
8. Cruz-Ramírez, M., Hervás-Martínez, C., Sánchez-Monedero, J., Gutiérrez, P.A.: Metrics to guide a multi-objective evolutionary algorithm for ordinal classification. Neurocomputing **135**, 21–31 (2014)
9. Fürnkranz, J., Hüllermeier, E., Vanderlooy, S.: Binary decomposition methods for multipartite ranking. In: Buntine, W., Grobelnik, M., Mladenić, D., Shawe-Taylor, J. (eds.) ECML PKDD 2009. LNCS, vol. 5781, pp. 359–374. Springer, Heidelberg (2009). doi:10.1007/978-3-642-04180-8_41
10. Galar, M., Fernández, A., Barrenechea, E., Bustince, H., Herrera, F.: A review on ensembles for the class imbalance problem: bagging-, boosting-, and hybrid-based approaches. IEEE Trans. Syst. Man Cybern. Part C Appl. Rev. **42**(4), 463–484 (2012)
11. Gutiérrez, P.A., Pérez-Ortiz, M., Sánchez-Monedero, J., Fernandez-Navarro, F., Hervás-Martínez, C.: Ordinal regression methods: survey and experimental study. IEEE Trans. Knowl. Data Eng. **28**(1), 127–146 (2016)
12. He, H., Garcia, E.A.: Learning from imbalanced data. IEEE Trans. Knowl. Data Eng. **9**(21), 1263–1284 (2009)
13. Herbrich, R., Graepel, T., Obermayer, K.: Support vector learning for ordinal regression. In: International Conference on Artificial Neural Networks, pp. 97–102 (1999)
14. Herbrich, R., Graepel, T., Obermayer, K.: Large margin rank boundaries for ordinal regression. Advances in Large Margin Classifiers, pp. 115–132. MIT Press, Cambridge (2000)
15. Hsu, C.W., Lin, C.J.: A comparison of methods for multi-class support vector machines. IEEE Trans. Neural Netw. **13**(2), 415–425 (2002)
16. Japkowicz, N., Stephen, S.: The class imbalance problem: a systematic study. Intell. Data Anal. **6**(5), 429–449 (2002)
17. McCullagh, P.: Regression models for ordinal data. J. Roy. Stat. Soc. **42**(2), 109–142 (1980)
18. Pérez-Ortiz, M., Cruz-Ramírez, M., Ayllón-Terán, M., Heaton, N., Ciria, R., Hervás-Martínez, C.: An organ allocation system for liver transplantation based on ordinal regression. Appl. Soft Comput. **14**(Part A), 88–98 (2014)
19. Pérez-Ortiz, M., Gutiérrez, P., Hervás-Martínez, C., Yao, X.: Graph-based approaches for over-sampling in the context of ordinal regression. IEEE Trans. Knowl. Data Eng. **27**(5), 1233–1245 (2015)
20. Pérez-Ortiz, M., Gutiérrez, P.A., Tino, P., Hervás-Martínez, C.: Oversampling the minority class in the feature space. IEEE Trans. Neural Netw. Learn. Syst. **27**(9), 1947–1961 (2016)

21. Rajaram, S., Agarwal, S.: Generalization bounds for k-partite ranking. In: Proceedings of the Seventeenth Annual Conference on Neural Information Processing Systems (NIPS2005), pp. 28–23 (2005)
22. Tang, Y., Zhang, Y.Q., Chawla, N.V., Krasser, S.: SVMs modeling for highly imbalanced classification. IEEE Trans. Syst. Man Cybern. B Cybern. **39**(1), 281–288 (2009)
23. Torgo, L., Ribeiro, R.P., Pfahringer, B., Branco, P.: SMOTE for regression. In: Correia, L., Reis, L.P., Cascalho, J. (eds.) EPIA 2013. LNCS (LNAI), vol. 8154, pp. 378–389. Springer, Heidelberg (2013). doi:10.1007/978-3-642-40669-0_33
24. Zheng, Z., Chen, K., Sun, G., Zha, H.: A regression framework for learning ranking functions using relative relevance judgments. In: Proceedings of the 30th Annual International ACM SIGIR Conference on Research and Development in Information Retrieval, pp. 287–294. ACM (2007)

Combining Ranking with Traditional Methods for Ordinal Class Imbalance

Ricardo Cruz[1]([✉]), Kelwin Fernandes[1,2], Joaquim F. Pinto Costa[3],
María Pérez Ortiz[4], and Jaime S. Cardoso[1,2]

[1] INESC TEC, Porto, Portugal
{rpcruz,kafc,jaime.cardoso}@inesctec.pt
[2] Faculty of Engineering, University of Porto, Porto, Portugal
[3] Faculty of Sciences, University of Porto, Porto, Portugal
jpcosta@fc.up.pt
[4] Universidad Loyola Andalucía, Córdoba, Spain
mariaperez@uloyola.es

Abstract. In classification problems, a dataset is said to be imbalanced when the distribution of the target variable is very unequal. Classes contribute unequally to the decision boundary, and special metrics are used to evaluate these datasets. In previous work, we presented pairwise ranking as a method for binary imbalanced classification, and extended to the ordinal case using weights. In this work, we extend ordinal classification using traditional balancing methods. A comparison is made against traditional and ordinal SVMs, in which the ranking adaption proposed is found to be competitive.

Keywords: Ordinal classification · Class imbalance · Ranking · SVM

1 Introduction

Ordinal classification, also known as ordinal regression, is a subset of multiclass classification problems where the target variable has an ordinal scale, and so it is possible to establish an order between any two classes. Often, it is desirable to punish more an error incurred from misclassification of an observation as an adjacent class than an error when the observation is misclassified as a more disparate classes. The extra ordinal constrains can, and have been used, to produce models that specifically optimize for these ordinal metrics.

Classification datasets, which feature a disproportion in the distribution of observations in each class, are said to be class imbalance. Traditional methods favor too much the majority classes. Furthermore, traditional metrics such as accuracy can produce apparently good results for models which consider only the majority class, and special metrics have been devised for the purpose of evaluating models applied to class imbalance problems. Much literature exists in the topic, but only one attempt has been made using ranking, by the same authors, in a binary [1] and ordinal [2] classification context.

© Springer International Publishing AG 2017
I. Rojas et al. (Eds.): IWANN 2017, Part II, LNCS 10306, pp. 538–548, 2017.
DOI: 10.1007/978-3-319-59147-6_46

Pairwise scoring rankers are an attractive family of models, since the problem is solved in the space of the differences between classes, and so is inherently balanced for the binary case. However, when transporting the problem to a multi-class ordinal context, the imbalance problem arises again between pairs of classes. In this work, approaches of tackling this imbalance are proposed and evaluated.

The work is divided as follows. Section 2 overviews some existing methods for ordinal classification. Section 3 overviews some methods for class imbalance. Section 4 details our proposal of combining ranking with traditional methods for ordinal class imbalance. Section 5 provides empirical experiments and results. Section 6 concludes the work.

2   Ordinal Classification

Many ordinal classifiers currently exist. oSVM [3] takes advantage of the fact that the decision boundaries are necessarily parallel in a well-formed ordinal problem. It transforms the original ordinal problem into a binary problem by increasing the number of dimensions, after which the multiple decision boundaries can be recreated.

SVOR [4] encompasses SVORIM and SVOREM which differ on how the constraints are defined. The idea is to find $k-1$ parallel discriminant hyperplanes in order to properly separate the data into ordered classes by modelling ranks as intervals [4].

Herbrich et al. [5] addresses ordinal classification using pairs in the space of differences. Let $C_1 < C_2 < \cdots < C_K$ be the K classes involved. Let $S_k = \{x_n^{(k)}\}$ be the set of N_k samples from C_k, with $N = \sum_{k=1}^{K} N_k$. Construct the differences $x_{mn}^{(k\ell)} = x_m^{(k)} - x_n^{(\ell)}$ with $C_k < C_\ell$. Like in the binary setting, solve the binary classification problem in the set of the differences $\{(x_{mn}^{(k\ell)}, +1), (-x_{mn}^{(k\ell)}, -1)\}$, where $+1$ and -1 are the labels of the samples $x_{mn}^{(k\ell)}$ and $-x_{mn}^{(k\ell)}$, respectively.

An issue with this approach arises when one of the classes is strongly misrepresented when compared with the others. The data from each class C_k is involved in $N_k(N-N_k)$ points in the set of the differences. If $N_k \ll N_\ell$ then also $N_k(N-N_k) \ll N_\ell(N-N_\ell)$. For instance, if $N_1 = 10$ and $N_2 = N_3 = 100$, then the data from C_1 is contributing to 2000 elements in the new space, while the data from C_2 or C_3 is contributing to 11000. So, the new learning problem will be dominated by the samples from C_2 and C_3 and it is likely that C_1 will be poorly estimated.

Traditional one-vs-rest or one-vs-all ensembles can also be used for ordinal classes, even if they do not take order in consideration. However, they do not take advantage of that extra information, and do not optimize for ordinal metrics. Furthermore, they may produce models whose decision boundaries make little sense in an ordinal context; for instance, decision boundaries should not cross in an ordinal context [6].

3   Traditional Methods for Class Imbalance

Several methods have been proposed in tackling class imbalance, which usually involve:

(a) Pre-processing;
(b) Training with costs;
(c) Ensembles.

(a) Pre-processing usually involves a mix of undersampling the majority class and creating new synthetic examples of the minority class [7,8]. *(b) Training with costs* involves the use of a cost matrix so that the cost of misclassifying a class is inversely proportional to its frequency, and therefore the estimation algorithm minimize an weighted loss function, rather than the original imbalanced loss function, so that the minority class contributes more to the loss than it would otherwise. *(c) Ensembles* by which each model within the ensemble is trained with balanced subsets of the data, coupled with the previous preprocessing techniques [9].

Some strategies of tackling multiclass problems such as one-vs-rest exacerbate the imbalance problem by training. Given K classes with N_k observations, this strategy solves the problem using an ensemble of binary classifiers, training each classifier i with N_i positively labeled data against the rest $N - N_i$ negatively labeled data. This creates an imbalance problem, even if all classes are equally represented [10].

Another strategy is known as one-vs-one, whereby each classifier i is trained K using N_i positively labeled data and N_j, $\forall j \neq i$, negatively labeled data, resulting in an ensemble of $K(K-1)$ classifiers. Even this strategy is not optimal for ordinal datasets because it generates decision boundaries that make little sense in the context of ordinal data due to the fact the decision boundaries are not parallel [6].

Work already exists to adapt these methods to be used with ordinal classifiers in imbalance situations [8]. In the rest of the work, we propose using these methods as-is combined with ranking.

4   Combining Ranking with Traditional Methods

Class imbalance has been previously addressed using pairwise scoring ranking in [1]. Ordinal classification seems like a natural extension because, unlike ordinary multi-class problems, any two classes can be compared, as in ranking applications.

4.1   Ranking for Binary Class Imbalance

In pairwise scoring ranking, a function $f\colon X \to \mathbb{R}$ is constructed so that $f(\mathbf{x}_i) > f(\mathbf{x}_j)$ when $\mathbf{x}_i \succ \mathbf{x}_j$ for every pair of observations $(\mathbf{x}_i, \mathbf{x}_j)$, where $\succ$ means "preferred".

In the case here considered, as in [5], a base estimator is trained using the space of differences. Consider two classes, $\mathcal{C}_1$ and $\mathcal{C}_2$, with a set $\mathcal{S}_1$ of N_1 examples from $\mathcal{C}_1$ and a set $\mathcal{S}_2$ with N_2 examples from $\mathcal{C}_2$. Construct all $N_1 \times N_2$ pairs $\mathbf{x}_{ij} = \mathbf{x}_i - \mathbf{x}_j$ with $\mathbf{x}_i \in \mathcal{S}_1$ and $\mathbf{x}_j \in \mathcal{S}_2$. Solve the binary classification problem using an ordinary SVM estimator in the set of the differences

$$\{(\mathbf{x}_{ij}, +1), (-\mathbf{x}_{ij}, -1) \mid \mathbf{x}_{ij} = \mathbf{x}_i - \mathbf{x}_j\},$$

where $+1$ and -1 are the labels of the samples $\mathbf{x}_{ij}$ and $-\mathbf{x}_{ij}$, respectively.

The big families of rankers are pointwise, pairwise and listwise. We focus on pairwise and, in particular, scoring pairwise rankers in order to produce a function $f\colon X \to \mathbb{R}$ so that we can afterwards build a threshold to convert back the ranking score to classes.

4.2 Ranking for Ordinal Class Imbalance

We have already suggested an initial ordinal class adaption of ranking in previous work [2]. Consider all K-tuples $(\mathbf{x}^{(1)}, \mathbf{x}^{(2)}, \dots, \mathbf{x}^{(K)})$, with $\mathbf{x}^{(k)} \in \mathcal{S}_k$. There are $\prod_{k=1}^{K} N_k$ of such K-tuples. Generate all pairwise differences between ordered elements in the K-tuple: $\mathbf{x}^{(k)} - \mathbf{x}^{(\ell)}$. There are $K(K-1)$ pairs built from a K-tuple. Like before, learn a binary classifier from the $\frac{K(K-1)}{2} \prod_{k=1}^{K} N_k$ pairs positively labeled, and the corresponding symmetric differences negatively labeled.

Note that in this case, each class is present in exactly the same number of elements in the new space: $(K-1) \prod_{k=1}^{K} N_k$ times. The imbalance binary case presented initially is a special case of this formulation, obtained by setting $K = 2$.

This approach is however repeating the pairs multiple times. A pairwise difference could be constructed with pairs $(\mathcal{C}_\ell, \mathcal{C}_m)$, repeated $(\prod_{k=1}^{K} N_k)/N_\ell N_m$ times. This would, however, be impractical. Several alternatives will be considered.

4.3 Balanced Difference Pairs

In order to balance the pairs of differences, conventional approaches from class imbalance are used, as mentioned in Sect. 3:

(a) Pre-processing;
(b) Training with costs;
(c) Ensembles.

Classes are balanced using the median of the distribution, so that, for every class k, $N_k' = \tilde{N}$, where $\tilde{N}$ is the median. The difference in the number of observations for each class is $\Delta N = \tilde{N} - N_k$. Therefore, the training sampling is oversampled if $\Delta N > 0$ and undersampled if $\Delta N < 0$.

For *(a) Pre-processing*, SMOTE [7] and MSMOTE [11] are evaluated. These oversampling techniques work by creating new synthetic examples based on the minority classes using an average of the nearest neighbors. While the original

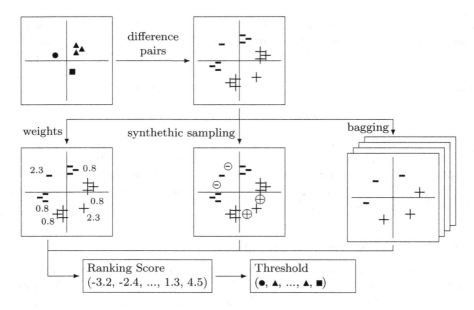

Fig. 1. Balanced ranking training.

paper [7] regards only binary problems, it is easily generalized to multiclass problems. MSMOTE is an extension that adds heuristics to identify points as outliers and refraining from oversampling using those.

(b) Training with costs involves using a cost matrix for each pair in the difference space defined to be inversely proportional to the frequency of each pair so that each pair contributes equally to the decision boundary.

When using *(c) Ensembles*, several strategies have been proposed which make use of either oversampling of the minority class or undersampling of the majority class so that each classifier is trained using balanced data [9]. Some work makes use of boosting where, in addition to fixing the imbalance problem, each classifier is trained to emphasize the most problematic cases. Boosting seems to muddle the causes of the balancing performance gain, and we suggest using bagging.

For the full picture of the method here proposed, with these several approaches, see Fig. 1.

4.4 Threshold for Ordinal Classes

After building the decision boundary from the difference pairs, the resulting score from the pairwise scoring ranker needs then to be transformed back to classes. Based on the training data, we obtain a ranking score s_i for each observation $\mathbf{x}_i$, which can be ordered, assuming that the score represents the order of y_i.

We here define the proposed threshold strategy recursively. Let s_i be the ordered score of observation i and k_i be the true class, we search the threshold left-to-right by invoking the function min_error with initial parameters $(s_0, k_0, 0)$.

$$\text{min\_error}(s_i, k_i, \hat{k}) = \begin{cases} \varepsilon_{k_i \hat{k}}, & \text{when } i = N, \\ \min\left\{\varepsilon_{k_i \hat{k}} + \text{min\_error}(s_{i+1}, k_{i+1}, \hat{k}), \ \text{min\_error}(s_i, k_i, \hat{k} + 1)\right\} \end{cases}$$

where $\varepsilon = \left[\varepsilon_{k\hat{k}}\right]$ is a cost matrix. Informally, min_error tests whether, at any given time, it is less costly to continue assuming observation i to be of class $\hat{k}$ or if it less costly to make a threshold and start assuming observations are now $\hat{k}+1$. Notice observations are ordered by the score and classes are ordinal.

For the cost matrix, absolute costs have been used, $\varepsilon_{k\hat{k}} = \{|k - \hat{k}| \mid \forall k, \hat{k}\}$.

5   Experiments

Pairwise scoring ranking has already been experimented within a binary classification context [1]. It has been found that, in a multi-class context, the difference space is no longer balanced. In this section, we experiment with the different balancing approaches previously discussed.

The proposed method is contrasted against state-of-the-art methods: One-vs-Rest SVM, One-vs-Rest SVM with a balanced cost matrix, SVOR [4], and oSVM [3]. All of which also use an SVM as the base estimator.

Each model is cross-validated by grid-search with $C \in \{10^{-3}, 10^{-2}, \ldots, 10^3\}$ using k-fold with $k = 3$. Final scores are obtained by 30-fold validation, using the same folds from [8].

Both linear and RBF kernels are tested. The proposed model with linear kernel is implemented by ourselves, while the RBF kernel version uses SVM[rank] by Thorsten Joachims[1]. His version has been modified to allow setting weights for each pair of differences. All implementations from our work including the dataset folds are made publicly available[2]. Python and scikit-learn were used.

5.1   Evaluation Metrics

Typically, in binary imbalance problems, special balanced metrics are used. The most popular are F_1 and G-mean. But these metrics are only well-established for binary settings. For ordinal classification, Mean Absolute Error (MAE) is widely used,

$$\text{MAE} = \frac{1}{N} \sum_i |k_i - \hat{k}_i|.$$

But this metric suffers from two problems. First, it treats an ordinal variable as a cardinal variable. Second, the metric is sensible to the per-class distribution of the magnitude of the errors, and is therefore not suitable for class imbalance. Like Pérez-Ortiz et al. [8], since the datasets are imbalance, we will contrast our imbalance ranking approach against conventional methods by using the Maximum Mean Absolute Error (MMAE) metric proposed by [12]. MMAE is defined as

$$\text{MMAE} = \max\{\text{MAE}_k \mid k = 1, \ldots, K\}.$$

---

[1] https://www.cs.cornell.edu/people/tj/svm_light/svm_rank.html.
[2] http://vcmi.inescporto.pt/reproducible_research/iwann2017/OrdinalImbalance/.

Table 1. Datasets used in the experiments.

Dataset	N	#vars	K	IR
balance-scale	625	4	3	0.170
car	1728	21	4	0.054
contact-lenses	24	6	3	0.267
cooling	768	8	8	0.066
diabetes5	43	2	5	0.091
diabetes10	43	2	10	0.167
newthyroid	215	5	3	0.200
pyrim5	74	27	5	0.250
pyrim10	74	27	10	0.143
squash-stored	52	51	3	0.348
squash-unstored	52	52	3	0.167
stock10	950	9	10	0.131
toy	300	2	5	0.356
triazines5	186	60	5	0.081
triazines10	186	60	10	0.040

5.2 Data

The datasets used come from real problems. The ordinal classification datasets are extracted from the benchmark repositories UCI [13] and mldata.org [14]. Some were originally regression problems converted into ordinal classification, and were obtained from the website of Chu and Ghahramani [15].

Datasets from [8] were used for the experiments, see Table 1. Here, the Imbalance Ratio (IR) metric represents $IR = \frac{\min_k N_k}{\max_k N_k}$, i.e. the ratio between the number of elements of the minority class to that of the majority class. $IR \in [0, 1]$, ranging from very imbalance to balanced, respectively. This provides a sense of the imbalance in each dataset.

Tables are ordered alphabetically. For performance reasons, not all datasets are used for the RBF kernel.

5.3 Models

The SVM models tested are **WRank**, **BRank**, **SRank**, and **MSRank**, which correspond to the proposed ranking method balancing difference pairs through weights, bagging and oversampling through SMOTE and MSMOTE, respectively. **OvR** and **OvR/w** are traditional SVM without and with balanced weights. **SVOREX** and **SVORIM** are from [4], and **oSVM** from [3].

These models are compared using linear and RBF SVM kernels. Two tables are presented for each kernel using the metrics discussed above: MAE and MMAE. The average of these metrics for each dataset is exhibited for the 30-fold validation. The best scores are presented in bold. Also in bold are scores which are statistically identical to the best score, using a paired difference Student's *t*-test with a 95% confidence level.

5.4   Results

Linear kernel results are presented in Table 2 for the aforementioned MAE and MMAE metrics. RBF kernel results are presented in Table 3.

Table 2. Results for SVMs with Linear kernel.

MAE

Dataset	WRank	BRank	SRank	MSRank	OvR	OvR/w	SVOREX	SVORIM	oSVM
balance-scale	0.12	0.52	0.11	1.00	0.20	0.19	**0.11**	**0.11**	0.12
car	0.09	0.11	0.09	1.07	0.12	0.09	0.14	0.12	**0.08**
contact-lenses	0.42	0.48	**0.38**	**0.39**	0.42	**0.44**	0.51	0.54	**0.42**
cooling	**0.41**	1.05	1.13	1.28	0.44	0.55	0.48	0.50	0.49
diabetes5	**0.64**	**0.67**	0.74	0.77	0.72	0.95	0.84	**0.67**	0.85
diabetes10	**1.68**	**1.72**	1.77	1.77	2.06	2.15	1.81	**1.69**	2.41
newthyroid	0.04	0.18	0.05	1.00	0.04	**0.03**	0.04	0.04	**0.03**
pyrim5	**0.58**	0.99	1.16	1.16	**0.58**	0.70	1.08	0.99	0.65
pyrim10	1.34	**1.26**	1.29	1.33	1.52	1.49	2.89	**1.32**	1.50
squash-stored	0.46	0.86	1.11	1.13	0.47	0.47	**0.41**	0.44	**0.38**
squash-unstored	**0.27**	**0.27**	**0.27**	**0.26**	0.30	**0.30**	**0.26**	0.26	0.33
stock10	0.64	0.66	0.67	1.02	**0.42**	0.42	0.68	0.63	0.70
toy	**0.84**	0.93	**0.87**	1.21	1.02	1.01	1.13	0.95	0.96
triazines5	0.70	1.31	0.98	1.08	0.69	**0.67**	**0.67**	**0.67**	**0.70**
triazines10	1.40	1.95	2.01	1.97	**1.33**	1.51	**1.37**	1.39	1.45
Average	0.64	0.86	0.84	1.10	0.69	0.73	0.83	0.69	0.74
Deviation	0.48	0.52	0.58	0.42	0.55	0.58	0.73	0.48	0.62
Winner	40%	26%	26%	13%	26%	26%	33%	33%	33%

MMAE

Dataset	WRank	BRank	SRank	MSRank	OvR	OvR/w	SVOREX	SVORIM	oSVM
balance-scale	0.21	1.01	0.15	1.10	1.00	0.96	**0.17**	**0.14**	0.21
car	0.47	1.06	**0.28**	1.15	0.77	**0.29**	1.36	1.01	0.49
contact-lenses	**0.81**	1.20	**0.78**	**0.76**	**0.88**	**0.73**	0.97	1.04	**0.82**
cooling	2.32	**1.72**	**1.73**	1.81	2.24	**1.80**	2.98	2.88	2.07
diabetes5	**1.15**	**1.17**	1.20	1.23	1.48	1.52	1.51	1.30	1.43
diabetes10	**3.09**	**3.16**	3.12	3.26	3.82	3.98	3.47	**2.94**	4.33
newthyroid	0.14	1.00	**0.09**	1.04	0.16	0.13	0.14	0.14	**0.13**
pyrim5	**1.40**	1.83	2.26	2.31	**1.30**	1.62	3.00	2.00	1.87
pyrim10	**3.80**	3.91	3.84	3.75	3.86	3.84	6.37	4.33	4.18
squash-stored	0.83	1.23	1.42	1.46	1.06	0.94	0.76	0.83	**0.66**
squash-unstored	0.57	1.00	0.55	0.52	0.76	0.80	**0.46**	0.46	**0.57**
stock10	1.02	1.04	0.97	1.64	1.03	**0.85**	1.30	1.05	1.29
toy	**1.79**	2.17	**1.67**	2.49	1.92	**1.57**	3.00	2.00	1.82
triazines5	2.77	2.11	**1.73**	2.52	2.99	2.94	3.00	3.00	2.79
triazines10	6.14	**4.46**	**4.58**	**4.55**	6.58	6.35	7.00	6.83	6.80
Average	1.77	1.87	1.63	1.97	1.99	1.89	2.37	2.00	1.96
Deviation	1.58	1.08	1.30	1.12	1.63	1.64	2.02	1.74	1.81
Winner	40%	33%	66%	26%	20%	40%	13%	13%	26%

Table 3. Results for SVMs with RBF kernel.

MAE

Dataset	WRank	BRank	SRank	MSRank	OvR	OvR/w	SVOREX	SVORIM
balance-scale	0.11	0.92	0.14	0.15	0.14	0.18	0.11	**0.05**
car	0.13	0.19	0.25	0.24	**0.12**	0.21	0.41	0.41
contact-lenses	0.52	0.51	0.45	0.46	0.47	**0.33**	1.31	0.98
cooling	0.62	0.61	0.63	0.63	**0.54**	0.62	0.58	**0.55**
diabetes5	**0.65**	P0.68	**0.64**	**0.65**	**0.65**	**0.67**	0.72	**0.69**
diabetes10	**1.35**	1.53	1.38	**1.37**	1.75	1.78	1.62	1.56
newthyroid	0.29	0.31	0.29	0.29	0.25	0.23	**0.16**	0.16
pyrim5	**0.47**	0.64	0.56	0.60	1.08	1.10	1.08	0.99
pyrim10	**1.02**	1.18	**1.03**	**1.06**	2.73	2.18	2.88	2.00
squash-stored	**0.57**	0.57	0.57	0.57	0.73	0.57	0.73	0.57
squash-unstored	0.54	0.54	0.54	0.54	**0.44**	0.44	0.49	0.50
stock10	1.35	1.38	1.33	1.30	0.18	**0.17**	0.27	0.26
toy	**0.03**	0.12	**0.03**	**0.04**	0.91	0.66	1.08	0.95
triazines5	**0.68**	1.10	0.88	0.87	**0.67**	1.18	0.67	0.67
triazines10	**1.28**	1.76	1.67	1.64	1.37	2.45	1.37	1.37
Average	0.64	0.80	0.69	0.69	0.80	0.85	0.90	0.78
Deviation	0.42	0.48	0.46	0.45	0.68	0.72	0.69	0.52
Winner	53%	6%	20%	26%	33%	20%	6%	20%

MMAE

Dataset	WRank	BRank	SRank	MSRank	OvR	OvR/w	SVOREX	SVORIM
balance-scale	0.20	1.79	0.19	0.19	1.00	0.24	1.00	**0.13**
car	1.98	2.00	1.00	1.01	1.13	**0.27**	3.00	3.00
contact-lenses	1.27	1.97	1.23	1.22	0.88	**0.53**	1.82	1.32
cooling	1.26	1.22	**0.99**	**0.98**	1.77	**1.01**	2.00	2.00
diabetes5	**1.90**	**1.97**	**1.97**	**1.88**	2.00	2.00	2.00	2.00
diabetes10	3.57	3.77	3.63	3.57	4.27	4.27	**3.06**	**2.84**
newthyroid	1.00	1.03	1.00	1.00	1.00	0.97	**0.64**	0.64
pyrim5	**1.15**	1.78	**1.14**	**1.22**	3.00	2.90	3.00	2.00
pyrim10	**2.65**	3.65	2.82	2.82	6.63	5.67	6.30	4.93
squash-stored	**1.00**	1.00	1.00	1.00	2.00	1.00	2.00	1.00
squash-unstored	**1.00**	1.00	1.00	1.00	1.00	1.00	1.00	1.00
stock10	3.71	4.84	3.27	3.16	0.66	**0.45**	1.14	1.11
toy	**0.10**	1.00	**0.10**	**0.10**	1.83	1.15	2.80	2.00
triazines5	2.66	**1.93**	**1.97**	**1.91**	3.00	2.82	3.00	3.00
triazines10	5.83	**4.73**	**4.99**	**4.69**	7.00	5.66	7.00	7.00
Average	1.95	2.25	1.75	1.72	2.48	1.99	2.65	2.26
Deviation	1.47	1.29	1.32	1.25	1.95	1.81	1.76	1.71
Winner	40%	20%	40%	40%	0%	26%	13%	13%

Using absolute costs in the thresholds performs usually better, albeit inverse frequency costs offers some competitiveness, especially in weighted models using MMAE. Interestingly, the RBF kernel is more stable across the two metrics.

Weighted difference pairs seem to provide the best results, albeit sometimes surpassed by SMOTE. MSMOTE is not found to be a competitive variant of SMOTE; rarely performing better and suffering from higher validation deviation for the linear kernel. These results are similar to the binary experiments from [1].

Bagging with models trained using undersampled and simple oversample of pairs did not offer much compelling results.

One-vs-Rest SVM is sometimes competitive, possibly due to the fact that its decision boundaries are not parallel constrained like traditional ordinal models.

6   Conclusion

Four traditional approaches are used in improving imbalance datasets metrics: pre-processing, using cost matrices, post-processing and ensembles, and often combinations of these. In a previous work, we have suggested ranking as an unexplored alternative to imbalance problems [1], in particular pairwise scoring ranking. Pairwise ranking models use an underlying estimator training in the space of difference pairs, therefore a necessarily balanced dataset in the binary case.

In a follow-up, the ordinal case was addressed [2]. However, it was verified that the imbalance problem was co-occurring in the new space. In this work, the new space is balanced through traditional balancing approaches, with the application of weights being generally superior.

Acknowledgment. This work was funded by the Project "NanoSTIMA: Macro-to-Nano Human Sensing: Towards Integrated Multimodal Health Monitoring and Analytics/NORTE-01-0145-FEDER-000016" financed by the North Portugal Regional Operational Programme (NORTE 2020), under the PORTUGAL 2020 Partnership Agreement, and through the European Regional Development Fund (ERDF), and also by Fundação para a Ciência e a Tecnologia (FCT) within PhD grant numbers SFRH/BD/122248/2016 and SFRH/BD/93012/2013.

References

1. Cruz, R., Fernandes, K., Cardoso, J.S., Pinto Costa, J.F.: Tackling class imbalance with ranking. In: International Joint Conference on Neural Networks (IJCNN). IEEE (2016)
2. Cruz, R., Fernandes, K., Pinto Costa, J.F., Perez Ortiz, M., Cardoso, J.S.: Ordinal class imbalance with ranking. In: Rojas, I., et al. (eds.) IWANN 2017, Part II. LNCS, vol. 10306, pp. 538–548. Springer, Cham (2017)
3. Cardoso, J.S., Costa, J.F.: Learning to classify ordinal data: the data replication method. J. Mach. Learn. Res. 8(Jul), 1393–1429 (2007)
4. Chu, W., Sathiya Keerthi, S.: New approaches to support vector ordinal regression. In: Proceedings of the 22nd International Conference on Machine Learning, pp. 145–152. ACM (2005)
5. Herbrich, R., Graepel, T., Obermayer, K.: Support vector learning for ordinal regression. In: Ninth International Conference on Artificial Neural Networks, ICANN 1999, (Conf. Publ. No. 470), vol. 1, pp. 97–102. IET (1999)
6. Pinto Costa, J.F., Sousa, R., Cardoso, J.S.: An all-at-once unimodal SVM approach for ordinal classification. In: Ninth International Conference on Machine Learning and Applications (ICMLA), pp. 59–64. IEEE (2010)

7. Chawla, N.V., Bowyer, K.W., Hall, L.O., Kegelmeyer, W.P.: SMOTE: synthetic minority over-sampling technique. J. Artif. Intell. Res. **16**, 321–357 (2002)
8. Pérez-Ortiz, M., Gutiérrez, P.A., Hervás-Martínez, C., Yao, X.: Graph-based approaches for over-sampling in the context of ordinal regression. IEEE Trans. Knowl. Data Eng. **27**(5), 1233–1245 (2015)
9. Liu, X.-Y., Jianxin, W., Zhou, Z.-H.: Exploratory undersampling for class imbalance learning. IEEE Trans. Syst. Man Cybern. **39**(2), 539–550 (2009)
10. Sahare, M., Gupta, H.: A review of multi-class classification for imbalanced data. Int. J. Adv. Comput. Res. **2**(5), 160–164 (2012)
11. Hu, S., Liang, Y., Ma, L., He, Y.: MSMOTE: improving classification performance when training data is imbalanced. In: 2nd International Workshop on Computer Science and Engineering, WCSE 2009, vol. 2, pp. 13–17 (2009)
12. Cruz-Ramírez, M., Hervás-Martínez, C., Sánchez-Monedero, J., Gutiérrez, P.A.: Metrics to guide a multi-objective evolutionary algorithm for ordinal classification. Neurocomputing **135**, 21–31 (2014)
13. Lichman, M.: UCI Machine Learning Repository (2013). http://archive.ics.uci.edu/ml
14. PASCAL. Pascal (pattern analysis, statistical modelling and computational learning) machine learning benchmarks repository (2011). http://mldata.org/
15. Chu, W., Ghahramani, Z.: Gaussian processes for ordinal regression. J. Mach. Learn. Res. **6**(Jul), 1019–1041 (2005)

Constraining Type II Error: Building Intentionally Biased Classifiers

Ricardo Cruz[1(✉)], Kelwin Fernandes[1,2], Joaquim F. Pinto Costa[3],
and Jaime S. Cardoso[1,2]

[1] INESC TEC, Porto, Portugal
{rpcruz,kafc,jaime.cardoso}@inesctec.pt
[2] Faculty of Engineering, University of Porto, Porto, Portugal
[3] Faculty of Sciences, University of Porto, Porto, Portugal
jpcosta@fc.up.pt

Abstract. In many applications, false positives (type I error) and false
negatives (type II) have different impact. In medicine, it is not considered
as bad to falsely diagnosticate someone healthy as sick (false positive)
as it is to diagnosticate someone sick as healthy (false negative). But we
are also willing to accept some rate of false negatives errors in order to
make the classification task possible at all. Where the line is drawn is
subjective and prone to controversy. Usually, this compromise is given
by a cost matrix where an exchange rate between errors is defined. For
many reasons, however, it might not be natural to think of this trade-off
in terms of relative costs. We explore novel learning paradigms where
this trade-off can be given in the form of the amount of false negatives
we are willing to tolerate. The classifier then tries to minimize false positives while keeping false negatives within the acceptable bound. Here we
consider classifiers based on kernel density estimation, gradient descent
modifications and applying a threshold to classifying and ranking scores.

Keywords: Classification costs · Cost matrix · Imbalance data · Kernel
density estimation · Ranking · Gradient descent · Neural networks

1 Introduction

A common problem in medical and financial decision-making is that when classifying an observation, different classification costs must be considered for different
classification errors. These costs are real, but expressing them in the form of a
number is messy, hard and prone to controversy. For instance, it is not obvious
how bad it is to fail to diagnose cancer versus being too careful and subjecting
the patient to undue biopsies.

Metrics based on minimizing the expected cost (or maximizing the expected
benefit) are not intuitive for human evaluators, such as physicians. Instead,
humans tend to instead avoid things like misclassifying positives beyond an
acceptable rate of false negatives. It is therefore important to hold our models

© Springer International Publishing AG 2017
I. Rojas et al. (Eds.): IWANN 2017, Part II, LNCS 10306, pp. 549–560, 2017.
DOI: 10.1007/978-3-319-59147-6_47

to the same standard if they are intended to be realistic contenders (or complements) to an existing human evaluator. A machine learning metric based on a false negative threshold for training and evaluating models can potentially be more intuitive and provide a higher sense of trustworthiness which is important for machine learning penetration in such fields as medicine.

This situation is further aggravated by the fact that too commonly these problems suffer from class imbalance; that is, there are typically too few observations of the more severe positive class, which further taints training toward the majority class [1].

The problem we aim to tackle is best illustrated in broad strokes by the following optimization problem:

$$\text{Maximize TNR}$$
$$\text{subject to FNR} \leq \rho.$$

Here, TNR refers to the true negatives rate, also known as specificity, and FNR is the false negatives rate, and is equal to $1-$ sensitivity. Throughout the manuscript, we will be using the user-defined parameter ρ to represent the user-acceptable FNR, and $\widehat{\rho}$ for the empirical FNR, estimated from the data sample.

As is common in the literature, and without loss of generality, we take the positive class $(+)$ to be the minority class, and assume this is the class whose classification errors we aim to control (FNR $\leq \rho$).

2   State of the Art

Imputing costs via a cost matrix is the *de facto* approach for tackling false classification trade-offs. In the most common case, when the correct decision has null cost, then the cost matrix has only one degree of freedom,

$$\begin{pmatrix} 0 & c_p/c_n \\ 1 & 0 \end{pmatrix}$$

These costs are then taken into account by the model through:

- **pre-processing** by changing the priors: either by stratification techniques (oversampling and undersampling) or by synthetically creating new observations [2] or even changing the class labels as done by MetaCost [3];
- **the training algorithm** may in some cases be made sensible to the misclassification costs as well. However, adding cost-sensitivity to the training algorithm is not always straight-forward and is sometimes cumbersome. Taking SVMs as an example, suggestions have been made to introduce costs in the feature space transformation by changing the kernel function [4], introducing different penalties for the positive and negative SVM slack variable ξ [5], among other approaches. Furthermore, such penalties might saturate the decision boundary as shown in Fig. 1;
- **post-processing** techniques involve using class posterior probabilities or the distance to the decision boundary, usually in the context of a ROC curve.

Training

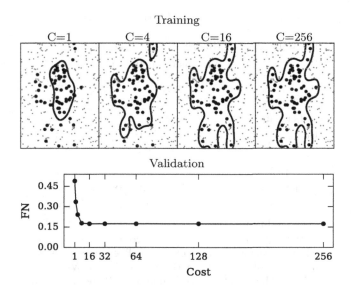

Fig. 1. SVM trained with several costs in a noisy synthetic sample. After a while, there are no gains in $\hat{\rho}$ in the validation sample.

All these solutions, however, are based on a relative trade-off between FNR and FPR, false negative and positive rates respectively. None of the approaches offers a means to define an absolute trade-off.

3   Proposal

Several proposals are explored in the following sections. The goal is to minimize true negatives while keeping false negatives under a user-specified threshold.

More concretely, if y is the endogenous label, we want to ensure $P(\hat{y} = - \mid y = +) \leq \rho$ (or, equivalently, $P(\hat{y} = + \mid y = +) \geq 1 - \rho$) in order to keep this type of error within a reasonable bound while maximizing specificity, $P(\hat{y} = - \mid y = -)$. The used-defined parameter ρ corresponds to an absolute trade-off in terms of false negatives, in contrast to the orthodox approach of using relative trade-offs between false negatives and false positives.

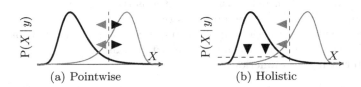

(a) Pointwise   (b) Holistic

Fig. 2. Comparing the current pointwise methodology to the proposed one.

Table 1. Datasets used for the experiments.

Dataset	Minority	N	Features	IR
Breast-cancer	Wisconsin	699	9	0.345
Car	Good	1728	6	0.040
German	–	1000	24	0.300
Haberman	–	306	3	0.265
Heart	–	270	13	0.444
Sonar	–	208	60	0.466
Transfusion	–	748	4	0.238
Vehicle	Van	846	18	0.235
Vowel	1	990	13	0.091
Winequality-red	7,8	1599	11	0.136

Acknowledgments: Datasets come courtesy of the UCI Machine Learning repository [6]. The breast cancer dataset was obtained from the University of Wisconsin Hospitals, Madison from Dr. William H. Wolberg [7]. The vehicle dataset is originally from the Turing Institute, Glasgow, Scotland. Wine-quality is originally from [8].

One way to consider this change in methodology is to consider that current approaches expand the decision boundary until $c_p P(\hat{y} = + \mid y = -) = c_n P(\hat{y} = - \mid y = +)$, see Fig. 2a, while our approach considers expanding the decision boundary of one class until the total error rate in the other class is controlled, see Fig. 2b.

Most of the approaches here presented can be described within the following general framework:

(a) a function $f \colon \mathbb{R}^n \to \mathbb{R}$ ranks how confident we are in that an observation $\mathbf{x}$ is positive, usually in the form of a probability $P(y = + \mid \mathbf{x})$;
(b) apply a threshold t such that $P(f(x) < t) = \rho$ or, in other words, find the ρ quantile of $f(x)$ for the training data.

The strategies hereon considered are:

1. Kernel density estimation
   (a) Multi-variate kernel density estimation (one-class)
   (b) Cascade of 1D kernel density estimations (two-classes)
2. Applying a threshold to the model's scores
3. Modifying the loss function, applied to gradient descent estimation
   (a) Adding a FNR term to the loss on batch training
   (b) Alternating loss function.

Along the exposition, empirical experiments are provided. These are the results of stratified k-fold with $k = 5$. For each dataset, we show scores for the training and testing set using the previously defined metrics. The datasets used are summarized in Table 1. IR is the imbalance ratio, IR $= N_+/N$, where N_+ and N are the number of positive and total observations. All empirical tests are targeting $\rho = 0.05$. The *Minority* column shows which class is being considered positive, if the dataset is multi-class.

All data and code used in the cis available from: http://vcmi.inescporto.pt/ reproducible_research/iwann2017/Type2Error/. Implementations used Python, SciPy, scikit-learn, and TensorFlow.

4   Kernel Density Estimation

4.1   Two-Class Approach

To include information of both classes, instead of multivariate KDE, we here propose each variable of the positive class to be modeled independently in multiple univariate KDE models. The negative class then guides the selection of a threshold for each model.

As an initial approach, we will be using as a model the kernel density estimation (KDE), also known as the Parzen-Rosenblatt window method.

KDE is a non-parametric way to estimate the probability density function of a random variable. Each observation is modeled by a kernel, all sharing the same parameters. Usually a Gaussian kernel is used (as we did), so that, in essence, the ensuing model is a Gaussian mixture. In the univariate case, for a given point x, the unknown function is thus being modeled as

$$p(x) = \frac{1}{nh} \sum_{k=1}^{n} K\left(\frac{x - x_k}{h}\right),$$

where K is the Gaussian kernel (the Gaussian density function for $N(0,1)$), and $h > 0$ is a smoothing hyperparameter. We have used the so-called Scott's rule [9], $h = n^{-\frac{1}{d+4}}$. n and d are the number of observations and variables, respectively.

4.2   One-Class Approach

As an initial approach, we have used multivariate KDE to model positive observations ($p_+(x_i) = p(x_i \,|\, y_i = +)$). This is a one-class model in that the model is based exclusively on positive observations. We then apply a threshold t such that

$$\hat{y}_i = \begin{cases} +, & \text{if } p_+(x_i) \geq t \\ -, & \text{if } p_+(x_i) < t. \end{cases}$$

The problem then becomes how to estimate this threshold t such that $P(p_+(x_i) < t) = \rho$. This corresponds to the ρ-quantile, so that $t = p_+(x_i)$ for $i = \lfloor \rho n_+ \rfloor$ assuming observations are ordered by $p_+(x_i)$. We have used floor approximations to ensure a ceiling on FNR.

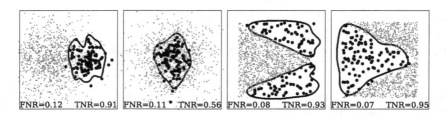

Fig. 3. One-class approach: synthetic example.

Illustrative results can be found in Fig. 3 based on synthetic samples. The first two correspond to partial and complete overlap of Gaussians while the other two correspond to a triangle inside a square. The dark and big dots correspond to observations of the positive class, while light and small dots to observations of the negative class. There is a 1–10 imbalance ratio in the illustration.

The two first columns of Table 2 exhibit empirical results for the aforementioned datasets. First column, *One-Class+*, refers to the description offered of modeling the positive class and using the positive class for the threshold, while the second column, *One-Class−*, refers to modeling the negative class while using the positive class for the threshold. Clearly, overfitting from using the same

Table 2. Evaluating KDE-based models. Train and test scores in white and gray, respectively, and $\rho = 0.05$

	One-Class +		One-Class -		Two-Class	
	FNR	TNR	FNR	TNR	FNR	TNR
breast-cancer-wisconsin	0.05	1.00	0.05	1.00	0.02	0.89
	0.98	1.00	0.06	0.97	0.03	0.88
car-good	0.04	1.00	0.04	1.00	0.00	0.93
	1.00	1.00	0.19	0.27	0.00	0.93
german	0.05	1.00	0.05	1.00	0.03	0.07
	1.00	1.00	0.06	0.07	0.06	0.07
haberman	0.05	0.11	0.05	0.14	0.01	0.04
	0.36	0.37	0.00	0.00	0.01	0.00
heart	0.04	1.00	0.04	1.00	0.02	0.13
	1.00	1.00	0.06	0.36	0.07	0.08
sonar	0.00	1.00	0.05	1.00	0.05	0.40
	1.00	1.00	0.05	0.04	0.15	0.32
transfusion	–	–	0.05	0.11	0.03	0.15
	–	–	0.07	0.03	0.07	0.15
vehicle-van	0.05	1.00	0.05	1.00	0.01	0.67
	1.00	1.00	0.06	0.66	0.02	0.67
vowel-1	0.04	1.00	0.04	1.00	0.00	0.58
	1.00	1.00	0.00	0.00	0.10	0.60
winequality-red-7,8	0.05	0.95	0.05	0.17	0.02	0.31
	0.78	0.96	0.04	0.08	0.10	0.30

positive class data for both modeling and threshold, allied with class imbalance, is preventing the results from being carried to the testset; the importance of using both classes for controlling FNR is shown to be paramount.

More concretely, at each iteration j, every variable k is modeled, $p_j(x_i^k)$. Here we relax the notation so that $p = p_+$, so that, for each variable x_i^k of the positive class, we consider one of the following thresholds, which correspond to the left and right tail of the distribution,

(1) Left

$$\hat{y}_i = \begin{cases} - & \text{if } p_j(x_i^k) < t_j \\ p_{j+1}(x_i^k) & \text{otherwise} \end{cases}$$

(2) Right

$$\hat{y}_i = \begin{cases} - & \text{if } p_j(x_i^k) > 1 - t_j \\ p_{j+1}(x_i^k) & \text{otherwise.} \end{cases}$$

In essence, at each iteration, a filter is created that clears some observations as negative. For the remaining observations, $\hat{y}_i = +$, when no observations exist below the threshold or fitting is no longer possible. For simplicity, the threshold at each iteration is fixed at $t_j = \rho/2^j$. This heuristic is chosen based on the geometric series, $\lim_{n \to \infty} \sum_{j=1}^{n} \rho/2^j = \rho$, so that $\hat{\rho} \leq \rho$, for the training set.

At each iteration j we choose: $w_j = (\text{variable} k, \text{binop}, \text{threshold } t)$ with binop $= (1)$ Left $\vee$ (2) Right. With the threshold fixed, there are yet two parameters to estimate. These two parameters are chosen by a greedy criterion based on whichever maximizes the highest TNR at the iteration, $\arg\max_{w_j} p_{-,j}$.

Illustrative results can be found in Fig. 4. This model can be seen as a cascade ensemble, as is exemplified by Fig. 5, though the final order by which the models are applied is not relevant.

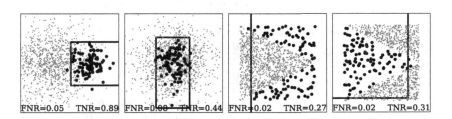

Fig. 4. Two-class approach: synthetic example.

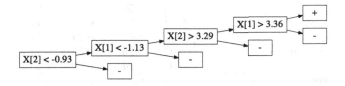

Fig. 5. Example of the cascade.

5 Scoring Threshold

A model is trained to produce scores representing the likelihood that the observation belongs to either class. A threshold is then used to find a quantile such that the FNR is contained. More concretely:

1. train the model, a model such that $f\colon \mathbb{R}^n \to \mathbb{R}$,
2. get the scores for the positive training data, $s_i = f(\mathbf{X}_i)$, $\forall i\ y_i = 1$,
3. choose a threshold t in the desired quantile, $\mathrm{P}(s_i < t) = \rho$.

A scoring function can be estimated for such models as SVM by using the distance to the hyperplane. More elegantly, rankers may also be built on top of SVM and other models so that the output is a score representing order. There is vast literature on this: we will here consider RankSVM, and compare it against SVM with linear kernel. The penalty term used was $C = 1$.

Using ranking for classification has already been used in [10] with good results for class imbalance. This is essentially the same approach, using a different threshold function.

5.1 Ranking Threshold

One possible family of methods to tackle this problem is pairwise scoring ranking models. In contrast to classifiers, which not always make it easy to apply

Table 3. Performance of threshold approach using SVM-based models ($\rho = 0.05$).

	Linear SVM		RankSVM	
	FNR	TNR	FNR	TNR
breast-cancer-wisconsin	0.05	0.98	0.05	0.98
	0.09	0.98	0.07	0.98
car-good	0.04	0.90	0.04	0.91
	0.11	0.80	0.17	0.83
german	0.05	0.39	0.05	0.41
	0.09	0.38	0.08	0.40
haberman	0.05	0.12	0.05	0.13
	0.08	0.11	0.08	0.12
heart	0.04	0.60	0.04	0.64
	0.06	0.59	0.10	0.59
sonar	0.04	0.99	0.04	0.77
	0.42	0.69	0.34	0.51
transfusion	0.05	0.26	0.05	0.27
	0.09	0.24	0.09	0.23
vehicle-van	0.05	0.99	0.05	0.99
	0.07	0.98	0.07	0.98
vowel-1	0.04	0.68	0.04	0.76
	0.26	0.68	0.27	0.76
winequality-red-7,8	0.04	0.56	0.04	0.57
	0.07	0.54	0.07	0.56

a threshold, this family of methods rank the observations making a threshold straightforward to apply (Table 3).

In ranking, observation $\mathbf{x}_i$ is compared with another observation $\mathbf{x}_j$, and we are interested in predicting whether $\mathbf{x}_i \succ \mathbf{x}_j$, meaning $\mathbf{x}_i$ is "preferred" to $\mathbf{x}_j$. In the particular case of pairwise scoring ranking, each observation $\mathbf{x}_i$ is compared against all others $\mathbf{x}_j$, and if $\mathbf{x}_i \succ \mathbf{x}_j$, then we train a function f so that $f(\mathbf{x}_i) > f(\mathbf{x}_j)$, with $f \colon X \to \mathbb{R}$ [11].

In this work, we have compared linear SVM with RankSVM [12]. In RankSVM, data is transformed into the space of differences, so the original dataset $\mathbf{X}$ becomes $\mathbf{X}'$, where $\mathbf{x}'_{ij} = \mathbf{x}_i - \mathbf{x}_j$ and $\mathbf{x}'_{ji} = \mathbf{x}_j - \mathbf{x}_i$, for all pairs (i, j) such that $y_i \neq y_j$, with $y'_{ij} = y_i$ and $y'_{ji} = y_j$. When it comes to training, RankSVM makes use of a linear SVM as a base estimator to classify observations within the space of differences, where the decision rule $\mathbf{w} \cdot (\mathbf{x}_i - \mathbf{x}_j) > 0$ can be transformed into a scoring function since $\mathbf{w} \cdot (\mathbf{x}_i - \mathbf{x}_j) > 0 \Leftrightarrow \mathbf{w} \cdot \mathbf{x}_i > \mathbf{w} \cdot \mathbf{x}_j \Leftrightarrow s(\mathbf{x}_i) > s(\mathbf{x}_j)$.

6 Gradient Descent Training

Several modifications to neural networks were considered. Neural networks seemed particularly enticing because they are traditionally trained using gradient descent. In the particular case of batch gradient descent, the gradient is computed using the entire training set, allowing us to estimate the FNR, and redirect the training process to keep $\hat{\rho} \leq \rho$.

Here we shall consider the following approaches:

1. Adding a term to the loss function;
2. Switching between loss functions.

6.1 Modifying the Loss Function

The term $\hat{\rho} = \frac{1}{N_1} \sum_i (1 - \hat{y}_i) y_i$ is considered to be an approximation to the false negative rate, so that

$$\text{Loss} = \frac{1}{2} \sum (y - \hat{y})^2 + \boxed{\exp(\rho - \hat{\rho}) - 1,}$$

where $\hat{\rho} = \frac{1}{N_1} \sum_i (1 - \hat{y}_i) y_i$ ($\approx$FNR). Of course, $\hat{y}_i \in [0, 1]$, but $y_i \in \{0, 1\}$.

The proposed loss function term is plotted in Fig. 6. It is inspired in the exponential utility function from microeconomics. False negatives can be seen as pollution: we do not want zero pollution (false negatives) because pollution is the byproduct of things we do want (in this case, true negatives), so we are willing to tolerate some pollution as a trade-off. But the more pollution there is, the less willing we are to tolerate it.

Some results are reproduced in Table 4. Here, desired $\rho = 0.05$. There is a big discrepancy between the training and evaluation results; solutions could encompass such things as regularization or evaluating $\hat{\rho}$ in a different sample while training. TensorFlow was used, with 10 hidden neurons. The table contrasts the previous approach (column *Threshold*) with this approach.

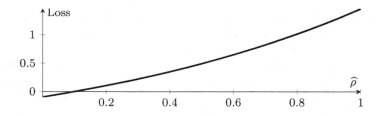

Fig. 6. Loss function term: $\exp(\widehat{\rho} - \rho) - 1$, here with $\rho = 0.10$.

Table 4. Results for the modified and alternating loss function ($\rho = 0.05$).

	Threshold		Loss Factor		Alternating	
	FNR	TNR	FNR	TNR	FNR	TNR
breast-cancer-wisconsin	0.05	0.98	0.00	0.95	0.02	0.98
	0.06	0.97	0.00	0.95	0.05	0.97
car-good	0.04	0.85	0.00	0.85	0.03	0.92
	0.07	0.85	0.00	0.85	0.14	0.86
german	0.05	0.30	0.10	0.51	0.03	0.61
	0.05	0.28	0.13	0.49	0.19	0.55
haberman	0.05	0.12	0.00	0.00	0.03	0.06
	0.06	0.11	0.00	0.00	0.09	0.12
heart	0.04	0.47	0.07	0.70	0.02	0.88
	0.06	0.47	0.10	0.64	0.22	0.74
sonar	0.04	0.57	0.03	0.62	0.00	1.00
	0.07	0.49	0.07	0.57	0.41	0.62
transfusion	0.05	0.26	0.00	0.00	0.04	0.21
	0.06	0.26	0.00	0.00	0.06	0.20
vehicle-van	0.05	0.56	0.00	0.63	0.02	0.99
	0.06	0.56	0.01	0.63	0.05	0.98
vowel-1	0.04	0.61	0.01	0.73	0.05	0.93
	0.04	0.60	0.03	0.72	0.51	0.85
winequality-red-7,8	0.04	0.48	0.08	0.61	0.02	0.41
	0.06	0.48	0.08	0.60	0.03	0.40

6.2   Alternating Loss Function

The optimization search within the neural network is changed so that at each epoch the loss function alternates:

1: **if** $\widehat{\rho} < \rho$ **then**
2:     controlled ← true
3: **if** controlled = true **and** $\hat{\rho} < \hat{\rho} + \sigma$ **then**
4:     compute gradient using positive cases only
5: **else**
6:     controlled ← false
7:     compute gradient using entire sample

Some results are presented in Table 4, contrasting with the threshold and previous section approaches. Worth noticing that too often the number of iterations in *controlling* reached close to 1, which indicated little computation was spent on specificity.

7   Discussion and Future Work

One major difficulty shows to be keeping the training in tandem with the validation results. Solutions could encompass (a) aggressive regularization strategies, (b) evaluating $\widehat{\rho}$ in a different sample while training, and (c) using a smaller desired TNR value $\rho' = \eta\rho$ with $0 < \eta < 1$ and obtained by cross-validation to ensure desired TNR is controlled.

Modeling observations using a density function lack expressiveness of more powerful models, but this lacks of expressiveness can also help in ensuring train-test FNR balance. A simple post-processing threshold method have proven simple and maybe acceptable, especially when used in tandeum with a scoring pairwise ranker. Most importantly, modifying the loss function of gradient descent has shown the most promising.

Possibly, rule induction could provide fruitful models for FNR-constraining. Models such as PNrule offer an interesting framework [13], related to our two-class KDE (Sect. 4.1). Positive and Negative rules are constructed, each with an associated level of PN or FN rate, which are then applied in sequence. Firstly, Positive rules are applied to reject examples; only then Negative rules are applied on top of the rejected examples, optimizing recall and precision in separate.

In multiclass scenarios, this new formulation to classification would be highly specific to the application. Suggestions would be to use a different ρ for each minority class or aggregate minority classes and then subclassify among them.

8   Conclusion

We have started by proposing a new learning problem: instead of defining a relative trade-off using cost matrices, we suggest it might be useful in some cases for learning algorithms to allow defining an absolute trade-off in the form of false negative threshold. In the terminology we have used, we try to maximize specificity (true negatives) while keeping Type II errors, false negatives, within a certain bound.

We suggest several learning estimators to be used in conjunction with traditional models which could help with this objective. These comprised: kernel density estimation, scoring thresholds and gradient descent learning. Scoring thresholds have shown to be simple and effective, albeit iterative methods like gradient descent are more malleable and may offer grounds for improvement.

Two problems and conclusions are arrived. (I) It is not easy to improve on a simple scoring threshold when considering specificity. (II) A big difficulty arises in keeping FNR at bay when using the estimated model on validation data.

Acknowledgment. This work was funded by the Project "NanoSTIMA: Macro-to-Nano Human Sensing: Towards Integrated Multimodal Health Monitoring and Analytics/NORTE-01-0145-FEDER-000016" financed by the North Portugal Regional Operational Programme (NORTE 2020), under the PORTUGAL 2020 Partnership Agreement, and through the European Regional Development Fund (ERDF), and also by Fundação para a Ciência e a Tecnologia (FCT) within PhD grant numbers SFRH/BD/122248/2016 and SFRH/BD/93012/2013.

References

1. Bessa, S., Domingues, I., Cardoso, J.S., Passarinho, P., Cardoso, P., Rodrigues, V., Lage, F.: Normal breast identification in screening mammography: a study on 18 000 images. In: IEEE International Conference on Bioinformatics and Biomedicine (BIBM), pp. 325–330. IEEE (2014)
2. Chawla, N.V., Bowyer, K.W., Hall, L.O., Philip Kegelmeyer, W.: SMOTE: synthetic minority over-sampling technique. J. Artif. Intell. Res. **16**, 321–357 (2002)
3. Domingos, P.: MetaCost: a general method for making classifiers cost-sensitive. In: Proceedings of the Fifth International Conference on Knowledge Discovery, vol. 55, pp. 155–164 (1999)
4. Gang, W., Chang, E.: Class-boundary alignment for imbalanced dataset learning. Twent. Int. Conf. Mach. Learn. (ICML) **1**, 49–56 (2003)
5. Bach, F.R.: Considering cost asymmetry in learning classifiers. JMLR **7**, 1713–1741 (2006)
6. Lichman, M.: UCI Machine Learning Repository (2013)
7. Mangasarian, O.L., Street, W.N., Wolberg, W.H.: Breast cancer diagnosis and prognosis via linear programming. Oper. Res. **43**(4), 570–577 (1995)
8. Cortez, P., Cerdeira, A., Almeida, F., Matos, T., Reis, J.: Modeling wine preferences by data mining from physicochemical properties. Decis. Support Syst. **47**(4), 547–553 (2009)
9. Scott, D.W.: Multivariate Density Estimation. Wiley Series in Probability and Statistics. Wiley, Hoboken (1992)
10. Cruz, R., Fernandes, K., Cardoso, J.S., Costa, J.F.P.: Tackling class imbalance with ranking. In: International Joint Conference on Neural Networks (IJCNN). IEEE (2016)
11. Li, H.: Learning to Rank for Information Retrieval and Natural Language Processing, vol. 4 (2011)
12. Herbrich, R., Graepel, T., Obermayer, K.: Support vector learning for ordinal regression a risk formulation for ordinal regression. In: Proceedings of the Ninth International Conference on Artificial Neural Networks, pp. 97–102 (1999)
13. Agarwal, R., Joshi, M.V.: Pnrule: a new framework for learning classifier models in data mining (a case-study in network intrusion detection). In: Proceedings of the SIAM International Conference on Data Mining, pp. 1–17. SIAM (2001)

Surveillance and Rescue Systems and Algorithms for Unmanned Aerial Vehicles

Pedestrian Detection for UAVs Using Cascade Classifiers and Saliency Maps

Wilbert G. Aguilar[1,3,4(✉)], Marco A. Luna[2], Julio F. Moya[2],
Vanessa Abad[5], Hugo Ruiz[1,6], Humberto Parra[1,7], and Cecilio Angulo[4]

[1] Dep. Seguridad y Defensa,
Universidad de las Fuerzas Armadas ESPE, Sangolquí, Ecuador
wgaguilar@espe.edu.ec
[2] Dep. Eléctrica y Electrónica,
Universidad de las Fuerzas Armadas ESPE, Sangolquí, Ecuador
[3] CICTE Research Center,
Universidad de las Fuerzas Armadas ESPE, Sangolquí, Ecuador
[4] GREC Research Group, Universitat Politècnica de Catalunya,
Barcelona, Spain
[5] Universitat de Barcelona, Barcelona, Spain
[6] PLM Research Center, Purdue University, West Lafayette, IN, USA
[7] Universidad Politécnica de Madrid, Madrid, Spain

Abstract. In this paper, we proposed algorithm and dataset for pedestrian detection focused on applications with micro multi rotors UAV (Unmanned Aerial Vehicles). For training dataset we capture images from surveillance cameras at different angles and altitudes. We propose a method based on HAAR-LBP (Local Binary Patterns) cascade classifiers with Adaboost (Adaptive Boosting) training and, additionally we combine cascade classifiers with saliency maps for improving the performance of the pedestrian detector. We evaluate our dataset by the implementation of the HOG (Histogram of oriented gradients) algorithm with Adaboost training and, finally, algorithm performance is compared with other approaches from the state of art. The results shows that our dataset is better for pedestrian detection in UAVs, HAAR-LBP have better characteristics than HAAR like features and the use of saliency maps improves the performance of detectors due to the elimination of false positives in the image.

Keywords: HAAR · HOG · LBP · Saliency maps · People detection · Cascade classifiers · UAVs

1 Introduction

In the field of computer vision, there are several applications in object detection, one of these applications is pedestrian detection, used on surveillance [1, 2], robotics [3, 4, 5, 6], navigation [7, 8, 9], driver assistance systems particularly in PPSs(pedestrian protection systems) [10, 11], and others. In the state of art multiple feature extraction algorithms working with machine learning and datasets have been created to deal with this problem.

© Springer International Publishing AG 2017
I. Rojas et al. (Eds.): IWANN 2017, Part II, LNCS 10306, pp. 563–574, 2017.
DOI: 10.1007/978-3-319-59147-6_48

Developments in computer vision have been introduced for UAVs [12, 13, 14]. Pedestrian detection can be used with UAVs taking into consideration that they have a complex dynamic and altitude variation that adding extra challenges to the detection [15, 16]. Conventional classifiers fail when altitude increases generating more false positives.

Our proposal for pedestrian detection in UAVs considers the altitude and introduces the CICTE-PeopleDetection dataset with images captured from surveillance cameras. We use two trained algorithms: The first one based on a combination of the feature extraction methods HAAR-LBP, and the second one based on HOG. Both algorithms use cascade classifiers with Adaboost training. In addition we propose an algorithm that merges Saliency Maps algorithm presented on [17] with cascade classifier to provide detection robustness. Our proposal is evaluated in images captured from UAVs in different scenarios.

This paper is organized as follows: Sect. 2 describes the related work on pedestrian detection. Next, our proposal for pedestrian detection, the creation of dataset and the algorithm are described in Sect. 3. In Sect. 4 we present the experimental results, followed by the summary. Finally conclusions and future works are presented in Sect. 5.

2 Related Works

In the literature, several research groups have created different datasets and methods for pedestrian detection. INRIA was introduced on [18], with training based on Histograms of Oriented Gradients (HOG). Widely used datasets are Caltech Pedestrian Dataset [19] and KITTI [20], due to they are comparatively large and challenging. According to [21, 22] there are two types of datasets: photo datasets and video datasets. Photo datasets like MIT [23], CVC [11], NICTA [24] aboard the classification problem: train binary classification algorithms. Video datasets as ETH [25], TUD-Brussels [26] or Dalmier (DB) [27] are focused on detection problem: design and test full image detection systems and human locomotion modeling.

Two important algorithms have been developed for pedestrian detection and object detection in general: Haar-like features [28] by Viola and Jones, and Dalal and Triggs algorithm called HOG [18]. Both algorithms have generated over 40 new approaches [21]. Several methods for pedestrian detection includes feature extraction algorithms: HAAR [28], HOG [18, 29], HOG-HAAR [30] and HOG-LBP [31]; working with machine learning approaches based on SVMs [18, 32] or Adaboost [11, 27].

The applications of pedestrian detection in UAVs are manifold: Human safety [33], rescue and monitoring missions [34, 35], track people systems [32, 36], and others. One of the challenges of pedestrian detection in UAVs is the camera perspective variations that deform the images. In [37, 38], they use thermal imagery combined with cascade classifiers to perform the detection. Few papers like [35] works on altitudes around five meters. In this paper, authors propose post-disaster victims detection with cascade classifier methods. In UAVs, the use of saliency maps is widely used to object and motion detection in aerial images [35, 39]. Works like [34] use saliency maps to detect people reducing the search space, choosing randomly bounding boxes to detect people inside saliency region and treating separately all detection windows; they fuse the results using mean-shift procedure applied in flights from 10 to 40 m of altitude.

3 Our Approach

3.1 Dataset Creation

One of the reasons for introducing our dataset is the requirement to detect people from UAV cameras. The main problem in pedestrian detection is the high altitude, where people images have deformation of their characteristics. The main difference of CICTE-PeopleDetection with previous photo datasets is the location and perspective of the cameras that emulate the onboard camera perspective of the UAV. We use surveillance cameras for photo dataset creation due to UAVs video captures are stable and comparable with fixed cameras. There are approximately 100 cameras (we can not specify the exactly number of cameras for security reasons) with D1 resolution located in the University between 2.3 m and 5 m of height looking down as shown in Fig. 1.

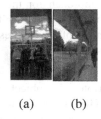

(a)   (b)

Fig. 1. Location of the cameras in the campus (a) 5 m height. (b) 2.3 m

For training we need positive and negative images. Positive images are the images that contain the object to be detected, in our case pedestrians. Negative images are frames without pedestrians. Our dataset has 3900 positive images and 1212 negative images. The positives images were captured in the Universidad de las Fuerzas Armadas ESPE during the day and the night in different scenarios, and contain entire and partial occluded people samples.

3.2 Training Process

Our approach consists in the combination of two algorithms for extraction of the feature set: Local Binary Patterns (LBP) and Haar-like features. We use Adaptive Boosting (AdaBoost) as training algorithm and a combination of Haar-LBP features due to them are algorithms of low computation time. To create our Haar-LBP algorithm we divided the all images in 70% for training and the other 30% for testing, after that we use the algorithm with a UAV images in different scenarios. Additionally, we train a HOG cascade classifier and compare it with Opencv HOG to validate our Dataset. The training processes are shown Fig. 2.

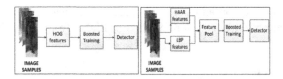

Fig. 2. Pedestrian detection training. (a) HOG features with Adaboost. (b) Haar-LBP features with Adaboost

The methods used for training the cascade classifiers are described as follows:

Local Binary Patterns (LBP)
This feature extractor was presented in [40] as a texture descriptor for object detection, and compares a central pixel with the neighbours. The window to be examined is separated into cells of 16 × 16 pixels. 8 neighbours are considered for each pixel inside the cell, the central pixel value is the threshold. A value of 1 is assigned if the neighbour is greater or equal to the central pixel, otherwise the value is 0.

Haar-like Features
Viola and Jones uses a statistical approach for the tracking and detection problem, describing the ratio between light and dark areas within a defined kernel. This algorithm is robust regarding to noise and lighting changes. The method uses simple feature sets similar to Haar basis functions [28, 41].

Histogram of Oriented Gradients (HOG)
This algorithm is a feature descriptor for object detection focused on pedestrian detection and introduced in [18]. The image window is separated into smaller parts called cells. For each cell, we accumulate a local 1-D histogram of gradient orientations of the pixels in the cell. Each cell is discretized into angular bins according to the gradient orientation and each pixel of the cell contributes with a gradient weight to its corresponding angular bin. The adjacent cells are grouped in special regions called blocks and the normalized group of histograms represents the block histogram.

Adaboost
Adaboost is a machine learning algorithm [42] that initially keeps uniform distribution of weights in each training sample. In the first iteration the algorithm trains a weak classifier using a feature extraction methods or mix of them achieving a higher recognition performance for the training samples. In the second iteration, the training samples, misclassified by the first weak classifier, receive higher weights. The new selected feature extraction methods should be focused in these misclassified samples.

3.3   People Detection Algorithm

In order to get a better performance of the classifier we implement a combination of cascade classifier with saliency maps, an algorithm presented in [17]. The purpose of saliency maps is to locate prominent areas at every location in the visual field. The areas with high saliency correspond to objects or places they are most likely to be found, and the areas with lower saliency are associated to background [43]. The saliency maps algorithms are deduced by convolving the function f by an isotropic bi-dimensional Gaussian function [44]:

$$S(X) = f(X)G_\sigma(X) \tag{1}$$

where σ is the standard deviation of the Gaussian function. The standard deviation depends on the experimental setup (size of the screen and viewing distance). To eliminate the false positives in the image we obtain the salient region; we consider a threshold from the salient map and we create a mask where values greater than threshold will belong to salient map. Additionally, this region was dilated to give it robustness. This algorithm is shown in Fig. 3.

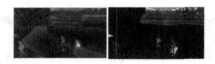

Fig. 3. Saliency maps algorithm. (a) Saliency map (b) Saliency region.

Once it has been obtained the salient region, our algorithm proposes take as true positives only the cascade classifier detections inside this region. For this reason we take the salient region as Region of Interest (ROI). To determinate if a detection bounding box is inside the salient region, we compute the center point of the bounding box with the formulas:

$$x_m = x + \left(\frac{w}{2}\right); \quad y_m = y + \left(\frac{h}{2}\right) \tag{2}$$

where x and y are the horizontal and vertical coordinates of the top left of the bounding box, x_m and y_m are the coordinates of the central point and w, h are the width and height. We take the center point as reference to avoid false positives that could have small parts of their bounding box in salient regions. Unlike other methods presented in the literature [34], we use our own algorithm for combination of cascade classifier with saliency maps. Our proposal is presented graphically in Fig. 4.

The results of the application of this algorithm are presented in the Sect. 4.

4   Results and Discussion

4.1   Dataset and Training Evaluation

The metric of evaluation for our approach is based on the sensitivity (true positive rate-TPR) and the miss rate (False negative rate-FNR). Defined as follows:

$$TPR = \frac{TP}{TP + FN} * 100\%; \quad FNR = \frac{FN}{TP + FN} * 100\% \tag{3}$$

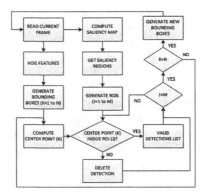

Fig. 4. Algorithm for people detection using HOG cascade classifier and saliency Maps

For the dataset evaluation we have trained the cascade classifier based on HOG features and compared this classifier with the OpenCV HOG cascade classifier. We tested the cascade classifier with videos captured from UAVs. Experimental results are presented in Table 1.

Table 1. Dataset training performance

Algorithm	Sensitivity (%)	Miss rate (%)
HOG-CICTE	**55.71%**	**44.28%**
HOG [45]	35.71%	64.28%

In this table, two cascade classifiers are compared: HOG-CICTE PeopleDetection and a HOG cascade classifier with the Adaboost training from the OpenCV library. Result shows our approach has better performance, the miss rate of our proposal is 20% lower than the conventional classifier miss rate, and the sensitivity is higher. ROC curves for comparing both algorithms are presented in Fig. 5.

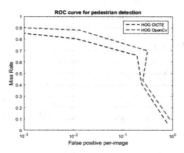

Fig. 5. Comparison of ROC curves for HOG from OpenCV cascade classifier and HOG-CICTE cascade classifier trained for pedestrian detection in UAVs.

In Fig. 5, HOG-CICTE classifier has a better performance that HOG from OpenCV cascade classifier in videos captured from UAVs.

4.2 Algorithm Evaluation

For the algorithm evaluation we are using 3 scenarios with 3 different altitudes. We compare HAAR-LBP features and HOG features (trained with CICTE-PeopleDetection) respect to other cascade classifiers. Results are presented in Table 2.

Table 2. Cascade classifiers performance

Algorithm	Altitude (m)	Sensitivity (%)	Miss rate (%)
HAAR-LBP	2	10.66%	89.34%
	3	7.18%	92.82%
	4	5.43%	94.57%
HAAR [28]	2	3.27%	96.73%
	3	2.28%	97.72%
	4	1.68%	98.32%
LBP [46]	2	41.98%	58.02%
	3	33.18%	66.82%
	4	23.09%	76.91%
HOG CICTE	2	**67.23%**	**32.76%**
	3	63.72%	36.27%
	4	60.44%	39.56%

In Table 2, the combination of HAAR-LBP features has low sensitivity compared with the other methods; however the proposal is higher that HAAR features. With altitude increasing, sensitivity decrease in all cascade classifiers. Performance curves are presented in the Fig. 6.

In the Fig. 6, the performance of the HAAR-LBP features algorithm is better than HAAR individually applied. HAAR-LBP features generate a lower rate of false positives. True positive rate of HAAR-LBP features is higher than HAAR features but

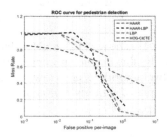

Fig. 6. Comparison of ROC curves for different approaches.

lower than LBP. Nevertheless, the HOG-CICTE cascade classifier still has the best performance due to its higher true positives rate and lower false positives rate.

4.3   Cascade Classifier-Saliency Maps Combination

Based on the results of performance we choose HOG CICTE cascade classifier to implement our algorithm. Graphical results are shown in the Fig. 7 and video results are provided by: https://www.youtube.com/watch?v=KN_hVgp1_t4

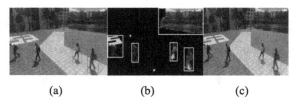

(a)   (b)   (c)

Fig. 7. Combination of Saliency Maps and Cascade classifier. (a) Cascade classifier result (b) saliency regions (c) final Result

As we can see in the Fig. 7, the use of saliency region helps to reject the false positives in the images. For the evaluation we take an additional metric of evaluation that is precision or positive predictive value (PPV), given by:

$$PPV = \frac{TP}{TP + FP} * 100\%$$

where TP are the true positive values and FP are the false positives. The results of precision of the detector with the application of the saliency region algorithm (SR) are shown in the Table 3.

Table 3. Comparison of precision between algorithms

Algorithm	Precision (%)
HOG-CICTE	72.23%
HOG-CICTE +SR	**92.1%**

In Table 3, the application of saliency region algorithm improves the precision of detection in 20% approximately, this denote an improvement in the performance too. The performance curves of two algorithms are shown in the Fig. 8.

The Fig. 8 shows that the use of saliency region algorithm improves the detection performance eliminating false positives.

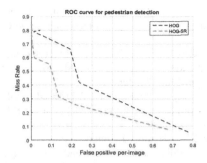

Fig. 8. ROC curves for HOG and HOG-SR

5 Conclusions and Future Work

Our proposal for pedestrian detection based on HOG features has higher performance that OpenCV HOG respect to sensitivity and miss rate (with an improvement of 20%), as shown in the Table 1 and Fig. 5, because the images used for training emulate UAVs perspective.

In order to improve the HAAR algorithm performance we combine two algorithms (HAAR and LBP). The sensitivity increased and the miss rate decreased as shows Table 2 and Fig. 6; however the performance is lower in comparison with HOG-CICTE and LBP algorithm. When the altitude increased from 2 to 4 meters, the sensitivity decreased in the four algorithms. Comparing HAAR-LBP and HAAR, HAAR-LBP has a better performance even in the altitude of 4 m.

The use of saliency maps improves the performance detectors, saliency map helps to eliminate background regions even in mobile cameras like UAVs, and these regions may contain objects that confuse the classifier that is important to decrease the number of false positives.

In the future is necessary to improve the detection. We will train new classifiers with images captured from UAVs, taking into consideration other human body parts like face, head, shoulders, etc. In addition, a robust of detection could be used for many applications like people tracking or people avoidance systems.

Acknowledgement. This work is part of the projects VisualNavDrone 2016-PIC-024 and MultiNavCar 2016-PIC-025, from the Universidad de las Fuerzas Armadas ESPE, directed by Dr. Wilbert G. Aguilar.

References

1. Torresan, H.: Advanced surveillance systems: combining video and thermal imagery for pedestrian detection. In: Proceedings of SPIE, pp. 506–515 (2004)
2. Zhang, L.Z.L., Wu, B.W.B., Nevatia, R.: Pedestrian detection in infrared images based on local shape features. In: 2007 IEEE Conference on Computer Vision and Pattern Recognition, pp. 1–8 (2007)

3. Aguilar, W.G., Angulo, C., Costa, R., Molina, L.: Control autónomo de cuadricopteros para seguimiento de trayectorias. In: Memorias del IX Congreso de Ciencia y Tecnología ESPE 2014 (2014)

4. Aguilar, W.G., Angulo, C.: Compensación de los Efectos Generados en la Imagen por el Control de Navegación del Robot Aibo ERS 7. In: Memorias del VII Congreso de Ciencia y Tecnolgia, ESPE 2012, pp. 165–170, June 2012

5. Jafari, O.H., Mitzel, D., Leibe, B.: Real-time RGB-D based people detection and tracking for mobile robots and head-worn cameras. In: Proceedings of IEEE International Conference on Robotics and Automation, pp. 5636–5643, April 2016

6. Kobilarov, M., Sukhatme, G., Hyams, J., Batavia, P.: People tracking and following with mobile robot using an omnidirectional camera and a laser. In: 2006 IEEE International Conference on Robotics and Automation 2006 ICRA, pp. 557–562, May 2006

7. Aguilar, W.G., Casaliglla, V., Pólit, J.: Obstacle avoidance based-visual navigation for micro aerial vehicles. Electronics 6(1), 10 (2017)

8. Cabras, P., Rosell, J., Pérez, A., Aguilar, W.G., Rosell, A.: Haptic-based navigation for the virtual bronchoscopy. In: 18th IFAC World Congress, Milano, Italy (2011)

9. Aguilar, W.G., Morales, S.: 3D environment mapping using the Kinect V2 and path planning based on RRT algorithms. Electronics 5(4), 70 (2016)

10. Gavrila, D.M.: Pedestrian detection from a moving vehicle. In: Proceedings of the 6th European Conference on Computer Vision, vol. 1843, pp. 37–49 (2000)

11. Gerónimo, D., López, A.M., Sappa, A.D., Graf, T.: Survey of pedestrian detection for advanced driver assistance systems. IEEE Trans. Pattern Anal. Mach. Intell. 32(7), 1239–1258 (2010)

12. Aguilar, W.G., Angulo, C.: Real-time video stabilization without phantom movements for micro aerial vehicles. EURASIP J. Image Video Process. 1, 1–13 (2014)

13. Aguilar, W.G., Angulo, C.: Real-time model-based video stabilization for microaerial vehicles. Neural Process. Lett. 43(2), 459–477 (2016)

14. Aguilar, W.G., Angulo, C.: Robust video stabilization based on motion intention for low-cost micro aerial vehicles. In: 2014 11th International Multi-Conference on Systems, Signals Devices (SSD), pp. 1–6 (2014)

15. Rudol, P., Doherty, P.: Human Body Detection and Geolocalization for UAV Search and Rescue Missions Using Color and Thermal Imagery

16. Aguilar, W.G., Luna, M.A., Moya, J.F., Abad, V., Parra, H., Ruiz, H.: Pedestrian detection for UAVs using cascade classifiers with meanshift. In: 2017 IEEE 11th International Conference on Semantic Computing (ICSC), pp. 509–514 (2017)

17. Itti, L., Koch, C., Niebur, E.: A model of saliency-based visual attention for rapid scene analysis. IEEE Trans. Pattern Anal. 20, 1254–1259 (1998)

18. Dalal, N., Triggs, W.: Histograms of oriented gradients for human detection. In: 2005 IEEE Computer Society Conference on Computer Vision Pattern Recognition CVPR 2005, vol. 1, no. 3, pp. 886–893 (2004)

19. Dollár, P., Wojek, C., Schiele, B., Perona, P.: Pedestrian detection: a benchmark. In: 2009 IEEE Computer Society Conference on Computer Vision and Pattern Recognition Workshops, CVPR Workshops 2009, pp. 304–311 (2009)

20. Geiger, A., Lenz, P., Urtasun, R.: Are we ready for autonomous driving? The KITTI vision benchmark suite. In: Proceedings of the IEEE Computer Society Conference on Computer Vision and Pattern Recognition, pp. 3354–3361 (2012)

21. Benenson, R., Omran, M., Hosang, J., Schiele, B.: Ten years of pedestrian detection, what have we learned? In: Proceedings of Computer Vision-ECCV 2014 Workshop, pp. 613–627 (2014)

22. Dollár, P., Wojek, C., Schiele, B., Perona, P.: Pedestrian detection: an evaluation of the state of the art. IEEE Trans. Pattern Anal. Mach. Intell. **34**(4), 743–761 (2012)

23. Papageorgiou, C., Poggio, T.: Trainable system for object detection. Int. J. Comput. Vis. **38** (1), 15–33 (2000)

24. Overett, G., Petersson, L., Brewer, N., Andersson, L., Pettersson, N.: A new pedestrian dataset for supervised learning. In: Proceedings of IEEE Intelligent Vechiles Symposium, pp. 373–378 (2008)

25. Ess, A., Leibe, B., Schindler, K., van Gool, L.: Robust multiperson tracking from a mobile platform. IEEE Trans. Pattern Anal. Mach. Intell. **31**(10), 1831–1846 (2009)

26. Wojek, C., Walk, S., Schiele, B.: Multi-Cue onboard pedestrian detection. In: 2009 IEEE Computer Society Confernce on Computer Vision Pattern Recognition Workshops CVPR Workshops 2009, pp. 794–801 (2009)

27. Enzweiler, M., Gavrila, D.M.: Monocular pedestrian detection: survey and experiments. IEEE Trans. Pattern Anal. Mach. Intell. **31**(12), 2179–2195 (2009)

28. Viola, P., Jones, M.: Rapid object detection using a boosted cascade of simple features. In: Conference on Computer Vision Pattern Recognition, pp. 1–9 (2001)

29. Zhu, Q., Avidan, S., Yeh, M.C., Cheng, K.T.: Fast human detection using a cascade of histograms of oriented gradients. IEEE Conf. Comput. Vis. Pattern Recognit. **2**, 1491–1498 (2006)

30. Wojek, C., Schiele, B.: A performance evaluation of single and multi-feature people detection. In: Pattern Recognition Symposium, pp. 82–91 (2008)

31. Wang, X., Han, T.X., Yan, S.: An HOG-LBP human detector with partial occlusion handling. In: 2009 IEEE 12th International Conference on Computer Vision ICCV, pp. 32–39 (2009)

32. Imamura, Y., Okamoto, S., Lee, J.H.: Human tracking by a multi-rotor drone using HOG features and linear SVM on images captured by a monocular camera. In: Proceedings of the International MultiConference of Engineers and Computer Scientists, vol. 1, pp. 8–13 (2016)

33. Lioulemes, A., Galatas, G., Metsis, V., Mariottini, G.L., Makedon, F.: Safety challenges in using AR. drone to collaborate with humans in indoor environments. In: Proceedings of 7th International Conference on Pervasive Technologies Related to Assistive Environments, p. 33 (2014)

34. Blondel, P., Potelle, A., Pegard, C., Lozano, R.: Human detection in uncluttered environments: from ground to UAV view. In: 2014 13th International Conference on Control Automation Robotics and Vision ICARCV 2014, pp. 76–81 (1997)

35. Andriluka, M., Schnitzspan, P., Meyer, J., Kohlbrecher, S., Petersen, K., Von Stryk, O., Roth, S., Schiele, B.: Vision based victim detection from unmanned aerial vehicles. In: 2010 IEEE/RSJ International Conference on Intelligent Robot and System (IROS), pp. 1740–1747, October 2010

36. De Smedt, F., Hulens, D., Goedeme, T.: On-board real-time tracking of pedestrians on a UAV. In: IEEE Computer Society Conference on Computer Vision and Pattern Recognition Workshop, pp. 1–8, October 2015

37. Rudol, P., Doherty, P., Science, I.: Human body detection and geolocalization for UAV search and rescue missions using color and thermal imagery. In: 2008 IEEE Aerospace Conference, pp. 1–8 (2008)

38. Gąszczak, A., Breckon, T.P., Han, J.: Real-time people and vehicle detection from UAV imagery. In: IS&T/SPIE Electron. Imaging, pp. 8–11, January 2011

39. Siam, M., Elhelw, M.: Robust autonomous visual detection and tracking of moving targets in UAV imagery. In: Proceedings of International Conference on Signal Process (ICSP), vol. 2, pp. 1060–1066, December (2012)

40. Wang, L., He, D.: Texture classification using texture spectrum. Pattern Recognit. **23**, 905–910 (1990)
41. Papageorgiou, C.P. Oren, M.: A general framework for object detection. In: IEEE International Conference on Computer Vision, pp. 555–562, January 1998
42. Schapire, R.E., Singer, Y.: Improved boosting algorithms using confidence-rated predictions. Mach. Learn. **37**(3), 297–336 (1999)
43. Moosmann, F., Larlus, D., Jurie, F.: Learning saliency maps for object categorization. In: International Workshop on the Representation and Use of Prior Knowledge in Vision (2006)
44. Le Meur, O., Baccino, T.: Methods for comparing scanpaths and saliency maps: strengths and weaknesses. Behav. Res. Methods **45**, 251–266 (2013)
45. Dalal, N., Triggs, B.: Histograms of oriented gradients for human detection. In: 2005 IEEE Computer Society Conference on Computer Vision Pattern Recognition, vol. 1, pp. 886–893 (2005)
46. Ojala, T., Pietikäinen, M., Mäenpää, T.: A generalized local binary pattern operator for multiresolution gray scale and rotation invariant texture classification. Adv. Pattern Recognit. **2013**, 399–408 (2001)

Obstacle Avoidance for Flight Safety on Unmanned Aerial Vehicles

Wilbert G. Aguilar[1,2,3(✉)], Verónica P. Casaliglla[2], José L. Pólit[2],
Vanessa Abad[4], and Hugo Ruiz[1,5]

[1] Dep. Seguridad y Defensa,
Universidad de las Fuerzas Armadas ESPE, Sangolquí, Ecuador
`wgaguilar@espe.edu.ec`
[2] CICTE Research Center,
Universidad de las Fuerzas Armadas ESPE, Sangolquí, Ecuador
[3] GREC Research Group, Universitat Politècnica de Catalunya,
Barcelona, Spain
[4] Universitat de Barcelona, Barcelona, Spain
[5] PLM Research Center, Purdue University, West Lafayette, IN, USA

Abstract. In this paper, we propose an obstacle avoidance system for UAVs using a monocular camera. For detecting obstacles, the system compares the image obtained in real-time from the UAV with a database of obstacles that must be avoided. In our proposal, we include the feature point detector Speeded Up Robust Features (SURF) for fast obstacle detection and a control law, with a defined obstacle as target. The system was tested in real-time on a micro aerial vehicle (MAV), to detect and avoid obstacles on unknown environment, and compared with related works.

Keywords: UAVs · Obstacle detection · SURF · Control system · Obstacle avoidance

1 Introduction

Unmanned Aerial Vehicles (UAV) are applied on several applications like mapping, journalism, transport, rescue military applications and environments where a human cannot access [1]. Kendoul [2] classified UAVs into five categories according to the size and payload: Full-scale, Medium-scale, Small-scale, Mini, and Micro air vehicles (MAVs).

Most of the commercial MAVs depend on the skill of the pilot, the robustness of the communication system and sensors on board the vehicle. Sometimes there is difficult to handle the device due to loss of visibility or because the global position system (GPS) is not available. Autonomous system is an alternative for solving this issue. The autonomous systems include motion planning [3, 4], path tracking, obstacle avoidance, target detection and other areas [5, 6]. These systems require sensing, state estimation, perception, and knowledge of the situation. The perception is used to detect and avoid obstacles in real time, recognize and tracking objects and environmental mapping [7].

© Springer International Publishing AG 2017
I. Rojas et al. (Eds.): IWANN 2017, Part II, LNCS 10306, pp. 575–584, 2017.
DOI: 10.1007/978-3-319-59147-6_49

Our system works in a low-cost UAV with a monocular camera, uses a perception method based on feature points for obstacle detection and proportional control for obstacle avoidance, and no depends on other sensors.

2 Related Works

Research groups on robotics have proposed different techniques for obstacle avoidance, based on sensors like LIDAR [8–10] and Kinect [11, 12] that show robustness to identify obstacles, but these sensors involves an additional cost for the any UAV. There are several vision systems based on optical flow like [13], where authors propose a system for controlling an ultra-light airplanes with translatory optic flow that avoid obstacles and keep distance from ground and ceiling. Others approach are autonomous collision avoidance systems for navigation within houses or indoor environments using optical flow, micro-sensors and neural networks [14, 15]. A simulation of a navigation system with optical flow for rotary-wing UAV to avoid lateral and frontal collisions for a 3D urban environment is presented in [16]. Some of Optical flow problems is that the method cannot detect frontal obstacles [17].

In spite of the use of Scale Invariant Feature Transform (SIFT), to recognize collisions by analyzing the change in scale and location between two images [18, 19], is stable for some works but is not recommended due to the low speed. We use Speeded-Up Robust Features (SURF) to detect obstacles. There are several application of SURF like face detection [20], target tracking [21, 22], simple visual navigation [23], and some works with UAVs. One of these is [24] that uses a bang-bang control. Our work propose a real time known obstacle detection algorithm based on feature points, and an offline modeling of the MAV for designing a controller for fixed obstacle avoidance in unknown environment.

3 Our Approach

3.1 Obstacle Detection

We use two images, one is located in a database that contains obstacles and other is captured with the onboard camera. In order to find correspondence between these images, feature point detection, description and matching are used. Additionally we calculate the obstacle area and mass center to be used as target in the controller.

Feature Point

For feature point detection, there are several works in the literature [25, 26, 27], but Oriented FAST and Rotated BRIEF (ORB) [28], Fast Retina Keypoint (FREAK) [29], Binary Robust Invariant Scalable Keypoints (BRISK) [30], Scale Invariant Feature Transform [31], and Speeded Up Robust Feature (SURF) [32], are widely used algorithms [33, 34, 35, 36]. In our proposal, we use SURF because its the computational cost is lower without reducing robustness [37].

The algorithm SURF uses a basic Hessian matrix approximation that has a high performance regarding the accuracy [38] and integral images for reducing computational time [39]. The scale-space function uses an image pyramid smoothing with a Gaussian Filter for finding edges and blobs on different scales, and sub-sampling to obtain the next higher level of the pyramid. The feature point location is estimated with a combination of hessian matrix and scale-space function. SURF descriptor determines the distribution of the pixel intensity within a neighbour region for each detected feature point. The method used a Haar wavelet to decrease the computation time and increase the robustness, Haar wavelets are simple filters that determine the gradient at image, these filters are considerate how block based methods to calculate a directional derivatives of the image intensity [40].

There is necessary to find correspondence between the image from the database and the image captured with the UAV. We compare the feature points with the same type of contrast, achieving a lower computational cost without reducing the descriptor performance. Random Sample Consensus (RANSAC) [41, 42] discards the set of the pairs points out of the model, Fig. 1.

Fig. 1. Matching between image from database and real-time image at UAV

Obstacle area and mass center

The obstacle area is inversely proportional to the distance between the obstacle and the UAV, however the camera perspective warps the obstacle geometry. There is required to compensate the perspective warping of the obstacle using a geometry transform. The affine transformation is widely used for motion compensation in [43–45]. This transformation is mathematically expressed as:

$$H = \begin{bmatrix} s\cos(\phi) & -s\sin(\phi) & t_y \\ s\sin(\phi) & s\cos(\phi) & t_x \\ 0 & 0 & 1 \end{bmatrix} \tag{1}$$

where ϕ is rotation roll, s is the scale, t_x and t_y are the translation in x and y between the current and the last frame. We use the affine transformation to compensate the warping

H. Additionally, we calculate the area defined by the compensated vertex of the image from UAV. The fact that the rectangle is a regular figure, we estimate the mass center by the average value of the x and y coordinates of each vertex.

3.2   Obstacle Avoidance

The platform used in the experimentation is the Bebop Drone 1. The control system of the MAV manipulates four different control action: pitch, roll, yaw, and altitude. In our model we use control actions as input (Fig. 2a) and velocities as outputs (Fig. 2b). We proposed two motions in the plane x and y for the avoid system, in the x-axis the motion is uniform, i.e. the linear speed x is constant. The motion in the y-axis depends of the obstacle location, so the control law will be applied on this axis.

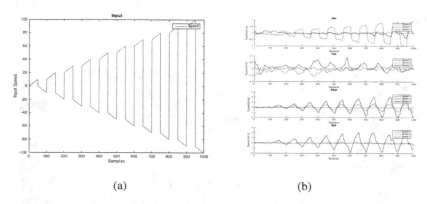

(a)   (b)

Fig. 2. System behavior to different input values (a) Speed input, (b) speed-output in x-axis, y-axis and z-axis

Based on the low level control system of Bebop and the acquired dataset, we have considered that the mathematical relation between angles and motion in axis and can be represented by a static non linear model combined with a dynamic linear model. The models for each motion axis can be decoupled and define by the correlations: the motion in y-axis can be controlled by roll, and the motion in x-axis can be controlled by pitch. There is necessary to estimate the mathematical model between roll control and the linear speed in the y-axis. The non linear part of the model can be estimated as a polynomial regression for the angular inputs of angles and the stationary speed values in the axis. Nevertheless, these speed values trends to saturation levels. We estimated the motion model in y-axis that depends only on the roll input the mathematical short model:

$$G(s) = \frac{K}{T_s + 1} \qquad (2)$$

Controller design

Our consideration of controller design is which after detecting the obstacle the UAV should recovery the path. Also, we must define if the obstacle is on the trajectory, the position error, and the area of the obstacle. For bebop drone we capture 640×480 pixels for each image of visual field, where the obstacles are defined by the mass center as $M_c(y, z)$ in the visual field image. The position error $e(t)$ is the difference between the obstacle mass center and the center $C(y, z)$ of the image (visual field): $e(t) = C(y, z) - M_c(y, z)$.

If $e(t)$ the obstacle is located in the right site of the drone otherwise the obstacle is in the left side. Maximum and minimum values of error are -320 and 320 for y-axis.

The obtained area is proportional to the distance between the onboard camera and the obstacle. In the Fig. 3 there is graphically explained the relation between the image plane and the obstacle length: $\frac{L_1}{IP_1} = \frac{L_1}{IP_1}$ and $\frac{L_2}{IP_2} = \frac{L_2}{IP_2}$.

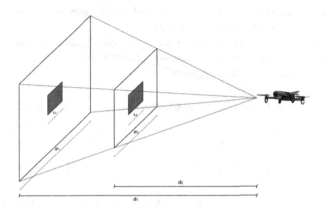

Fig. 3. Relation between the area and distance.

If $L_{P2} > L_{P1}$, the image plane I_{P2} increase otherwise I_{P2} decrease. The relation between image plane and the distance between UAV and obstacle is $\frac{I_{P1}}{I_{P1}} = \frac{d_1}{d_2}$.

Most of control systems are focused on reducing to zero the error between the output and set point. However, for the avoidance system, the controller increases the difference between the obstacle mass center and the visual field center. If the position error is low, i.e. the obstacle is close to the path center, the control system must send a higher speed signal to the motor controllers in order to keep distance respect to obstacles. In the other hand, when the position error is high, the vehicle is far from the path center reducing the collision probability. The control law depends on the location of the obstacle, is positive when the error is $e(t) < 0$ and negative when $e(t) > 0$. This means that the controller output $u(t)$ is inversely proportional to the error $e(t)$. We obtained a control law with a proportional gain K_p and a bias P defined as:

$$u(t) = K_p * e(t) + P \tag{3}$$

Our algorithm starts when the obstacle area is greater than limit area, this means that the vehicle is closer to the obstacle. The limit area value is experimentally obtained defining the dimensions of obstacles in pixels at a specific distance. The output of the control law moves the vehicle away from the center of the obstacle avoiding it. If the error is greater than zero, the UAV moves to the left side, otherwise, the vehicle moves to the right side. The average speed obtained from the control signal is used for recovering the original path.

4   Results and Discussion

We performed an experimental comparison between our autonomous algorithm, the algorithm proposed by [24], and the tele-operation of two persons with different experience levels, using the same UAV.

The metrics of evaluation are: time to complete the path, maximum speed, minimum speed, average speed, distance and battery.

Table 1 presents the average values of the successfully flights for one fixed and two mobile obstacles, and the Table 2 a resume of them. The normalized values were obtained from the official Android App of Parrot.

Table 1. Results of one fixed and two mobile obstacles

Control	Max speed [m/s]	Min speed [m/s]	Ave. speed [m/s]	Distance [m]	Time [s]	Battery [%]
Autonomous algorithm	0,760	0,104	0,235	11,759	48,271	13,888
Bang-Bang	0,942	0,108	0,304	16,997	51,800	13,184
Tele-operator with experience	2,189	0,118	0,915	15,859	20,490	11,390
Tele-operator without experience	2,910	0,103	0,767	22,040	34,038	12,419

Experimental results showed that the autonomous system has lower ratio of battery usage and travelled distance, due to the system has not high variations of speed. The tele-operators used the less time to complete the path, for the ease of avoiding one obstacle, but if we compared with bang-bang system the time is similar.

Based on the experimental results, our proposal has a better performance than others controllers because the travelled distance and the time are lower, and additionally the stable speed allows successfully flights. In spite of that use of battery and the average time required to complete the path is lower for teleoperators, the number of unsuccessful flights ratio is higher. When the number of obstacles is higher the performance of human tele-operators decrease, as consequence of fatigue, in the other hand our proposal keeps this performance. Unlike the Bang-Bang controller, our proposal includes a path recovery system in order to return to the original trajectory.

Table 2. Results of flights with different controls

Control	Number of flights	Successful flights	Unsuccessful flights	Ratio [%]
Autonomous algorithm	20	16	4	80
Bang-Bang	20	12	8	60
Tele-operator with experience	20	13	7	65
Tele-operator without experience	20	11	9	55

The importance of the path recovery system is evident when increases the number of obstacles, for example: for two or more obstacles, the bang- bang controller avoids the first obstacle and loses the path for the next obstacles, ending in other location. Our approach finishes in the correct goal.

5 Conclusions and Future Work

On this paper, we experimentally tested the optimal and robust performance of our system including obstacle detection and avoidance. Flexibility and energy efficiency are important features for autonomous navigation of UAVs. In our approach, flexibility is given by the effectiveness responding to unspecified number of fixed and mobile obstacles in unknown positions. SURF obtains matching between the image from the database and captured frame without incrementing the computational cost. Our proportional controller for obstacle avoidance between the start and goal point is optimal, faster and with higher performance than a bang-bang controller, human controllers with and without experience.

Acknowledgement. This work is part of the projects VisualNavDrone 2016-PIC-024 and MultiNavCar 2016-PIC-025, from the Universidad de las Fuerzas Armadas ESPE, directed by Dr. Wilbert G. Aguilar.

References

1. Barrientos, A., del Cerro, J., Gutierrez, P., San Martin, R., Martinez, A., Rossi, C.: Vehiculos aereos no tripulados para uso civil. tecnologia y aplicaciones. Universidad politecnica de Madrid, Madrid (2007)
2. Kendoul, F.: Survey of advances in guidance, navigation, and control of unmanned rotorcraft systems. J. Field Robot. **29**(2), 315–378 (2012)
3. Aguilar, W.G., Morales, S.: 3D environment mapping using the Kinect V2 and path planning based on RRT algorithms. Electronics **5**(4), 70 (2016)
4. Cabras, P., Rosell, J., Pérez, A., Aguilar, W.G., Rosell, A.: Haptic-based navigation for the virtual bronchoscopy. In: 18th IFAC World Congress, Milano, Italy (2011)

5. Aguilar, W.G., Angulo, C., Costa, R., Molina, L.: Control autónomo de cuadricópteros para seguimiento de trayectorias. In: IX Congreso de Ciencia y Tecnología ESPE, Sangolquí, Ecuador (2014)
6. Ortega, D.V., Bueno, J.A.G.-C., Merino, R.V., Sanz, S.B., Correas, A.H., Campo, D.R.: Pilotos de dron (RPAS). Ediciones Paraninfo, SA (2005)
7. Beyeler, A., Zufferey, J.-C., Floreano, D.: Vision-based control of near-obstacle flight. Auton. Robots **27**(3), 201–219 (2009)
8. Scherer, S., Singh, S., Chamberlain, L., Elgersma, M.: Flying fast and low among obstacles: methodology and experiments. Int. J. Robot. Res. **27**(5), 549–574 (2008)
9. Wurm, K.M., Kummerle, R., Stachniss, C., Burgard, W.: Improving robot navigation in structured outdoor environments by identifying vegetation from laser data. In: 2009 IEEE/RSJ International Conference on Intelligent Robots and Systems. IEEE (2009)
10. Merz, T., Kendoul, F.: Beyond visual range obstacle avoidance and infrastructure inspection by an autonomous helicopter. In: 2011 IEEE/RSJ International Conference on Intelligent Robots and Systems. IEEE, pp. 4953–4960 (2011)
11. Bachrach, A., Prentice, S., He, R., Roy, N.: Range–robust autonomous navigation in GPS-denied environments. J. Field Robot. **28**(5), 644–666 (2011)
12. Lange, S., Sünderhauf, N., Neubert, P., Drews, S., Protzel, P.: Autonomous corridor flight of a UAV using a low-cost and light-weight RGB-D camera. In: Rückert, U., Joaquin, S., Felix, W. (eds.) Advances in Autonomous Mini Robots, pp. 183–192. Springer, Heidelberg (2012). doi:10.1007/978-3-642-27482-4_19
13. Beyeler, A., Zufferey, J.-C., Floreano, D.: 3D vision-based navigation for indoor microflyers. In: Proceedings 2007 IEEE International Conference on Robotics and Automation, pp. 1336–1341. IEEE (2007)
14. Oh, P.Y., Green, W.E., Barrows, G.: Neural nets and optic flow for autonomous micro-air-vehicle navigation. In: ASME 2004 International Mechanical Engineering Congress and Exposition. American Society of Mechanical Engineers (2004)
15. Zufferey, J.-C., Floreano, D.: Fly-inspired visual steering of an ultralight indoor aircraft. IEEE Trans. Robot. **22**(1), 137–146 (2006)
16. Muratet, L., Doncieux, S., Meyer, J.-A.: A biomimetic reactive navigation system using the optical flow for a rotary-wing UAV in urban environment. In: Proceedings of the International Session on Robotics (2004)
17. Merrell, P.C., Lee, D.-J., Beard, R.W.: Obstacle avoidance for unmanned air vehicles using optical flow probability distributions. In: Optics East. International Society for Optics and Photonics, pp. 13–22 (2004)
18. Alenya, G., Negre, A., Crowley, J.L.: A comparison of three methods for measure of time to contact. In: 2009 IEEE/RSJ International Conference on Intelligent Robots and Systems, pp. 4565–4570. IEEE (2009)
19. Chavez, A., Gustafson, D.: Vision-based obstacle avoidance using SIFT features. In: Bebis, G., et al. (eds.) ISVC 2009. LNCS, vol. 5876, pp. 550–557. Springer, Heidelberg (2009). doi:10.1007/978-3-642-10520-3_52
20. Kim, D., Dahyot, R.: Face components detection using SURF descriptors and SVMS. In: International Machine Vision and Image Processing Conference, IMVIP 2008, pp. 51–56. IEEE (2008)
21. Chu, D.M., Smeulders, A.W.M.: Color invariant SURF in discriminative object tracking. In: Kutulakos, K.N. (ed.) ECCV 2010. LNCS, vol. 6554, pp. 62–75. Springer, Heidelberg (2012). doi:10.1007/978-3-642-35740-4_6
22. He, W., Yamashita, T., Lu, H., Lao, S.: Surf tracking. In: 2009 IEEE 12th International Conference on Computer Vision, pp. 1586–1592. IEEE (2009)

23. Krajník, T., Nitsche, M., Pedre, S., Preucil, L., Mejail, M.E.: A simple visual navigation system for an UAV. In: 2012 9th International Multi-Conference on Systems, Signals and Devices (SSD), pp. 1–6. IEEE (2012)

24. Mori, T., Scherer, S.: First results in detecting and avoiding frontal obstacles from a monocular camera for micro unmanned aerial vehicles. In: 2013 IEEE International Conference on Robotics and Automation (ICRA), pp. 1750–1757. IEEE (2013)

25. Canny, J.: A computational approach to edge detection. IEEE Trans. Pattern Anal. Mach. Intell. **6**, 679–698 (1986)

26. Harris, C., Stephens, M.: A combined corner and edge detector. In: Alvey Vision Conference, vol. 15, p. 50. Citeseer (1988)

27. Miksik, O., Mikolajczyk, K.: Evaluation of local detectors and descriptors for fast feature matching. In: 2012 21st International Conference on Pattern Recognition (ICPR), pp. 2681–2684. IEEE (2012)

28. Rublee, E., Rabaud, V., Konolige, K., Bradski, G.: ORB: an efficient alternative to SIFT or SURF. In: 2011 International Conference on Computer Vision, pp. 2564–2571. IEEE (2011)

29. Alahi, A., Ortiz, R., Vandergheynst, P.: FREAK: fast retina keypoint. In: 2012 IEEE conference on Computer Vision and Pattern Recognition (CVPR). IEEE (2012)

30. Leutenegger, S., Chli, M., Siegwart, R.Y.: BRISK: binary robust invariant scalable keypoints. In: 2011 International Conference on Computer Vision. IEEE (2011)

31. Lowe, D.G.: Object recognition from local scale-invariant features. In: The Proceedings of the Seventh IEEE International Conference on Computer Vision, vol. 2, pp. 1150–1157. IEEE (1999)

32. Bay, H., Ess, A., Tuytelaars, T., Van Gool, L.: Speeded-up robust features (SURF). Comput. Vis. Image Underst. **110**(3), 346–359 (2008)

33. Aguilar, W.G., Angulo, C.: Estabilización de vídeo en micro vehículos aéreos y su aplicación en la detección de caras. In: IX Congreso de Ciencia y Tecnología ESPE, Sangolquí, Ecuador (2014)

34. Aguilar, W.G., Angulo, C.: Estabilización robusta de vídeo basada en diferencia de nivel de gris. In: VIII Congreso de Ciencia y Tecnología ESPE, Sangolquí, Ecuador (2013)

35. Aguilar, W.G., Angulo, C.: Compensación y aprendizaje de efectos generados en la imagen durante el desplazamiento de un robot. In: X Simposio CEA de Ingeniería de Control, Barcelona, Spain (2012)

36. Aguilar, W.G., Angulo, C.: Compensación de los efectos generados en la imagen por el control de navegación del robot Aibo ERS 7. In: VII Congreso de Ciencia y Tecnología ESPE, Sangolquí, Ecuador (2012)

37. Juan, L., Gwun, O.: A comparison of SIFT, PCA-sift and SURF. Int. J. Image Process. (IJIP) **3**(4), 143–152 (2009)

38. Viola, P., Jones, M.: Rapid object detection using a boosted cascade of simple features. In: Proceedings of the 2001 IEEE Computer Society Conference on Computer Vision and Pattern Recognition, CVPR 2001, vol. 1, pp. 511–518. IEEE (2001)

39. Huang, D.-S., Jo, K.-H., Hussain, A.: Intelligent Computing Theories and Methodologies: 11th International Conference, ICIC 2015, Fuzhou, China, 20–23 August 2015, vol. 9226. Springer, Heidelberg (2015)

40. Torr, P.H., Zisserman, A.: MLESAC: a new robust estimator with application to estimating image geometry. Comput. Vis. Image Underst. **78**(1), 138–156 (2000)

41. Derpanis, K.G.: Overview of the RANSAC algorithm. Image Rochester N. Y. **4**(1), 2–3 (2010)

42. Fischler, M.A., Bolles, R.C.: Random sample consensus: a paradigm for model fitting with applications to image analysis and automated cartography. Commun. ACM **24**(6), 381–395 (1981)

43. Aguilar, W.G., Angulo, C.: Real-time model-based video stabilization for microaerial vehicles. Neural Process. Lett. **43**(2), 459–477 (2016)
44. Aguilar, W.G., Angulo, C.: Real-time video stabilization without phantom movements for micro aerial vehicles. EURASIP J. Image Video Process. **1**, 1–13 (2014)
45. Aguilar, W.G., Angulo, C.: Robust video stabilization based on motion intention for low-cost micro aerial vehicles. In: 11th International Multi-conference on Systems, Signals & Devices (SSD), Barcelona, Spain (2014)

RRT* GL Based Optimal Path Planning for Real-Time Navigation of UAVs

Wilbert G. Aguilar[1,2,3(✉)], Stephanie Morales[2], Hugo Ruiz[1,4],
and Vanessa Abad[5]

[1] Dep. Seguridad y Defensa, Universidad de las Fuerzas Armadas ESPE,
Sangolquí, Ecuador
wgaguilar@espe.edu.ec
[2] CICTE Research Center, Universidad de las Fuerzas Armadas ESPE,
Sangolquí, Ecuador
[3] GREC Research Group, Universitat Politècnica de Catalunya,
Barcelona, Spain
[4] PLM Research Center, Purdue University, West Lafayette, IN, USA
[5] Universitat de Barcelona, Barcelona, Spain

Abstract. In this paper, we propose a path planning system for autonomous navigation of unmanned aerial vehicle based on a Rapidly-exploring Random Trees (RRT) combination of RRT* Goal and Limit. The system includes a point cloud obtained from the vehicle workspace with a RGB-D sensor, an identification module for interest regions and obstacles of the environment, and a collision-free path planner based on RRT for a safe and optimal navigation of vehicles in 3D spaces.

Keywords: Path planning · RRT · Point cloud registration · 3D modeling · Mobile robotics · RGB-D segmentation · Computational geometry

1 Introduction

UAVs (Unmanned Aerial Vehicle) have a growing interest due to their multiple applications in several areas like surveillance [1], filming [2–4], search and rescue, military functions [5], etc. However, different issues must be solved before their application in non-controlled environments as energy efficiency and autonomous navigation [6]. Nowadays, commercial UAVs require human intervention with long training times, and includes the variable expertise of the pilot for controlling the UAV. Autonomous navigation [7] is an alternative for solving this issue and depends on perception and motion estimation algorithms to obtain the workspace and the pose of the UAV in the workspace. Additionally, path planning and optimization are required for connecting the start point with the goal point in the workspace.

In our system, we use a low cost RGB-D sensor, located in a zenith perspective, that generates a 3D binary occupancy grid map as the workspace. A RRT based path planner obtains a solution path between start and goal point in the workspace [8, 9]. These points are recognized by morphological segmentation [10, 11], and the RGB-D coordinate frame is calibrated respect to the workspace coordinate frame using a registration technique. In the path planning literature, sample-based methods are widely used because their

© Springer International Publishing AG 2017
I. Rojas et al. (Eds.): IWANN 2017, Part II, LNCS 10306, pp. 585–595, 2017.
DOI: 10.1007/978-3-319-59147-6_50

effectiveness and low computational cost on high dimensional spaces. These methods use a representative configuration space and build a collision-free roadmap connecting points sampled from the obstacle free space. RRT is an incremental sample-based method commonly used for fast trajectory search. There are several versions of the RRT developed to improve the solution path cost, so we compared the RRT with the RRT* algorithm and two additional variations using the path cost as evaluation metrics.

2 Related Works

There are different techniques for UAV navigation like 3D perception and mapping based on RGB-D cameras [12], laser range scanners [13] and stereo cameras [14]. The Kinect was released on November 2010 as an RGB-D commercial camera, and after that, several applications on robotics appeared such as on [15, 16].

Segmentation and object recognition are fundamental algorithms for vision applications. Recently, the growing interest on 3D environments for UAV navigation has increased the research on 3D segmentation. In [17] a color and depth segmentation is proposed to detect objects of an indoor environment by finding boundaries between regions based on the norm of the pixel intensity. A novel technique for depth segmentation uses the divergence of a 2D vector field to extract 3D object boundaries [18].

Works on the literature for planning the path to connect target with goal. The Ant Colony Algorithm based on the behavior of ants searching for food is used on many research projects due to its global optimization performance in multi goal applications [19]. Others use neural networks to give an effective path planning and obstacle avoidance solution that leads the UAV to the target position [20, 21]. Some strategies use behavior-based navigation and fuzzy logic approaches. The image provides the data for fuzzy logic rules that guide the UAV to the safest and most traversable terrain region [22].

In this paper we use a probabilistic path planning algorithm. Several optimizations have been developed on this area like the RRT-Connect approach [23] that uses a simple greedy heuristic. Other variants have been presented in [24, 25]. Our contribution is focused on the overall system of an optimal UAV path planner, comparing the resulting paths of two variants of the RRT algorithm.

3 Our Workspace

A path planning algorithm requires a workspace, the initial and goal positions, and a robot, in our case the UAV. Our proposed system has three steps: environment perception, detection of target points and path planning. In this section we talk about first and second steps.

The workspace is obtained by using the proposal of Lars Ivar Hatledal [26] and a RGB-D sensor (Kinect [27]). According to [26], each pixel size has an estimated value of 6 mm for width and 3.38 mm for height, giving the approximate resolution of our point cloud in x and y axis. Based on [26] we obtain the colored point cloud in millimeters from the RGB-D sensor. We align the point cloud relative to the Kinect to the workspace coordinate frame, previous to image segmentation. A technique to solve this issue is point cloud registration. The objective of registration is the matching of two sets of points by estimation of the transform that maps one point set to the other [28, 29]. A widely used registration method is the Iterative Closest Point (ICP) due to simplicity and low computational complexity. ICP assigns correspondence based on the nearest distance criteria and the least-squares rigid transformation giving the best alignment of the two point sets [30, 31].

Assuming that the point set $\{m_i\}$ and $\{d_i\}$, $i = 1\ldots N$ are matched and related, R is a 3×3. rotation matrix, T is a 3D translation vector and V_i is a noise vector [32, 33], for obtaining the optimal transformation that maps the set $\{m_i\}$ to $\{d_i\}$, we minimize the least-square error (1).

$$\Sigma^2 = \sum_{i=1}^{N} \left\| d_i - \hat{R}m_i - \hat{T}^2 \right\| \tag{1}$$

ICP uses an outlier filter before the transformation matrix calculation in order to solve the incorrect correspondence issue. This is because the least-squares solution is not optimal if there are many outliers on the dataset. After obtaining a suitable UAV coordinate frame, we segment the image into regions. Each region is defined as a homogenous group of connected pixels respect to a selected interest property such as color intensity or texture [34, 35]. In our case, color threshold [36, 37] is an effective approach for segmenting, but the input image is generated from the point cloud, so each pixel has a corresponding xyz coordinate. The threshold values [38] are locally obtained by selecting a target region pixel.

A binary mask is obtained and undesired noise and holes are removed by morphological operations (Dilatation, Erosion, Opening and Close). The dilation of a matrix A with a structuring element B is defined as (2), and removes holes smaller or equal than the structuring element [39]. Erosion removes noise smaller than the structuring element, and is defined by (3). The opening of an image A with a structuring element B is (4), and its effect is to remove edges and some of the foreground bright pixels. Close (5) operation keeps the background regions with a similar size of the structuring element, and removes all other pixels of the background. Based on theses morphological operators, we obtain a reliable estimation of the interest regions.

$$A \oplus B = \bigcup_{b \in B} A_b \tag{2}$$

$$A \ominus B = \bigcap_{b \in B} A_{-b} \tag{3}$$

$$A \circ B = (A \ominus B) \oplus B \tag{4}$$

$$A \cdot B = (A \oplus B) \ominus B \tag{5}$$

The next step is the representation where adjacent pixels of the same class are connected to form spatial sets $s_1 \ldots s_m$. For the representation of our binary image we use the blob detection method of Peter Corke [40]. Blob is a contiguous spatial region of pixels. Property like area, perimeter, circularity, moments or centroid coordinates are used to identify these regions. We identify our interest regions filtering the blobs based on their position and area in order to simplify the problem. Knowing the boundaries of the workspace and the minimum and maximum size of the target circles, we obtained the interest blobs and their centroid coordinates.

4   RRT Algorithms

RRT is a widely used algorithm in the literature for solving the motion planning because provides an efficient solution for the travelling salesman problem. However, there are several variation of the RRT proposed to optimized the solution path, i.e., a collision free-path that starts in X_{init}. and ends in X_{goal}. The optimization problem is focused on a reliable path with minimal cost. We introduce the standard RRT algorithm and the RRT* variations tested in our system.

4.1   RRT Standard

Using the notation from [23], be: X is a bounded connected open subset of $\mathbb{R}^d$ where $d \in \mathbb{N}, d \geq 2$. $G = (V, E)$ is a graph composed by a set of vertexes V and edges E. X_{obs}. and X_{goal}, subsets of X., are the obstacle region and the goal region respectively. The obstacle free space $X \backslash X_{obs}$ denoted as X_{free} and the initial state x_{init} is an element of X_{free}. A pathn X_{free} should be a collision-free path. The RRT standard algorithm starts in the initial state. Then the graph incrementally grows by sampling a random ate x_{rand} from X_{free}, and connects $x_{nearest}$ with this x_{rand}. The Algorithms 1 and 2 present the RRT standard and the standard extended.

Algorithm 1. RRT standard

---

1. $V \leftarrow \{x_{init}\}; E = \emptyset; i = 0;$

2. $while\ i < N\ do$

3. $G \leftarrow (V, E);$

4. $x_{rand} \leftarrow Sample(i);$

5. $i \leftarrow i + 1$

6. $(V, E) \leftarrow Extend(G, x_{rand})$

Algorithm 2. Extend Function of the RRT algorithm

---

1. $V' \leftarrow V; E = E';$

2. $x_{nearest} \leftarrow Nearest(G, x);$

3. $x_{new} \leftarrow Steer(x_{nearest}, x);$

4. $if\ ObstacleFree(x_{nearest}, x_{new})\ then$

5. $V' = V' \cup \{x_{new}\};$

6. $E' = E' \cup \{(x_{nearest}, x_{new})\};$

7.   return $G' = (V', E')$

4.2   Rrt*

The RRT* algorithm has two additional optimization than the standard RRT. The first one is done after the addition of the new vertex to the tree. The connection between the new and each returned vertex with lowest cost functions is saved and the others are deleted. The second optimization procedure, if the cost function from x_{init} to each near vertex through x_{new} is less, a rewire procedure is executed. Both optimization procedures are included in the RRT* shown on the Algorithm 3.

Algorithm 3. Extended Function of the RRT* algorithm

---

1. $V' \leftarrow V; E = E';$
2. $x_{nearest} \leftarrow Nearest(G, x);$
3. $x_{new} \leftarrow Steer(x_{nearest}, x);$
4. $if\ ObstacleFree(x_{nearest}, x_{new})\ then$
5. $V' = V' \cup \{x_{new}\};$
6. $x_{min} \leftarrow x_{nearest};$
7. $X_{near} \leftarrow NearVertices(G, x_{new});$
8. $for\ all\ x_{near} \in X_{near}\ do$
9. $if\ ObstacleFree(x_{near}, x_{new})\ then$
10. $c' \leftarrow Cost(x_{near}) + Cost\big(Line(x_{near}, x_{new})\big);$
11. $if\ c' < Cost(x_{near})\ then$
12. $x_{min} \leftarrow x_{near};\ //Choose\ new\ parent\ for\ x_{new}$
13. $E' = E' \cup \{(x_{min}, x_{new})\};$
14. $for\ all\ x_{near} \in X_{near} \backslash \{x_{min}\}\ do$
15. $if\ ObstacleFree(x_{near}, x_{new})\ then$
16. $c' \leftarrow Cost(x_{new}) + Cost\big(Line(x_{near}, x_{new})\big);$
17. $if\ c' < Cost(x_{near})\ then$
18. $x_{parent} \leftarrow Parent(x_{near});\ //Rewire$
19. $E' \leftarrow E' \backslash \{(x_{parent}, x_{near})\};$
20. $E' \leftarrow E' \cup \{(x_{new}, x_{near})\};$
21.   return $G' = (V', E')$

4.3   RRT* Goal and RRT* Limits

In order to decrease the time necessary for finding a reliable solution, accelerating the rate of convergence and optimization, we proved two variations of the RRT*. The modifications are on the Sample function, changing the probability of the generated random node. We have called these version RRT* Goal (Algorithm 4), RRT* Limits (Algorithm 5) and the combination of both RRT*GL.

Algorithm 4. Sample Function of the RRT* Goal algorithm

1.	$if\ (i\ MOD\ 2 = 0)\ OR\ (feasible\ Path\ is\ TRUE)$
2.	$\quad x_{rand} \leftarrow random(X_{free});$
3.	$else$
4.	$\quad x_{rand} \leftarrow X_{goal};$
5.	$return\ x_{rand}$

Algorithm 5. Sample Function of the RRT* Limits algorithm

1.	$if\ feasible\ Path\ is\ TRUE$
2.	$\quad minL = min(V)$
3.	$\quad maxL = max(V)$
4.	$\quad x_{rand} \leftarrow random(X_{free}, [minL, maxL]);$
5.	$else$
6.	$\quad x_{rand} \leftarrow random(X_{free});$
7.	$return\ x_{rand}$

For the RRT* Goal, we guide the exploration of the tree in the way of the goal region. We give a probability of 50% for sampling the goal region. The other 50% returns an identically distributed sample from X_{free}. Once the reliable path is found, we increase the probability of the random sample in a specific space for the RRT* Limits. This accelerates the rate of convergence to an optimal low cost path. We obtain the uniform distributed random sample by delimiting the random range to the minimum and maximum coordinates (x, y, z) of the found path. This increases the density of random samples at the path surroundings, avoiding unnecessary optimizations for other sections of the tree.

5   Results and Discussion

The experiment was divided into three parts: (a) the alignment of the workspace point cloud of the Kinect to the UAV coordinate frame, (b) the target point recognition and, (c) the path planning algorithm with the variations.

The method used for the point cloud alignment was registration of two data sets. We rotated the Kinect for testing the transformation matrix effect on the original point cloud. Our system obtains an aligned point cloud with high accuracy, where we present information of eight workspace boundary circles (landmarks), their centroid coordinates estimated from the Kinect coordinate frame, the real coordinates on the workspace, the calibrated coordinates after registration and the error percentage of each circle centroid coordinate. Because the circles are located on the same plane, we can calculate the average error percentage as the error of the workspace plane on the robot coordinate frame related to the Kinect pose by:

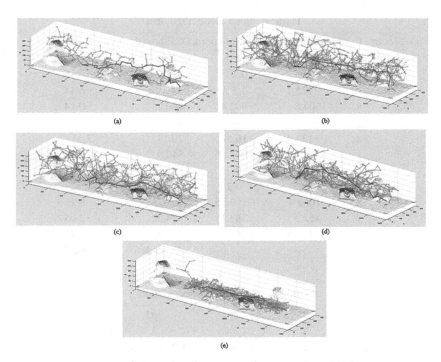

Fig. 1. Comparison between algorithms in a simulation example. (a) The RRT standard; (b) the RRT*; (c) the RRT* Goal; (d) the RRT* Limits and (e) the RRT* GL.

$$E_x\% = \frac{calibrated_x - real_x}{delta_x} \times 100 \qquad (6)$$

Where $delta_x$ is the difference between the maximum and minimum value of the workspace seen by the Kinect on the x coordinate (Table 1).

For the target point recognition, we present the location on millimeters of five circle centroids recognized with the algorithm on the xy plane and the corresponding error percentage. The error was also calculated with respect to the robot workspace seen by the Kinect (Table 2).

For the testing of the path planning approaches, each algorithm was tested ten times with 1000 (maximum number of iteration) for obtaining the mean cost, number of segments in the resulting path, the time for locating the target point and the execution time. These comparative results are shown on the following Table 3, and Fig. 1.

RRT* Goal algorithm decreases the computational cost for finding a reliable solution, and the RRT* Limits algorithm increases the density of the tree branches around the path. The combination of both methods produces a faster and optimal path without increases the number of iterations.

Table 1. Point cloud alignment

Vertex	Coord.	Kinect coord.	Real coord.	Calibrated coord.	Error %
1	x	0.7	−1.5	−0.91	0.04
	y	79.9	0.1	−0.52	−0.16
	z	−125.7	2.3	−4.14	−6.44
2	x	604.8	602.41	600.2	−0.14
	y	467.8	394.37	390.1	−1.02
	z	−134.4	16.58	20.11	3.53
3	x	942.5	942.56	940.9	−0.10
	y	65.1	−9.72	−14.4	−1.11
	z	−88.8	30.25	31.97	1.72
4	x	1603.4	1598.5	1596	−0.15
	y	452.5	383.1	374.5	−2.05
	z	−133.5	21.54	21.08	−0.46
				Average	**−0.27**

Table 2. Target points recognition

	p1		p2		p3	
	X	Y	X	Y	X	Y
Real	252.2	300.1	497.9	151.2	988.52	208.37
Calculated	249.6	299	497	152.8	985.3	205.7
Error %	0.16	0.26	0.06	0.38	0.20	0.64

Table 3. Comparative results between RRT algorithms and our proposal.

Obstacles	Algorithm	# Iterations	Total cost	# Segments	Goal time	Total time
1	RRT	225	1501.8	31	4.94	4.94
	RRT*	1000	1403.4	23	8.3	180.25
	RRT* Goal	1000	1312.3	26	0.81	161.24
	RRT* Limits	1000	1366.3	26	28.54	191.2
	RRT* GL	1000	1299.1	26	0.86	198.01
2	RRT	190.5	1853.65	38	4.015	4.015
	RRT*	1000	1398.4	24	2.4	188.61
	RRT* Goal	1000	1366.6	26.5	1.285	158.26
	RRT* Limits	1000	1425.4	30	7.22	185.17
	RRT* GL	1000	1332.2	27	1.5	187.46
3	RRT	264	1324.5	27.5	5.53	5.53
	RRT*	1000	1137.2	20	2.95	165.3
	RRT* Goal	1000	1029.3	21	1.35	189.73
	RRT* Limits	1000	1073.8	25	10.84	220.16
	RRT* GL	1000	1001.7	21	1.61	270.71

6   Conclusions and Future Work

Our system was experimentally tested, and has shown flexibility and energy efficiency with a combined process of point clouds alignment, target point detection and path planning. Flexibility is given by the automatic alignment of the point cloud to the robot coordinate frame based on ICP, and target point detection based on color threshold technique for obtaining the goal region centroid. This approach showed a high performance with RGB-D devices. Several variations of the RRT algorithm were tested in order to compare the results respect to autonomy and energy efficient for navigation UAVs. The RRT* Goal reduces the time required for finding a reliable solution. RRT* Limits improves the optimization by increasing the density of the tree branches on the path region. We have selected RRT* GL that achieves a fast and optimal path between the start and goal point. As future works, we will work with onboard RGB-D devices for UAVs, reactive navigation, and our combination RRT* GL in order to achieve an autonomous navigation system for the aerial vehicle.

Acknowledgement. This work is part of the projects VisualNavDrone 2016-PIC-024 and MultiNavCar 2016-PIC-025, from the Universidad de las Fuerzas Armadas ESPE, directed by Dr. Wilbert G. Aguilar.

References

1. Aguilar, W.G., Angulo, C.: Estabilización de vídeo en micro vehículos aéreos y su aplicación en la detección de caras. In: IX Congreso de Ciencia y Tecnología ESPE, Sangolquí, Ecuador (2014)
2. Aguilar, W.G., Angulo, C.: Real-time model-based video stabilization for microaerial vehicles. Neural Process. Lett. **43**(2), 459–477 (2016)
3. Aguilar, W.G., Angulo, C.: Real-time video stabilization without phantom movements for micro aerial vehicles. EURASIP J. Image Video Process. **1**, 1–13 (2014)
4. Aguilar, W.G., Angulo, C.: Robust video stabilization based on motion intention for low-cost micro aerial vehicles. In: 11th International Multi-Conference on Systems, Signals & Devices (SSD), Barcelona, Spain (2014)
5. Koren, Y.: Robotics for Engineers. McGraw-Hill, New York (1998)
6. Gonzalez, R., Safabakhsh, R.: Computer vision techniques for industrial applications and robot control. Computer (2006)
7. Aguilar, W.G., Angulo, C., Costa, R., Molina, L.: Control autónomo de cuadricópteros para seguimiento de trayectorias. In: IX Congreso de Ciencia y Tecnología ESPE, Sangolquí, Ecuador (2014)
8. Vasishth, O., Gigras, Y.: Path planning problem. Int. J. Comput. Appl. **104**(2) (2014)
9. Cabras, P., Rosell, J., Pérez, A., Aguilar, W.G., Rosell, A.: Haptic-based navigation for the virtual bronchoscopy. In: 18th IFAC World Congress, Milano, Italy (2011)
10. Bruce, J., Balch, T., Veloso, M.: Fast and inexpensive color image segmentation for interactive robots. In: IEEE/RSJ International Conference on intelligent Robots and Systems (2000)
11. Heinz, K., Hanson, W.: Interactive 3D segmentation of MRI and CT volumes using morphological operations. J. Comput. Assist. Tomogr. **16**(2), 285–294 (1992)

12. Henry, P., Krainin, P., Herbst, E., Ren, X., Fox, D.: RGB-D mapping: using depth cameras for dense 3D modeling of indoor environments. In: The 12th International Symposium on Experimental Robotics (ISER) (2010)

13. Thrun, S., Burgard, W., Fox, D.: A real-time algorithm for mobile robot mapping with applications to multi-robot and 3D mapping. In: IEEE International Conference on Robotics and Automation, San Francisco (2000)

14. Gutmann, J.-S., Fukuchi, M., Fujita, M.: 3D perception and environment map generation for humanoid robot navigation. Int. J. Robot. Res. **27**, 1117–1134 (2008)

15. Oliver, A., Kang, S., Wunsche, B., MacDonald, B.: Using the Kinect as a navigation sensor for mobile robotics. In: Conference on Image and Vision Computing, New Zealand (2012)

16. Benavidez, P., Jamshidi, M.: Mobile robot navigation and target tracking system. In: The 6th International Conference on System of Systems Engineering, Albuquerque (2011)

17. Rao, D., Le, Q., Phoka, T., Quigley, M., Sudsang, A., Ng, A.Y.: Grasping novel objects with depth segmentation. In: IEEE/RSJ International Conference on Intelligent Robots and Systems (IROS), Taipei (2010)

18. Ali Shah, S.A., Bennamoun, M., Boussaid, F.: A novel algorithm for efficient depth segmentation using low resolution (Kinect) images. In: IEEE 10th Conference on Industrial Electronics and Applications (ICIEA), Auckland (2015)

19. Liu, J., Yang, J., Liu, H., Tian, X., Gao, M.: An improved ant colony algorithm for robot path planning. Soft Computing, **1**(11) (2016)

20. Glasius, R., Komoda, A., Gielen, S.C.A.M.: Neural network dynamics for path planning and obstacle avoidance. Neural Netw. **8**(1), 125–133 (2000)

21. Xin, D., Hua-hua, C., Wei-kang, G.: Neural network and genetic algorithm based global path planning in a static environment. J. Zhejiang Univ. Sci. A **6**(6), 549–554 (2005, 2006)

22. Seraji, H., Howard, A.: Behavior-based robot navigation on challenging terrain: a fuzzy logic approach. IEEE Trans. Robot. Autom. **18**(3), 308–321 (2002)

23. Kuffner, J.J., LaValle, S.M.: RRT-connect: an efficient approach to single-query path planning. In: IEEE International Conference on Robotics and Automation, San Francisco (2000)

24. Devaurs, D., Siméon, T., Cortés, J.: Efficient sampling-based approaches to optimal path planning in complex cost spaces. In: Akin, H.L., Amato, N.M., Isler, Volkan, Stappen, A.F. (eds.) Algorithmic Foundations of Robotics XI. STAR, vol. 107, pp. 143–159. Springer, Cham (2015). doi:10.1007/978-3-319-16595-0_9

25. Gammell, J.D., Srinivasa, S., Barfoot, T.: Informed RRT*: optimal sampling-based path planning focused via direct sampling of an admissible ellipsoidal heuristic. In: IEEE/RSJ International Conference on Intelligent Robots and Systems (IROS 2014) (2014)

26. Hatledal, L.I.: Kinect V2 SDK 2.0 – Colored point clouds, 15 August 2015. http://laht.info/kinect-v2-colored-point-clouds/

27. Fankhauser, P., Bloesch, M., Rodriguez, D., Kaestner, R., Hutter, M., Siegwart, R.: Kinect v2 for mobile robot navigation: evaluation and modeling. In: 2015 International Conference on Advanced Robotics (ICAR), Istanbul (2015)

28. Myronenko, A., Song, X.: Point set registration: coherent point drift. IEEE Trans. Pattern Anal. Mach. Intell. **32**(12), 2262–2275 (2010)

29. Karaman, S., Frazzoli, E.: Incremental sampling-based algorithms for optimal motion planning. Int. J. Robot. Res. **30**, 846–894 (2010)

30. Lachat, E., Hélene, M., Tania, L., Pierre, G.: Assessment and calibration of a RGB-D camera (Kinect v2 Sensor) towards a potential use for close-range 3D modeling. Remote Sens. **7**(10), 13070–13097 (2015)

31. Pagliari, D., Pinto, L.: Calibration of kinect for Xbox one and comparison between the two generations of microsoft sensors. Sensors **15**(11), 27569–27589 (2015)

32. Eggert, D.W., Lorusso, A., Fisher, R.B.: Estimating 3-D rigid body transformations: a comparison of four major algorithms. Mach. Vis. Appl. **9**, 272–290 (1997)
33. Sreedhar, K., Panlal, B.: Enhancement of images using morphological transformations. Int. J. Comput. Sci. Inf. Technol. (IJCSIT) **4**(1), 33–50 (2012)
34. Aguilar, W.G., Angulo, C.: Compensación y aprendizaje de efectos generados en la imagen durante el desplazamiento de un robot. In: X Simposio CEA de Ingeniería de Control, Barcelona, Spain (2012)
35. Aguilar, W.G., Angulo, C.: Compensación de los efectos generados en la imagen por el control de navegación del robot Aibo ERS 7. In: VII Congreso de Ciencia y Tecnología ESPE, Sangolquí, Ecuador (2012)
36. Navon, E., Miller, O., Averbuch, A.: Color image segmentation based on adaptive local thresholds. Image Vis. Comput. **23**, 69–85 (2005)
37. Aguilar, W.G., Angulo, C.: Estabilización robusta de vídeo basada en diferencia de nivel de gris. In: VIII Congreso de Ciencia y Tecnología ESPE, Sangolquí, Ecuador (2013)
38. Sahoo, P.K., Soltani, S., Wong, A.K.C.: A survey of thresholding techniques*. Comput. Vis. Graph. Image Process. **41**, 233–260 (1988)
39. The MathWorks, Inc.: pcregrigid documentation (2015). http://www.mathworks.com/help/vision/ref/pcregrigid.html. Accessed 24 Febrero 2016
40. Corke, P.I.: Robotics, Vision and Control: Fundamental Algorithms in Matlab. Springer, Heidelberg (2011)

Visual SLAM with a RGB-D Camera on a Quadrotor UAV Using on-Board Processing

Wilbert G. Aguilar[1,3,4]($\boxtimes$), Guillermo A. Rodríguez[2,3],
Leandro Álvarez[2,3], Sebastián Sandoval[2,3], Fernando Quisaguano[2,3],
and Alex Limaico[2,3]

[1] Dep. Seguridad y Defensa,
Universidad de las Fuerzas Armadas ESPE, Sangolquí, Ecuador
wgaguilar@espe.edu.ec
[2] Dep. Eléctrica y Electrónica,
Universidad de las Fuerzas Armadas ESPE, Sangolquí, Ecuador
[3] CICTE Research Center,
Universidad de las Fuerzas Armadas ESPE, Sangolquí, Ecuador
[4] GREC Research Group,
Universitat Politècnica de Catalunya, Barcelona, Spain

Abstract. In this article, we present a high accuracy system for real-time localization and mapping using a RGB-D camera. With the use the RGB-D sensor Microsoft Kinect and the small and powerful computer Intel Stick Core M3 Processor, our system can run the computation and sensing required for SLAM on-board the UAV, removing the dependence on unreliable wireless communication. We use visual odometry, loop closure and graph optimization to achieve this purpose. Our approach is able to perform accurate and efficient on-board SLAM by analyzing data and maps generated on several tests of the system.

Keywords: SLAM · RGB-D · Loop closure detection · Graph optimization · Visual odometry · RANSAC · UAVs

1 Introduction

One of the most desired goals of the field of mobile robotics is to create a truly autonomous robot [1–4], which could be placed in an unknown environment and obtain the location without any previous knowledge of relative position. One solution to this problem is SLAM (Simultaneous Localization and Mapping). Over the last few years, many approaches to this solution have been developed, with bundle adjustment [5–7] or filtering-based estimation [8, 9]. These methods provide approximations of the solution, but the process for obtaining relevant reference points gives up important information that compromises the accuracy of the entire system.

In recent years, interest in RGB-D cameras like Microsoft Kinect or ASUS Xtion Pro has grown because of their ability to provide real-time color images and depth maps. New approaches for SLAM combine the scale information of 3D depth sensing

© Springer International Publishing AG 2017
I. Rojas et al. (Eds.): IWANN 2017, Part II, LNCS 10306, pp. 596–606, 2017.
DOI: 10.1007/978-3-319-59147-6_51

with the visual information of the cameras to create accurate 3D environment maps. Some methods use dense visual odometry [10–12], showing accurate results on the main problematic, but making use of a laptop computer to process the visual data. This represents a considerable complication when implementing the methods on UAVs, which cannot support high processing. This inconvenience can be solved using onboard micro computers.

Our approach consists in an onboard SLAM for a UAV, using the dense color and depth images obtained from RGB-D cameras [13]. We based our work on contributions made in loop closure detection [14] and graph-based SLAM [15]. Our system is able to perform SLAM on an UAV using an Intel Stick M3 computer, optimizing communication resources and simultaneously processing the algorithms in real-time.

2 Related Works

The SLAM problem has been a discussed topic throughout the history of computer vision [16–20] and robotics [21–25]. This is because of its vantages for mounting an autonomous robotic system. Visual SLAM systems [26, 27] are applied to extract interest points from the camera images [28–30], simplifying data association. For online loop closure detection, bag-of-words [31] approach is used [32, 33]. The bag-of-words approach consists in representing each image by visual words taken from a vocabulary. Graph pose optimization approaches [34, 35] can be employed to reduce odometry errors using poses and link transformations inside each map and between the maps.

Different sensor methods are based on SLAM, including 2D scanners [36, 37], monocular cameras [38, 39], and recently RGB-D sensors such as the Microsoft Kinect [40] or the Asus Xtion Pro Live. SLAM approaches for flight systems include vision guided MAV (Micro Air Vehicle) to avoid obstacles using vision guidance and on-board processing [41], but are unable to create a map and accurate localization of the robot. In [42], the autonomous mapping and localization system in outdoor environments performs the hard processing work off-board in a ground station, taking away the real time capabilities from the system.

3 Our Approach

In our work, the basic structure of the map is a graph with nodes and links. These nodes store important information, like the odometry poses for every location taken in the map, the visual information of the RGB, depth images of the Kinect, and the visual words used for loop closure detection. As for the links, they save the rigid geometrical transformations within nodes. Close links are added among the current and the previous nodes with their corresponding odometry transformation. Loop closure links are added when a loop closure is detected within the current node and one from the same or previous map. Our contribution consists in the combined use of visual odometry, loop closure detection [10] and optimization to run real time SLAM in an Intel Stick M3.

3.1 Visual Odometry

We use visual odometry to estimate the trajectory of the RGB-D sensor from a specific region within the image. This method is useful for the pose estimation problem which will be linked directly to the measurements given by the RGB-D sensor via a non-linear model. This model is responsible for the 3D geometric configuration of the current environment. The Fig. 1 shows the visual odometry performed by the system. For a robust visual odometry model, the approach from [43] is used. This method defines a RGB-D sensor with a color brightness function $I(p,t)$ and a depth function $D(p,t)$, where $p = (u, v)$ are pixel coordinates within the image acquired at time t. After defining a series of motion models and mathematical transformations, a non-linear least square cost function

$$C(x) = \sum_{P^* \in \mathcal{R}^*} \left(\mathcal{I}\left(w\left(P^*; T(x)\hat{T}\right)\right) - \mathcal{I}^*(P^*)\right)^2 \tag{1}$$

is obtained, where $P^* = \{p, D\} \in \mathbb{R}^{nx3}$ and are the 3D points associated with the depth image and the image points p, the current image $\mathcal{I} = \{I, D\}$ which is the set containing both brightness and depth, w the motion model that defines the 3D geometric deformation of a structured light RGB-D camera, $T(x)$ the incremental pose to be estimated, $\hat{T}$ the estimated pose of the current image and $\mathcal{I}^*$ the reference image. By minimizing the cost function (1), the pose and trajectory of the camera can be estimated. The minimization algorithm from [43] estimates these parameters.

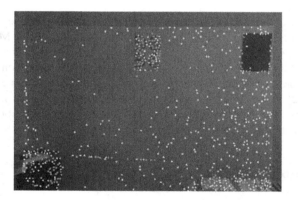

Fig. 1. Visual odometry performed by the system

3.2 Loop Closure Detection

Our approach uses a Bayesian filter to evaluate loop closure hypotheses over all previous images, based on the method described in [14]. The loop closure detector uses a bag-of-words (visual words, which are SURF features quantized to an incremental visual dictionary) approach to determinate the likelihood that a new image comes from a previous or new location. When a loop closure hypothesis is accepted, a new constraint is added to the map graph, following by a graph optimizer that minimizes the errors in the map. A loop closure is detected when a pre-defined threshold H is reached by the loop closure hypothesis. We use the map memory management approach of [15] to limit the number of locations used for loop closure detection and graph optimization algorithms, thus respecting the real-time limitations on large-scale environments.

The visual words are extracted from the RGB image. This image is registered with the depth image, knowing that a 3D position can be computed using the calibration matrix and the depth information provided by the depth image for each point in the RGB image. The RANSAC algorithm [44] uses the 3D visual word matches to compute the rigid transformation between the corresponding images when a loop closure is detected. The loop closure is accepted and a link with this transformation between the loop closure hypothesis node and the current node is added to the graph, only in the case of a minimum of I inliers are found (Figs. 2 and 3).

Fig. 2. Loop closure detection performed by the system

Fig. 3. Results of the graph optimization stage of the system.

3.3   Graph Optimization

We use the tree based parametrization [22] to describe an efficient configuration of the nodes in the graph. It can construct a graph with the given trajectory of the UAV. The pose and link transformations are used as the limitations. Errors produced by the visual odometry estimation can be propagated to all the links when a loop closure is found, and correcting the map at the same time. In order to decrease the computational cost of the algorithms, we are not using the approach proposed in [14]. Instead, we will use a more straightforward method, using the tree based parametrization algorithm [22] to create a tree from the map graph with only one map. By this procedure the tree of the algorithm will only have one root, removing the requirement of a robust memory management for the system.

4   Results and Discussion

The metric of evaluation used to obtain the error on the distance estimation is the percent error of measurements taken at different instances, defined as follows:

$$Error\% = \frac{|Experimental\ value - Reference\ value|}{Reference\ value} * 100\% \qquad (2)$$

For obtaining the error on the pose estimation, the absolute error was used and defined as follows:

$$Error\% = |Experimental\ value - Reference\ value| \qquad (3)$$

We test the difference in the pose referred to an initial reference point taking two different measurements, one before (Initial) and another after (Final) performing the mapping trajectory, in different environments and using two luminous intensities. We use a light bulb as a source of high brightness, and the LED flash of a common smartphone as a source of low brightness. We will test our approach in two different indoor environments and an outdoor environment. The poses were measured in meters. Video results are provided on https://www.youtube.com/watch?v=UQgHNN7g-xk.

4.1   Indoor Environment 1

The first mapping environment is a small 2,72 m × 3,64 m office, as shown in Fig. 4. This location was used to test the mapping capabilities of the system in a small environment. As mentioned earlier, we used a reference point to obtain the values of the poses before and after the mapping process.

To obtain the values of the poses we took the average value of five measurements acquired by our system. The results of these measurements at different light variations and the error calculated is shown in the following table.

In Table 1, we can appreciate that the error in the measurements of the pose is less than 4 cm when mapping with a constant light source, but increases considerably with

Fig. 4. Mapping of a small environment for pose error measurement. The left image shows the mapping performed with high brightness, the right image with low brightness.

Table 1. Pose error at different instances in a small environment

Pose	High brightness			Low brightness		
	Initial	Final	Error	Initial	Final	Error
X	−0,0038	−0,0211	0,0173	−0,0022	0,0475	0,0497
Y	−0,0528	−0,0798	0,027	−0,0093	0,1794	0,1887
Z	−0,0018	−0,0225	0,0207	−0,0018	−0,1779	0,1761
Roll	0,0069	−0,0043	0,0112	0,0015	−0,3087	0,3102
Pitch	−0,0015	0,0112	0,0127	−0,0009	0,0714	0,0723
Yaw	0,0297	0,0531	0,0234	0,0055	0,1893	0,1838

a low light source. Although the error in the measurements is high, the mapping was performed with success, showing the robustness of our system for different light conditions.

4.2   Indoor Environment 2

The next mapping environment used is a medium 4,80 m × 5,24 m office, as shown in Fig. 5. Similar to the last scenery, this location was used to test the mapping capabilities of the system in a medium size environment.

Fig. 5. Mapping of a medium environment for pose error measurement. The left image shows the mapping performed with high brightness, the right image with low brightness.

Same as the last environment, we took the average value of five measurements of pose acquired by our system. The results were tested at different light variations and the error calculated is shown in the following table.

In Table 2, we can appreciate that the error in the measurements of the pose is less than 5 cm when mapping with high brightness, but increases greatly when doing so with low brightness. Same as the last environment, the error in low brightness measurements is high and the system had difficulties correlating different reference points. Nevertheless, the map was performed with success.

Table 2. Pose error at different instances in a medium environment

Pose	High brightness			Low brightness		
	Initial	Final	Error	Initial	Final	Error
X	0,0168	0,0628	0,046	−0,0012	0,0934	0,0946
Y	−0,0415	−0,0629	0,0214	−0,0063	0,1943	0,2006
Z	−0,0063	−0,0487	0,0424	−0,0023	−0,1584	0,1561
Roll	−0,0008	0,0231	0,0239	0,0074	−0,2352	0,2426
Pitch	−0,0016	0,0189	0,0205	−0,0011	0,1043	0,1054
Yaw	0,0141	−0,0183	0,0324	0,0037	0,0342	0,0305

4.3   Outdoor Environment

The next mapping environment used is a 12,50 m outside wall of the offices, as shown in Fig. 6. This location was used to test the mapping capabilities of the system in a large size outdoor environment.

Fig. 6. Mapping of a large environment for pose error measurement. The image shows the mapping of a large outside wall and performed during the day.

Following the procedure used in the former environment, we get the average value of five measurements of pose during the day. Because of the low level of brightness available, mapping could not be performed at night, and values could not be obtained. The results are shown in the following table.

In Table 3, we can appreciate that the error in the measurements of the pose is less than 10 cm when mapping in the daylight. Due to the lack of light, the mapping could not be performed at night. This shows our system is limited to outdoor environments in the daylight.

Table 3. Pose error at different instances in a large environment

Pose	Daylight		
	Initial	Final	Error
X	−0,0022	0,1457	0,1479
Y	−0,0227	0,0712	0,0939
Z	−0,0021	0,0854	0,0875
Roll	0,0015	0,0658	0,0643
Pitch	−0,0024	−0,0987	0,0963
Yaw	0,0227	0,1118	0,0891

4.4 Distance Estimation Error

To estimate the error in the distance measured by the device, we take a similar approach to the previous tests. Based on a known distance to a reference point, we calculate the error at different instances of the mapping process. The results are shown in Table 4. The measured error is 2.506% after mapping. This shows that our system is accurate enough to perform SLAM on board.

Table 4. Distance taken at different times and the error in the measurements

Real distance (m)	Distance before mapping (m)	% Error before mapping	Distance after mapping (m)	% Error after mapping
2,139	2,153	0,650	2,194	2,506

4.5 CPU Usage and Power Consumption

Using the tools provided by the operating system, we measure the use of each CPU from the computer, which is shown in Table 5.

Table 5. CPU usage

CPU number	No processing (%)	Performing SLAM (%)	Increment (%)
1	9,4	38,4	29
2	6,0	43,8	37,8
3	9,9	35,0	25,1
4	12,0	44,2	32,2

We carried out the same procedure to obtain the power consumption of the system before and after mapping, shown in Table 6. Even though there is notable increase in the CPU usage and power consumption, the computer works better than expected and is able to run real-time SLAM without lagging.

Table 6. Power consumption

No processing (W)	Performing SLAM (W)	Increment (W)
5,48	8,20	2,72

5   Conclusions

Our proposal for on-board SLAM can estimate the UAV position with high accuracy and run on an Intel Stick core M3 computer in real time, removing the dependence on unreliable wireless communication. As shown in Fig. 6, our system can perform mapping in real time with a small difference in the pose, as shown in Table 1. The system does not behave in the same way while doing the mapping with a weaker light source. In order to obtain the best results from the system, the mapping needs to be performed in the day.

Our approach uses the loop closure detection algorithm to relate reference points in different instances in time. We have shown that the system is capable of updating the map in order to adapt to the environment, which is suitable for UAV applications.

Acknowledgement. This work is part of the projects 2016-PIC-024 and 2016-PIC-025, from the Universidad de las Fuerzas Armadas ESPE, directed by Dr. Wilbert G. Aguilar.

References

1. Aguilar, W.G., Verónica, C., José, P.: Obstacle avoidance based-visual navigation for micro aerial vehicles. Electronics **6**(1), 10 (2017)
2. Aguilar, W.G., Morales, S.: 3D environment mapping using the kinect V2 and path planning based on RRT algorithms. Electronics **5**(4), 70 (2016)
3. Cabras, P., Rosell, J., Pérez, A., Aguilar, W.G., Rosell, A.: Haptic-based navigation for the virtual bronchoscopy. In: 18th IFAC World Congress, Milano, Italy (2011)
4. Aguilar, W.G., Angulo, C., Costa, R., Molina, L.: Control autónomo de cuadricópteros para seguimiento de trayectorias. In: IX Congreso de Ciencia y Tecnología ESPE, Sangolquí, Ecuador (2014)
5. Huang, A.S., Bachrach, A., Henry, P., Krainin, M., Maturana, D., Fox, D., Roy, N.: Visual odometry and mapping for autonomous flight using an RGB-D camera. In: International Symposium on Robotics Research (ISRR) (2011)
6. Mur-Artal, R., Montiel, J., Tardo, J.: ORB-SLAM: a versatile and accurate monocular SLAM system. IEEE Trans. Robot. **31**(5), 1147–1163 (2015)
7. Triggs, B., McLauchlan, P.F., Hartley, R.I., Fitzgibbon, A.W.: Bundle adjustment — a modern synthesis. In: Triggs, B., Zisserman, A., Szeliski, R. (eds.) IWVA 1999. LNCS, vol. 1883, pp. 298–372. Springer, Heidelberg (2000). doi:10.1007/3-540-44480-7_21

8. Engel, J., Schöps, T., Cremers, D.: LSD-SLAM: large-scale direct monocular SLAM. In: Fleet, D., Pajdla, T., Schiele, B., Tuytelaars, T. (eds.) ECCV 2014. LNCS, vol. 8690, pp. 834–849. Springer, Cham (2014). doi:10.1007/978-3-319-10605-2_54

9. Kohlbrecher, S., von Stryk, O., Meyer, J., Klingauf, U.: A flexible and scalable SLAM system with full 3D motion estimation. In: IEEE International Symposium on Safety, Security, and Rescue Robotics (2011)

10. Whelan, T., Kaess, M., Leonard, J., McDonald, J.: Deformation based loop closure for large scale dense RGB-D SLAM. In: IEEE/RSJ International Conference on Intelligent Robots and Systems (2013)

11. Nister, D., Naroditsky, O., Bergen, J.: Visual odometry. In: Computer Vision and Pattern Recognition, pp. 652–659 (2004)

12. Konolige, K., Agrawal, M., Sola, J.: Large-scale visual odometry for rough terrain. In: International Symposium Robotics Research (2007)

13. Kerl, C., Sturm, J., Cremers, D.: Dense visual slam for RGB-D cameras. In: Proceedings of the International Conference on Intelligent Robot Systems (IROS) (2013)

14. Michaud, M., Labbe, F.: Appearance-based loop closure detection for online large-scale and long-term operation. IEEE Trans. Robot. 29(3), 734–745 (2013)

15. Michaud, M., Labbe, F.: Online global loop closure detection for large-scale multi-session graph-based SLAM. In: Proceedings of IEEE/RSJ International Conference on Intelligent Robots and Systems, pp. 2661–2666 (2014)

16. Stühmer, J., Gumhold, S., Cremers, D.: Real-time dense geometry from a handheld camera. In: Goesele, M., Roth, S., Kuijper, A., Schiele, B., Schindler, K. (eds.) DAGM 2010. LNCS, vol. 6376, pp. 11–20. Springer, Heidelberg (2010). doi:10.1007/978-3-642-15986-2_2

17. Aguilar, W.G., Angulo, C.: Estabilización de vídeo en micro vehículos aéreos y su aplicación en la detección de caras. In: IX Congreso de Ciencia y Tecnología ESPE, Sangolquí, Ecuador (2014)

18. Aguilar, W.G., Angulo, C.: Estabilización robusta de vídeo basada en diferencia de nivel de gris. In: VIII Congreso de Ciencia y Tecnología ESPE, Sangolquí, Ecuador (2013)

19. Aguilar, W.G., Angulo, C.: Compensación y aprendizaje de efectos generados en la imagen durante el desplazamiento de un robot. In: X Simposio CEA de Ingeniería de Control, Barcelona, Spain (2012)

20. Aguilar, W.G., Angulo, C.: Compensación de los efectos generados en la imagen por el control de navegación del robot Aibo ERS 7. In: VII Congreso de Ciencia y Tecnología ESPE, Sangolquí, Ecuador (2012)

21. Thrun, S.: Robotic mapping: a survey. In: Lakemeyer, G., Neberl, B. (eds.) Exploring Artificial Intelligence in the New Millennium. Morgan Kaufmann, San Francisco (2003)

22. Grisetti, G., Grzonka, S., Stachniss, C., Pfaff, P., Burgard, W.: Efficient estimation of accurate maximum likelihood maps in 3D. In: Proceedings of the International Conference on Intelligent Robots and Systems (IROS) (2007)

23. Dellaert, F.: Square root SAM. In: Proceedings of Robotics: Science and Systems (RSS), pp. 177–184 (2005)

24. Kaess, M., Ranganathan, A., Dellaert, F.: iSAM: incremental smoothing and mapping. IEEE Trans. Robot. 24(6), 1365–1378 (2008)

25. Jin, H., Favaro, P., Soatto, S.: Real-time 3-D motion and structure of point features: front-end system for vision-based control and interaction. In: IEEE Conference on Computer Vision and Pattern Recognition (CVPR) (2000)

26. Murray, G., Klein, D.: Parallel tracking and mapping for small AR workspaces. In: Proceedings of IEEE and ACM International Symposium on Mixed and Augmented Reality (ISMAR) (2007)

27. Strasdat, H., Montiel, J.M., Davison, A.: Scale drift-aware large scale monocular SLAM. In: Proceedings of Robotics: Science and Systems (2010)
28. Aguilar, W.G., Angulo, C.: Real-time model-based video stabilization for microaerial vehicles. Neural Process. Lett. **43**(2), 459–477 (2016)
29. Aguilar, W.G., Angulo, C.: Real-time video stabilization without phantom movements for micro aerial vehicles. EURASIP J. Image Video Process. **1**, 1–13 (2014)
30. Aguilar, W.G., Angulo, C.: Robust video stabilization based on motion intention for low-cost micro aerial vehicles. In: 11th International Multi-Conference on Systems, Signals & Devices (SSD), Barcelona, Spain (2014)
31. Zisserman, J., Sivic, A.: Video Google: a text retrieval approach to object matching in videos. In: Proceedings of 9th International Conference on Computer Vision, pp. 1470–1478 (2003)
32. Botterill, T., Mills, S., Green, R.: Bag-of-words-driven, single-camera simultaneous localization and mapping. J. Field Robot. **28**(2), 204–226 (2011)
33. Konolige, K., Bowman, J., Chen, J., Mihelich, P., Calonder, M., Lepetit, V., Fua, P.: View-based maps. Int. J. Robot. Res. **29**(8), 941–957 (2010)
34. Christensen, H.I., Folkesson, J.: Closing the loop with graphical SLAM. IEEE Trans. Robot. **23**(4), 731–741 (2007)
35. Johannsson, H., Kaess, M., Fallon, M., Leonard, J.J.: Temporally scalable visual SLAM using a reduced pose graph. In: RSS Workshop on Long-term Operation of Autonomous Robotic Systems in Changing Environments (2012)
36. Montemerlo, M., Thrun, S., Koller, D., Wegbreit, B.: FastSLAM: a factored solution to the simultaneous localization and mapping problem. In: Proceeding of the National Conference on Artificial Intelligence (AAAI) (2012)
37. Grisetti, G., Stachniss, C., Burgard, W.: Improved techniques for grid mapping with Rao-Blackwellized particle filters. IEEE Trans. Robot. (T-RO) **23**, 34–46 (2007)
38. Weiss, S., Scaramuzza, D., Siegwart, R.: Monocular-SLAM-based navigation for autonomous micro helicopters in GPS-denied environments. J. Field Robot. **28**(6), 854–874 (2011)
39. Tardos, R., Mur-Artal, J.D.: Visual-inertial monocular SLAM with map reuse. IEEE Robot. Autom. Lett. **2**, 796–803 (2016)
40. Engelhard, N., Endres, F., Hess, J., Sturm, J., Burgard, W.: Realtime 3D visual SLAM with a hand-held RGB-D camera. In: RGB-D Workshop on 3D Perception in Robotics at the European Robotics Forum (2011)
41. Heng, L., Meier, L., Tanskanen, P., Fraundorfer, F., Pollefeys, M.: Autonomous obstacle avoidance and maneuvering on a vision-guided MAV using on-board processing. In: Proceedings of ICRA, pp. 2472–2477 (2011)
42. Fraundorfer, F., Heng, L., Honegger, D., Lee, G., Meier, L., Tanskanen, P., Pollefeys, M.: Vision-based autonomous mapping and exploration using a quadrotor MAV. In: Intelligent Robots and Systems (IROS) (2012)
43. Audras, C., Comport, A., Meilland, M., Rives, P.: Real-time dense appearance-based SLAM for RGB-D sensors. In: Australasian Conference on Robotics and Automation (2011)
44. Bolles, M.A., Fischler, R.C.: Random sample consensus: a paradigm for model fitting with apphcatlons to image analysis and automated cartography. Commun. ACM **24**, 381–395 (1981)

End-User Development for Social Robotics

An End-User Interface to Generate Homeostatic Behavior for NAO Robot in Robot-Assisted Social Therapies

Hoang-Long Cao[1,3]($\boxtimes$), Albert De Beir[1,3], Pablo Gómez Esteban[1,3],
Ramona Simut[2], Greet Van de Perre[1,3], Dirk Lefeber[1,3],
and Bram Vanderborght[1,3]

[1] Robotics & Multibody Mechanics Research Group,
Vrije Universiteit Brussel, Pleinlaan 2, 1050 Brussels, Belgium
Hoang.Long.Cao@vub.be
[2] Department of Clinical and Life Span Psychology Group,
Vrije Universiteit Brussel, Pleinlaan 2, 1050 Brussels, Belgium
[3] Flanders Make, Strategic Research Centre Manufacturing Industry,
Lommel, Belgium
http://www.dream2020.eu

Abstract. Homeostatic drive theory is a popular approach for decision-making of robot behavior in social robotic research. It is potentially to be used in social therapies. To increase the involvement of end-users in the robot's control, we present an end-user interface allowing the therapists to generate homeostatic behavior for NAO robot in social skills training for children. We demonstrate the system by two interactions in which the robot homeostatic behavior is adapted to children's behavior. The result shows that the system provides a practical solution for therapists to implement interaction scenarios to robot behavior.

Keywords: Homeostasis · Robot behavior · End-user development · NAO · Robot-Assisted Therapy

1 Introduction

Social robots have been widely used in robot-assisted therapy (RAT) e.g. elderly care [8], social therapies for children with autism spectrum disorder [3,10,14, 24] and diabetes [5,9]. The advantages using robots in these domains can be explained by the social simplicity, predictability, and responsiveness of the robot behavior [22]. In many studies, robot behavior is manually controlled by a human operator using the Wizard of Oz approach [11,20,21]. This approach allows developers to set up robot behavior for human-robot interaction scenario quickly. However, it requires the operator a significant amount of time to manually select robot actions and can not totally focus on the human-robot interaction events especially the human behavior. Therefore, there is a trend – and a challenge – to go beyond the WoZ approach to increase the level of autonomy and reduce direct human operations [19,23].

© Springer International Publishing AG 2017
I. Rojas et al. (Eds.): IWANN 2017, Part II, LNCS 10306, pp. 609–619, 2017.
DOI: 10.1007/978-3-319-59147-6_52

To adopt this challenge, a number of behavior generation architectures have been developed enabling the robot the select actions more autonomously. Decision-making systems of these architectures are implemented following different approaches inspired from psychology or neurobiology. One of the most widely used solutions for decision-making is the homeostatic drive theory, in which the agent's actions are selected to maintain the internal needs within certain ranges [6]. The internal needs vary from physiological, e.g. hunger, thirst, and nutrient appetites, to cognitive and social necessities, e.g. being cared [2,18]. When one or several needs are not satiated, a motivational state is created, named as "drive" to trigger appropriate correction actions. Hence, the agent is able to adapt to environment changes and its behavior is understandable. Many studies have applied this theory to generate robot behavior, e.g. Aibo [1] Kismet [7], Maggie [18], iCub [25]; and it is potential to be utilized in robot-assisted social therapies.

A collaboration of developers and end-users is beneficial to improve the applicability of robot-assisted social therapy system. Consequently, intuitive tools for end-users should be designed to increase their involvement in the robot's control [4]. Robot behavior generation systems using homeostatic drive approach are also subjected to these recommendations. In this paper, we present an end-user interface to generate homeostatic behavior for robots in social therapies. We select a case of training social skills for children with NAO robot, and the primary group of end-users is therapists. Therefore, the terms "end-user" and "therapist" are used interchangeably throughout this paper. With a graphical user interface, the end-users can define needs, drives, actions, and the relationships among these objects. Based on these relationships, robot actions are autonomously selected according to the interaction events (e.g. children's behavior, performance, attention, emotion) analyzed using a Kinect 2.0. We follow the supervised-autonomy approach which allows the therapist to intervene the robot operation at any time e.g. correcting perceived stimuli or robot actions, manually activating missed stimuli. We conducted two interactions with children playing games with NAO robot to demonstrate the usability of the end-user interface.

This paper is organized as follows. Section 2 describes the homeostatic drive modeled used in the end-user interface and how it is used to generate robot's homeostatic behavior. Section 3 presents a scenario of using the end-user interface for children-robot interaction. Finally a conclusion will be given.

2 The End-User Interface for Homeostatic Behavior Generation

In this section, we present a homeostatic drive model that is used to develop the end-user interface and how the end-user can configure the robot behavior (NAO) in social therapies with children by a graphical user interface (GUI). The process of behavior generation and robot operation is supervised by the therapist i.e. the human decision has the highest priority.

2.1 Homeostatic Drive Model

For homeostatic drive models used in human-robot interaction studies, the term "need" is expanded to cognitive and social necessities. The working principle of homeostasis is illustrated in Fig. 1 in which a robot has a number of needs which are requited to be kept within a saturated (equilibrium) region limited by a lower bound and an upper bound. When a need value decreases below the lower bound to the under-saturated region, a drive is generated to look for actions that bring the need value back to the saturated region. For example, when the robot does not have enough attention from human, there is a drive to look for actions such as saying "Hello" or approaching the human to get more attention. Once the attention level is sufficient, the drive is satiated and the robot continues to look for actions satiating other drives (if they exist). Similarly, when a need value exceeds the upper bound to the over-saturated region, a drive is generated to decrease this need value.

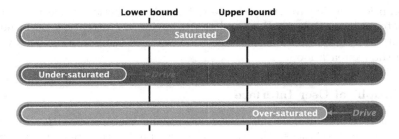

Fig. 1. The working principle of the homeostatic drive theory: the need is maintained within a specific range limited by a lower bound and an upper bound i.e. the saturated region. If going into the under-saturated or over-saturated regions, a drive is generated to trigger appropriate actions bringing the need value back to the saturated region.

In our model, the relationships between needs and drives are expanded to many-to-many as can be seen in Fig. 2. A drive can be triggered by different needs with influence factors. For each iteration, all drives values are calculated by the following Eq. 1:

$$D_j(t) = \frac{1}{n} \sum_i^n \gamma_{ij} \Delta_{N_i}(t) \tag{1}$$

where $D_j(t)$ is the value of drive j at time t, $\Delta_{N_i}(t)$ is the deviation of need i from its upper/lower bound at time t, γ_{ij} is the influence factor of need i on drive j, and n is the total number of needs.

The drive with the highest value is considered to be the most urgent one to be satiated. The selected drive triggers its "drive satiator" which includes a set of actions that might bring the need values back to the saturated regions. A specific action from a drive satiator is selected based on the pre-defined priorities

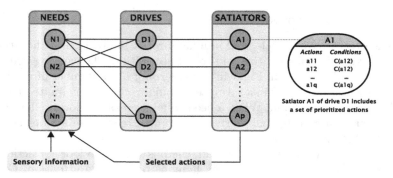

Fig. 2. The homeostatic drive model used in the end-user programming tool. The selected actions are sent to the robot. These actions and interaction events are then used to update the need values.

and conditions, which are set by the user. The selected action is sent to the robot (NAO) to perform. Robot actions and user behavior (e.g. attention, performance, emotion) analyzed from data of a Kinect 2.0 are used to update the need values, closing the feedback loop.

2.2   Graphical User Interface

We developed an end-user interface allowing the therapists to develop robot behavior using the homeostatic drive model presented in the previous subsection. The software architecture is depicted in Fig. 3 which includes three components: *Needs and stimuli management*, *Drives and satiators management*, and *Actuation and Sensory*. For the ease of configuring these components, three windows were developed as shown in Fig. 4. Since the aim is to use this interface with children and NAO robot for social therapies, the software is connected with a NAO and a Kinect 2.0.

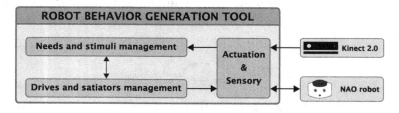

Fig. 3. Software architecture of the end-user interface to generate robot's homeostatic behavior. The system is connected with a NAO robot and a Kinect 2.0.

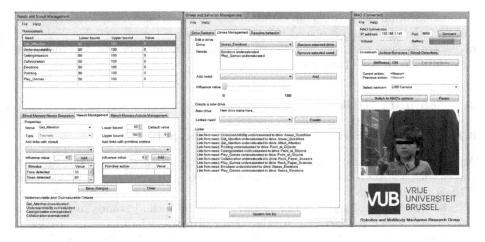

Fig. 4. The graphical user interface is composed of three windows: *Needs and stimuli management*, *Drives and satiators management*, and *Actuation and Sensory*. Not only managing the robot behavior, the therapist is also allowed to intervene the robot behavior during the robot operation.

Needs and Stimuli Management. This window is used to define needs. A need is defined with a lower and an upper bound from a range from 0% to 100%. Influenced elements of each need (e.g. actions, perceived stimuli) and how much each element affects the need value are also defined here.

Drive and Satiators Management. This window allows the therapist to configure the relationships between drives and needs i.e. how much each need influences a particular drive. Next, the user creates drive satiators when drives are selected. A drive satiator is a list of prioritized actions with conditions. When a drive satiator is executed, the software searches for the most appropriate action by the following rule: *the action with the highest priority and all of its conditions are satisfied*. In case more than one actions fulfill the selection rule, the software randomly selects on of them. The user creates drive satiators using a list of available actions and a list of conditions.

Actuation and Sensory. This window sets up connections with the NAO robot (wireless) and the Kinect 2.0 (USB 3.0). On the one hand, selected actions from the software are sent to the robot to perform. On the other hand, information gathered from the Kinect is sent back to the software as perceived stimuli (e.g. the child's performance, attention, emotions).

Following the supervised-autonomy approach, the human therapist controls the whole process of behavior generation and robot operation. The numerical values of needs and drives can be adjusted during the interaction to deviate the robot behavior. Regarding the sensory input, perceived stimuli can be corrected, or missed stimuli can be manually activated. And importantly, the selected action can be interrupted and replaced by any action from a list of available actions if necessary. In such way, although the robot actions are autonomous, the therapist stays fully in charge of the robot behavior.

3 Usability Test

3.1 Interaction Scenario and Experimental Setup

An interaction scenario is selected by a therapist in which a child plays different games with a NAO robot namely "Point at objects", "Rock Paper Scissors", and "Guess emotions".

The experimental setup is illustrated in Fig. 5. A child stands in front of a NAO robot which is placed on a table. Four objects (a hammer, an orange, a glass and a flute) are placed on the table for the game "Point at objects". The child is asked to remain standing in front of the robot at approximately one meter. The therapist sits behind the computer running the software during the interaction for supervision purpose. A Kinect 2.0 is used to detect the child's response and emotions. External cameras are used to record the interaction for video analysis. Robot behavior (e.g. need values, drive values, selected drives, selected actions) is also recorded to evaluate the system. During the interaction, a vocal instruction is given to the child in case he or she has difficulties of understanding the robot behavior e.g. speak louder. The interaction lasts about 15 min.

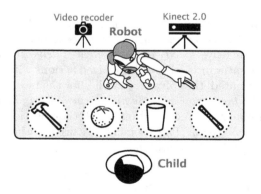

Fig. 5. The experimental setup. A child stands in front of a NAO robot to play different games. Four objects (a hammer, an orange, a glass, and a flute) are used as materials of the games. Different types of cameras are used to provide interaction data to the software as well as for analysis and evaluation purposes (Color figure online).

3.2   Using the Interface to Generate Robot's Homeostatic Behavior

To create the homeostatic behavior for the robot in this case, the therapist needs to define needs, drives and satiators. The drive–need relationships is visualized in Fig. 6. Seven needs are defined as follows.

Fig. 6. Need–drive–satiator relationships of the interaction scenario. The connections and influences among these objects are configured using the graphical user interface.

Get attention: a need of having attention
Understandability: a need of being understood by the child
Categorization: a need of teaching how to categorize objects.
Collaboration a need of teaching how to collaborate in a task or in a game.
Emotions: a need of teaching how to recognize emotions.
Pointing: a need of teaching pointing.
Play games: a need of having engagement and interest in the games

The first two needs (Get attention, Understandability) are social needs required to initiate an interaction between the child and the robot. The four following ones (Categorization, Collaboration, Emotions, Pointing) are connected to training activities in social therapies for children. The last need (Play games) ensures the child is engaged in the activities. From the needs, five drives are defined with their satiators.

Answer questions. This drive is triggered when the child does not understand what to do or does not answer to the robot. Its satiator selects an appropriate answer depending on the child's response.

Attract attention. This drive is triggered when the child is not sufficiently attentive and stops looking at the robot. Its satiator randomly selects an action to attract attention. For example, the robot waves and says "Look at me".

Game: Point at objects. In this game, the robot points at an object and asks questions. The game is inspired by the works of joint attention [16]. The questions include the color, the usage or the material of the pointed object. The child can answer this question by saying "Yes" or "No".

Game: Rock Paper Scissors. In this game, the robot and the child play the game rock–paper–scissors. The robot provides feedback after each turn. For example: "You did rock. I did scissors. You won!".

Game: Guess emotions. In this game, an emotional behavior is performed by the robot. The child is requested to recognize the corresponding emotion. The emotion library is inspired by the work of Haring *et al.* [15].

3.3 Interactions with Children

We conducted two interactions with children of six years old (a boy and a girl) to demonstrate the working of the system. The session was organized at a primary school following the above-mentioned experimental setup (Fig. 7). A written informed consent was signed by each participant legal tutor prior to the experiment. To analyze the system, we present two subsets of the two interactions.

Fig. 7. Child-robot interaction. The robot is explaining the game to the child.

Figure 8 shows the evolution of drives that leads to the robot actions in two cases, in which the robot behavior is adapted to each child. The bullets indicate the selected drives at each iteration. In the first case (a girl), the robot mainly stays at the game "Guess emotions" except one time going to the "Answer questions" drive. This can be explained by the girl's emotion. She smiles a lot during the interaction and most of the time the detected emotion is happy (56.3% *happy*, 28.1% *neutral*, 15.6% *unknown*). Since there is a link between the happy

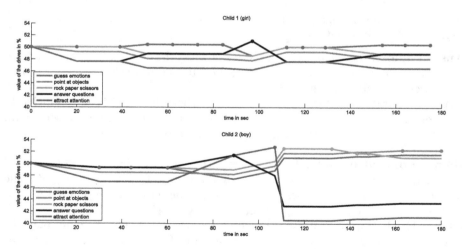

Fig. 8. Evolution of drives in two subsets of interactions. Bullets represent the selected drives. In the first case, the girl enjoys the interaction, the drive of playing the game "Guess emotions" is selected most of the time. In the second case, the boy's emotion is mainly neutral. The system switches drives to seek for a more suitable game.

emotion (perceived stimulus) and the game, the system understands that the child enjoys the game and there is a need to keep playing this game. On the contrary, in the second case (a boy), the robot switches among the drives many times. The boy seems to be shy and his facial expression is mainly neutral during the interaction (96% *neutral*, 4% *unknown*). Therefore, there is no need to stay long in any game, but instead, to switch to another game.

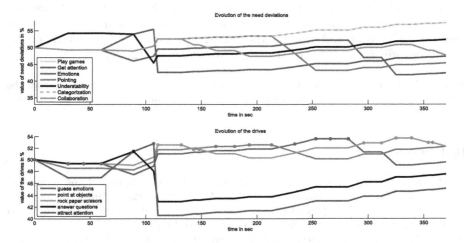

Fig. 9. Evolutions of need deviations and drives in the second interaction (a boy). The modification of the internal needs (upper part of the figure) results from the interaction between the robot and child. These needs influence the drive selections.

To further analyzing the system, we look at evolutions of need deviations and drives in the second interaction (the boy) shown in Fig. 9. At the beginning, since all needs are set at the same values, all drives value are equal and a random game is selected. When the boy does not understand the instruction and sometimes does not look at the robot (loses attention), the need "Understandability" and "Get attention" deviate. Since these needs are linked to the drives "Answer questions" and "Attract attention, these drives are selected. When the child has enough attention at the robot, the need "Understandability" and "Get attention" are satisfied. Hence, the drive to play the game "Rock Paper Scissors" is selected. However, the child's emotion indicates that he does not enjoy the selected game (but still has attention at the robot). After that, the needs for social activities vary and lead to the variations of the corresponding drives. Consequently, the selected game is regularly changed.

4   Conclusion and Future Work

In this paper, we present an end-user interface to generate homeostatic behavior for NAO robot in social therapies with children. With a graphical user interface, the therapist is able to set up the needs, drives, satiators and their relationships. During the interaction, the robot autonomously selects actions following the configuration of the homeostatic drive system. This allows the human-robot interaction to go beyond the WoZ setup and reduce repetitive works. We demonstrated the working of the system by conducting two interactions with children. The result shows that the system provides a practical solution for the therapist to implement an interaction scenario to robot behavior. We suggest that the numbers of needs and drives should not be limited to the level that the need–drive relationships are still intuitive to the users. Future work includes implementing a drag-and-drop interface using Boxology [17]; and improving the emotion expression of NAO by using eyebrows plug-in [12,13].

Acknowledgement. The work leading to these results has received funding from the European Commission 7th Framework Program as a part of the project DREAM grant no. 611391.

References

1. Arkin, R.C., Fujita, M., Takagi, T., Hasegawa, R.: An ethological and emotional basis for human-robot interaction. Robot. Auton. Syst. **42**(3), 191–201 (2003)
2. Bach, J.: Principles of Synthetic Intelligence PSI: An Architecture of Motivated Cognition, vol. 4. Oxford University Press, Oxford (2009)
3. Barakova, E.I., Bajracharya, P., Willemsen, M., Lourens, T., Huskens, B.: Long-term LEGO therapy with humanoid robot for children with ASD. Expert Syst. **32**(6), 698–709 (2015)
4. Barakova, E.I., Gillesen, J.C., Huskens, B.E., Lourens, T.: End-user programming architecture facilitates the uptake of robots in social therapies. Robot. Auton. Syst. **61**(7), 704–713 (2013)
5. Belpaeme, T., Baxter, P.E., Read, R., Wood, R., Cuayáhuitl, H., Kiefer, B., Racioppa, S., Kruijff-Korbayová, I., Athanasopoulos, G., Enescu, V., et al.: Multimodal child-robot interaction: building social bonds. J. Hum. Robot Interact. **1**(2), 33–53 (2012)
6. Berridge, K.C.: Motivation concepts in behavioral neuroscience. Physiol. Behav. **81**(2), 179–209 (2004)
7. Breazeal, C.: Designing Sociable Robots. MIT press, Cambridge (2004)
8. Broekens, J., Heerink, M., Rosendal, H.: Assistive social robots in elderly care: a review. Gerontechnology **8**(2), 94–103 (2009)
9. Cañamero, L., Lewis, M.: Making new "new AI" friends: designing a social robot for diabetic children from an embodied AI perspective. Int. J. Soc. Robot. **8**(4), 523–537 (2016)
10. Cao, H.-L., et al.: Probolino: a portable low-cost social device for home-based autism therapy. In: Tapus, A., André, E., Martin, J.C., Ferland, F., Ammi, M. (eds.) Social Robotics. LNCS, vol. 9388, pp. 93–102. Springer, Cham (2015)
11. Dautenhahn, K.: Methodology and themes of human-robot interaction: a growing research field. Int. J. Adv. Rob. Syst. **4**, 15 (2007)
12. De Beir, A., Cao, H.-L., Esteban, P.G., Van de Perre, G., Lefeber, D., Vanderborght, B.: Enhancing emotional facial expressiveness on NAO. Int. J. Soc. Robot. **8**(4), 513–521 (2016)

13. De Beir, A., Cao, H.-L., Esteban, P.G., Van de Perre, G., Vanderborght, B.: Enhancing NAO expression of emotions using pluggable eyebrows. In: New Friends International Conference (2015)
14. Diehl, J.J., Crowell, C.R., Villano, M., Wier, K., Tang, K., Riek, L.D.: Clinical applications of robots in autism spectrum disorder diagnosis and treatment. In: Patel, V.B., Preedy, V.R., Martin, C.R. (eds.) Comprehensive Guide to Autism, pp. 411–422. Springer, Heidelberg (2014)
15. Haring, M., Bee, N., Andre, E.: Creation and evaluation of emotion expression with body movement, sound and eye color for humanoid robots. In: RO-MAN, pp. 204–209. IEEE (2011)
16. Kozima, H., Michalowski, M.P., Nakagawa, C.: Keepon. Int. J. Soc. Robot. 1(1), 3–18 (2009)
17. Lemaignan, S.: Boxology (2017). https://github.com/severin-lemaignan/boxology
18. Malfaz, M., Castro-González, Á., Barber, R., Salichs, M.A.: A biologically inspired architecture for an autonomous and social robot. IEEE Trans. Auton. Ment. Dev. 3(3), 232–246 (2011)
19. Rabbitt, S.M., Kazdin, A.E., Scassellati, B.: Integrating socially assistive robotics into mental healthcare interventions: applications and recommendations for expanded use. Clin. Psychol. Rev. 35, 35–46 (2014)
20. Riek, L.D.: Wizard of Oz studies in HRI: a systematic review and new reporting guidelines. J. Hum. Robot Interact. 1(1), 119–136 (2012)
21. Riek, L.D.: Robotics technology in mental health care. In: Luxton, D. (ed.) Artificial Intelligence in Behavioral and Mental Health Care, p. 185. Elsevier, New York (2015)
22. Robins, B., Dautenhahn, K., Te Boerkhorst, R., Billard, A.: Robots as assistive technology-does appearance matter? In: The 13th IEEE International Workshop on Robot and Human Interactive Communication, pp. 277–282. IEEE (2004)
23. Thill, S., Pop, C.A., Belpaeme, T., Ziemke, T., Vanderborght, B.: Robot-assisted therapy for autism spectrum disorders with (partially) autonomous control: challenges and outlook. Paladyn 3(4), 209–217 (2012)
24. Vanderborght, B., Simut, R., Saldien, J., Pop, C., Rusu, A.S., Pintea, S., Lefeber, D., David, D.O.: Using the social robot probo as a social story telling agent for children with ASD. Interact. Stud. 13(3), 348–372 (2012)
25. Vouloutsi, V., Lallée, S., Verschure, P.F.M.J.: Modulating behaviors using allostatic control. In: Lepora, N.F., Mura, A., Krapp, H.G., Verschure, P.F.M.J., Prescott, T.J. (eds.) Living Machines 2013. LNCS, vol. 8064, pp. 287–298. Springer, Heidelberg (2013). doi:10.1007/978-3-642-39802-5_25

Graphical Programming Interface for Enabling Non-technical Professionals to Program Robots and Internet-of-Things Devices

Igor Zubrycki[✉], Marcin Kolesiński, and Grzegorz Granosik

Lodz University of Technology, Łódź, Poland
igor.zubrycki@dokt.p.lodz.pl
http://robotyka.p.lodz.pl

Abstract. This paper presents a graphical programming interface that enables non-technical professionals to program robots. Increasingly, robots are used by non-technical persons, such as service workers, therapists or marketers and graphical programming enables them to adjust robots to the situational needs through intuitive but expressive ways.

We present our implementation of graphical environment for programming a set of internet-of-things (IoT) devices and robots for therapists of autism. It is based on Robot Operating System (ROS) and Snap, and is called Robokol. Compared to previous solutions our system is easily extensible to new devices, with an interface enabling plug-and-play device discoverability and uses nonproprietary, well-known tools.

We detail two use cases of our interface, one where therapists of autism create sense-act loops for sensory therapy employing robot-like or IoT devices, and second where Robokol is used to create interfaces for Wizard-of-Oz scenario with robots.

1 Introduction

In our project Robokol, we were focused on making a tool for non-technical users to program a heterogeneous and possibly variable set of robots and other devices. The need for such software arose when we have been creating a system of therapeutic devices for use in autism therapy. Therapists – persons that would be using such system every day, wanted more control and flexibility than could be possible through just setting some knobs or switches on the equipment. They wanted to control the devices logic (e.g. set appropriate actions for a particular child or automatically change the intensity of the stimuli with the session progress) as well as to combine devices to create settings that would be interesting and helpful. Such system, however, needs to be simple to use and, overall, the programming cannot be time-consuming because it would not be the priority task of therapists workday.

Such cases, when non-technical persons would benefit from programming robots, are arriving more and more frequently, as robots come to service industries (hotel, restaurant, entertainment robots) where employees need to program

© Springer International Publishing AG 2017
I. Rojas et al. (Eds.): IWANN 2017, Part II, LNCS 10306, pp. 620–631, 2017.
DOI: 10.1007/978-3-319-59147-6_53

"business logic" of their particular workplace, or homes where household members or caregivers would want to give appropriate behaviour for their robots. Moreover, in the future, users may want to program groups of different robots to fit their needs.

Our solution combines Snap – a JavaScript-based version of popular graphical programming language Scratch – and ROS (Robot Operating System) being the most popular robotic framework and toolchain often used when creating software for robotic needs. In our solution robots introduce themselves to the programming environment, facilitating discoverability (one of the system's functionalities). We have also enabled sending binary files through the interface to the robots, in order to simplify the creation of multimedia robot behaviors without the need to learn other tools (i.e. to manage network files or discs) by the users. Our philosophy was to take the best features of Scratch, such as ease of use, ease of understanding of a system's state, intuitiveness, and to expand them to the multi robots domain, without increasing complexity on the user's side. Also, we wanted to use as much as possible open, non-proprietary and popular tools to ease a future code maintenance and changes.

To explain the value of our system to the robotic community we provide two use cases: rapid prototyping of robot behaviors for Wizard-of-Oz robot experiments and use of Robokol to program interactive therapy using two therapeutic devices – Ono Robot and Sensory Sleeve.

2   Review

Simplifying robot programming for persons that have been not explicitly trained in programming robots or programming in general was studied in several contexts before.

A number of authors studied simplifying the programming of industrial robots or manipulators. Simplification was done by means of programming by demonstration, where an industrial robot was guided through a trajectory, guided with sensory feedback or such trajectory was presented by using programmer's hand or some demonstration device [3]. Additionally, the operator could receive feedback through augmented reality or various auditory or visual devices [4]. In general, simplification of programming was focused on how to ease teaching the movement of several degrees of freedom robots with additional feedback about robots limitations and singularities. Some of the other approaches are using higher-level primitives via task-oriented robot programming languages [18].

The topic of simplification the programming of social and therapeutic robots was also studied. Here, the focus was on programming sequences of (social) actions to create meaningful behavior such as robots telling a story or recognizing objects and reacting to them. A fuzzy method based programming language, capable of a natural language input for programming therapeutic robots was described in [2]. Creators of *Social Robotic Toolkit* used an alternative, tangible, programming method, where a sequence of commands or conditional commands can be drawn on paper and presented to the robot vision system [11].

Tangible programming, in a form of connecting together physical blocks, was used to create programs for robots, without the need to program on the computer [10].

Graphical programming is increasingly used to program robots and internet-of-things devices, mainly as it provides a more effective way to facilitate comprehension of fundamental programming concepts needed to create useful programs [19]. This results in better understanding and, in the case of non-technical users, gives a way to understand and program without large time investment. *CustomPrograms* is a Blockly based software environment for programming Savioke Relay robots [7]. Particularly, authors focused on enabling hotel staff and business people to program robots behavior. *Interaction Composer* is a visual programming environment for creating social human-robot interactions based on state flows, used to program the behavior of shopping mall robots, robotic wheelchairs or helper robots [5]. Also, several educational robots use a graphical approach to teach programming, most notably *Choregraphe* for NAO and Pepper robots [15], *EV3 Programmer App* for programming LEGO Mindstorms [17] and *Scratch for Arduino* for programming Arduino based robots [1]. Compared to previous solutions, our environment is ROS-centric and provides a way to introduce new robots or components to the system with no need for users to manually insert new device specifications. Open roberta [8] is a blockly based (a Scratch derivate) programming environment with goals similar to ours.

There were also previous attempts to introduce graphical programming to ROS-based systems. *ROS Commander* is an expert user interface for constructing robot behaviors out of Hierarchical Finite State Machines [13]. *ROS from Scratch* and *ROS-Scratch* are two projects interfacing the most popular graphical programming language – Scratch – to ROS [16]. Both of these projects (as of 2016) are no longer supported (*ROS from Scratch* was available for Ros Electric, while the current version is ROS Jade, and *ROS-Scratch* worked with Scratch 1.4, while the current version is Scratch 2.0). Our project, based on JavaScript version of Scratch called Snap, uses ROS-bridge to interface with a larger set of ROS versions. Also, in contrast to previous solutions, which supported only ROS-topics for communication, our system offers higher level abstractions, and ability for each new device to introduce their own communication protocol and available functionalities.

Our solution is highly connected to previous attempts to communicate and control a heterogeneous group of robots. *PyRo* is an open source design of controllers and simulators for robots, with Bluetooth based communication [6]. It provides a master-slave protocol with each device (robot) presenting its own specification that can be used when programming this device (robot) using python language. *Buzz* is another language suitable for programming robot swarms; it is based on Lua and C, allows for creating software for a single robot or groups of robots, and facilitates information sharing between robots [14]. Compared to these solutions, ours uses graphical programming language instead of text-based ones but shares similar features – communication transparency and ease of creating multi-robot scenarios.

3 Description of Robokol Graphical Programming Environment

3.1 Specification of Needs

We have conducted a need-finding studies on a group of autism therapists as a part of the Roboterapia project in which goal was to create technologies helpful in autism therapy [22]. During these studies therapists stressed the need to control and ability to change the details of the behaviour of therapeutic devices. This led us to design a set of programmable devices for sensory therapy and presented us with a need for a programming environment that could be used by nontechnical people to work with robots or other programmable devices [20].

In this section, we present Robokol, a graphical programming environment capable of programming different multisensory devices. We did not aim to create a new programming environment, rather, the current solutions (listed in Table 1) did not have features necessary for our case, which led us to modify Snap environment for multi-device and multi-system scenario all keeping the ease of use needed in end-user development.

First, the interface must allow the user to program in an *intuitive* way. That is, it can be used for programming particular tasks without technical preparation or reading documentation. An extended study on syntactic and semantic issues in introductory programming and usability of programming interfaces from technological and psychological perspective can be found in [9]. According to this analysis, a *usable* interface, which is easy to learn for beginners, is the one with several features such as closeness of mapping, consistency, low error-proneness, lack of hard mental operations and expressiveness. Additionally, to ease the usage without a prior knowledge, the system should feature *discoverability*, that is providing a clear way to find and understand functions of the elements in the system without explicit training. These functions must be on the right abstraction level.

Table 1. Graphical programming environments overview. Abbreviations used in columns: PHT - Allows for programming heterogeneous groups of robots FL - Free license MP - Multi-platform AS - Active support or community FIL - Fast, intuitive to learn EX - Built on free and extensible technologies P&P - Plug and Play

	PHT	FL	MP	AS	FIL	EX	P&P	Comment
Scratch for Arduino	No	Yes	No	No	Yes	Yes	No	For Arduino based devices only
EV3 Programmer App	No	No	No	Yes	Yes	No	No	For Lego products
Interaction Composer	Yes	No	No	No	No	Yes	Yes	Focused on dialog behaviors, built on top of Eclipse
ROS commander	Yes	Yes	Yes	No	No	Yes	?	Difficult UI
ROS from Scratch, ROS-Scratch	Yes	Yes	Yes	No	Yes	Yes	No	For old Scratch versions only
Choregraphe	No	No	Yes	Yes	Yes	?	No	For Aldebaran products only
Open Roberta	Yes	Yes	Yes	Yes	Yes	Yes	Yes	Similar solution to our

Secondly, as programming interface will be used to program multiple devices, the system must be capable of handling communication between these devices, easily detect them and setup. Users may want to attach additional devices/ functionalities while the system is running, therefore, such action should be *plug and play* (not require to set the system up for a particular device but rather the device should introduce itself and deliver its own description). This is to allow spontaneity of the use where the whole system would have Lego-like (the toy set) capability to reconfigure.

Thirdly, as the user may need to program/modify programs using a range of devices (tablets, PC's, smartphones) it should be available on various machines with different operating systems. Preferably, with no need for installation, because users may not want to or not be capable of doing this procedure (administrative reasons or lack of knowledge).

Lastly, it should be based on standard tools, that are actively developed (preferably by a large community). As developing robot architecture is a big task, and changes in robotics are rapid, large portions of necessary functionality could be imported from other, open-source projects.

3.2   Architecture and Implementation of the System

On the highest abstraction level, our system that has the capability of being programmed through the graphical programming interface consists of:

- A data exchange server, that manages communication between devices, keeps the device addresses (IP, ports) and a list of each device functions
- Programming interface client, which provides a way to program any device that provides a list of programming affordances to it (such as list of sensors and its possible conditions, possible actuation methods etc.)
- One or more nodes, each providing its own description of programming affordances as well as a description of possible setup choices (i.e. Linux setup files, folder to which multimedia files can be uploaded etc.). Such nodes can have physical form (e.g. a robot) as well as virtual form, such as an AI helper agent or some internet API client.

ROS Based Implementation of the System with Graphical Programming Availability. To implement the elements and functionality of system described in the section above, we have modified browser-based graphical programming interface Snap! [12], to integrate it with ROS (Robot Operating System), being responsible for system setup and communication in our system. Elements of our implementation are presented in Fig. 1. We have called the modified Snap programming interface with additional protocol specification Robokol.

Data exchange server is a computer running ROS core services, rosbridge server that enables the browser to communicate with the ROS through HTML5 WebSockets and JSON data. Each system node has a running ROS process. Nodes can run on physically separate computers (robots) connected together through WiFi or Ethernet.

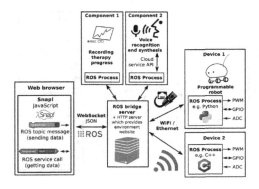

Fig. 1. Implementation of the system with graphical programming availability

Additional HTTP service on the communication server provides a Snap! website for the local network, making possible to control robotized environment from any connected computer with a web browser on it. A computer that opens the Snap! website becomes a programming interface client via ROS network. Through the website, the user can create or execute programs that will control the nodes.

For a device to be connected to the system, it has to be connected with ROS and publish to a specific topic with a message describing its own functionality. A message has to be in a form of XML data block consisting of: device name, icon, description as well as all programming affordances that are added by the device when connected to the system in a form of Snap blocks. Currently, the affordances have a form of Snap XML blocks (as explained in Snap reference manual [12]) and are automatically imported to the programming interface. Detailed example of such XML block is provided in Subsect. 3.2.

The process of searching newly added device is triggered from programming interface via "search for connected devices" button, which sends a trigger message to a particular topic. Each device publishes also its own description to the system. A special, safety mode can be triggered from the graphical interface in case of emergency, stopping all critical activity. Example graph of the ROS network is presented in Fig. 2.

From the Robokol interface, the user can also send any files directly to one of the devices. The user first selects the destination device and then uploads files to the browser-based interface, the Snap program then wraps the file into a message, with an additional header specifying the file name and function, and sends them to listening node on the device.

Other services available on the device are available on the Robokol interface through the blocks specified via the XML file available on the device.

Block Definition Files. New block or set of blocks can be added to the Robokol interface by providing an adequate message to the listener node. Blocks added

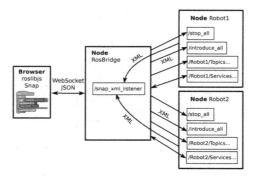

Fig. 2. Graph of ROS organization convention in the proposed system

to the interface conform to Snap's color and shape coding, that is, different colors are used for commands related to motion, device appearance or sensing (conditional) behavior. Snap's sprite – an icon representing the device – animates when the state of the device is changed, to facilitate better understanding. Also, similarly to the Snap!, the user can see the state of particular value (i.e. sensor reading) by moving the mouse pointer over the sensing block. This enables user to understand system's state easily.

As an example block for controlling Ono robot is presented in Fig. 5. It uses sub-blocks for communicating with ROS, such as "string ROS message" block for sending a message for a particular topic. This is, however, invisible to the user in normal mode, and only visible in developer mode. An XML file with definitions of blocks that would be sent to add the device affordances is listed in Listing 1.1.

Listing 1.1. Example of basic XML device definition

```xml
<sprites app="Snap! 4.0, http://snap.berkeley.edu" version="1">
    <sprite name="Ono" idx="2" x="0" y="0" heading="90" scale="1"
        rotation="1" draggable="true" costume="1" color="0,0,0" pen="tip" id="2">
        <costumes>
            <list id="1">
                <item>
                    <costume name="Ono" center-x="39.5" center-y="50"
                        image="data:image/png;base64dataFormat" id="4"/>
                </item>
            </list>
        </costumes>
        <sounds><list id="5"></list></sounds>
        <variables></variables>
    <blocks>
        <block-definition s="emocja %'emotion'" type="command" category="looks">
        <header/><code/>
            <inputs>
                <input type="%txt" readonly="true">
                    <options>
                        alert
                        excited
                        happy
                        bored
                        sad
                    </options>
                </input>
            </inputs>
            <script>
                <block s="rosStringMessage"><l>%%/onoEmotion</l>
                    <block var="emotion"/>
                </block>
```

```
            </script>
          </block-definition>
        </blocks>
        <scripts></scripts>
      </sprite>
</sprites>
```

4   Use Cases

In order to show how the Robokol can be used in robot programming, we present and explain in detail use cases of our system.

4.1   Using the System to Create the Sensory Therapy Scenario

In the first of presented scenarios, the Robokol was used to program a set of two devices, social robot Ono and Sensory Sleeve (see photograph in Fig. 3), which is a therapeutic device developed for sensory therapy [21]. The goal was to create an interaction supporting cognitive processes in an autistic child: if the child puts the hand into the Sensory Sleeve, Ono robot would react with a face expression change.

Both devices had special Snap blocks ascribed to them and stored on the devices. In the case of Ono there were blocks connected to face expressions and emotions (based on circumplex emotion circle). In the case of Sensor Sleeve, there were blocks connected to proximity sensing, lighting up and vibrations.

The program, shown in Fig. 4 reacts to the depth to which the hand is put into the device. Readings from four proximity sensors of the Sensor Sleeve are used to change the behavior of both the Ono robot and the Sensor Sleeve. Therapist, using the system, can change the behavior (i.e. Ono robot expression, or light) online.

Such use case – Robokol used as a way to prototype an interaction between a child and a device was tested in several settings. In one, described in, we have used Robocol with virtual devices and performed a series of tests to check the

Fig. 3. Robot Ono (yellow) and Sensory Sleeve (black) used to create an interactive scenario for use in sensory therapy. Video recording of the interaction is presented in: https://youtu.be/Mr_JedZMaJ0 (Color figure online)

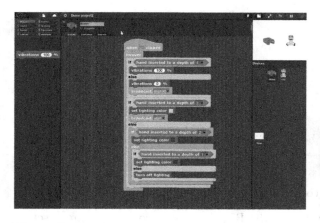

Fig. 5. Definition of set emotion block visible from snaps editor in developer and user modes

Fig. 4. Demo of our interface, showing a program for two devices: Ono robot and Sensory Sleeve.

ability of therapists of autism to create programs (involving loops, conditional commands and use of multimedia). In second, a "sensor box" device was programmed by two therapists throughout a group therapy session. In this setting therapists managed to change the program "online" while using the Ono robot and "sensory box" to interact with autistic child. In third, therapists participated in a trade fair for medical devices, where they presented the project. Throughout three days of the trade fair, therapists created a series of demonstration programs for the trade audience. Currently, some of devices created in Roboterapia project, are used in Navicula centre for autism therapy, where Robokol programming system is used to program them.

4.2   Rapid Prototyping of Wizard-of-Oz Robot Control Interface

The Ono robot was used in an experiment where it was interacting with a human, describing or commenting (through voice and face expressions) emotions that it was recognizing. As the main goal was to investigate the human-robot interaction and not (at this stage) develop a robust emotion recognition and dialogue system for the robot, it was decided that the robot will be teleoperated. This required an interface for robot control, that would enable selecting appropriate voice and gestural signals. Also, as the subject of the experiment was in the different room, robot's operator used wifi network to connect to the robot and ROS system.

The Robokol was used to create a set of blocks, each consisting of voice command (with a voice synthesized in separate ROS node, based on Ivona voice synthesiser) and face expression. The operator clicked on the appropriate set of blocks which ran the voice synthesiser and Ono programs. In order to facilitate fast responses, an additional program was running in parallel, where the robot reacted with "It tickles" whenever it was touched on the head. A screen of the program used in the experiment is shown in Fig. 6.

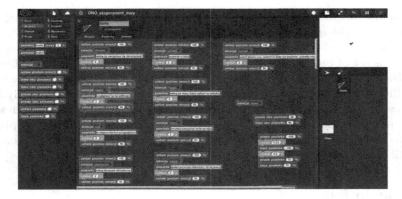

Fig. 6. Robokol used to rapidly create an interface for ROS connected robot (Ono) in Wizard-of-Oz scenario (interface in The Polish language).

5   Discussion

We have presented an idea, architecture, and implementation of the Robokol programming interface. It is our solution to the lack of features necessary for our use case, of other graphical programming systems (see Table 1).

The basis of our programming system is a programming interface derived from Scratch called Snap. It is easy to use and provides all the features necessary for an *intuitive* system as discussed in Sect. 3.1, in particular: the state understanding, discoverability, and puzzle like block behavior that eases the construction of valid programs.

The system is available to the end user via a web page, that can be accessed from all devices in the same network (smartphones, pc's). The web technologies used, require only a moderately capable computer – we have validated that the program can run on various devices from the therapy centre, such as Windows Vista/7 based laptops from around 2008–10. However, as currently the program itself runs in the browser's JavaScript machine, this limits the system capabilities. Namely, program execution is limited by the computational power of the browser's host which can lead to poor performance in larger, more computationally demanding programs. Also, as data from ROS must be available for program logic, transport of the data through WiFi network can introduce delays.

These issues could be accommodated if parts of the program were compiled and binaries run on a server or directly on the controlled device, as in [5]. However, understanding of programs state could be inhibited as it would require a debugger like capability from the system.

Our system has plug-and-play device discoverability, where each device carries its own description, available in programming interface, as presented in Sect. 3.2. This provides an easy way, for users not acquainted with the new device's features, to know what functions are available and how to use them. This allows also for spontaneous use of various devices – in our test therapists not accommodated with system modified programs on-the-go when interacting with children to provide more interesting therapy.

By using ROS, different devices and software programs can create a heterogeneous set of programmable nodes, available in the programming interface. Initialization of each device can be also done through the Robokol because it has the ability to send text and multimedia files (such as robot voices) directly from the browser. This further eases the development, as data transport and setup is transparent to the users. Additional nodes were connected during the project (such as voice generation when using Robokol in Wizard-of-oz type of scenario) but still using simple to use graphical programming interface.

6 Conclusion

To conclude, our system provides an extension of Snap's graphical program to heterogeneous robotic systems, with a focus on user-friendliness and usability. As presented in use cases, it can be utilized in different scenarios, to give end users capability to program robots without extensive training or to rapidly prototype robot behaviors, for human-robot interaction studies.

In our project Roboterapia, we focus on giving therapists of autism new tools for using modern technologies, with the ability to program and connect together different functionalities being a very important part.

In future work, we hope to extend the functionality of the Robokol to a larger set of robot behaviors, such as setting automatic reporting or AI. Also, as stated in the discussion, compiling some of the blocks and running them on the devices could bring the ability to program real-time control loops.

References

1. Hardware that can connect to scratch (2015). http://wiki.scratch.mit.edu/wiki/Hardware_That_Can_Connect_to_Scratch
2. Arent, K., Kabala, M., Wnuk, M.: Programowanie i konstrukcja kulistego robota spolecznego wspomagajacego terapie dzieci autystycznych. Ph.D. thesis, Politechnika Wroclawska (2005)
3. Billard, A., Calinon, S., Dillmann, R., Schaal, S.: Robot programming by demonstration. In: Siciliano, B., Khatib, O. (eds.) Springer Handbook of Robotics, pp. 1371–1394. Springer, Heidelberg (2008)
4. Chong, J., Ong, S., Nee, A., Youcef-Youmi, K.: Robot programming using augmented reality: an interactive method for planning collision-free paths. Robot. Comput. Integr. Manufact. **25**(3), 689–701 (2009)
5. Glas, D.F., Kanda, T., Ishiguro, H.: Human-robot interaction design using interaction composer: eight years of lessons learned. In: The Eleventh ACM/IEEE International Conference on Human Robot Interation, HRI 2016, pp. 303–310. IEEE Press, Piscataway (2016). http://dl.acm.org/citation.cfm?id=2906831.2906884
6. Granosik, G., Stasiak, K.: Pyrobotics managing the family of heterogeneous robots. In: 13th IEEE IFAC International Conference on Methods and Models in Automation and Robotics, pp. 965–969. IEEE (2007)

7. Huang, J., Lau, T., Cakmak, M.: Design and evaluation of a rapid programming system for service robots. In: The Eleventh ACM/IEEE International Conference on Human Robot Interation, HRI 2016, pp. 295–302. IEEE Press, Piscataway (2016). http://dl.acm.org/citation.cfm?id=2906831.2906883

8. Jost, B., Ketterl, M., Budde, R., Leimbach, T.: Graphical programming environments for educational robots: open roberta - yet another one? In: 2014 IEEE International Symposium on Multimedia, pp. 381–386, December 2014

9. McIver, L.: Syntactic and Semantic Issues in Introductory Programming Education. Monash Univesity, Clayton (2001)

10. McNerney, T.S.: Tangible programming bricks: an approach to making programming accessible to everyone. Ph.D. thesis, Massachusetts Institute of Technology (1999)

11. Gordon, M., Ackermann, E., Breazeal, C.: Social robot toolkit: Tangible programming for young children. HRI 15 Extended Abstracts, Portland, OR, USA, March 2015

12. Mnig, J.: Snap! (build your own blocks) 4.0. (2016). http://snap.berkeley.edu/

13. Nguyen, H., Ciocarlie, M., Hsiao, K., Kemp, C.C.: ROS commander (ROSCo): behavior creation for home robots. In: 2013 IEEE International Conference on Robotics and Automation (ICRA), pp. 467–474. IEEE (2013)

14. Pinciroli, C., Lee-Brown, A., Beltrame, G.: Buzz: An extensible programming language for self-organizing heterogeneous robot swarms. CoRR abs/1507.05946 (2015). http://arxiv.org/abs/1507.05946

15. Pot, E., Monceaux, J., Gelin, R., Maisonnier, B.: Choregraphe: a graphical tool for humanoid robot programming. In: The 18th IEEE International Symposium on Robot and Human Interactive Communication, RO-MAN 2009, pp. 46–51. IEEE (2009)

16. Thomas, B.: ROS Scratch: Enabling Block-Based Robotics. 1 edn. (2011). http://robotics.cs.brown.edu/test/slides/brian-final.pdf

17. Valk, L.: LEGO MINDSTORMS EV3 Discovery Book: A Beginner's Guide to Building and Programming Robots. No Starch Press, San Francisco (2014)

18. Wahl, F.M., Thomas, U.: Robot programming-from simple moves to complex robot tasks. Institute for Robotics and Process Control, Technical University of Brawnschweig (2002)

19. Werner, L., Campe, S., Denner, J.: Children learning computer science concepts via alice game-programming. In: Proceedings of the 43rd ACM Technical Symposium on Computer Science Education, pp. 427–432. ACM (2012)

20. Zubrycki, I., Kolesiski, M., Granosik, G.: A participatory design for enhancing the work environment of therapists of disabled children. In: 2016 25th IEEE International Symposium on Robot and Human Interactive Communication (RO-MAN), pp. 781–786, August 2016

21. Zubrycki, I., Granosik, G.: Designing an interactive device for sensory therapy. In: The Eleventh ACM/IEEE International Conference on Human Robot Interation, HRI 2016, pp. 545–546. IEEE Press, Piscataway (2016). http://dl.acm.org/citation.cfm?id=2906831.2906969

22. Zubrycki, I., Granosik, G.: Understanding therapists' needs and attitudes towards robotic support. The roboterapia project. Int. J. Soc. Robot. 8(4), 553–563 (2016). http://dx.doi.org/10.1007/s12369-016-0372-9

Biomimetic Navigation Using CBR

Jose Manuel Peula$^{(\boxtimes)}$, Joaquín Ballesteros, Cristina Urdiales,
and Francisco Sandoval

Department of Tecnología Electrónica, University of Málaga, Málaga, Spain
{peula,jballesteros,acurdiales,fsandoval}@uma.es

Abstract. Biologically inspired design consists in the creation of technological systems using as starting or inspirational point biological systems. Indeed, it has been used widely in robotics in different areas, such as mechanics, coordination or navigation. For example, in robot navigation, biomimetic algorithms can be specially useful in certain circumstances, such as when a robot needs to interacts closely with users. Using biomimetic navigation robot movements would be more similar to human ones but maintaining some basic navigation factors such as the safety. It is important in systems such as assistive systems in which human and robot control can be switch or combined –depending on the kind of system– to obtain the final command. Thus, in these systems interaction is very close to the user and it is advisable to make robot commands as similar as possible to the user ones. Otherwise, the user could even reject robot assistance depending on the disagreement between user and robot commands to reach a destiny. This disagreement provokes user's frustration and stress and, in extreme, assistive system rejection. In this paper we propose a biomimetic navigation algorithm based on Case-Based Reasoning that learns from real traces –performed by volunteers– in order to achieve robot navigation as close as possible to the human one.

Keywords: Case-based reasoning · Biomimetic navigation · Learning

1 Introduction

Robotics have been highly influenced by biology since its very beginning. Indeed, biological inspired algorithms have been applied in almost all robotics areas such as coordination [13,15], mechanics and aerodynamics [25], robot navigation [9,19], mapping [11,14] or robot control [16]. Biology has always be a source of inspiration and, for example, in the area of robot navigation animal behaviors inspire researchers to find new approaches to problems in complex environments as in some cases biological systems and behaviors can be more efficient and refined than the equivalent robotic systems [9]. Moreover, there are specific application areas, such as assistive systems, in which a biomimetic navigation algorithm can be recommended for improving user acceptance [24] and, thus, system succeed.

Assistive systems, such as wheelchairs, walkers or rollators, usually make use of any kind of shared control to interact with the user and help him/her when

© Springer International Publishing AG 2017
I. Rojas et al. (Eds.): IWANN 2017, Part II, LNCS 10306, pp. 632–643, 2017.
DOI: 10.1007/978-3-319-59147-6_54

necessary. More common assistive systems switch control from user to robot at wish or automatically depending on the situation [5,17]. One specific kind of shared control is collaborative control which, instead of switching control, combines continuously user and robot commands so that both of them have an impact in the final executed command. Main advantage of this control is that users have part of the system control all the time what makes them to practice their driving skills improving rehabilitation and/or preventing loss of residual skills [22]. However, main drawback of this control is that robot and user commands need to be combined correctly to avoid user frustration and stress if he/she feels loss of system control. It happens, usually, when robot and user commands differs too much in non dangerous situations. The reason is that user trajectories and intermediate goals do not usually fit with robot ones, although both of them could be considered correct. In extreme, it can provoke that users reject the collaborative control assistance [23]. A possible solution to avoid it could be to substitute the habitual analytic algorithms used in collaborative controls –Potential Field Approach (PFA), Dynamic Window Approach (DWA), Vector Field Histogram (VFH), etc.– with a biomimetic one that will fit better with the user trajectories. Indeed, this approach was applied before in assistive wheelchairs with succeed at low navigation level [21].

In this work we want to apply the commented biomimetic navigation approach to assistive walker control at a higher navigation level. According to the biological navigation hierarchy proposed in [6], biomimetic navigation can be divided in local navigation behaviors and way-finding behaviors. In this hierarchy, local navigation behaviors require the recognition of only one location –the goal– while way-finding ones involve the recognition of other places besides the goal –for example the starting point– and the representation of relations between these places. As our objective is to imitate the way users move from one point to another taking into account possible obstacles in the environment, way-finding behaviors fits better with the biomimetic navigation level searched. Specifically between the three possible way-finding levels [6], the one that best suits with our approach is the recognition-triggered response behavior. This behavior consists in connecting two locations by a local navigation method or, in other words, associating the start location –that can be defined by a sensory pattern– and an action –to reach the goal– [7].

This behavior could be implemented via an analytical algorithm or via learning. However, it could not be easy to find an analytical algorithm that fits correctly a biomimetic human navigation. Thus, we have choose a learning approach. Between the different machine learning techniques one of the most typical ones is Learning from demonstration (LfD) [2], that consists in associating the pair state/action making possible learning from examples provided by a supervisor or from data gathered. In our case, we will use as training data real traces gathered from volunteers. Thus, the navigation algorithm could be more flexible and would be easier to perform different specific responses –comparing to the use of an analytic option– depending on the data used for the training. LfD comprehends high variety of machine learning strategies [1] such as decision-tree learning, evolutionary learning, Bayesian learning or case-based learning. Each one has its pros and cons and can be better for some kind of applications. However, we have chosen CBR technique because the way it works results very intuitive for humans, as it is similar to the way we reason and because it fits perfectly with the way recognition-triggered response behavior works, as both consist of associating a state/action. Moreover, we have chosen CBR instead of other popular techniques such as Artificial Neural Networks [18] because in CBR technique knowledge can be stored in a plain database, what have some advantages: (i) it makes possible to check why the system returns a specific response to a specific input, and (ii) it is possible to modify directly and at hand a specific response if needed.

Thus, this paper presents a biomimetic navigation behavior based on CBR. CBR will be trained with real traces performed by volunteers in a hospital environment. This navigation behavior could be an alternative to analytical algorithms usually applied in assistive system. Hence, the main goal of this biomimetic navigation algorithm is to achieve robot navigation as close as possible to the human one improving, when possible, user acceptance of assistive systems such as collaborative control one. This paper is structured as follows. First of all, biomimetic navigation approach is explained in Sect. 2. Then, Sect. 3 explains how user data is gathered and which information is available. After that, Sect. 4 describes CBR structure and parameters. Finally, Sect. 5 presents the comparison of the proposed system with an analytical one and Sect. 6 presents our conclusions and future work.

2   Biomimetic Navigation System

The objective of the biomimetic navigation algorithm proposed is to imitate, as closer as possible, the way an user moves when using a walker. As the navigation algorithm is a recognition-triggered response behavior, in order to work it is necessary to know [7]: (i) the starting point, that can be described by the obstacles configuration, and (ii) an approximate goal or destiny to reach. Then, the biomimetic navigation behavior should be able to determine short-term trajectory that an user would perform to reach that goal from that initial position. Thus, it is necessary to use any algorithm to determine local-partial destinies that will be one of the inputs for the biomimetic behavior. Hence, our global system will consist in a two layer structure. The upper layer will be a global planner –we will use the well known A^* algorithm– that will propose to the lower layer –the biomimetic recognition-triggered response behavior– the next destiny to reach. However, it is important to take into account that destinies proposed by A^* algorithm to biomimetic behavior are only proposals, because the biomimetic algorithm takes into account for its output path the proposed destiny as well as the surrounded obstacles.

Figure 1 shows an example of how the global navigation system works. First of all, the initial position and global destiny are set –Fig. 1a–. Then, A^* algorithm is launched and intermediate destinies to reach the global destiny are calculated

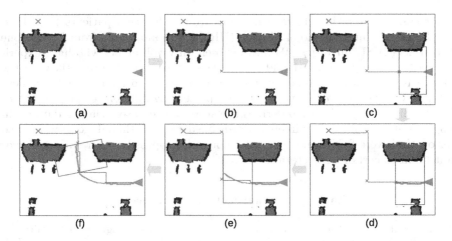

Fig. 1. Different stages during system work: (a) Initial position and destiny. (b) A^* launch and intermediate destinies set. (c) Calculaton of local destiny for local window. (d) Biomimetic navigation application. (e) Local destiny reached. Repeats (c) and (d). (f) Goal reached differs from local destiny proposed. Re-launch A^* from current position and repeat (c) and (d) steps.

–Fig. 1b–. After that, a local window, that determines the range of operation of the biomimetic recognition-triggered response behavior, is set and local destiny is estimated from A^* intermediate destinies –Fig. 1c–. This local destiny and the obstacle configuration are the input for the biomimetic behavior that returns the user path that best fits in that specific situation. After applying the path proposed by the biomimetic behavior, if the local destiny is reached, as in Fig. 1d, local window is moved to the end of the current path and a new local destiny is estimated –according to prior A^* intermediate destinies–. However, if local destiny is not reached –as in Fig. 1e– because user movement in that obstacle configuration did not arrive exactly at the local destiny proposed, A^* algorithm is launched again considering the new local destiny reached as the initial position –Fig. 1f–. Thus, A^* do not force the biomimetic behavior to follow the trajectory calculated, as it recalculates the proposed destinies if reached goals differs from the proposed one. A^* algorithm is only used for local destinies proposal.

As commented before, the biomimetic behavior has been implemented as a CBR system. Thus, it is necessary a set of steps in order to get it ready to work. These steps are: (i) Data gathering and processing, (ii) CBR configuration and training using data gathered and, (iii) to launch CBR in standalone mode reproducing the trajectories trained. This will be described in the following sections.

3 Data Gathering

In order to imitate the way users move, CBR training phase needs real traces to learn the appropriate response to the different situations it could face.

In our case, data gathered are real traces from volunteers patients from Hospital Regional Universitario of Malaga (HRU)[1] and inpatients from Fondazione Santa Lucia (FSL) in Rome. We have chosen only volunteers who have experience with rollators –to avoid cold start issues– and were able to walk with the aid of a rollator. Total number of volunteers were 11 –6 females and 5 males– with an average age over 69 years old and presenting different cognitive and/or physical disabilities. Table 1 describes the state of each user. During data gathering volunteers could move freely towards fixed goals. Although goals were fixed a priori they were free to move in order not to limit possible user trajectories or limit them to certain trajectories.

Table 1. Description of the 11 volunteers.

User	Gender	Age	Physical disability	Cognitive disability
1	Female	77	Prosthetic femur fracture (Left)	-
2	Men	86	Intertrochanteric hip fracture (Left)	Psychological distress
3	Female	82	Intertrochanteric fracture femur (Left)	-
4	Men	59	Above knee amputation (Left)	-
5	Female	83	Total hip replacement (Left)	-
6	Men	30	Polytraumatism (Left)	-
7	Men	59	Above knee amputation (Left)	-
8	Female	71	-	Mild Parkinson
9	Male	71	-	Isquemia (Left)
10	Female	73	-	Mild Parkinson
11	Female	78	-	Mild dementia

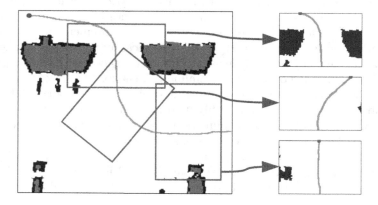

Fig. 2. Extraction of local trajectories from global one. (Color figure online)

---

[1] Collaboration under the framework New technologies in rehabilitation: walking aids: a pilot study with robotic walker.

Raw data gathered consists of a global map of the obstacles –shown in a probability grid map–, the volunteers' trajectories and the global destiny. Figure 2 shows an example of the trajectories obtained. After raw data gathering, each global trajectory is divided in local snapshots that represent short-term local trajectories of each volunteer. Snapshots are stored when trajectory curvature or obstacle configuration change significantly. These snapshots are stored as 40×60 pixels images –grid map– that represent the obstacles –red–, the destiny –blue– and the volunteer trajectories –green–. Figure 2 shows an example of how each global trajectory is split into snapshots representing short-term local trajectories.

4   Case-Based Reasoning

CBR is a learning technique that solves a problem by retrieving and adapting past experiences that become the proposed solutions for the current problem [12]. In CBR, the key concept is the assumption that *similar problems have similar solutions*. Hence, if CBR learns solutions to the more common and habitual problems it could have solutions for most of them in the future. Knowledge learnt is stored in a CBR database which is form by cases, that are the joining of the parameters that describes a problem and its corresponding solution.

A CBR system normally consists of four steps: *retrieve, reuse, revise* and *retain*. In our case, we will active only *retrieve* and *reuse*, that consist in retrieving and reusing from the database the case that best fits and solves current problem. The other two steps –revise and retain– consists in modifying and adapting cases to the current problem if retrieved case does not solve correctly faced problem. Thus, this steps have been disabled for maintaining original users responses and avoiding to *contaminate* them through adaptation.

CBR has been used in the last 20 years in a wide variety of areas such as classification and diagnosis in medical applications [4,20], robot coordination [3] or robot navigation [8]. In our case it will be used for implementing a biomimetic recognition-triggered response navigation algorithm. Thus, first of all it is necessary to define the CBR database structure and CBR case, that will determine the parameters that described the problem and its solution and how they are stored. After that, we have to choose the search and distance algorithms that will use CBR system. CBR configuration will be described in the following sections.

4.1   CBR Database and Case Structure

CBR database structure can be flat structure or hierarchical one. A flat structure retrieves the cases that match best with the input case and adding new cases is cheap and easy. However, retrieval time can be very high. On the other hand, a hierarchical structure allow lower retrieval time than flat structure, but keeping the structure in optimal conditions requires a significant consume of time. In our case, we will use a plain structure for comfort and, in case of problems with time retrieval we could apply a clustering algorithm to the database or, in extreme, change to a hierarchical structure if necessary.

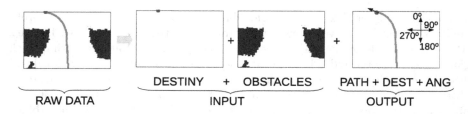

Fig. 3. Extraction of the input and output case parameters from raw data gathered.

Respect to the case definition, a case is form by a pair problem/solution, described by their corresponding parameters. As CBR is used to implement a biomimetic navigation behavior, input and output is determined by the behavior requirements. Thus, input should be form by the proposed destiny to reach respect to the origin position and by the configuration of obstacles in the environment –that determine the start position respect to them–. The output, on the other hand, will be form by the path that the user perform in this environment to reach or to approximate to the goal, the destiny finally reached and the angle of the user when arriving at the destiny. Figure 3 shows orientation of the angle at destiny point and how snapshots extracted from raw data gathered are converted into the CBR case parameters. Table 2 describes in more detail the CBR parameters and their structure in the case.

Table 2. CBR case inputs and outputs

CBR case inputs		CBR case outputs		
(In_1, In_2)	$In_3...In_{2402}$	$(Out_1, Out_2)...(Out_{19}, Out_{20})$	(Out_{21}, Out_{22})	Out_{23}
$(Goal_x, Goal_y)$	$Pix_1...Pix_{2400}$	$(Path_1^x, Path_1^x)...(Path_{10}^x, Path_{10}^x)$	$(Dest_x, Dest_y)$	Ang

Being $(Goal_x, Goal_y)$ the coordinates x and y of the goal proposed to reach, $Pix_1...Pix_{2400}$ the pixels of the grid map that describes the obstacles in the environment, $(Path_i^x, Path_i^x)$ the coordinates x and y of the i-th position of the path, $(Dest_x, Dest_y)$ the coordinates x and y of the destiny reached and Ang the angle of the user at destiny position.

4.2   CBR Search and Distance Algorithms

Other parameters that are important to determine in order to configure the CBR system are the search algorithm and the distance used for the retrieval step. In our case, search algorithm used is the common nearest-neighbour (NN) algorithm [10] and distance used is an Euclidean distance.

5   Experiments and Results

This section briefs results from several simulations of the biomimetic navigation behavior. Map used for the tests has been the one extracted from the environment where the real data was gathered. Indeed, for the tests we have chosen a specific part of the map that includes right and left turns and forward movements in order to reach the global destiny. The objective of the area selected was to obtain the higher variety of possible movements in the tests.

Once selected the test map, and as commented in Sect. 2, A^* algorithm is launched to determine intermediate destinies to arrive at the global destiny. Then, local destinies are calculated and sent to the CBR. Each time CBR receives an input case –form by local destiny and obstacles in the environment– CBR searches in the database the user path that best fits for this obstacle configuration and destiny, and checks if this path provokes a collision. If path return by the CBR is safe, it is used and applied for simulated walker user advance. Else, we check if following more similar CBR case retrieved is safe. CBR system checks till the 10 more similar cases. If none of them is safe for the current input simulation is stopped.

In order to check that biomimetic navigation algorithm returns paths more similar to the user than an analytical one, the same tests have been done with the analytical A^* algorithm. Both responses –analytical and biomimetic one– have been compared with the original path perform by each user in that situation. Figure 4 shows the results of some simulations in which it can be seen Fig. 4a and b that, although the three paths are not really close, but biomimetic path –green– tend to be more close to the user path –blue– than the analytical algorithm –red–. It is important to take into account that, although objective is to imitate user path as close as possible, it is not always possible, as it can be seen in Fig. 4c, in which user takes a different route to reach the goal and biomimetic and analytical paths can be barely compared to the user one. This kind of problems have been quite exceptional but at the level the biomimetic algorithm have been perform these problems could not be solved.

Table 3 sums up the results of the tests with 11 users. This table shows the error between the user path and the analytical or biomimetic ones and the mean and the standard deviation error. Similarity measure has been made using Dynamic Time Warping (DTM) to minimize possible errors in the paths comparison. Data in the table have been divided depending on the kind of movement (forward movement and turn movement). It can be observed that path return by the CBR –biomimetic– algorithm is, in general, more similar –lower error– to the user than analytical one. Indeed, it improves the A^* algorithm results in a 52.74% in forward movement and in a 85.93% in turns movement. As it was expected, CBR improves the turns movements because the A^* prioritizes the shortest path –which is not always similar the human one–. However, there are cases in which difference between biomimetic and analytical path is not so high and, indeed, there are even cases in which analytical path is more similar to the user one, such as user 3 and 7 –in forward move– and user 2 and 5 –in turn move–. This indicates that, although biomimetic algorithm could be closer

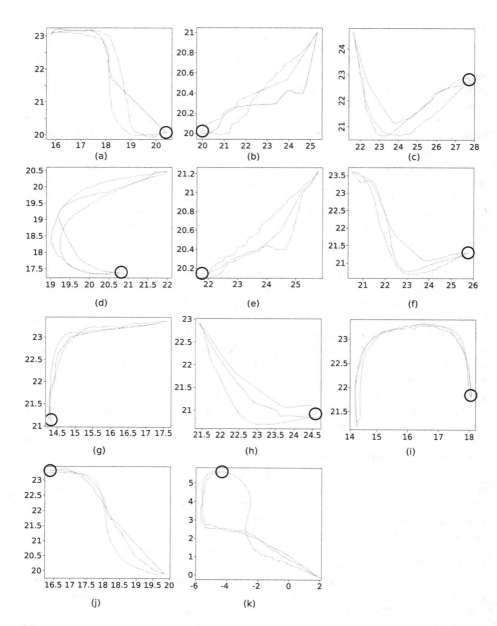

Fig. 4. Different tests showing user (green), biomimetic (blue) and analytical (red) path. The circle marks the goal. (Color figure online)

Table 3. Error path between A^* algorithm-user and CBR algorithm-user for forward and turn movements.

	Forward movement		Turn movement	
	A^*	CBR	A^*	CBR
User 1	77,70	56,99	155,83	82,02
User 2	75,57	30,12	74,97	87,95
User 3	107,35	108,73	126,84	115,44
User 4	51,25	47,08	209,51	134,66
User 5	146,71	130,11	113,33	116,75
User 6	252,70	139,71	89,46	63,97
User 7	41,12	82,36	908,52	307,03
User 8	153,38	46,10	294,88	114,65
User 9	50,71	9,59	104,98	62,78
User 10	51,21	13,67	120,55	115,66
User 11	59,79	34,42	92,67	31,58
Mean	97,04	63,53	208,32	112,04
Std	64,55	45,39	240,80	71,60

to user one, and mean error value and standard deviation are lower but, logically, a generic biomimetic algorithm cannot imitate correctly all kind of users. In any case, results show that, in general, the biomimetic navigation algorithm approach tends to be more similar to user trajectories than the analytical one.

6   Conclusions and Future Work

This work has presented a LfD approach to achieve a biomimetic navigation behavior for its use in assistive systems using collaborative control. A biomimetic navigation system can be useful in this assistive systems for improving user acceptance of the system, as commands and trajectories returned by the biomimetic system should be more similar to the user ones than an analytical one, reducing possible feeling of loss of control.

Our biomimetic navigation algorithm have been implemented using CBR. It has been trained using real traces from volunteers at an hospital environment. These traces have been processed for creating the CBR case, formed by the proposed destiny to reach and the obstacle configuration –the input– and the path performed by the user, the real destiny reached and the angle of the user at destiny –the output–. As the biomimetic algorithm needs as input a proposed destiny, the global navigation system consist of this biomimetic algorithm and an upper layer that calculates intermediates destinies to reach the global goal. In our case the upper layer was an A^* algorithm.

The biomimetic navigation algorithm has been tested using simulation in the map extracted from the hospital where the data was gathered. Trajectories performed by the biomimetic algorithm have been compared with an analytical one $-A^*-$ and it has been checked that biomimetic trajectories are more similar to the ones performed by a user. However, results are still preliminary and there are cases in which biomimetic system does not respond properly.

Future work will focus in improving the system for solving problems found. One possible solution could be adding data from more volunteers in order to train CBR system in more situations. We will also work to improve the upper layer of the biomimetic algorithm, for reducing its impact in the resulting trajectory. Finally, we also want to check that, using this biomimetic algorithm, collaborative control acceptance really improves.

Acknowledgements. This work has been partially supported by the Spanish Ministerio de Educacion y Ciencia (MEC), Project. TEC2011-29106, Project no. TEC2014-56256-C2-1-P, by the Junta de Andalucia project No. TIC-7839, Hospital Regional Universitario of Malaga and Fondazione Santa Lucia of Rome.

References

1. Abramson, M., Mittu, R.: Learning and coordination: an overview. In: 2011 International Conference on Collaboration Technologies and Systems (CTS), pp. 343–350 (2011)
2. Argall, B., Chernova, S., Veloso, M., Browning, B.: A survey of robot learning from demonstration. Robot. Auton. Syst. **57**(5), 469–483 (2009)
3. Auslander, B., Apker, T., Aha, D.W.: Case-based parameter selection for plans: coordinating autonomous vehicle teams. In: Lamontagne, L., Plaza, E. (eds.) ICCBR 2014. LNCS, vol. 8765, pp. 32–47. Springer, Cham (2014). doi:10.1007/978-3-319-11209-1_4
4. Begum, S., Ahmed, M., Funk, P., Xiong, N., Folke, M.: Case-based reasoning systems in the health sciences: a survey of recent trends and developments. IEEE Trans. Syst. Man Cybern. Part C Appl. Rev. **41**(4), 421–434 (2011)
5. Bruemmer, D.J., Few, D.A., Boring, R.L., Marble, J.L., Walton, M.C., Nielsen, C.W.: Shared understanding for collaborative control. IEEE Trans. Syst. Man Cybern. Part A Syst. Hum. **35**(4), 494–504 (2005)
6. Franz, M.O., Mallot, H.A.: Biomimetic robot navigation. Robot. Auton. Syst. **30**(53), 133–153 (2000)
7. Franz, M.O., Stürzl, W., Hübner, W., Mallot, H.A.: A robot system for biomimetic navigation-from snapshots to metric embeddings of view graphs. In: Jefferies, M.E., Yeap, W.K. (eds.) Robotics and Cognitive Approaches to Spatial Mapping, vol. 38, pp. 297–314. Springer, Heidelberg (2007)
8. Hodál, J., Dvořák, J.: Using case-based reasoning for mobile robot path planning. Eng. Mech. **15**(3), 181–191 (2008)
9. Hoy, M., Matveev, A.S., Savkin, A.V.: Algorithms for collision-free navigation of mobile robots in complex cluttered environments: a survey. Robotica **33**(3), 463–497 (2015)
10. Kruusmaa, M.: Global navigation in dynamic environments using case-based reasoning. Auton. Robots **14**(1), 71–91 (2003)
11. Li, Y., Li, S., Ge, Y.: A biologically inspired solution to simultaneous localization and consistent mapping in dynamic environments. Neurocomputing **104**, 170–179 (2013)

12. Mantaras, L.D., McSherry, D., Bridge, D., Leake, D., Smyth, B., Craw, S., Faltings, B., Maher, M.L., Cox, M.T., Forbus, K., et al.: Retrieval, reuse, revision and retention in case-based reasoning. Knowl. Eng. Rev. **20**(3), 215–240 (2005)
13. Masár, M., Budinská, I.: Robot coordination based on biologically inspired methods. Adv. Mater. Res. **664**, 891–896 (2003). Trans Tech Publications
14. Milford, M.J., Wyeth, G.F.: Mapping a suburb with a single camera using a biologically inspired slam system. IEEE Trans. Robot. **24**(5), 1038–1053 (2008)
15. Mondada, F., Gambardella, A.L., Floreano, A.D., Dorigo, A.M.: Swarm-bots: physical interactions in collective robotics. Robot. Autom. Mag. **12**(2), 21–28 (2005)
16. Nichols, E., McDaid, L.J., Siddique, N.: Biologically inspired snn for robot control. IEEE Trans. Cybern. **43**(1), 115–128 (2013)
17. Parikh, S.P., Grassi, V., Kumar, V., Okamoto, J.J.: Usability study of a control framework for an intelligent wheelchair. In: Proceedings of the 2005 IEEE International Conference on Robotics and Automation, ICRA 2005, pp. 4745–4750. IEEE (2005)
18. Saeedi, S., Paull, L., Trentini, M., Li, H.: Neural network-based multiple robot simultaneous localization and mapping. IEEE Trans. Neural Netw. **22**(12), 2376–2387 (2011)
19. Savkin, A.V., Wang, C.: A simple biologically inspired algorithm for collision-free navigation of a unicycle-like robot in dynamic environments with moving obstacles. Robotica **31**(6), 993–1001 (2013)
20. Sharaf-El-Deen, D.A., Moawad, I.F., Khalifa, M.: A new hybrid case-based reasoning approach for medical diagnosis systems. J. Med. Syst. **38**(2), 9 (2014)
21. Urdiales, C.: Collaborative Assistive Robot for Mobility Enhancement (CARMEN): The Bare Necessities Assisted Wheelchair Navigation and Beyond. Springer, Berlin (2012)
22. Urdiales, C., Perez, E.J., Peinado, G., Fdez-Carmona, M., Peula, J.M., Annicchiarico, R., Sandoval, F., Caltagirone, C.: On the construction of a skill-based wheelchair navigation profile. IEEE Trans. Neural Syst. Rehabil. Eng. **21**(6), 917–927 (2013)
23. Urdiales, C., Peula, J.M., Fdez-Carmona, M., Barrué, C., Pérez, E.J., Sánchez-Tato, I., Del Toro, J., Galluppi, F., Cortés, U., Annichiaricco, R., et al.: A new multi-criteria optimization strategy for shared control in wheelchair assisted navigation. Auton. Robots **30**(2), 179–197 (2011)
24. Urdiales, C., Peula, J.M., Fernández-Carmona, M., Sandoval, F.: Learning-based adaptation for personalized mobility assistance. In: Delany, S.J., Ontañón, S. (eds.) ICCBR 2013. LNCS, vol. 7969, pp. 329–342. Springer, Heidelberg (2013). doi:10.1007/978-3-642-39056-2_24
25. Ward, T.A., Rezadad, M., Fearday, C.J., Viyapuri, R.: A review of biomimetic air vehicle research: 1984–2014. Int. J. Micro Air Veh. **7**(3), 375–394 (2015)

A Pseudo-3D Vision-Based Dual Approach for Machine-Awareness in Indoor Environment Combining Multi-resolution Visual Information

Hossam Fraihat, Kurosh Madani[(✉)], and Christophe Sabourin

Signals, Images, and Intelligent Systems Laboratory (LISSI/EA 3956),
Senart-FB Institute of Technology, Université Paris-Est,
36-37 rue Charpak, 77127 Lieusaint, France
fraihathossam@yahoo.fr, {madani,sabourin}@u-pec.fr

Abstract. In this paper we describe a pseudo-3D vision-based dual approach for Machine-Awareness in indoor environment. The so-called duality is provided by color and depth cameras of Kinect system, which presents an appealing potential for 3D robots vision. Placing the human-machine (including human-robot) interaction as a primary outcome of the intended visual Machine-Awareness in investigated system, we aspire proffering the machine the autonomy in awareness about its surrounding environment. Combining pseudo-3D vision, and salient objects' detection algorithms, the investigated approach seeks an autonomous detection of relevant items in 3D environment. The pseudo-3D perception allows reducing computational complexity inherent to the 3D vision context into a 2D computational task by processing 3D visual information within a 2D-images' framework. The statistical foundation of the investigated approach proffers it a solid and comprehensive theoretical basis, holding out a bottom-up nature making the issued system unconstrained regarding prior hypothesis. We provide experimental results validating the proposed system.

Keywords: Pseudo-3D vision · Visual saliency · Autonomous detection · Multi-resolution · Fusion

1 Introduction

In order to assist human in their daily tasks and to evolve with them and beside them within their surrounding environment, robots ought to come by other autonomy regarding their perceptual skills, their awareness about the surrounding environment in order to build their knowledge about it [1]. This means that robots have to be able to extract adequate information from surrounding environment namely from data provided by their sensors (as visual sensors, auditory sensors, etc.). The recent decades' advances relating sensors, linking signal and image processing techniques and vis-à-vis automation technologies allow to qualify nowadays' robots as robotic entities (bodies) integrating quite plenty multi-resolution sensory devices and thus reaching a tempting

© Springer International Publishing AG 2017
I. Rojas et al. (Eds.): IWANN 2017, Part II, LNCS 10306, pp. 644–654, 2017.
DOI: 10.1007/978-3-319-59147-6_55

level of providing data pertaining their surrounding environment [2]. In fact, if nearly all of the commercially available robotic bodies embed 2D cameras, most of them (especially those dedicated to cooperate with humans) are equipped with auditory and 3D visual sensors.

Awareness embodies an eclectic scope of notions. Relating human beings, it is defined as the ability to perceive, to feel, or to be conscious of events, objects, thoughts, emotions, or sensory patterns (Wikipedia). However, the same source (providing the cited definition) emphasizes the fact that in such level of consciousness, sense-data can be processed without necessarily implying understanding or interpretation. A rougher notion of "awareness" relates it to the state or quality of being awake of something. The biological psychology defines the "awareness" as a human's or an animal's perception and the transpired cognitive reaction to a condition or event. If the cited definitions supply the conceptual notion of "awareness", they only show how challenging is making a robot fitting human's awareness skills. In fact, a lot remains to be accomplished before reaching such machine's autonomy, however, recent technological advances accomplished in sensors' area as well as in computational intelligence allow to prospect significant innovations toward the expected target. According to what has been stated, Machine Awareness (MA) or Artificial Awareness (AA) could be defined as the machine's ability to perceive and to identify objects, events or actions within its surrounding environment through sensory patterns, however without necessarily implying construal analysis of the perceived reality.

In the same way, the recent decade has been a token of numerous progresses in computer vision techniques and visual sensors offering appealing potential to look at the above-mentioned dilemma within innovative slants. In fact, on the one hand, numerous image processing techniques with reduced computational complexity have been designed and on the other hand, a number of new combined visual sensors with appealing features and accessible prices have been presented as standard market products. "Kinect", a Microsoft product which has been initially designed for Xbox play station in 2008, is a typical example of such combined low-priced standard-market visual sensor [3]. It provides a pseudo-3-D visual capture of surrounding environment by providing the depth (in meters) using an infra-red device and an color image using a standard camera [4].

In this research we are especially focusing 3D indoor environment through visual perception. Combining pseudo-3D vision, and salient objects' detection (and recognition) algorithms, the investigated approach seeks an autonomous detection of relevant items in 3D environment. The pseudo-3D perception allows reducing computational complexity inherent to the 3D vision context into a 2D computational task by processing 3D visual information within a 2D-images' framework. On the other hand, the statistical foundation of the investigated approach proffers it a solid and comprehensive theoretical basis. Moreover, the aforementioned statistical foundation holds out an appealing bottom-up nature of the investigated saliency detection making it unconstrained regarding prior hypothesis relating the potentially salient items. Finally, its fusion-based construction of saliency features provides it benefiting from multi-resolution visual information boosting the accuracy of the investigated approach.

It is pertinent to note that the robot's awareness about its surrounding environment is a foremost requirement for accurate and contextual interaction of the robot with the user. In this context, the saliency detection is one of the foremost skills for robot's autonomy in visual perception of the environment in which it evolves and in which it is supposed to interact with users. The above-mentioned features of the investigated approach proffer a robot equipped with such a system an enhanced awareness about the 3D surrounding environment and thus, open appealing perspectives for an accurate and realistic interaction with the human.

The rest of this paper is organized as follows: Sect. 2 provides a brief overview of state-of-art of different fields relating the presented work. Section 3 introduces the proposed approach. Section 4 presents the validation and experimental results. Finally, the last section will conclude the paper sketching perspective of the presented work.

2   State-of-Art and Accomplishments in Related Fields

Microsoft's Kinect is a cheap pseudo-3D system which directly provides color and depth information [3–5]. Even if its field of view is limited [6] (about 60° vertical and 40° horizontal) and the data is noisy, its ever-increasing usage in many domains as medical, robotics, home automation, holograms' creation, has been appreciable during recent years [7–12]. It is pertinent to emphasize the diversity of Kinect's applications, especially over recent years, covering as well its typical areas of applicability such as scene modeling [7] or objects' recognition ([8, 9]), as dealing with more way-out fields such as robot control [10], biomedical [11] or forensic [12] applications. Kinect provides spatial depth coordinates between 0.6 to 4 meters and thus could swathe spatial features within the expected area. It is pertinent to notice that Kinect's spatial coverage gap fits with the human's typical indoor living space making this low-cost device an appealing pseudo-3D sensor for our purpose [6]. This fact and the previously highlighted potentiality of Kinect's applicability render our motivation to consider this pseudo-3D sensor as visual sensor of the investigated MA system.

Visual saliency is described as a perceptual quality that makes a region of image stand out relative to its surroundings and to capture attention of the observer [13]. Although there exist biologically based computational approaches to visual saliency, most of the existing works do not claim to be biologically plausible: instead, they use purely computational techniques to achieve the goal. One of the first works using visual saliency in image processing has been published by Itti et al. [14]. Authors use a biologically plausible approach based on a center-surround contrast calculation using "Difference of Gaussians". Published more recently, other common techniques of visual saliency calculation include graph-based random walk [15], center-surround feature distances [16], multi-scale contrast, center-surround histogram and color spatial distribution or features of color and luminance [17]. A less common approach is described in [18]. It uses content-sensitive hyper-graph representation and partitioning instead of using traditional features and parameters commonly considered in images. Finally, in their recent works, authors of [18, 19] have investigated an intelligent autonomous 2D salient vision system for humanoid robots' autonomous knowledge acquisition as well as a powerful segmentation approach taking advantage from RGB representation in spherical coordinates [21].

A number of recent works have investigated pseudo-3D based saliency detection, highlighting the pertinence of depth information in effective enhancement of salient items' detection in 3D environment [22–25]. if all of those works converge toward the pertinent impact of depth information on enhancement of the detection accuracy, however, all of them are based on prior restrictive hypothesis relating either the visual field or the potential salient objects' locations within the scenery [26, 27]. In fact, often the target salient items in those works are located within the central area of the image. In the same way, the peripheral regions of the scenery are frequently excluded (or not considered). Finally, all of the above-mentioned works deal with "unique" salient object's detection: a very restrictive prior hypothesis regarding the reality of usual complexity of the surrounding landscapes. Nevertheless, the above-mentioned research works constitute an appealing source of information relating this new field by providing the state-of-art and resources as benchmarking data (provided by [23]).

3   Proposed Pseudo-3D-Based Saliency Detection Approach

As it has been already mentioned in introductory section, the investigated approach takes its first advantage from a 2D processing of visual information relating 3D reality by jointly using color and depth information provided by the Kinect. Figure 1 illustrates the block diagram of the overall architecture of proposed visual Machine-Awareness system. Apart of the pseudo-3D sensor, namely Kinect, our system includes four main operational streams. The first one involves the RGB image provided by Kinect and consists of a set of processing tasks resulting in surrounding environment's chromatic and luminance saliency detection. The resulted image, called "Color-Luminance Saliency Map" (CLSM), carries items' chromatic and luminance relevancies. CLSM results from computation of two elementary saliency maps relating overall saliency (i.e. Global Saliency Map) and local saliency (i.e. Local Saliency Map) of objects [19–21]. The second one involves the so-called Depth image (provided by the same pseudo-3D imaging module). It extracts saliency feature regarding the third dimension of the landscape, dealing with the items' relevance regarding their locations in space. The resulted image, called "Depth Saliency Map" (DSM).

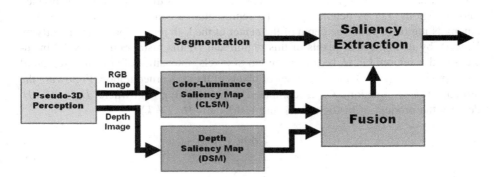

Fig. 1. Block diagram of operational stream of the proposed system.

The third unit performs segmentation. In principle, any image segmentation technique may be used for performing this task. However, the segmented items' quality impacts the detected objects' contour and thus, influences the segmented outline's quality. We have used the technique described in [21] offering an appealing balance between segmentation quality and computational time. The fourth main operation is devoted to the nonlinear fusion of so-called CLSM and DSM. The result is an image, called "Final Saliency Map" (FSM), leading to the saliency extraction operation. Finally, the last chief-operation relates the saliency extraction mask, obtained from FSM and segmented image.

3.1   Color-Luminance Saliency Map

As described in our farther works ([19–21]), two kinds of saliency maps, representing two different saliency levels, contribute to computation of CLSM: "Global Saliency Map" (GSM) and "Local Saliency Map" (LSM). The GSM handles overall sway of items' strikingness dealing with their overall contrast within the image. LSM conveys persnickety bends of those items coping with their local center-surround contrast in the considered image. It is obtained on the basis of center-surround histograms' concept, involving statistical properties of two centered windows (over each pixel) sliding alongside whole the image. CLSM, the constituents of which are denoted $M_{CLSM}(x)$, results from a conditional fusion involving GSM and LSM. It is computed using the relation (1). $x \in N^2$ denotes 2D-pixel position. $M_G(x)$ and $M_L(x)$ represent constituents of GSM and LSM, respectively.

$$M_{CLSM}(x) = \begin{cases} M_L(x) & \text{if } M_L(x) > M_G(x) \\ \sqrt{M_G(x)M_L(x)} & \text{otherwise} \end{cases} \tag{1}$$

3.2   Depth Saliency Map

Computation of DSM is inspired from research work presented in [22] and involves statistical properties of regions of depth image. This means that depth image is divided into N regions. Let us denote by C_k the representative contrast of k-th region (k-th among N regions), by Z_k the depth of the center of the k-th region (provided directly by Kinect), by n_k number of pixels of this region and H_k the histogram of pixels' intensities of this region. Let H_j be the histogram of pixels' intensities of any other region of the depth image $(j \neq k)$ and n_j number of pixels of the concerned region. Following the aforementioned notation, C_k is computed using the Eq. (2), where $D_{kj} = H_k \cdot H_j$ denotes the product of histogram H_k and histograms H_j $(j \neq k)$.

$$C_k = \frac{2 Z_k n_k \sum\limits_{j \neq k} D_{kj}}{\sum\limits_{j=1}^{N} n_j} \tag{2}$$

The saliency of the k-th region of the DSM, denoted as S_k, is then obtained from the Eq. (3), where $C_{max} = Max(C_k)$ represents the highest representative contrast.

$$S_k = 1 - \frac{C_k}{C_{max}} \qquad (3)$$

Figure 2 gives examples of CLSM (c) and DSM (d), respectively.

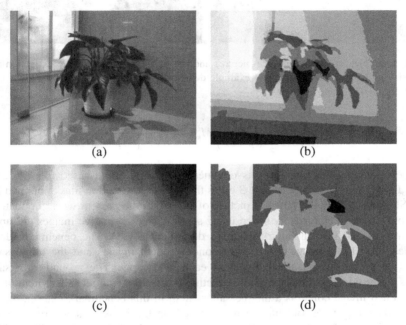

(a)   (b)

(c)   (d)

Fig. 2. Examples of CLSM and DSM obtained from RGB and depth images provided by Kinect: (a) RGB image, (b) corresponding depth image, (c) CLSM and (d) DSM.

3.3   Final Saliency Map and Saliency Extraction Mask

As previously mentioned, FSM results from nonlinear fusion of CLSM and DSM. The fusion is performed region-by-region (for all N regions), involving corresponding areas in RGB and depth images. In this way, the resulted FSM, which constituents of its k-th region are denoted as $M_k^{FSM}(x)$, is computed according to the Eq. (4), where $M_k^{CLSM}(x)$ is the constituent of k-th region in CLSM corresponding to a pixel located at coordinate x, $M_k^{DSM}(x)$ is the constituent DSM in the same region and corresponding to the same pixel. As example of FSM, Fig. 3 gives the obtained final map issued from CLSM and DSM depicted in Fig. 2.

Fig. 3. Examples of DSM (left-side picture) and the corresponding saliency extraction mask (right-side picture) obtained using the example depicted in Fig. 2.

$$M_k^{FSM}(x) = \left(1 - \frac{1}{1 + e^{-\frac{c_k}{c_{max}}}}\right) M_k^{CLSM}(x) + \frac{1}{1 + e^{-\frac{c_k}{c_{max}}}} M_k^{DSM}(x) \qquad (4)$$

The saliency extraction mask is obtained from a threshold-based filtering of the FSM. The right-side picture of Fig. 3 gives the saliency extraction mask obtained from the DSM depicted by the left-side picture of Fig. 3. In the present work the matching threshold has been empirically stated using set of 22 representative images (22 among 1000 images) of the benchmark database described in [23]. The benchmark dataset includes 1000 RGB images with corresponding depth images and the target salient items (i.e. ground-truth salient object in each landscape). The ground-truth salient objects are provided as binary masks extracting precisely the target salient item of each image. Each image contains a unique salient item in this benchmark dataset.

The matching threshold has been computed on the basis of standard F-measure criterion given by Eq. (5), where P denotes the "precision", R stands for "Recall", TP for number of True-Positive samples (samples correctly matched), FP for number of True-Negative samples (sample incorrectly matched) and FN for number of False-Negative samples (samples not belonging to the correct category correctly matched). The value of threshold maximizing F-measure has been retained (i.e. 70).

$$F_{Measure} = \frac{(1 + 0.5)P \times R}{0.5P + R}$$
$$\text{with} \qquad (5)$$
$$P = \frac{TP}{TP + FP} \quad and \quad R = \frac{TP}{TP + FN}$$

In other words, if the target salient object of an image is correctly detected (i.e. figures in the set of potentially salient items) then the sample is considered as TP else it is qualified as FP. In fact, contrary to the other techniques (which presuppose that the unique relevant object is located within the central area of the inspected landscape), the presented approach is able to detect any potentially salient item of the landscape, without any restrictive hypothesis.

4 Experimental Results and Validation

As mentioned in previous section, the validation has been performed using the benchmark database provided by [23]. However, we have extended the experimental validation using another benchmark database provided by [24] containing 135 images. As the first database, this second benchmark database provides, for each image, the RGB image, the corresponding depth image and the target ground-truth salient object. The assessment has been performed using F-measure and another measure known as "Area Under Curve" (AUC) standard, measuring the likeness within the frame of "classification task" [28].

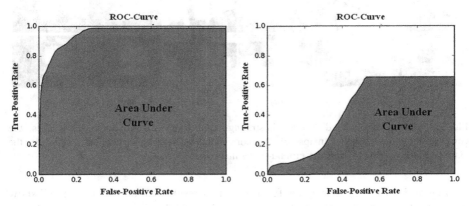

Fig. 4. Examples of ROC-curve and the corresponding areas under curve for an accurate classification matching perfectly the ground-true (left-side) and an appalling classification far from matching the ground-true (right-side).

AUC measure, computes the evaluation score as the area under ROC-curve (see [29]), following the frame of a classification task categorizing pixels as belonging either to the class of "salient pixels" or to the class of "background pixels". ROC-curve (Receiver Operating Characteristic curve) is a diagram of classification rates relating the pixels' classification as "True Positive" (TP) and "False Positive" (FP) patterns. In other words, ROC-curve represent TP rate versus FP rate. The subtracted score, conformably to AUC-based standard, is the area under curve (e.g. the integral) of the so-called ROC-curve. The highest attainable score is 1 while, a uniformly random saliency map conducts to the score of 0.5 meaning that such a measure evaluates the assessed technique's quality versus the random process: scores reaching values below 0.5 are interpreted as "worse than random process". Figure 4 gives two examples of ROC-curves with the corresponding areas under curve. The first one (left-side diagram) corresponds to an accurate classification matching perfectly the ground-true while the second (right-side diagram) corresponds to an appalling classification far from matching the ground-true.

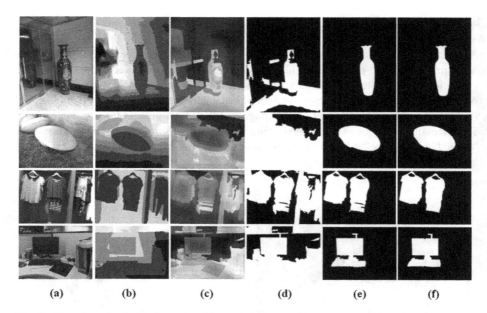

(a)	(b)	(c)	(d)	(e)	(f)

Fig. 5. Samples of obtained results, illustrating input RGB (a) and depth (b) images, the corresponding final saliency maps (c), the corresponding saliency extraction masks (d), the extracted target items (e) and the ground-true salient shapes.

Figure 5 illustrates four examples of obtained results. Columns "a" and "b" show input RGB and depth images. The corresponding final saliency maps and saliency extraction masks are shown in columns "c" and "d". Finally, columns "e" and "f" of this figure depict the extracted target items and the ground-true salient shapes, respectively. The assessment of the both two benchmark datasets leaded to the scores summarized in Table 1. Referring to samples depicted in Fig. 5, as well as to whole outcomes resulted from both benchmark databases, can note the very good accuracy of target salient items' detection. This fact is confirmed by numerical scores issued from AUC and F-measure metrics. The comparable high scores (0.86 for the first database and 0.84 for the second database) highlight an alluring shape of the ROC-curves for both databases. This means that the proposed approach leads to outstanding TP rates for both databases. On the other hand, it is pertinent to note that the proposed technique is able to detect other potentially saliencies within the inspected landscapes.

Table 1. Percentage of correct detections of object over testing image set using Viola-Jones detection framework.

Benchmark database	Number of images	AUC score	F-measure
Database 1 ([23])	1000	0.84	0.86
Database 2 ([24])	135	0.86	0.88

5   Conclusion and Further Work

We have presented a pseudo-3D vision-based dual approach for Machine-Awareness in indoor environment. Combining pseudo-3D vision and salient objects' detection algorithms, the investigated approach seeks an autonomous detection of relevant items in 3D environment. The pseudo-3D perception allows reducing computational complexity inherent to the 3D vision context into a 2D computational task by processing 3D visual information within a 2D-images' framework. On the other hand, the statistical foundation of the investigated approach proffers it a solid and comprehensive theoretical basis. Moreover, the aforementioned statistical foundation holds out an appealing bottom-up nature of the investigated saliency detection making it unconstrained regarding prior hypothesis relating the potentially salient items. Finally, its fusion-based construction of saliency features provides it benefiting from multi-resolution visual information boosting the accuracy of the investigated approach.

Performed involving two benchmark datasets, the assessment of the proposed concept and the issued system resulted in high scores (0.86 for the first database and 0.84 for the second database) highlighting very good accuracy of detected saliency.

The short-term perspective will focus on integration of an adaptive threshold-based strategy for building the saliency extraction masks.

References

1. de Greeff, J., Delaunay, F., Belpaeme, T.: Human-robot interaction in concept acquisition: a computational model. In: Proceedings of International Conference on Development & Learning, pp. 1–6 (2009)
2. Araki, T., Nakamura, T., Nagai, T., Funakoshi, K., Nakano, M., Iwahashi, N.: Autonomous acquisition of multimodal information for online object concept formation by robots. In: Proceedings of IEEE/IROS, pp. 1540–1547 (2011)
3. Kinect camera http://www.xbox.com/en-US/kinect/default.htm
4. Han, J., Shao, L., Xu, D., Shotton, J.: Enhanced computer vision with microsoft kinect sensor: a review. IEEE Trans. Cybern. 43(5), 1318–1334 (2013)
5. Zhang, Z.: Microsoft kinect sensor and its effect. IEEE Multimedia Mag. 19(2), 4–10 (2012)
6. Khoshelham, K., Elberink, S.O.: Accuracy and resolution of kinect depth data for indoor mapping applications. Sensors 12(2), 1437–1454 (2012)
7. Camplani, M., Mantecon, T., Salgad, L.: Depth-color fusion strategy for 3-D scene modeling with kinect. IEEE Trans. Cybern. 43(6), 1560–1571 (2013)
8. Lloyd, R., Closkey, S.M.: Recognition of 3D package shapes for single camera metrology. In: Proceedings of IEEE Winter Conference on Applications of Computer Vision (IEEE-WACV 2014), pp. 99–106 (2014)
9. Skalski, A., Machura, B.: Metrological analysis of microsoft kinect in the context of object localization. J. Metril. Measure. Syst. 22(4), 469–478 (2015)
10. Zolkiewski, S., Pioskowik, D.: Robot control and online programming by human gestures using a kinect motion sensor. In: Rocha, Á., Correia, A.M., Tan, F.B., Stroetmann, K.A. (eds.) New Perspectives in Information Systems and Technologies, Volume 1. AISC, vol. 275, pp. 593–604. Springer, Cham (2014). doi:10.1007/978-3-319-05951-8_56

11. Stone, E.E., Skubic, M.: Fall detection in homes of older adults using the microsoft kinect. IEEE J. Biomed. Health Inf. **19**(1), 290–301 (2015)

12. González-Jorge, H., Zancajo, S., González-Aguilera, D., Arias, P.: Application of kinect gaming sensor in forensic science. J. Forensic Sci. **60**(1), 206–211 (2015)

13. Achanta, R., Hemami, S., Estrada, F., Susstrunk, S.: Frequency-tuned salient region detection. In: Proceedings of IEEE International Conference on Computer Vision and Pattern Recognition, pp. 1597–1604 (2009)

14. Itti, L., Koch, C., Niebur, E.: A model of saliency-based visual attention for rapid scene analysis. IEEE Trans. Pattern Anal. Mach. Intell. **20**, 1254–1259 (1998)

15. Harel, J., Koch, C., Perona, P.: Graph-based visual saliency. Adv. Neural. Inf. Process. Syst. **19**, 545–552 (2007)

16. Achanta, R., Estrada, F., Wils, P., Süsstrunk, S.: Salient region detection and segmentation. In: Gasteratos, A., Vincze, M., Tsotsos, J.K. (eds.) ICVS 2008. LNCS, vol. 5008, pp. 66–75. Springer, Heidelberg (2008). doi:10.1007/978-3-540-79547-6_7

17. Liu, T., Yuan, Z., Sun, J., Wang, J., Zheng, N., Tang, X., Shum, H.-Y.: Learning to detect a salient object. IEEE Trans. Pattern Anal. Mach. Intell. **33**(2), 353–367 (2001)

18. Liang, Z., Chi, Z., Fu, H., Feng, D.: Salient object detection using content-sensitive hypergraph representation and partitioning. Pattern Rec. **45**(11), 3886–3901 (2012)

19. Ramik, D.M., Sabourin, C., Madani, K.: Hybrid salient object extraction approach with automatic estimation of visual attention scale. In: Proceedings of IEEE SITIS, pp. 438–445 (2011)

20. Ramik, D.M., Sabourin, C., Moreno, R., Madani, K.: A machine learning based intelligent vision system for autonomous object detection and recognition. J. Appl. Intell. **40**(2), 358–375 (2014)

21. Moreno, R., Ramik, D.M., Graña, M., Madani, K.: Image segmentation on the spherical coordinate representation of the RGB color space. IET Image Proc. **6**(9), 1275–1283 (2012)

22. Desingh, K., Madhava Krishna, K., Rajan, D., Jawahar, C.V.: Depth really matters: improving visual salient region detection with depth. In: Proceedings of 24th British Machine Vision Conference (BMVC 2013), Bristol, UK, pp. 98.1–98.11 (2013)

23. Peng, H., Li, B., Xiong, W., Hu, W., Ji, R.: RGBD salient object detection: a benchmark and algorithms. In: Fleet, D., Pajdla, T., Schiele, B., Tuytelaars, T. (eds.) ECCV 2014. LNCS, vol. 8691, pp. 92–109. Springer, Cham (2014). doi:10.1007/978-3-319-10578-9_7

24. Cheng, Y., Fu, H., Wei, X., Xiao, J., Cao, X.: Depth enhanced saliency detection method. In: Proceedings of International Conference on Internet Multimedia Computing and Service (ICIMCS 2014), Xiamen, China, pp. 23–27 (2014)

25. Tang, Y., Tong, R., Tang, M., Zhang, Y.: Depth incorporating with color improves salient object detection. Vis. Comput. **32**(1), 111–121 (2016)

26. Bertasius, G., Park, H.S., Shi, J.: Exploiting egocentric object prior for 3D saliency detection (2015). Preprint arXiv:1511.02682, arxiv.org

27. Cheng, M.M., Mitra, N.J., Huang, X.: Global contrast based salient region detection. IEEE Trans. Pattern Anal. Mach. Intell. **37**(3), 569–582 (2015)

28. Riche, N., Duvinage, M., Mancas, M., Gosselin, B., Dutoit, T.: Saliency and human fixations: state-of-the-art and study of comparison metrics. In: Proceedings of IEEE ICCV, pp. 1153–1160 (2013)

29. Fawcett, T.: An introduction to ROC analysis. Pattern Recogn. Lett. **27**, 861–874 (2006)

Artificial Intelligence and Games

Analysis of the Protocols Used to Assess Virtual Players in Multi-player Computer Games

Cindy Even[1,2]([✉]), Anne-Gwenn Bosser[1], and Cédric Buche[1]

[1] Lab-STICC, ENIB, CERV, 25 Rue Claude Chappe, 29280 Plouzané, France
{even,bosser,buche}@enib.fr
[2] Virtualys, 41 Rue Yves Collet, 29200 Brest, France

Abstract. Recently, the development of believable agents has gained a lot of interest and many solutions have been proposed by the research community to implement such bots. However, in order to make advances in this field, a generic and rigorous evaluation that would allow the comparison of new systems against existing ones is needed. This paper provides a summary of the existing believability assessments. Seven features characterising the protocols are identified. After a comprehensive analysis, recommendations and prospects for improvement are provided.

Keywords: Believability · Evaluation · Benchmarking · Turing test · Non player characters · Virtual players · Bots · Multi-player games

1 Introduction

Computer games can be populated by human players represented by their avatar as well as computer-controlled players, also known as Non-Player Characters NPCs or bots. They may have different roles in the game such as acting as traders, providing services, quests or clues to the human players. In multi-player games, a special type of bot - that we call "virtual player"- can be used instead of human players. Their role is to play the game as a human player would. They are necessary for players who want to practice before facing human opponents, players who do not have the possibility to connect with other players, or to fill in spots on a server when there are too few human players. The popularity of a video game (and therefore its commercial success) is linked to the quality of these bots. For example, an unbeatable bot would be frustrating to play against while a predictable one would be boring. Indeed, according to Livingstone [1], modern video games do not require unbeatable AI but believable AI. Also, recent experimental results [2] show that believable bots increase users' enjoyment. Different approaches have been adopted for the development of believable bots, such as systems based on connectionist models [3,4], production systems [5,6] or probabilistic models [7–9] - to mention just a few. Generally, the proposed systems are not assessed, and when they are, the results obtained can not be compared as different protocols have been used. However, in order to make advances in

© Springer International Publishing AG 2017
I. Rojas et al. (Eds.): IWANN 2017, Part II, LNCS 10306, pp. 657–668, 2017.
DOI: 10.1007/978-3-319-59147-6_56

this field, many authors [8,10,11] pointed out the need of a generic and rigorous evaluation that would allow the comparison of new systems against existing ones. The evaluation of AI in games research has been identified as one of the main challenges in game AI research [12]. In this paper, we review evaluation techniques for assessing the believability of virtual players and we provide a comprehensive analysis of the evaluation features. We conclude by suggesting prospects for improvement.

2   Assessing Believability

Authors have worked on criteria-based assessment methods [13,14] where the believability of bots is ranked by the amount of criteria they meet. Even though such lists can be interesting, it can be difficult to take all the items into account during the assessment [15]. These lists are rather intended to provide a roadmap for the design of human-like bots. However, the notion of believability being highly personal, subjective assessments are a more common approach to measuring believability.

A way of evaluating AI is to organise competitions. According to Togelius [16], the advantage of competitions is that they provide fair, transparent and reusable benchmarks. In recent years we have seen the emergence of competition oriented toward the implementation of human-like (or believable) opponents such as the 2 K Botprize competition [17] or the Turing Test track of the Mario AI Championship [18]. The BotPrize is particularly interesting as it has evolved significantly over the years. It is a variant of the Turing test [19] which uses the "Deathmatch" game-type mode of the video game *Unreal Tournament 2004 (UT2004)*. For the first two editions [17], each human judge played against a human confederate and a bot. At the end of each round, the judges were asked to evaluate the two opponents on a rating scale and to record their observations. For the next edition, a new protocol was implemented [20], in order to make the judging process part of the game. A weapon in the game had two firing modes that could be used to tag an opponent as being human or bot. Both bots and humans were equipped with the judging gun and could vote. This modification to the system introduced a bias in the evaluation process as the game-play was adversely affected. Whilst players previously had to move quickly in order to not present an easy target, in the new competition players are tempted to stop and observe their opponents to make a judgement [21]. Furthermore, judges may be inclined to attempt to communicate through movements and shooting patterns [6]. This kind of behaviour would not naturally occur in normal game-play. For the last edition[1], the novelty was the addition of a third-person believability assessment (i.e. the judges observe the game).

Third person assessment was also used in [22]. Videos were recorded where an expert player played against bots and human players with different levels.

---

[1] Human-Like Bots Competition, presented at the IEEE CIG conference by Raúl Arrabales: http://www.slideshare.net/array2001/arrabales-bot-prize2014v2.

After watching videos, judges were asked to evaluate human-likeness on a 7-point Likert scale. A similar approach was used in [5,8] but with a different FPS game (Quake II). The protocol's characteristics of these player believability assessments can be found in Table 1 along with relevant references regarding *player believability* (the belief that a character is controlled by a human player [23,24]), and *character believability* (the belief that the character itself is real [24]) assessments.

As we can see from the descriptions below, the protocols used in the past for the assessment of virtual player's believability have characteristics that vary significantly. The process of judging the behaviours of a bot is, by nature, a subjective process [1,10,11] as it depends on the perceptions of the people playing or watching the game. Having no obvious physical attributes or features that can be measured, the only solution for measuring the believability of bots that can be considered is the use of a questionnaire [10]. In some cases, the players fill the questionnaire after playing the game for a few minutes, in other cases they vote during the game. The judgement can be done by the players or by observers, and different types of questionnaires are used such as ranking or comparison. In the next section we propose to analyse the characteristics of the protocols collected in Table 1.

3 Assessment's Characteristics Analysis

3.1 Application

The application used for the evaluation process can be pre-existing or developed specially for the test. The implementation of a sample game can be necessary when no open-source games are available [27] but it needs to be well-thought-out in order to not introduce bias unintentionally.

There are many advantages when choosing a pre-existing video game. According to Tencé et al. [23], the game needs to be a multi-player game (indeed, the role of virtual players is to be played against) offering a lot of interaction between the players. Action, role playing, adventure and sport games meet these criteria. Adventure and sport games tend to be difficult to modify and in particular, they rarely offer the possibility to add customised bots. The main draw-back of role playing games is that they rely in large part on communication and natural language which is not what we intend to evaluate here. Similarly, in order to not impact the assessment, all "chat" options should be disabled [17]. Action games, especially FPSs, are often a good choice. For the BotPrize contest, Hingston [17] chose UT2004 because it is affordable, readily available, customizable, bots and humans can play together and do not need to be collocated, and it is easy to interface a bot to the game. Togelius et al. [24] argued that FPS are not suitable for believability assessments as players encounter their opponents for only a few seconds and in the middle of a chaotic situation. For this reason they preferred to use the single player game *Infinite Mario Bros* which does not meet the criteria of being a multi-player game.

Table 1. Comparison of the existing experiments

Reference	Application	1st or 3rd person	assessment	Duration	No. of judges	Judges' level novice	medium	expert	Information given	Subjective assessment type binary	comparison	scale	comments	How
[5]	Quake II Deathmatch	3rd	16 x 1 video candidate's view	3 min	8	✓		✓	A	✓		✓ 1 to 10	✓	n/a
[10]	Simulation of a bar	3rd	2 simulations global view	as long as needed	13	✓	✓		B		✓ 2 choices	✓ 1 to 5		pen & paper
[11]	Pong game	3rd	video global view	n/a	n/a		n/a		A				✓	n/a
[8]	Quake II Deathmatch	3rd	15 x 3 videos 1st person view	20 sec	20	✓	✓	✓	A		✓ 4 choices	✓ 1 to 5	✓	on-line
[25]	CoPeFoot		1st	n/a	48	✓		✓	C	✓				pen & paper
[17] (BotPrize v1)	UT2004 Deathmatch		1st	10 min	5		✓	✓	A			✓ 1 to 5	✓	n/a
[20] (BotPrize v2)	UT2004 Deathmatch		1st	n/a	7		✓	✓	A	✓				in-game
[4]	UT2004 Deathmatch	3rd	10 x 1 video 3rd person view	15 min	3	✓	✓	✓	n/a	✓				in-game
		1st		1 min	12		✓		n/a	✓				crowdsourcing platform
[22]	UT2004 Capture The Flag	3rd	1 x 4 videos 1st person view	n/a	10		n/a		n/a			✓ 1 to 7		n/a
[18]	Infinite Mario Bros	3rd	2 videos global view	1 min	73		n/a		n/a		✓ 4 choices			on-line
[26]	Everyday life of the Darug people	3rd	14 x 2 videos 1st person view	n/a	43		n/a		B		✓ 3 choices	✓ 1 to 5		on-line

Character believability assessment.

A Judges are told that there is a mix of bots and humans.
B Judges know the nature of each entity.
C Judges are given no information.

3.2   1st or 3rd Person Assessment

Believability assessment may consider both first person and third person reports. In first person assessment, the judge has two simultaneous roles: to play the game, and to judge opponents. On the other hand, in third person assessment, the judge is only a spectator observing the game being played.

In [4,5,18] the authors argued that assessing believability from a first-person perspective might be distracting since the judge has to pay attention both to the game experience and to the behaviour of the other players for the evaluation. Livingstone answered in his paper [1] with: *"in game development the aim is to satisfy the needs of the players of a game and not those of watchers"*. However, even if computer games are primarily designed for the players, video game spectating has recently become a popular activity [28,29]. In Cheung et al. paper [28], the authors report that there are some spectators that actually prefer to watch professionals playing rather than playing the game themselves.

First person assessment is possible only with applications that can be played by at least two players simultaneously. The third person assessment however, can be used with any application. When performing a third person assessment, judges are asked to give their judgement after watching a video of the game previously recorded. To reduce the subjectivity and the guesswork, Gorman et al. [8] suggested to show more than one video to the judges in order for them to have a basis for comparison. They also pointed out the risk of introducing a bias when selecting videos for the assessment. The person in charge of the selection might pick parts of the video that could influence the responses.

When recording the videos, different points of view can be used. In some cases the application does not offer many possibilities. The Pong game for example, can only be played with a global view, representing the tennis table and the two paddles. In other video games such as FPSs, it is possible to choose between the first and third person view. Therefore, videos can be recorded from the confederate's or the candidate's first or third person view.

Confederate's 1st or 3rd person view. The confederate's 1st person view is most commonly used for assessing the believability of bots. This might be due to the fact that it is easily recorded during game play, particularly during a first person assessment. These points of view allow us to capture the game as if the judges were in-play. The main drawback of these points of view is that a considerable portion of recording can not be used. Indeed, all the moments when the confederate is in the environment without facing the candidate are useless and need to be cut from the video.

Candidate's 1st person view. When using the candidate's first person view, the judges have less resources to evaluate the entity: for instance, they can not see its movements.

Candidate's 3rd person view. This solution has never been used in our knowledge. Yet it could be especially interesting since it would capture both the perception and the actions of the candidates. This could allow a better understanding of the decisions made by the candidate. Moreover, it would not require cuts in the recording as even the time when the candidates are alone in the environment could be used for the judgement, which would be time saving and would reduce the risk of introducing the aforementioned bias when selecting videos for the assessment.

3.3 Duration

The duration of video and game play varies greatly from one experience to another, going from 20 s to as long as the judge desires. It might depend on the nature of the game but most of the time, the choice of the experiment's duration relies on the organisers' opinion [18,24] and is never justified. In their experiment, Soni et al. [2] tried to examine the role of predictability by using two different bots during their assessment. Unfortunately, the subjects did not notice any difference between the two bots. The authors hypothesised that the experiment was too short and that longer sessions could give the judges enough time to make a distinction. The observation of Paritosh et al. [30] regarding the Loebner competition[2] (the first formal instantiation of a Turing Test) is similar. They argue that the test is too short (only few minutes) to allow any depth in the judgement. Even if it is important to allow enough time for the judges to make a judgement, the assessment can not be too long as it can induce inattention or mistakes due to judges' boredom or fatigue [31].

3.4 Number of Judges

The assessment being of subjective nature, it seems important to collect a significantly large number of judgements in order to cancel out the biases introduced by that type of assessment [32]. The use of on-line surveys eases the collection and treatment of results. For their experiment, Llargues Asensio et al. [4] used a crowd-sourcing platform for mobile devices that allows to conduct a video-based poll experiment where the users can vote at the end of each video clip.

3.5 Judges' and Confederates' Expertise

The level of the judges is sometimes taken into account for the experiment. As it has been noticed by Mac Namee [10], the experience of players in video games can introduce a difference between the subjects. In general, for an experienced player it will be quicker and easier to recognise a bot than for a novice player. For example, in Laird et al.'s paper [5], only the expert player made no mistake in differentiating between bots and humans. Novice players might not fully know the rules of the game or the available actions which could make the whole experience too confusing and they would not be capable to sensibly evaluate the players' behaviours [1].

Another interesting element that has been taken into account in [5,18,22] is the level of the confederates. They have a major role in the assessment as

---

[2] http://www.loebner.net/Prizef/loebner-prize.html.

their behaviours directly influence the judges' evaluation. For example, a high-performing expert-player confederate could easily be mistaken for a bot by non-expert players [6]. On the contrary, novice-players confederates who are still learning how to play the game and how to use the controls might be mistaken with a weak bot by expert players. Confederates should be provided with sufficient time for gaining control over the game rules and commands before starting the evaluation. Hingston [17] avoided these potential problems by choosing confederates who were all of a reasonable level of experience, i.e. neither expert nor novice.

3.6 Information Given to the Judges

As we can see in Table 1, judges can be given different information before starting the experiment. Most of the time, they are informed that they will see a combination of bots and human players (A in Table 1). In other cases, (B) they know the nature of the entity they are evaluating. Finally, (C) judges are not informed as to the purpose of the experiment. For instance, in [25], judges were invited to play a football video game, where all the players had a number. After a given time, the game was paused and they were given a table and the following instructions: "Cross the box corresponding to the two players controlled by humans in the simulation, if and only if you are confident in your answer. If in doubt, write nothing". The analysis of the results revealed that judges were considerably better at distinguishing bots from human players after the first attempt.

In two other experiments, half of the participants were informed that the other character in the game would be controlled by another person, while the other half were informed that it would be controlled by a computer (AI). In fact, for all the participants, the character were controlled by a computer in [33], and by a human in [34]. In the first experiment, the participants who played against the character that they believed to be human-controlled, reported stronger experiences of presence, flow, and enjoyment. And in the second experiment, the participants exhibited greater physiological arousal and reported greater presence and likeability when the character was introduced as being human controlled rather than computer controlled. These results demonstrate that the information given to the judges can significantly alter their judgement.

3.7 Subjective Assessment Types

When assessing players' believability in a game, players are asked to give their opinion [24]. Their answer can have the form of a free response or of forced data retrieved through questionnaires.

Free response answers can contain much richer information but they are also much harder to analyse appropriately. Sometimes judges have the opportunity to give a free response in the form of comments [17]. These comments can be useful for identifying areas for improvement for the bots implementation but are generally not used for evaluation.

On the other hand, by using a questionnaire, subjects are constrained to choose between some specific items, yielding data that is easier to analyse. Different types of forced questionnaires can be identified [24]:

- *Binary*: Subjects can answer by *Yes* or *No* to a simple question (e.g. *is this player a bot?*, or, *is this bot believable?*).
- *Scale*: Judges are asked to rate the humanness of the players' behaviour or to choose an answer within a list (e.g. [8] *1: Human, 2: Probably Human, 3: Don't Know, 4: Probably Artificial, 5: Artificial*).
- *Comparison*: Subjects are asked to compare two or more players (e.g. *did player A or B act more like a human player?*).

With ranking questionnaires, it is not possible to analyse the interpretation of the rating categories across subjects [35]. To minimise the subjective notion of scaling and allow a fairer comparison between the subjects' answers, comparison and boolean questions can be used [24]. But as mentioned by Hingston [17], a binary choice might have the effect of forcing the subjects to "toss a coin" if they are unable to choose an answer. In an effort to reduce subjectivity, in [1,10] subjects were not asked to rate believability, instead, they were asked to compare two players and say which was more believable or acted more like a human player. The choice items may be presented in different ways, for instance, the subjects can choose between 2 solutions (*player A* or *player B*). They can also be offered more options such as *there is no difference*, or *both equally* and *none of them*, following the 4 alternative forced choice (4-AFC) protocol proposed by Yannakakis and Hallam [36].

4   Discussion

When studying the protocols used in the past to assess virtual players' believability, we identified some characteristics that varied significantly from one assessment to another, giving results that can not be correlated.

Application. First of all, different types of games were used such as FPS, sport or platform games. The main criterion when choosing the game is that it needs to be a multi-player game where one can face virtual players. The second criterion, which restricts significantly the range of games that can be considered, is that it has to be possible to interface a bot.

1st *or* 3rd *person assessment.* Even when the types of games used in the assessments were similar, judges had different roles. They were either part of the game (first person assessment), with the ability to interact with the candidates but also with the risk of modifying the game-play. Or they were spectators (third person assessment), assessing a game in which they were not involved. The recent interest for game spectating can be an additional argument in favour of this choice. For this type of assessment, the judges watch videos of the game. These videos can be recorded using different points of view. The most commonly used is the confederate's first person view but a solution that seems to have potential and needs to be tested is the candidate's third person point of view.

Duration. The duration of the assessment is another characteristic that can vary significantly. Judges might give a random answer if they do not have enough time to evaluate a bot. In order to avoid this situation it seems important to define a minimum assessment duration.

Number of judges. As the notion of believability is very subjective, it is important to collect a large number of judgements. The use of an on-line questionnaire or crowd-sourcing platform seems unavoidable as they can allow for the collection of more data that would give more accurate results. In order for the protocol to be rigorous, a minimum number of participants must be defined.

Judges' and confederates' expertise. The judges' and confederates' level of experience is sometimes taken into account. In general, we recommend training novices before involving them in the roles of judge or confederate as they need to know the rules, the commands and to have experimented with the game. Otherwise, confederates could easily be mistaken with weak bots and judges could be too confused to be able to make a judgement. It would be interesting to study the influence of the judges' level on the results when the number of judges is high.

Information given to the judges. As we saw in Sect. 3.6, recent experiments have shown the influence of the information given to the judges on their judgement. This part of the assessment protocol needs to be carefully designed in order to avoid introducing a bias. When conducting a first person assessment, the gameplay might be modified if the judges know the aim of the assessment. The only way to avoid this is to keep the question secret and to ask the player only at the end of the game, whether he thought he was playing against a human player or a bot. Of course, the player could be asked only once. During a third person assessment, the best solution seems to be keeping the nature of the candidate secret and telling the judges that they would see a mix of bots and human players, so that they have no prejudices.

Subjective assessment types. Finally, different types of questionnaire have been used (binary, scale or comparison) to collect the judges' opinions, giving data that can not be compared from one assessment to another. Regardless of the type of questionnaire, the question(s) as well as the offered solutions will have to be adapted according to the type of assessment (first or third person) and the information previously given to the judges.

5 Conclusion and Future Work

Virtual players play a major role in the success of video games. A new challenge is to develop believable bots that could blend in among human players. Over the years, different approaches have been used for the implementation of such bots. However most of the time, these bots were either not evaluated, or they were evaluated using different protocols. Yet, in order to make improvements in the development of believable bots, a generic and rigorous evaluation needs to be set up, that would allow the comparison between new systems and existing ones. According to Clark et al. [37], *"standardised tests are an effective and practical assessment of many aspects of machine intelligence, and should be part of any comprehensive measure of AI progress"*. Although the evaluation of bots' performance can be performed through objective measures (comparing score or time spent to complete a level), the evaluation of bots' believability is complex due to its subjective aspect.

In this paper we analysed the protocols previously used to assess the believability of virtual players. We identified seven features that characterise the assessments and which vary significantly from one to another. When designing a new protocol, these features need to be chosen carefully in order to not introduce a bias into the evaluation. After an in-depth analysis of these protocols, we gave recommendations for the features that are well established. In order for the protocol to be rigorous and reusable, other features still need further study and testing to be determined.

References

1. Livingstone, D.: Turing's test and believable AI in games. Comput. Entertainment **4**(1), 6 (2006)
2. Soni, B., Hingston, P.: Bots trained to play like a human are more fun. In: IEEE International Joint Conference on Neural Networks (IEEE World Congress on Computational Intelligence), pp. 363–369, June 2008
3. van Hoorn, N., Togelius, J., Wierstra, D., Schmidhuber, J.: Robust player imitation using multiobjective evolution. In: 2009 IEEE Congress on Evolutionary Computation, pp. 652–659. IEEE, May 2009
4. Asensio Llargues, J.M., Peralta, J., Arrabales, R., Bedia, M.G., Cortez, P., Peña, A.L.: Artificial Intelligence approaches for the generation and assessment of believable human-like behaviour in virtual characters. Expert Syst. Appl. **41**(16), 7281–7290 (2014)
5. Laird, J.E., Duchi, J.C.: Creating human-like synthetic characters with multiple skill levels: a case study using the Soar Quakebot. Papers from 2001 AAAI Spring Symposium, Artificial Intelligence and Interactive Entertainment I, pp. 54–58 (2001)
6. Polceanu, M.: Mirrorbot: using human-inspired mirroring behavior to pass a turing test. In: IEEE Conference on Computational Intelligence in Games (CIG 2013), pp. 1–8. IEEE (2013)
7. Le Hy, R., Arrigoni, A., Bessière, P., Lebeltel, O.: Teaching Bayesian behaviours to video game characters. Robot. Auton. Syst. **47**(2–3), 177–185 (2004)
8. Gorman, B., Thurau, C., Bauckhage, C., Humphrys, M.: Believability testing and bayesian imitation in interactive computer games. From Anim. Animats **9**(1), 655–666 (2006)
9. Tencé, F., Gaubert, L., Soler, J., Loor, P., Buche, C.: CHAMELEON: online learning for believable behaviors based on humans imitation in computer games. Comput. Animat. Virtual Worlds (CAVW) **24**(5), 477–496 (2013)
10. Mac Namee, B.: Proactive Persistent Agents: Using Situational Intelligence to Create Support Characters in Character-Centric Computer Games. Ph.D thesis, University of Dublin, Trinity College, August 2004

11. McGlinchey, S., Livingstone, D.: What believability testing can tell us. In: Proceedings of the International Conference on Computer Games: Artificial Intelligence, Design, and Education, pp. 273–277 (2004)
12. Lucas, S.M., Mateas, M., Preuss, M., Spronck, P., Togelius, J.: Artificial and computational intelligence in games (Dagstuhl Seminar 12191). Dagstuhl Rep. **2**(5), 43–70 (2012)
13. Hinkkanen, T., Kurhila, J., Pasanen, T.A.: Framework for evaluating believability of non-player characters in games. In: AI and Machine Consciousness (2008)
14. Arrabales, R., Ledezma, A., Sanchis, A.: ConsScale: a pragmatic scale for measuring the level of consciousness in artificial agents. J. Conscious. Stud. **17**(3–1), 131–164 (2010)
15. Arrabales, R., Ledezma, A., Sanchis, A.: ConsScale FPS: cognitive integration for improved believability in computer game bots. In: Hingston, P. (ed.) Believable Bots, pp. 193–214. Springer, Heidelberg (2012)
16. Togelius, J.: How to run a successful game-based AI competition. IEEE Trans. Comput. Intell. AI Games **8**(1), 95–100 (2016)
17. Hingston, P.: A turing test for computer game bots. IEEE Trans. Comput. Intell. AI Games **1**(3), 169–186 (2009)
18. Shaker, N., Togelius, J., Yannakakis, G.N., Poovanna, L., Ethiraj, V.S., Johansson, S.J., Reynolds, R.G., Heether, L.K., Schumann, T., Gallagher, M.: The turing test track of the 2012 Mario AI Championship: entries and evaluation. In: IEEE Conference on Computational Intelligence in Games (CIG 2013), pp. 1–8. IEEE (2013)
19. Turing, A.M.: Computing machinery and intelligence. Mind **59**(236), 433–460 (1950)
20. Hingston, P.: A new design for a Turing Test for Bots. In: Proceedings of the 2010 IEEE Conference on Computational Intelligence and Games, pp. 345–350. IEEE, August 2010
21. Thawonmas, R., Murakami, S., Sato, T.: Believable judge bot that learns to select tactics and judge opponents. In: IEEE Conference on Computational Intelligence and Games (CIG 2011), pp. 345–349 (2011)
22. Acampora, G., Loia, V., Vitiello, A.: Improving game bot behaviours through timed emotional intelligence. Knowl.-Based Syst. **34**, 97–113 (2012)
23. Tencé, F., Buche, C., De Loor, P., Marc, O.: The challenge of believability in video games: definitions, agents models and imitation learning. In Mao, W., Vermeersch, L. (eds.) 2nd Asian Conference on Simulation and AI in Computer Games (GAMEON-ASIA 2010), Eurosis, pp. 38–45 (2010)
24. Togelius, J., Yannakakis, G.N., Karakovskiy, S., Shaker, N.: Assessing believability. In: Hingston, P. (ed.) Believable Bots: Can Computers Play Like People?, pp. 215–230. Springer, Heidelberg (2012)
25. Bossard, C., Benard, R., De Loor, P., Kermarrec, G., Tisseau, J.: An exploratory evaluation of virtual football player's believability. In: Proceedings of 11th Virtual Reality International Conference (VRIC 2009), pp. 171–172 (2009)
26. Bogdanovych, A., Trescak, T., Simoff, S.: What makes virtual agents believable? Connection Sci. **28**(1), 83–108 (2016)
27. Bernacchia, M., Hoshino, J.: AI platform for supporting believable combat in role-playing games. In: Proceedings of the 19th Game Programming Workshop in Japan, pp. 139–144 (2014)
28. Cheung, G., Huang, J.: Starcraft from the stands: understanding the game spectator. In: Proceedings of the SIGCHI Conference on Human Factors in Computing Systems, pp. 763–772 (2011)

29. Kaytoue, M., Silva, A., Cerf, L.: Watch me playing, i am a professional: a first study on video game live streaming. In: Proceedings of the 21st International Conference Companion on World Wide Web, June 2009, pp. 1181–1188 (2012)

30. Paritosh, P., Marcus, G.: Toward a comprehension challenge, using crowdsourcing as a tool. AI Mag. **37**(1), 23–30 (2016)

31. Brace, I.: Questionnaire Design: How to Plan, Structure and Write Survey Material for Effective Market Research. Kogan Page Publishers (2008)

32. Hyman, H.H., National Opinion Research Center: Interviewing in social research. A research project of the National Opinion Research Center. University of Chicago Press, Chicago (1954)

33. Weibel, D., Wissmath, B., Habegger, S., Steiner, Y., Groner, R.: Playing online games against computer- vs. human-controlled opponents: effects on presence, flow, and enjoyment. Comput. Hum. Behav. **24**(5), 2274–2291 (2008)

34. Lim, S., Reeves, B.: Computer agents versus avatars: responses to interactive game characters controlled by a computer or other player. Int. J. Hum. Comput. Stud. **68**(1–2), 57–68 (2010)

35. Friedman, H.H., Amoo, T.: Rating the rating scales. J. Market. Manag. **9**(3), 114–123 (1999)

36. Yannakakis, G.N., Hallam, J.: Real-time game adaptation for optimizing player satisfaction. IEEE Trans. Comput. Intell. AI Games **1**(2), 121–133 (2009)

37. Clark, P., Etzioni, O.: My computer is an honor student but how intelligent is it? standardized tests as a measure of AI. AI Mag. **37**(1), 5–12 (2016)

The *Long Path of Frustration*: A Case Study with *Dead by Daylight*

Pablo Delatorre[1](✉), Carlos León[2], Alberto Salguero[1], and Cristina Mateo-Gil[1]

[1] Universidad de Cádiz, Cádiz, Spain
{pablo.delatorre,alberto.salguero}@uca.es
cristina.mateogil@alum.uca.es
[2] Universidad Complutense de Madrid, Madrid, Spain
cleon@ucm.es

Abstract. Playability is a key factor in video-games. From a narrative standpoint, the play process is usually designed as sequences of episodes triggered by the player's motivations, which unfold along a sense of suspense-relief. Suspense, as a factor on engagement, has a strong impact on the narrative of video-games: when it decreases, so does the engagement. This is a common pattern when players are aware that losing is unavoidable. As we point out, many players disconnect from the game in this situation. In this paper we evaluate how suspense affects playability, to analyse how the lack of uncertainty due to the knowledge of the rules may degrade *Dead by Daylight* game players experience when they are bound to fail. We have observed that players acknowledging that there are no chances to win tend to leave the game. Results also reveal that suspense is modulated by the player's knowledge of the game.

1 Introduction

Narrative-based games may lead to situations where it is practically impossible to overcome a challenge unless the character is ready enough to face them, fact that not always the free plot is able to guarantee. The improvements that players incorporated to their characters can do nothing but prolong the encounter, that inexorably will end with a defeat. When players are aware that the situation is lost but not still finish, they may only watch impotently as the outcome progressively approaches (for example, waiting a number of turns in role playing games or the character's life bar is completely drained in other action games). This way, if skills are insufficient to win but not to defeat instantly, game experience may become boring. Clearly about to fail, no more suspense is perceived beyond the chance of "touching" the enemy or emerging unscathed front any assault.

In these situations, players experience a cognitive and emotional state that we call the player's *long path of frustration*. The interesting aspect of this effect is that it is not produced because the player is playing bad with respect to the antagonist, or due to chance, but above all else due to his certainty regarding the rigidity of the game rules. Thus, while the immediate effect is: (a) the certainty that it is not possible to do any useful action in the time this *long path of*

© Springer International Publishing AG 2017
I. Rojas et al. (Eds.): IWANN 2017, Part II, LNCS 10306, pp. 669–680, 2017.
DOI: 10.1007/978-3-319-59147-6_57

frustration lasts; the secondary effect is (b) the tendency to break the suspension of disbelief (individual's willingness to accept the world of the character as "real" [11, p. 515]) as a defensive mechanism for such frustration. Consequently, a third effect is (c) the feeling that the time in which this *long path of frustration* occurs is a waste, because the player is not in control [10, p. 13].

This *long path of frustration* becomes especially important in the field of suspense video-games as narrative discourse. Effect (a) implies that chances of plot twist are limited in comparison with the design of classical discourses such as cinema or literature. These limitations are mainly based on the "narrative paradox" or how to reconcile the needs of a participant user rather than a spectator with the idea of narrative coherence [1, p. 35]. In a film or a book, a group of unprepared characters could face a potential stronger antagonist, and the audience might still have the hope that the author would come to a solution that includes in the story an "investor effect" that helps them overcome. Since the germ of suspense is the approach to a crucial outcome, until the fatality becomes irreversible the emotion persists along with the hope of avoiding it [18, pp. 134–139]. However, in the case of video-games such irreversibility occurs when it is known that the rule system can only lead to the fatal denouement. Because there is nothing the player can do to avoid this, the time between this certainty and the arrival of that outcome is a long path of frustration for the player.

As referred by aforementioned effect (b), there is a tendency for the players to "quit the game" if they are detected, since they can not break the rules. Studying the behaviour of *YouTubers'* gameplays, from the moment they begin their *long path of frustration* and without having finished the game, they tend to start to make references to the video-game itself as an interactive system, instead of describing their experiences in the context of the game. They talk about how wrong they has played and the unfairness of the game, usually questioning other characters' strategies. There are some players who simply and directly disconnect from the game.

As an example, in *Shadow Tactics*[1] [12], a real-time infiltration game, players tend to reload from the last checkpoint when they are discovered, for example, although it is still possible (but more difficult, as a penalty part included in the game design) to hide again and try to find another way to solve the situation. Just as another example, in *Metal Gear Solid V: The Phantom Pain* [9] enemies will be provided with flash-lights and night-vision goggles if they detect our character during a nocturnal incursion. Being detected makes players reload the game at the point they have not still found.

This strategy supposes a "time-travelling" to a previous state when the player has again the control of a more solvable situation. Otherwise, the existence of new enemies or skills in the scene will require much more time studying possible ways of solving it. Because experience with the game indicates that players are not likely going to solve the new situation [2, p. 97], the effect (c) appears because

---

[1] German Developer Award 2016: Best Game Design, Best PC/Console Game and Best German Game.

they perceive that the effort employed in solving it does not worth it. They feel they are wasting their time.

At present, our research about suspense effects and their computational modelling has led us to the review of suspense in different narrative discourses and interactive techniques such as video-games. In particular, we have found the same *long path of frustration* and its resultant disconnecting trend among all the players while analysing suspense responses in cooperative/competitive games like *Dead by Daylight* [3] or *Friday the 13th* [8], where there is no artificial intelligence and all the characters are handled by human players.

Moreover, in cooperative games, the consequences of those disconnections due to the conviction of "waiting for nothing" are not only limited to the relation of the frustrated player with the video game, but it also affects to the rest of the participants: although the player's character could not actively participate in the game, could still be part of the antagonist's universe and, therefore, could cause a distracting effect. Consequently, other players may take advantage of this situation because they have more time to perform other actions. Furthermore, a disconnection of the player may even lead to the premature termination of the game for all the players if he is hosting the P2P game server.

On that basis, the present work is based on the following hypothesis:

Knowledge of rules influences cognitive and emotional aspects related to the player disinterest.

Although there exist some proposals in the literature to increase the playability of video games, to our best knowledge there are no authors who deeply analyse how suspense influence it nor, alternatively, its conservation through the increase of the domain of solutions. In this respect, in this article a study is carried out to verify the existence of the *long path of frustration* effect and a design proposal is provided to take into account this effect. In Sect. 2, a preliminary analysis through the review of gameplays published in *YouTube* is presented. Based on this information, an experiment that relates the player's safety to the rigidity of the rules of the game and the perception of suspense is presented in Sect. 3. Finally, Sects. 4 and 5 are intended to discuss the proposal and conclude this paper, respectively.

2   Preliminary Analysis

In order to design a focused experiment providing insight about how players react to the decrease of suspense in video-games, we first ran a preliminary analysing of gameplays in which this happens. The analysing was run against *Dead by Daylight*, a suspenseful video-game in which a killer chases, wounds and hangs the characters to death in a closed scenario. The skill of the game controlling the killer and the specific circumstances were also recorded. Although relative influential, other aspects like the environment, experience as a victim or other character features were discarded in order to keep the study focused and to provide a well scoped analysis.

Dead by Daylight is a match-based, multi-player game which takes place at night, in an outdoor map with three fundamental elements: electricity generators, hooks and exit doors. Four players (the victims), starting with a full blood/life bar, cooperate with each other (more or less intentionally) trying to switch on five generators. When the fifth generator is switched on, the exit doors are activated and can be opened after a short time. On the other hand, the player who acts as the killer (only one) must try to kill all the players, hitting them with his or her weapon[2]. In case of knocking a victim down, which usually requires two strokes, the killer can wait to bleed themselves out or hang them on one of the hooks. The victim is finally eaten by a monster called *the Entity* (who is only showed its spider-like legs) after a short time in the hook. While the killer drags a victim to one of the hooks, the victim may struggle to free himself. The farther the hook, the more possibilities for the victim to escape. Otherwise, once on the hook the blood bar decreases gradually. Then, the victim goes through three different phases: (1) if the blood bar is above the half value and it is the first time the victim is hung, he or she can escape by himself (although the probability is low) or can be unhooked by another player; (2) if the blood bar is under the half or it is the second time the victim is hung, *the Entity* appears and the player must fight against it pressing the space key until being rescued by another player or the blood bar is exhausted; (3) if the player stops fighting, the blood bar is exhausted or the victim is hung for the third time, *the Entity* kills the player.

2.1 Method

Gameplays were collected as follows: first, recent games were chosen over older ones to increase the probability to analyse the behaviour of experienced players. Secondly, in order to avoid subsequent video editions that may have cut part of the content, only live videos that were published directly from the game were selected, which we also reviewed. Taking into account these two criteria, a search on *YouTube* was made using the phrase "Dead by Daylight Livestream", and the results were sorted by publication date. As an additional condition, those games where players communicated among other players through microphones and headphones were discarded to avoid additional advantages beyond the pre-established rules, which may decrease suspense. We finally obtained a set $C1$ (control group, without any particularity) of 26 videos (a total of 164 games) from 24 different *YouTubers* ($mean_{C1} = 6.31$ games per video, $stdev_{C1} = 4.05$), all published between December 31, 2016 and January 2, 2017.

We also collected gameplays of expert players playing as killers in order to analyse if the chance of disconnection depends on the killer experience and the resources the have acquired (players gather experience points each game, serving these for improving characters' abilities). For this reason, *Dead by Daylight* play lists published by twenty *YouTubers* were examined in order to choose the one with more videos and experience. The user chosen was GENuINE993, with 129

---

[2] Victims know when the killer is approaching because of a increasing heartbeat sound, urging them to run away.

videos of four games each and more than 300 h in *Dead by Daylight* on Steam. Although this player does not publish live matches (only recordings), they were all checked to verify that there was no post-production beyond adding an introduction and the face of the player to the video. From this set $C2$ (games with GENuINE993 as experienced killer in which no victim was a subscriber), the 30 most recent videos or a total of 71 games as killer were reviewed ($mean_{C2} = 2.37$ games per video, $stdev_{C2} = 0.63$). Other 49 games with subscribers, coming from 24 videos, where also collected in a new set $C3$ (games as experienced killer, being all victims obtained randomly from his subscribers; $mean_{C3} = 2.04$ games per video, $stdev_{C3} = 0.04$). This was done in order to check differences between playing with or without subscribers. Games took place between November 11 and December 31, 2016.

The analysis of each video consisted on visualizing the summary screen of each individual game, which shows the disconnections of players by the symbol of a cable unplugged[3]. When disconnections were found, conditions of the game prior to that disconnections were reviewed. Disconnections were classified according to the circumstances of the player at the time of his or her disconnection: *in Danger* in the case of threat (about to be hung, about to die [blood bar under 25%], lying on the ground, pursued by the killer or injured less than ten seconds before); and *unknown* in case of the disconnection occurs in other circumstances, which also includes reasons beyond the player's control as network slowdowns, main server failures or game host outages.

2.2   Results

From the set of games $C1$ (control group in which capabilities of characters were not considered), 14 *in Danger* disconnections and 24 *unknown* disconnections were registered. Taking this into account, *in Danger* disconnections occurred in 8.54% of the analysed gameplays. On the other hand, in the game set $C2$ (games with an experienced killer) the number of *in Danger* disconnections did not change (affecting 19.72% of games), and *unknown* disconnections decreased to 5. Finally, there were no disconnections *in Danger* in the set of games $C3$ (games with an experienced killer, who plays with his subscribers as victims). This may be explained because the satisfaction the player experiment when sharing a video with subscribers is higher than the objective of the game, so the player tries to prolong the experience. Anyway, 3 *unknown* disconnections were still registered in this set. Table 1 shows the results obtained for each group.

A discrepancy is observed between the higher number of *unknown* disconnections of the group $C1$ with respect to the other two groups. The reason is that this type of disconnections, which not infrequently occurs at the beginning of the matches, are not removed from the live gameplays in $C1$, in contrast to the gameplays of the chosen *YouTuber*. Therefore, it is likely that the total number

---

[3] In case the *YouTuber* ignored this screen, the video progress bar was displaced to the last moments of the game, manually checking the status of each player.

Table 1. Percent of disconnections per games, group and type of disconnection

Disconnection/Group	$C1$	$C2$	$C3$
in Danger	8.54%	19.72%	0.00%
unknown	14.63%	7.04%	6.12%

of *in Danger* disconnections of the groups $C2$ and $C3$ will also be greater in the games played but these are probably not published.

Obtained results suggest the existence of a relation between the escape possibilities and players disconnections. On average, potentially almost one in eleven games has at least one player who quits because of a *in Danger* situation, while in case of an experienced killer this is as high as one in five.

This preliminary analysis provides useful insight, though far from yielding concluding results. Obtaining videos from *YouTube* is straightforward but, given the lack of control of how videos are produced, a real evaluation must be performed. Next section details how this has been tackled for the current research.

3   Experiment

Once we obtained insight about the relation between possibilities of escaping and players disconnections, we carried out an experiment trying to confirm our hypothesis about how knowledge of rules influences cognitive and emotional aspects which to the player's disinterest. A set of eight scenes of *Dead by Daylight* were screened to the participants in the experiment. For each scene and participant, we collected a number of questions about the perceived suspense, the perceived chances for the character to escape and player own enjoyment.

A total of thirty seven undergraduate students ($N = 37$; seven women, thirty men) from the University of Cadiz, with ages ranging from 19 to 40 years ($mean = 23.70$, $stdev = 4.32$), voluntarily took part in this experiment. A total of 2923 data records were collected. Participants were divided in three groups: Group A ($N_A = 12$, all men), composed by participants with previous experience in the game (at least 50 gameplay hours); Group B ($N_B = 11$; three women, eight men), and group C ($N_C = 14$; four women, ten men), both randomly composed by participants without previous knowledge about the game.

We decided to stick to canonical narrative structures in order to approximate general accepted notions of what a narrative is. Therefore, scenes were chosen to evoke suspense through a similar narrative plot keeping the Freytag's curve [6]. Therefore, selected sequences were focused in one specific victim (per sequence) and they contained the following stages: (1) *pre-exposition* stage: the victim is busy doing something (searching, repairing a generator, breaking a trap...); (2) *exposition* stage: the victim stops performing her current task and starts to walk; (3) *rising action* stage: the killer gradually approaches, discovers and ultimately knocks the victim over; (4) *pre-climax* stage: the killer holds the victim up and goes to the closest hook; (5) *climax* the victim is about to escape by his own means, he is being unhooked by a partner or she is about to die; (6) *falling action*: the victim runs away; and (7) *denouement* stage: the heartbeat that indicates the killer proximity is not heard anymore. Stages 6 and 7 are replaced by the pre and post game scene of death in the case of the victim finally dies.

For the elaboration of the sequences, *Dead by Daylight* gameplays have been obtained from *YouTube*. In this case, unlike the analysis presented in Sect. 2, these recordings were just used as basic material, and the in-game reactions were not measured. They were only meant to serve as a basis for providing post-processed videos to the evaluators.

We set four requirements for the chosen gameplays: they must contain the aforementioned stages, they must not have comments (whether spoken or written), they must not have conversations between participants and they must have sufficient video quality (720 p minimum) so that the resolution did not influence the viewers' perception when screening in a standard projector. To do this, we searched on *YouTube* using the expression "dead by daylight no-comment", studying the first page results, measuring a total of twenty. Among all the candidates met the referred criteria, `Lizzy Beam` was the selected user, due to the largest number of videos on his *Dead by Daylight* playlist, counting 214 videos.

Videos were reviewed from the most modern to the oldest, choosing gameplays where any victim is pulled to a hook. The scenes were selected by taken into account they have to take place in the same scenario and with the same killer[4], in order to avoid variability in the perceived suspense due to the scenario.

Since some game events are more likely than others, candidate gameplays were filtered proportionally in terms of these events (the percentage was kept and the amount was normalized). Finally, eight videos were selected: the victim escaping from the hook by himself (*EBH*); the victim saving another player from the hook (*SPA*); the victim being unhooked by another player (three different episodes, *SBP1*, *SBP2* and *SBP3*, with different characters); and the victim dying in the hook even starting from maximum level of blood (three different episodes, *DFM1*, *DFM2* and *DFM3*, also with different characters).

Selected gameplays were then post-processed with Adobe Premiere Pro CC v11.0 in order to obtain a homogeneous aspect, normalize sound volume and re-synchronize audio and video. Each gameplay was trimmed to extract the part of the scene corresponding with our Freytag's curve based design of stages. In that way, influence of non-related sub-episodes is avoided (it would make the analysis more complex and the results less conclusive). The final duration of the sequences varies between 1'23" and 3'31" (*mean* = 2'16", *stdev* = 34"). All these sequences were later edited to obtain a second version of each one in which left and right lower zones of the screen were covered, in order to hide any information about the state of the group or the skills of the characters.

Finally, videos were randomized before to start the experiment. The order was: *SPA*, *SBP1*, *DFM1*, *EBH*, *SBP2*, *SBP3*, *DFM2*, *DFM3*. The same order is used for the three groups.

---

[4] In the current version, *Dead by Daylight* has six different killers and scenarios, for a total of thirty six different combinations.

3.1   Method

Separately, each group was brought in a classroom. Participants were randomly placed, keeping an empty table between each pair. After a demographic survey, another questionnaire was given to each participant in which they had to answer the following three questions: *how much suspense does the situation generate?*; *what hope do you think the character has to escape?*; and *what degree of enjoyment are you experiencing?*. The responses to questions are given in a 4-likert scale with the following values: *none, low, high* and *very high*, corresponding to values ranging from 1 to 4, respectively (numbers at left, in Fig. 1).

The experiment was carried out by group A (participants with previous experience in *Dead by Daylight*), then for the group B and finally for the group C, both without any experience or knowledge about the game. To differentiate these both groups and prior to play the video sequences, the rules of the game were clarified for the group B, emphasizing the circumstances in which a character cannot escape. This clarification was not made for the group C and, in addition, the screened videos were those edited to hide the state of the group or the skills of the characters. Therefore, group C had almost no knowledge about the game rules besides intuitive assumptions about the scene it represents.

For each video sequence, the participant's emotion was evaluated during three different situations: state *0*, once the victim is captured and seized by the killer, being still able to escape by struggling; state *1*, when the victim is hung on the hook but *the Entity* has not yet appeared, being able to escape by jump only if it is the first time in the hook (otherwise, being unhooked by another player is the only way out); and state *2*, when *the Entity* appears and the only possibility is to be saved by another player, if any is available for. The screening was paused when the character entered a new state so participants could fill in the answers to the questions for the current instant. The screening was resumed after five seconds. Once a complete scene was ended, a new one began. The experiment finished after all the eight video sequences were played.

3.2   Results

Obtained values evidence clear differences between scenes where chance to escape is possible against those in which death is unavoidable. Regarding *suspense*, the first case shows differences among groups tend to be small and even non-significant ($\chi_0^2 = 29.994$, $p < 0.000$, $\chi_1^2 = 9.993$, $p < 0.007$, $\chi_2^2 = 3.018$, $p < 0.3$), around *high* for all states. On the other hand, differences are remarkable when escaping is not possible ($\chi_0^2 = 54.882$, $\chi_1^2 = 56.815$, $\chi_2^2 = 56.997$, $p < 0.000$). Specifically, group A reports a *none/low* value of *suspense*, group B values decrease from *low/high* in state *0* to *none/low* in state *1*, and group C increases its *suspense* from *high* in state *0* to *very high* in states *1* and *2*.

Similarly, reports about *hope* do show a weak and barely significant difference among groups in the case of chances ($\chi_0^2 = 22.112$, $p < 0.000$, $\chi_1^2 = 0.756$, $p < 0.7$, $\chi_2^2 = 5.856$, $p < 0.06$), with *low* average values in all the states for groups A and B, and some fluctuation from *high* to *low* for group C. Nevertheless, if there is not escape, *hope* values for groups A and B are similar to their respective *suspense* values, while group C does not show differences in *hope* with respect to the case that escaping is possible ($\chi_0^2 = 23.043$, $p < 0.000$, $\chi_1^2 = 10.616$, $p < 0.005$, $\chi_2^2 = 23.471$, $p < 0.06$).

Lastly, in the case of chance *amusement* decreases from *high* to *low* for group A, *high* in average for group B and from *high* to *very high* for group C, increasing the difference of reported values state by state ($\chi_0^2 = 5.5794$, $p < 0.07$, $\chi_1^2 = 35.856$, $p < 0.000$, $\chi_2^2 = 59.855$, $p < 0.000$). On the other hand, if evasion is not possible, reported *amusement* for group A is *low* (just in state *0*) or *none*, group B maintains a *high* value for all three states, and group C goes from *high* to mainly *very high* ($\chi_0^2 = 28.291$, $\chi_1^2 = 37.773$, $\chi_2^2 = 53.802$, $p < 0.000$).

In contrast to *suspense* and *hope*, *amusement* presents a significant difference between groups A and B. We may attribute this to the fact that participants of group B had never watched a *Dead by Daylight* gameplay: watching character deaths, even when expected, can be a new (and seemingly funny) experience.

Figure 1 illustrates the evolution in each state per group of *suspense*, *hope* and *amusement*, respectively, grouping by chance to escape.

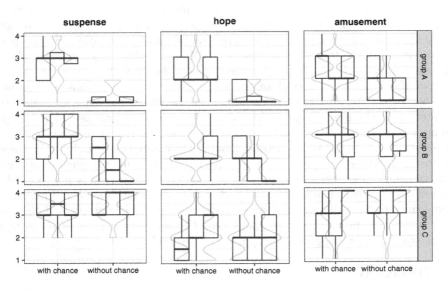

Fig. 1. Reported states evolution per group/chance (states 0, 1 and 2)

With respect to relations between reported emotions, correlation between *suspense* and *amusement* is significant and expected in all three states ($\rho_0 = 0.272$, $\rho_1 = 0.485$, $\rho_2 = 0.547$, $p < 0.001$), although the weak dependency in the initial state can be observed when the intentions of the killer are still unknown[5].

---

[5] Killer can abandon to the victim on the floor, bleeding to death. This is an usual strategy when there are not hook near around.

A relation between *suspense* and *hope* is also observed in the states *0* and *1*, being moderate/weak ($\rho_0 = -0.302$, $\rho_1 = -0.188$, $p < 0.001$). In addition, an inversion occurs in the state *2*, becoming a weak uphill ($\rho_2 = 0.232$, $p < 0.001$).

Weak correlations between *suspense* and *hope* and the inversion in the state *2*, may be explained when effects of knowledge about the victim fate are reviewed. As expected, when there is confidence about the chance to escape, *suspense* in all states is up to moderately downhill relation to *hope* for group *A* ($\rho_0 = -0.534$, $p < 0.001$, $\rho_1 = -0.659$, $p < 0.001$, $\rho_2 = -0.445$, $p < 0.01$), group *B* ($\rho_0 = -0.383$, $p < 0.01$, $\rho_1 = -0.687$, $p < 0.001$, $\rho_1 = -0.491$, $p < 0.01$) and group *C* ($\rho_0 = -0.619$, $p < 0.001$, $\rho_1 = -0.311$, $p < 0.01$, $\rho_2 = -0.231$, $p < 0.2$). An exception can be observed in group *C*, where the relation between *suspense* and *hope* significantly decreases, until a non-significant weak relation is reached.

Paradoxically, the situation is reversed when the victim has no chances and the participant has experienced in the game. In that case, reported *suspense* is not related to the low *hope* for surviving. Thus, for group *A*, correlation between both variables is equally moderate, but close to the limit of significance as opposed to the case where chances exist ($\rho_0 = 0.476$, $p < 0.05$, $\rho_1 = 0.415$, $p < 0.05$, $\rho_2 = 0.361$, $p < 0.1$). For its part, group *B* shows an oscillating non-significant relationship ($\rho_0 = 0.396$, $p < 0.1$, $\rho_1 = -0.297$, $p < 0.1$, $\rho_2 = -0.149$, $p < 0.6$). However, there is still an increasing downhill correlation for group *C* ($\rho_0 = -0.373$, $p < 0.05$, $\rho_1 = -0.466$, $p < 0.05$, $\rho_2 = -0.669$, $p < 0.001$), where the value grows as the states approximate to the death of the character.

Summarising, the observed values are in line with the proposed hypothesis; that is, the knowledge of rules influences the emotional aspects related to disinterest. If there is any possibility to escape, all three groups report a *high* value of *suspense* for all the states. Otherwise, group *A* reports a *none/low suspense* which gradually is reached by group *B* too. Emotion is only maintained by group *C*, where *very high* value is reported. This is mainly consistent with reported *hope* (*low*) when escape is still possible. If surviving is not possible, *suspense* and *hope* get alike for groups *A* and *B*, reaching *none/low*. Finally, in comparison to the other groups, *amusement* is lower for the group *A*, falling to *none* when it is impossible to escape; group *B*'s *amusement* is reported as *high* for all states. Likewise, group *C* reported a *very high* level for states *1* y *2*.

4   Discussion

Despite these results, some games based on the systematic death of the character have been very successful. A well-known example is *Demon Souls* [7] and similar games of *From Software*'s, in which the character can easily die [16, p. 17]. While some authors relate this difficulty to a discursive context of ubiquitous self-surveillance [15, p. 2], criticism about the excessive difficult can be easily found on the Internet. In our opinion, this criticism is related to the same emotional mechanism that provokes the *long path of frustration*. Moreover, we defend that too much and continuous deaths directly affect to the suspension of disbelief, which is a key point for player engagement.

In addition, the experiment has only been run with a single type of killer (a male masked Jason Voorhees style taken from classical slasher movies [14]) and environment (in a forest, as archetype for Gothic games [13, p. 103]). The literature, however, reports on emotional differences when the gender, environment and the aspect of characters vary [4,5,14]. Accordingly, differences could be expected due to participants' gender. This limit is aggravated by the fact that the percentage of female participants is low, which does not yield sufficient data to discard the influence of genre. Furthermore, it was not possible to find a woman with experience in *Dead by Daylight* among the available participants.

We need also to emphasize that the effect of suspense in experimental prototypes and practical narrative are often not the same. Stories developed through research projects generally create much shorter and less intense narrative experience than films, novels or story-centered commercial games do [17, p. 338]. We have taken this effect into account to interpret the results adequately.

5   Conclusions and Future Work

We have introduced the idea of the *long path of frustration* as an emotional state of the player. This state arises during the period between the certainty that current events lead irremediably towards the character's fail and the failure itself, due to the lack of uncertainty because of his knowledge of the game rules.

A preliminary analysis based on published gameplays of the game *Dead by Daylight* has evidenced that between 8% to 20% of gameplays end up with disconnections, presumably to try to avoid this frustration. The harder the antagonist plays, the more disconnections are observed. After this analysis, visualization of gameplays were compared, taking into account the previous experience of the participants and their knowledge about the rules.

Results show that experience and knowledge significantly affect the perception of suspense, the hope that the player will achieve his goal and, finally, the audience amusement. The conclusions of this study reflect that the audience has more fun when rules are unknown. In these cases, when viewers notice the impossibility of overcoming the situation, suspense and amusement grows significantly, in contrast with the cases of experimented audience. The experiment also reveals a relation between this decrease and the number of disconnections.

Based on the evidences and the objective of this study, we are working on a proposal of game design that takes the *long path of frustration* into account as an inherent part of playability. We aim to reduce its emotional effect and, consequently, players' disengagement.

Acknowledgements. This work has been supported by the Andalusian Government under the University of Cadiz programme for Researching and Innovation in Education 2015/2016 (SOL-201500054211-TRA); by the IDiLyCo project (TIN2015-66655-R) funded by the Spanish Ministry of Economy, Industry and Competitiveness; and by the projects WHIM 611560 and PROSECCO 600653 funded by the European Commission, Framework Program 7, the ICT theme, and the Future and Emerging Technologies FET program.

References

1. Aylett, R.: Emergent narrative, social immersion and "storification". In: Proceedings of the 1st International Workshop on Narrative and Interactive Learning Environments, pp. 35–44 (2000)
2. Bakkes, S., Spronck, P., van den Herik, J.: Rapid and reliable adaptation of video game AI. IEEE Trans. Comput. Intell. AI Games 1(2), 93–104 (2009)
3. Behaviour Interactive. Dead by Daylight (2016). http://www.deadbydaylight.com/
4. Belmi, P., Neale, M.: Mirror, mirror on the wall, who's the fairest of them all? thinking that one is attractive increases the tendency to support inequality. Organ. Behav. Hum. Decis. Process. 124(2), 133–149 (2014)
5. Clover, C.J.: Her body, himself: Gender in the slasher film. In: Her body, Himself, chap. 6, pp. 91–133 (1987)
6. Freytag, G.: Freytag's Technique of the Drama: An Exposition of Dramatic Composition and Art. Scott, Foresman, Chicago (1894)
7. From Software. Demon's Souls (2009). http://www.jp.playstation.com/scej/title/demons-souls/
8. Gun Media, IllFonic. Fryday the 13th: the game (2017). http://f13game.com/
9. Kojima Productions. Metal Gear V: the Phantom Pain (2015). https://www.konami.com/mg/mgs5/
10. Korhonen, H., Koivisto, E.M.: Playability heuristics for mobile games. In: Proceedings of the 8th Conference on Human-Computer Interaction with Mobile Devices and Services, pp. 9–16. ACM (2006)
11. Lewis, M.L., Weber, R., Bowman, N.D.: "They may be pixels, but they're MY pixels:" Developing a metric of character attachment in role-playing video games. CyberPsychol. Behav. 11(4), 515–518 (2008)
12. Mimimi Productions. Shadow Tactics: Blades of the Shogun (2016). http://www.mimimi-productions.de/game/shadow-tactics/
13. Müller, I.: Gaming after dark-visual patterns and their significance for atmosphere and emotional experience in video games (2011)
14. Sapolsky, B.S., Molitor, F., Luque, S.: Sex and violence in slasher films: Reexamining the assumptions. Journalism Mass Commun. Q. 80(1), 28–38 (2003)
15. van Nuenen, T.: Playing the panopticon: procedural surveillance in dark souls. Games Cult. 11(5), 510–527 (2016)
16. Vaz, B.: La convergencia ludo-narratológica en el diseño de videojuegos de rol: las mecánicas jugables de Dark Souls contra el diagrama de flujo de Mass Effect. Razón y Palabra 92(diciembre-marzo), 1–28 (2015)
17. Wei, H.: Structuring narrative interaction: what we can learn from *heavy rain*. In: Si, M., Thue, D., André, E., Lester, J.C., Tanenbaum, J., Zammitto, V. (eds.) ICIDS 2011. LNCS, vol. 7069, pp. 338–341. Springer, Heidelberg (2011). doi:10.1007/978-3-642-25289-1_43
18. Zillman, D.: Anatomy of suspense. In: The Entertainment Functions of Television, pp. 133–161. Psychology Press (1980)

Optimising Humanness: Designing the Best Human-Like Bot for Unreal Tournament 2004

Antonio M. Mora[1](✉), Álvaro Gutiérrez-Rodríguez[2],
and Antonio J. Fernández-Leiva[2]

[1] Departamento de Teoría de la Señal, Telemática y Comunicaciones,
ETSIIT-CITIC, Universidad de Granada, Granada, Spain
amorag@ugr.es
[2] Departamento de Lenguajes y Ciencias de la Computación,
Universidad de Málaga, Málaga, Spain
alvarogutirodri@hotmail.com, afdez@lcc.uma.es

Abstract. This paper presents multiple hybridizations of the two best bots on the BotPrize 2014 competition, which sought for the best human-like bot playing the First Person Shooter game Unreal Tournament 2004. To this aim the participants were evaluated using a Turing test in the game. The work considers MirrorBot (the winner) and NizorBot (the second) codes and combines them in two different approaches, aiming to obtain a bot able to show the best behaviour overall. There is also an evolutionary version on MirrorBot, which has been optimized by means of a Genetic Algorithm. The new and the original bots have been tested in a new, open, and public Turing test whose results show that the evolutionary version of MirrorBot apparently improves the original bot, and also that one of the novel approaches gets a good humanness level.

1 Introduction

Most of modern videogames are designed to provoke intense feelings on the player. To this end, many times they include non-player characters (NPCs) who try to empathize with the human, showing human-like behaviors and feelings. Thus, they implement internally an Artificial Intelligence (AI) engine focused on their *humanness*, i.e. including a set of rules to guide their actions similar to those that a real human would follow. It is also called *believability* of the NPCs.

Thus, the final aim is that those characters would be able to pass a Turing test [1] inside the game, which would mean that the human player could lead to think that they are 'real humans'. In this line, some years ago the *2K Botprize Competition* arose. It had as objective to find the best human-like NPC (or Bot) for the First Person Shooter (FPS) game Unreal Tournament™ 2004 [2], also known as UT2K4. That game, in its DeathMatch mode (combats between two or more bots trying to defeat the opponents an survive), was considered as the scenario for evaluating the bots in order to pass an adapted version of the Turing Test inside the game.

© Springer International Publishing AG 2017
I. Rojas et al. (Eds.): IWANN 2017, Part II, LNCS 10306, pp. 681–693, 2017.
DOI: 10.1007/978-3-319-59147-6_58

In the last edition of the competition, 2014, the two first Bots obtained very good humanness levels. They were named *MirrorBot* [3] and *NizorBot* [4].

This paper continues a previous work [5] in which those bots were analyzed and their advantages and weaknesses were identified. In that study several ways for optimizing and combining those bots were suggested, so, in the present paper some of them have been addressed. Thus, here we describe two different hybrid approaches which combine the best parts of every bot with different features from the other one. Moreover, an evolutionary-based improvement of the winner of the competition, MirrorBot, is also presented.

The bots have been tested using an open/public and online Turing Test, in which anonymous people have judged the humanness of the bots in many different videos showing battles.

2   State of the Art

The objective of creating human-like or believable characters in videogames is to show a life-like behaviour, including features such as personality, emotions, empathy or almost-real movements. There are normally interactions between the controlled characters in the game, so, the aim is then to show the illusion that these virtual players are controlled by a human [6]. This is a very important issue in current games, because this can enhance, for instance, the immersion of the player in the game and thus, his/her satisfaction [7].

However, evaluating the level of humanness that a virtual player exhibits is quite difficult, as it is normally a subjective measure. Thus, a way to evaluate this 'believability' can be the Turing test [1], or, currently, its adaptation to the scope of videogames [6] such as the 2K Botprize Competition (see Sect. 3).

Nevertheless, modelling a credible human-like behaviour is a very hard task, since it is not as simply as following a predefined set of states or mathematical formulae. So, the usual solutions [8] are focused on 'simulating' typically human actions, such as medium or high-level effectiveness in playing, make somehow unexpected mistakes from time to time, take different decisions even in the same conditions (with a stochastic factor), and show any kind of 'emotion'.

FPS games are one of the most considered scenarios for Turing test in videogames. Thus, there have been several proposals of human-like agents in this scope. From the SOAR Bot for Quake, presented by Laird [9] in 2000, which modelled a human-like behaviour through the so-called cognitive architecture. Choi et al. in [10] improved that architecture in an autonomous agent for the game Urban Combat. Their enhanced version was able to use knowledge and learn, by means of memories, such as skill or prioritized list of goals the agent should attempt to achieve.

However, the most extended environment for human-like bots in FPSs has been Unreal Tournament^{TM} 2004 game (UT2K4). Several proposals in this scope have applied a variation of Evolutionary Algorithms (EAs). For instance [11] which implemented evolution and co-evolution techniques, or [12] in which an evolutionary rule-based system was used. Schrum et al. [13] considered the combination of EAs with Artificial Neural Networks (ANNs) in order to learn to play as a human by imitating players' traces. Or the proposal by Soni and Hingston [14] which tried to imitate human's behaviour by means of ANNs.

Finally, the authors presented in [15] a bot which modelled the behaviour of an expert Spanish human player. It included a two-level FSM, in which the main states define the high level behaviour of the bot (such as attack o retreat), meanwhile the secondary states (or substates) can modify this behaviour in order to meet immediate or necessary objectives (such as taking health packages or a powerful weapon that is close to the bot's position). This approach was improved through a parameter-tuning made by means of an EA.

As stated before, this paper tries go a step further to the creation of the best human-like bot through the hybridization of two of the best bots in the last BotPrize Competition, and also enhancing the winner of that competition applying an evolutionary method.

3   Botprize Competition: A Turing Test for Bots

This test is a variation of the classical Turing Test in which a human judge who looks and interacts with a virtual world (a game), must to distinguish if the other actors (players) in the game are humans or bots. This test was proposed in order to advance in the fields of Artificial and Computational Intelligence in videogames, since a bot which can pass this test could be considered as excellent, and could increase the quality of the game from the players' point of view. Moreover the test tries to prove that the problem of AI for videogames is far from being solved.

The Turing Test for bots is focused on a multiplayer game in which the bot has to cooperate or fight against other players (humans or bots), making the same decisions that a human would take. This was transformed into an international competition with the features:

- The Deathmatch mode is considered in rounds of 10 min.
- There will be (ideally) three players: a human player, a bot and a judge.
- The bot must 'simulate' to be more human than the human player and both of them receive an independent mark.
- The three players cannot be distinguished from 'outside' (even with a random name and appearance).
- Bots cannot have omniscient powers as in other games. They can just react to the same stimuli (caught by means of sensors) than a human player.

In 2008 it was held the first 2K BotPrize competition (BotPrize from now on), in which UT2K4 was considered as the 'world' for this test. The participants should create their human-like bots using an external library to interact with the game by means of TCP connections (Pogamut [16]).

In the first editions of Botprize (2008 to 2011) the marks of the bots were not able to overcome to any of the human players. Anyway, the maximum humanness score for the human players was just 41.4%. This demonstrates the limitations of the test (or the competition), since even appraise a human behaviour is a quite complex task.

The first two bots in 2014 edition of Botprize were MirrotBot [3] (which also won 2012 edition) created by Mihai Polceanu; and a proposal by J.L. Jiménez and two of the authors of this study, *NizorBot* [4].

In the original competition, a number of judges that participated directly in the matches were responsible with the evaluation of humanness of bots; this means a First Person Assessment (FPA). In the edition of 2014, a Third Person Assessment (TPA) was also included by means of the participation of (external) judges via a crowdsourcing platform. The humanness (H) was evaluated according to the following formula:

$$H = (FPA * FP_{wf}) + (TPA * TP_{wf}) \tag{1}$$

where FP_{wf} and TP_{wf} are weighting factors (ranging in $[0.0, 1.0]$) for FPA and TPA respectively. For the 2014 edition, $FP_{wf} = TP_{wf} = 0.5$.

The results of 2014 Competition are plotted in Fig. 1. As the results figures show, MirrorBot was very close to completely pass the Turing test proposed in that edition, which was a new and harder evaluation system, it does no reach the value for being consider human (i.e., 0.5) although is relatively close to it. NizorBot showed also a very good performance finishing in the second position, obtaining a humanity factor relatively close to be considered as human.

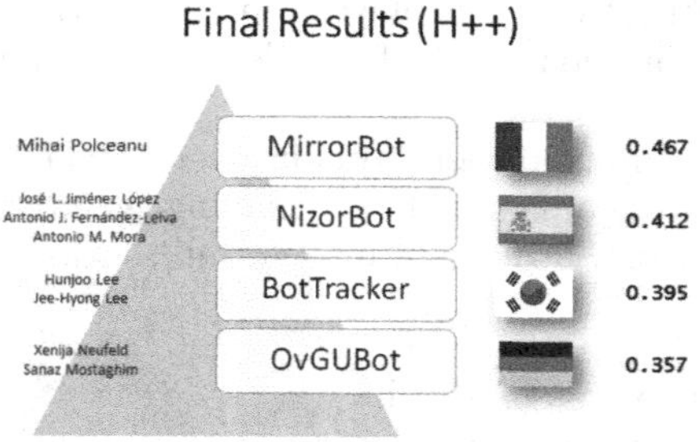

BotName	FPA	TPA	H++
Xenija	0.17139763	0.8235294	0.4974635
MirrorBot	0.20164771	0.7333333	0.4674905
Player	0.19328127	0.6315789	0.4124301
tmchojo	0.17757519	0.6470588	0.4123170
NizorBot	0.11821633	0.7058824	0.4120493
BotTracker	0.20070203	0.5909091	0.3958056
CCBot	0.06214746	0.7058824	0.3840149
Juan_CVC	0.12372294	0.6190476	0.3713853
OvGUBot	0.10545765	0.6086957	0.3570767
ADANN	0.08351664	0.4761905	0.2798536

Fig. 1. Results of Botprize 2014. Yellow cells are the best competitors whereas blue ones where humans. (Color figure online)

4 MirrorBot and NizorBot

This section presents the two considered bots in our study as the basis of the created hybridizations.

4.1 MirrorBot

MirrorBot [3] was developed in 2012, specifically for the 2K BotPrize competition, and was submitted again for the 2014 edition, which it won. It is based in two main behavioural modules:

– *Default module*: used frequently to navigate through the map, gathering items and weapons, shooting to enemies and avoiding their attacks. It is composed by several submodules. The three main ones let the bot *aiming* automatically to enemies computing trajectories or to a point in the path to simulate anticipation; *navigating* applying a modified version of the standard graph navigation which adds some kind of 'noise' or distortion to the movement, in order to hide bot-like displacement; and finally, *shooting* to the most appropriate enemy taking into account its weapon type, splash damage and distance to it.

– *Mirroring module*: which is only activated when an enemy is considered as unaggressive (every opponent is considered like this by default, until it shoots to the bot). The reason is that the enemy is probably a (human) judge inside the game (FPA). When the mirroring behavior is activated for a target, MirrorBot will begin recording all observable low-level actions of the opponent: aim, movement, fire, jumping, crouching and weapon choice. These are stored as frames in a sequence, which are to be replayed by MirrorBot itself. The orientation is inverted and movement maintains a constant distance to the target. Additionally a delay is introduced in the sequence in order to cheat the judge looking at this bot.

4.2   NizorBot

NizorBot [4] was based on the idea shown in the aforementioned ExpertBot [15], which modelled the behavior of an expert human player using a two-level finite state machine (FSMs).

ExpertBot was formed by two layers: The first one is the *cognitive layer*, in charge of controlling the FSM taking into consideration the environmental stimuli (perceived by sensors). It decides the transitions between states and substates using the expert system and the knowledge database. The second one is the *reactive layer*, which does not perform any kind of reasoning, and just reacts immediately to events during the match.

NizorBot is an implementation over ExpertBot by applying an Interactive Evolutionary Algorithm (IEA) [17], in which human experts guide the optimization of the bot's parameters in order to obtain a human-like bot. The basic idea is to let the experts rule out those candidate solutions (i.e., individuals) that perform *subjectively* worse than others from the point of view of humanness. More specifically, every **individual** in the IEA is a chromosome with 26 genes, divided into 6 blocks of information. Each block represents the behavior of a specific feature of the bot: distance, weapon selection, weapon priority, profile, risk, time.

The **fitness function** to evaluate the individuals is a combination of enemy kills (frags), number of own deaths, and the damages dealt and received by the bot. The function rewards the individuals with a positive balance (more frags than deaths) and a high number of frags. In addition, individuals which deal a high amount of damage to enemies are also rewarded, even if they have not got a good balance.

The *evaluation of an individual* consists of setting the values of the chromosome in the NizorBot AI engine, then a 1 vs 1 combat is launched between this and a standard UT2K4 bot at its maximum difficulty level. Once the time defined for the match is finished, the summary of the individual (bot) performance regarding these values is considered for the fitness computation.

Regarding the genetic operators considered, a *probability roulette wheel* has been used as **selection mechanism**, with *5 elitism*. **Uniform crossover** operator is applied, so that every gene of a descendent has the same probability of belonging to each one of the parents.

The **interaction of the game expert** has been conducted at some specific points of the evolution, where the expert should conduct a TPA (watching a video of the bot) and identify those specific features (e.g. distance selection, weapon selection, etc.) that they consider more human-like in the bot. Then, the gene blocks associated to the selected features are 'blocked' so that they are not altered by the genetic operators during the evolution. This affects the rest of the population when this individual combines and spreads its genetic information. This interaction guides the search to find more human-like individuals.

5   Hybrid Bots

Two bots have been proposed in this paper as hybrid approaches of MirrorBot and NizorBot. Regarding this one, we have used the parameter setting of the best individual obtained in our previous work [4], after the whole interactive evolutionary process.

Each hybrid bot combines the best part of one of the two reference bots with a complementary part from the other. These are namely:

- **MIRZorBot**: This bot combines the best parts of MirrorBot, namely, the navigation module (target selection, pathfinding, aiming and movement), and the mirroring ability (mirroring module). As these have been considered as the key of MirrorBot's human-like behaviour, after a deep analysis. The aiming and movement seem to be conditioned by the human perception in the game. The aiming is not always perfect. The mirroring module adds an unexpected - but close-to-natural - behaviour.

  These modules and submodules have been included inside the NizorBot's FSM, which also adds the expert weapon selection system - very proficient as it was designed by a human expert taking into account many weapon-related parameters -. Moreover, the division into primary and secondary states seems to be very close to the actual human's priorities in the game.

- **NIZRorBot**: This bot uses almost all the internal structure of NizorBot, however it includes the navigation module of MirrorBot, as it was the weakest part of the initial ExpertBot [15]. Thus, NIZRorBot makes use of the navigation module with all the 'tricks' that MirrorBot implements, including an improved implementation of Pogamut's Navigation Mesh (with optimal obstacle avoidance) and an own RayCasting system based in 24 rays: 16 for

the detection of horizontal collisions (direct obstacles, items, weapons, enemies), and 8 for vertical collisions (45° for the detection of non-floor, holes or
falls).

6   Evolutionary MirrorBot

In addition to the two hybrid proposals, here we present an improvement of the
initial MirrorBot, *EVOMirBot*. It is based on a parameter tuning or optimization
by means of a classic Genetic Algorithm (GA) [18].

The aim is to enhance the overall behaviour of the original bot, in order
to show a more 'offensive profile' (i.e. being more aggressive), since sometimes
it just waits for the opponent's actions and do not react properly, i.e. as a
human would do (just moves around the rival). To this end, the hand-coded
parameters on which it depends the behaviour of this bot, have been identified
and 'extracted', in order to compose a chromosome or individual for the GA.

Every individual in the GA is a chromosome with 12 genes, namely:

- *Gene 1* (initial value = 8136): time devoted to imitate, once a potential
  human player has been identified for mirroring. It is measured in milliseconds
  in the range [1000,10000].
- *Gene 2* (initial value = 2982): time to consider a target enemy as lost.
  Milliseconds in the range [1000,9000].
- *Gene 3* (initial value = 7): aggressiveness level of the enemy. It the value
  is greater than this gen, the opponent is discarded for mirroring (it is too
  aggressive for being a human). Value in the range [1,10].
- *Gene 4* (initial value = 115): imitation delay when the mirroring module
  is reproducing the recorded movements of the opponent. This value is very
  important to 'guide' the observer's impression regarding the behaviour the
  bot is showing. Value in milliseconds in the range [0,500].
- *Gene 5* (initial value = 842): last time the opponent to be imitated was
  seen. This value represents the end of the imitation flow. It is measured in
  milliseconds and in the range [0,1000].
- *Gene 6* (initial value = 5): voting value to consider an opponent to be imitated. If it receives more than this number, it will become mirrored.
  Value in the range [1,7].
- *Gene 7* (initial value = 2000): average distance between the bot and the
  candidate rival to be imitated. It is adjusted in order to observe the other bot
  without being attacked by it. Value in the range [1200,2000].
- *Gene 8* (initial value = 8211): time considering an opponent as 'nemesis', i.e.
  the bot will attack it as soon as it is detected. Value in milliseconds in the
  range [1000,9000].
- *Gene 9* (initial value = 2142): time to forget the list of nemeses. Value in
  milliseconds in the range [1000,5000].
- *Gene 10* (initial value = 300): time for uncontrolled (or pseudo-random)
  shooting. This is done to show an unexpected behaviour from time to time.
  Value in milliseconds in the range [100,500].

- *Gene 11* (initial value = 3500): distance considered as far from the enemy. Value in [500,5000].
- *Gene 12* (initial value = 600): distance considered as short from the enemy. Value in [100,800].

The list shows the initial values that MirrorBot had set, as a reference. As it can be seen, the ranges have been defined so a great variation of MirrorBot could be obtained through evolution.

The *fitness function* is defined as:

$$f(fr, d, dmgG, dmgT) = ((fr * 50) - (d * 5)) + (dmgG - dmgT/10) \quad (2)$$

where fr is the number of enemy kills the bot has obtained (frags), d is the number of own deads, $dmgG$ is the total damage produced by the bot, and $dmgT$ is the total damage it has received. As in NizorBot, this function rewards a lot individuals with positive balances, i.e. more kills than deads and more produced than received damage, aiming to obtain the aforementioned 'offensive profile'.

The *evaluation of an individual* is done setting the values of the chromosome in the MirrorBot's AI code and then running a 1 vs 1 Deathmatch against NizorBot in UT2K4 during 1 min. Once the battle finishes the fitness is computed considering the performance of the bot.

A *probability roulette wheel* has been used as selection mechanism, considering the fitness value as a proportion of this probability. In addition, a *Stationary* replacement policy has been conducted, so just the worse individual is replaced every generation. Finally, *uniform crossover* has been applied, and the mutation generates a random value in the corresponding interval of the parameter.

7   Experiments and Results

This section analyzes the obtained results, first regarding the evolutionary approach of MirrorBot, and by means of a Third Person Assessment Turing Test.

7.1   Evolutionary Optimization

In this experiment the so-called EVOMirBot has been obtained as an optimization of MirrorBot. The parameter setting has been: 30 individuals, 50 generations, 1/12 of mutation probability. The evaluation has been conducted as a 1-minute 1 vs 1 combat against NizorBot, always in the map DM-TrainingDay (frequently used in the UT2K4 competitions). 10 runs have been conducted.

The evolution of the fitness for all the runs is plotted in Fig. 2.

As it can be seen, there is an improvement tendency on the average best fitness along generations. However it is a bit 'slight', due to the *noisy* nature of the problem [19], i.e. an individual can be valued as good in one combat, but the same bot yield very bad results in another match. This happens due to the high pseudo-stochastic component present in these battles, since the results do

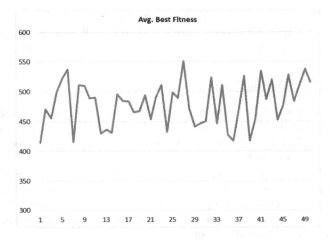

Fig. 2. Evolution of the average of the best fitness per generation considering 10 runs.

not depend completely on our bot, but also on the enemy's actions which we cannot control.

Thus, we selected as the definitive EVOMirBot one individual from the last generations. It was not the one with the highest fitness value, but a bot which showed the most human-like behaviour from our point of view (we checked several combats of each candidate bot). The obtained optimized values for the parameters of this bot are presented in Table 1.

Table 1. Gene values for the final EVOMirBot

Gn1	Gn2	Gn3	Gn4	Gn5	Gn6	Gn7	Gn8	Gn9	Gn10	Gn11	Gn12
6231	5364	1	25	612	2	1560	3194	2880	447	3398	245

Looking at the results, we can remark that the aggressiveness of the Mirror-Bot has been increased overall in EVOMirBot. For instance *Gene 2* (changed from 2982 to 5364) means that the bot will 'remember' for longer time its nemesis, *Gene 8* (value 3194 instead of 8211) will affect the time the bot considers another one as its nemesis. *Gene 12* (changed to 245 from 600) will turn the measure for short distances, which will lead to a closer combat style.

Regarding the imitation ability of MirrorBot, its has been enhanced or better adapted for the game. Thus, for instance the value for *Gene 1* is much lower than the original (6231 instead of 8136), which means than the bot will be less time imitating the opponent. *Gene 3* new value (1 instead of 7) will lead to a lower imitation rate, but the opponents, considered as bots will be attacked more frequently, which is more recommended in this game. The new value of *Gene 5* (612 instead of 842) will mean that the imitation will start sooner. *Gene 6* (value 2 instead of 6) will affect the probability of choosing candidates for imitation. The new value of *Gene 10* (447 instead of 300) will lead to EVOMirBot to conduct more random or imprecise shoots, which probably will increase the perception of an inaccurate player which would be more likely to be a human.

7.2 Open TPA Turing Test

In this experiment the three new bots (MIRZorBot, NIZRorBot, and EVOMir-Bot) and the originals NizorBot and MirrorBot have been evaluated in an open Turing Test based in a Third Person Assessment (TPA) on the website http://1-dot-proyecto-tfg.appspot.com/.

To this end 7 different videos, of 20 s each, have been recorded and presented to the 'voluntary judges' (whoever has accessed the web). Every video shows a short combat stage between two players, being each of them one of the bots or a human. Thus, every bot has participated in two videos: one against another bot and one against a human player.

After a video, the judge must decide about the humanness of the contenders, with the options: *(a) player 1 is human, (b) player 2 is human, (c) both are humans, (d) none is human, (e) not sure, (f) wrong test.*

The test has been open during three weeks, and 61 judges have participated. We have considered the votes for every bot as human, and also the votes to both of them. The rest are omitted in the computation of humanness level for every bot. This value has been calculated as the number of votes received divided by 122, which is the maximum number of votes that a bot could receive (61 votes × 2 videos in which the bot is present). The results are shown in Table 2.

Table 2. TPA Turing Test results.

Bot	Votes as human	Humanness
NizorBot	57	46.72
EVOMirBot	53	43.44
MIRZorBot	51	41.80
MirrorBot	47	38.52
NIZRorBot	46	37.70

It can be seen high voting values for all the bots, which is a good sign of their human-like behaviour.

The most remarkable fact that we can see in these results is the (apparent) improvement that EVOMirBot have meant with respect to MirrorBot. The first has obtained the second best results in the test, just behind NizorBot, which has 'won'. MIRZorBot has also obtained good results, however NIZRorBot has been the worse. The reason for poor performance is probably that the raycasting system or MirrorBot has been in conflict with the target selection method of the original NizorBot, which has meant a non-proper movement behaviour.

However, looking at the whole figures in the results, we think that there is place for improvement in the videos, such as their duration (maybe too short), or the point of view which is not probably the best to evaluate the opponent. But in this kind of open test, it is very important to reach an accurate number of videos with an accurate duration, in order to avoid tiredness or disappointment in the judges.

8   Conclusions and Future Work

This paper has presented three different approaches for human-like bots for the First Person Shooter Unreal Tournament 2004. All of them have been obtained as variation/enhancement of the two first bots in the last 2014 edition of the 2K Botprize Competition (a Turing test for bots): MirrorBot and NizorBot.

These are *MIRZorBot* (based on MirrorBot with some components of Nizor-Bot), *NIZRorBot* (structure of NizorBot with movement module of MirrorBot), and *EVOMirBot* (evolutionary optimization of MirrorBot).

In the results we have firstly analyzed the obtained improvement of MirrorBot by means of a Genetic Algorithm, paying attention to the new values for the parameters and their influence on the bot's behaviour, getting, in summary, a more aggressive bot.

Then, an open and online Turing Test has been conducted as a Third Person Assessment, so the voluntary judges have revised some videos of the bots fighting in the game and decided about who is the human (if there is any in the match). The results of this test yield two main conclusions: EVOMirBot seems to be a real improvement of MirrorBot, and MIRZorBot has obtained a very good humanness level. NIZRorBot has got a worse value, but the reason could be an incompatibility between one of the modules with a raycasting system, which we will solve in the near future.

Other future lines of work will be dealing with the noise in the evolutionary process (evaluation function), in order to get a better improvement progression. In addition, taking into account the voting results in the Turing test, there have been some 'not sure' or 'wrong test' votes, which lead us to think that the videos must be improved maybe better focusing on every bot, with a longer duration or reducing the vote to just one bot per video.

Acknowledgements. This work has been supported by MINECO project EPHE MECH (TIN2014-56494-C4-1-P, 3-P), and KNOWAVES (TEC2015-68752) (MICINN and FEDER), and Universidad de Málaga (Campus de Excelencia Internacional Andalucía Tech). The authors are very grateful to Mihai Polceanu and José L. Jiménez, authors respectively of MirrorBot and NizorBot, for providing us their source code and their support for the development of this work.

References

1. Turing, A.M.: Computing machinery and intelligence. Mind **59**(236), 433–460 (1950)
2. Unreal tournament (2014). http://www.unrealtournament.com/
3. Polceanu, M.: Mirrorbot: using human-inspired mirroring behavior to pass a turing test. In: 2013 IEEE Conference on Computational Intelligence in Games (CIG), pp. 1–8. IEEE (2013)
4. Jiménez, J.L., Mora, A.M., Fernández-Leiva, A.J.: Evolutionary interactive bot for the FPS unreal tournament 2004. In: Camacho, D., Gómez-Martín, M.A., González-Calero, P.A. (eds.) Proceedings 2nd Congreso de la Sociedad Española para las Ciencias del Videojuego, Barcelona, Spain, 24 June 2015. CEUR Workshop Proceedings, vol. 1394, pp. 46–57. CEUR-WS.org (2015)
5. Polceanu, M., Mora, A.M., Jiménez, J.L., Buche, C., Leiva, A.J.F.: The believability gene in virtual bots. In: Proceedings of the Twenty-Ninth International Florida Artificial Intelligence Research Society Conference, FLAIRS 2016, Key Largo, Florida, 16–18 May 2016, pp. 346–349. AAAI Press (2016)
6. Livingstone, D.: Turing's test and believable AI in games. Comput. Entertainment **4**(1), 6 (2006)
7. Soni, B., Hingston, P.: Bots trained to play like a human are more fun. In: IEEE International Joint Conference on Neural Networks, IJCNN 2008 (IEEE World Congress on Computational Intelligence), pp. 363–369 (2008)
8. Yannakakis, G., Togelius, J.: A panorama of artificial and computational intelligence in games. IEEE Trans. Comput. Intell. AI Games **7**(4), 317–335 (2015)
9. Laird, J.E.: It knows what you're going to do: adding anticipation to a quakebot. In: AAAI 2000 Spring Symposium Series: Artificial Intelligence and Interactive Entertainment SS-00-02 (2000)
10. Choi, D., Könik, T., Nejati, N., Park, C., Langley, P.: A believable agent for first-person shooter games. In: Proceedings of the Third Artificial Intelligence and Interactive Digital Entertainment Conference, Stanford, California, 6–8 June 2007, pp. 71–73. The AAAI Press, USA (2007)
11. Priesterjahn, S., Kramer, O., Weimer, A., Goebels, A.: Evolution of human-competitive agents in modern computer games. In: IEEE World Congress on Computational Intelligence 2006 (WCCI 2006), pp. 777–784 (2006)
12. Small, R., Bates-Congdon, C.: Agent Smith: towards an evolutionary rule-based agent for interactive dynamic games. In: IEEE Congress on Evolutionary Computation 2009 (CEC 2009), pp. 660–666 (2009)
13. Schrum, J., Karpov, I., Miikkulainen, R.: Ut2: human-like behavior via neuroevolution of combat behavior and replay of human traces. In: 2011 IEEE Conference on Computational Intelligence and Games (CIG), pp. 329–336 (2011)
14. Soni, B., Hingston, P.: Bots trained to play like a human are more fun. In: IEEE International Joint Conference on Neural Networks, IJCNN 2008, pp. 363–369 (2008)
15. Mora, A.M., Aisa, F., García-Sánchez, P., Castillo, P.Á., Guervós, J.J.M.: Modelling a human-like bot in a first person shooter game. IJCICG **6**(1), 21–37 (2015)
16. Pogamut - virtual characters made easy (2014). http://pogamut.cuni.cz/main/
17. Takagi, H.: Interactive evolutionary computation: fusion of the capabilities of EC optimization and human evaluation. Proc. IEEE **9**, 1275–1296 (2001)

18. Goldberg, D.E.: Genetic Algorithms in Search, Optimization and Machine Learning. Addison Wesley, Redwood City (1989)
19. Mora, A.M., Fernández-Ares, A., Merelo, J.J., García-Sánchez, P., Fernandes, C.M.: Effect of noisy fitness in real-time strategy games player behaviour optimisation using evolutionary algorithms. J. Comput. Sci. Technol. **27**(5), 1007–1023 (2012)

Combining Neural Networks for Controlling Non-player Characters in Games

Ismael Sagredo-Olivenza[✉], Pedro Pablo Gómez-Martín,
Marco Antonio Gómez-Martín, and Pedro Antonio González-Calero

Dep. Ingeniería del Software e Inteligencia Artificial,
Universidad Complutense de Madrid, Madrid, Spain
{isagredo,pedrop,marcoa,pagoncal}@ucm.es

Abstract. Creating the behavior for non-player characters in video games is a complex task that requires the collaboration among programmers and game designers.

Usually game designers are only allowed to change certain parameters of the behavior, while programmers write new code whenever the behavior intended by designers cannot be achieved by just parameter tweaking. This becomes a time-consuming process that requires several iterations of designers testing the solution provided by programmers, followed by additional changes in the requirements that programmers must again re-implement.

In this paper, we present an approach for creating the behavior of non-player characters in video games that gives more power to the game designer by combining program by demonstration and behavior trees. Our approach is able to build some parts of a behavior tree with the observed data in a previous training phase.

1 Introduction

Game development is a multidisciplinary task that involves expertise from different areas with different knowledge and motivations, mainly artists, programmers and game designers. Our work specifically concentrates on the creation of behaviors for non-player characters (NPCs) from the point of view of the relationship between programmers and designers. In this context, to create a character behavior mainly two roles are involved: *programmers*, that write the code to control the character and *designers* that specify the way the character must behave in order to make the game fun to play.

Game AI development is a hard problem. Having characters that behave as intended is a challenge for both programmers and designers. On the one hand, game programmers deal with low-level algorithms, perception systems or behavior representation. On the other hand, designers envision the way their characters behave, and they represent this behavior using high level descriptions, usually in natural language [1]. Designers must have in mind the player experience and whether the final behavior are enjoyable to play with or not.

Supported by the Spanish Ministry of Science and Education (TIN2014-55006-R).

I. Rojas et al. (Eds.): IWANN 2017, Part II, LNCS 10306, pp. 694–705, 2017.
DOI: 10.1007/978-3-319-59147-6_59

In order to successfully accommodate both sides of the same problem, programmers and designers must work together. As designers do not usually have programming skills, implementation is mainly under programmers sole responsibility. Developing NPC AI is an iterative process where designers specify behaviors, programmers develop them, and designers test them.

In order to alleviate the process, programmers develop tools for designers that expose some parameters of their code and let designers to tweak them. Their goal is to ease the task of behavior development in such a way that designers, which usually are not good at programming, have enough control to fulfill their goals in terms of game interaction, while at the same time they do not have the power to break the system, introducing hard to find bugs.

Rule-based systems, finite state machines (FSMs) and scripting languages are examples of technologies used in game development for designing game AI. Over the last decade, however, behavior trees (BTs) [2, 3] have gained momentum. BTs have replaced FSMs for orchestrating non-trivial NPCs behavior in professional game development. Although BTs can be created through visual interfaces without writing code, they involve concepts that can be difficult for a game designer to fully grasp, such as parallel execution versus iteration, and for that reason they are primarily a tool for programmers. One of the goals of our work is to bring the expressive power of BTs to game designers, by letting them demonstrate the intended behavior and then editing the resulting BT.

In previous work, we have already tested the differences between programmers and designers when building BTs [4]. We could observe that although designers without programming skills can design some high level behaviors, the development time was higher and the quality was lower than those designed by programmers. In order to facilitate this process to designers, we propose to use Program by demonstration (which can also be found as *Learning from Demonstration* or *Learning by Imitation* in the literature) to let designers generate an NPC behavior just by playing the game. This process is accomplished in two steps: first game traces are recorded with the designer controlling the NPC to be programmed, and then those traces are processed using machine learning techniques.

The main contribution of our work is to provide a framework for NPC AI that can easily integrate a machine learning model induced from the traces with hand-coded knowledge represented in BTs. We have extended behavior trees with a new type of node, *Trained-query nodes* (TQN) which run a machine learning model inside of a BT.

In this paper we demonstrate how two different machine learning techniques, namely neural networks and k-nearest neighbor, can produce good results when learning simple behaviors but are not good enough for complex ones. We show how we can generate complex behaviors by combining models trained with simple behaviors which are integrated through hand-coded nodes in a BT.

The rest of the paper runs as follows. Next Section describes some related work on applying program by demonstration to generate NPC AI. Section 3 describes trained-query nodes, our extension to behavior trees that allow the combination of program by demonstration with hand-coded knowledge. Section 4 presents our experiments with different combinations of machine learning and behavior trees. Finally Sect. 5 presents some conclusions and future work.

2   Related Work

Significant work has been done under the label of *Program by demonstration* or *learning from demonstration* [5]. Program by demonstration (PbD) is a sub-field of machine learning that studies how to learn to perform a task by observing the behavior of an expert while doing the task that the system has to learn. In the context of game creation and the creation of behavior for NPCs, the key idea is to let designers play the game controlling the character that they want to design, and use the information gathered on these plays to automatically build the behavior of the character. We can use neural networks to learn a model from the traces, or we can use lazy machine learning techniques such as Case-Based Reasoning (CBR) or k-Nearest Neighbors (k-NN).

Programming by demonstration has been used in the literature for building AIs in multiple contexts, such us the one reported by Floyd et al. [6] that use it for training agents to play in RoboCup Simulation League by observing the play of other teams. Another example is provided by Ontañon et al. [7] that describes its use in a real time strategy game (RTS) named Darmok 2. They managed to learn plans while designers play the game and afterward, they used these plans to make automatic decisions, controlling the game with an autonomous agent using Case-Based Reasoning. Another use of program by demonstration and CBR was the one presented by Rubin and Watson [8] in the creation of a Poker-playing agent, and the AI created by Jaidee et al. [9] for moving and coordinating multiple agents in Wargus, a RTS game created for research purposes.

PbD has been combined with other techniques such as Hidden Markov Models [10] used to learn a model of the behavior of agents in real-time strategy games (RTS). In this direction, genetic programming is used to create the behavioral engine of bots to play RTS games in [11]. [12] contains a comparative study between PbD and other techniques employed to imitate human plays such as Dynamic Scripting [13] and Neuro-Evolutionary [14] using *Super Mario Bros* as a testbed.

A relevant feature of our approach is the use of behavior trees, which are *de facto* standard for representing complex AI in game industry. Our goal is to extend a technique already in use in industry. Also connecting with BTs is the work presented in [15] where an approach to generating BTs by demonstration using genetic algorithms is presented. The main difference with the approach presented here is that their goal is to generate the full behavior automatically from the traces, while we intend to integrate a learned model within a BT that can be further tweaked. This distinction is significant because we provide more control to the designer, something important in game industry where designers usually prefer controlled predictable behaviors instead of emergent unpredictable ones.

3 Trained Query Nodes

As explained before, designers envision and decide the behaviors of the NPCs, and programmers makes those ideas real. This requires an iterative process where implemented behaviors are refined until designers' intentions are fulfilled.

But this is not always easy. Designers and programmers usually have opposite interests and concerns, that affect even the way they communicate. A way to reduce communication errors and development time consists on easing the way behaviors are specified. Designers have a plethora of alternatives to create them without the tedious need of writing code, mainly using visual editors to create FSMs, BTs or data flow charts. However, the use of these behavior editors requires certain technical knowledge that not all designers possess.

To solve this problem, we present Trained-Behavior Trees, and approach that intends to help non-technical designers create behaviors in a more intuitive way, just by playing the game as the NPC would do. To get this, we propose a new internal node from the BTs named *Trained Query Node* (TQN). This node is an evolution of Query Node described in [16] where the ontology description is replaced by a model learned from traces generated by the designer. The node is able to observe the designer in a pure game session where he simulates the intended behavior of the NPC, and at the same time it takes snapshots of the state of the game and the task that the player is performing. The variables recorded in the training session are a subset of the variables recorded in the blackboard of the BT (a blackboard is the structure commonly used in BTs for sharing variables between nodes that contains, among others, perception information). This subset is selected by the designer and is serialized together with the task executed. The system is taking samples and storing them periodically while the designer is controlling the target character of the training.

The knowledge acquired in this training session, is later used at run-time when the player executes the game. The Trained Node selects one of the tasks from the available ones, using the knowledge acquired in the training session. Thanks to that, the designer can create behaviors without programming.

The approach is based on the existence of a set of basic tasks that the NPC is able to perform and that are provided by programmers. The complex behaviors created using the Trained Query Nodes use as primitive actions that basic tasks. Moreover, this task collection is also the set of options designers have available in the training session.

In essence, we propose the following methodology: the designer starts with the definition of the behavior of the game NPCs using very high level specifications, for example, natural language, tables, rules or diagrams. Using this description, designers and programmers identify and select the basic tasks of the NPC. These tasks are implemented by the programmers and later used by designers as *bricks* when building the behaviors using the BT editor. If the designer reaches a dead end and is not able to express himself using BTs, he can try to generate one by demonstration placing a Trained Query Node on the BT. After the training phase, designers may test the behavior learned in an usual game session to check if it fits his expectations and repeat the training phase or fine-tune the behavior. In that game session, the Trained Query Node retrieves the task more appropriately to execute taking into account the current state of the world. The methods we can use to infer this behavior can be several, but in the experiment nowadays the system can use a k-NN algorithm and a neural network.

On the one hand, we can choose a k-NN [17] as the strategy used by the TQN to select the action to perform considering the state of the world. This algorithm will select the most similar cases to the current state, comparing the environment parameters that were considered more relevant during sampling with the current ones. The current criteria available to select is the most popular task among the k selected, when $k > 1$ and the similarity measure used is a weighted Euclidean distance among the parameters recorded while training.

On the other hand, if we choose the neural network, the system implements a multilayer perceptron [18] with one hidden layer. In order to facilitate the network setup and the training process, the system only asks to the designer some parameters: the learning rate, the momentum of the inertia and the number of the hidden neurons. The system encodes all input values as real values between 0 and 1 and the output is encoded with as many neurons as possible outputs there, being the neuron with more amount of activation that will indicate the task to select. When the network training begins, the system shows the training and validation error, so that the designer can stop the training when it suffers from over-adapt to the training data. To do this the designer must provide a second set of data to validate the training. This second set can be generated as if it was a another training.

As the system is integrated into a BT, we can put the Trained-Query Node anywhere in BT or even put more than one. As we will show in the experiment described in the next section, having more than one TQN and training them separately brings great flexibility to the designer.

4 Experiment

The aim of the experiment is validating our mixed approach, where we embed program by demonstration into Behavior Trees, the behaviour modeling tool widely used in the game industry. Our initial hypothesis is that program by demonstration can help designers to make behaviors without programming, but when the behavior is complex, program by demonstration is not enough to achieve quality behaviors. Our mixed approach should achieve better results than using only program by demonstration.

In order to validate our hypothesis, we have carried out a series of experiments in a real domain, using an in-house developed game called TowoT as a testbed. TowoT is a tower defense game that combines tower defense mechanics with third-person action elements. With the aim of simplify the test environment, in the experiment the player can not place towers, because all of them are already put in place in the scene. The main NPC to train is the TowoT, a robotic character that accompanies the player. This character is slower than the player, but its firepower is bigger and from time to time, he need to recharge its energy in a charging station.

The mini-map shown on the lower, right-hand corner of Fig. 1 presents the charging station as a petrol pump icon, where TowoT must occasionally be recharged. The yellow spot shown at the bottom marks the position of the spaceship, that plays the role of the *base* or *core* present in virtually all tower defense games. The oil extractor tower points the location of the *refinery*, another strategic point that the player must defend.

Fig. 1. Training interface of TowoT (Color figure online)

4.1 Experimental Setup

In the experiments, the game was played following 4 strategies from the point of view of the Towot NPC. They are described below:

- Defensive: the Towot defends the core and the player (Jacob) attacks to enemies near their spawn points placed in left and right upper corner and protect the refinery. Trying to keep the Towot always with enough energy. The smart thing is to recharge the Towot between enemy waves.
- Offensive: the Towot defends the refinery, which is a position in the forefront trying to the robot can destroy the mostly enemies before they reach the core where the player is waiting for them. Of course, the Towot must keep its batteries recharged to be effective. The player can also attack the two enemy generators at the bottom of the map.
- Mixed strategy: the player can choose if each wave must attack or defend and Towot must choose the opposite strategy that chosen by the player.
- Complex: the TowoT always defend the refinery, unless Jacob (player) or the core is in danger (be surrounded by enemies). In that case, the Towot defends in order of priority: first the core and then Jacob.

Figure 2 shows the state of the mini-map in particular moments of some of them, where the small red icons represent the enemies. First wave comes from the generator in the upper-left corner of the map, second wave from the generator in the bottom-left, and third wave simultaneously spawns enemies from upper-left, upper-right and bottom-right. Enemies randomly select which element to attack but with certain probabilities. The target most likely to be attacked is the core, the next is the mining tower and finally the charging station.

Fig. 2. The three different enemy waves configured in the experiment. (Color figure online)

During training phase, TQN saves a subset of the behavior tree parameters, selected by designer in advance. We recorded one for use as training and another 10 for use as validation for each strategy. The BT used for training is shown in Fig. 3 where there is only a perception node to inject the parameters in the BT blackboard and the Trained Query Node which record the game and then will execute the learned behavior. The games recorded were different within the limits that describes each strategy. For example, in the mixed strategy, sometimes Jacob attacks at first and third wave and defend in the second one and in other games, Jacob defend at first and thirds waves and attack in the second one. In this way, we try to verify whether the different algorithms have a good capacity of abstraction. In each trace recoded, the number of the examples was about 2,000–3,500, depending on the game duration. Recordings are performed every 100 ms and include the parameters of the Table 1 and the task that the Towot was performed at each time.

As we said in the Sect. 3, TQN is able to execute different algorithms to achieve the behavior model. For this experiment, we have used k-Nearest Neighbor (k-NN) and a multilayer perceptron. On one hand, k-NN was configured with $k = 10$ and the algorithm selects the most repeated tasks among the k more similar. The k value was selected because in previous experiments, this value obtained a good results. The similarity measure used was the *weighted Euclidean distance* (weights are shown in the Table 1) and all values are normalized between 0 and 1. On the other hand, the multilayer perceptron used the *backpropagation* with inertia as learning algorithm, with only one hides layer, eight inputs and four outputs (one for every possible task to be done by Towot)

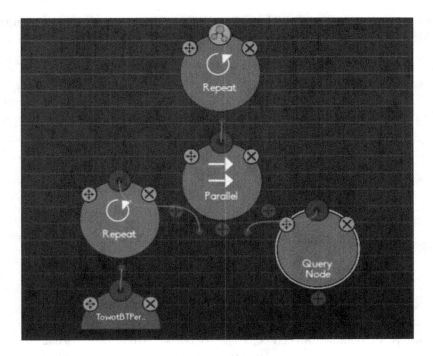

Fig. 3. BT used to training the Towot behavior

and the output neuron with more activation value determines the task to be selected. The *Sigmoid* function was the activation function and all values were normalized between 0 and 1. We perform different trainings of the multilayer perceptron for each strategy, with different number of hidden neurons and learning rate. The *momentum* was always 0.1.

For training the network, we use one of the recorded games of each strategy as training data and another as validation and it is the designer who, looking at training and validation error, must stop the network when it detects over-fitting. In the mixed strategy, the training data was the union of both defend and attack traces.

4.2   Experimental Results

Once trained the network, we use the ten plays recorded to give examples to the model and compare the task output of the network with the task recorded that was executed by the Towot in this environment. We made the same for the model built with k-NN. The results shown in Table 2 are the average and standard deviation of the hit rate of each algorithm in each strategy with k-NN with and without weights and neural network. Furthermore, in the neural network also shows the number of hidden neurons and the learning rate used to achieve it.

Table 1. Attributes recorded in the training with its ranges and weights (used in the k-NN)

Attribute	Range	Weight in k-NN
TowotEnergy	[0, 100]	2
RefineryRiskLevel	[0, 4]	1
CoreRiskLevel	[0, 4]	1
JacobRiskLevel	[0, 4]	1
CoreLife	[0, 1000]	0,25
RefineryLife	[0, 500]	0,25
JacobLife	[0, 500]	0,25
JacobDistanceToCore	[0, 130]	1

Table 2. Summary of experiment results

	Defensive	Offensive	Union 1 & 2	Complex
k-NN avg	0,939	0,9422	0,7468	0,7833
k-NN S.Desv	0,01	0,004	0,08	0,032
k-NN no Weights avg	0,91	0,945	0,676	0,779
k-NN no Weights S.Desv	0,006	0,007	0,054	0,032
Neural avg	0,9418	0,9465	0,78385	0,7536
Neural S.Desv	0,011	0,009	0,063	0,071
Hidden neurons	10	14	21	10
learning rate	0,1	0,15	0,2	0,1

The results show a similar performance between k-NN and neural network (except if k-NN does not use the weights where performance is significantly lower). Somewhat better in some cases using k-NN (in the fourth strategy, for example) and somewhat better using the perceptron (in the third strategy). But more interesting is to compare among different strategies. In the first and second strategy, both models have a high hit rates about 93 to 95%, but, the third and fourth strategy the hit rates are significantly lower. In none of the cases the differences are statistically significant (p-value = 0.05)

Thanks to our hybrid approach between program by demonstration and behavior trees, a possible solution to improve the hit rate in the third strategy is to use a few BT nodes to select between the defend and attack strategy. So we created a new behavior tree with two TQNs, one of then trained with a neural network with the defensive examples and another trained with offensive examples, and a priority selector that select one or another according to the player behavior. The new BT is shown in Fig. 4.

Those nodes are simulated in the experiments, creating two network models trained with defensive and offensive traces (the same trained in the strategy 1 and

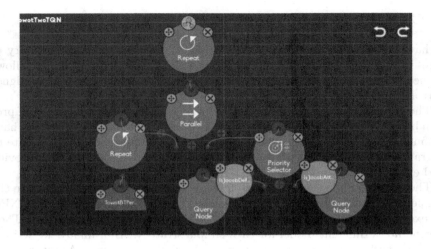

Fig. 4. BT where was merged both behaviors: Defensive and offensive

2 in the Table 2) and encoding the equivalent of the priority node by programming. As the Table 3 shows, the hit rate was increased noticeably, approaching to 84% on average. This difference is statistically significant with p-value = 0.05.

Table 3. Hit rate among k-NN, Neural network and hybrid PdB & BT.

Third strategy	k-NN	Neural	Hybrid PdB & BT
Union 1 & 2 Mean	0,7468	0,78385	0,839
Union 1 & 2 S.Desv	0,08	0,063	0,07

By testing the behaviors in the game, the mixed approach generally behaves as expected. The neural networks version had better results in terms of gaming experience than k-NN, because the latter had more oscillations. That is, sometimes when it should defend the refinery, the Towot move to defend for a few seconds the core and vice-versa. In the 10 training sessions that were carried out, with k-NN seven suffered from certain erratic behavior. With neural network only three of them suffer this problem and when using our hybrid approach joining BTs and PbD, only on one occasion the Towot tried for a few seconds to descend to defend the core when it should not.

With these results, we can conclude that our hybrid approach is a very good tool to create behaviors without programming. At the same time, we also give a lot of freedom so that designers can use different techniques to generate the behavior model. For example, we allow to divide the more complex behaviors into simpler sub-behaviors to train and learn by PbD, then to merge them by programming using BTs. This way many behaviors can be created by designers in an autonomous way.

5   Conclusions and Future Work

We have presented an extension to the behavior trees used in game industry to represent complex NPC behavior. Such extension, trained-query nodes, allows for the integration of a model learned from the traces generated by the designer controlling the NPC.

The main advantage of our approach is that it allows the combination of program by demonstration with hand-coded portions of the behavior trees. We have demonstrated in our experiments that this combination is specially adequate to program complex behaviors where program by demonstration can not provide good enough results by itself.

There is an implicit methodology for applying our approach to generate the complete behavior of an NPC. First, we identify simple strategies that the NPC can execute and generate an independent set of traces for every strategy. Then we generate a different model for each simple strategy. And, finally, we write a BT nodes with conditions to choose one of the simple strategies. As future work we plan to further formalize and test this methodology.

We have also tested the use of two different machine learning techniques within our combined framework: k-NN and neural networks. The first conclusion is that we have not found in our experiments any statistically significant difference between them. Nevertheless, there are differences regarding the ease of use and runtime efficiency.

k-NN is largely dependent on a good weight configuration to obtain good results and this configuration usually depends on the designer's knowledge about the representation of the information in the game level, which in general we can not assume. As can be seen in Table 2 when no additional knowledge is used to determine the weights by making all of them equal, k-NN ability to make good predictions is significantly reduced. Furthermore, k-NN is CPU expensive because it must process a large number of examples at runtime. Optimizations can be used such as k-d trees [19] to partition the set of examples and speed-up the retrieval process, and the retrieval process can be distributed along a number of game frames, thus reducing frame load. Nevertheless, k-NN is computationally more expensive than neural networks at runtime, what becomes crucial for games, where performance is crucial.

Neural networks have also demonstrated in our experiments a better tolerance to input noise than k-NN. However, the configuration and training of a neural network requires some technical knowledge that a game designer does not typically has and may be difficult for him to get. For that reason, we plan as future work to explore the automatic configuration of the neural network learning process, which can be typically accomplished through genetic algorithms.

References

1. Rasmussen, J.: Are behavior trees a thing of the past (2016). http://www.gamasutra.com/blogs/JakobRasmussen/20160427/271188/Are_Behavior_Trees_a_Thing_of_the_Past.php

2. Isla, D.: Halo 3 - building a better battle. In: Game Developers Conference (2008)
3. Isla, D.: Handling complexity in the Halo 2 AI. In: Game Developers Conference (2005)
4. Sagredo-Olivenza, I., Gómez-Martín, M.A., González-Calero, P.A.: Supporting the collaboration between programmers and designers building game AI. In: Chorianopoulos, K., Divitini, M., Hauge, J.B., Jaccheri, L., Malaka, R. (eds.) ICEC 2015. LNCS, vol. 9353, pp. 496–501. Springer, Cham (2015). doi:10.1007/978-3-319-24589-8_46
5. Ontañón, S., Mishra, K., Sugandh, N., Ram, A.: On-line case-based planning. Comput. Intell. **26**(1), 84–119 (2010)
6. Floyd, M.W., Esfandiari, B., Lam, K.: A case-based reasoning approach to imitating robocup players. In: Proceedings of the Twenty-First International Florida Artificial Intelligence Research Society Conference, Florida, 15–17 May 2008, pp. 251–256. AAAI Press, Coconut Grove (2008)
7. Ontanón, S., Ram, A.: Case-based reasoning and user-generated artificial intelligence for real-time strategy games. In: González-Calero, P.A., Gómez-Martín, M.A. (eds.) Artificial Intelligence for Computer Games, pp. 103–124. Springer, New York (2011)
8. Rubin, J., Watson, I.D.: On combining decisions from multiple expert imitators for performance. In: IJCAI-11 (2011)
9. Jaidee, U., Muñoz-Avila, H., Aha, D.W.: Case-based goal-driven coordination of multiple learning agents. In: Delany, S.J., Ontañón, S. (eds.) ICCBR 2013. LNCS, vol. 7969, pp. 164–178. Springer, Heidelberg (2013). doi:10.1007/978-3-642-39056-2_12
10. Dereszynski, E.W., Hostetler, J., Fern, A., Dietterich, T.G., Hoang, T.T., Udarbe, M.: Learning probabilistic behavior models in real-time strategy games. In: AIIDE (2011)
11. Fernández-Ares, A., Mora, A., García-Sánchez, P., Castillo, P., Merelo, J.: Analysing the influence of the fitness function on genetically programmed bots for a real-time strategy game. Entertainment Comput. **18**, 15–29 (2017)
12. Ortega, J., Shaker, N., Togelius, J., Yannakakis, G.N.: Imitating human playing styles in super mario bros. Entertainment Comput. **4**(2), 93–104 (2013)
13. Spronck, P., Ponsen, M., Sprinkhuizen-Kuyper, I., Postma, E.: Adaptive game AI with dynamic scripting. Mach. Learn. **63**(3), 217–248 (2006)
14. Floreano, D., Dürr, P., Mattiussi, C.: Neuroevolution: from architectures to learning. Evol. Intel. **1**(1), 47–62 (2008)
15. Robertson, G., Watson, I.: Building behavior trees from observations in real-time strategy games. In: v Innovations in Intelligent SysTems and Applications (INISTA), pp. 1–7. IEEE (2015)
16. Flórez-Puga, G., Gómez-Martín, M.A., Gómez-Martín, P.P., Díaz-Agudo, B., González-Calero, P.A.: Query enabled behaviour trees. IEEE Trans. Comput. Intell. AI Games **1**(4), 298–308 (2009)
17. Silverman, B.W., Jones, M.C.: E. Fix and J.L. Hodges (1951): an important contribution to nonparametric discriminant analysis and density estimation: commentary on fix and Hodges. Int. Stat. Rev./Revue Internationale de Statistique **57**(3), 233–238 (1989)
18. Rosenblatt, F.: Principles of neurodynamics. Perceptrons and the theory of brain mechanisms. Technical report, DTIC Document (1961)
19. Bentley, J.L.: Multidimensional binary search trees used for associative searching. Commun. ACM **18**(9), 509–517 (1975)

Supervised, Non-supervised, Reinforcement and Statistical Algorithms

A Classification System to Assess Low Back Muscle Endurance and Activity Using mHealth Technologies

Ignacio Diaz-Reyes[1]([✉]), Miguel Damas[1], Jose Antonio Moral-Munoz[2], and Oresti Banos[3]

[1] Department of Computer Architecture and Computer Technology,
University of Granada, Granada, Spain
idiaz@mdurance.es, mdamas@ugr.es
[2] Department of Nursing and Physiotherapy, University of Cadiz, Cadiz, Spain
joseantonio.moral@uca.es
[3] Telemedicine Group, Center for Telematics and Information Technology,
University of Twente, Enschede, Netherlands
o.banoslegran@utwente.nl

Abstract. Low back pain remains a major cause of absenteeism in the world. In addition to its socio-economic impact, the age at which the first symptoms appear is decreasing. Consequently, there are more experts who start incorporating prevention plans for the lumbar area in their work routines. In addition, the continued market growth of wearable sensors and the potential opened up by wearable technology allows experts to obtain a precise feedback from improvements in their patients in a daily basis. For this reason, this work wants to continue with the development and verification of the usefulness of mDurance, a novel mobile health system aimed at supporting specialists in the functional assessment of trunk endurance and muscle activity by using wearable and mobile devices. This work presents an extension of this system to classify low back muscle activity in the low back. mDurance has been tested into a professional football team. Clustering and data mining are applied in a new dataset of endurance and muscle activity data collected through mDurance. In addition, these results are cross-related with a questionnaire created to evaluate how the football players perceive themselves physically and mentally. The results show a clear correlation between the perception participants have about their low back endurance and the objective measurements conducted through mDurance. The results obtained through mDurance and the football players answers show a 68.3% of accuracy and 83.8% of specificity in the first approach to build a classifier to assess low back muscle endurance and activity using mDurance system.

1 Introduction

Low back pain (LBP) is considered an extremely common health problem, and the major cause of activity limitation and work absence in the world. In addition,

© Springer International Publishing AG 2017
I. Rojas et al. (Eds.): IWANN 2017, Part II, LNCS 10306, pp. 709–721, 2017.
DOI: 10.1007/978-3-319-59147-6_60

it continues to decrease the age at which the first signs of fatigue or discomfort in the low back begins to appear [1]. Thus, experts are beginning to make plans to improve the endurance and to discharge these important muscles in daily activity. According to Kim [2], electromyography (EMG) is based on the study of muscle activity through observation and analysis of the electrical signals monitored during voluntary or involuntary muscle contractions. EMG is useful to study muscular function during sport activities, biofeedback training, daily living or detect pathological states of the musculoskeletal systems. In order to assess all those functions, EMG offers useful information about timing of muscular activity and its relative intensity. Fernandez [3] reports that muscle fatigue is manifested as a reduction in the ability to keep a certain level of strength in a sustained contraction or as the inability to achieve a level of initial strength in intermittent contractions and it is accompanied by changes in muscle electrical activity. Muscle fatigue has central and peripheral components. The first are manifested as an inability to realize an induced activity for the development of this activity. The second stands out as a muscle inability to produce a certain level of strength. In applications of electrical stimulation it is important to assess the development of muscle fatigue to prevent deterioration of the mechanical behavior of the muscle being stimulated.

Muscle fatigue can be evaluated by EMG signals. The effect of fatigue on muscle mechanics is well represented by various metrics, such as the mean frequency (MNF), median frequency (MDF), the root mean square (RMS), the average rectified value (ARV) and the maximum voluntary muscle contraction (MVC) [2,4,5]. This information is very interesting to compare the evolution of the muscle strength among sessions, as well as to measure the effectiveness of potential treatments. In the time domain the RMS of the EMG signal is considered the most reliable parameter. An increase of the RMS with advancing fatigue has been reported in many studies [6,7]. RMS is not affected by the cancellation due to motor unit action potential train superposition, which may affect other processing techniques involving rectification [7]. In [2] it is shown that muscle fatigue is accompanied with an increase in the RMS and ARV and a decrease in the MNF.

On the other hand, the design and implementation of wearable EMG systems for health monitoring has got lot of attention throughout the world, specially in the sport, the physiotherapy and scientific community during the last years. These wearable sensors allow experts to obtain a precise feedback from improvements in their patients daily [8]. The studies that use surface EMG in sport sciences are mostly related to determination of the mechanism of contraction and relaxation of muscles while also dealing with evolution of injuries. The data obtained from these studies can be used in the following areas [9]: the evaluation of the technical development, the establishment of the suitable exercise programs and the follow up of the development of the athletes. Techniques that take full advantage of the extraordinary amount of data that such sensors/systems can gather are lacking. Several analysis provide a very large data volume coming from EMG register and physical examinations. The analysis and treatment of these data is difficult and time consuming.

For this reason, mDurance was presented as a novel mobile health system aimed at supporting specialists in the functional assessment of trunk endurance and muscle activity by using wearable and mobile devices [10,11]. The present study aims to analyze the endurance and the muscle activity of the low back during an endurance muscle test execution by mDurance. This work studies the effects of the muscle fatigue of a football team. In addition, this study is intended to be the starting point for the development of a new classification system of low back muscle activity, by applying clustering and data mining in the new data generated by mDurance.

The design of this new classification system can be applied in several investigations, including the functional evaluation of muscular processes and verification of the response of rehabilitation therapy in sports medicine, occupational medicine and physical medicine and rehabilitation, the study and ergonomic analysis of the workload and the prevention of muscular fatigue in activities of human occupation. The rest of the paper is structured as follows. Section 2 presents a description of the main features of the study. The fundamental results and discussion about this work are outlined in Sect. 3. Final conclusions and remarks are summarized in Sect. 4.

2 Methodology

2.1 Subjects

Fifteen professional soccer players from 19 to 32 years old (mean $\pm$ SD; age 24.05 ± 3.32 years old; height 178.07 ± 5.21 cm; weight 74.39 ± 5.19 kg) were recruited to be evaluated by one external physical therapist using mDurance. Before performing the evaluation, the volunteers were informed about the research aims, risks and benefits of participation. All subjects were tested during the 2016/2017 Spanish competitive soccer preseason.

2.2 Instrumentation

For this work, mDurance was used to measure the low back endurance and muscle activity. The mDurance system consists of a wearable inertial sensor to track the patient trunk posture and a portable electromyography sensor to seamlessly measure the electrical activity produced by the trunk muscles. All the information registered through these sensors is intelligently managed by a mobile or tablet application by mHealthDroid [12]. In addition, all data is stored in a new storage system in the cloud, formed by a back-end service and API Rest application. The back-end platform is here used for data engine can apply data mining techniques and applications.

2.3 Test Procedure

Different tests are available to assess the trunk endurance in people with or without LBP. This kind of tests are performed by a specialist, and they normally consist of the measurement of the time a person can hold a specific posture involving the trunk muscles. During the execution of the test, the health professional has to control the patient position, and with the help of mDurance, the specialist decides when the test ends, according to some established termination criteria. The results obtained for a given patient help experts determine their status and muscular capacity, as well as their ability to hold a posture normally related to daily living activities. To assess the low back stabilization, mDurance uses a functional trunk endurance test widely used: the static trunk extensor endurance test (STEET), also known as Sorensen test [13] (see Fig. 1).

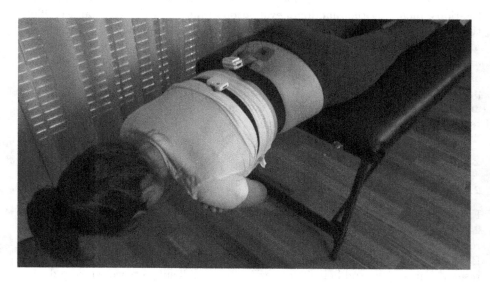

Fig. 1. STEET procedure using mDurance.

In the STEET, the subject has to maintain a horizontal unsupported posture with the upper body extending beyond the edge of the bench. Special remarks are that two chances are given to the individual to execute the STEET. The position is held up to a maximum of 240 s. A detailed description of this test, including posture, procedure and finalization criteria, is available in our previous work [10].

The STEET test was executed once per week to minimize the fatigue effects on the performance of the volunteers. The test was realized both before and after the training sessions up to a maximum of three times, in order to get a higher variety in the results for reducing the muscle fatigue intervals of each player. In both cases, each player filled up a questionnaire (see Fig. 2) every time that he realized the test. This questionnaire was created and approved by the physical therapists who evaluated the players. The main goal is to evaluate how the players are perceived themselves physically and psychologically and to check which is their condition and to improve their performance. This evaluation is carried out by comparing the responses obtained from the questionnaires filled in by the players and the results obtained through mDurance for each test. The complete procedure was explained to the subjects before performing the sessions, assuring the full understanding of their phases.

```
Name      _____

Date      _____

Have you played today?
     a)  No
     b)  Yes

Which is your condition today?
     a)  Ok
     b)  Fatigue
     c)  Discomfort
     d)  Injury
```

ID		YES	NO
Q1	Do you feel low back pain at this moment?		
Q2	Have you performed any activity that may have caused you low back fatigue?		
Q3	How many hours do you think you spent sitting today?		
Q4	Do you feel rested at this moment?		

Fig. 2. Part of the questionnaire used for the physical and mental player assessment.

2.4   Dataset Analysis

The dataset consists of the final results obtained through mDurance and the results obtained through the questionnaires. This dataset represents how the endurance and muscle fatigue affect the soccer players performance according to several features. They are divided into two blocks: Questionnaire and mDurance results.

Questionnaire results:

– *Player's condition.* This represents the health condition of each player in the exact moment of realizing the test. The condition can take the following values: "Ok", "Fatigue", "Discomfort", "Injury" depending on the player was in perfect condition, or he had done any diary activity which could cause muscle fatigue, or the player suffered some low back pain during the test or the player was injured in that date.

- *Has_Played.* This indicates if the players who were going to do the test had played or not. *Q1, Q2 and Q4* represent the different yes/no questions in the questionnaire. They describe both physical and mental status of the players for a particular date. *Q3* describes the number of hours that each player was sitting that day.

mDurance results:

- *Total time.* It indicates the endurance total time in second, which the player lasted in the STEET test.
- *Ratio.* It refers to a value calculated by the total time and an average value. It can be consulted in our previous work [10].
- *Ratio_label.* It has two levels: "Up" and "Down": The values are classified according to if the ratio is higher than 1 or lower than 1. In the first case, it means that the player is above the average and he is below in the opposite case.
- *RMS, ARV and MVC.* They describe the absolute muscle fatigue values measured in the end of the test by the application. They are expressed in mV.
- *RMS_per_second, ARV_per_second.* Another way to compare the muscle activity among the players is to create some new attributes, which allow to check that players had a major muscle activity. Thus, these variables are calculated dividing the final RMS and ARV values over the total time obtained in the test. They are expressed in mV/s.
- *Endurance label.* It describes the three possible endurance states that are registered by the mDurance application. They are: "bad shape", "good shape", "excellent shape" according to the time duration in the test.
- *Fatigue label.* One of the difficulties that this work has encountered is to get to categorize the results of muscle activity obtained by the application. We could not find any bibliography which allowed to label the muscle activity values. For this reason, we classify these results in three levels: "minimum fatigue", "medium fatigue" and "maximum fatigue". Each fatigue level is used to represent the minimum, medium and maximum value obtained in the three tests executed for each football player. For example, if a football player got 0.4, 0.5 and 0.6 mV for each RMS result obtained for each test, the first value would be classified as "Minimum fatigue", the second value as "Medium fatigue" and the third value as "Maximum fatigue".
- *Fatigue per second label.* Finally, it is also decided to create a new EMG range to categorize the results obtained by the RMS and ARV per second, in order to classify every player based on the same unit. The values are classified again in three levels: "minimum", "medium" and "maximum" fatigue per second.

3   Results and Discussion

3.1   Muscle Activity Analysis

Most important attribute to detect the muscle fatigue is the RMS, so it is plotted with respect to the endurance and fatigue labels. Figure (see Fig. 3) shows that

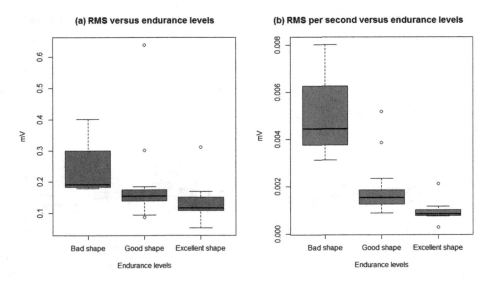

Fig. 3. Endurance levels with respect to: (left) RMS values and (right) RMS per second values.

RMS values are greater when the players endurance is worse. It indicates that many of the players who realized the tests and that lasted a short time obtained a higher RMS than the players who lasted more longer value. Thus, this is a clear indicator of the existence of muscle fatigue in these players. In addition, it can be observed that between the "good" and the "excellent" time classification, most of tests which had an "excellent" time classification obtained the lower RMS values. This supports the previous conclusion. In the figure (see Fig. 4) is compared how the RMS and RMS per second values are distributed for the three fatigue classifiers. It shows that the maximum, mean and minimum values of the "Maximum" group are bigger than these same values in the "Medium" and "Minimum" group. In addition, "Medium" values are bigger than these in the "Minimum" group. However, the "Medium" group takes values both of the "Maximum" group for the RMS values and the "Maximum" group for the RMS/s values. This can make difficult the supervised classification task.

3.2   Clustering Analysis

In the clustering theory, the first is to select an approximation functions. The choice of function determines the future results and the clustering quality. Moreover, it is important to select the attributes that are going to participate in the clustering process. There are two very differentiable variables sets: the quantitative attributes describing the final results obtained for each test and the qualitative attributes, which represent both the questionnaire results as the endurance and muscle activity. The most important variables are selected through Chi-squared methods [14] and the correlation feature selection method (CFS) [15]

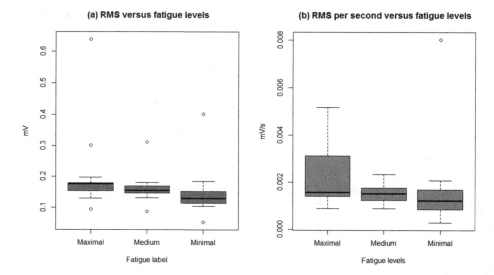

Fig. 4. Muscle activity levels with respect to: (left) RMS values and (right) RMS per second values.

were used. Finally, the selected attributes used in clustering were: RMS, RMS_sec representing to the quantitative variables and Condition, Q1, Q4, Ratio_label, Endurance_label and Fatigue_perSecond_label.

The hierarchical clustering approach is a method of group analysis which seeks to build a hierarchy of groups. Most algorithms are of the agglomerative type, i.e., partitions start with as many groups as items and at each step are united together. This approach could give a global idea about how many groups should be considered. The Ward method was selected [16], because it offered the best results. Thus, the distance matrix of the data was calculated, because these algorithms are based on making transformations about this matrix decreasing its size. To select the best group, it is necessary to use the silhouette coefficient of the possible partitions. The silhouette coefficient [17] is a measure that combines the cohesion and separation. It compares the distance means among instances of a same group and the distance means among instances of different groups. The results obtained are showed in Table 1.

When two groups are only considered (K = 2), the cohesion and the separation are the largest with respect to three (K = 3) and four (K = 4) groups. However, this happens because the second group has few units with respect to the first group. Thus, this group is not a good option. If the groups are compared with three and four groups respectively, it is observed that the Silhouette coefficients are very similar (0.3418, 0.3456). However, it is not worth to take four groups because one of these groups just has one item. Thus, it was decided to take the three groups option.

Table 1. Units per group and Silhouette coefficient for all the possible groups in all the possible pruning (k = 2, 3 and 4).

	Size	Ind. Silhouette width	Mean Silhouette width
K = 2			
Group 1	37	0.6022	0.5825
Group 2	4	0.3999	
K = 3			
Group 1	16	0.2648	0.3418
Group 2	21	0.4036	
Group 3	4	0.3251	
K = 4			
Group 1	16	0.2648	0.3456
Group 2	21	0.4023	
Group 3	1	0.0000	
Group 4	3	0.4932	

Figure 5 shows that players of the second group that have a major RMS value per second will have a major probability of being classified into "Maximum" or "Medium" muscle activity. In addition, all players labeled with the worst endurance needed more muscle activity to maintain the same load as others labeled with better endurance. The 71.42% the players classified with the best endurance needed the minimum muscle activity. Another aspect to note is the clustering of these instances according to the endurance labels. In Fig. 5, it is also demonstrated the accuracy of the classification used by mDurance to label the RMS value in function of the player endurance. It is seen that the worst endurance results obtained in the tests correspond with the instances of the third group, labeled as "Maximum muscle activity" instances. This aspect is extended to the second and first group, where most of the instances are grouped as good and excellent shape and "Minimum muscle activity".

3.3   Classification Supervised Analysis

Once it has been able to determine the number of groups where to classify any new RMS value obtained in the execution of a new STEET test, it is time to build a classifier and evaluate its performance. The supervised classification allows classifying a new instance in a certain groups of the considered ones.

Cross-validation is a model validation technique for assessing how the results of a statistical analysis will generalize to an independent data set. Leave-p-out cross-validation (LpO CV) involves using p observations as the validation set and the remaining observations as the training set [18]. This is repeated to split the original sample into a validation set of p observations and a training set. In this work, it is used Leave-one-out cross-validation (LOOCV) because there are not a

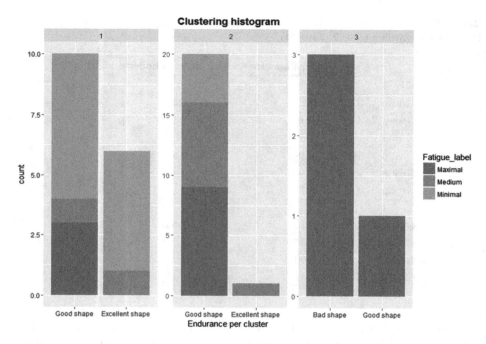

Fig. 5. Representation of the three groups obtained by Ward methods in histograms. The groups are represented by the results obtained through mDurance and the questionnaires.

large number of samples. This is a particular case of leave-p-out cross-validation with p = 1. Classification algorithms can be grouped into two large families attending the goodness of the results. *Individual classifiers* are learning a single model, which can be of various kinds: KNN [19], SVM Linear [20] and SVM Radial [21]. *Ensemble classifiers* are algorithms that combine several individual classifiers, decision trees normally. For this study was used classification trees (CTree) [22].

In Table 2, these groups are compared on equal terms. The same variables used for clustering are used in this section.

Table 2. Comparison between the main classifiers mentioned above.

Method	Accuracy	Mean bal. accuracy	Mean sensitivity	Mean specificity
KNN (k = 9)	0.671	0.699	0.652	0.826
SVM linear	0.658	0.733	0.640	0.825
SVM radial	**0.683**	**0.756**	**0.674**	**0.838**
CTree	0.610	0.694	0.586	0.801

It may be concluded that the individual classifiers clearly outweigh the ensembles classifiers, and SVM Radial seems to be the most appropriate. In this paper, with 68.3% accuracy, it is shown that a classifier could be constructed taking into account the results obtained through mDurance and the answers obtained through a simple questionnaire. In addition, this could increase until a 75.6% if it is measured the mean of all the accuracies obtained for each fold in LOOCV. It should be noted that the mean specificity, which identifies the percentage of non-fatigued people who are correctly identified is the best result obtained in this study with a 83.8%. On the contrary, the mean sensitivity, which measures the percentage of fatigued people who are correctly identified as having is the lowest result.

4   Conclusion

It has been shown a study about the behavior of the muscles of the lumbar area on a professional football team. This work intends to be the starting point for the development of a new classification system of muscle activity in the low back muscle, by applying clustering and data mining in the new data collected through mDurance results and the answers to the questionnaire (see Fig. 2). This questionnaire aims to evaluate how the players are perceived themselves physically and mentally, and to check what are their conditions tò can improve their performance. The first groups of what could be a new unsupervised classification system of the lumbar muscle activity have been grouped. The results claim that groups correspond to the values of minimum-medium and medium-maximum muscle activity, further demonstrating that muscle activity shown by the players was in most cases consistent with their perception. Once it was possible to group instances into three groups, they were applied the most famous supervised classification techniques, both individual and ensembles, in order to test the accuracy of the classification of new instances. It must be concluded that the highest accuracy, specificity and sensibility were obtained with a single classifier, such as SVM Radial, which obtained a 68.3%, 67.4% and 83.8% in precision and sensitivity and specificity respectively.

The design of this new classification system can be applied in several investigations: functional evaluation and verification of the rehabilitation therapies and the prevention of muscular fatigue in activities of human occupation.

Acknowledgements. This work was supported by the Spanish Ministry of Economy and Competitiveness (MINECO) Projects TIN2015-71873-R and TIN2015-67020-P together with the European Fund for Regional Development (FEDER). The authors want to especially thank all of the volunteers who participated in the experiments and Cristian Rivera Peregrina for your help.

References

1. Woolf, A.D., Pfleger, B.: Burden of major musculoskeletal conditions. Bull. World Health Organ. **81**(9), 646–656 (2003)

2. Kim, G., Ahad, M.A., Ferdjallah, M., Harris, G.F.: Correlation of muscle fatigue indices between intramuscular and surface EMG signals. In: Proceedings SoutheastCon, pp. 378–382, IEEE (2007)
3. Fernández, J.M., Acevedo, R.C., Tabernig, C.B.; Influencia de la fatiga muscular en la señal electromiográfica de músculos estimulados eléctricamente. Revista EIA (7), 111–119 (2007)
4. Phinyomark, A., Phukpattaranont, P., Limsakul, C.: Feature reduction and selection for EMG signal classification. Expert Syst. Appl. 39(8), 7420–7431 (2012)
5. Phinyomark, A., Thongpanja, S., Hu, H., Phukpattaranont, P., Limsakul, C.: The usefulness of mean and median frequencies in electromyography analysis, pp. 195. INTECH Open Access Publisher (2012)
6. Siemienski, A., Kebel, A., Klajner, P.: Fatigue independent amplitude-frequency correlations in EMG signals. arXiv preprint physics/0611120 (2006)
7. Al Zaman, A., Sharmin, T., Mohammad Ashraf Ali Khan, and Ferdjallah, M.: Muscle fatigue analysis in young adults at different MVC levels using EMG metrics. In: Proceedings SoutheastCon, pp. 390–394. IEEE (2007)
8. Pantelopoulos, A., Bourbakis, N.G.: A survey on wearable sensor-based systems for health monitoring, prognosis. IEEE Trans. Syst. Man Cybern. Part C (Appl. Rev.) 40(1), 1–12 (2010)
9. Türker, H., Sözen, H.: Surface electromyography in sports and exercise. In: Electrodiagnosis in New Frontiers Clinical Research, p. 181 (2013)
10. Banos, O., Moral-Munoz, J.A., Diaz-Reyes, I., Arroyo-Morales, M., Damas, M., Herrera-Viedma, E., Hong, C.S., Lee, S., Pomares, H., Rojas, I., et al.: mDurance: a novel mobile health system to support trunk endurance assessment. Sensors 15(6), 13159–13183 (2015)
11. Banos, O., Bang, J.H., Kang, D.U., Hong, C.S., Lee, S., Munoz, J.A.M., Damas, M., Reyes, I.D., Pomares, H., Rojas, I., Villalonga, C., Morales, M.A.: Facilitating trunk endurance assessment by means of mobile health technologies. In; ACM International Joint Conference on Pervasive and Ubiquitous Computing, UbiComp, pp. 149–152. ACM (2015)
12. Banos, O., Villalonga, C., Garcia, R., Saez, A., Damas, M., Holgado-Terriza, J.A., Lee, S., Pomares, H., Rojas, I.: Design, implementation and validation of a novel open framework for agile development of mobile health applications. Biomed. Eng. Online 14(2), S6 (2015)
13. Biering-Sorensen, F.: Physical measurements as risk indicators for low-back trouble over a one-year period. Spine 9(2), 106–119 (1984)
14. Higgins, J.E., Koch, G.G.: Variable selection and generalized chi-square analysis of categorical data applied to a large cross-sectional occupational health survey. Int. Stat. Rev./Revue Internationale de Statistique 45, 38–51 (1977)
15. Hall, M.A.: Correlation-based feature selection for discrete and numeric class machine learning. In: Proceedings of the Seventeenth International Conference on Machine Learning, pp. 359–366. Morgan Kaufmann Publishers Inc. (2000)
16. Batagelj, V.: Generalized ward and related clustering problems. In: Classification and Related Methods of Data Analysis, pp. 67–74 (1988)
17. Rousseeuw, P.J.: Silhouettes: a graphical aid to the interpretation and validation of cluster analysis. J. Comput. Appl. Math. 20, 53–65 (1987)
18. Arlot, S., Celisse, A., et al.: A survey of cross-validation procedures for model selection. Stat. Surv. 4, 40–79 (2010)
19. Murugappan, M.: Electromyogram signal based human emotion classification using KNN and LDA. In: IEEE International Conference on System Engineering and Technology (ICSET), pp. 106–110 (2011)

20. Burges, C.J.C.: A tutorial on support vector machines for pattern recognition. Data Min. Knowl. Disc. **2**(2), 121–167 (1998)
21. Cristianini, N., Shawe-Taylor, J.: An Introduction to Support Vector Machines and Other Kernel-Based Learning Methods. Cambridge University Press, Cambridge (2000)
22. Kaur, A., Sharma, S.: Human activity recognition using ensemble modelling. In: Unal, A., Nayak, M., Mishra, D.K., Singh, D., Joshi, A. (eds.) SmartCom 2016. CCIS, vol. 628, pp. 292–298. Springer, Singapore (2016). doi:10.1007/978-981-10-3433-6_35

Probabilistic Leverage Scores for Parallelized Unsupervised Feature Selection

Bruno Ordozgoiti$^{(\boxtimes)}$, Sandra Gómez Canaval, and Alberto Mozo

Department of Computer Systems, Universidad Politécnica de Madrid,
Madrid, Spain
bruno.ordozgoiti@upm.es

Abstract. Dimensionality reduction is often crucial for the application of machine learning and data mining. Feature selection methods can be employed for this purpose, with the advantage of preserving interpretability. There exist unsupervised feature selection methods based on matrix factorization algorithms, which can help choose the most informative features in terms of approximation error. Randomized methods have been proposed recently to provide better theoretical guarantees and better approximation errors than their deterministic counterparts, but their computational costs can be significant when dealing with big, high dimensional data sets. Some existing randomized and deterministic approaches require the computation of the singular value decomposition in $O(mn\min(m,n))$ time (for m samples and n features) for providing leverage scores. This compromises their applicability to domains of even moderately high dimensionality. In this paper we propose the use of Probabilistic PCA to compute the leverage scores in $O(mnk)$ time, enabling the applicability of some of these randomized methods to large, high-dimensional data sets. We show that using this approach, we can rapidly provide an approximation of the leverage scores that is works well in this context. In addition, we offer a parallelized version over the emerging Resilient Distributed Datasets paradigm (RDD) on Apache Spark, making it horizontally scalable for enormous numbers of data instances. We validate the performance of our approach on different data sets comprised of real-world and synthetic data.

Keywords: Machine learning · Feature selection · Distributed computing

1 Introduction

Over the last few years, machine learning and data mining have proved to be useful for different applications such as automated classification, forecasting and anomaly detection. However, high dimensional data, which arises fequently in numerous domains, pose certain challenges. The first and perhaps the most

The research leading to these results has received funding from the European Union under the FP7 grant agreement n. 619633 (project ONTIC) and H2020 grant agreement n. 671625 (project CogNet).

© Springer International Publishing AG 2017
I. Rojas et al. (Eds.): IWANN 2017, Part II, LNCS 10306, pp. 722–733, 2017.
DOI: 10.1007/978-3-319-59147-6_61

important one is the curse of dimensionality, which makes learning more difficult. Second, a high number of dimensions can sometimes have a significant impact on performance. There exist well-established techniques to fight these problems, such as principal component analysis (PCA), which transforms the data into a new, lower dimensional subspace which captures the majority of the information of the original records in some sense. This representation, however, can be difficult to interpret for domain experts. To overcome this issue, feature selection methods can be employed. These techniques eliminate redundant or uninformative features in favor of the most relevant ones, which results in a lower dimensional and hence more manageable data set. The size of today's data sets makes these techniques especially attractive, since their use can improve running times, but it can also be a hurdle for their application. Even when these algorithms depend linearly on the number of samples they can become unusable in the presence of billions of records. In a context where the amount of data generated and stored every day is growing exponentially, applying traditional machine learning and data mining techniques to derive useful conclusions from data is becoming increasingly difficult.

In order to cope with massive data sets, two approaches can be adopted: scaling vertically - i.e. increasing the power of available machines - and scaling horizontally - i.e. increasing the number of available machines. The latter is often preferred due to its lower costs and better flexibility. This preference can also be explained by the emergence of large-scale data processing platforms during the last decade. In 2004, MapReduce [2] introduced a shared-nothing, fault-tolerant and resilient distributed computation architecture that allows the parallel processing without the usual impediments of distributed computing. As MapReduce enables high horizontal scalability, Hadoop MapReduce, the open source implementation of this paradigm, has been adopted widely by both the research and the industrial communities. However, MapReduce does not provide an efficient way to implement algorithms that repeatedly access the same data, because temporary data is written and read from disk instead of keeping it in RAM. To cope with this shortcoming, Apache Spark has emerged as a novel parallel computing platform. It is based on the Resilient Distributed Datasets paradigm (RDDs) [15], a new abstraction designed to provide distribution and fault tolerance efficiently. RDDs allow the programmer to cache data in main memory. Therefore, well designed algorithms using this abstraction clearly outperform their MapReduce counterparts.

Contribution. In this paper we propose **PPICS** (Probabilistic Parallelized Independent Column Selection), a new unsupervised feature selection algorithm designed to address the previously discussed scalability and high dimensionality issues. Our algorithm (1) computes approximate leverage scores (see Sect. 3) in $O(mn\rho)$ (for ρ computed principal components) based on probabilistic principal components analysis (PPCA), increasing the dimensionality to which it is applicable. In addition, it is parallelized over the RDD abstraction on Apache Spark, making it horizontally scalable for enormous numbers of data instances and achieving a total execution time that depends linearly on the number of worker nodes in the cluster.

The rest of the paper is structured as follows: Sect. 2 describes the related work on parallel feature selection algorithms. Section 3 presents an overview of PPICS algorithm and in Sect. 4 we describe the detail of parallel PPICS implementation on Apache Spark using the RDD abstraction. In Sect. 5 we present experimental results and conclusions in Sect. 6.

2   Related Work

Over the last few years, the problem of selecting a subset of columns to obtain low-rank approximations of large matrices has received considerable attention.

In [5], Frieze et al. studied the idea of choosing a small number of matrix entries to efficiently obtain a low-rank approximation. This proposal is built upon in [3], where the authors propose an iterative adaptive sampling method that favors data points lying far from the currently sampled subspace.

A key idea for choosing a column subset was proposed by Jolliffe in [7], often known as *leverage scores*. The scores for a column is equal to the sum of squares of the corresponding row of the matrix containing the leading right singular vectors. A thorough analysis of this method is provided in [10] along with a deterministic column selection algorithm. The algorithm proposed in [1] views these scores as probabilities, which are used to sample various candidates among which the best one is chosen *a posteriori*. This method allows for strong theoretical guarantees and is very effective in practice.

Other feature selection algorithms, both supervised and unsupervised, have been proposed recently to deal explicitly with the enormous size of today's data sets, exploiting the benefits of parallel computing. These approaches are based on concepts as varied as mutual information [12], rough set theory [6] and logistic regression [11]. An unsupervised method for Message Passing Interface (MPI) is described in [16] and one for MapReduce is provided in [4]. Recently, the authors proposed in [8] a parallelized hybrid algorithm, similar to the one described in [1], that needs to calculate the singular value decomposition (SVD) of the data matrix. An efficient algorithm, as well as methods to deal with large matrices, is described by the authors in [9].

The leverage scores approach, though effective, requires the computation of the SVD, which takes $O(mn\min(m,n))$ time. Nowadays, where data sets are often extremely large and high-dimensional, this can result in prohibitively costly computations, rendering the cited methods unusable. In order to overcome this drawback we propose a two-fold approach. First, in order to eliminate the quadratic dependence on the number of columns we propose a parallelized version of the expectation-maximization algorithm for PPCA that iterates in $O(mn\rho)$ time (for ρ computed principal components). Second, in order to enable the processing of huge numbers of samples, we proposed a parallelized approach for choosing the best candidate among an arbitrarily large sample in an efficient, scalable manner.

3   PPICS Algorithm Overview

This section provides a description of the main ideas behind our algorithm.

Notation. We denote sets by uppercase letters (S, Θ). Uppercase bold letters $(\mathbf{A}, \mathbf{\Sigma})$ denote matrices, while lowercase bold letters $(\mathbf{x})$ denote vectors. $\mathbf{A}_{(i,:)}$ and $\mathbf{A}_{(:,j)}$ represent the i-th row (understood as a column vector) and the j-th column of $\mathbf{A}$ respectively. $\mathbf{A}_{(\Theta,:)}$ represents the rows of $\mathbf{A}$ whose indices are the elements of the set Θ (we use analogous notation for columns). $\mathbf{A}_{(:i,j)}$ denotes the first i elements of the j-th column of $\mathbf{A}$. By RDD(Θ) we denote a Resilient Distributed Dataset whose entries are the elements of the set Θ.

The parallelized algorithm we propose provides an approximate solution to the Column Subset Selection Problem (CSSP). The CSSP provides an interesting framework for developing unsupervised feature selection methods, and can be defined as follows:

Definition 1. *Column Subset Selection Problem. Given a matrix* $\mathbf{A} \in \mathbb{R}^{m \times n}$ *and a positive integer k, pick k columns of $\mathbf{A}$ forming a matrix* $\mathbf{C} \in \mathbb{R}^{m \times k}$ *such that the residual*

$$\|\mathbf{A} - \mathbf{C}\mathbf{C}^{+}\mathbf{A}\|_F$$

is minimized over all possible $\binom{n}{k}$ choices for the matrix $\mathbf{C}$. $\mathbf{C}^{+}$ denotes Moore-Penrose generalized inverse of $\mathbf{C}$. and $\|\mathbf{A}\|_F = \sqrt{\sum_i a_{ii}^2}$ is the Frobenius norm.

The CSSP is believed to be NP-Hard. Therefore, no known efficient algorithm can find the solution efficiently with guarantees. In order to provide performance guarantees, some of these approaches rely on random sampling of possible candidates. However, this requires computing a suitable distribution, and testing the adequacy of all candidates to choose the best one, imposing significant computational costs for large data sets. We propose a parallelized hybrid algorithm, using efficient methods both for computing the sampling probabilities and for determining the best candidate.

Inspired by [1], our algorithm works in two stages: a randomized one, where we randomly sample $\gamma \in \mathbb{N}$ candidate column subsets using a judicious probability distribution, and a deterministic one that chooses the best candidate among the randomly sampled subsets. For the random selection in the first phase we propose the discrete probability distribution $\pi = (p_1, p_2, \ldots, p_n)$ where the probability associated with column i is

$$p_i \propto \| (\mathbf{\Sigma}\mathbf{V}^T)_{(:\rho,i)} \|_2^2 \tag{1}$$

where $\mathbf{A} = \mathbf{U}\mathbf{\Sigma}\mathbf{V}^T$ is the singular value decomposition of $\mathbf{A}$. By increasing ρ we approach a uniform distribution (for normalized columns), which is interesting if we have enough resources to process many candidates. Intuitively, high values in this distribution correspond to columns that are well aligned with a singular vector and that, because of the orthogonality of $\mathbf{V}$, are almost orthogonal to the space spanned by the rest of the columns. By multiplying by $\mathbf{\Sigma}$ we favor the leading singular vectors. This also means that it is not necessary to compute all vectors. In practice, the top-k or a slightly superior number will suffice.

We sample $k \log k$ columns from this distribution and run a rank-revealing QR factorization on them, keeping only the k first elements from the resulting permutation. These factorizations discard sets of (nearly) linearly dependent columns by permuting them to the right of the matrix.

The deterministic phase requires the computation of the residual error on the Frobenius norm for all candidate choices in order the choose the one that minimizes it. We define the residual error δ_i corresponding to candidate column choice i, for $1 \leq i \leq \gamma$ as

$$\delta_i = \|\mathbf{A} - \mathbf{C}_i\mathbf{C}_i^+\mathbf{A}\|_F \tag{2}$$

Note that for any real $m \times k$ matrix C, with $k \leq n$,

$$\mathbf{C}^+\mathbf{A} = \underset{\text{rank}X=k}{\arg\min}\|\mathbf{A} - \mathbf{CX}\|_F \tag{3}$$

i.e. $\mathbf{C}^+\mathbf{A}$ contains the coefficients of the set of linear combinations of the columns of $\mathbf{C}$ that best approximate matrix $\mathbf{A}$.

The Moore-Penrose generalized inverse of $\mathbf{C}$ can be calculated as follows. If $\mathbf{C} = \mathbf{U}\mathbf{\Sigma}\mathbf{V}^T$ is the singular value decomposition of $\mathbf{C}$, then $\mathbf{C}^+ = \mathbf{V}\mathbf{\Sigma}^+\mathbf{U}^T$. Of course, the computation of the singular value decomposition for all our γ candidate choices would be very costly. For that reason, we follow the approach derived in [8], which merges all the sampled columns into one matrix, to significantly decrease the necessary computation time. This way we can compute the solution to (3) for all column choices with only one large matrix product.

The computation of the probability distribution shown in Eq. (1) has certain drawbacks in terms of computational complexity, since the computation of the right singular vectors requires $O(mn \min(m, n))$ time. To overcome this issue, we propose the use of probabilistic principal component analysis.

3.1   Probabilistic Principal Component Analysis

Principal component analysis (PCA) [14] is a well-known technique for dimensionality reduction. In essence, PCA provides the k orthonormal axes that maximize the variance under projection of our m observations. These k axes equal the k dominant eigenvectors of the covariance matrix $\mathbf{S} = \Sigma_n(\mathbf{x_n} - \boldsymbol{\mu})(\mathbf{x_n} - \boldsymbol{\mu})^T/m$, and consequently the right singular vectors of a colum-wise mean-centered dataset. Additionally, the variance of the projection onto each of the axes is proportional to the corresponding singular value. Therefore, computing the principal components of our data is sufficient to obtain the column sampling probability distribution. The computation of this subspace, like the SVD, requires $O(mn \min(m, n))$ operations. However, a probabilistic interpretation of PCA [13] enables the derivation of an expectation-maximization algorithm whose iterations run in $O(mn\rho)$ each. A parallelized version of this algorithm over RDDs is given in Algorithm 2 as part of the proposed solution.

Algorithm 1. PPICS

1: **procedure** PPICS$(\mathbf{A}, k)$
2:     $c \leftarrow k \ln k$; rows $\leftarrow$ RDD$(\{\mathbf{A}_{(i,:)} | 1 \leq i \leq m\})$
3:     $(\mathbf{V}, \sigma(\mathbf{A})) \leftarrow$ computePPCA(rows, k); $\mathbf{\Sigma} \leftarrow$ diag$(\sigma(\mathbf{A}))$
4:     **for** $i = 1, \ldots, n$ **do**
        $p_i \leftarrow \| (\mathbf{\Sigma}\mathbf{V}^T)_{(:,\rho,i)} \|_2^2$
5:     **end for**
6:     $s \leftarrow \sum_i p_i$
7:     $\boldsymbol{\pi} = (p_1/s, \ldots, p_n/s)$
8:     **for** $i = 1, \ldots, \gamma$ **do**
9:         $\tilde{\Omega}_i \sim \mathcal{W}(c, \mathbf{1}_n, \boldsymbol{\pi})$; $\Omega_i \leftarrow$ RRQR$_k\left((\mathbf{\Sigma}\mathbf{V}^T)_{(:,\tilde{\Omega}_i)}\right)$
10:    **end for**
11:    $\Theta \leftarrow \bigcup_i \Omega_i$; $T \leftarrow \{\Omega_i | 1 \leq i \leq \gamma\}$
12:    $\mathbf{Q}^T\mathbf{A} \leftarrow$ rows.**map**{ $\mathbf{x} =>$
13:        $\mathbf{q} \leftarrow \mathbf{x}_{(\Theta)}$; output $\mathbf{q}\mathbf{x}^T$
14:    }.reduce$\{(\mathbf{X}, \mathbf{Y}) => \mathbf{X} + \mathbf{Y}\}$
15:    $P \leftarrow \emptyset$
16:    **for** $i = 1, \ldots, \gamma$ **do**
        $\mathbf{C}_i^T\mathbf{A} \leftarrow (\mathbf{Q}^T\mathbf{A})_{(\Omega_i',:)}$; $\mathbf{C}_i^T\mathbf{C}_i \leftarrow (\mathbf{Q}^T\mathbf{A})_{(\Omega_i',\Omega_i')}$
17:        $P \leftarrow P \cup \{(\mathbf{C}_i^T\mathbf{C}_i)^{-1}\mathbf{C}_i^T\mathbf{A}\}$
18:    **end for**
        broadcast(P); broadcast(T)
19:    $\delta \leftarrow$ rows.**map**{ $\mathbf{x} =>$
20:        **for** $i = 1, \ldots, \gamma$ **do**
21:            $\mathbf{d}_i \leftarrow \mathbf{x} - \mathbf{x}_{(\Omega_i)}^T \mathbf{C}_i^T\mathbf{A}$

22:        **end for**
23:        output $(\|\mathbf{d}_1\|_2^2, \ldots, \|\mathbf{d}_\gamma\|_2^2)$
24:    }.reduce$\{(\mathbf{x}, \mathbf{y}) => \mathbf{x} + \mathbf{y}\}$
25:    output Ω_i with $i = \arg\min_j \delta_j$
26: **end procedure**

---

4 PPICS Detailed Algorithm Description

This section provides a detailed description of PPICS, the proposed algorithm. Algorithm 1 provides a pseudo-code representation of PPICS, while Algorithm 2 contains the parallel expectation-maximization (EM) algorithm for PPCA, which enables the computation of the sampling probabilities.

Algorithm 1 takes as input the data matrix $\mathbf{A}$ and the number of columns to be kept, k. We treat $\mathbf{A}$ as an RDD whose elements are its rows. We then call subroutine *computePPCA*. This subroutine first builds an RDD[1] whose elements are 3-tuples containing each $\mathbf{x}_i$, $\mathbf{x}_i\mathbf{x}_i^T$ and each data point $\mathbf{x}$ (line 5). We use

---

[1] If the cluster has enough memory, we recommend caching this RDD, since it is used in two different operations.

this RDD to compute the matrices that form $\tilde{\mathbf{W}}$ (line 8) and to compute the new value of σ^2. We suggest the reader to refer to [13] for details. At convergence, we obtain the estimated principal subspace from the maximum likelihood estimator of $\mathbf{W}$:

$$\mathbf{W}_{ML} = \mathbf{U}_k(\mathbf{\Lambda}_k - \sigma^2\mathbf{I}_k)^{1/2}\mathbf{R}$$

where the columns of $\mathbf{U}_k$ are eigenvectors of the data covariance matrix and the diagonal of $\mathbf{\Lambda}_k$ contains its eigenvalues. We can find $\mathbf{R}^T$ as the eigenvectors of the $k \times k$ matrix $\mathbf{W}^T\mathbf{W}$, project $\mathbf{W}$ and orthogonalize to obtain the final estimation (line 14). The eigenvalues are also easily obtained from $\mathbf{R}^T$ (line 14).

The algorithm continues with the computation of the sampling probabilities, buidling vector $\boldsymbol{\pi} = (p_1/s, \ldots, p_n/s)$ where p_i Eq. (1) and s is a normalizing factor. We then proceed to sample γ random selections of c columns. We set $c = k \ln k$, although other small factors of k might work well too. We set γ to $\min(4\sqrt{\frac{n}{3}-k}, 100)$, although it can also be varied slightly with little impact on the performance of the algorithm. The rationale behind this choice is that as k approaches the (loose) numerical rank of the data matrix (which we estimate to be approximately of $n/3$ in our experiments), the random choices will resemble each other more and more, thus requiring fewer samples to be drawn. The drawn choices $\tilde{\Omega}_i$ are actually samples from a multivariate instance of Wallenius' noncentral hypergeometric distribution with parameters $c, \mathbf{1}_n, \boldsymbol{\pi}$.

In order to reduce these samples to exactly k columns we then perform a rank-revealing QR factorization of the γ matrices $(\mathbf{\Sigma}_k\mathbf{V}_k^T)_{(:,\tilde{\Omega}_i)}$ (of dimension $k \times (k \ln k)$) and keep the first k columns of the resulting permutation. In line 9 we use RRQR_k to denote a function that returns the k corresponding indices.

We can now define two sets: (1) Θ which contains all the indices of the columns that are included in at least one of the Ω_i's, and (2) T, whose elements are each of the Ω_i's. We then compute $\mathbf{Q}^T\mathbf{A}$ (line 12), from which we can retrieve $\mathbf{C}_i^+\mathbf{A}$ for all i (line 17) by taking the corresponding rows of $\mathbf{Q}^T\mathbf{A}$, indexed by Ω_i'. These are the coefficient matrices we use to rebuild our original data using the selected columns as a basis.

The only thing remaining to do is to compute the residual norms. We do this in parallel by broadcasting the indices of the column choices and the approximating linear maps $\mathbf{C}_i^+\mathbf{A}$ for $1 \leq i \leq \gamma$ to compute the residual for each rebuilt data sample, then reducing to sum up all the computed residuals (lines 19–24). We output the column choice whose residual is minimum.

Finally, we note that for all computations, the input matrix $\mathbf{A}$ can be replaced with $\mathbf{\Sigma}\mathbf{V}^T$, which can be beneficial if $m \gg n$.

5   Experiments

In order to validate the performance of our algorithm we have carried out a set of experiments of three types: accuracy, horizontal scalability and scalability with respect to the number of features. The results show that our algorithm achieves approximation errors close to those of state-of-the-art randomized proposals. We

---

Algorithm 2. Parallel EM for PPCA

1: **procedure** COMPUTEPPCA(rows, k)

2:     $\mathbf{W} \leftarrow \begin{pmatrix} \mathbf{I}_\rho \\ \mathbf{0} \end{pmatrix}$; $\sigma^2 \leftarrow 1$

3:     **while** not converged **do**

4:         $\mathbf{M} \leftarrow \mathbf{W}^T\mathbf{W} + \sigma^2\mathbf{I}$; broadcast($\mathbf{M}^{-1}$)

5:         preFactors $\leftarrow$ rows.**map**$\{\mathbf{x} =>$

6:             $\langle \mathbf{x} \rangle \leftarrow \mathbf{M}^{-1}\mathbf{W}^T(\mathbf{x} - \boldsymbol{\mu})$; $\langle \mathbf{x}\mathbf{x}^T \rangle \leftarrow \sigma^2\mathbf{M}^{-1} + \langle \mathbf{x} \rangle\langle \mathbf{x} \rangle^T$

7:             output $(\langle \mathbf{x} \rangle, \langle \mathbf{x}\mathbf{x}^T \rangle, \mathbf{x})$ $\}$

8:         $(\mathbf{X}, \mathbf{Y}) \leftarrow$ preFactors.**map**$\{(\langle \mathbf{x} \rangle, \langle \mathbf{x}\mathbf{x}^T \rangle, \mathbf{x}) => (\mathbf{x}\langle \mathbf{x} \rangle^T, \langle \mathbf{x}\mathbf{x}^T \rangle))\}$.**reduce**$(+)$

9:         $\tilde{\mathbf{W}} \leftarrow \mathbf{X}\mathbf{Y}^{-1}$

10:        $\tilde{\sigma}^2 \leftarrow$ preFactors.**map**$\{(\langle \mathbf{x} \rangle, \langle \mathbf{x}\mathbf{x}^T \rangle, \mathbf{x}) =>$

11:            $\left(\|\mathbf{x} - \boldsymbol{\mu}\|^2 - 2\langle \mathbf{x} \rangle^T\tilde{\mathbf{W}}(\mathbf{x} - \boldsymbol{\mu}) + \mathrm{tr}(\langle \mathbf{x}\mathbf{x}^T \rangle\tilde{\mathbf{W}}^T\tilde{\mathbf{W}}) \right)\}$.**reduce**$(\frac{1}{Nd}+)$

12:            $\mathbf{W} \leftarrow \tilde{\mathbf{W}}$; $\sigma^2 \leftarrow \tilde{\sigma}^2$

13:        **end while**

14:        $\mathbf{U}, \boldsymbol{\Sigma}, \mathbf{V} \leftarrow svd(\mathbf{W}^T\mathbf{W})$; $\mathbf{Q}, \mathbf{R} \leftarrow$ qr($\mathbf{W}\mathbf{V}^T$); $\lambda \leftarrow (\mathbf{V}\mathbf{W}^T\mathbf{W}\mathbf{V}^T + \sigma^2)^{1/2}$

15:        output $(\mathbf{Q}, \lambda)$

16: **end procedure**

---

also show how our algorithm benefits greatly from RDD-based parallelization, achieving a significant speedup when the number of worker nodes in the cluster are increased.

The experiments have been run on a cluster of 10 worker nodes equipped with an Intel quad-core processor and 4 GB of RAM each, connected over an ethernet switch with a capacity of 100 Mbps on each link. Our algorithm, as well as the ones we have compared it with, have been parallelized over the RDD abstraction and implemented using Scala 2.10.4 on Apache Spark 1.4.1, and the HDFS distributed file system implemented on Hadoop 2.6 has been employed for data storage and access.

The following algorithms have been implemented and compared:

- **PPICS**: The algorithm presented in this paper. We set the number of iterations for the expectation-maximization to 3, as we have observed this value to provide a good balance between approximation error and running time.
- **Two-stage**: This is the two-stage algorithm presented in [1], which is a very good representative of randomized approaches. We implement it with a slight variation. Specifically, the column sampling in the random phase has been done as described in Sect. 4 rather than independently and with replacement as suggested by the authors. The reason is that the latter approach could result in poor samples being drawn. Since the authors do not propose a method for choosing the best candidate subset, we use the second phase of **PPICS** for this purpose.
- **Uniform**: This is similar to PPICS, although we use a uniform probability distribution instead of the one described in Eq. (1). This allows us to assess

the effectiveness of using our proposed distribution and the leverage scores approximated by PPCA instead of simply sampling columns uniformly.
- **Deterministic**: This algorithm is based on the deterministic approach described in [10], which consists in choosing the k columns with the best leverage scores obtained from the top-k right singular vectors. This is a standard approach for deterministic column subset selection, regularly used throughout the literature.

We have used the following data sets to validate our results:

- **ONTS**[2]: These data consist of an Internet traffic trace collected from the core network of a medium-sized Spanish ISP, as part of the FP7 project ONTIC[3], funded by the European Union. The captured packets have been aggregated as TCP session flows, containing 105 standard network flow features such as average RTT, transmitted data, number of packets with payload, number of ACK packets, etc. We have tested the algorithms on a small subset of the ONTS totalling 8,000,000 rows. This data set has been used to evaluate both the accuracy and the running times of our algorithm.
- **Synthetic**: We have synthetically built matrices with entries generated uniformly at random. These matrices are all of 20,000 rows and a number of columns ranging from 100 to 3,200. This data set has been used to validate only the better scalability of our approach with respect to the number of features.

5.1 Accuracy

First of all, we run a set of experiments to test the accuracy of our approach in terms of the approximation error. We use a 2,000,000-row sample of the **ONTS** data set with varying values of k, ranging from 3 to 25 (out of the full 105 feature set). As shown in Fig. 1, our algorithm achieves errors comparable to those of **Two-stage** and consistently outperforms **Uniform** and **Deterministic**. The relatively good performance of **Uniform** also indicates the benefits to be gained from applying the approach for choosing the best random sample, since even with a uniform random sampling distribution a relatively good feature subset can be obtained efficiently. It is interesting to point out, however, that nothing prevents **Uniform** from sampling rank-defficient column subsets if they exist in the data set, often resulting in unsuitable candidate subsets. This was never a problem in our experiments with **PPICS** or **Two-stage**.

5.2 Horizontal Scalability

Next, we run experiments to assess the running time of our algorithm in different scenarios. First, we use 1, 2, 4 and all the 8 million rows of the ONTS data set. As expected, Fig. 2 reveals that **Uniform** and **Deterministic** are the fastest,

---

[2] http://ict-ontic.eu/index.php/onts-data/onts-request-access.
[3] http://ict-ontic.eu/.

since they do not need to compute SVD and the selection of the best candidate respectively. The fact that the running time of **PPICS** increases linearly with respect to the number of rows means that larger data sets can be easily processed by an equivalently larger number of worker nodes. To assess the benefit of parallelization, we run the algorithms on a fixed sample of 2,000,000 rows using 1, 2, 4, 6, 8 and all 10 worker nodes. Figure 3 shows that the running time of **PPICS**, **Two-stage** and **Uniform** decreases significantly when more worker nodes are added. Since this data set is not too large, the point of diminishing returns is reached soon, with the overhead of network I/O and cluster management becoming dominant. Also, note that given the relatively small size of n, working on $\mathbf{\Sigma V}^T$ instead of $\mathbf{A}$ would likely have improved the running times.

5.3 Scalability with Respect to the Number of Columns

Finally, we run experiments using matrices filled with random values with different numbers of columns (100, 200, 400, 800, 1600, 3200), and a fixed $k = 5$. This serves the purpose of showing how the algorithms behave when the value of n increases. As shown in Fig. 4, **PPICS** outperforms both **Two-stage** and **Uniform**. In the case of **Two-stage** this is to be expected because of its quadratic

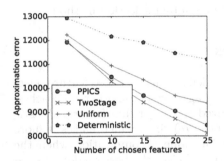

Fig. 1. Approximation error.

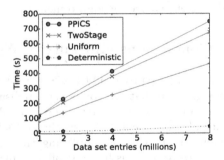

Fig. 2. Time for different data set sizes

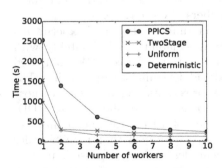

Fig. 3. Time with respect number of workers

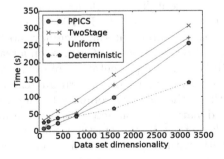

Fig. 4. Time for varying dimensionality

dependence on n, a value that is more significant in this data set than in the previous one. The fact that this quadratic tendency is not observed is due to the dominance of the second phase of the algorithm (the choice of the best candidate). In the case of **Uniform**, it performs worse than **PPICS** due to the larger size of matrix **Q** resulting from uniform sampling. The performance penalties incurred by the computation of the SVD, which requires $O(mn \min(m, n))$ operations, are clearly visible in the case of **Deterministic**, which does not need a computationally demanding second phase. The complexity of the column sampling phase in **PPICS** ($O(mn\rho)$) seems to manifest itself in its faster running times than **TwoStage**, although again the second phase seems to mask this result. However, at this dimensionality the tendency of **PPICS** and the extent of its benefits with respect to **Two-stage** are not clear, so further experiments would be interesting.

Based on the results shown in this section, we can conclude that **PPICS** (1) achieves approximation errors comparable to those of state-of-the-art randomized methods (2) scales linearly with respect to the number of data samples, making it possible to handle increasing data set sizes by simply increasing the number of worker nodes and (3) is faster than methods that exactly compute the leverage scores when n is large. Unlike the rest of the evaluated algorithms, **PPICS** performs well with respect to these three criteria, making it a reliable and efficient choice for high-dimensional column subset selection method.

6   Conclusions and Future Work

In this paper we proposed PPICS, a novel parallelized unsupervised feature selection algorithm that overcomes certain disadvantages present in other approaches. It computes leverage scores using a parallelized EM algorithm for probabilistic principal components analysis that runs in $O(mn\rho)$ time, which makes it more efficient than methods employing the singular value decomposition when the number of columns is large. It offers a method to select the best randomly sampled candidate column subset in an efficient manner. The algorithm has been parallelized over the emerging Resilient Distributed Datasets abstraction, making it horizontally scalable for enormous numbers of data samples.

In order to validate PPICS we carried out different types of experiments to measure three criteria: accuracy, horizontal scalability and scalability with respect to the number of features, on a real and synthetic datasets. We compared PPICS against representatives of randomized and deterministic approaches. The results showed that our algorithm achieves approximation errors close to those methods that exactly compute the leverage scores. We also showed how RDD-based parallelization improves the performance of PPICS as the number of worker nodes in the cluster is increased. Also, we verified that PPICS scales linearly with respect to the number of data samples, making it possible to handle enormous data set sizes by simply increasing the number of worker nodes. Also, PPICS depends linearly on the number of features in the input data set.

In the future we plan to study the behavior of PPICS on larger data sets. We also plan to carry out a thorough study of how probabilistic PCA can be leveraged for column sampling and to try to establish approximation guarantees.

References

1. Boutsidis, C., Mahoney, M.W., Drineas, P.: An improved approximation algorithm for the column subset selection problem. In: Proceedings of the Twentieth Annual ACM-SIAM Symposium on Discrete Algorithms, pp. 968–977. Society for Industrial and Applied Mathematics (2009)
2. Dean, J., Ghemawat, S.: Mapreduce: simplified data processing on large clusters. Commun. ACM **51**(1), 107–113 (2008)
3. Deshpande, A., Rademacher, L., Vempala, S., Wang, G.: Matrix approximation and projective clustering via volume sampling. In: Proceedings of the Seventeenth Annual ACM-SIAM Symposium on Discrete Algorithms, pp. 1117–1126. Society for Industrial and Applied Mathematics (2006)
4. Farahat, A.K., Elgohary, A., Ghodsi, A., Kamel, M.S.: Distributed column subset selection on mapreduce. In: IEEE 13th International Conference on Data Mining (ICDM), pp. 171–180. IEEE (2013)
5. Frieze, A., Kannan, R., Vempala, S.: Fast monte-carlo algorithms for finding low-rank approximations. J. ACM (JACM) **51**(6), 1025–1041 (2004)
6. He, Q., Cheng, X., Zhuang, F., Shi, Z.: Parallel feature selection using positive approximation based on mapreduce. In: 11th International Conference on Fuzzy Systems and Knowledge Discovery (FSKD), pp. 397–402. IEEE (2014)
7. Jolliffe, I.T.: Discarding variables in a principal component analysis. I: artificial data. Appl. Stat. **21**, 160–173 (1972)
8. Ordozgoiti, B., Canaval, S.G., Mozo, A.: Parallelized unsupervised feature selection for large-scale network traffic analysis. In: Proceedings of the ESANN (2016)
9. Ordozgoiti, B., Canaval, S.G., Mozo, A.: A fast iterative algorithm for improved unsupervised feature selection. In: IEEE 16th International Conference on Data Mining (ICDM), pp. 390–399. IEEE (2016)
10. Papailiopoulos, D., Kyrillidis, A., Boutsidis, C.: Provable deterministic leverage score sampling. In: Proceedings of the 20th ACM SIGKDD International Conference on Knowledge Discovery and Data Mining, pp. 997–1006. ACM (2014)
11. Singh, S., Kubica, J., Larsen, S., Sorokina, D.: Parallel large scale feature selection for logistic regression. In: SDM, pp. 1172–1183. SIAM (2009)
12. Sun, Z., Li, Z.: Data intensive parallel feature selection method study. In: International Joint Conference on Neural Networks (IJCNN), pp. 2256–2262. IEEE (2014)
13. Tipping, M.E., Bishop, C.M.: Probabilistic principal component analysis. J. Roy. Stat. Soc. Ser. B **61**(3), 611–622 (1999)
14. Wold, S., Esbensen, K., Geladi, P.: Principal component analysis. Chemometr. Intell. Lab. Syst. **2**(1), 37–52 (1987)
15. Zaharia, M., Chowdhury, M., Das, T., Dave, A., Ma, J., McCauley, M., Franklin, M.J., Shenker, S., Stoica, I.: Resilient distributed datasets: a fault-tolerant abstraction for in-memory cluster computing. In: Proceedings of the 9th USENIX Conference on Networked Systems Design and Implementation, p. 2. USENIX Assoc. (2012)
16. Zhao, Z., Zhang, R., Cox, J., Duling, D., Sarle, W.: Massively parallel feature selection: an approach based on variance preservation. Mach. Learn. **92**(1), 195–220 (2013)

General Noise SVRs and Uncertainty Intervals

Jesus Prada[(✉)] and Jose Ramon Dorronsoro

Universidad Autónoma de Madrid, Cantoblanco, Madrid, Spain
jesus.prada@estudiante.uam.es, jose.dorronsoro@uam.es

Abstract. Building uncertainty estimates is still an open problem for most machine learning regression models. On the other hand, general noise–dependent cost functions have been recently proposed for Support Vector Regression, SVR, which should be more effective when applied to regression problems whose underlying noise distribution follows the one assumed for the cost function. Taking this into account, we first propose a framework that combines general noise SVR models trained by Naive Online R Minimization Algorithm, NORMA, optimization with uncertainty interval estimates for their predictions. We then provide the theoretical details required to implement this framework for several noise distributions and carry out experiments, whose results show an improvement over the ones obtained by classical ϵ-SVR and also support the hypothesis that the model and error intervals with the noise distribution assumption closest to the real one yield the best results. Finally, and in accordance with the principle of reproducible research, we make the implementations developed and the datasets employed in the experiments publicly and easily available.

Keywords: Support Vector Regression · Uncertainty intervals · General noise model · NORMA · Noise distribution · Maximum likelihood · Reproducible research

1 Introduction

Support vector regression, SVR, machines are one of the most powerful machine learning techniques and has been widely used with excellent results in many real-world problems such as stock market [1], wind energy [2] or solar radiation [3] forecasting. SVR usually employ the ϵ–insensitive loss function, which is justified [4] under the assumption that the noise in the data is additive and Gaussian. This paper presents a method to train SVR models suited for any noise distribution and give concrete implementations for several noise distributions, namely zero-mean and general Laplace, zero-mean and general Gaussian, and the Beta, Weibull and Marshall-Olkin Generalized Exponential distributions. We will call the resulting models *general noise SVR*, which we build using NORMA [5] as the optimization method. Moreover, extending our previous work in [6], we also propose a method to compute uncertainty intervals for the model predictions. Thus, the main contributions of this paper are:

© Springer International Publishing AG 2017
I. Rojas et al. (Eds.): IWANN 2017, Part II, LNCS 10306, pp. 734–746, 2017.
DOI: 10.1007/978-3-319-59147-6_62

- We propose a novel framework based on NORMA to train general noise SVR models and to estimate error intervals for their predictions.
- We give the theoretical background as well as explicit formulations to implement this framework for the following distributions: zero-mean and general Laplace, zero-mean and general Gaussian, Beta, Weibull and Marshall-Olkin Generalized Exponential, MOGE.
- We extend the uncertainty intervals formulations described in [6], adding to these the ones corresponding to MOGE noise distribution assumption.
- Experiments are carried out showing that the proposed models can outperform classical SVR over artificially created datasets when an accurate noise distribution supposition is chosen. Best results correspond to the noise assumption closest to the real one present in the data.

Moreover, in accordance with the principle of reproducible research, all our implementations are publicly accessible via CRAN libraries written in R; similarly, the artificial datasets used are available through Github.

The rest of this paper is organized as follows. Section 2 briefly reviews the prior theoretical background for classical ϵ-SVR formulation, general noise SVR models, NORMA optimization and uncertainty intervals for SVR. In Sect. 3 there is an in-depth description of the proposed approach combining general noise models, NORMA optimization and error interval estimations. Section 4 contains some implementation details, the experiments over artificial datasets and their analysis. The paper ends with a short section on conclusions and pointers for further work.

2   Theoretical Background

2.1 ϵ-SVR

The classical linear Support Vector Regression cost function is

$$\ell_S(w, b) = \sum_p [y^p - w \cdot x^p - b]_\epsilon + \frac{1}{C} \|w\|_2^2 \tag{1}$$

where we use ℓ_2 regularization and the ϵ-insensitive loss function, or ILF, $\ell(y, \hat{y}) = [y - \hat{y}]_\epsilon = \max\{|y - \hat{y}| - \epsilon, 0\}$, that defines an ϵ-wide, penalty-free "error tube" around the model. The optimal w^*, b^* in SVR are found rewriting (1) as a constrained minimization problem which is then transformed using Lagrangian theory into a much simpler dual problem, the one actually being solved; see [7]. The optimal w^*, b^* are then obtained from the dual solution through the KKT equations. It turns out that the dual problem only involves patterns through their dot products and a natural extension to improve on a purely linear model is to apply the kernel trick [8]. It replaces the initial dot products $x \cdot x'$ with the values $k(x, x')$ of a positive definite kernel k that can be written as $k(x, x') = \phi(x) \cdot \phi(x')$, where the x are mapped through an implicitly defined map $\phi(x)$ into a Hilbert space. We thus arrive to a kernel–based formulation of the dual problem, which is then usually solved using the well known SMO algorithm [9]. Since we can write the optimal model vector W^* as $W^* = \sum \alpha_p^* \phi(x^p)$, the final non linear model becomes

$$f(x) = b^* + W^* \cdot \phi(x) = b^* + \sum \alpha_p^* \phi(x^p) \cdot \phi(x) = b^* + \sum_{\alpha_p^* > 0} \alpha_p^* k(x^p, x). \quad (2)$$

The x^p for which $|\alpha_p^*| > 0$ are the Support Vectors and the Gaussian kernel $k(x, x') = \phi(x) \cdot \phi(x') = e^{-\gamma \|x - x'\|^2}$ is usually chosen.

2.2   General Noise SVR

In [7] an SVR formulation was proposed with a general cost function. Its primal problem is to minimize

$$\min_{\beta, \beta_0, \xi_i, \hat{\xi}_i} \quad \frac{1}{2} \|\beta\|^2 + C \sum_{i=1}^{N} (c_i(\xi_i) + c_i(\hat{\xi}_i))$$

$$\text{subject to} \quad \xi_i, \hat{\xi}_i \geq 0, \ i = 1, ...N \quad (3)$$

$$f(x_i) - y_i \leq \epsilon_i + \xi_i, \ i = 1, ...N$$

$$y_i - f(x_i) \leq \epsilon_i^* + \hat{\xi}_i, \ i = 1, ...N,$$

where ϵ_i, ϵ_i^* are chosen such that $c(\xi, y_i, y_i + \xi) = 0, \ \forall \ \xi \in [-\epsilon_i^*, \epsilon_i]$, $c_i(\xi_i) = c(\xi, y_i, y_i + \epsilon_i + \xi)$ and $c_i(\hat{\xi}_i) = c(\xi, y_i, y_i - \epsilon_i^* - \hat{\xi})$. Note that c_i is the ϵ-clipped version of the original loss function.

2.3   NORMA

Based on classical stochastic gradient descent, *NORMA* is an optimization method that focus on the *instantaneous regularized risk*

$$R_{inst,\lambda}[f_t, x_t, y_t] := l(f_t(x_t), y_t) + \frac{\lambda}{2} \|f_t\|^2, \quad (4)$$

where $\lambda \geq 0$ is the regularization parameter and l is a given loss function. NORMA performs gradient descent with respect to $R_{inst,\lambda}$ with the updates

$$f_{t+1} = f_t - \eta_t \partial_f R_{inst,\lambda}[f_t, x_t, y_t], \quad (5)$$

where ∂_f is the gradient with respect to f and $\eta_t > 0$ the learning rate, which usually is chosen to be constant, i.e., $\eta_t = \eta$.

To derive $\partial_f R_{inst,\lambda}[f_t, x_t, y_t]$ we work independently with the two terms in (4). For the first one, $\partial_f l(f_t(x_t), y_t)$, proceeding as in [5] and working on a *reproducing kernel Hilbert space* $\mathbb{H}$ with a kernel k, the reproducing property $< f, k(x, \cdot) >_{\mathbb{H}} = f(x), \forall x \in \mathbb{X}$, yields

$$\partial_f l(f_t(x_t), y_t) = l'(f_t(x_t), y_t) \ k(x_t, \cdot),$$

with $l'(x, y) = \partial_x l(x, y)$.

Moreover, we also have $\partial_f \frac{\lambda}{2} ||f_t||^2 = \lambda f_t$ for the second term. Plugging these equalities into (5) we get

$$f_{t+1} = f_t - \eta_t(l'(f_t(x_t), y_t) \, k(x_t, \cdot) + \lambda f_t) = (1 - \eta_t \lambda)f_t - \eta_t l'(f_t(x_t), y_t) \, k(x_t, \cdot),$$

where $\eta_t < \frac{1}{\lambda}$ must hold for the algorithm to work properly. Writing the previous f_t as a kernel expansion [10]

$$f_{t+1} = (1 - \eta_t \lambda) \sum_{i=1}^{t-1} \alpha_i k(x_i, x) - \eta_t l'(f_t(x_t), y_t) k(x_t, \cdot), \tag{6}$$

defining

$$\hat{\alpha}_i := (1 - \eta_t \lambda)\alpha_i \text{ for } i < t, \ \hat{\alpha}_t := -\eta_t l'(f_t(x_t), y_t), \tag{7}$$

and plugging (7) into (6) yields

$$f_{t+1} = \sum_{i=1}^{t} \hat{\alpha}_i k(x_i, x). \tag{8}$$

NORMA optimization iterates on (7) and (8) starting at an initial hypothesis f_1 usually taken as $f_1 = 0$. As showed in [5] an extra update rule to include the possibility of the existence of an offset b term for the function f can be added as $b_{t+1} = b_t - \eta_t \, l'(f_t(x_t), y_t)$.

2.4 Uncertainty Intervals for SVR

Classical SVR does not give probability intervals to address the uncertainty in the predictions and, in fact, error interval estimation for SVR has received a somewhat limited attention in the literature. In [11], a rather direct and simple approach to obtain these estimates for SVR is presented. This method assumes prediction errors to follow a specific probability distribution that, in turn, is used to define probability intervals for them. In particular, the idea is to model the distribution of prediction errors, Ψ, based on a set of out-of-sample residuals $\{\psi_i\}_{i=1}^{l}$ obtained by k-fold cross-validation, CV, over the training data.

In [11] it is assumed that the conditional distribution of y given x depends on x only through $\hat{f}(x)$. In theory, the distribution of Ψ may depend on the input x and therefore the width of the predictive interval may vary from one example x to another. However, it is claimed in [11] that despite of this independence, same width intervals can be justified if we consider the probability to be taken over all possible input values.

Two noise families, zero–mean Gaussian and Laplace, are considered in [11] and their parameters are fitted by maximum likelihood estimation, MLE. Denoting the log–likelihood by l and assuming prediction errors, $\{\psi_i\}_{i=1}^{N}$, to be independent, MLE is reduced to maximize $\sum_{i=1}^{N} \log f(\psi_i | \theta)$. Since both noise distributions are zero–mean and symmetric, we can obtain an error interval $[a, b]$ for a predefined probability $1 - 2s$ simply by setting $a = -p_s$ and $b = p_s$, where p_s is the the upper s–th percentile. As mentioned, we get the same interval for each test instance x.

738   J. Prada and J.R. Dorronsoro

3 Proposed Model

3.1 General Noise SVR Using NORMA

It is well known that the usual cost functions for regression are chosen according to the model noise being assumed and this is also the case for the standard classical SVR [4]. However, in some real-world applications the model's noise may follow other distributions. Because of this, here we shall consider the Laplace and Gaussian distributions [11] but also the Beta [12] and Weibull [13] distributions, relevant in wind energy forecasting, and the Marshall-Olkin Generalized Exponential distribution, MOGE [14], which arises in healthcare regression problems. The MOGE density is

$$P(\psi_i) = \begin{cases} \frac{\alpha\lambda\theta e^{-\lambda x}(1-e^{-\lambda x})^{\alpha-1}}{(\theta+(1-\theta)(1-e^{-\lambda x})^\alpha)^2} & , \psi_i > 0 \\ 0 & , \psi_i \le 0 \end{cases} , \; \alpha > 0, \; \lambda > 0, \; \theta > 0 \qquad (9)$$

This suggests that building SVR models for general noise formulation may be of interest for some regression problems. As mentioned, such a general noise formulation for SVR has been proposed in [7], arriving to a dual problem that allows to work with general loss functions. However, for most noise densities other than Gaussian or Laplace, their associated cost functions make the dual problem very difficult to solve and standard SVM optimizers such as SMO can no longer be applied. However, NORMA has a fairly simple formulation and implementation, and can be used in a straightforward manner for regression problems under general loss functions; furthermore, the kernel trick makes its extension from linear to non-linear models largely direct. Thus, NORMA is perfectly suited for our goal and is the optimization method we will use in this work.

To apply NORMA we must first compute the loss functions associated to general noise. Following a similar approach as described in [13], for a concrete error distribution $P(\psi_i)$, we take its minus logarithm $l(\psi_i) = -\log P(\psi_i)$ as the associated loss function as it is the optimal choice in a maximum likelihood sense; In Table 1 we give the loss functions for the distributions we consider in this paper and their corresponding derivatives.[1]

Now we can apply NORMA's iterations by plugging these derivatives into NORMA's update rule described in (7). As stated before, NORMA is based on stochastic gradient descent and [15] proves asymptotic convergence to a stationary point in the non-convex case; however, this point is not guaranteed to be a global minimum. We address this issue in Sect. 4. To apply the preceding to the computation of uncertainty intervals two steps are required: parameter estimation via maximum likelihood and probability intervals computation. We describe both next.

---

[1] Technically, at $\psi = 0$ the loss functions corresponding to Laplace distributions are non-differentiable. This case corresponds to predictions with no error, so we take as a proxy for the derivative at this point the value $l'(0) = 0$.

3.2 Parameter Estimation

In [6] we gave the computations required for parameter estimation via MLE for all distributions considered in this work except for the MOGE distribution. Values for all of them are shown in Table 2. For the MOGE case, writing $\Psi = (\psi_1, \ldots, \psi_N)$ for the vector of sample errors, the log-likelihood is

$$l(\Psi|\lambda, \alpha, \theta) = \sum_{i=1}^{N} \log \frac{\alpha\lambda\theta e^{-\lambda\psi_i}(1 - e^{-\lambda\psi_i})^{\alpha-1}}{(\theta + (1 - \theta)(1 - e^{-\lambda\psi_i})^\alpha)^2}$$

$$= N(\log\alpha + \log\lambda + \log\theta) - \lambda\sum_{i=1}^{N}\psi_i + (\alpha - 1)\sum_{i=1}^{N}\log(1 - e^{-\lambda\psi_i})$$

$$-2\sum_{i=1}^{N}\log\left(\theta + (1 - \theta)(1 - e^{-\lambda\psi_i})^\alpha\right) \tag{10}$$

Computing the derivatives of $l(\Psi|\lambda, \alpha, \theta)$ with respect to λ, α and θ, setting them to zero and then rewriting the factor of the resulting equations dependent of $\{\lambda, \alpha, \theta\}$ as $F_1(\lambda, \alpha, \theta), F_2(\lambda, \alpha, \theta)$ and $F_3(\lambda, \alpha, \theta)$ respectively, we get

$$\sum_{i=1}^{N}\psi_i = \frac{N}{\lambda} + (\alpha - 1)\sum_{i=1}^{N}\frac{\psi_i e^{-\lambda\psi_i}}{1 - e^{-\lambda\psi_i}} - 2\alpha(1 - \theta)\sum_{i=1}^{N}\frac{\psi_i e^{-\lambda\psi_i}(1 - e^{-\lambda\psi_i})^{\alpha-1}}{\theta + (1 - \theta)(1 - e^{-\lambda x_i})^\alpha} = F_1(\lambda, \alpha, \theta)$$

$$0 = \frac{N}{\alpha} - \sum_{i=1}^{N}\log(1 - e^{-\lambda\psi_i}) + 2\theta\sum_{i=1}^{N}\frac{\log(1 - e^{-\lambda\psi_i})}{\theta + (1 - \theta)(1 - e^{-\lambda\psi_i})^\alpha} = F_2(\lambda, \alpha, \theta)$$

$$0 = \frac{N}{\theta} - 2\sum_{i=1}^{N}\frac{1 - (1 - e^{-\lambda\psi_i})^\alpha}{\theta + (1 - \theta)(1 - e^{-\lambda\psi_i})^\alpha} = F_3(\lambda, \alpha, \theta) \tag{11}$$

Iterative methods may be employed for the numerical solution of the Eq. (11). Newton-Raphson's method, involving the linearization of F_1, F_2, F_3 in the neighborhood of the root, leads to the following iterative scheme:

$$\sum_{i=1}^{N}\psi_i = F_1(\lambda_j, \alpha_j, \theta_j) + (\lambda_{j+1} - \lambda_j)(\frac{\partial F_1}{\partial\lambda})_{(\lambda_j,\alpha_j,\theta_j)}$$

$$+ (\alpha_{j+1} - \alpha_j)(\frac{\partial F_1}{\partial\alpha})_{(\lambda_j,\alpha_j,\theta_j)} + (\theta_{j+1} - \theta_j)(\frac{\partial F_1}{\partial\theta})_{(\lambda_j,\alpha_j,\theta_j)} \tag{12}$$

$$0 = F_2(\lambda_j, \alpha_j, \theta_j) + (\lambda_{j+1} - \lambda_j)(\frac{\partial F_2}{\partial\lambda})_{(\lambda_j,\alpha_j,\theta_j)}$$

$$+ (\alpha_{j+1} - \alpha_j)(\frac{\partial F_2}{\partial\alpha})_{(\lambda_j,\alpha_j,\theta_j)} + (\theta_{j+1} - \theta_j)(\frac{\partial F_2}{\partial\theta})_{(\lambda_j,\alpha_j,\theta_j)} \tag{13}$$

$$0 = F_3(\lambda_j, \alpha_j, \theta_j) + (\lambda_{j+1} - \lambda_j)(\frac{\partial F_3}{\partial\lambda})_{(\lambda_j,\alpha_j,\theta_j)}$$

$$+ (\alpha_{j+1} - \alpha_j)(\frac{\partial F_3}{\partial\alpha})_{(\lambda_j,\alpha_j,\theta_j)} + (\theta_{j+1} - \theta_j)(\frac{\partial F_3}{\partial\theta})_{(\lambda_j,\alpha_j,\theta_j)} \tag{14}$$

Table 1. Loss functions corresponding to several distributions and their derivatives.

Error distr.	Loss function	Derivative
0–mean Laplace	$l(\psi_i) = \frac{\|\psi_i\|}{\sigma}$	$l'(\psi_i) = \begin{cases} \frac{1}{\sigma} & , \psi_i > 0 \\ 0 & , \psi_i = 0 \\ -\frac{1}{\sigma} & , \psi_i < 0 \end{cases}$
Laplace	$l(\psi_i) = \frac{\|\psi_i - \mu\|}{\sigma}$	$l'(\psi_i) = \begin{cases} \frac{1}{\sigma} & , \psi_i - \mu > 0 \\ 0 & , \psi_i - \mu = 0 \\ -\frac{1}{\sigma} & , \psi_i - \mu < 0 \end{cases}$
0–mean Gaussian	$l(\psi_i) = \frac{\psi_i^2}{2\sigma^2}$	$l'(\psi_i) = \frac{\psi_i}{\sigma^2}$
Gaussian	$l(\psi_i) = \frac{(\psi_i - \mu)^2}{2\sigma^2}$	$l'(\psi_i) = \frac{\psi_i - \mu}{\sigma^2}$
Beta	$l(\psi_i) = (1 - \alpha) \log \psi_i + (1 - \beta) \log (1 - \psi_i)$	$l'(\psi_i) = \frac{1-\alpha}{\psi_i} - \frac{1-\beta}{1-\psi_i}$
Weibull	$l(\psi_i) = \begin{cases} (1 - \kappa) \log \psi_i + (\frac{\psi_i}{\lambda})^\kappa & , \psi_i > 0 \\ 0 & , \psi_i \leq 0 \end{cases}$	$l'(\psi_i) = \begin{cases} \frac{1-\kappa}{\psi_i} + \frac{\kappa}{\lambda}(\frac{\psi_i}{\lambda})^{(\kappa-1)} & , \psi_i > 0 \\ 0 & , \psi_i \leq 0 \end{cases}$
MOGE	$l(\psi_i) = \begin{cases} 2\log(\theta + (1 - \theta) \\ (1 - e^{-\lambda\psi_i})^\alpha) + \lambda\psi_i \\ +(1 - \alpha) \log (1 - e^{-\lambda\psi_i}) & , \psi_i > 0 \\ 0 & , \psi_i \leq 0 \end{cases}$	$l'(\psi_i) = \begin{cases} \lambda(1 + e^{-\lambda\psi_i} 2\alpha(1 - \theta) \\ (1 - e^{-\lambda\psi_i})^{\alpha-1} + \\ \frac{1-\alpha}{1-e^{-\lambda\psi_i}}]) & , \psi_i > 0 \\ 0 & , \psi_i \leq 0 \end{cases}$

In this work initial values $(\lambda_0, \alpha_0, \theta_0)$ are chosen empirically through cross-validation over the datasets used in Sect. 4.

3.3   Uncertainty Intervals

For all the considered distributions, we compute the prediction error interval $[a, b]$ for a pre-specified probability $1 - 2s$ as follows:

1. **Zero Mean Laplace and Gaussian:** For a zero-mean symmetric variable with density $p(z)$, we can obtain the upper s–th percentile, p_s, just by solving $1 - s = \int_{-\infty}^{p_s} p(z)dz$. The prediction error interval is $(-p_s, p_s)$ in this case.
2. **Non-Zero Mean Laplace and Gaussian:** The percentile p_s is determined as before. However, as in this case the distribution is centered at μ and not zero, the prediction error interval is $(\mu - (p_s - \mu), \mu + (p_s - \mu))$.
3. **Beta, Weibull and MOGE:** For a beta distribution we have $z \geq 0$, so it holds that $1 - s = \int_0^{p_s} p(z)dz$ and therefore the prediction error interval is $(0, p_s)$. For the Weibull and MOGE distributions only the case $z \geq 0$ is relevant, so we determine the prediction error interval the same way.

As stated before, the conditional distribution $y|x$ is assumed to depend on x only through the prediction value $\hat{f}(x)$ and, therefore, these intervals are the same for each test instance. In [6] we propose to use clustering methods, empirical and analytical such as k-means, on the data and build different intervals for each cluster of points to lessen the impact of this drawback. Testing these techniques in the experiments carried out is a topic of further research for this paper.

Table 2. Estimated parameters via MLE corresponding to several distributions

Distribution	MLE parameters
Zero-mean Laplace	$\hat{\sigma} = \frac{\sum_{i=1}^{n}\lvert\psi_i\rvert}{n}$
Laplace	$\hat{\sigma} = \frac{\sum_{i=1}^{n}\lvert\psi_i - \mu\rvert}{n}$, $\mu = m_{\psi_i}$. m_{ψ_i} is the median of $\{\psi_i\}_{i=1}^{n}$
Zero-mean Gaussian	$\hat{\sigma} = \frac{\sum_{i=1}^{n}\psi_i^2}{n}$
Gaussian	$\hat{\sigma} = \frac{\sum_{i=1}^{n}(\psi_i - \mu)^2}{n}$, $\mu = \sum_{i=1}^{n}\frac{\psi_i}{n}$
Beta	$\frac{\sum_{i=1}^{n}\log\psi_i}{n} = F_1(\alpha_j,\beta_j) + (\alpha_{j+1} - \alpha_j)\left(\frac{\partial F_1}{\partial\alpha}\right)_{(\alpha_j,\beta_j)} + (\beta_{j+1} - \beta_j)\left(\frac{\partial F_1}{\partial\beta}\right)_{(\alpha_j,\beta_j)}$, $\frac{\sum_{i=1}^{n}\log(1-\psi_i)}{n} = F_2(\alpha_j,\beta_j) + (\alpha_{j+1} - \alpha_j)\left(\frac{\partial F_2}{\partial\alpha}\right)_{(\alpha_j,\beta_j)} + (\beta_{j+1} - \beta_j)\left(\frac{\partial F_2}{\partial\beta}\right)_{(\alpha_j,\beta_j)}$
Weibull	$\lambda = \left(\frac{1}{n}\sum_{i=1}^{n}\psi_i^{\kappa}\right)^{\frac{1}{\kappa}}$, $\frac{\sum_{i=1}^{n}\log\psi_i}{n} = G(\kappa_j) + (\kappa_{j+1} - \kappa_j)\left(\frac{\partial G}{\partial\kappa}\right)_{(\kappa_j)}$

4 Experiments

4.1 Implementation

We have implemented the proposed models as well as the software used in our experiments in R. Two R packages have been developed for this purpose:

- **errint:** Builds and analyzes error intervals for a particular model predictions assuming different distributions for noise in the data.
- **NORMA:** Builds general noise SVR models using NORMA optimization.

Both are freely available via CRAN[2]. For standard ϵ-SVR models we used the R wrapper of the popular library LIBSVM [16], available in CRAN package *e1071*.

4.2 Artificial Datasets

We created several artificial datasets consisting of $2,000$ instances following the expression

$$y_i = 3x_i + x_i^2 + \delta_i, \ i = 1,2,\ldots,2,000. \tag{15}$$

with δ_i random noise. We consider eight types of datasets, each one with a different noise distribution; Five datasets are built for each of these types and the mean of experiment results over them is computed to contemplate possible deviations on results obtained. 70% of each dataset is used for training and validation and 30% for testing. The concrete distributions are[3]

1. **Fixed:** $\delta_i = 0$.
2. **Zero Laplace:** δ_i extracted from a Laplace with $\mu = 0$ and $\sigma \in [0.1, 10]$.

---

[2] https://cran.r-project.org/web/packages/available_packages_by_name.html.

[3] Distribution parameters like σ are randomly computed among the selected interval only once for each dataset, i.e. they remain constant for all δ_i extractions of a dataset.

3. **General Laplace:** Same as the previous case but this time $|\mu| \in [10^2, 10^3]$.
4. **Zero Gaussian:** δ_i extracted from a Gaussian with $\mu = 0$ and $\sigma^2 \in [0.1, 10]$.
5. **General Gaussian:** Same as the previous case but this time $|\mu| \in [10^2, 10^3]$.
6. **Beta:** δ_i extracted from a Beta distribution with $\alpha, \beta \in [1, 100]$.
7. **Weibull:** δ_i extracted from a Weibull distribution with $\kappa, \lambda \in [0.1, 10]$.
8. **MOGE:** δ_i extracted from a MOGE distribution with $\lambda, \alpha, \theta \in [0.1, 1]$.

These artificial datasets are available through Github[4].

4.3   Metrics and Parameter Selection

For model comparison and CV we use the relative mean absolute error

$$RelMAE = \frac{1}{N} \sum_{i=1}^{N} \frac{|\hat{f}(x_i) - y_i|}{|y_i|} \tag{16}$$

To test the accuracy of the error intervals for a pre-specified probability $1 - 2s$, we compare the percentage of test prediction errors, ψ_i^{test}, lying in their prediction intervals $[a, b]$ with the expected number, $(1 - 2s)\%$, i.e.:

$$per_{err} = |\{\# \ of \ \psi_i^{test} \in [a, b]\} - (1 - 2s)| \tag{17}$$

We choose here an absolute error as accuracy measure over one with weights for positive or negative errors because preference towards a positive or negative mistake is problem-dependent and here we opt to use a more general measure.

Regarding parameter selection, ϵ-SVR parameters, $\{C, \epsilon, \gamma\}$, are selected using CV over a standard grid search. For general noise SVR models using loss functions other than ILF, parameters are selected applying the MLE formulas showed in Table 2. These parameters are also used to build the corresponding error intervals.

4.4   Experiment I. SMO ϵ-SVR vs. NORMA ϵ-SVR

Before proceeding to the general noise models and uncertainty intervals experiments, we have checked that the NORMA and SMO optimizers build ϵ-SVR models with highly comparable errors on our artificial datasets. The results of this experiment can be seen in Table 3, which shows that training an ϵ-SVR model using SMO or NORMA optimization yields very similar results. Thus, our NORMA implementation works as expected and, therefore, we can consider only NORMA–based ϵ-SVRs in the remaining experiments.

---

[4] https://github.com/jesuspradaalonso/IWANN_2017

Table 3. Experiment I results. RelMAE comparison for SMO vs NORMA ϵ-SVR.

Dataset	RelMAE SMO	RelMAE NORMA	Difference
Fixed	2.84e$-$06	3.56e$-$06	$-$0.72e$-$06
Zero Laplace	0.63	0.62	$+$0.01
General Laplace	6.38e$-$06	6.33e$-$06	$+$0.05e$-$06
Zero Gaussian	0.25	0.30	$-$0.05
General Gaussian	1.57e$-$05	1.48e$-$05	$+$0.9e$-$05
Beta	0.01	0.01	0
Weibull	0.02	0.02	0
MOGE	0.05	0.06	$-$0.01

4.5 Experiment II. General Noise SVR

The goal of this experiment is to test, for the loss functions in Table 1 plus the ILF, whether the best general noise SVR model is the one that uses the loss function which corresponds to the real data noise distribution.

As mentioned in Sect. 3.1, using non-convex loss functions could lead to local minima. We use two mechanisms to deal with this problem. The first one is to constrain the parameters of the chosen distribution to be outside the set of parameters which cause the loss function to be non-convex. The second one, less restrictive and more general, is to apply the optimization algorithm several times from different initial points and keep the best solution to the optimization problem as our final function. We use both approaches and show the results of the model that gives the best results.

These results are given in Table 4, which shows that, for all datasets tested, the best model is the one with the loss function corresponding to the data noise distribution; in particular, it beats the model built with the classical ILF loss, which seems to confirm our initial hypothesis and the usefulness of our approach to build general noise SVR models.

It is also interesting to note that for the general Gaussian dataset, the second best model is the one using ILF as loss function, a result that fits in well with the statements in [4] that the use of the ILF is justified under the assumption that the noise is additive and Gaussian with random mean and variance.

Table 4. Experiment II results. RelMAE for each noise distribution–loss pair.

Dataset	ILF	Z-Lap	G-Lap	Z-Gau	G-Gau	Beta	Weib	MOGE
Fixed	3.27e$-$06	2.57e$-$06	9.38e$-$05	**1.01e$-$06**	8.35e$-$05	8.23e$-$06	3.59e$-$06	4.61e$-$06
Zero Laplace	0.63	**0.13**	0.17	0.48	0.53	0.99	0.66	0.85
General Laplace	7.01e$-$04	1.05e$-$04	**7.74e$-$06**	9.37e$-$04	1.93e$-$05	4.13e$-$03	1.49e$-$04	1.01e$-$04
Zero Gaussian	0.35	0.37	0.42	**0.07**	0.11	0.59	0.55	0.80
General Gaussian	3.52e$-$05	1.08e$-$04	9.86e$-$05	5.07e$-$05	**9.18e$-$06**	4.11e$-$03	7.37e$-$04	6.82e$-$04
Beta	0.01	0.11	0.10	0.13	0.11	**8.93e$-$03**	0.11	0.11
Weibull	0.03	0.19	0.17	0.07	0.05	0.28	**6.37e$-$03**	0.02
MOGE	0.09	0.51	0.47	0.26	0.19	0.49	0.12	**0.01**

Table 5. Experiment III results. per_{err} for each noise distribution–loss pair.

Dataset	Z-Lap	G-Lap	Z-Gau	G-Gau	Beta	Weib	MOGE
Fixed	**1.0**	**1.0**	**1.0**	1.2	2.2	3.0	3.8
Zero Laplace	**0.3**	**0.3**	1.1	1.3	1.0	1.6	1.7
General Laplace	1.0	**0.4**	1.8	1.4	1.2	1.4	2.0
Zero Gaussian	2.4	2.5	**0.2**	**0.2**	1.4	0.8	1.6
General Gaussian	3.3	2.3	1.1	**0.1**	3.8	1.1	1.8
Beta	2.9	2.9	1.4	1.1	**0.2**	1.1	1.3
Weibull	3.9	3.4	1.5	1.3	2.3	**0.3**	1.2
MOGE	3.6	3.2	2.6	2.4	2.8	1.4	**0.8**

4.6 Experiment III. Uncertainty Intervals

This experiment is analogous to the previous one but for uncertainty intervals computation. Its aim is to test if, among all distributions considered, the error intervals obtaining the best accuracy are the ones built using the noise distribution assumption corresponding to the real noise in the data.

Results can be found in Table 5. The experiment is carried out two times, one with $s = 0.1$, i.e. intervals should contain 80% of the errors, and the second with $s = 0.05$, i.e., 90% intervals. Both results are averaged to obtain the final error. We point out that these results are quite similar to the ones reported in [6] and, again, seem to confirm our starting hypothesis that for a dataset with a given noise, the best error intervals are the ones built using its associated loss function.

We list here some factors contributing to moderate differences between performances of the zero-mean and general versions of Laplace and Gaussian approaches in these experiments. First, variations in the mean of the validation population with respect to the test one. Moreover, although theoretically SVR models should have zero-mean errors in practice this is not always achieved and an error bias appears. Finally, the bias term, b, is not always perfectly estimated.

5   Conclusions and Further Work

In this paper we propose a method to build general noise SVR models using NORMA optimization and an approach to compute error intervals for them, and in general for any regression model. Both techniques rely on the assumption that data noise follows a concrete distribution. We give mathematical framework for their implementation under several noise distributions and the concrete details for MOGE density. We have made public via CRAN the R implementation of the proposed models and error intervals. Furthermore, artificial data created for our experiments can be freely accessed via Github to facilitate reproducibility.

Our experiments on artificial datasets show that best results are achieved when the noise assumption matches the real noise distribution in the data, as we expected. We also show that the proposed model achieves better results, both in terms of regression and intervals accuracy, than ϵ–SVR for these synthetic sets.

A first line of further work is to test the proposed models in real-world problems to verify that the conclusions drawn for artificial data persist. A second line is to consider other error distributions such as the Cauchy or Logistic ones. We also want to improve on the x–independent error intervals, for which a possibility is to identify different clusters and build different uncertainty models on each of them, in a similar manner as in [6]. Finally, similar techniques may be applied to other regression methods such as ensemble or Deep Learning frameworks.

Acknowledgments. With partial support from Spain's grants TIN2013-42351-P, TIN2016-76406-P, TIN2015-70308-REDT and S2013/ICE-2845 CASI-CAM-CM. Work supported also by project FACIL–Ayudas Fundación BBVA a Equipos de Investigación Científica 2016, and the UAM–ADIC Chair for Data Science and Machine Learning. We also acknowledge the use of the facilities of Centro de Computación Científica (CCC) at UAM.

References

1. Yang, H., Chan, L., King, I.: Support vector machine regression for volatile stock market prediction. In: Yin, H., Allinson, N., Freeman, R., Keane, J., Hubbard, S. (eds.) IDEAL 2002. LNCS, vol. 2412, pp. 391–396. Springer, Heidelberg (2002). doi:10.1007/3-540-45675-9_58

2. Kramer, O., Gieseke, F.: Short-term wind energy forecasting using support vector regression. In: Corchado, E., Snášel, V., Sedano, J., Hassanien, A.E., Calvo, J.L., Ślęzak, D. (eds.) SOCO 2011. AISC, vol. 87, pp. 271–280. Springer, Heidelberg (2011). doi:10.1007/978-3-642-19644-7_29

3. Gala, Y., Fernández, Á., Díaz, J., Dorronsoro, J.R.: Support vector forecasting of solar radiation values. In: Pan, J.-S., Polycarpou, M.M., Woźniak, M., Carvalho, A.C.P.L.F., Quintián, H., Corchado, E. (eds.) HAIS 2013. LNCS, vol. 8073, pp. 51–60. Springer, Heidelberg (2013). doi:10.1007/978-3-642-40846-5_6

4. Pontil, M., Mukherjee, S., Girosi, F.: On the noise model of support vector machines regression. In: Arimura, H., Jain, S., Sharma, A. (eds.) ALT 2000. LNCS, vol. 1968, pp. 316–324. Springer, Heidelberg (2000). doi:10.1007/3-540-40992-0_24

5. Kivinen, J., Smola, A.J., Williamson, R.C.: Online learning with kernels. IEEE Trans. Sig. Process. **52**, 2165–2176 (2004). IEEE

6. Prada, J., Dorronsoro, J.R.: SVRs and uncertainty estimates in wind energy prediction. In: Rojas, I., Joya, G., Catala, A. (eds.) IWANN 2015. LNCS, vol. 9095, pp. 564–577. Springer, Cham (2015). doi:10.1007/978-3-319-19222-2_47

7. Schölkopf, B., Smola, A.J.: Learning with Kernels: Support Vector Machines, Regularization, Optimization, and Beyond. MIT Press, Cambridge (2002)

8. Minh, H.Q., Niyogi, P., Yao, Y.: Mercer's theorem, feature maps, and smoothing. In: Lugosi, G., Simon, H.U. (eds.) COLT 2006. LNCS, vol. 4005, pp. 154–168. Springer, Heidelberg (2006). doi:10.1007/11776420_14

9. Platt, J., et al.: Fast training of support vector machines using sequential minimal optimization. In: Advances in Kernel Methods-Support Vector Learning, Cambridge, MA, vol. 3 (1999)

10. Schölkopf, B., Herbrich, R., Smola, A.J.: A generalized representer theorem. In: Helmbold, D., Williamson, B. (eds.) COLT 2001. LNCS, vol. 2111, pp. 416–426. Springer, Heidelberg (2001). doi:10.1007/3-540-44581-1_27

11. Lin, C., Weng, R.: Simple Probabilistic Predictions for Support Vector Regression. National Taiwan University, Taipei (2004)

12. Bludszuweit, H., Domínguez-Navarro, J.A., Llombart, A.: Statistical analysis of wind power forecast error. IEEE Trans. Power Syst. **23**, 983–991 (2008). IEEE

13. Hu, Q., Zhang, S., Xie, Z., et al.: Noise model based ν-support vector regression with its application to short-term wind speed forecasting. Neural Netw. **57**, 1–11 (2014). Elsevier

14. Klein, J.P., Keiding, N., Kamby, C.: Semiparametric Marshall-Olkin models applied to the occurrence of metastases at multiple sites after breast cancer. Biometrics **45**, 1073–1086 (1989). JSTOR

15. Kushner, H.J., Clark, D.S.: Stochastic Approximation Methods for Constrained and Unconstrained Systems. Springer, New York (2012)

16. Chang, C., Lin, C.: LIBSVM: A library for support vector machines. ACM Trans. Intell. Syst. Technol. **2**, 1–27 (2011

Towards Visual Training Set Generation Framework

Jan Hůla$^{(\boxtimes)}$, Irina Perfilieva, and Ali Ahsan Muhummad Muzaheed

University of Ostrava, 701 03 Ostrava, Czech Republic
{jan.hula,irina.pefilieva}@osu.cz
http://www.osu.cz, http://irafm.osu.cz

Abstract. Performance of trained computer vision algorithms is largely dependent on amounts of data, on which it is trained. Creating large labeled datasets is very expensive, and therefore many researchers use synthetically generated images with automatic annotations. To this purpose we have created a general framework, which allows researchers to generate practically infinite amount of images from a set of 3D models, textures and material settings. We leverage Voxel Cone Tracing technology implemented by NVIDIA to render photorealistic images in realtime without any kind of precomputation. We have build this framework with two use cases in mind: (i) for real world applications, where a database with synthetically generated images could compensate for small or non existent datasets, and (ii) for empirical testing of theoretical ideas by creating training sets with known inner structure.

1 Introduction

We introduce a framework for generating training data for supervised learning algorithms in computer vision. Currently, the most successful algorithms for computer vision tasks are very data hungry. They require large amount of training images, to be able to achieve low test errors. Creation of large training sets like ImageNet [RDS+15], which contain millions of labeled images is often considered as one of the main factors influencing the progress in machine learning. The creation of such large training sets is prohibitively expensive and for tasks like optical flow, where one would need to label every pixel, it is practically impossible. Some researches try to avoid this bottleneck by creating synthetic datasets, where the training examples are generated along with the labels [dSGCP16, SQLG15, GAGM15, XVL+13, IMS+16]. Because computer graphics can today create imagery, which has high degree of photorealism in reasonable time, it offers cheap way how to generate practically infinite amount of annotated images from limited amount of 3D models. Different configurations of 3D models, with different textures and materials applied to them, with different lighting and viewing conditions create combinatorially very large number of possible scenes. Due to the complete control over the rendering algorithm, all kinds of ground truth annotations could be generated along with the photorealistic images. These, beside other things, include: semantic labels,

I. Rojas et al. (Eds.): IWANN 2017, Part II, LNCS 10306, pp. 747–758, 2017.
DOI: 10.1007/978-3-319-59147-6_63

edges and depth maps (See Fig. 1). From these, other kinds of ground truth data could be calculated [MHLD16]. Unlike many of the previous attempts [MHLD16, HPB+15, RSM+16], that were concentrated on creation of specific datasets, we instead create a general framework, which allows to generate data according to the needs of the researcher. Our main contribution does not lie at bringing some new knowledge to the community, but at building a framework, which we hope will be useful for researchers in computer vision. In this paper we describe technical details, that were made to meet this purpose.

Fig. 1. Example of ground truth images. Top left: surface normals, Top right: Object ID, Bottom left: surface edges, Bottom right: depth map.

2   Problem Statement

Our aim is to create a framework which, given a database of 3D models, textures and scene descriptions, generates large amount of images together with ground truth annotations in reasonable time (e.g. days). In order to be sure about the results, we constrain ourselves to images of indoor scenes, which are much easier to make photorealistic than for example images of organic nature. To achieve this goal, two main subtasks should be identified. Assembling 3D scenes from individual models and rendering images from these scenes. Two main criteria, which influence our solution, are photorealism of the rendered images and the speed of rendering. We also develop this framework with two use case scenarios in mind:

- As a tool for real world applications, which compensates for small or non existent training sets.
- As an experimental framework, which generates data according to the need of the experiment. Because with computer graphics we could control almost every detail of the image, we could conduct truly controlled experiments.

More generally we try to accommodate our framework to the changing needs for different datasets and to prevent researchers from wasting their efforts by creating one shot datasets.

3   Related Work

Many researchers in the field of machine learning successfully leveraged synthetic datasets in the past. [GVZ16, WWCN12] generated large datasets of synthetic images with text to train OCR systems with great accuracy. [dSGCP16] trained a classifier for action recognition from rendered videos containing procedurally guided models of humans. [SQLG15] outperform state of the art view estimation methods on PASCAL 3D+ benchmark with convolutional neural network (CNN) trained on rendered objects. [GAGM15] use synthetic images of objects to train CNN, which replaces objects in RGB-D scenes with 3D models from a database. [PSAS14, AR15] study the invariances and features of CNN in experiments done on synthetically generated images. [HPB+15, MHLD16] assemble indoor scenes using models from ShapeNET database [CFG+15] in a same fashion as we do, and render large amount of RGB-D images with ground truth annotations from these scenes. Synthetic images are also used a lot for the task of pedestrian detection [XVL+13, HNBKK15, VLM+14]. [IMS+16] use CNN to estimate optical flow. Due to unavailability of training data for this task, they generate their own using computer graphics. [LVVG16] uses synthetic images to learn to decompose shading from albedo in real images. [RSM+16] produce virtual environment with the aim to train algorithms for self driving cars. [SDLK17] does the same but for drones. [RVRK16] figured out a method, how to leverage computer games for generation of labeled images. [QY16] released a plugin for Unreal Engine, which allows to render images from premade scenes. [LGF16, KWR+16, BCP+16] use game environments to train reinforcement learning agents. [SPT+16, BSD+16] make the synthetic images more realistic by using adversarial network to change the distribution of pixels. They then show, that thus modified images increase the accuracy of the classifier trained on them (compared to classifier trained on the original synthetic images). The main difference in our work is, that we are creating a general framework for rendering datasets instead of creating one concrete dataset or studying effects of learning from synthetic data.

4   Effectiveness of Synthetic Data Sets

As mentioned earlier, there are two use cases which could benefit from synthetic datasets. For real world applications, synthetic datasets could serve as data augmentation technique. It is well known, that data augmentation is very useful technique, which could besides other thing reduce overfitting. In computer vision data augmentation is often done by introducing noise and different kinds of occlusion, mirroring the image horizontally or cropping the image at different positions, rotations and scales. Thus, from one image one creates many others. All these transformations are very limited and obviously do not capture too much of variations which arise in real world images.

On the other hand, when generating images synthetically, one could model any variation one wants. Every dimension (e.g. texture, lighting parameters, camera parameters) of variation creates many possible images, which could be generated, combinatorially giving rise to enormous pool of possible images, which could be very tedious to label manually. Because in computer graphics we have complete information about every pixel of the generated image, we get the labels for free. Lot of previous work which was described in last section has benefited from this fact.

The second use case for synthetic datasets is empirical testing of theoretical ideas. Here we believe our framework brings much larger gains and it was also our main motivation. As an analogy, one could imagine that the framework for generating synthetic images would in computer vision play similar role as synthetic biology plays in molecular biology. Synthetic biology constitutes great experimental framework, which allows researchers to test their theories quickly and interpret the results of the experiment with more certainty. For example when conducting experiments with synthetic cell, there is much less unknown variables involved than in experiments with biological cell. Similarly there is a lot of unknowns in natural images, which makes the interpretation of machine learning algorithms in computer vision very hard.

With training set generation framework one could generate dataset tailored for the experiment in mind. As an illustration example, one could test how the order of training examples affects the convergence of learning, the idea behind curriculum learning [BLCW09]. With such framework it is very easy to generate the training images in the order of some predefined criterion, and thus test the idea of curriculum learning. If, as in our case, the generation is realtime, the learning process could be made dynamic by establishing two way communication between the learner and the teacher. By teacher we here mean the score function together with the generation framework. The teacher could generate images on the fly, taking into account the behaviour of the learner.

5   Componentization

Instead of creating one monolithic piece of software, we have divided our framework into two separate components, which could be used independently. These two components can be used for following tasks:

- Stochastic scene generation: generating empty rooms with walls, windows and doors, generating furniture layouts, creating lighting of a scene

– Rendering: assigning materials to objects, creating camera settings and rendering images

Tasks in stochastic scene generation are specific to indoor scenes, the rest could be used for all other projects. This componentization offers easier reuse. Researchers could easily exchange any component with their own, if they adopt our interface.

6   Stochastic Scene Generation

For generating empty rooms and placing lights we use very simple heuristics, because we do not have any hints, why modeling them in more complex ways would benefit the learning. For empty room we first sample the type of room (e.g. bathroom, kitchen) which influences the size of the room and also the distribution of types of furniture. We set the dimensions of the room according to the distribution for that type of room. Next we sample count of windows and doors to use for that type of room and also their placement. The models of doors and windows are sampled from the database. We use open sourced modeling program Blender to model the rooms and save them to .obj files.

When creating lighting for the scene we sample number of lights according to the type of room and place the lights randomly below the ceiling and to objects (e.g. lamps) which have special sockets indicating the placement of light. We utilize IES profiles of lights, which are freely available and create realistically looking light effects (Fig. 2).

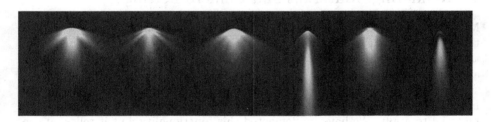

Fig. 2. Examples of various ies profiles of lights.

For generating furniture layouts we utilize factor graphs, kind of graphical model which allows us to specify probable configurations of objects by defining local constraints (factors) between them. In our case, these constraints encode information like distance of objects from each other (objects should not overlap, chairs should be near a table), distance of objects from walls (objects should not intersect with the wall, objects like wardrobe should be placed near the wall) and other relationships or ergonomic factors. For more involved set of factors see [YYT+11]. The value in which these constraints are unsatisfied in some configuration is used as a penalty for the probability of that configuration. The probability of a configuration is equal to the product of local factors:

$$P(X) \sim \prod_{i=1}^{n} f_i$$

where X is a global configuration of variables and f_i is a local factor, whose arguments are variables, on which this factor depends.

One way to get a sample from this probability distribution is to use Markov Chain Monte Carlo algorithm (MCMC), where one creates Markov chain whose stationary distribution is equal to the desired probability distribution. After certain number of steps in this Markov chain, one can get correlated samples, whose probability is given by the desired distribution. Pure MCMC algorithm is very slow, therefore other variations of it are used in practice (e.g. simulated annealing, parallel tempering). For more complete description of how factor graphs and MCMC are used for stochastic scene generation see [YYT+11, YYW+12]. For our proof of concept we have used Locally annealed Reversible Jump MCMC developed by [YYW+12], mainly because it can generate layouts, where the count of objects is not fixed.

When synthesizing furniture layouts from databases of models like ShapeNET [CFG+15], there arises a problem with inconsistent metric units. We have used same heuristics as [MHLD16], who leveraged distribution of height of object categories, which is encoded in a dataset called SUN RGD-B [SLX15]. For drawbacks of this heuristic see [MHLD16].

7   Rendering, Materials and Camera Settings

Most of our attention during the process of designing this framework went to the rendering part, because it is most reusable. This part consist of three main tasks: assigning materials and textures to models, setting the camera parameters and rendering the images.

For assignment of materials it would be great to use semantic labels (information about the type of material) of 3D model, in order to choose texture and material to apply. We have implemented this functionality, but it relies on specific details, which may not be present. Concretely on the type of material of the model (or its subpart) and real world scale of its UV maps and textures from which we choose. Given the information about material type, we could assign a material and texture, which have the properties fulfilling the properties of the material type. To this purpose we have collected set of material settings, which imitate some real world materials. As a concrete example, if the model would have material type *wood*, then we could choose from different material settings, which imitate different types of wood (e.g. polished, brushed) and for this again different types of textures could be used (e.g. oak, beech, maple). This would create very large number of possible object appearances.

Unfortunately, the database of models we currently use (ShapeNET [CFG+15]) does not contain semantic labels for materials types or information about the scale of UV maps. Therefore, in these cases, we load the materials, which were created by the author of the model. We hope for better standardization, so that curated collections of models and textures can be used for creating very large number of possible appearances of objects.

The most critical part of our framework was the rendering engine. As mentioned previously, the main criterion affecting our framework was photorealism of images and speed of rendering. One could rightly ask, how is the accuracy of a trained algorithm affected by the photorealism of the training set, when it is tested on real images. It could possibly be the case, that the algorithm trained on large sample of artificially looking images would give similar accuracy as if it was trained on photorealistic images. Results of [SPT+16] show, that it is not so, as one would intuitively expect. They have trained a kind of adversarial network [GPAM+14], which takes as input the synthetic image and as an output it produces semantically same image, but whose pixel distribution is much closer to that of natural images (they look more realistic). They then show that discriminative algorithm trained on thus modified images achieves significant improvement over the same algorithm trained on the original synthetic images. This is our justification for the effort to reach high photorealism.

There is obviously a trade off between the photorealism of rendered images and speed of rendering. On one side, in game industry there exist rendering engines, which produce dozens of images per second, but one could definitely recognize that they are not real. On the other side, rendering engines developed for special effects industry produce images, which are often indistinguishable from real images, but take hours to render. In the following paragraphs we describe rationale behind decisions we made when choosing a rendering technology.

We base our decisions on the number of images we would like to generate. This number is specific to the use case. For small experiments we would like to generate thousands of images, but for real world applications the number of generated images would be ideally around one million and more. If the image would take 10 s in average to render, then million images would render for almost 4 months. This could be possibly justifiable for some large projects, or if the resulting dataset would be reused many times by some community, but for our purposes this is too prohibitive. Also for small experiments we would like to be able to iterate the generation many times as new ideas appear or some deficiencies are discovered. Generation of 10 000 images with rendering time 10 s per image would take almost whole day.

The rendering technology for creating highly realistic images is often based on some form of path tracing, which simulates how the light rays bounce around the scene. It is often implemented in progressive form, which means that the image is being continuously updated as the simulation progresses. One can stop the simulation, when the quality of the image is satisfying. This allows the user to make a trade off between quality and speed. Unfortunately, for indoor scenes, the quality of the images rendered under 10 s is very poor. One could utilize few tricks like caching the information needed to calculate the radiance values (photon mapping) and reusing this cache across many frames from the same scene and thus amortizing the cost of the calculation, but the results are still unsatisfying and it works only for static scenes. For few examples see [MHLD16]. They managed to achieve render times below 3 s on average, but the resulting images look noisy with visible splotches from the under sampled photon map. The problem with noise could be partially remedied by denoising algorithms tailored specially for Monte Carlo type renderers [BRM+16]. These algorithms could improve the quality of the image significantly, but their run times is in tens of seconds. They are used to denoise images, which normally take many minutes to render, so there the run time of denoising is negligible. For our purposes their run time is unacceptable. Possibly in few years the path traced and similar renderers will be producing images of satisfying quality in seconds or even less, but for our project we have finally started to search for something else.

Eventually we have decided to render the images with game engine, which could render many frames per second. When neglecting the loading time of the scenes and saving the images, we could render around 15 high resolution frames per second on NVIDIA GTX980Ti graphics card. Concrete frame rates depends on lighting setup and few other settings. Thus, neglecting the cost of loading the models, we could render one million images in less than a day.

Concretely, we have chosen Unreal Engine 4 (UE), which is currently among the leaders in terms of photorealism and at the same time it is open sourced. The photorealism of UE is constantly getting better and its maintenance is almost guaranteed, thus we get improvements for free. The main hindrance for us was the fact, that until recently the photorealism in game engines was achieved by light maps, which are precomputed during the creation of the game environment. These light maps often take few minutes to precompute, which would make whole engine unusable for our purposes.

Fortunately for us, new kind of technology was recently developed for real-time graphics, which achieves very realistically looking images without the need of precomputing lightmaps. It is called Voxel Cone Tracing and it computes Indirect illumination by dynamically dividing 3D scene to voxels, which approximate the original geometry [CNS+11]. Researchers from NVIDIA have implemented this algorithm under codename VXGI and they also released a version of UE, which uses their VXGI technology. We are using this version of UE for rendering part of our framework. UE also provides setting, which enables to mimic real world camera. It contains settings for exposure, depth of field, lens distortions, motion blur and dozen other settings. The code for loading a geometry, applying materials, creating lights and rendering the images is packaged in a plugin, which could be installed into UE. We have reused lot of work from [QY16], who released a plugin called UnrealCV, which allows to call commands in UE from programming language Python. This enables researchers to use our commands without learning about the architecture of UE. At the same time it allows to combine our generation framework with machine learning libraries like TensorFlow [AAB+16]. The framework could generate images at the time of learning, although at this point this would make the learning prohibitively slow. We will be releasing the code for whole project after the presentation of this paper (Fig. 3).

Fig. 3. Comparison of scene rendered using UE (left) and the same scene rendered using off the shelf path tracing renderer (right). The images were rendered in same resolution ($1600 \times 1024\,\mathrm{px}$). The left image was rendered under $100\,\mathrm{ms}$, the right took around $20\,\mathrm{min}$ to render.

8   Use Case Example

We provide simple use case example of how our framework simplified the creation of testing data for geometrical edge detection algorithm. By the term geometrical edge we mean every edge which was created by discontinuities in the geometry (as opposed to discontinuities in light or texture).

We were faced with the problem of creating a small benchmark for a competition in geometrical edge detection. Creating ground truth annotations for natural images manually would be very time consuming. At the same time, we wanted also to create easier version of the benchmark, which would not contain textures. This was very easy to create with our framework. We have reused the scenes from [HPB+15] and rendered test images with ground truth annotations in matter of seconds. Unfortunately some images required manual correction, but

Fig. 4. Example of images from benchmark in geometrical edge detection

it still spared us a lot of manual work. See Fig. 4 for example pair of images. All images are placed on website of the competition: http://irafm.osu.cz/edge2017.

9   Conclusion

We have described an implementation of a framework, which allows to generate large collection of realistically looking images with ground truth annotations in small amount of time. It is achieved by exploiting combinatorially large set of possible configurations of different models, materials, lighting and viewing conditions. We have developed our framework with two use cases in mind. For augmenting a training set in order to achieve higher accuracy in real world applications and also for generating datasets in a very controlled manner, which is very useful for testing theoretical ideas empirically. As a main bottleneck we regard the lack of databases with high quality models and material annotations. Our future work will focus in this direction. We will also focus on photorealism of the generated images, mainly by incorporating the work of [SPT+16] in adversarial image modification. Lastly we will incorporate more recent algorithms for stochastic scene generation.

References

[AAB+16] Abadi, M., Agarwal, A., Barham, P., Brevdo, E., Chen, Z., Citro, C., Corrado, G.S., Davis, A., Dean, J., Devin, M., et al.: Tensorflow: Large-scale machine learning on heterogeneous distributed systems. arXiv preprint arXiv:1603.04467 (2016)

[AR15] Aubry, M., Russell, B.C.: Understanding deep features with computer-generated imagery. In: Proceedings of the IEEE International Conference on Computer Vision, pp. 2875–2883 (2015)

[BCP+16] Brockman, G., Cheung, V., Pettersson, L., Schneider, J., Schulman, J., Tang, J., Zaremba, W.: Openai gym. arXiv preprint arXiv:1606.01540 (2016)

[BLCW09] Bengio, Y., Louradour, J., Collobert, R., Weston, J.: Curriculum learning. In: Proceedings of the 26th Annual International Conference on Machine Learning, pp. 41–48. ACM (2009)

[BRM+16] Bitterli, B., Rousselle, F., Moon, B., Iglesias-Guitián, J.A., Adler, D., Mitchell, K., Jarosz, W., Novák, J.: Nonlinearly weighted first-order regression for denoising monte carlo renderings. Comput. Graph. Forum **35**, 107–117 (2016). Wiley Online Library

[BSD+16] Bousmalis, K., Silberman, N., Dohan, D., Erhan, D., Krishnan, D.: Unsupervised pixel-level domain adaptation with generative adversarial networks. arXiv preprint arXiv:1612.05424 (2016)

[CFG+15] Chang, A.X., Funkhouser, T., Guibas, L., Hanrahan, P., Huang, Q., Li, Z., Savarese, S., Savva, M., Song, S., Hao, S., et al.: Shapenet: An information-rich 3d model repository. arXiv preprint arXiv:1512.03012 (2015)

[CNS+11] Crassin, C., Neyret, F., Sainz, M., Green, S., Eisemann, E.: Interactive indirect illumination using voxel cone tracing. Comput. Graph. Forum **30**, 1921–1930 (2011). Wiley Online Library

[dSGCP16] de Souza, C.R., Gaidon, A., Cabon, Y., López Peóa, A.M.: Procedural generation of videos to train deep action recognition networks. arXiv preprint arXiv:1612.00881 (2016)

[GAGM15] Gupta, S., Arbeláez, P., Girshick, R., Malik, J.: Aligning 3D models to RGB-D images of cluttered scenes. In: Proceedings of the IEEE Conference on Computer Vision and Pattern Recognition, pp. 4731–4740 (2015)

[GPAM+14] Goodfellow, I., Pouget-Abadie, J., Mirza, M., Xu, B., Warde-Farley, D., Ozair, S., Courville, A., Bengio, Y.: Generative adversarial nets. In: Advances in Neural Information Processing Systems, pp. 2672–2680 (2014)

[GVZ16] Gupta, A., Vedaldi, A., Zisserman, A.: Synthetic data for text localisation in natural images. In: Proceedings of the IEEE Conference on Computer Vision and Pattern Recognition, pp. 2315–2324 (2016)

[HNBKK15] Hattori, H., Boddeti, V.N., Kitani, K.M., Kanade, T.: Learning scene-specific pedestrian detectors without real data. In: Proceedings of the IEEE Conference on Computer Vision and Pattern Recognition, pp. 3819–3827 (2015)

[HPB+15] Handa, A., Patraucean, V., Badrinarayanan, V., Stent, S., Cipolla, R.: Scenenet: Understanding real world indoor scenes with synthetic data. arXiv preprint arXiv:1511.07041 (2015)

[IMS+16] Ilg, E., Mayer, N., Saikia, T., Keuper, M., Dosovitskiy, A., Brox, T.: Flownet 2.0: Evolution of optical flow estimation with deep networks. arXiv preprint arXiv:1612.01925 (2016)

[KWR+16] Kempka, M., Wydmuch, M., Runc, G., Toczek, J., Jaśkowski, W.: Vizdoom: A doom-based AI research platform for visual reinforcement learning. arXiv preprint arXiv:1605.02097 (2016)

[LGF16] Lerer, A., Gross, S., Fergus, R.: Learning physical intuition of block towers by example. arXiv preprint arXiv:1603.01312 (2016)

[LVVG16] Lettry, L., Vanhoey, K., Van Gool, L.: Darn: a deep adversarial residual network for intrinsic image decomposition. arXiv preprint arXiv:1612.07899 (2016)

[MHLD16] McCormac, J., Handa, A., Leutenegger, S., Davison, A.J.: Scenenet RGB-D: 5m photorealistic images of synthetic indoor trajectories with ground truth. arXiv preprint arXiv:1612.05079 (2016)

[PSAS14] Peng, X., Sun, B., Ali, K., Saenko, K.: Exploring invariances in deep convolutional neural networks using synthetic images. CoRR, abs/1412.7122 **2**(4) (2014)

[QY16] Qiu, W., Yuille, A.: UnrealCV: connecting computer vision to unreal engine. In: Hua, G., Jégou, H. (eds.) ECCV 2016. LNCS, vol. 9915, pp. 909–916. Springer, Cham (2016). doi:10.1007/978-3-319-49409-8_75

[RDS+15] Russakovsky, O., Deng, J., Hao, S., Krause, J., Satheesh, S., Ma, S., Huang, Z., Karpathy, A., Khosla, A., Bernstein, M., et al.: Imagenet large scale visual recognition challenge. Int. J. Comput. Vis. **115**(3), 211–252 (2015)

[RSM+16] Ros, G., Sellart, L., Materzynska, J., Vazquez, D., Lopez, A.M.: The synthia dataset: a large collection of synthetic images for semantic segmentation of urban scenes. In: Proceedings of the IEEE Conference on Computer Vision and Pattern Recognition, pp. 3234–3243 (2016)

[RVRK16] Richter, S.R., Vineet, V., Roth, S., Koltun, V.: Playing for data: ground truth from computer games. In: Leibe, B., Matas, J., Sebe, N., Welling, M. (eds.) ECCV 2016. LNCS, vol. 9906, pp. 102–118. Springer, Cham (2016). doi:10.1007/978-3-319-46475-6_7

[SDLK17] Shah, S., Dey, D., Lovett, C., Kapoor, A.: Aerial informatics and robotics platform. Technical report MSR-TR-9, Microsoft Research (2017)

[SLX15] Song, S., Lichtenberg, S.P., Xiao, J.: Sun RGB-D: a RGB-D scene understanding benchmark suite. In: Proceedings of the IEEE Conference on Computer Vision and Pattern Recognition, pp. 567–576 (2015)

[SPT+16] Shrivastava, A., Pfister, T., Tuzel, O., Susskind, J., Wang, W., Webb, R.: Learning from simulated and unsupervised images through adversarial training. arXiv preprint arXiv:1612.07828 (2016)

[SQLG15] Su, H., Qi, C.R., Li, Y., Guibas, L.J.: Render for CNN: viewpoint estimation in images using CNNs trained with rendered 3D model views. In: Proceedings of the IEEE International Conference on Computer Vision, pp. 2686–2694 (2015)

[VLM+14] Vazquez, D., Lopez, A.M., Marin, J., Ponsa, D., Geronimo, D.: Virtual and real world adaptation for pedestrian detection. IEEE Trans. Pattern Anal. Mach. Intell. **36**(4), 797–809 (2014)

[WWCN12] Wang, T., Wu, D.J., Coates, A., Ng, A.Y.: End-to-end text recognition with convolutional neural networks. In: 21st International Conference on Pattern Recognition (ICPR), pp. 3304–3308. IEEE (2012)

[XVL+13] Xu, J., Vázquez, D., López, A.M., Marin, J., Ponsa, D.: Learning a multi-view part-based model in virtual world for pedestrian detection. In: IEEE Intelligent Vehicles Symposium (IV), pp. 467–472. IEEE (2013)

[YYT+11] Yu, L.F., Yeung, S.K., Tang, C.K., Terzopoulos, D., Chan, T.F., Osher, S.J.: Make it home: automatic optimization of furniture arrangement (2011)

[YYW+12] Yeh, Y.-T., Yang, L., Watson, M., Goodman, N.D., Hanrahan, P.: Synthesizing open worlds with constraints using locally annealed reversible jump MCMC. ACM Trans. Graph. (TOG) **31**(4), 56 (2012)

Author Index

Printed in the United States
By Bookmasters